Handbook of Research on Collaborative Learning Using Concept Mapping

Patricia Lupion Torres
Pontifícia Universidade Católica do Paraná, Brazil

Rita de Cássia Veiga Marriott
University of Birmingham, UK

Information Science REFERENCE

INFORMATION SCIENCE REFERENCE
Hershey · New York

Director of Editorial Content: Kristin Klinger
Senior Managing Editor: Jamie Snavely
Assistant Managing Editor: Michael Brehm
Publishing Assistant: Sean Woznicki
Typesetter: Michael Brehm
Cover Design: Lisa Tosheff
Printed at: Yurchak Printing Inc.

Published in the United States of America by
Information Science Reference (an imprint of IGI Global)
701 E. Chocolate Avenue
Hershey PA 17033
Tel: 717-533-8845
Fax: 717-533-8661
E-mail: cust@igi-global.com
Web site: http://www.igi-global.com/reference

Library of Congress Cataloging-in-Publication Data

Handbook of research on collaborative learning using concept mapping / Rita de Cassia Veiga Marriott and Patricia Lupion Torres, editors.

p. cm.

Includes bibliographical references and index.

Summary: "This book presents innovative educational and learning models that meet current complex educational demands"--Provided by publisher.

ISBN 978-1-59904-992-2 (hardcover) -- ISBN 978-1-59904-993-9 (ebook) 1. Team learning approach in education--Handbooks, manuals, etc. 2. Concept mapping--Computer-aided design--Handbooks, manuals, etc. 3. Computer-assisted instruction--Handbooks, manuals, etc. I. Marriott, Rita de Cassia Veiga. II. Torres, Patrmcia Lupion.

LB1032.H367 2009

371.3'6--dc22

2009001057

British Cataloguing in Publication Data
A Cataloguing in Publication record for this book is available from the British Library.

Editorial Advisory Board

List of Reviewers

List of Contributors

Table of Contents

Section 3
The Use of Concept Mapping and Collaborative Learning at Pre-School, Primary and Secondary Schools

Detailed Table of Contents

This chapter aims a method of collaborative learning for distance education that was defended in a PhD thesis and used in several subsequent research studies. The proposed method, plus the quantitative and qualitative analysis were applied to students on a postgraduate course in education. The steps taken to carry out the research were: new bibliographic review; restructuration, implementation and follow-up of the methodological proposal of an on-line lab; preparation and application of questionnaires; analysis and discussion of the data obtained. The authors worked with a non-probabilistic convenience sample.

In this chapter, Marriott proposes the use of collaborative learning and concept mapping activities for teaching foreign languages to higher-intermediate and advanced language students. She introduces LAPLI (The Language Learning Lab), a blended methodology of integrative CALL (computer assisted language learning) and the Internet. By using the tools available in a collaborative VLE (virtual learning environment) in this student-centred approach, students not only develop their reading, writing and communicative skills but are also encouraged to learn meaningfully and to develop their creativity, responsibility and social skills while working individually and collaboratively throughout the activities.

The authors discuss how interactive assessment, combined with cognitive mapping techniques and the use of online portfolios, can contribute to the debate about new pedagogical and learning assessment practices in the context of interactive online information and comunication technologies.

This chapter presents the contributions drawn from the study exploring the use of inquiry maps in academic research for eliciting thinking skills. This work also points out the potential collaborative learning environments (CLEs) have to enable students to learn different mapping techniques and to help them share ways in which they can apply inquiry maps to elaborate their scientific projects.

This chapter covers a description of a theoretical framework that adopts concept maps as a tool to enhance collaborative learning in virtual teams. It was developed by drawing on studies in the literature and our own research in online collaborative learning in coursework and pre-service teacher education programs. The framework is underpinned by socio-constructivist learning theories for collaborative learning with online technologies. The use of a concept map to illustrate the framework is presented and the pedagogical benefits are highlighted. The framework is applicable to all higher education courses in promoting collaborative virtual team learning.

This chapter describes a study that used the community of learning and inquiry and concept maps as strategies to facilitate individual construction of knowledge in an asynchronous online course. Six factors influenced the concept map creation, which in turn affected individual construction of knowledge:

In the chapter "Mapping Concepts with Fisherfolk", Denis Hellebrandt starts out by presenting a conceptual background about complex systems, followed by an account of the use of concept mapping in a case study of small-scale fisheries in southern Brazil. Denis' chapter ends with a reflection on his current experience and on the application of the technique in similar projects.

The chapter presents two complementary studies, a research project and an innovation project, that highlight the differences between the non-regulative and the regulative use of concept mapping, and the differences between individual and collaborative concept mapping. The results demonstrate the advantages of regulative and collaborative use.

Some recent experiments related with the collaborative use of concept maps to physics teaching are presented in this chapter. The first is a study of how a team of teachers designed learning sequences using three-dimensional maps. In the second, concept maps are constructed and then collaboratively re-constructed by various groups of students.

This chapter focuses on the use of concept mapping, broadly defined to include both graphical and textual representations, for assessment in collaborative learning contexts. Several tools developed by the authors integrate concept mapping as the primary means of assessing progress of learning in complex and problem-solving domains. This chapter presents an overview of the theoretical and empirical foundation for these assessment tools, and discusses their applicability to collaborative learning environments (CLE).

Section 3
The Use of Concept Mapping and Collaborative Learning at Pre-School, Primary and Secondary Schools

This chapter reports on a case study investigating the use of an authoring tool for preschool concept mapping. Beginning with a literature review of the field, it continues by explaining why collaborative learning is close to passive with current mapping approaches used in preschool. The development of a tool designed to address these issues is described. The study's characteristics and results on how the tool was used for organizing, retaining, and sharing knowledge are presented. Finally, an analysis of the results demonstrates that tools which enhance children's control over the process of map building can promote child autonomy, active participation, and transform teachers into partners.

This work shows a research to adapt concept maps to the sociocognitive competence of 5 year-olds participating in an early childhood classroom, understood to be a social context for learning. In this study the authors intend to demonstrate the sociocognitive benefits experienced by young children, specifically 5 year-old children, as they participate in the production of consensual concept maps.

This chapter describes the fundamentals of collaborative learning and the use of concept maps with different individuals in a hospital. The study also aimed to involve health professionals and health educators in this process and thus gather information to develop a Learning Object.

Concept mapping and collaborative learning are discussed as literacy strategies in this ethnographic case study. This paper emphasizes the use of concept mapping as a formative assessment strategy in the content area of marine ecology in high school science classrooms. Concept mapping integrated with collaborative learning was used to engage students to construct and re-construct their understanding of a complex scientific concept, the energy cycle. The results showed that students benefited from the combination of collaborative learning and concept maps to focus their writing on key ideas, to organize their ideas, and include specific details. Most importantly, initial concept maps and revisions provided the teacher with evidence of student learning in the form of formative assessment products, to guide teachers' focused feedback and clarify specific ideas for re-teaching, as well as students' self-assessment.

Foreword

It is our privilege to have this opportunity to introduce to a wide audience this important first effort to present the good work that is being done with the use of the concept mapping tool to facilitate collaborative learning. Patricia Torres and Rita Marriott have solicited and edited chapters from scholars who have been using the concept map tool in their work. The chapters cover work with a wide range of ages, from pre-school to adult education, and for a wide range of subject matter fields. The chapters provide illustrations for the improvement of curriculums, facilitation of learning, new methods of assessment, community learning, metacognitive learning, and enhanced self-concept.

There is a growing worldwide recognition that globalization is placing new demands on individuals and societies. More than ever, we need individuals who are flexible and creative in their thinking and who have the self confidence needed to be creative and competitive. It is also widely recognized, as is well illustrated in these chapters, that we need to move learners from traditional rote learning methods to more effective meaningful learning methods.

Traditionally, most learning has taken place in settings and with methods that are directed toward individual learners. Collaboration in learning in extreme cases had been regarded as "cheating". The consequence of these methods has not only limited the development of social and cognitive skills but has also lost the important advantages that collaborative learning confers to individual learners in these areas. The dialogue between learners that is encouraged with collaborative learning mirrors more closely the kind of skills that are needed in the real world exchanges for most occupations. Moreover, there is a need for increasing collaboration between individuals in different countries with widely differing cultures. Collaborative learning can help to build the skills needed with this increasing of diversity in work environments.

Another important change that has occurred in the last decade is the explosive development of the World Wide Web and the range of resources available to learners from the WWW. The resources available go beyond those in traditional texts and include various kinds visual media. These resources can enhance work by research groups, as is illustrated in some of the chapters. If we want to build individual and team problem solving skills, the data presented show that collaborative learning using concept maps can be highly effective.

Alberto J. Cañas
Associate Director of the Institute for Human and Machine Cognition, USA

Joseph D. Novak
Professor Emeritus, Cornell University
Senior Research Scientist at the Florida Institute for Human and Machine Cognition, USA

Alberto J. Cañas, *BSc, computer systems engineering (Instituto Tecnológico de Monterrey, México, 1975); master of mathematics in computer science (University of Waterloo, Canada, 1981); PhD management sciences (University of Waterloo, Canada, 1985); founding director, computer science department, (Instituto Tecnológico de Costa Rica, 1976-1979); assistant professor, Department of Computer Science, (Tulane University, 1985-1986); assistant professor, director of the Department of Computer Services and director of the information systems department (Instituto Centroamericano de Administración de Empresas [INCAE], Costa Rica, 1986-1987); visiting professor (Centro de Investigaciones en Computación, Instituto Tecnológico de Costa Rica, 1987-1989); director (IBM Latin America Education Research Center [CLIE], Costa Rica, 1988-1989); director, institutional computing department, University of West Florida (1996-1997); visiting professor, (School of Business, Stellenbosch University, South Africa, 2003); visiting professor (Department of Pedagogy and Psychology, Unversidad Pública de Navarra, Spain, 2003-2004); associate professor, (Deparment of Computer Science, University of West Florida, 1990-2005); co-founder and associate director, (Institute for Human and Machine Cognition [IHMC], Pensacola, FL, 1990). For many years, Dr. Cañas has been involved in the use of computers in education, with particular interest in understanding the pedagogical aspects of using technology, and leveraging on his computer science background to come up with innovative solutions. He is interested not only in the theoretical aspects, but also in the implementation details and scalability of computers in education efforts. He has been a consultant to presidents of Costa Rica and Panama in the large scale introduction of computers into the public school systems, resulting in the creation of the Omar Dengo Foundation in Costa Rica and the Conéctate al Conocimiento Project in Panama. He directed the Quorum Project while at the University of West Florida, a joint effort with IBM Latin America that led to the creation of a computer network that allowed thousand of students in schools throughout seven countries in the Americas to have their own email address and work on collaborative projects before Internet arrived in those countries. At IHMC, with the support of NASA and the US Department of Defense, and the Government of Panama, he has led the development of CmapTools, a software suite to represent, visualize and share knowledge models that is used by students and professionals in over 150 countries.*

Joseph D. Novak *completed his graduate studies at the University of Minnesota in 1958; after which, Dr. Novak taught biology at Kansas State at Emporia, and Purdue University. From 1967 to 1995, he was professor of education and biological sciences at Cornell University where his research focused on human learning, educational studies and knowledge creation. He is currently Professor Emeritus, Cornell University and senior research scientist at the Florida Institute for Human and Machine Cognition, Pensacola, Florida. He is author or coauthor of 29 books and more than 150 book chapters and papers in professional books and journals. He has consulted with more than 400 schools, universities and corporations, including work with Procter and Gamble, and NSA. His recent book, Learning, Creating, and Using Knowledge: Concept Maps as Facilitative Tools in Schools and Corporations (LEA, 1998) has been translated into 5 foreign languages. Dr. Novak is listed in Who's Who in America, and other lists, and has received a number of awards and honors including a honorary doctorate from The University of Comahue in 1998 in Nuquen, Argentina, The Public University of Navarra in 2002 in Pamplona, Spain and University of Urbino, Urbino, Italy in 2006. He received the first award for contributions to science education from the Council of Scientific Society Presidents. His current research work includes studies on student's ideas on learning and epistemology, and methods of applying educational ideas and tools (such as concept mapping for knowledge archiving and utilization) and has developed a new model for education in corporate settings, schools, universities and distance learning. He is married with three children and 2 grandchildren.*

Preface

Currently, one of the major discussions taking place concerns the lack of a clear coherent definition, by the different actors involved in classroom and distance learning, of the pedagogical approach that each of them defends and follows in terms of theory and practice.When those involved in the educational process are asked about the pedagogical approach that they defend, the answers given nearly always lead to the defence of *critical teaching*. Interactivity, cooperation, collaboration, dialogue, exchange, knowledge production and group communication are constantly stressed in speeches and projects, although few programmes actually incorporate pedagogical strategies that guarantee such practices. These terms are frequently used without any corresponding commitment to them in practice. Very often, the pedagogical method used involves transmitting information under the guise of a new approach based on the use of new communications and information technologies to disseminate information. (Bochniak and Torres, 2003)

The methods used by current e-learning programs rely very little on interactivity. In this form of transmissive pedagogy, the student "receives" the content by means of a continuous monologue in which he does not participate, which gives him little opportunity to express himself and which he can neither modify nor even criticize. Dialogue and group communication, together with interaction, are highly valued in current educational thinking, although little used. Interactive exchanges between learners form the basis of methods currently advocated because they promote meaningful learning. Such learning only takes place when the student participates actively in the process of building his own knowledge (Henri, 1992).

Even institutions which develop distance-learning programmes based on the use of new technologies in an integrated-media model still face challenges related to the virtualization of teaching.

Faced with this new knowledge society and with so many socioeconomic and technological pressures arising from a globalised world, methodological solutions need to be reviewed. At such a unique conjuncture, with unprecedented resources, the opportunity of making learning situations more flexible and of advancing, enriching and socializing them must not be missed.

Educators are seeking new methodologies to respond to the demand for personal education and knowledge acquisition and production.

A study by Laister & Kober (2008) identified a number of reasons for the success of Collaborative Learning (CL) as a teaching approach. Of these reasons, the following stand out:

- the effectiveness of both short-term learning, in terms of the subject / material being studied, and long-term learning, in terms of cognitive skills and self-esteem (both considered by many theoreticians as necessary conditions for the development of both independence and skills needed for long-term learning).

- when Collaborative Learning is compared with individual and competitive learning scenarios, it can be seen to help students perform better by increasing their ability to resolve problems and helping develop personality traits that will be of benefit to them in both their academic and professional lives.
- Collaborative Learning empowers the individual and gives him/her the skills to live a more independent, collaborative and pleasant life.

Collaborative Learning thus offers the possibility of methodological innovation, as it is "based on the notion that knowledge construction is basically a social event, and adequate collaboration is particularly important for learning complex knowledge and higher order cognitive skills." (Lehtinen et al., 2001, p. 24).

Therefore, it seems clear to us that Collaborative Learning meets the demands of the knowledge and information society, in which the 'information explosion' and the demands for greater and more varied supply make it necessary to convert new knowledge produced by science and technology into educational resources and content as a matter of urgency.

The technological advances that have taken place in this new knowledge society have made it much easier both to access and to disseminate this knowledge. Nevertheless, in spite of the extraordinary advances in communications and in all areas of knowledge, the enormous amount of information available has given rise to concerns and worries among teachers the world over regarding how to understand and learn about the information that is being disseminated and use it to construct knowledge.

Many institutions and organisations, which also face the same pedagogical challenges and seek ever-greater educational excellence and quality, believe they have found a technique that may help them in their search: the construction of concept maps.

A concept map is a graphical representation of concepts and propositions. For Boxtel, it "represents the main concepts and relationships within a domain. It is a network in which the nodes represent concepts, the lines linking the nodes represent relationships, and the labels on the lines represent the nature of the relationships." (2002)

Concept mapping is a powerful learning and teaching technique. Based on Meaningful Learning, one of Ausubel's most important principle in his theory of Cognitive Learning (Cañas and Novak, 2008), learners learn meaningfully by anchoring new concepts and propositions to ones they already know.

In the light of the above and considering our experience in the development, research and evaluation of methodologies that encompass the latest educational trends, we felt that a publication aimed at educators which relates the innovative collaborative learning methodology to the technique for constructing concept maps would be of great benefit and relevance.

Therefore, in the first two chapters in section I, we present the methodologies involving the use of concept mapping and collaborative learning developed by Torres (2002) in her Doctoral research and by Marriott (2004) in her Masters research which inspired the production of this handbook.

The subsequent chapters and sections interweave theory and practice. They focus on authors' innovative experiences in the face-to-face, blended and distance learning modes of delivery, at four educational levels, namely pre-school, primary, secondary and higher education, reporting on the complex challenges and demands of education in the digital age.

Thus the *Handbook of Research on Collaborative Learning using Concept Mapping* can easily meet the aims it set out for. How? With accounts that skillfully blend theory with practice and which allow for the adaptation of the methodologies utilized into other knowledge domains.

Patricia Lupion Torres and Rita de Cássia Veiga Marriott
Editors

REFERENCES

Bochniak, R., & Torres, P. L. (2003). A atualidade de Montessori: Da sala ambiente aos ambientes virtuais, *Anais do III Congresso Paranaense de Educação*, Curitiba.

Boxtel, C. V. (2002). *Collaborative concept mapping: provoking and supporting meaningful discourse.* Retrieved December 17, 2008, from http://findarticles.com/p/articles/mi_m0NQM/is_1_41/ai_90190488/print?tag=artBody;col1

Cañas, Alberto, & Novak, Joseph. (2008). Concept Mapping using CmapTools to Enhance Meaningful Learning. In, A. Okada, S. Buckingham Shum & T. Sherborne (Eds) *Knowledge Cartography - Software Tools and Mapping Techniques.* London: Springer-Verlag.

Henri, France. (1992). Formation à distance et téléconférence assistée par ordinateur: Interativité, quasi-interativité, ou monologue? *CADE: Journal of Distance Education/Revue de l'enseignement à distance,* 7,1. Retrieved March 28, 2008, from http://cade.athabascau.ca/vol7.1/07_henri_9html

Laister, J. & Kober, S. *Social aspects of collaborative learning in virtual learning environments.* Retrieved July, 2008 from http://comma.doc.ic.ac.uk/inverse/papers/patras/19.htm

Lehtinen, E., Hakkarainen, K., Lipponen, L., Rahikainen M., & Muukkonen, H. (2001). *Computer Supported Collaborative Learning: A Review.* Retrieved December 17, 2008, from http://www.comlab.hut.fi/opetus/205/etatehtava1.pdf

Marriott, R. de Cassia Veiga. (2004) *Do LOLA - Laboratório on-line de Aprendizagem ao LAPLI - Laboratório de Aprendizagem de línguas: uma proposta metodológica para o ensino semi-presencial em ambiente virtual.* Curitiba. Dissertação de Mestrado do Programa de Pós-graduação em Educação da Pontifícia Universidade Católica do Paraná - PUCPR.

Torres, P. L. (2002) *Laboratório On line de Aprendizagem: uma proposta crítica de aprendizagem colaborativa para a educação.* Florianópolis. Tese de Doutorado do Programa de Pós Graduação em Engenharia de Produção da Universidade Federal de Santa Catarina.

Acknowledgment

To the many authors who contributed to this publication for their commitment and enthusiasm throughout the publication process.

To the team of reviewers who attentively read every article and made skilful suggestions to the betterment of the whole volume.

To the members of the Editorial Advisory Board for their dedication and availability during the key stages of the handbook.

To John Novak and Albert Cañas for sharing their distinctive theory.

To Soeur Cristina, Principal of Sion College, Curitiba, Brazil, our heartfelt gratitude for continuing to provide theoretical and practical guidance on our journey through the area of education.

To the valuable support of the distance learning Department from PUCPR.

To our husbands Gilberto and John, with whom we share our lives, dreams, children and principally our love, for their caring support throughout the whole process.

To our children Tattiana, Patrick, Nicholas and Rebecka for allowing us the freedom to work on this book.

And to our mothers Maria Helena Ribas and Maria de Lourdes Silva Veiga who continually encourage us, for their dedication and love, and for sharing their life experiences.

This handbook is dedicated to the memory of our maternal grandparents Marina and Gustavo Ribas and Maria Luisa and Jose Patrocinio da Silva who through their example made us who we are today.

Patricia Lupion Torres and Rita de Cássia Veiga Marriott
Editors

Section 1
The Use of Concept Mapping and Collaborative Learning in E-Learning

Chapter 1
LOLA:
A Collaborative Learning Approach Using Concept Maps

Patricia Lupion Torres
Pontifícia Universidade Católica do Paraná, Brazil

ABSTRACT

The Online Learning Laboratory, LOLA, is a collaborative learning methodology for classroom-based and distance learning that was defended in a PhD thesis and used in several subsequent research studies. In this chapter the author describes the first results of a case study undertaken with students on a postgraduate course in education. A further activity (the construction of a concept map) was added to the six activities in LOLA, and its inclusion was the object of this study. The methodological approach used in this research was a case study with an exploratory objective. To answer the questions raised in the study, a new review of the literature related to the following subjects was carried out: restructuring, implementing and following up the online learning laboratory methodology using concept maps; drawing up and applying a questionnaire; and analysis and discussion of the data obtained. A non-probabilistic convenience sample was used. Their analysis and preliminary considerations focus on the data gathered from the students.

INTRODUCTION

At the same time as the virtualization of teaching represents a challenge for education in the 21st century, the use of Virtual Learning Environments, whether to support classroom-based teaching or develop distance-learning courses, also represents a possible response to the need for continuing teacher education.

For Siqueira (2005, p. 190) "the Internet is not a passing fashion, fad or craze, as some of its more radical critics believe. It has come to stay. And it is evolving at incredible speed." While the Internet and virtualization of teaching are an established fact, it should be borne in mind that the mere use of information and communication technologies in the teaching-learning process does not in itself guarantee a quality process. In this scenario, a need therefore arises to seek new educational models that

DOI: 10.4018/978-1-59904-992-2.ch001

are able to meet the demands of the educational community.

This need led to the development of the Collaborative Learning methodology known as the Online Learning Laboratory (LOLA), the aim of which is to ensure that the subject develops an independent, critical attitude to the production of knowledge. This concept was developed in a doctoral thesis in 2002, and in the six years since then it has been used in various postgraduate classes in education and applied in a number of later research studies. As with any innovation in the field of learning, particularly in the context of virtual education, the processes involved need to be constantly reevaluated to ensure the quality of the training itself. In the case of LOLA, this was no different. In each class, experiences are reconstructed and suggestions from students and teachers are added, as there is still much to be done to develop a pedagogical approach that meets the needs of this new model of virtual education based on collective knowledge construction.

In this paper we describe research undertaken with master and doctoral students in the PUCPR Postgraduate Program, in which a further activity (the construction of a concept map) was added to the six activities in LOLA. The inclusion of this new activity was monitored and evaluated and resulted in a new study, which is described in this paper.

COLLABORATIVE LEARNING IN A VIRTUAL LEARNING ENVIRONMENT

Cooperative and collaborative approaches are frequently confused both in terms of what they mean and the concepts involved. This failure to distinguish between them may be due to the similarity between some of the principles that characterize them. Both approaches are based on active learning involving the sharing of experiences between the actors in the teaching-learning process; the combining of individual and social skills to produce knowledge; the taking of responsibility for one's own learning and for that of others; and respect for the diversity and heterogeneous nature of the members of the group (Matthews, Cooper, & Hawkes, 2004).

In spite of this and other similarities, there are fundamental differences between the two approaches. They differ in terms of the role, authority and degree of intervention of the teacher; the hierarchicalization of the teaching-learning process; the construction and assimilation of knowledge; the way in which the groups are organized; and individual and group responsibility for the activities (Matthews, Cooper, & Hawkes, 2004).

The idea of cooperation in the teaching-learning process is known to date back to classical antiquity, but it is only in modern times that it has come to acquire greater importance. Under the aegis of the Newtonian-Cartesian paradigm, cooperation has lent itself since the 18th and 19th centuries to professional education and the preparation of students for life in society. Because it leads to more structured, controlled activities through a more hierarchical process, cooperation has very often been used as a technique for group work.

In the beginning of the 20th century, with the emergence of the movement known as the New School, the concept of cooperation gained strength. Various educators in this movement, such as Profit, Freinet, Cousinet and Ferrière, started to include cooperative activities in their methods. Particularly worthy of note is the work carried out with cooperative groups by Dewey, who proposed that citizenship and democracy should be exercised by means of work in groups and interaction with fellow students.

As early as the first half of the 20th century, some researchers, such as Koffka, Lewin, Deutsch, Piaget and Vygotsky, were introducing ideas related to cooperation and collaboration into their theories. Later, Suchman and Bruner carried out research into inquiry learning, and Skinner investigated programmed learning and behavior modification, corroborating the discussions about cooperation, collaboration and competition. It is also worth mentioning the pioneering:

Johnson brothers (David and Roger), of the University of Minnesota, and Robert Slavin, of John Hopkins University. Others who can also be considered pioneers are Elliot Aronson (University of California, Stanford), Spencer Kagan (University of California, Riverside), Noreen Webb (University of California, Los Angeles) and Sholomo Sharan (University of Tel-Aviv, Israel). (Freitas & Freitas 2002, p.13)

More recently, with the ever-more intensive use of information and communication technologies in education, the characteristics associated with the hierarchicalization of the teaching-learning process and the style and degree of intervention of the teacher in cooperative learning have proved to be less well suited to an approach in line with the paradigm of complexity. Indeed, these resources favor nonlinear, interactive communication, which ends up significantly changing the acquisition and appropriation of knowledge. Thus, it seems quite clear to us that the choice of a collaborative learning approach is much better suited to exploration of the range of possibilities offered by online education.

Collaborative learning is a model for the Internet, which promotes interactivity and collaboration between peers and the exchange of information between individuals in a group (Cord, 2001, p.1). Interactivity is one of the most important characteristics of communication and one of the determining factors in learning (Henri, 2002, p.3).

LOLA: A COLLABORATIVE LEARNING SPACE USING CONCEPT MAPS

The purpose of the Online Laboratory is to promote learning in a collaborative framework using planned and interactive activities. The activities in LOLA, whether individual or group ones, are aimed at encouraging knowledge construction. It is during the process of managing the activities in LOLA that the groups organize themselves and prepare a work plan that is defined and negotiated collectively.

The aim in the Online Laboratory is to provide an environment characterized by an active and interactive collaborative methodology, which is ensured through the use of technologies that allow online communication. Silva & Santos (2007, p. 28-29) pointed out that to construct an interactive approach, it is necessary to allow multiple experiments and expressions; to provide an assembly of network connections to allow multiple occurrences; to provoke situations that generate creative restlessness; to engineer hypertext routes; and to mobilize the experience of knowledge.

The Online Learning Laboratory, LOLA, is implemented by developing the seven fundamental activities it embraces: discussing texts chosen by the teacher; inserting links and proposing connections with systematized knowledge; questioning existing knowledge; replying to questions that have been drawn up; preparing concept maps; producing new knowledge; and evaluating the procedures that were used.

The seven activities were carried out in the classroom or in the EUREKA virtual learning environment. According to Silva (2003, p.62), "the virtual environment is the online classroom."

LOLA SUPPLEMENTED WITH CONCEPT MAPS: SOME CONSIDERATIONS ABOUT THE EXPERIENCE GAINED

The seven activities in LOLA are developed in a classroom or virtual learning environment to achieve the theoretical and practical aims of a critical, creative and reflexive education. The activities are as follows:

1. The first LOLA activity, Commented Reading, involves reading material related to the theoretical basis for the subject proposed by the teacher. Each group or student is responsible for reading a text introduced by the teacher and, after discussing it using the chat tool, must prepare a commentary on it using the chat (synchronous) and/or forum (asynchronous) tools to write it. Once the comments on the text that was read have been prepared, it must be published and can be the subject of new comments by other teams. During this collective process of critical analysis of the contents of the text, the students are practicing the management of heterogeneity.

2. In the second exercise, Inserting Links, the students must carry out the bibliographical research, or research into the "state of the art" of important topics related to the subject matter. Any additional information obtained must be made available to all the students so that they have a range of views on the same subject. By delegating the act of looking for texts on the Internet and publishing them with comments, the student has to select course content and becomes responsible for his/her own learning. Torres (2007, p.14) emphasizes that material chosen by students for teachers and students is just as useful as that chosen by teachers. Teachers and students thus establish a true partnership that

Figure 1. Links

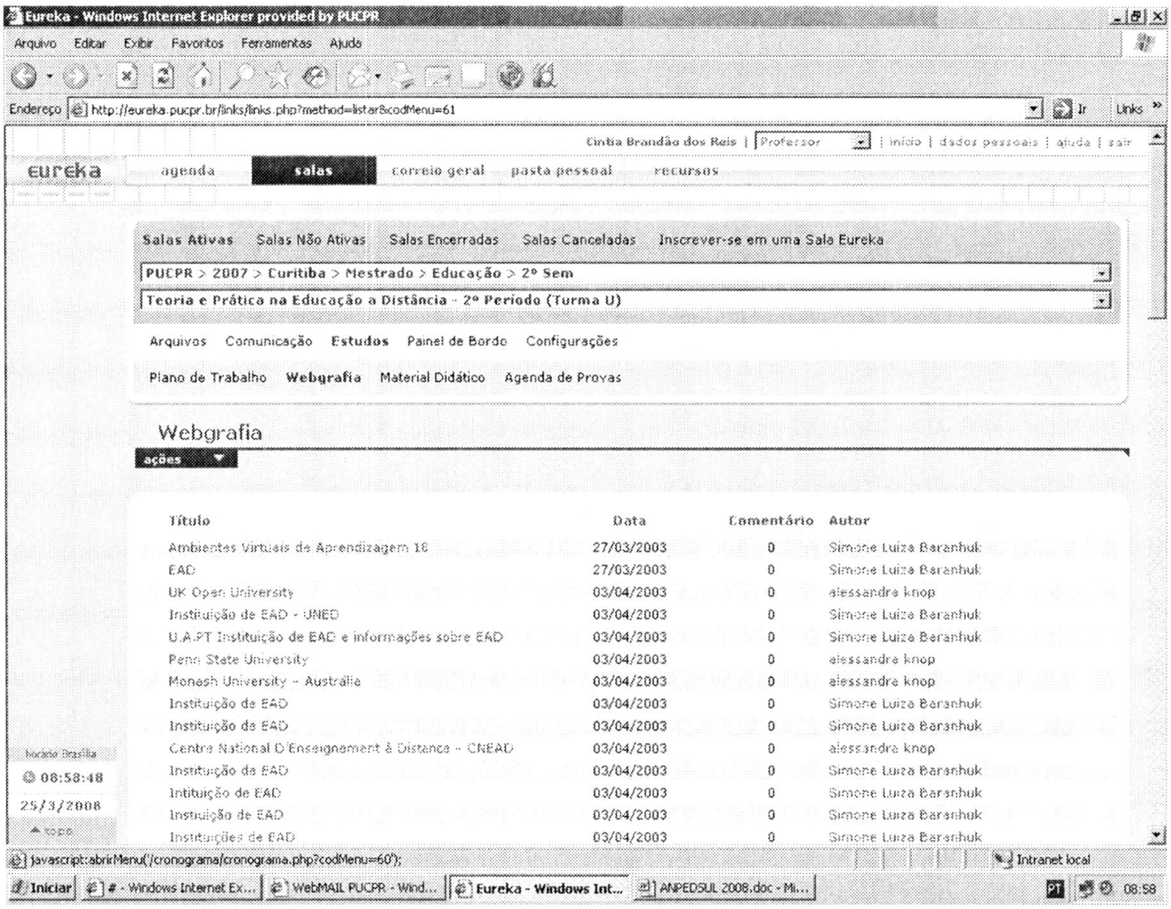

helps them to play a collaborative, active and primarily reflective role in the process of knowledge acquisition and production. The links to content matter chosen by students for students were published using the EUREKA tool known as Links along with comments made after a critical reading (Figure 1). When students accessed their fellow students' texts they were expected to publish new comments in the virtual environment.

3. The third exercise is that of Questioning existing knowledge and is fundamental to LOLA. It aims to develop the skills of questioning knowledge. Working at times individually (QuestInd) and at times in small groups (QuestGru), the objective is for the students themselves to prepare questions on the text they have read. For Torres (2007, p. 15) the teacher's role is to provide guidance. He or she must encourage the students to go beyond formulating questions that are merely factual or conceptual and encourage them to try to develop more complex questions related to interpretation, comparison of uses, analyses, syntheses and evaluations. The activities of questioning existing knowledge and answering questions were carried out with the Forum tool in the VLE (Figure 2).

4. The fourth exercise is that of answering the questions that were drawn up. In this activity the students must answer questions prepared

Figure 2. Forum

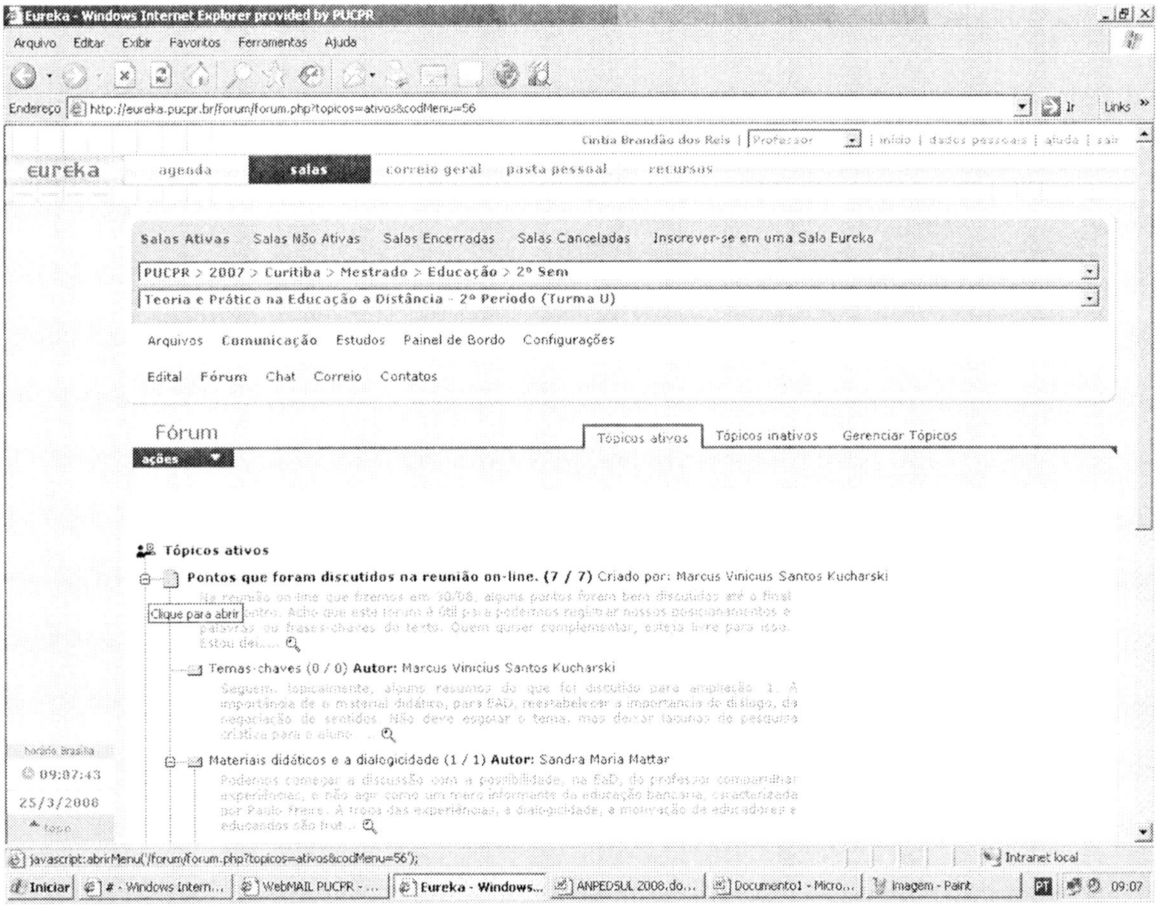

by their peers. By working individually (RespInd), the students establish their own criteria both for choosing the question and for the content and length of their answer. However, when the questions are answered in groups (RespGru), agreement about the possible choices must be negotiated and the social skills required when working together must be developed.

5. The fifth exercise is that of preparing concept maps using texts recommended by the teacher or by the students as the starting point. According to Molina, Ontoria, and Gómez (2006, p. 108) "A concept map is a schematic resource for presenting a set of conceptual meanings included in a structure of propositions. Concept maps provide a schematic summary of what has been learnt, arranged hierarchically". The preparation of the comments in the first two activities, Commented Reading and Inserting Links, prepares the students for the activity Preparing a Collective Text. After carrying out the first five activities, the students have to prepare a concept map that is a synthesis of the texts the team has worked on (synthesis here is understood to mean not merely a summary of someone else's ideas but rather the students' own individual and singular productions), thus making them work as authors in the production of knowledge, as "a map represents the knowledge of the person who created it at a particular time" (Marriott & Torres, 2006, p.

Figure 3. Concept maps produced by student Raquel

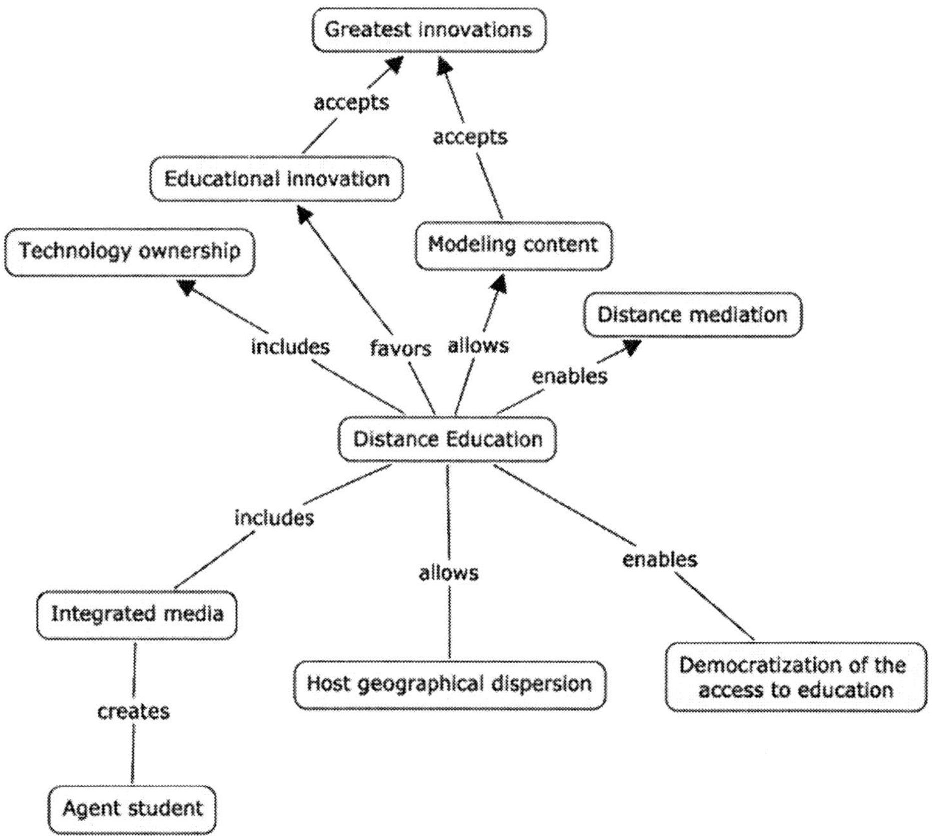

11). For illustrative purposes, two concept maps produced by the students are shown (Figure 3 and Figure 4).

6. The sixth exercise is that of producing new knowledge. The first stage of the exercise consists of the dialog between all the group members to choose the subject about which they will produce an article. During the second stage, the students start the actual production of the text. Each member writes their own contributions, which are forwarded to other members, who in turn complete them and refute or add ideas. It is in this interactive process of producing the text, when the dialog between all those involved allows them to reciprocally modify it, that knowledge is produced. The texts that were created during the activity involving production of new knowledge were also sent to the other participants by email and published in the portfolios in the Content tool (Figure 5). For Leite and Fernandes (2002, p. 60), a portfolio is a pedagogical procedure whose aim is to encourage active learning and promote the development of cognitive, social and affective competencies. According to the same authors, a portfolio makes students aware of their responsibility for their own or their peers' learning process.
7. The seventh exercise is that of evaluating the procedures used. The cycle of activities finishes with an evaluation by the students themselves of their performance. Armed with the portfolios, the students discuss

Figure 4. Concept maps produced by student Marcus

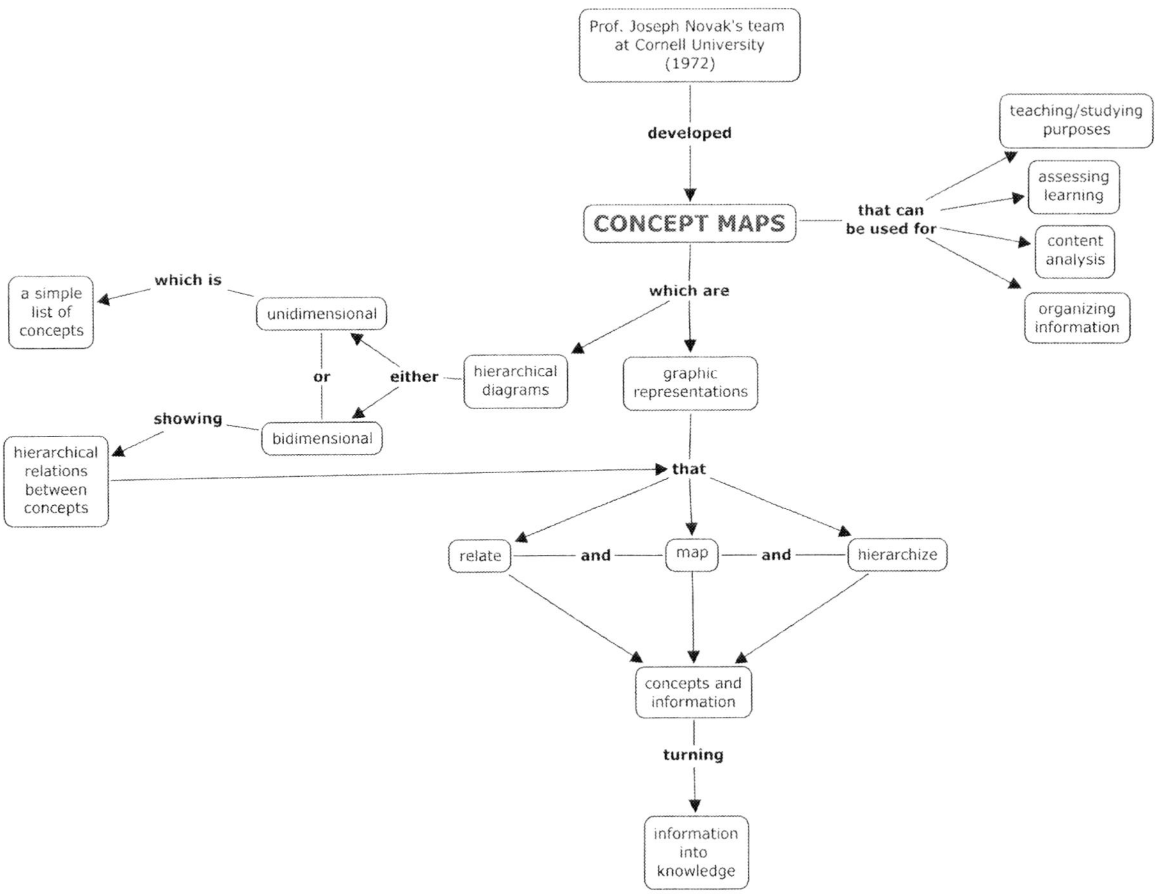

issues such as the questioning of existing knowledge, the answers that were drawn up, the connections to existing knowledge proposed, the production of the collective text and even the students' ability to evaluate these very activities.

It should be noted that all the LOLA activities, including those that were carried out in the classroom, were published in the EUREKA environment at the Pontifical Catholic University of Paraná, enabling a virtual portfolio with all the individual and group work by the students to be created. EUREKA is a web-based environment that allows collaborative activities to be developed.

In this environment communications for the whole group were sent using the info tool (Figure 6), and e-mail is used for communication between individuals.

SOME BRIEF CONSIDERATIONS ON THIS STUDY

Having given a brief bibliographic review of the subject of this study and a short description of the activities in the online learning laboratory, we now present a number of considerations about LOLA that are the fruit of one of many studies undertaken to improve this methodology for classroom-based and distance education.

Figure 5. Content

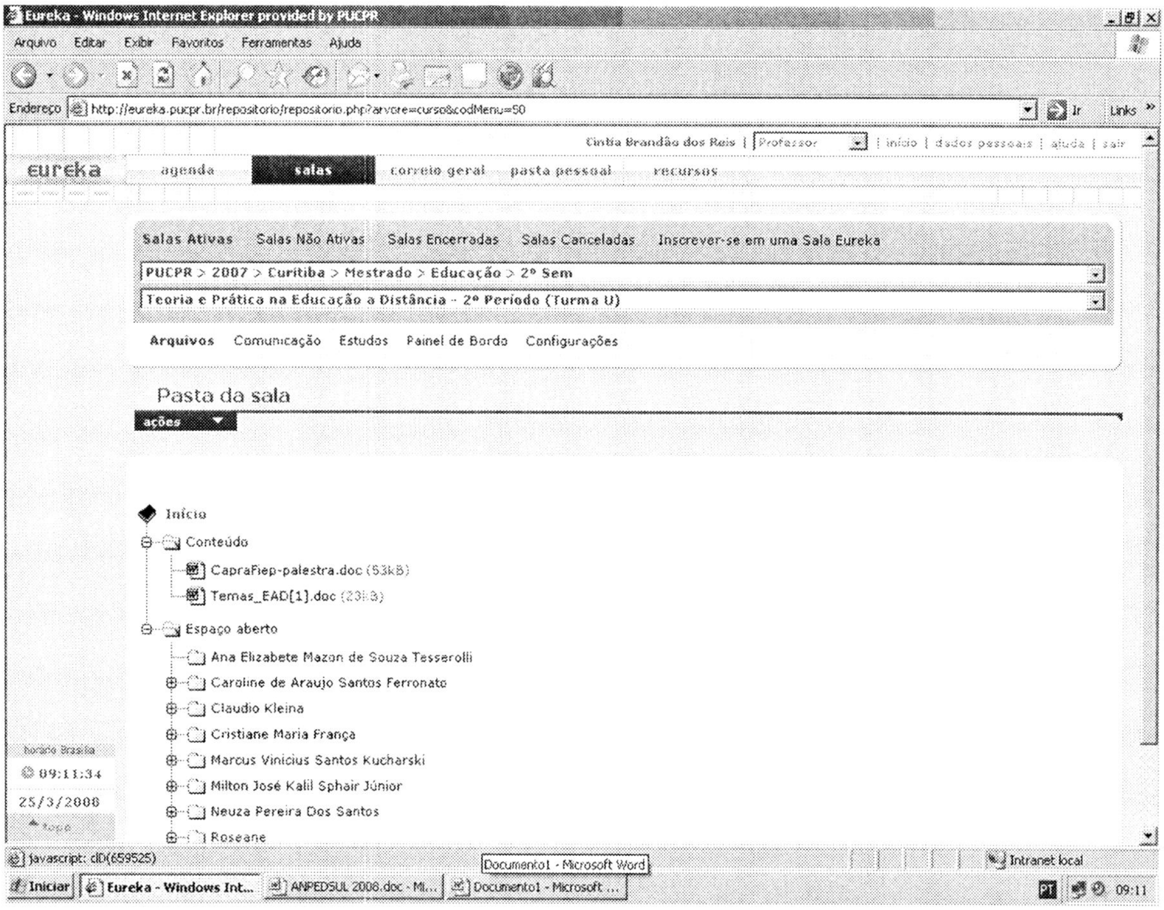

Figure 6. Notice Board

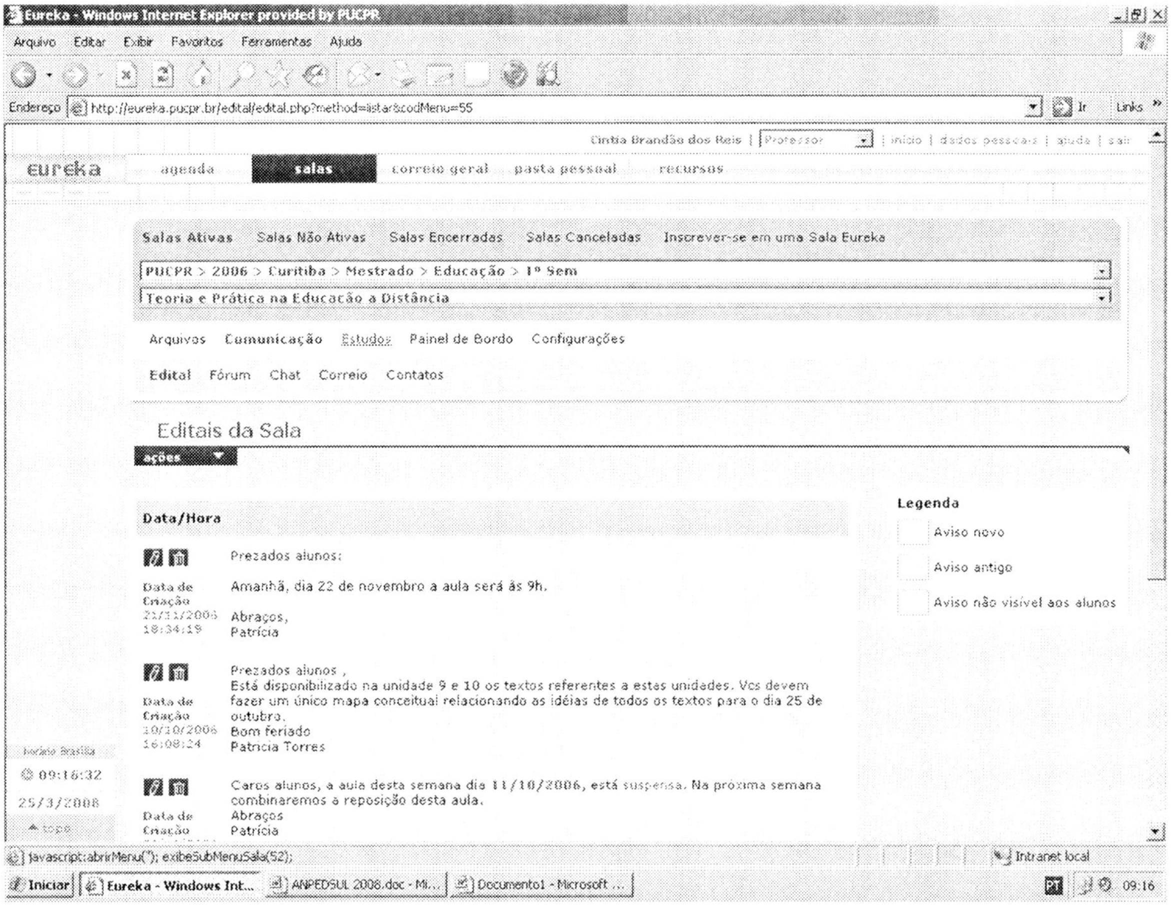

This study was carried out with students on the Masters and PhD courses in Education at the Pontifical Catholic University of Paraná. Most of these students are teachers who work in primary or secondary education or lecturers in higher education. Thus, the data that were collected relate to a homogeneous population made up of the students in the group under observation.

We used the methodology known as Descriptive Research, as according to Knupfer this combines elements of quantitative and qualitative research and in some cases also includes ethnographic aspects. In this method nothing is inferred, but everything that is happening is described accurately. Knupfer also states that this methodology has gained recognition and acceptance in studies in the area of educational technologies (in Gomes, 2003, p.258).

It was also defined that this research would make use of a type of descriptive study known as a case study, as according to Lüdke & André (1986, p.17) this helps focus the investigation on "what is unique about it or particular to it, even if it subsequently becomes clear that there are certain similarities with other cases or situations".

Gil (2002, p.54) points out that the results of a case study are very often "presented in open form, i.e., as hypotheses rather than conclusions", allowing the process to be more flexible and any corrections that are needed for new investigations to be made.

The data were collected by means of a questionnaire distributed to all the students, who responded voluntarily. The qualitative analysis was based on the students' experience as teachers and relevant data and observations to complement and provide greater depth to the interpretation of the data from the questionnaires.

Based on the data gathered at various stages during the study, the preliminary conclusions described below can be drawn.

FUTURE TRENDS

The experience gained from the use of concept maps in the online learning laboratory is highly significant and well suited to the educational demands of a new technology-based, globalized society. It undoubtedly represents another step forward in the development of educational approaches based on innovative teaching-learning concepts supported by the use of new information and communication technologies. The author therefore suggests that this research be continued to improve this model so that it may be used in other levels and forms of education.

SOME PRELIMINARY CONCLUSIONS

The use of virtual learning environments and the Internet as tools in the teaching-learning process is a recent experience that is still in the process of being discovered. In consequence, suitable methodologies for these new educational environments need to be developed. Recent studies indicate that many researchers recognize that interaction, whether it be in the form of collaborative or cooperative methodologies, or a combination of both, is an essential element. It is through interaction that one can combat a sense of isolation, maximize learning and develop aspects such as responsibility and independence. It is also through interaction that students can be prepared for better educational and social performance and for the work of researching, analyzing, negotiating and writing and the very production of knowledge.

In LOLA the seven activities proposed promoted student involvement in the collaborative methodology used in the subjects. This involvement meant that in addition to producing knowledge, the students sought to face their discomfort in challenging interactive situations and develop personal and social relationship skills. It was also observed that the fact that students had to behave as researchers, the need for them to change roles, the changing from individual work to work in small groups, the constant preparation of questions and comments, and the self-assessment in relation to the whole process meant, according to the students, that they matured in terms of their teaching-learning process. In addition, and still from the perspective of the students, the work in LOLA led to most of them adopting a more proactive, responsible and independent attitude.

With regard to the addition of the activity involving the building of concept maps, most of the students felt that the use of this technique helped to understand the texts chosen by the teacher and peers, to organize ideas and thoughts, to structure reasoning, to produce new knowledge, to reflect on subjects, to systematize concepts and to organize the study.

It is worth noting that new procedures open doors to a new era of methodologies and challenges for the teaching-learning process supported by the new technologies. In an age when information is at everybody's finger tips, it is important to know how to look for, evaluate, discuss and produce this information.

REFERENCES

Cord, B. (2001). *Internet et pédagogie: état des lieux*. Retrieved from http://wwwadm.admp6.jussieu.fr/fp/uaginternetetp/definition_travail_collaboratif.htm

Freitas, L. V., & Freitas, C. V. (2002). *Aprendizagem Colaborativa*. Porto, Portugal: Edições Asa.

Gil, A. C. (2002). *Como elaborar projetos de pesquisa* (4th ed.). São Paulo Brazil: Atlas.

Gomes, P. (2003a). Internet e universidade: o EUREKA da PUCPR, 35.000 usuários depois. In C. Maia (Ed.), *Experiências em educação a distância no Brasil: Reflexões atuais, em tempo real*. São Paulo, Brazil: Anhembi Morumbi.

Gomes, P. (2003b). O MATICE da PUCPR: Uso da Internet no Ensino de arquitetura. In *CREAD: Congresso de Educação a Distância* Mercosul 2003, 7., 2003, Florianópolis. Aprendizagem e Trabalho: Políticas e Tecnologias, Anais, Florianópolis.

Henri, F. (2002). Formation à distance et téléconférence assistée par ordinateur: Interativité, quasi-interativité, ou monologue? *CADE: Journal of Distance Education*, *7*(1), 1–4.

Leite, C., & Fernandes, P. (2002). *Avaliação das aprendizagens dos alunos*. Porto, Portugal: Asa.

Lüdke, M., & André, M. E. D. (1986). *A. Pesquisa em educação: abordagens qualitativas*. São Paulo, Brazil: EPU.

Marriott, R. C. V., & Torres, P. L. (2006). *Tecnologias educacionais e educação ambiental: uso de mapas conceituais no ensino e na aprendizagem*. Curitiba: FAEP.

Matthews, R. S., Cooper, J. L., & Hawkes, P. (2004). Building bridges between cooperative and collaborative learning. *Change*, *27*, 35–40.

Molina, A., Ontoria, A., & Gómez, J. P. R. (2006). *Potencializar a capacidade de aprender e pensar*. São Paulo, Brazil: Madras.

Silva, M. (Ed.). (2003). *Educação on-line*. São Paulo, Brazil: Loyola.

Silva, M., & Santos, E. (2007). A Pedagogia da Transmissão e a Sala de Aula Interativa. In P. L. Torres (Ed.), *Algumas Vias para Entretecer o Pensar e o Agir*. Curitiba, Brazil: SENAR-PR.

Siqueira, E. (2005). *2015: como viveremos*. São Paulo, Brazil: Saraiva.

Torres, P. L. (2007a). A Trama do Conhecimento. In P. L. Torres (Ed.), *Alguns Fios para Entretecer o Pensar e o Agir*. Curitiba, Brazil: SENAR-PR.

Torres, P. L. (2007b). Sugerindo Caminhos para Explorar a Trama do Conhecimento no Programa Agrinho. In P. L. Torres (Ed.), *Alguns Fios para Entretecer o Pensar e o Agir*. Curitiba, Brazil: SENAR-PR.

KEY TERMS AND DEFINITIONS

Collective Knowledge Construction: Knowledge constructed collectively by students engaged in collaborative or cooperative activities.

Collaborative Learning: Collaborative learning is a student-centered methodology for learning. By working in groups, students' knowledge is constructed by their participation and interaction. This is achieved by becoming involved in activities aimed toward a common goal. It solidifies socialization not only "by" learning but also "in" learning.

Eureka: PUCPR's collaborative virtual learning environment developed since 1995 in partnership with Siemens. In 1999 it became fully functional and started being used by the university as a support for its online education. It now has more than 30,000 students in its nearly 8,000 classes.

Interaction: The process that emerges from the participation of all learners that interact among themselves by means of information, points of view, queries, and ideas that occur in a learning environment.

Interactivity: Process that emerges from the participation of all learners that interact among themselves by an active dialogue, a constant exchange of information, points of view, queries, and ideas that occur in a learning environment.

LOLA: Online Learning Laboratory, LOLA, is a collaborative learning methodology for classroom-based and distance learning that was defended in a PhD thesis and used in several subsequent research studies.

Chapter 2
Collaborative Learning and Concept Mapping for Language Teaching

Rita de Cássia Veiga Marriott
University of Birmingham, UK

ABSTRACT

This chapter outlines how collaborative learning and concept mapping have been incorporated and implemented within a blended foreign language course. Focusing on these two approaches, it introduces the reader to LAPLI – The Language Learning Lab: a methodology of integrative CALL using the Internet. The aim in LAPLI's 12 activities is to challenge high-intermediate and advanced language students to go beyond their limitations and be more active and responsible for their own learning. Students, based on authentic material selected by themselves, work individually and collaboratively throughout its activities. They are stimulated to develop fluency and accuracy in the foreign language, focusing on the development of their reading and writing skills, but also promoting their oral and social skills. Some feedback from the students is presented. The chapter concludes with a few considerations on the challenges of life-long education.

INTRODUCTION

The use of collaborative learning and concept mapping activities in a language learning methodology is not what most language teachers would expect to come across. So why propose a methodology involving these approaches?

By and large, in foreign language courses for under-graduate students of all levels, teachers follow an adopted course book. This course book provides a pre-planned path, with pre-planned presentation material and exercises, aiming at achieving carefully selected and graded linguistic outcomes. The lessons are usually structured in the same way: the teacher has to cover parts "1", "2" and "3", on page "X", unit "Y" in classes "A" to "Z" and the students are asked to read, write, listen, watch, repeat and do the written and oral exercises in parts "1", "2" and "3", on page "X", unit "Y" in classes "A" to "Z". The teacher "teaches" and the students "learn". The

DOI: 10.4018/978-1-59904-992-2.ch002

teacher, at the front of the class, asks students to open their books to page "K", presents the new material and asks students to practise. Students, in turn, open their books to page "K" and either work individually, in pairs (usually with their best friend or a colleague who invariably sits next to him/her in every class) or in groups (with usually the same mates). So, why offer alternatives to this peaceful learning environment?

By following the adopted book, probably written for an international market, teachers do what the author of the book suggests, not necessarily covering the subjects, grammar and social skills their group needs.

The teaching of languages to higher-intermediate and advanced level students can make use of a distinctive methodology. Most of these students are, or will be, language teachers and need to be motivated to continue learning the foreign language, practice the acquired knowledge and develop teaching and researching skills. They need to be challenged to practice their language skills and to expand their vocabulary. Besides, as it is the case with pre-service students, they need to be prepared for the job market to fulfil their professional and personal ambitions. Therefore they need to practise the target language in subjects that are interesting and relevant to them, in which they feel encouraged to communicate and contribute with (new) ideas, making decisions, accepting other people's opinion, supporting and refusing contributions.

For this to happen, it is necessary to change the current paradigm, centred on the teacher, to one which is centred on the student and which promotes responsibility, critical analysis and autonomy. As Behrens (2000) says, it is important to:

[a.] gradually reduce the number of theoretical lessons, increasing the time available to do research, to access databases, to give support in the construction of activities and the students' own texts; [b.] encourage students' development in both well planned individual and collective group work activities with defined responsibilities; [c.] organise differentiated activities, events that require creativity, challenging projects that provoke cross-referencing, dialogue with authors and own production; and [d.] promote the use of electronic devices, of IT, of multimedia and telecommunications with all the available resources of the school campus. (p.121-2).

Teachers in general and language teachers in particular can maximise the advantages of collaborative learning. In collaborative learning, learners solve problems together and "learning is expected to happen as a side-effect of problem solving, measured by the elicitation of new knowledge or by the improvement of problem solving performance." (Dillenbourg, 1999, p. 4). However, to maximise the possibilities of collaborative learning to occur, Dillenbourg (1999) posits that some types of interaction need to happen which he classifies in four categories: a) to set up initial conditions (for instance, how many students will there be in each group, who will chose group members, the teacher or the students themselves; should the members of the group share the same point of view or not); b) to over-specify the 'collaboration' contract with a *scenario* based on *roles* (e.g., asking students to play different roles or giving members access to different data); c) to scaffold productive interactions by encompassing interaction rules in the medium (such as asking all members of the group to express their opinion at each stage); and d) to monitor and regulate the interactions (by the teacher assuming the role of a 'facilitator' instead of a 'tutor'. The teacher does not provide the right answer or reveals who is correct – s/he only directs the group work by providing 'hints' to the students, interfering with their work as little as possible (p. 6). For Kirschner, Strijbos & Kreijns (2005, p. 31), teachers should make students aware of their responsibilities when working in groups. Each group member needs to be accountable for his/her own work ('individual accountability' in Slavin, 1980), for the group is, on the one hand, dependable on the effort put in by each individual and, on the other hand, responsible for each member's achievement ('positive interdependence' in Johnson, 1981). This interdependence provides the context for promotive interaction ('promotive interaction' in Johnson & Johnson, 1996) to occur, which is the group's support and encouragement to each of its members in order to achieve the group's goal. The development of collaborative learning (CL) activities can assist students in the achievement of the desirable aforementioned skills.

Complementary to this approach, concept mapping (CMing) can also be very enriching for the students. Concept mapping is widely known as a tool to facilitate meaningful learning. It is a powerful learning technique which helps students reflect on their learning process, on the structure of knowledge and on its production; in other words, on meta-knowledge (Novak and Gowin, 1999). Whether they include pictures or not, concepts are written in boxes which are organized in a hierarchical form – from top to bottom or from the centre to the sides – and are linked to each other by lines which include linkage words (such as prepositions, verbs, linking words) that express the relationship between these concepts. CMing stimulates the development of both sides of the brain, the analytical and the creative; it encourages judgement, reflexion, revision, critical thinking and knowledge construction.

In this chapter, the benefits of collaborative learning and concept mapping to language learning are outlined. On this basis, LAPLI – The Language Learning Lab (*Laboratório de Aprendizagem de Linguas* in Portuguese)1 is presented, a blended methodology for language learning in which these two approaches play an important part. The chapter is concluded by considering the role of students in the learning process and the demands of life-long education.

COLLABORATIVE LEARNING AND FOREIGN LANGUAGE TEACHING

Collaborative learning (CL) represents a significant change in the typical classroom where the teacher is at the centre of the process. In CL, teachers do not regard themselves as the experts. Both teachers and students interact and combine their intellectual efforts to attempt to understand, explore and solve issues, generate ideas and create a product together. Students are involved in activities designed to explore course material instead of having it presented to them by the teacher. Gokhale (1995) points out the potential that CL has in the development of critical thinking "through discussion, clarification of ideas, and evaluation of others' ideas." However, Gokhale also remarks that for this to happen, "…the instructor must view teaching as a process of developing and enhancing students' ability to learn" (1995).

CL in distance learning methodologies can reduce the feeling of loneliness. When a group of students starts collaborating and interacting they feel a sense of belonging which encourages learning. In a study conducted by Laister & Kober (2005), the authors observe that "models of collaborative learning are very effective as means of learning" both subject-matter and soft skills that are seen as necessary for long-term successful learning, such as cognitive skills and self-esteem. They also state that compared to individual learning scenarios, collaborative learning takes students to a higher achievement level, increases their problem-solving abilities, "offers cognitive advantages to learners and also has a positive influence in enhancing the personality traits that are beneficial for future learning or future autonomous or co-operative learning and working" (2005).

The benefits of CL, therefore, not only target the learning of content being studied at present but also contribute to the development of students as citizens, polishing up their personal skills which can in turn bring them social and professional benefits.

CL activities in the foreign language (FL) classroom make students more active and responsible participants in the learning process. By engaging in activities not primarily focused on the development of language skills, students use the FL as a tool to verbalise opinions, clarify ideas, ask questions, defend points of view, all in the target language. CL activities encourage students to use the language effectively in meaningful contexts. Teachers, as facilitators - and not presenters - ask questions to help students think of the steps needed to solve the problem or create a product, as well as "hint, prompt, help to structure activity [and] mediate" (Pilkington, 2007).

A significant section of CL is collaborative writing (CW). CW is seen by Lowry, Curtis & Lowry (2004) to be of great importance in the foreseeable future as "increasing globalization magnifies the need for collaborative work, and the Internet magnifies the ability to collaborate" (p. 67). In spite the fact that Lowry et al (2004) view CW as more effective than single-author writing, they point out some of the challenges faced when working collaboratively on a piece of writing, of which I highlight: writing processes generate strong emotions; groups can revise their papers indefinitely; it is challenging for all to converge toward a common goal or even use a common language; and that success is difficult to predict and guarantee. (p. 70). For them, although CW is a dynamic and non-linear process, it still follows a linear progress with a starting and an ending point with uncertainties in the middle. To lessen unpredictability, exchanges of outlines and drafts are suggested. Their dynamic approach to CW entails six activities, namely brainstorming, outlining, drafting, reviewing, editing and revising (p. 83), with writers assuming roles such as writer, consultant, editor, reviewer, scribe and facilitator.

The exchange of roles and negotiation that takes place in some collaborative approaches led some authors to refer to students as *actors* in the process (Dillenbourg, 2000; Torres, 2002; Marriott, 2004). Faced with challenges, debates, questionings, different points of view and work styles, the *actors* experience diversity and responsibility for shared authorship and value freedom when work alternates between individual and collaborative activities.

CONCEPT MAPPING FOR LANGUAGE LEARNING / TEACHING

Although concept maps have been primarily used in the teaching and learning of sciences and mathematics, research has been developed to investigate their benefits in nearly all domains of studies, among all age groups all over the world.

In the field of languages, Dias (1996) reports of a study conducted with false beginners English language learners in the Technical School at UFMG in Belo Horizonte. In this study, she found that the construction of concept maps contributes positively to process, retain and recover the information read in the text, fostering a more responsible and autonomous posture in students (Dias, 1996). Bahr & Dansereau (2004) have been using bilingual knowledge maps (BIK-maps) for the acquisition of foreign language vocabulary as a way of going beyond rote memorization. By writing the concepts in the boxes both in the mother tongue and in the target language (hence the name bilingual knowledge maps) they have found that "BIK-map students recall more vocabulary items than do list learners" (p. 61) and suggest that they can be used as a study strategy as well.

The implementation of concept mapping into the foreign language classroom can promote a significant change in the teaching methodology and in the students' and teacher's level of participation in the learning process. As noted by Telebinezhad (2007) "Students maximize their learning by using concept mapping in their essay writing; hence they feel more independent and feel more responsibility for their own learning." Telebinezhad describes some benefits in his experiment with a group of

English Language Proficiency students, stating that "concept mapping [...] helped students attend to writing tasks, and control their learning more effectively. It helped students facilitate their learning by organizing key concepts into visual representation. They simply represented visually their understanding of ideas and their relationships. This created a much more tangible evidence of the quality of both the learning process and concept understanding." (2007)

CMing can help break away from teacher-centered approaches and the "cut-and-paste" philosophy (i.e., when students simply copy and paste bits of information into their document, sometimes not even reading it properly or failing to acknowledge their source). By encouraging students to engage in CMing activities, they are provided with an opportunity to become more active and responsible for their own learning. They are invited to practise analyzing information on hierarchical levels of detail and, from a linguistic point of view, to think about how concepts can be linked together to convey one's own ideas. This exercise of linking the concepts together make mappers think harder on the relationship of those concepts and thus illustrates better his/her knowledge at that particular time. Due to the linguistic and cognitive affordances of concept mapping, I have chosen it as a mapping tool for language learning above all other types of mapping techniques.

By incorporating concept mapping and collaborative learning activities to LAPLI, our aim is to challenge students to develop their social, collaborative, creative, and oral and written linguistic skills while they grow together to become more autonomous learners.

WHAT IS LAPLI AND WHO IS IT FOR?

LAPLI is a methodology for practising and developing language skills, based on educational theories, principles and practices mentioned in the next section. It makes use of Computer Mediated Communication (CMC), a collaborative virtual learning environment (VLE) and the Internet. It is a language development course designed for students of an intermediate level and above (pre-service students/teachers taking Language degree courses, language courses in private/state schools, teacher training/refreshment courses) who not only want to further their studies in the target language, developing fluency and accuracy in reading and writing skills, practising oral skills and increasing vocabulary and knowledge in areas of interest, but also want to develop teaching and researching skills using the resources of the Internet and those of a VLE.

This methodology has been developed based on the activities proposed by LOLA – The Online Learning Lab, a methodology created for distance learning by Torres, in 2002. LOLA was in the spotlight in 2003 for being awarded 1st place by the Brazilian Association for Distance Learning (ABED) and by EMBRATEL, a Brazilian telephone communications company, in the category "Research" (Torres & Marriott, 2004). It is made up of six activities based on Bochniak's (1998) "Pedagogy of Research" which comprises three exercises for face-to-face teaching. One of the main objectives of these three methodologies is to overcome the learning/teaching model of reproduction and accumulation of knowledge by developing students' responsibility and autonomy. All twelve activities that form LAPLI are explained later in this chapter.

LAPLI is a carefully planned course whose framework is provided by the teacher, but whose material is produced by the students themselves. When developing the activities either face-to-face or via e-learning, students construct knowledge by constantly exchanging information, points of view, questioning, solutions to problems and evaluations.

The collaboration amongst the individuals and group members allows for a coherent and unique production of the class, as all work is shared and published in the VLE. This collaboration is not meant to create standardisation, but it respects the students as individuals, who, in their heterogeneity, produce and grow together. In the management of this heterogeneity, students handle conflicts, propose alternatives, review concepts and ideas, discuss positions, share authorship, negotiate and are responsible for their own actions.

LAPLI's activities give meaning to the action of the group whilst at the same time energise it. They challenge the students to become more autonomous and each of them to become a researcher. Moreover, they are planned to have a spiral movement, a succession of balance and counterbalance, being open to what is new, reasoning between old and new meanings. Their objective is to promote creativity, questioning, paradoxes and discussions, all to enhance the production of knowledge. To handle these activities, members organise themselves, interact, discuss ideas and positions, share roles and define subtasks. The educational and linguistic theories that support LAPLI's activities are presented below.

METHODOLOGIES INVOLVED IN LAPLI

LAPLI promotes a shift from rote-mode to meaningful-mode learning through task-based learning activities in which students use language for a communicative purpose, i.e. they do things and exchange meaning in order to achieve an outcome (Willis, 1996). They develop their reading and writing skills by being exposed to a rich authentic linguistic input in a low anxiety situation. When selecting, reading the texts and engaging in LAPLI's activities, students activate their previous knowledge and background information (Brown's Schema Theory) and practise skimming, scanning, intensive and extensive reading (Brown, 1994), develop cognitive and meta-cognitive strategies, prediction and inference (Nicholls, 2001; Bastos, 1998), as well as get involved in process writing (Brown, 1994) with strategies like – brainstorming, planning, outlining, writing, re-writing, editing, peer-editing, proof-reading and feedback.

LAPLI's activities stimulate cognitive development and the development of both language acquisition and fluency (in Part 1) and language accuracy (in Part 2). They provide an opportunity for reflection (individual work) and collaborative learning (group work), encouraging students to learn to do, learn to know, learn to live together and learn to be (Delors, 1998). By working collaboratively to accomplish tasks, learners gain more knowledge through the help of colleagues with different sets of skills (Vygotsky's social interaction and ZPD - Zone of Proximal Development, as cited in Moreira, 1999). They get involved in natural learning and discovery-based learning when offered opportunities to question, to explore the resources of the Internet and to practise their initiative. Meaningful learning and constructivism are stimulated in activities such as concept mapping, working with texts selected by themselves and in their area of interest, and the construction of lexical lists and group articles.

LAPLI falls into Warschauer's third phase of CALL named as Integrative CALL and the Internet, which uses the computer as a tool and enables the student to use and understand the target language, making use of word processors, spelling and grammar checkers, desk-top presentation programs and concordancers, using computer mediated communication (CMC) (Warschauer, 1986). Being connected to the Internet allows students to access authentic material and to publish their own material.

LAPLI'S ACTIVITIES

The Languages Learning Lab – LAPLI - is made up of 12 activities which are developed in the following sections of a collaborative virtual learning environment (VLE): Message Board

Figure 1. LAPLI's activities and their origin

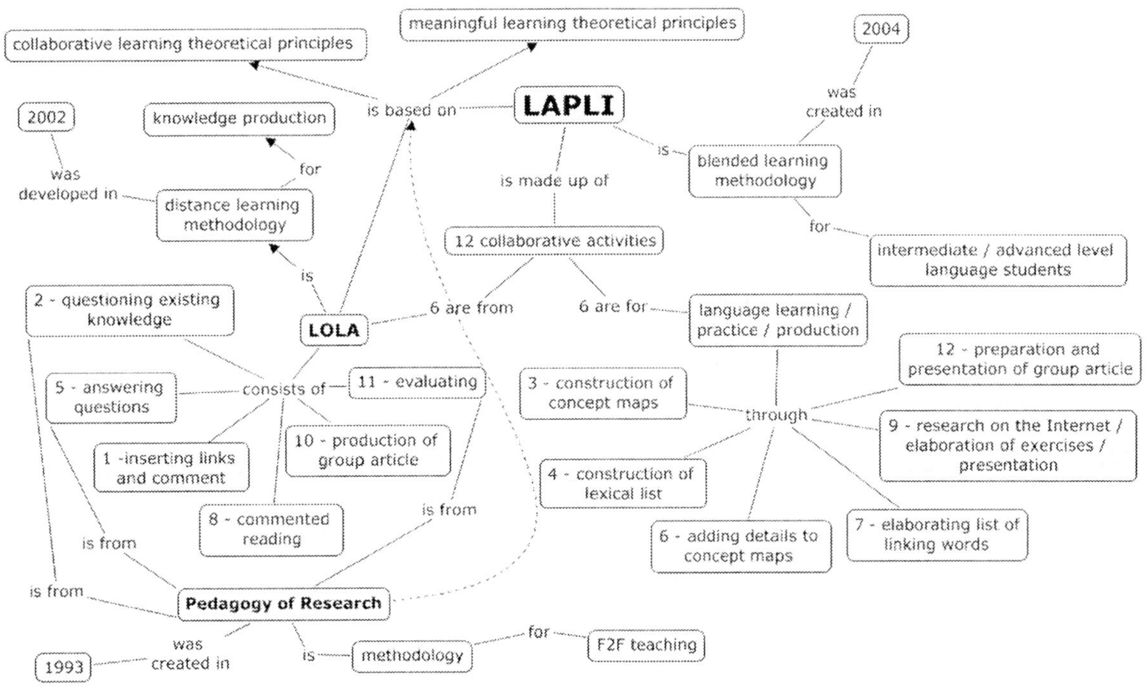

(where instructions about the class and important information about the course are published by the teacher); Info (information about course participants); Chat (for synchronous communication); Mail (for asynchronous communication, allowing messages to be forwarded or replied to and for attachments to be included); Content (where both students and teacher can upload classroom work and reference material); Forum (allowing students several levels of asynchronous reply); and Links (where participants contribute with their favourite Links / webliography for the course). In the concept map (Figure 1), LAPLI's 12 activities and their origin are illustrated.

Activities 1 to 8 make up Part 1, when students develop their fluency and communicative skills – Message Oriented – whereas activities 9, 10 and 12 form Part 2, focusing on accuracy – Language Oriented. Activity 11, if possible, is carried out in the students' mother tongue.

The activities are either developed individually, in the first cycle, or collaboratively, in the second cycle. The balance between them allows students to experience both situations, as they are completely different and equally enriching. Table 1 shows how the activities are split between individual and group work.

When students work individually, in cycle 1, they reflect to make their choices and do not have to express themselves verbally to justify or defend a point of view. They simply follow their own criteria. On the other hand, in cycle 2 (when they form new work groups based on similar interests that arise from those specific activities), they need to work collaboratively, sharing their choices and negotiating with the other members, thus exercising their fluency (oral or written), linguistic, social and communicative skills. It is important that students are made aware at the beginning of the course that they will be working in different groups in cycle 2, and that it is important they do this for the development of their communicative, social, negotiation and management skills.

Table 1.

Activity	First Cycle	Second Cycle
1. Link + C	Individual	Group 1
2. Q	Individual	Group 2
3. CM	Individual	Group 2
4. Lex List	Individual	Group 2
5. A	Individual	Group 1 or Group 3
6. CM+	Individual	Group 1 or Group 3
7. LW	Individual	Group 1 or Group 3
8. CR	Individual	Group 4
9. GT	Individual	Group 5
10a. Ind_R 10b. Ind_RGR 10c. Group_Art	10a. Individual 10b. Individual 10c. Group 1	10a. Individual 10b. Individual 10c. Group 1
11. Evaluation	Individual: in the Forum Class—Face-to-face or videoconference	Individual: in the Forum Class—Face-to-face or videoconference
12. Presentation	----------	Group 1

Table 1 Legend:
Individual: Individual work
Group 1: Group based on common theme formed in cycle 1 activity 10
Class: Everybody together
Groups 2, 3, 4, & 5: Newly formed work groups based on common themes of interest

LAPLI uses the resources of a collaborative VLE and can be delivered fully distance learning, face-to-face or in a hybrid mode. My recommendation is that it be implemented in a hybrid mode (50% distance learning and 50% F2F) as it maximises students learning allowing for F2F contact and the development of real life social skills and relationships, and contributes to fostering autonomy, self-trust, responsibility and collaboration. It encompasses the best of both worlds (Murphy, 2002; Young, 2002; Matthews, 2002; Spannier, 2003) and enhances "time management, critical thinking skills, problem solving skills, computer skills – increasing opportunities for academic and professional success – promote[ing] self-directed learning" (*Teaching-Learning Centre*, n.d.)

The twelve activities in LAPLI are introduced below, with special focus being given to the concept mapping and group work tasks.

Activity 1: Inserting Links and Comment (Link + C)

As researchers, students search on the Internet for a text of their interest (that will be the basis for the broadening of their lexical, syntactical, semantical, textual and specific/general knowledge) and publish it in the VLE with a comment on why they have chosen this text and what it has to offer. The purpose of this activity is to: 1. motivate the comprehension of authentic texts in the target language and to stimulate the broadening of their personal vocabulary; 2. develop their critical sense when we give them the responsibility for the selection of topics and texts; 3. offer opportunity to the students to establish connections between the learning of the foreign language and their field or subject of interest; 4. encourage students to practise skimming a text (reading quickly for main ideas) in the process for the selection of their link / text to publish in the VLE; and 5. get

Figure 2. Inserting link and comment

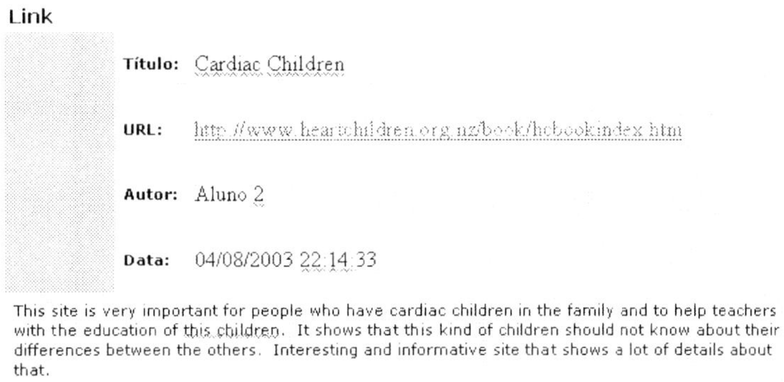

students to practise scanning (reading quickly for specific information) in order to write their comments. In their comments, students produce new knowledge while practising their communicative competence when 1. they use their own words; 2. they motivate their colleagues' visit ("I hope you like it"); and 3. exercise their critical sense when they demonstrate their opinion with phrases like *"Interesting and informative site"*, as can be seen in Figure 2.

The next three activities - Questioning, Construction of Concept Maps and Construction of Lexical Lists - form a mini-sequence of collaborative activities. Students choose a topic / text / link that has not been published by themselves (which I will refer to as 'secondary text') and start working individually on the activities. They go back to work on their main text (which I will refer to as 'primary text') in activities 5, 6, 7 and 10. In the second cycle, students need to form a different group (Group 2) to collaborate in this mini-sequence of activities.

Activity 2: Questioning (Q)

In this activity, students are again encouraged to play an active role, using the foreign language as a tool to practise interrogative structures motivated by the content / subject. They choose to work on a primary text of a fellow colleague (their secondary text) and formulate two or three questions on it. These can be either questions concerning "reference to form" evaluating students' understanding of the text, thus leading them to practise scanning, or "reference to function" encouraging students to think and express their ideas (Widdowson, 1991), leading them to practise top-down reading strategies (using context and existing knowledge to predict meaning). The latter type of questions are stimulated.

Traditionally, it is the teacher's responsibility to ask questions concerning the content of the subject being studied. The students' role, on the other hand, is to ask clarifying questions to the teacher and, as mentioned by Shor and Freire (1986):

Students are excluded from searching, from the activity, from the demand. Answers are given to them to memorize. Knowledge is passed to them like a corpse of information – a corpse of knowledge – and not as a live connection to their own reality. Hour after hour, year after year, knowledge is nothing but a monotonous task from an official syllabus. (p. 14-15)

In LAPLI, it is the students' responsibility to elaborate content questions to each other, making them an active and reflective participant in the process. As in LOLA, when students need to form new groups while developing the fluency activities, they have a chance to experience different roles in each group. At school, students usually "elect" their work group and do not want to work in another group, establishing interpersonal links amongst themselves which are accepted by the institute without any questioning or analysis of their implications. This permission impoverishes the process, for each member of the group "specialises" in a role: the spokesperson, the writer, the researcher, the one that prepares the presentation and so on. When the groups are constantly altered, a change of roles is automatically promoted as participants have unique personalities. This, in turn, challenges them to position themselves in more precise ways and develop different linguistic and social skills.

Activity 3: Construction of Concept Maps (CM)

The construction of Concept Maps is a powerful learning technique used to represent and organise information visually. To make a concept map in LAPLI, students have to read the secondary text carefully, employing skimming, scanning, top-down and bottom-up (working out meaning from individual words) reading techniques, identifying key words and supporting ideas, to create a summary of the main ideas of the text. This exercise promotes the understanding of micro and macro structures in the text, helping them in the identification of the framework of the text and how the ideas are related, knowledge which is useful when writing their own texts later on. By using key words in the reconstruction of the meaning of the text, students deepen their active vocabulary and develop their grammar. They also practise word transformation to construct the map, taking their proficiency in the foreign language one step higher.

Concept maps can be produced using pen and paper, a word processor with drawing facility or can be created with CmapTools (a piece of software available online with free download for pedagogical purposes). The use of CmapTools is recommended since it is user-friendly and it fa-

Figure 3. "Blood donating"

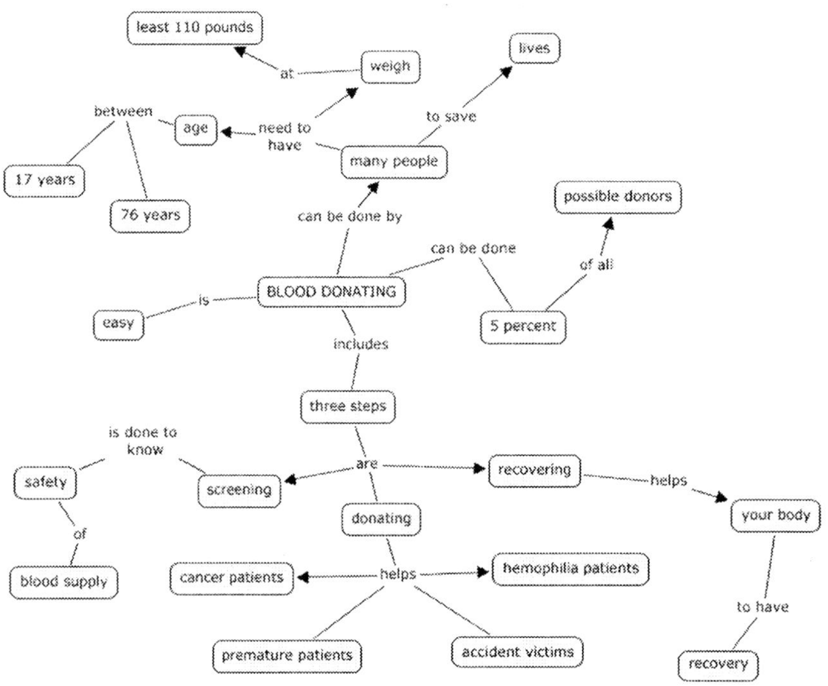

cilitates the dragging, deleting and/or re-writing of concepts. It allows the work to be done at a distance and promotes collaborative learning. Students can work with each other online, collaborating, interfering, modifying, doing research, inserting annotations, documents, links and pictures and setting up discussion threads. By engaging in the construction of concept maps, students think about both the content and the form. The organised visual representation of content also helps in the transfer of knowledge from the short-term memory to the long-term memory, anchoring new concepts to previously acquired ones.

The CM illustrated in Figure 3 was drawn individually in cycle 1 on one of the topics/text chosen by a fellow colleague. As can be seen on his map, it displays key ideas and a good understanding of how the concepts are related, but at this stage no cross links were identified.

To create a map, students need time, concentration and understanding of the concepts being exposed or discussed in the text. For those who are only used to "ticking" the correct answer or to answering unchallenging questions, this activity seems too difficult, complicated and demanding. The building of concept maps requires a change of attitude. Once again, students are encouraged to become more active and responsible for their learning.

Activity 4: Construction of Lexical Lists (Lex List)

At this point, students list the vocabulary related to the theme, which gives cohesion to the text, preferably in alphabetical order, providing the meaning of the word if they feel the need for it. This thematic list of words can receive further contributions by other students working on the newly selected texts related to the same theme in the subsequent cycles, and provides a good resource for writing their individual and collective articles. The relationship between these concepts can be visualised in the corresponding concept maps.

This is the last activity in the mini-sequence of collaborative activities. From now on, students go back to work on the main (primary) text, which was chosen by themselves, and get involved in individual (cycle 1) or collaborative group work (cycle 2), with the objective of constructing the group article.

Activity 5: Answering (A)

This activity, which is linked to the "Questioning" activity, consists of giving students freedom to choose two or three of the questions created by their fellow colleagues in Activity 2. These can be about their research theme or not, although the tendency is to choose to answer the questions formulated based on their primary text, i.e., the text they themselves had selected in Activity 1. In cycle 2, if they find questions based on their main text, they get together in their Group 1 group to answer them. However, if no questions have been asked about their topics, they answer questions on another subject(s) (which should be different from the one they have just worked on) and form another group, Group 3, with the students interested in answering those questions. The teacher is informed about the formation of this group.

To elaborate these answers, students practise fluency in the target language, using and adapting the structures and vocabulary of the main / source text in their answers in a meaningful and communicative way. Good questions motivate more research, and students have the freedom to search for new information.

The mere activity of choosing which question to answer from the many options published, and the discussion - oral or written - on the criteria for the selection, is already a research activity, as the arguments used to accept or reject a question often constitute knowledge being constructed collectively, and sometimes are in themselves the answer to the question. According to Torres (2002), students select the questions based on two criteria: interest or usefulness (if they are interested in the theme or find it useful) and ease or difficulty (an easy question to answer or one that represents a challenge). This individual or group choice on which questions to answer and how detailed an answer to give makes them once again responsible for their learning. These choices, which result from the criteria established by the learners, imply an acceptance of the consequences when the "Evaluation" activity is undertaken. This emphasises the importance of choosing questions that promote more growth in the construction of knowledge (and in turn fosters the use of the target language in a contextualized way).

Activity 6: Adding Details to Concept Maps (CM+)

Here, students work on the concept map constructed by their colleagues based on their primary texts, (i.e., their own text/article selected in Activity 1). Their role is to add more information / details to this map, thus creating CM+. To achieve this, students have to refer back to the text to check which information has already been included in the map and which information could / should also be appended to make it more complete.

The aim of this activity is to develop students reviewing, critical, analytical and concentration skills by engaging them in a deeper and more careful reading of the main text in order not only to compare it with the information already mapped but also to add more details to the map to make it more complete. This exercise aids in the assimilation of new vocabulary and grammatical structures, taking students' proficiency in the target language a stage further, both in terms of their linguistic skills and specific subject matter. When in the second cycle, just as was explained in the previous activity, students either continue working with Group 1 or with the newly formed Group 3. Figure 4 shows students' map on "Blood Donating" after both contributions in cycle 2.

The topic of "Blood Donating" was chosen and developed in both cycles and new concepts were included. The difference in type highlight the additions made in the previous two stages (activity 6 [CM+] cycle 1 and activity 3 [CM] cycle 2). In the latest additions made in activity 6 [CM+] in cycle 2 (in italics, bold and underlined), it is possible to observe that the major contributions were the insertion of four cross links and the change from *three steps* to *five steps,* apart from the addition of more concepts/linking phrases.

To produce this map, students faced some challenges, for example the negotiation of precisely which concepts / linking words to use and how the cross links could be established. To come to a consensus, they used and practised not only their language skills but also their argumentative and persuasive abilities when debating on what to include and on how to express their thoughts in the map.

Activity 7: Elaborating List of Linking Words (LW)

To elaborate a list of linking words, students look back to the main text in order to detect and list the linking words used by the author. These are connectors used to illustrate a point of view, an opposing idea, an explanation or a sequence of events, linking words that establish coherence (linking / sequencing of ideas) and cohesion (making sense) within phrases, sentences and paragraphs in a text. This exercise not only helps them develop their discourse competence but also makes them more aware of the importance of properly using these connectors when writing their own texts later on.

Moreover, the act of listing words of the same category helps the transfer of vocabulary from the short-term to the long-term memory, and this organized anchoring facilitates recovery in future situations. (Dias, 1996). For this activity, students work individually in cycle 1, and in cycle 2, they

Figure 4. "Blood Donating" at the end of cycle 2

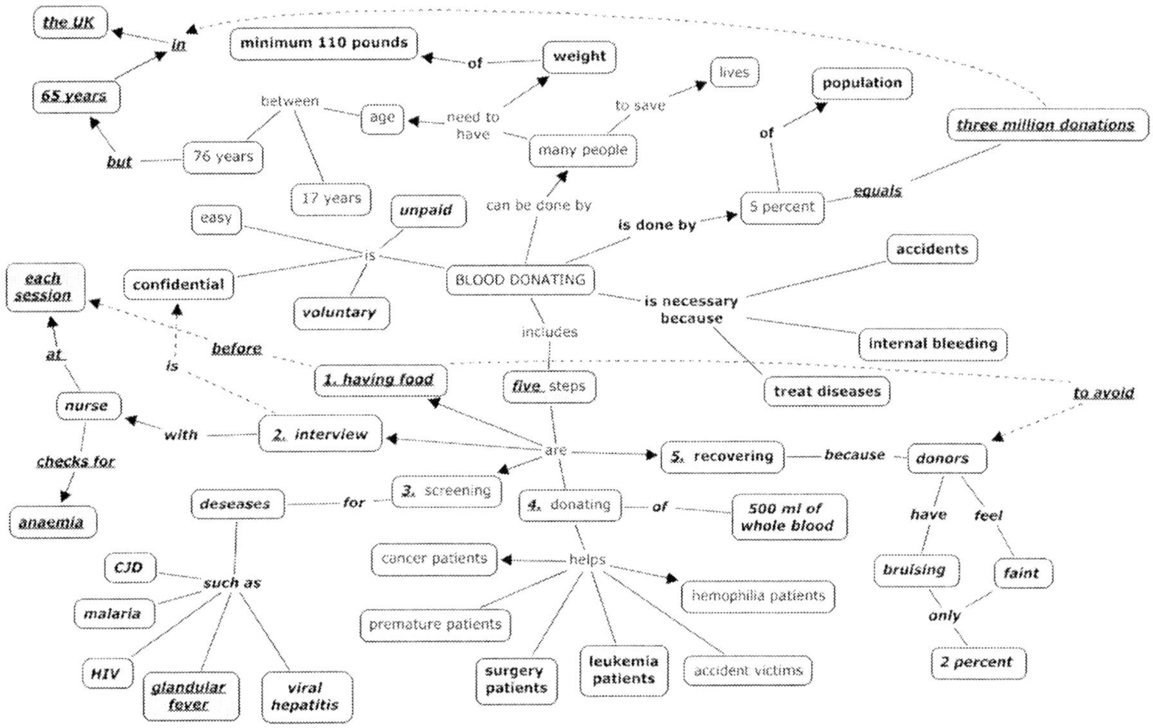

work with Group 1 or Group 3, as in the previous two activities.

Activity 8: Commented Reading (CR)

The goal of "Commented Reading" is to provide students with more time for oral practice, based on the reading of texts, selected and supplied by the teacher, that help them in the construction of their articles. This activity can be developed face-to-face or using resources like videoconferencing or a web camera. Students read all the texts provided but come to class or to the videoconference prepared to initiate the discussion on a particular text selected for/by him/herself (or for/by the group). In cycle 2, students work with other class members, according to their interests, forming group 4.

This activity is fundamental to the course, for it is the channel of communication the teacher has to deliver theoretical content on topics such as the new methodology / activities being implemented and on the construction of their articles.

Activity 9: Research on the Internet on Grammar Topics, Theoretical Explanation and Exercises, and Presentation of Findings / Exercises to Colleagues (GT)

This is the first activity in Part 2 – language oriented. Students do research on grammatical issues (with which they have encountered problems when expressing themselves in the earlier activities) and then prepare interactive, challenging and meaningful exercises, making use of the technology available (VLE and Internet). The explanation of the grammar point and the exercises (which should be targeted just above their peer group level of fluency) are presented, using software such as Microsoft PowerPoint, and should take no longer than 10 minutes. The answers are provided in a separate document. Students' work should be carefully prepared and should not contain any grammar, vocabulary or spelling mistakes. In cycle 2, students form yet another work group (Group 5) based on their linguistic difficulties or research interests.

Activity 10: Production of Group Article (Ind_R, Ind_RGR, & Group_Art)

This activity is made up of three distinguishable and complementary phases: Production of Individual Review (Ind_R), Production of Individual Reflection on Group Review (Ind_RGR) and Production of Group Article (Group_Art).

Students (who have in the previous exercises done research on subjects of personal interest related to the course – cycle 1) now exchange ideas, via Chat or Forum, to select the topic of their article. Group 1 is then formed according to the similarity of interests in their research topic. If they want to, they can (and are encouraged to) combine two topics, concentrating their research on more specialised themes such as Bilingualism and Deaf Children, exploring the linking threads between them.

Once the groups are formed, students refer back to the articles inserted in the Links section (either their main text or a secondary one – if their main text is completely different or cannot offer a different insight to the newly chosen topic), compile all the work done on them (questions, concept maps, answers, lexical list) and work on the first Individual Review (this task is always done individually), sending the finished product to the teacher by e-mail. Once feedback from the teacher is received, each individual produces an Ind_RGR by skilfully joining together all of the group's Ind_Rs, making his own review of the group's work.

The Group_Art is the group's chosen Ind_RGR (this could be the best marked review) which then receives new contributions from the other members of the group to make it even better or to make it complete. The second Group_Art (in cycle 2) always incorporates the output of the first Group_Art and the current Ind_RGRs.

At this stage, students are encouraged to consult online concordancers (like the "BNC – The British National Corpus" or "Just the Word") to help them express their ideas more clearly because, as Raimes (2002) said, "Writing is for discovery of learning, not just demonstration of learning" (p. 309).

For each one of the three steps mentioned above there is a different formative evaluation chart in which colours are used to highlight the ambiguities. Core aspects such as content (ideas developed), organisation (introduction, development and conclusion) and use of English (vocabulary, grammar and punctuation) are pointed out, taking into account the peculiarities of each phase (for example, the use of linking words when joining the Ind_Rs together).

In LAPLI, writing is understood as a process. Students are encouraged to edit their own work, sending it back to be (re)marked as many times as they want, but always making sure to publish the more updated version in the VLE so that it is available for the collaborative activities.

In this exercise, the individual and the collective mix at all stages. Their personal universe merges with a new universe discovered by the group in LAPLI. Each member of the group can interact with any of the colleagues, establishing a communications network. In this process, everybody assumes the role of writer, researcher, reviewer and critic.

Activity 11: Evaluating (Evaluation)

At this stage, students reflect on LAPLI's activities and issues such as their criteria for choosing the links, questions to answer, difficulties encountered during the process, and how they were overcome. This sharing and exchanging of information is done orally with the whole class (face-to-face or by videoconference, for group feedback) and individually, in writing, in the Forum (for individual feedback). If feasible, it takes place in the students' mother tongue at the end of each cycle, allowing them to express themselves fully.

Students reflect on the process, and not on the product, thinking about the methodology developed, the course itself, the interaction and collaboration amongst themselves, and the process of learning a foreign language. They discuss: the communication within the groups and the difficulties in managing the conflicts, made more difficult by the type of communication (mainly written); their commitment to success, made evident by the publication of their work; the time/space relationship determined by synchronous or asynchronous communication; the roles performed when working in (different) groups; and mainly the fact that everyone – individuals or groups – belongs to the same large group, a feeling emphasized by the fact that their work is published so that it can be used as a reference by any one of the members, and is thus owned by everyone.

Activity 12: Preparation and Presentation of Group Article (Only in the Second Cycle)

In this last activity, students create a 20-minute presentation in groups (Group 1) for the whole class (face-to-face or by videoconference), summarising their work in no more than 10 slides, work that is the product of the collaboration and insight of all the participants. This is a relaxed but formal situation, in which pupils show what they have learnt, demonstrating proficiency in the foreign language and incorporating the new vocabulary both in the written and oral delivery. When concluding their presentations, sometimes students decide to use concept maps as a means to summarise their findings.

As can be seen above, LAPLI's 12 activities have been carefully planned to develop students' language and social skills. As language learners in this student-centred approach, students are presented with the opportunity to act as researchers, responsible for the selection of content and production of knowledge, whereas the teacher's main role is both to guide them throughout the activities and to be a mediator in the face-to-face and the distance learning lessons. The alternation between individual and collaborative work, and between online and face-to-face encounters also encourages them to develop their initiative, critical sense and problem solving skills, which are deemed to be desirable and much sought after qualities in our new professionals and in lifelong education.

THE RESEARCH METHODOLOGY AND RESEARCH QUESTION

LAPLI was created with the objective of developing a methodology to promote collaborative and meaningful learning for use with students of a higher-intermediate level of proficiency and above in the foreign language. It focuses on improving reading and writing skills, but also caters for their speaking skills, by means of a VLE and the Internet, in a hybrid mode. The following methodological procedure was followed for its development: review of the literature; construction of a methodological proposal for CALL; implementation of the methodological proposal for CALL; monitoring of students during the process; adjustments; analyses of students' production; construction of a questionnaire for data collection; validation of the questionnaire; application of the questionnaire for data collection; analysis and discussion of results; and conclusion.

To develop this methodology, I worked with a group of pre-service teachers taking their fourth semester towards a degree (total of eight semesters) in English and Portuguese Languages at the Catholic University of Paraná in Brazil from August to December 2002, subject English Language Intermediate I, and adopted one type of

Descriptive Research called Case Study.

This study aimed to answer the following main research question:

How can the LOLA (The On-line Learning Lab) methodology contribute to the teaching of foreign languages with fluent students, giving emphasis to the reading and writing skills but not excluding the communicative skills, considering the use of both collaborative and meaningful learning as the basis of the production of knowledge?

More specifically, I aimed to:

- Investigate alternatives to current language teaching methodologies, placing the students in the centre of the process and giving special emphasis to collaborative learning, meaningful learning, learning by doing and the new technologies;
- Contribute to the teaching of languages via a proposal which involved collaborative learning in a virtual environment;
- Better prepare pre-service language teachers for the job market, leveraging their potential in terms of the use of the new technologies;
- Investigate the benefits of a blended modality course for the teaching of languages to fluent students.

Qualitative and quantitative data was collected to address the main questions and to investigate the issues raised. It was gathered by means of two questionnaires, with closed questions offering possibility of expansion of the choice selected, which were applied before the course started and at the end of the course. Moreover, data was also collected throughout the course, during the Evaluation Activity (activity 11) which was published in the Forum of the VLE. The results of some of the data gathered is explored in the next session.

RESULTS

Thirteen out of the fifteen students (pre-service teachers) who formed the group in which LAPLI was implemented believe that using the LAPLI methodology increases their world knowledge and improves their linguistic skills in the target language (87%). For 93% of them, the course develops reading comprehension. Both for the students who said that it "improved a lot" (33%) as well as for those who said that it "improved" (60%), it is evident that the need to read to accomplish the tasks during the semester contribute to this improvement: "The more you read, the easier it gets", "We had to read a lot", "I feel I understand the ideas in a text better now and I have assimilated some sentence construction" and "I've learnt to want to read". The research and selection of texts were also highlighted as important factors that helped increase their reading comprehension: "The research on the Internet helped." and "I learnt to look for and select good texts." (Marriott, 2004, p. 253)

93% of the pupils said that LAPLI develops their writing skills and helps them increase their vocabulary (87%) ("...because I read so many different texts, I learnt many new words.", "I had to consult dictionaries, very important to acquire vocabulary" (p. 254)). 93% of pre-service teachers believe that a VLE contributes to the acquisition of knowledge (33% of them said "significantly") and that it increases their knowledge of IT (in the beginning of the course, 20% said they were proficient in the use of computers. This percentage rose to 67% at the end of the course). The alternation between face-to-face and distance learning classes is said to be beneficial (87%, 60% said "very beneficial") and for 80% of them, the course should always be offered in a blended mode because it develops "autonomy", "self-trust" and "responsibility".

Both the collaborative activities and the concept mapping exercises were exciting new highlights to the course. However, for some students they proved a bit challenging as this was the first time they were exposed to these approaches.

47% of students said they favoured working collaboratively (p. 252). This belief was expressed in comments such as: "It is good to have other people's opinion on what we are researching. We see from another point of view."; "Two heads think better than one."; "I feel I work better in groups."; and "In groups we can exchange opinions and learn new things." (p. 252). Nevertheless, 14% of students felt the collaborative activities were "very difficult" (p. 251). This was justified by a) the need to wait for other students' contributions to continue their work; b) their dislike to depend on other people and have other people dependant on them; and c) the difficulty faced by some when dealing with their colleagues ("It's difficult to accept and be accepted by others"; "It's difficult to accept our own mistakes" and "I need to get better at accepting clashing opinions" (p. 251)).

As far as the concept mapping activities are concerned, students acknowledged their benefit to learning and the development of linguistic skills (33% thought they were "important" for learning and 27% found them to be "very important". (p. 256)). Some specific comments made by the students were "A concept map provides a summary, is objective and guides reading"; "They help organise the ideas."; "They help the visualisation of content."; "Now I pay a lot more attention to 'linking words'" (pages. 253-4), although some of them wished they had had a bit more time to develop their maps before concentrating in the next activity.

As is the case with some case studies, I have also encountered some difficulties in developing this research. The methodology proposed by LAPLI was very innovative to these students: up until now they were only having face-to-face encounters and were following a course book. The implementation of a student-centred approach, collaborative and meaningful learning activities, concept mapping exercises, a hybrid course, the use of computers and the Internet as well as a virtual learning environment in an English language course, all at once, demanded a lot more from them and this experience proved to be too much for some students. Initially there were 23 students in the course. However, after the first couple of weeks, this number dropped to 15. The reasons given by the students who gave up were basically five: lack of time to dedicate to studies as they worked full-time; difficulty in gaining access to a computer with Internet; lack of willingness to participate actively in the activities; lack of IT knowledge; and lower level of target language compared with the level of the group. These were acceptable reasons due to the circumstances but perhaps these particular students should have been informed of all these innovations before the course started and should have been given the freedom to join it or not.

Admittedly, time was the biggest issue in this course, both for the students to accomplish the activities to their satisfaction and for the teacher to give them support, guide them throughout the activities and to mark their work in a formative way. The students were not used to dedicating more than two hours a week to their English language course so some of them felt the difference. Nevertheless, nearly 87% of them said they would do the course again if they were allowed more time to develop the activities.

Nevertheless, based on these results, I now make sure students understand more precisely how the course is run and what their responsibility is in relationship to their own and their colleagues' learning throughout the activities. Also, more time has been allowed for the development of the collaborative and concept mapping activities, as was suggested by many of them. Developing online activities takes time. However, it is important to note that it is both the individual's and the group's responsibility to be on time, once delivery dates are agreed on.

FUTURE TRENDS

From the world scenario of constant changes resulting from the globalisation of economies and technological evolution – in which knowledge is the basis – one of the greatest challenges to education emerges: the access to lifelong education made easy to all segments of society. Even technologically prepared countries face difficulties when dealing with social transformation forces, and both developed and developing countries have been seeking to train their professionals so that they can respond to the demands and changes of the market, products and processes. To learn a second, third or even fourth language, which is now being required by some companies in today's globalised world, every professional needs and has the right to continue his studies, keeping up to date, as the knowledge learnt at school is nowadays nearly always insufficient and in many cases obsolete.

Companies also need to realise that it is through the ideas generated by these professionals that businesses grow, and that they have to be encouraged to feed on the knowledge available. All this can be achieved by motivating the workforce to do computer-assisted courses when they are at work in the company or at home. As Bill Gates (1995) says, "The highway [Internet] will alter the focus of education from the institution [school or company] to the individual" (p. 204).

At the same time as the world is becoming globalised, it is important to preserve peoples' cultural and linguistic identities. Incentives need to be given to the learning of foreign languages, ensuring linguistic diversity, for use both at the national and supranational levels, as well as catering for the demands of a society in evolution.

To meet these educational, cultural and company demands, it is necessary to develop new mechanisms and methodologies that allow for flexible formal and informal educational processes and increase the efficiency of communications.

LAPLI's methodology represents a step forward in language teaching methodologies. Nevertheless, more research needs to be carried out into its use in companies and the benefits of its practice in virtual reality worlds should be investigated. The linguistic barriers to progress need to be overcome and more opportunities of lifelong education need to be offered.

CONCLUSION

The use of collaborative learning and concept mapping for language learning are innovative techniques to the teaching of languages. LAPLI's methodological approach promotes a change in the role of the teacher and on the teaching process as a whole, promoting a shift from rote-mode to meaningful-mode learning. The teacher cannot be the centre of attention anymore and students need to develop an active and more responsible attitude towards their learning. Collaborative learning stimulates the development of autonomy and self-trust and can engage students and the teacher in qualitative interactive communication. On the other hand, concept mapping activities not only help students understand the meaning of a text more easily and summarise and represent the information more effectively but also "allow them to reflect on their experiences, build complete new meanings and develop strategies for creative and intelligent thinking." (Marriott & Torres, 2008)

LAPLI's activities are designed to make students break away from the "cut-and-paste" philosophy, helping them internalize the foreign language grammar and vocabulary and making them more aware of the macro and micro structures of a text. The mapping of texts, either as an individual or as a collaborative activity, is not a common practice in the teaching of languages. On the contrary, there are only a few publications about it. Dias (1996) mentions a study conducted with false beginners at the Federal University of Minas Gerais (UFMG) in Brazil, coming to the conclusion that "when they are constructed by the students themselves they positively contribute to the understanding of this text, its assimilation and the retrieval of information" (Dias, 1996, p. 158). In LAPLI, CMing is recognised as a valuable tool in the organisation of ideas and in the construction of articles, as reported by one of the students "The construction of maps summarizes the main concepts of the text that we are working on and this helps when we are constructing our own texts.". (Torres & Marriott, 2004, p. 608).

As LAPLI is a distinctive methodological proposal, students need to be given an explanation on how the course and the activities will develop. They also need to be given enough time for the development of the collaborative activities (i.e. doing research; constructing of concept maps, reviews and articles; preparation of presentations) so that they have time to make their choices, read the material, work with the foreign language and assimilate the language and content, and feel happy with what they produce. In the case of the Production of a Group Article, they need to be able to implement the changes suggested by the teacher and publish the new text in the VLE for the next collaborative task.

In LAPLI, students not only become more proficient in the foreign language and develop their analytical, critical, evaluative, social and collaborative skills but they are also encouraged to be researchers, readers, writers, editors and proofreaders. By fostering creativity, autonomy and responsibility for their learning we will be teaching students not only how to construct their knowledge but also how to keep on learning by themselves, useful skills throughout their academic life and professional careers. And this could be the single most important lesson that teachers could hope their students learn!

REFERENCES

Bahr, G. S., & Dansereau, D. F. (2004). Bilingual knowledge (BIK-) maps: Study strategy effects. In A. J. Cañas, J. D. Novak, & F. M. Gonzáles (Eds.), *Concept Maps: Theory, Methodology, Technology. Proceedings of the First International Conference on Concept Mapping* (Vol. 1). Pamplona, Spain: Universidad Pública de Navarra.

Bastos, H. M. (1998). A escrita no ensino de uma língua estrangeira: reflexão e prática. In V. L. M. Paiva (Ed.) *Ensino de Língua Inglesa: reflexões e expressões*. Campinas, Spain: Pontes.

Behrens, M. A. (2000). Projetos de aprendizagem colaborativa num paradigma emergente. In J. M. Moran, M. T. Masetto & M. A. Behrens (Eds.), *Novas Tecnologias e Mediação Pedagógica* (pp. 67-132). Campinas, Spain: Papirus.

Bochniak, R. (1998). *Questionar o conhecimento. A interdisciplinariedade na escola e fora dela.* (2nd Ed). São Paulo, Brazil: Loyola.

Brown, D. (1994). *Teaching by principles: An interactive approach to language pedagogy*. Englewood Cliffs, NJ: Prentice Hall Regents.

Delors, J. (1998). *Report to UNESCO of the International Commission on Education for the Twenty-first Century*. Retrieved March 22, 2003 from http://www.unesco.org/delors/

Dias, R. (1996). A Incorporação de estratégias ao ensino-aprendizagem de leitura em inglês como língua estrangeira. In V. L. M. de O. Paiva (Ed.) *Ensino de Língua Inglesa: Reflexões e Experiências* (pp. 137–160). Campinas, Spain: Pontes Editores.

Dillenbourg, P. (1999). What do you mean by collaborative learning? In P. Dillenbourg (Ed.) *Collaborative-learning: Cognitive and computational approaches*. Oxford, UK: Elsevier.

Dillenbourg, P. (2000). Learning in the new millennium: Building education strategies for schools. In *Workshop on Virtual Learning Environments*. Retrieved July, 2008 from http://tecfa,unige.ch/tecfa/publicat/dil-papers-2/Dil.7.5.18.pdf

Gates, B. (1995). *The Road Ahead*. New York: Viking.

Gokhale, A. A. (1995). *Collaborative Learning enhances critical thinking*. Retrieved May 16, 2008 from http://scholar.lib.vt.edu/ejournals/JTE/jte-v7n1/gokhale.jte-v7n1.html

Kirschner, P., Strijbos, J. W., & Kreijns, K. (2005). Designing integrated collaborative e-learning. In W. Jochems, J. van Merrienboer, & R. Koper (Eds.), *Integrated E-Learning: Implications for Pedagogy, Technology & Organization*. New York: RoutledgeFalmer.

Laister, J., & Kober, S. (2005). *Social aspects of collaborative learning in virtual learning environments*. Retrieved July, 2008 from http://comma.doc.ic.ac.uk/inverse/papers/patras/19.htm

Lowry, P. B., Curtis, A., & Lowry, M. R. (2004). Building a taxonomy and nomenclature of collaborative writing to improve interdisciplinary research and practice. *Journal of Business Communication, 41*(1), 66–99. doi:10.1177/0021943603259363

Marriott, R. C. V. (2004). *Do LOLA—Laboratório on-line de aprendizagem ao LAPLI—Laboratório de Aprendizagem de Línguas: Uma proposta metodológica para o ensino semi-presencial em ambiente virtual*. Masters in Education, Pontifícia Universidade Católica do Paraná (PUCPR), Brazil.

Marriott, R. C. V., & Torres, P. (2008). Enhancing collaborative and meaningful learning using concept mapping. In A. Okada, S. Buckingham Shum, & T. Sherborne (Eds.), *Knowledge Cartography: Software Tools and Mapping Techniques*. London: Springer.

Matthews, H. (2002). *Hybrid courses and the future of teaching and learning at UC*. Retrieved April 30, 2003 from http://www.uctltc.org/news/2002/12/matthews.html

Moreira, M. A. (1999). *Teorias de aprendizagem*. São Paulo, Brazil: EPU.

Murphy, P. (2002). *The hybrid strategy: Blending face-to-face with virtual instruction to improve large lecture courses*. Retrieved April 30, 2003 from http://www.uctltc.org/news/2002/12/feature.html

Nicholls, S. M. (2001). *Aspectos pedagógicos e metodológicos do ensino de inglês*. Maceió, Brazil: EDUFAL.

Pilkington, R. (2007). Collaborative and problem-based learning [Unpublished notes]. *Learning and ICT, Session 8*.

Raimes, A. (2002). Ten steps in planning a writing course and training teachers of writing. In J. C. Richards & W. A. Renandya (Eds.), *Methodology in language teaching: An anthology of current practice* (pp. 306-314). Cambridge, UK: Cambridge University Press.

Shor, I., & Freire, P. (1986). *Medo e ousadia: O cotidiano do professor*. Rio de Janeiro, Brazil: Paz e Terra.

Telebinezhad, M. R. (2007). The Effect of Explicit Teaching of Concept Mapping in Expository Writing on EFL Students' Self-regulation. *The Linguistics Journal 2*(1). Retrieved July 25, 2007 from http://www.linguistics-journal.com/April_2007_mrt&gmn.php

Torres, P. L. (2002). *Laboratório on-line de aprendizagem: Uma proposta crítica de aprendizagem colaborativa para a educação*. Doctoral dissertation, Universidade Federal de Santa Catarina (UFSC), Brazil.

Torres, P. L., & Marriott, R. C. V. (2004). The contribution of concept maps to LOLA: The online learning lab. *Proceedings of the First International Conference on Concept Mapping, Spain,* (pp. 603-610).

Warschauer, M. (1986). Computer-assisted language learning: An introduction. In Fotos S. (Ed.), *Multimedia language teaching*. Tokyo: Logos International. Retrieved July 01, 2004 from http://www.ict4lt.org/en/warschauer.htm

Widdowson, H. G. (1991). *O Ensino de Línguas para a Comunicação.* J. C. P. de Almeida Filho (Trans.). Campinas, Spain: Pontes.

Willis, J. (1996). *A Framework for Task-based Learning.* London: Longman

Young, J. (2002, March 22). 'Hybrid' Teaching Seeks to End the Divide Between Traditional and Online Instruction by blending approaches, colleges hope to save money and meet students' needs. *The Chronicle of Higher education Information Technology.* Retrieved April 30, 2003 from http://chronicle.com/free/v48/i28/28a03301.htm

KEY TERMS AND DEFINITIONS

Autonomy: ability to work independently, to be able to manage the learning process and to act thoughtfully when working on activities and exercises without anybody else's help.

Blended learning: The term blended learning (or hybrid learning) in LAPLI is used to refer to lessons which are delivered partially face-to-face (F2F) with students and teacher sharing the same real life classroom with access to computers and the Internet, and partially distance learning with students and teacher communicating synchronously and asynchronously, making use of technology-based materials and a collaborative virtual learning environment.

Collaborative Learning: Collaborative learning is a student-centred methodology which is used by some practitioners in e-learning. By working in groups, students are the subject of their construction of knowledge while at the same time they contribute to their peers learning. The construction of knowledge is achieved by the participation and interaction that occurs amongst all students when they get involved in activities that aim at a common goal. It solidifies socialisation not only by learning, but mainly in learning.

Collaborative Virtual Learning Environment: A collaborative VLE is student friendly and very easy to use and navigate. In a collaborative VLE, students have a high degree of autonomy and can perform tasks such as: communicating with all colleagues and tutor synchronously or asynchronously; suggesting interesting websites that can be accessed and commented by all at the click of the mouse; creating folders to manage their own work; uploading material in their or their colleague's folder; and initiating new topics for discussion in the Forum which allows for several layers of response.

Interaction: Process that emerges from the participation of all learners that interact amongst themselves by an active dialogue, a constant exchange of information, points of view, queries and ideas that occur in a learning environment.

LAPLI: The Language Learning Lab (*Laboratório de Aprendizagem de Línguas*) is a blended methodology for foreign language teaching / learning / acquisition created by Rita Marriott for her Masters in 2004. It makes use of technology that allows for synchronous and asynchronous communication, and a pedagogical methodology based on 1. students active participation in the learning process; 2. use of the foreign language as a tool to perform activities; 3. learning mediated by tutors/teachers; 4. collective construction of knowledge that emerges from exchanges between students; 5. interactivity amongst all involved in the process; 6. encouragement of the processes of expression and communication; 7. flexibilisation of roles in the communication process in order to allow for the collective construction of knowledge; 8. acceptance of the diversities and differences of students; 9. development of students' autonomy in the learning process; 10. value of freedom with responsibility; 11. respect to authorship; 12. value given to both the process and the product.

LOLA: The Online Learning Lab (*Laboratório On-line de Aprendizagem*) is a distance learning methodology created by Patricia Lupion Torres in 2002.

ENDNOTE

1 This methodology was first published in Panayiotis Zaphiris & Giorgos Zacharia (Eds) (2006) in *Computer Assisted Language Learning*, Hershey: Information Science Publishing. Re-published with permission.

Chapter 3
The Assessment of Interactive Learning:
The Contributions Made by Online Portfolios and Cognitive Mapping

Edméa Santos
Faculty of Education of State of Rio de Janeiro, Brazil

Marco Silva
Estácio de Sá University, Brazil

ABSTRACT

This chapter proposes the use of a communicational approach to rethink conceptual and methodological aspects of learning assessment in the context of interactive online information and communication technologies. The approach makes use of the digital online portfolio interface together with cognitive mapping techniques (mind maps and concept maps) as devices for assessing learning in online education. The examples described in the text are the result of pedagogical practice and research undertaken by the authors.

INTRODUCTION

The assessment of learning in the online classroom requires that a break be made with the traditional model of assessment that has developed over time in the physical classroom. If teachers do not want to underuse the potential of the digital online medium or repeat the same mistakes as those made by teachers using traditional assessment, they will have to discover new attitudes and new engagement strategies for their teaching and learning situations and so re-dimension the way they assess learning and their own performance.

DOI: 10.4018/978-1-59904-992-2.ch003

The traditional model of learning assessment is characterized by arbitrary procedures linked to the cumulative measurement of the results of specific tests, defined by the teacher, of the student's work and attitudes. The student's performance is measured in the same way as one measures length, quantity and volume, using scales and assigning numerical values to it.

According to Hoffman (2004a, p. 25), "assessment in schools is a laborious process involving the judgment of results." A practice consisting of "the recording of results about the student's performance during a particular period" or "final exams and the assignment of marks to classify the results." For her, a conscious or subconscious perception became established as part of a distressing collective practice based on historic bureaucratic demands and a deep-rooted authoritarian attitude.

This author became widely known in the academic and educational world for her involvement in the "assessment phenomenon." The conclusions she reaches in her research strongly suggest a preference for "mediatory assessment", her theoretical and practical approach for overcoming the deficiencies of the traditional assessment model from a perspective of "constructivist and liberating assessment." This approach is built on the principles of independence, dialogue, participation and collaboration in order to overcome the deficiencies associated with the arbitrary nature of the assessment-based model, which revolves around exams, grades, reports, repeat exams and failure.

"Giving a grade is assessment, sitting an exam is assessment, the record of the grades is called assessment." This definition suggested by the author shows the reductionist view that is taken of learning assessment in schools and universities. To assess is to judge the result of the learner's work after the learning has taken place. To assess is to issue a final appreciation dissociated from the learning process but intimately linked to the ghosts of control and authoritarianism that have historically marked education (Hoffman, 2004b; 2004c).

This model survives to the present day. It brings the stamp of authoritarianism and arbitrariness, which are incompatible with the emerging dynamics of learning, to the knowledge society and to "cyberculture" (Lemos, 2002a; 2002b).

Henceforth the perspective is one of interactivity supporting learning communities in online networks that value independence, dialogue, participation and collaboration. The old model of assessment will cease to occupy such a dominant position in the new sociotechnical context, allowing new learning and assessment practices more in tune with the dynamics of our time to be developed.

In this context, new pedagogical practices capable of resignifying teaching and learning processes, and hence learning assessment practices, must be developed. In this work we discuss how the theory of interactive assessment combined with cognitive mapping techniques (concept maps and mind maps) and the use of online portfolios can make a useful contribution to the debate.

THE USE OF INTERACTIVE ASSESSMENT IN THE ONLINE CLASSROOM

Teachers and students accustomed to the primacy of transmission in education and the mass media now have to develop alternatives to transmissive teaching and the associated learning assessment model. They have to develop their creative imagination to meet the new sociotechnical demands for independence, a multiplicity of connections, dialogue, collaboration and interactivity.

Teachers and students may fail to make full use of the potential of the digital environment and the Internet in the classroom if they do not understand the complex concept of interactivity. By this is meant the communication mode that is increasingly occupying a position of central importance both in the digital age and in cyberculture. The concept involves consciously making available a more communicational, deliberately complex medium prepared by the sender, giving the receiver the possibility of replying to the system of expression and dialoguing with it. Thus understood, interactivity represents a qualitative leap in relation to the mass communication mode that prevailed until the end of the 20th century and whose univocal logic is now under threat in a context in which it is expected that the restrictions imposed by passive reception will be overcome.

In-depth examination of interactivity leads to a number of different suggestions for approaches that can be used to remodel the teaching and learning practice known as transmissive pedagogy. With these suggestions, teacher and students can review their roles in the construction of learning and assessment, transforming the communication environment and making possible the transition from a model whose central element is a linear sequence that unidirectionally links steps, ages and packages of knowledge to a plural, decentralized model, the key to which is the coming together of text, hypertext and mapping.

However, before bringing together these suggestions, it is worth considering a number of points raised in the book "Sala de Aula Interativa" (Silva, 2000) that lead one to agree that there is a need for investment in interactive learning and assessment.

Firstly, the teacher will need to be aware that digital technologies have their own momentum, which is in tune with cyberculture and the communication profiles of learners, who learnt using a remote control unit and the univocal logic of the mass media and will from now on learn with a mouse, a touch sensitive screen and mobile three-dimensional windows that allow them to do much more than just watch.

The prevailing method of teaching revolves around the speaking-dictating of the master, who continues to be the guardian of culture and the transmitter of closed packages of information. He transmits so that what is deposited in the learners' minds can be filed, as if he wished to cement the belief in the receiver that the more knowledge one transmits, the more one appropriates it.

He remains attached to the transmissive model, which involves repeating information rather than mapping and constructing knowledge. He does not develop a way of communicating that can promote participation and dialogue as a *sine qua non* for learning and assessment. At best, he develops an attitude to communication that only respects participation and dialogue but does not promote them in a concrete fashion as part of learning and assessment.

Changing this attitude to communication requires specific techniques and at the same time a critical view of a sociotechnical change that is in progress. In tune with our times, the teacher allows learners to take part in mapping and constructing knowledge and communication itself, which is understood to be the collaborative effort of transmitting and receiving.

Unlike when he transmits to a massified receiver, the teacher learns with the dynamics of digital technologies and online connectivity and makes personalized, operational, collaborative communication available to the learner.

We teachers, who learnt with esteemed masters, must learn with the spirit of our times — indeed, we can learn with the work of the web designer. A site should not be watched the way one sits in front of a television screen; rather, a site should be manipulated, browsed, mapped and operated. It is important to be aware that implicit in the screen of a networked computer is the assumption that the individual will become immersed and participate/intervene — an experience that is rare in the mass media.

Note that many websites are still generally for looking at and plundering rather than for interacting with. We can conclude that there are specialists in the production of websites who are not making full use of the digital medium or cyberspace and, still envious of television, are just making online content available for internet users to watch and download. Indeed, we can conclude that the power of the culture of transmission is still felt not only in traditional education but also in the very latest media and that teachers and web designers need to be aware that in education and communication, interaction involves more than watching and carrying out specific tasks.

We need to be clear that interactivity is not just a passing fashion. And that even if we bear this in mind we run the risk of remaining constrained by transmission. Not through bad faith but through the weight of tradition of the speaking-dictating of the teacher, we run the risk in the name of interactivity of being satisfied with advertising ideology, marketing strategy, the generation of loyalty and the production of public opinion.

We run the risk of the term "interactivity" becoming hackneyed, as in a TV advertisement for a pair of trainers in which the product is described as "interactive". We can observe an increasing use of the adjective "interactive" to describe anything whose operation allows the user/consumer/spectator/receiver to participate, change actions and control events to some extent (computers and associated devices, toys, clothing, white goods, online banking, shows, the theater, advertising and marketing strategies, radio and TV programs etc.).

We will need to criticize the interactivity industry, which is showing vigorous growth and promising an interactive future. We cannot just be satisfied with interactive cars, games, refrigerators and microwaves. This represents increasing hackneying of the term "interactivity", seen as an excellent sales argument, the promise of an enriching dialog that helps to swallow the pill.

Despite this hackneying, we can see the emergence of interactivity over time as a new paradigm in communication. Transmission (sending that requires no more than passive reception) loses its power in the digital age and in cyberculture, when the overlap of at least two factors is increasingly apparent:

- Technological: The computer screen is not a space in which transmission occurs but one that can be penetrated and manipulated, with mobile windows that are open to multiple offline and online connections and allow their content to be interfered with and modified.
- Social: There is a new spectator, who is less passive when faced with a message that is more amenable to intervention and who learned with the TV remote control unit and video game joystick and is now learning with the mouse.

In fact, a new communication scenario is emerging with the overlap of technological, marketing and social scenarios. In this new scenario, a transition from the logic of distribution to the logic of communication is taking place, implying a radical change in the classical information scheme based on the unidirectional transmitter-message-receiver connection. Education benefits from this change. Its social function, of socializing citizens, benefits from the timely momentum provided by the new communication scenario We have a scenario that favors the following repositioning of the teacher's role:

- The teacher no longer transmits what is usually understood as a closed message. He offers the learner a range of items and possibilities for him to manipulate and use his operational creativity on.
- His message is no longer "transmitted", it is no longer a closed, paralyzed, immutable, untouchable, sacred world. It is an open world that can be modified in so far as it responds to the requests of the person consulting it.
- The learner is no longer in the classic position of reception. He is the new spectator invited to create freely. The teacher's message acquires new meaning with the student's intervention.

Education, in its social function of socializing and promoting participation and collaboration, benefits from this paradigmatic change in communication theory and practice. The message only acquires its full meaning with the intervention of the receiver, who becomes, to a certain extent, a creator. This is extremely welcome as a communication environment that will henceforth have an ever greater influence on educational systems and the associated assessment processes.

Faced with this situation, we teachers must prepare ourselves for the fact that our message can now be recomposed, reorganized and modified under the impact of learner interventions. We must prepare ourselves for closed transmission and assessment to lose their hegemony We will need to prepare ourselves to revitalize our profession when there is a change in the classic information scheme based on the unilateral or unidirectional transmitter-message-receiver connection.

We will need to be aware that in every learning situation interactive education and assessment must promote the free expression of the fundamentals of interactivity:

- Participation-intervention. The teacher assumes that there will be participation-intervention by the receiver. This participation is not limited to answering "yes" or "no"; it is much more than choosing a particular option, much more than asking a question. To participate is to interfere in the message, to collectively construct and map learning and communication.
- Bidirectionality-dialogue. To communicate presupposes that there is bidirectionality between teacher and learners. Communication is a joint production by everyone. Faced with the learning content and maps, the teacher is a potential learner, and the learner a potential teacher. The two poles encode and decode.
- Multiplicity-connectivity. The teacher provides the opportunity for constructing multiple maps of networks connecting the learning contents. He does not provide closed content; rather, he provides information in networks made up of connections that give the learner significant freedom to change, virtualize, simulate, associate and signify.

We have the opportunity afforded by the spirit of the time to create a new communication environment with learners in physical and virtual classrooms. In the digital online environment we can learn that to communicate is not simply to transmit but to provide multiple opportunities for the learner to intervene, as communication only takes place with his participation.

Interactive communication is a challenge for all of us who are used to the paradigm of transmission. It can in fact be divided into three parts. The teacher must at the same time be aware of hypertext, use it to enhance his pedagogical action without losing his authorship and, lastly, appreciate that it is not a question of invalidating the classical paradigm.

It is the teacher's responsibility to formulate more exactly the extent of the challenge:

- The teacher must work with learning content as he would work with hypertext, i.e., like a non-sequential piece of writing, an assembly of network connections that, by allowing and demanding a multiplicity of investigations, transforms reading into writing.

The teacher must realize that hypertext, networked constructions and mapping will enhance the figure of the teacher and his profession. From being a mere transmitter of knowledge, he will have to change his role to that of someone who formulates problems, provokes questioning, coordinates teams and systematizes experiences, as well as being the living memory of an education that values and enables dialogue and collaboration between learning participants instead of clinging to the pedagogy of transmission (Silva, 2000). Maps allow the teacher to develop course content in hypertext. Let us consider an example (Figure 1).

We need to understand that this change in the way that we communicate is in tune with the emergence of a new reader or new spectator. The reader is no longer a person who follows the pages of a book one at a time and in sequence, but

Figure 1. An example of course content using hypertext and maps integrating networked connections of texts, images, forums, chat rooms, activities for students, questions and comments

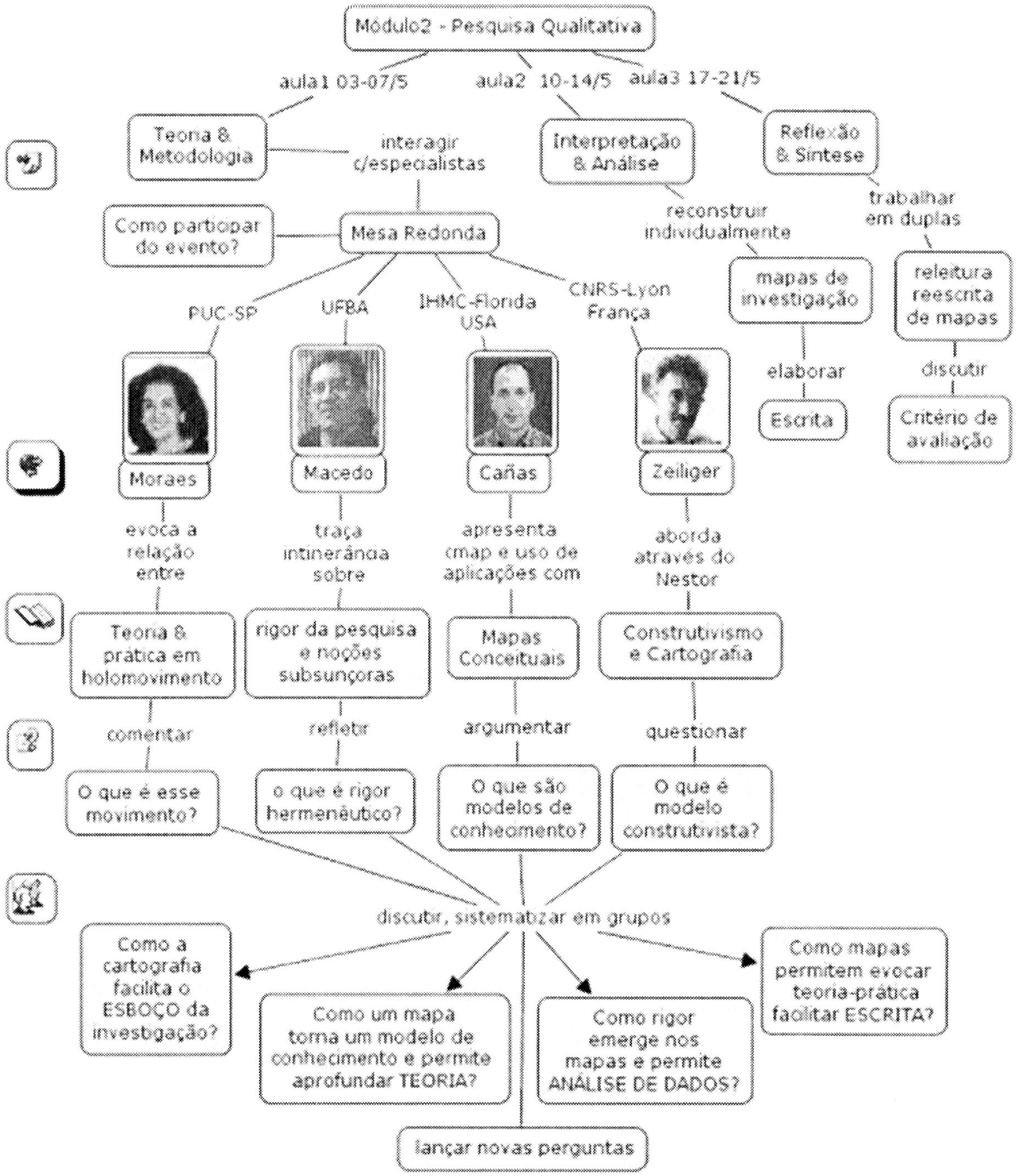

someone who jumps from one place to another, making his own map of his reading itinerary. He no longer submits to declamations that form part of a transmission, but, as he no longer considers himself merely a receiver, interferes, maps, manipulates, changes and thus reinvents the message.

We can observe from this distinction that intervening in the message is not the same as being a passive receiver of information. When the teacher realizes this, he can redesign his classroom and change the communicational principles on which he relies so that they are in tune with the spirit of the digital and cybercultural age. He modifies the model based on his speaking-dictating and provides the student with as wide a range as possible of opportunities for authorship, co-authorship, participation, manipulation and information so that the message can be exchanged, associated, formulated and modified.

We must not only transmit; rather, we must make available conditions that favor the collaborative construction of knowledge. We must not merely distribute knowledge that has been accumulated over time; rather, we must promote a communications environment that favors the critical appropriation of knowledge.

Making such conditions available in the classroom requires essentially three investments:

- Offering multiple items of information and mappings in images, sounds, texts etc., using digital technologies or not, but in the knowledge that when used interactively these can considerably enhance actions that result in knowledge.
- Facilitating, providing an opportunity for, preparing, weaving the threads of the web, developing and mapping multiple paths for connections and expressions so that learners can express themselves as they manipulate information and follow the routes that have been designed
- Stimulating each learner to contribute new information and to create and offer more and better routes mapped by him, so that he participates as co-author of the process (Silva, 2005).

Teacher and students build, map and browse a network rather than a route. The teacher defines a set of territories to be explored. Learning and assessment occur when learners are exploring — experiencing participation, collaboration, creation and co-creation — and not as a result of the teacher's declamations or speaking-dictating. This means that the teacher's classic posture in the classroom must be changed.

The teacher allows learners to experience knowledge by making webs and maps available, by creating possibilities for learners to become involved, by providing opportunities for them to create and acquire, and by stimulating intervention by them as co-authors of their actions.

In our role of interactive authorship we will be more than merely counselors or bridges between information and knowledge. We will be more than just partners, more than just facilitators. The terms "counselor", "partner" and "facilitator" are obvious and simplistic. Our role of interactive authorship requires a figure who systematizes experience, who provides an opportunity for learning, who makes the elements of the web available and weaves together with the learner, and who unveils new territories of knowledge to be mapped and maps these with the learner.

The teacher must provide possibilities for multiple experiments and expressions. This is exactly what the software designer — the web designer — can do: provide an assembly of network connections that allows multiple investigations. Thus, we teachers can change the way that we communicate both by using hypertext and by acting as systematizers of experiences.

From mere transmitters of knowledge, partners or counselors, we can become problem formulators, instigators of situations, architects and cartographers of routes; in short, knowledge-construction brokers in the live experiment that is the classroom.

In the online learning environment, we need to consider learning content and maps as open work in which immersion, browsing, exploration and polyphony can flow in the logic of completion, i.e., so that the learner can complete and not just contemplate. All the participants will be able to make a contribution to the connective and collaborative dynamic.

To achieve this, we have to pay attention to a number of essential details. We can share suggestions about how to conduct the process in order to ensure that interactive learning and assessment in the online classroom is effective.

We can create hypertext environments that have intertextuality — connections with other sites or documents; intratextuality — connections within the same document; multivocality, a multiplicity of points of view; usability — a simple environment that is easily browsed in an intuitive way, with transparent information; mixing — the integration of several languages (sounds, text, dynamic and static images, graphics, maps); and hypermedia — the integration of different types of media support that can accommodate new links and additions.

We can make synchronous (real-time) and asynchronous interactivity possible (the latter being communication at any time, when the transmitter and receiver do not need to be communicating at the same time).

We can create research activities that stimulate knowledge construction using problem situations, where the subject can contextualize local and global questions from his cultural universe.

We can create environments for interactive assessment, where knowledge is constructed and mapped in a negotiation process and where decision making is constantly used in the evolving re-signification of authorships and co-authorships.

We can make available and encourage ludic and artistic expression, using games, simulations and learning objects. We can provide opportunities for learning by making available experiments that we will perform jointly with learners.

We can mobilize contacts and relationships by mapping between the different fields of knowledge considered as an inter/transdisciplinary network and at the same time stimulate learners to participate creatively, taking into account their sensorial, motor, affective, cognitive and cultural makeup.

These are not a recipe for making a cake, but proposals to which additions can be made according to each appropriation, each situation and each occasion. For the purposes of this text, they are approaches to which the term "mediatory assessment" used by Hoffman can be applied. If it is to be brought up to date, the learning and assessment environment in the online classroom must be consciously prepared. Suitable tools and interfaces[1] will be required. We can, for example, use those suggested by Schlemmer (2005, p.141):

- "**Cooperative authorship of assessment methods, instruments and criteria:** this allows assessment methods, instruments and criteria to be defined collectively, inserted, consulted, mapped, changed and excluded;
- **Self-assessment:** this allows the individual assessment (development) process to be described, inserted, mapped and consulted according to established criteria;
- **Group assessment:** this allows the collective assessment (development) process to be described, inserted, mapped and consulted according to collective criteria;
- **Teacher assessment:** this allows the individual assessment (development) process to be described, inserted, mapped and consulted according to established criteria;
- **Qualitative historical records:** allow activities undertaken by each member of a particular community to be consulted, mapped and monitored;

- **Quantitative historical records:** generate reports with statistical data about the environment, communities and individual." (Silva, 2006).

These suggestions are also proposals to which additions can be made according to each appropriation, each situation and each occasion. Learning communities are invited to share other suggestions. All of this is favored by the present times, which support "mediatory assessment", interactivity and "citizen education" in tune with the spirit of the time. Before, we only had esteemed teachers such as Paulo Freire (1979 e 1996) inviting us to be more daring in learning and assessment. Henceforth we will have cyberculture, digital support, interactivity, learning communities, a new spectator and the online classroom.

THE PORTFOLIO AND THE CONCEPT MAP: THE INTERFACE, THE DEVICE AND THEIR CONTRIBUTIONS TO LEARNING ASSESSMENT IN ONLINE EDUCATION

We teachers and researchers, for whom the research process and pedagogical practice are interactive communicational learning spaces, need to include in our devices not only the possibility of access to and the transmission of various languages, but especially the sharing of these ideas. Interactive communication between content and authors must be guaranteed, as these may be separated by large distances and geographically dispersed. In the present study we will examine how an online portfolio and cognitive mapping techniques are devices that can lead to the emergence of new interactive assessment practice in online education. (Santos, 2003).

A portfolio is here considered to be an asynchronous communication interface, a device[2] for pedagogical research and practice and, above all, a device for interactive assessment. A portfolio can be combined with other digital interfaces, such as a blog, forum, web conference, map and chat room. In the case of blogs and maps, we can use them to reflect on the individual and collective learning process by sharing our personal and work-related dilemmas. The educational process is made up of this reflective and dialogic sharing. In the case of forums, web conferences and chat rooms, however, we can use them to share and debate different issues arising from the communicational organization of the teaching-learning process. Thus we can deepen and resignify our subsuming concepts[3]. In forums, maps and blogs we communicate asynchronously, and in chat rooms and web conferences the communication is synchronous. But how specific is the portfolio? Why should it be used in online education?

For researchers, teachers and students who are involved with interactive assessment practices, this process is often understood to be an event that does not separate itself from its products and authors. Interactive assessment is an interactive act in which teachers and students negotiate knowledge-production strategies that are constantly analyzed in activities to diagnose and plan new strategies and in decision making so that learning is in fact achieved. Thus, procedures and devices that illustrate the process in a meaningful way are required. To this end, we must not separate the process from its products, let alone from the strategies for their production.

Starting with these assumptions, it is important to consider the portfolio as a device that increases the visibility of each subject's production and mapping, allowing the subject to share his production with the whole subject-group, so that at the same time he is author and assessor not only of his own work and production but also of that of his peers. When we say that assessment is an interactive act, we believe that each subject who is assessed must also be a subject who can and must *assess*. A portfolio is thus an interactive device that allows:

- **Self-assessment:** the assessor is the author of the action, the production or the performance being assessed. The author is responsible for, and conscious of, his learning process.
- **Co-assessment:** the assessor is a *peer* in relation to the action, the production or the performance being assessed. This "peer" must in fact be the group that will intervene in the process in a cooperative, shared and global manner, adding value to the productions of all those involved.
- **Hetero-assessment:** the assessor is an actor (usually the teacher for the activity) with more experience in the study object to be assessed: someone who has a wide repertory and interacts with the whole group in a more deliberate and planned way. (Santos, 2005).

In the following section we describe a teaching and research case that aimed, through the use of an online portfolio and mapping techniques, to develop an interactive assessment experience.

CASE REPORT OF A RESEARCH EXPERIMENT INTO PEDAGOGICAL PRACTICE IN ONLINE EDUCATION

What we describe here is the result of pedagogical research and practice in online education based on participation in the course "The Use of Software in Qualitative Research" (www.cogeae.pucsp.br)[4]. The main aim of the course was to train researchers and provide the effective means to analyze, define, create and monitor online activities and support for classroom-based education involving interactive processes and technological interfaces so that the subject-group could develop related activities in their research and work environments. The learning objectives of this course are:

- To discuss important concepts relating to maps and qualitative research (their complexity and nonlinear dynamics);
- To learn about a number of examples of maps used in academic projects and to reflect on the benefits, difficulties, obstacles and new applications;
- To learn about and apply a number of software resources: Cmap and Nestor Web Cartographer[5];
- To prepare maps to facilitate the analysis and production of knowledge in qualitative research (Santos, 2005).

Based on these objectives, the aim of the course was to develop a virtual research community through the use of online activities and the development of a meaningful project to support theoretical reflection and professional practice. The virtual environment was thus intended to guide the process in a flexible interactive environment, in which the contents and activities were made available as the process progressed, primarily according to the group's interests and needs.

During the online course, each module resulted in various productions that challenged the researchers to map the specific stage of their research using cognitive mapping techniques (mind maps, concept maps and web maps[6]). Researchers were either in the pre-project preparation phase, reviewing the literature, analyzing the data, writing the final report or moving between the different stages. After all, the different stages in academic research and pedagogical practice are neither linear nor stationary processes. They form a network that organizes itself as a function of the movement between the researcher, the object being studied and the subjects and processes involved.

Against this background, we endeavored to mediate the construction of mind maps, concept maps and web maps that would help the researcher reflect on his research in a meaningful way within a suitable context. A number of productions were generated. However, only the more meaningful ones were included in the portfolio. A portfolio is not a repository where productions are accumulated at random. It is a special collection of the best productions, of production-syntheses, of productions that were previously analyzed and organized by the author by means of self-assessment.

By constructing the portfolio as part of an evolving process, the researcher can perceive himself as an author, and the consequent feeling of ownership means that the productions cease to be merely the means with which he is accountable to the teacher who is examining them. From the moment the student-researcher shows, and can see, his learning process, he feels prepared to communicate what he has learnt, and to defend his position on various topics in the course, thus helping him to maintain a dialog with other subjects involved, whether these be fellow students or teachers. Using a number of examples, we describe below how the portfolio was assembled in our study.

The first activity involving maps was to create an *introductory map*. Each researcher presented something about his personal and professional background in a graphical format. In this way we were able as a group to learn a little about the lives of each person taking part in the research and to map the context within which each researcher's scientific and research interest emerged. (see Figure 2)

After the participants had introduced themselves, we encouraged each researcher to build a mind map about the current state of their research. Mind maps "are a strategy developed by the psychologist Tony Buzan in the beginning of the 1970s in the book *Use your head*" (Okada, 2004, p. 4). This technique consists of expressing thinking about a knowledge object in an external medium. In the case of our online course, the knowledge object was each researcher-participant's research project. The main challenge of the exercise is to express in a diagram, in a medium outside the mind, ideas about each researcher's thinking.

Another aim of the technique is, given the organization of the mind and the difficulties involved in externalizing, to help unblock or "put out" ideas that were or are being mentally signified by the knowing subjects. The lack of visibility of the process of signification very often makes

Figure 2. Introductory map for teacher-researcher Solange Nogueira[7]

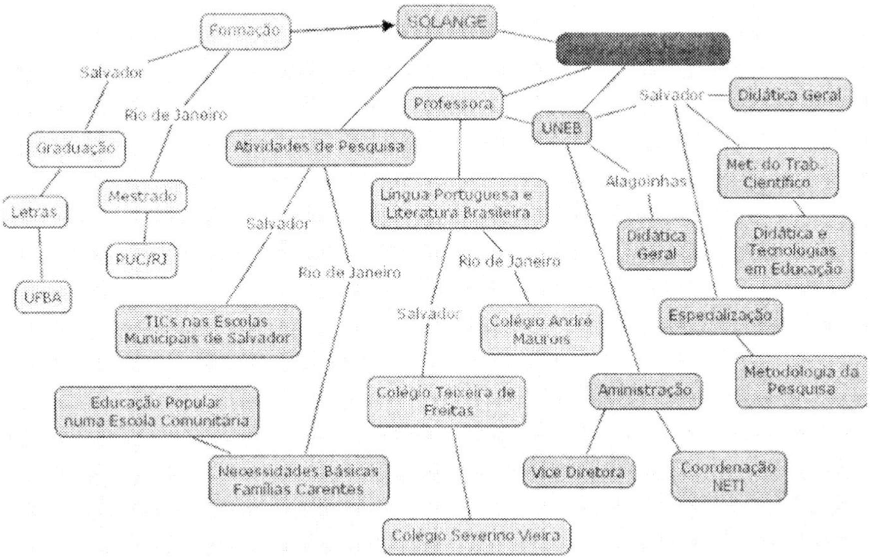

learning difficult because the subject does not exercise his capacity for reflecting on his actions, reading and experience. These processes form the complex movement of the act of researching. The technique of mind maps does not provide rigid models, let alone recipes, that have to be followed. According to Okada:

The organization of mind maps reflects how the mind processes thinking as ideas arise. In the beginning, few words appear, but as they are registered and linked, they stimulate others to appear... How can a mind map be constructed? The main subject is recorded in the center of the page with a word or picture that is clear and meaningful. From then on, various questions can be asked so that new ideas can be recorded: What? How? Why? What for? Where? When? Who? As the keywords arise, new links may appear and related issues emerge, allowing more and more links to be made. Always record ideas, trying not to write sentences but rather a keyword or concept. Meaningful symbols can be used and may include drawings, icons, figures, pointers... In addition, different colors can be chosen to bring different parts of the map together, to select and classify common or more important topics and so help with identification, particularly when the map becomes very big. (Okada, 2004, p. 5)

Let us look at an example taken from the teacher-researcher's portfolio (Figure 3).

Studies of cognitive mapping have shown that the more we manage to express ourselves with forms that are similar to our cognitive structures, the more visible will be the process of knowledge construction and the better and more varied will be the ways of resignifying learning and subsuming concepts (Santos, 2005).

Thinking is built with networks and associations. We produce new knowledge in a hypertextual network; we do not think linearly. New knowledge connects to knowledge that has already been constructed and can be updated and even refuted. Everything depends on our production of meanings, on how we signify, on how we make science.

Thus, concept maps are potent devices for new learning. With the aid of concept maps, expression of the thought network — cognitive mapping — can be externalized / internalized / externalized in a dynamic movement as a "hypertext portrait" of the mind.

Concept maps are diagrams that show relationships, links or associations between concepts. The way concept maps are organized depends uniquely and exclusively on the researcher's thinking, on how he structures his ideas based on his interaction with the object he is studying, whether it be in the theoretical or empirical field or at the interface. To build a concept map that expresses the cognitive mapping of the act of researching, interfaces must be used, and these can be either physical (paper and pencil) or digital (software).

Concept maps can be more easily constructed with software because of the flexibility and plasticity inherent to digital technologies. How are concept maps drawn? The main subject is recorded inside a rectangle on the screen in the form of a clear and meaningful concept. After this, the concepts that are directly related to the initial topic are written underneath in rectangles and arrows with descriptions drawn between the elements More specific concepts are always placed further down, and the structure gradually branches out like a tree. Concepts at the same hierarchical level are shown in parallel on the map. New arrows can appear, and there can be images as well as words. The choice of colors is left to the person building the map. (see Figure 4)

A number of points need to be borne in mind when preparing concept maps:

- The subject to be addressed must be chosen and the main objective defined;
- The initial concepts must be registered and a hierarchical sequence defined;

Figure 3. Mind map of Solange Nogueira's research project

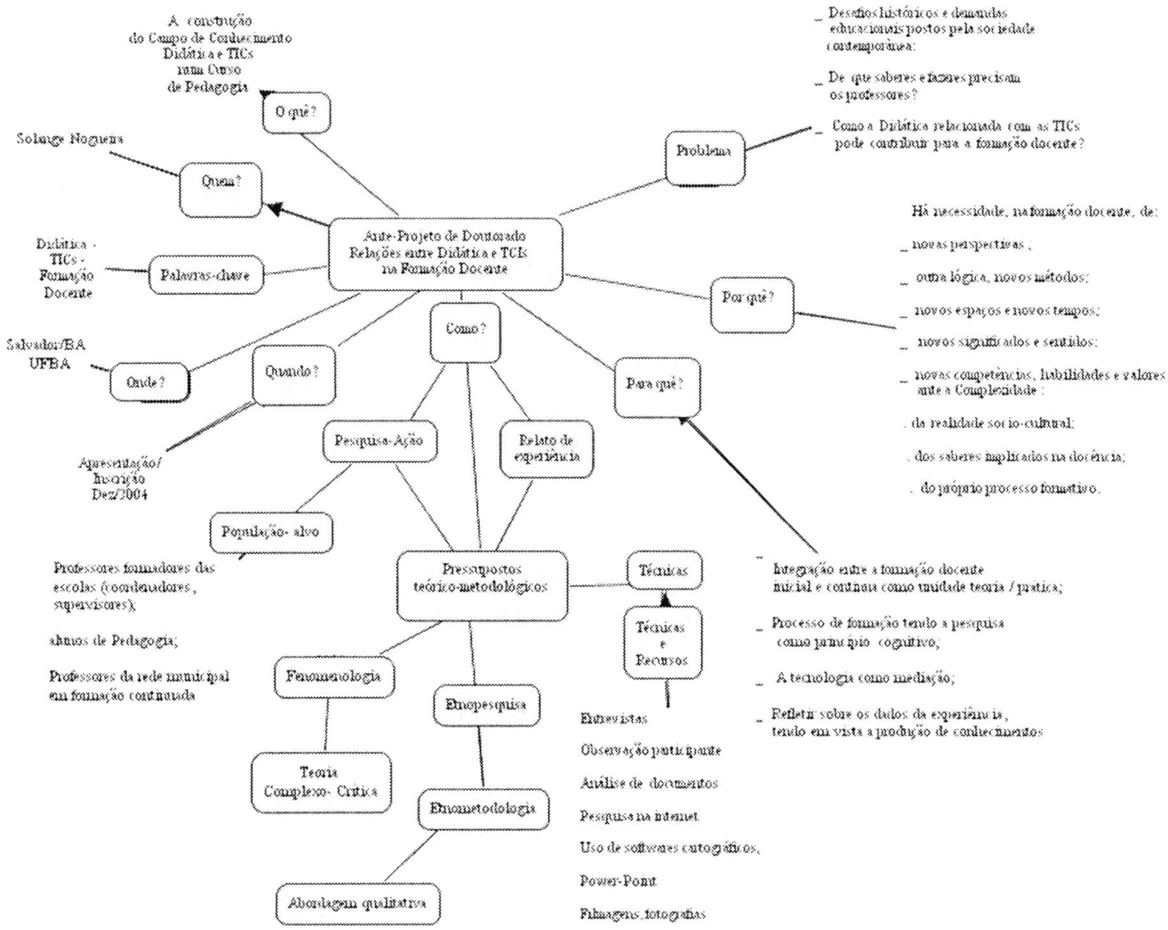

- The links between elements must be shown by means of lines: a) the purpose of the connection must be shown by recording this on the line; b) the meaning of the keywords must be researched so that they are as clear and precise in the context of the subject being addressed; c) feedback sessions should be allowed to obtain other people's opinions so that concepts can be reconsidered and the instrument being used can be reassessed in order to emphasize the most important points.

Learning implies not only additions but also changes to the cognitive structure. For meaningful learning it is important to remember that:

- Concepts registered at the different levels must represent the context in question;
- A hierarchical organization facilitates the assimilation of new concepts; new ideas and concepts must be "potentially meaningful" for the learner;
- By establishing links between new and existing concepts, cognitive structures will be able to assimilate more easily and remember meaning more clearly.

Figure 4. Concept map about fields of knowledge in teaching prepared by Solange Nogueira with Cmap Tools http (Taken from the teacher-researcher's portfolio)

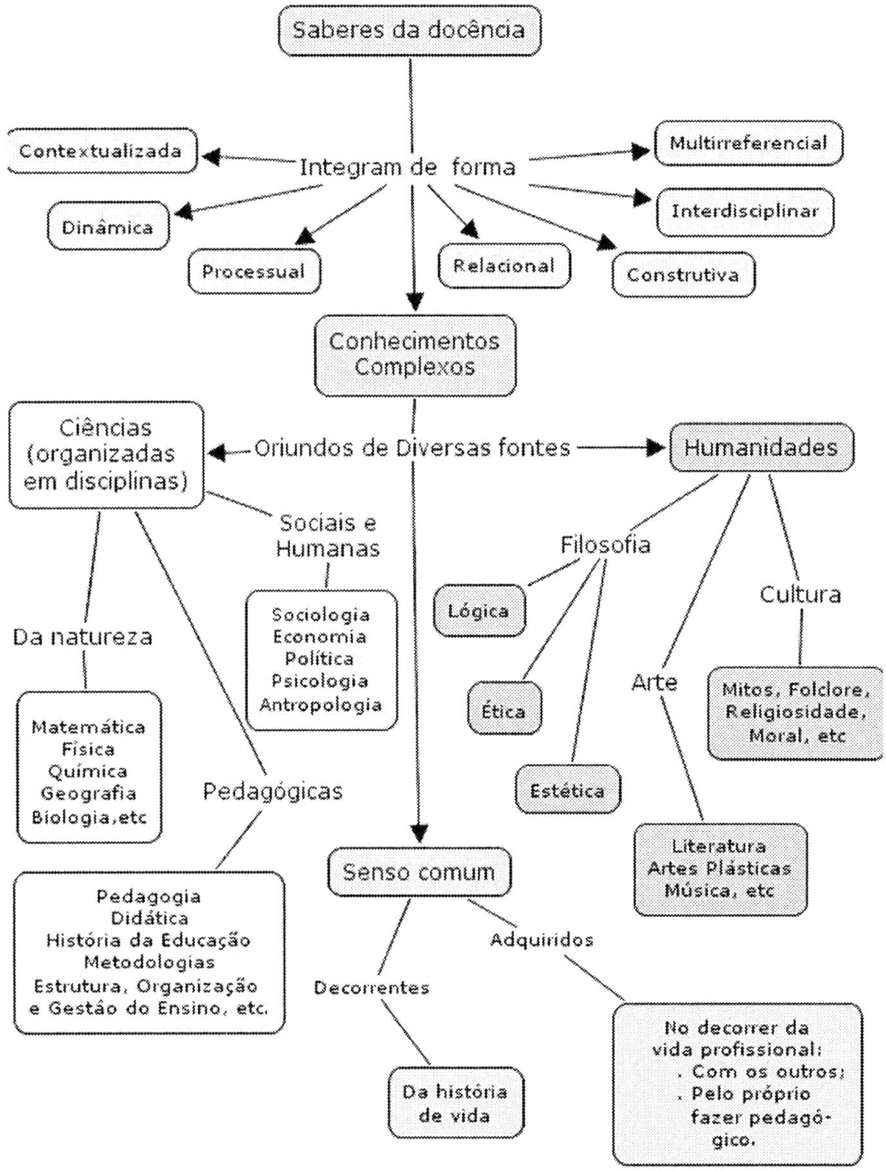

Mapping allows the researcher to visualize his cognitive map, enhancing knowledge construction. Hypertext mapping socialized in an online portfolio allowed everybody to learn from everybody else's work, as participants assessed others and assessed themselves throughout the process.

FINAL CONSIDERATIONS

The use of a portfolio together with the techniques of concept maps allowed each participant to consciously think about his thinking (metacognition) and about the path that he had traveled in his research-education.

Table 1.

Communicational principles of interactive assessment	Activities in the portfolio using cognitive maps	Benefits
Participation-intervention. Continuous interactive assessment requires a large number of tasks that provide an opportunity for the student to express himself. In an environment where freedom of expression is guaranteed, students have various opportunities to express their knowledge about a study or idea. Their various expressions and interventions must be observed by the teacher, who must pay close attention to their evolution in terms of building knowledge and participation. Bidirectionality-dialogue. The knowledge that the student develops is constructed as a co-creation in the relationship with others. His engagement in the learning process is the result of bidirectionality and dialogue among the students and teacher. The mediation of learning is collective and takes place as a result of challenging questioning activities rather than as a result of a new explanation by the teacher or a piece of work done individually by a student. Multiplicity-connectivity. Learning content is offered as "raw material" so that learners can, in the heterogeneity of the classroom and the diversity of thinking, doing and knowing, act in networks of connections, redo exercises, exchange ideas and in fact form a group with a view to extending their knowledge.	Subjects' introductory mind maps; Mind maps of the research projects, in which each researcher can express their authorship as the research develops, while having the collaboration of their colleagues (co-assessment) and teachers (hetero-assessment). Individual and collaborative concept maps of study texts used during the course (from a theoretical and methodological perspective). Participation by everyone (students and teachers) in the interactive part of the portfolios, where assessment and learning occur together. Everyone can assess everyone else and make comments on everyone's productions. In addition, each learner can reflect on and control his learning through self-assessment.	The possibility of including different kinds of productions and resources in different languages. The singularity and plurality of the productions associated with the development of an intentional pedagogical practice in the context of an online course. A dynamic assessment method, primarily because the teachers can monitor the process of change that occurs while the individual is learning about the specific course study object as well as the study process and study object of each piece of research individually. A feeling of belonging and authorship. Each researcher (learner and teacher) self-regulates their own learning process.

We were thus able as part of our study to practice the construction of portfolios that made the learning process apparent, so that each researcher and their peers (colleagues and teachers) could follow the learning process using the concept of interactive assessment. The fact that individual productions were visible, combined with the interaction with fellow researchers and their learning backgrounds, made putting learning assessment into practice a unique experience. In Table 1 a number of observations based on this study are presented.

We were thus able as part of our study to practice the construction of portfolios that made the learning process apparent, so that each researcher and their peers (colleagues and teachers) could follow the learning process using the concept of interactive assessment. The fact that individual productions were visible, combined with the interaction with fellow researchers and their learning backgrounds, made putting learning assessment into practice a concrete rehearsal for interactive assessment.

It should be stressed that even in the light of the above observations the challenge of interactive assessment is a constant one. Our approach to assessment was, and continues to be, based on the use of exams, even when we try out and have the opportunity to experience new assessment practices. This study also showed the difficulties we experience when we are faced with the challenge of co-assessment. Assessment very often continues to be unidirectional (assessment by the teacher of the student) and is rarely bidirectional between teacher and student. In short, an investment must be made in interactive measures involving hypertext that allow everyone to assess everyone in the co-creation of knowledge and learning.

REFERENCES

Ardoíno, J. (2003). *Para uma pedagogia socialista*. Brazil: Editora Plano.

Freire, P. (1979). *Conscientização: Teoria e prática da libertação*. São Paulo, Brazil: Moraes.

Freire, P. (1996). *Pedagogia da autonomia: Saberes necessários à prática educativa*. São Paulo, Brazil: Paz e Terra.

Hoffman, J. (2004a). *Avaliação: mito e desafio: Uma perspectiva construtivista*. Porto Alegre, Brazil: Mediação.

Hoffman, J. (2004b). *Avaliação mediadora: Uma prática em construção da pré-escola à universidade*. Porto Alegre, Brazil: Mediação.

Hoffman, J. (2004c). *Avaliar para promover: As setas do caminho*. Porto Alegre, Brazil: Mediação.

Lemos, A. (2002a). *Cibercultura: Tecnologia e vida social na cultura contemporânea*. Porto Alegre, Brazil: Sulina.

Lemos, A. (2002b). *Cultura das redes: Ciberensaios para o século XXI*. Salvador, Brazil: EDUFBA.

Moreira, M. A., & Masini, E. F. S. (1982). *A aprendizagem significativa: A teoria de David Ausubel*. São Paulo, Brazil: Moraes.

Morin, E. (1999). *Ciência com consciência*. Rio de Janeiro, Brazil: Bertrand Brasil.

Okada, A. (2004). Cartografia cognitiva: novos desafios e possibilidades. In A. Okada, E. Santos, and F. Almeida, (Eds.), *Curso online: Uso de software na pesquisa qualitativa*. PUS-SP, COGEAE. Retrieved March 2004 from http://cogeae.dialdata.com.br/soft/520/1/1/modulos/texto2.php

Santos, E. (2003). Dos. Articulação de saberes na educação *online*: por uma rede interdisciplinar e interativa de conhecimentos em ambientes virtuais de aprendizagem. In M. Silva, (org.). *Educação Online* (pp. 217-272). São Paulo, Brazil: Loyola.

Santos, E. (2003). Portfólio e cartografia cognitiva: dispositivos e interfaces para a prática da avaliação formativa em educação online. In E. Santos, & M. Silva, (Eds.) *Avaliação da aprendizagem em educação online* (pp. 316-331). São Paulo, Brazil: Loyola

Santos, E. (2005). *Educação Online: cibercultura e pesquisa-formação na prática docente*. Doctoral thesis, Faculdade de Educação. Universidade Federal da Bahia - FACED/UFBA, Salvador, Brazil.

Schlemmer, E. (2005). Ambiente virtual de aprendizagem (AVA): Uma proposta para a sociedade em rede na cultura da aprendizagem. In B. Valentini, Soares & M. S. Eliana (Eds.), *Ambientes virtuais: Compartilhando idéias e construindo cenários*. Caxias do Sul.

Silva, M. (2005). *Educatión interactiva: enseñança y aprendizaje presencial y on-line*. Barcelona, Spain: Gedisa.

Silva, M. (2006). *Sala de Aula Interatia* (4th ed.). Rio de Janeiro, Brazil: Quartet.

Silva, M., & Santos, E. (Eds.). (2006). *Avaliação da aprendizagem em educação online*. São Paulo, Brazil: Loyola.

Steven, J. (2001). *Cultura da interface: como o computador transforma nossa maneira de criar e comunicar.* Rio de Janeiro, Brazil: Jorge Zahar.

KEY TERMS AND DEFINITIONS

Concept Maps: Concept maps are diagrams that show relationships, links or associations between concepts.

Interactive Learning: Interactive communication is a challenge for all of us who are used to the paradigm of transmission. It can in fact be divided into three parts. The teacher must at the same time be aware of hypertext, use it to enhance his pedagogical action without losing his authorship and, lastly, appreciate that it is not a question of invalidating the classical paradigm.

Mind Maps: The organization of mind maps reflects how the mind processes thinking as ideas arise.

Portfolio: A portfolio is here considered to be an asynchronous communication interface, a device for pedagogical research and practice and, above all, a device for interactive assessment.

ENDNOTES

1 The term "tool" is often used incorrectly instead of the term "interface". In an industrial society, a tool is an individual instrument for making or manufacturing. An interface, in the sociotechnical context of the online computer, is the collective communication space between two or more geographically disperse human or infotechnical faces. It does more than mediate interaction or translate sensitivities between faces. This would be a "tool", a term that is unsuitable for expressing the sense of an interactive meeting place in "cyberspace" or a "parallel universe of ones and zeros" (Johnson, 2001:19).

2 The concept discussed here is inspired by Ardoino, for whom a device is "an organization of material and/or intellectual means forming part of a strategy for knowledge about an object" (ARDOINO, 2003, p. 80).

3 Subsuming concepts are the analytical categories that are the result of the analysis of and dialogic interpretation between theory and empiricism in a meaningful learning process. For Ausubel (in Moreira, 1982), meaningful learning is a dynamic process in which new information is anchored in relevant preexisting concepts in the learner's cognitive structure—a hierarchical structure of concepts that are abstractions of individuals' experiences—whenever a new concept is signified.

4 This course was initially set up to develop a field of research for two doctoral studies. The first related to online education and research-education, by the present authors, and the second, by Alexandra Okada, investigated the relationship between maps and academic research. The course is run by the COGEAE at PUC-SP.

5 The software for cognitive mapping can be accessed free of charge. Cmap Tools (http://www.uwf.com) was developed by the IHMC, University of West Florida, under the supervision of Dr. Alberto J. Cañas. The Nestor Web Cartographer software (http://www.gate.cnrs.fr/~zeiliger/nestor/nestor.htm) was developed at the National Center for Scientific Research in Lyon, France, by Romain Zeiliger.

6 In this study we will not discuss the concept of the web map. To learn more about this concept, visit the following site:

7 Solange Nogueira is associate professor of Didactics and Technology at the State University of Bahia, UNEB. During the course "The Use of Software in Qualitative Research", she created various maps that illustrate the structure of her action-research project. Her project can be seen in the webfolio in the following website.

Chapter 4
Eliciting Thinking Skills with Inquiry Maps in CLE

Alexandra Okada
The Open University, UK

ABSTRACT

The first aim of this chapter is to present the contributions drawn from the study exploring the use of inquiry maps in academic research for eliciting thinking skills. The second objective of this work is to highlight the potential collaborative learning environments (CLEs) have to enable students to learn different mapping techniques and to help them share ways in which they can apply inquiry maps to elaborate their scientific projects. While the study is informed by qualitative research methodology, it employs quantitative data to describe the fieldwork: an online course, which was organized by the author. The participants were lecturers and research students from different countries: Brazil, United Kingdom and Portugal. Findings indicate six kinds of inquiry maps that can be applied in academic research and may contribute to developing thinking skills such as, critical thinking, content thinking and creative thinking.

INTRODUCTION: INQUIRY MAPS FOR ACADEMIC RESEARCH IN CLE

Information literacy is a vital skill for research students in the digital age. Students need to know how to locate, evaluate and use information effectively in their academic courses and in their workplace. They also have to be able to structure the stages of their investigation, and integrate theory and data. Mapping software tools can help them construct meaning from the information selected through search engines, news feeds, course content and research literature.

DOI: 10.4018/978-1-59904-992-2.ch004

Knowledge Cartography (Okada, Buckingham Shum & Sherborne, 2008) is one of the most promising resources for these challenges. Through knowledge maps, learners can integrate information with graphical representations of key components and connections. Concept mapping helps students represent and visualize concepts that they know and do not know (Cañas & Novak, 2008). Mapmaking scaffolds different forms of reasoning about arguments (Van Gelder, 2002), engaging students in meaningful learning (Novak, 1998) and critical thinking (Jonassen, 2000; Jonassen, Beissner, & Yacci, 1993).

This chapter presents how mapping techniques and software tools (e.g. Cmap Tools, Nestor Web Cartographer, Compendium and Freemind) can be used by PhD students to connect knowledge during their research projects. In this study, we denominate "inquiry maps" as a range of six kinds of knowledge maps for developing academic research:

1. Research map for designing a research project.
2. Reference map for collecting references in the literature.
3. Reading map for selecting key ideas of papers' content.
4. Theory map for organising key concepts and definitions from the literature.
5. Fieldwork map for structuring key data from a corpus of documents.
6. Writing map for integrating key arguments for an essay.

The term "inquiry maps" is used in this work to denote graphical representations of knowledge during a research process. The thesis of this study is that these inquiry maps play an important role for eliciting thinking skills by helping researchers identify, connect and interpret key issues, ideas, concepts, data and arguments. Knowledge mapping software, in which learners can construct, examine and transform their thinking, acts as mediating inquiry tools. These tools for representational guidance mediate learning interactions and thinking by providing learners with means to represent emerging knowledge graphically (Suthers, 2003; Roschelle, 1994).

This work also describes a collaborative learning environment (CLE) that employed inquiry maps for research students and educators to learn software tools and apply mapping techniques to develop their research projects. Another purpose for this CLE was engaging participants in sharing their inquiry maps and improving their ways of mapping with peers. These collaborative interactions and feedback about the process of inquiry mapping might lead them to develop thinking skills and improve their inquiry projects. In the CLE analysed in this study, we used three kinds of maps application:

1. Personal map for participants introducing themselves in the CLE
2. Learning path map for participants accessing and visualising activities and content.
3. Portfolio map for participants accessing and visualising their individual and collective productions.

In order to explain each of the above map models, examples were selected and analysed from a CLE created during an online course – *Using Software for Qualitative Research*. This course was offered at the University of PUC-SP in Brazil from 2004 to 2005. The number of participants was 35 research students and 20 lecturers from Brazil, Portugal and The United Kingdom.

This study, thus, aims to address the following research questions:

- What are the contributions of applying inquiry maps to academic projects?
- What are the benefits of using CLEs with diverse mapping techniques for participants?
- What are the challenges of using inquiry maps to elicit thinking skills?

THEORETICAL PRINCIPLES: THINKING SKILLS THROUGH INQUIRY MAPS

The contemporary critical thinking movement, which started in the 60s as an antidote to reproductive and passive learning, argues that learners should not acquire knowledge just by memorising and repeating what texts and experts say. The simple process of "copying and pasting information" does not mean acquiring knowledge. Thus, a significant construction of knowledge involves interpretation, sensemaking and critical thinking (Jonassen, 2000).

At the same time, inquiry-based learning became an important approach to engage students in research projects. Its core purpose is to help learners act as critical thinkers for managing their own investigation rather than act as passive receivers of content. Freire (1967) points out that critical thinking is an important skill not only for apprehending meaning, knowledge and truth of the reality, but also for making decisions, implementing actions and improving results provoking changes. To be critical means thinking-acting-reflecting in order to make improvements (praxis). It also involves reading and writing the world - not only identifying words, but also understanding their meanings, reasons, consequences, aims, context, references and evidence.

Inquiry-based learning has been considered a complex process. Teachers need to provide learners with strategies, tools and guidance by helping them apply what they know and also new knowledge in problem-based activities (Edson, 1997; Hmelo-Silver, Duncan & Chinn, 2007). Inquiry-based learning requires students to develop several skills. Therefore in order to construct knowledge during their investigation, students must be able to:

- Formulate key questions.
- Select relevant information to address the main issues.
- Identify new knowledge and make sense in order to construct meanings.
- Choose appropriate methods of inquiry.
- Develop possible solutions and draw conclusions.
- Get feedback and points of view to evaluate the process and products.

Academic research is also framed as ongoing and complex process of raising significant questions, integrating relevant information and generating acceptable lines of reasoning grounded on scientific assumptions and bodies of knowledge (Veerman, 2003). A key capability to foster in research students to tackle this challenge is the ability (1) to map information, ideas and arguments; and (2) to manage the rich connections that emerge between them using a range of cartographic techniques. (Okada & Buckingham Shum, 2006).

Cartographic representation is one of the most ancient forms of communication and tools for thinking (Harley & Woodward, 1987). Map-making, which pre-dates both numbering systems and writing, has been used to represent not only geographic space, but knowledge areas as well. The earliest concept maps were found during Middle age to describe the nature of elements, concepts and meanings (Edson, 1997). During the early modern period, concept maps were used to organise and classify different areas of knowledge, curriculum and libraries.

However, the digitalisation of cartography and the widespread use of computer have led to rapid manipulation, transformation and reconstruction of graphical representations than ever before. The diversity of software tools (e.g. Cmap Tools, Freemind, Nestor Web Cartographer and Compendium), mapping techniques and tutorials available on the internet, allow individuals or groups to create high quality maps for representing and sharing knowledge. Knowledge cartography as a strategy *to capture, mediate, and improve constructive discourse* presents diverse mapping

Table 1. Genres of knowledge map knowledge cartography

Mapping techniques	Aims	Freeware Tools
Concept Map (Novak, 1998)	to organise concepts	Cmap Tools
Mind Map (Buzan, 1993)	to generate ideas	Freemind
Web Map (Okada & Zeiliger, 2003)	to collect web resources	Nestor Web Cartographer
Issue Map (Conklin, 2006);	to structure discussions	Compendium
Argument Map (Van Gelder 2002)	to develop argumentation	Compendium

techniques and software tools to map knowledge (see Table 1), which can be applied to learning and research (Okada, Buckingham Shum and Sherborne, 2008).

Well-designed maps are flexible sources of communication and tools for thinking because they help people use their minds to identify key elements and connections that may otherwise not be noticed without graphical visualisation (Dodge and Kitchin, 2001).

Weaving connections between nodes in the network is the most flexible way to bring ideas and information into locally coherent relationships with each other, knowing that there is always another viewpoint on the validity of these patterns. (Okada, Buckingham Shum, & Sherborne, 2008, p. ii)

Interpreting knowledge from maps help students visualise and identify important structures or steps around problem-solution such as: generalisation, enumerations, sequence, classification and, compare and contrast (Cook and Mayer, 1988). McTighe(1992: 183) points out that graphical representations "have proven to be effective tools for enhancing thinking and promoting meaningful learning by helping teachers and students to organise information, generate many ideas, represent abstract concepts, illustrate relationships, relate new information to prior knowledge, store and retrieve information, and assess thinking and learning".

Making maps helps learners make their thinking explicit and that this can support them in the metacognitive process of developing better thinking strategies. The inquiry pathways represented by inquiry maps provide learners with graphical representations for reflecting in action and reflecting about their own reflections during their research projects. Inquiry maps can be applied in several stages of a research project to make thinking visible by drawing out lines of reasoning. These inquiry pathways provide researchers with representational guidance to interpret and construct meanings by visualising key components and their connections.

Paul (1992) emphasises the importance of thinking about thinking. Eliciting thinking skills require students to think in order to improve their own thinking by skillfully taking charge of the structures inherent in thinking. He defines critical thinking as the capacity to question positions, arguments, assumptions and values in order to identify the real meaning. Thinking skills comprise the ability to formulate, analyse and assess problems. It also implies the study of assumptions, concepts, evidence, inferences, purposes and consequences.

In order to investigate the use of inquiry maps in academic research for eliciting thinking skills, this qualitative research draws on the work of Jonassen (2000) who points out some principles to elicit thinking skills in terms of:

- **Content/Basic Thinking:** It represents the ability to make sense of accepted information, declarative and explicit knowledge. It refers to the skill of interpret general

Figure 1. Eliciting thinking skills through inquiry maps (Okada, 2006)

Table 2.

Inquiry Maps	Thinking skills	Research Steps
Research Map	Problem solving Designing	Problematisation: map a brainstorm of questions in order to find key issues
Reference Map	Decision-making	Literature Review: map relevant sources of reference in order to select key literature to ground your ideas.
Reading Map	Evaluating	Interpretation: map the content of the selected papers to make sense of key concepts
Theory Map	Connecting	Conceptual Studies: map different approaches to integrate a key body of knowledge
Fieldwork map	Analysing Elaborating	Analysis: map your data based on an appropriate inquiry method to address the research questions
Writing map	Synthesising Imagining	Synthesis: map key components of the research process: issues, references, concepts, methods, data and findings in order to visualise key arguments and develop a coherent summary.

knowledge and common sense information. Content basic thinking requires learning and retrieving what has been learned.

- **Critical Thinking:** This represents the dynamic process of mapping knowledge in meaningful and usable ways though analysis, evaluation and connections. It integrates important skills such as evaluating the process by appropriate criteria analysing interrelationships among relevant elements mapped through connections and recognising gaps, vagueness and misunderstandings.
- **Creative Thinking:** It shows the ability to go beyond accepted knowledge to create and reconstruct new knowledge. Creative

thinking must be connected to content thinking and critical thinking in order to integrate existing knowledge with the skill of innovative thinking.

Figure 1 and Table 2 describe six kinds of inquiry maps which can be used to develop six stages of a research project. It also presents the types of thinking skills which were integrated by the author in the inquiry cycle.

FRAMEWORK FOR ASSESSMENT: INQUIRY MAPS IN CLE

Baker (2003) emphasises that collaborative problem-solving and argumentative discussions help students choose better problem solutions, develop thinking skills and co-elaborate deeper understanding. McTighe (1992:190) points out that the uses of graphical representations in CLE benefits students in at least four ways:

1. They provide a focal point for group discussions by offering a commons frame of reference for thinking
2. They provide a "group memory" or tangible product for the group's discussion
3. They encourage students to expand their own thinking by considering different points of view
4. They helps to articulate diverse lines of reasoning and helps to render the invisible process of thinking visible for all participants.

However, some scholars argue that maps constructed by learners are difficult to be understood by other learners (Mayer, 2003). Representational notations in maps manifest themselves as constraints, presenting limits on expressiveness, and on the sequence in which knowledge units can be expressed (Suthers, 2003). A map's content can be clear for some mappers, but sometimes can not be understood for some readers. "Maps can work well as a tool for one's own sense-making, but not necessarily as a tool for transmitting knowledge to someone else (Zimmer)" (Okada and Connolly, 2008:12).

There are several factors involved for creating well-designed maps such as domain expertise, fluency with the tools, familiarity with mapping techniques (Okada & Zeiliger, 2003) and structural patterns (Chen and Czerwinski, 1997).

The "*elicit*" principles developed by Okada (2004) during the online course aims to provide some guidelines for applying mapping tools and techniques to create well designed maps. Through some structural patterns, maps can be assessed and improved by offering a clear structure, easy to be understood according to six properties:

- **(E)**xplicit goals: Representing clear goals to be achieved with the map is an initial step for the inquiry pathway. This start point can be a question or problem. Making research aims explicit in the map helps researchers understand what they can use it for and start their research projects.
- **(L)**earning activities or research actions: Connecting learning activities in the map based on the research goals helps students plan next steps in their inquiry projects and visualise tasks that were not solved. This set of research actions will guide learners to achieve their aims during their navigation and mapping.
- **(I)**nteresting information: Integrating relevant information in the map will be useful for making sense of new concepts. Organising and connecting interesting information help researchers identify what they already know and what they want to discover.
- **(C)**lear connections: Describing connections through links, text, and icons can be meaningful for making the lines of reasoning explicit. Structuring connections without making the map confusing adds

more value by making the thinking process clear.

- **(I)**ntegrated overview: Getting the big picture in the map means be able to zoom out the map and visualise the connection among the most relevant points - familiar and unknown ones. Offering a big picture of the main topics through a simple map interface will allow researchers to see the key information initially and, through the pre-defined filtering or layering process, give them the opportunity to magnify or zoom into the areas that they want to explore.
- **(T)**rail as a marked route: Visualising and tracing learning pathways means to explore, discover, and ascertain significant steps. This trail with key steps will help learners to feel more confident in exploring, discovering, and ascertaining different pathways without becoming lost.

Another way to improve maps is the ability of assessing maps in group. Learners can improve their maps by getting collaborative feedback about what is not clear (Veerman, 2003).

Research students must be aware of questions such as: What is this map for? What am I trying to accomplish by using this map? What does this map show that I want to discover? What does this map show that I already know? What is missing in this map that I should include to make it clear? Is there anything that I could change in this map to make it more useful? What other situations and issues that this map can be useful for?

Table 3 shows some questions that teachers and researchers used in this course to assess collaboratively each kind of inquiry maps.

Table 3. Table for assessing inquiry maps (Okada, 2006)

Inquiry Maps	Rubrics	Some questions to assess content of maps
Research map	**Research-questions**	• Does your map present good research questions or aims?
	Research-description	• Is your research project well described through key questions: What? Why? For what? Who? Where? How? When?
	Research-knowledge	• Is your prior knowledge about the topic visible?
Reference Map	**Reference-relevance**	• Does your map indicate relevant references in the field?
	Reference-quantity	• Does your map show enough references to start your study?
	Reference -structure	• Are your references well organised by key concepts?
Reading Map	**Reading-summary**	• Does you map show a good summary of your text
	Reading-headings	• Are the key concepts to structure your reading well described? (theme, relevance, aims, concepts, analysis, claims, evidence, conclusion)
	Reading-understanding	• Does your map allow you to understand the content?
Theory Map	**Theory-relevance**	• Does your map present relevant theory?
	Theory-viewpoints	• Does your map integrate different viewpoints?
	Theory-meaning	• Does your map allow you visualize new meanings?
Fieldwork Map	**Data-relevance**	• Does your map present relevant data from your fieldwork?
	Data-quantity	• Does your map show enough data for your study?
	Data-structure	• Is your map well structure allow you find specific data quickly?
Writing Map	**Writing-structure**	• Does you map present a clear structure for you writing about your research?
	Writing-connections	• Does you map connect key-categories such as context, hypothesis, aims, background, methodology, findings, and considerations?
	Writing-understanding	• Does your map help you write your understanding about the topic?

Table 4. Table for evaluating inquiry maps (Okada, 2006)

Rubrics	Some questions to assess aesthetic characteristics of maps
Structure	Is the title of the map visible and clear? Are the components and their connections well organized? Are the relationships between objects well described? Is the map easy to be understood?
Illustration	Does the map offer a global picture of its content? Are the components relevant and clear? Does the map achieve its purpose? Does the map allow you to understand its content?
Layout	Is the design of the map clear? Are the text and images well organized in the map? Are the connections visible and easy to be identified? Does the map allow you to read and browse its content easily?

Another relevant step to improve knowledge mapping is evaluating aesthetic characteristics of a map. The content of an inquiry map can be better understood when researchers improve their maps' structure. Well structured maps can facilitate the process of inquiry as well. Table 4 presents the questions that teachers and researchers used in this course to help them evaluate structure, illustration and layout of their maps.

CASE STUDY: FOSTERING CRITICAL THINKING THROUGH INQUIRY MAPS IN CLE

In this section, some strategies about the uses of conceptual maps to foster critical thinking in CLEs are presented. In order to understand how this framework can be applied in online courses, I analyse some maps built by a community of researchers during the online course "*The uses software in qualitative research*" (USQR).

CLE: Our Aims

"*The uses of software in qualitative research*" (USQR) was an online course organized by the author at Pontificia Universidade Catolica in Sao Paulo Brasil between 2004 e 2005. The aim of this course was to apply mapping techniques in research projects.

In 2006, the author organized another course entitled "*Writing academic papers using maps*" (WAP) for all previous participants interested in some mapping techniques to write academic papers. In 2007, the author published the online course at OpenLearn Project for the Open Research Community in Collaborative Learning CoLearn to map open educational resources.

The learning outcomes of this online course were to:

- Understand concepts which underpin the uses of mapping for qualitative research.
- Apply mapping techniques in a research project to collect web resources (web mapping), generate new ideas (mind mapping), organise concepts (concept mapping) and structure arguments (argument mapping).
- Use different mapping software tools, depends on the context and interests, such as: Nestor Web Catographer, CMap Tools, FreeMind and Compendium.

The learning outcomes of the WAP course were to:

- Analyse key ideas during academic papers reading
- Integrate most relevant ideas from different sources
- Systematise key arguments for writing academic papers

CLE: Academic Actors and their Networks

The participants of this community were postgraduate students, researchers and lecturers. Table 5 shows participants who registered in the courses, completed all activities and at the end participated in the Emapbook. Some participants, due to had personal problems, had to leave the community before finishing the course. For instance, in USQR2 40% of the participants were busy finishing their thesis; therefore the participation was lower than other groups. Few students had problems with equipment, internet and software tool installation and were not able to participate.

Table 6 shows the participants of this research community organized by fields and professional activities, while Table 7 shows the participants organized by their cities. Most participants were PhD and MA students (20) from São Paulo (25).

In the CLE environment, the participants introduced themselves through concept maps using CMap tools to describe their personal and professional life. Box 1 and Figure 2 show the example of a MBA student´s concept map created to introduce herself in the USQR community. The text shows some information about Laura´s professional and personal life. The content in the map is different to the text. It shows how she

Table 5. Participants of online courses about the uses of maps for research 2004-2006

Participants	2004 USQR1	2004 USQR2	2005 USQR3	2005 USQR4	Total	2006 WAP
Registered	15	13	9	18	55	12
Active members	14	7	6	15	42	12
Emapbook authors	9	6	5	10	30	24

Table 6. Participants - Field

Participants	Qt	Field
Lecturers	7	Biology, History, English Language and Computer Science.
Scholars	5	Medicine, Psychology, Business, Marketing, Law.
Educators	7	Education, History, Social Science, Health.
MA students	11	Education, Business, Maths, Journalism, Social Science, Biology,
PhD students	8	Education, Economy, Social Science, Technology, Anthropology.
Research fellows	4	Economy, Architecture, Maths, Psychology,

Table 7. Participants - Location

Location	(Brazil)					(UK)	(Portugal)
	Sao Paulo	Rio de Janeiro	Bahia	Goias	Rio G. do Sul		
Quantity	25	5	3	3	2	2	2

Box 1.

My name is Laura, I graduated in Business at PUC / SP in 1974 and I then start the Masters in Business Management at FGV. Unfortunately, I have not finished this course. I have been a consultant since 1974. Moreover, I have developing with my own company in the areas of marketing and strategic planning since 1990 and most recently, knowledge management. I am a lecturer at ESPM - School of Marketing and I am taking a MBA course in knowledge management at PUC / SP. I am glad to share my life with a wonderful partner and twelve wonderful cats, which were rescued from the street. Protecting animals is one of my current activities, so if someone would like a pet, let me know.

Figure 2. Personal map for participants´ introduction to USQR with Cmap Tools

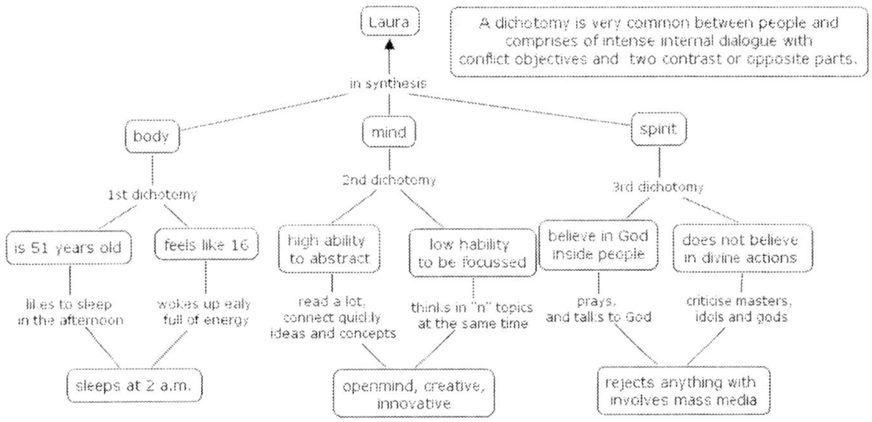

represents her reflection about herself. In this map, Laura points out three dichotomies in her life. She indicates some of her skills (e.g. ability to connect ideas and concepts quickly) and difficulties (e.g. low ability to be focused). In this example, it is possible to notice that introducing herself through concept map helped her reflect on and share her personal aspects as a researcher. The map reveals some aspects of her personality that are not described in the text.

When participants created and shared maps in the CLE, they started to know each other and themselves in different ways, which promoted more thinking and familiarity with the group. The research students mentioned that maps were useful to identify common interests and similarities between them, which helped their communication, and collaboration.

CLE: Interactions and Learning Activities

Learning activities were planned to make the participants explore collectively mapping techniques in their individual research projects and also to reflect and discuss about the uses of mapping techniques to develop academic projects.

The course was organized in two parts:

1. Mapping techniques and software tools.
2. Mapping techniques and qualitative research.

In the first part, the participants introduced themselves in the forum; they installed the software tools and explore different kinds of mapping technique such as concept maps, mind maps and webmaps. The second part, whose content was presented through a learning path map (*Figure 3*), focused on principles to support the uses of maps in qualitative research. Its first activity was a "round table", where four authors Moraes, Macedo, Canas and Zeiliger presented papers with discussion forums with questions related to authors' papers. In the second activity, the participants had to improve their research map based on teacher's feedback and colleagues' comments. In the third activity, they should work in groups in order to evaluate and improve their maps. Finally, they should write a paper with their maps (map-paper).

Through a learning path map, the participants were able to visualise a global picture of learning activities and identify groups to interact and give contributions. They were able to navigate through hyperlinks, select papers to read and choose groups to discuss based on their interests.

The participants exchanged ideas in the discussion forum about the papers, raised new questions and shared their reflections about new concepts (e.g. subsumer, theoretical and empirical research, models of knowledge and constructivism approach). Some participants logged the following in their blogs:

- *"The website of an online course is more attractive and objective through concept maps"*

Figure 3. Learning path map created in Cmap tools created by Okada (2004)

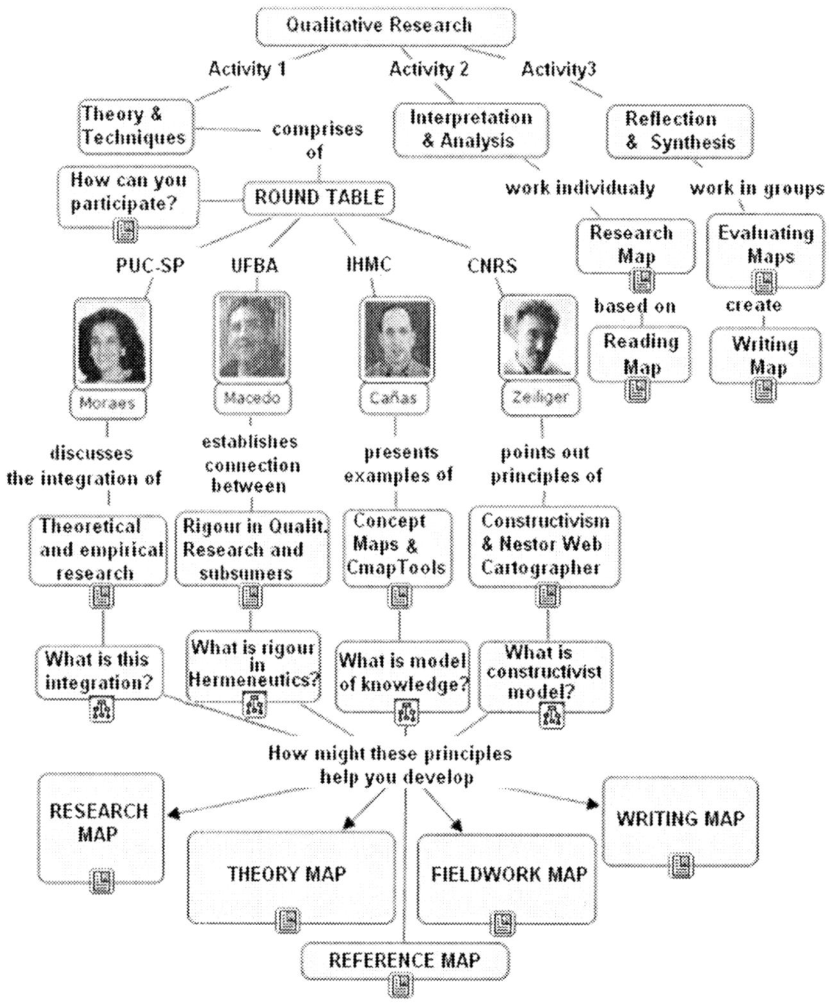

- *"Maps allow us to visualize different options and select what we are interested in"*
- *"Maps help us identify relations between concepts while we are browsing the content"*
- *"Through maps, it is easier to connect our reading to activities and learning goals"*

CLE: Individual and Collective Contributions of Participants and Groups

The online learning environment was created to promote a virtual community of research. The learning activities were planned to engage participants in collaborative learning. The technology used was a Moodle platform, in which several maps

Figure 4. Portfolio map about individual and collective productions in Nestor Web cartographer created by Okada (2005)

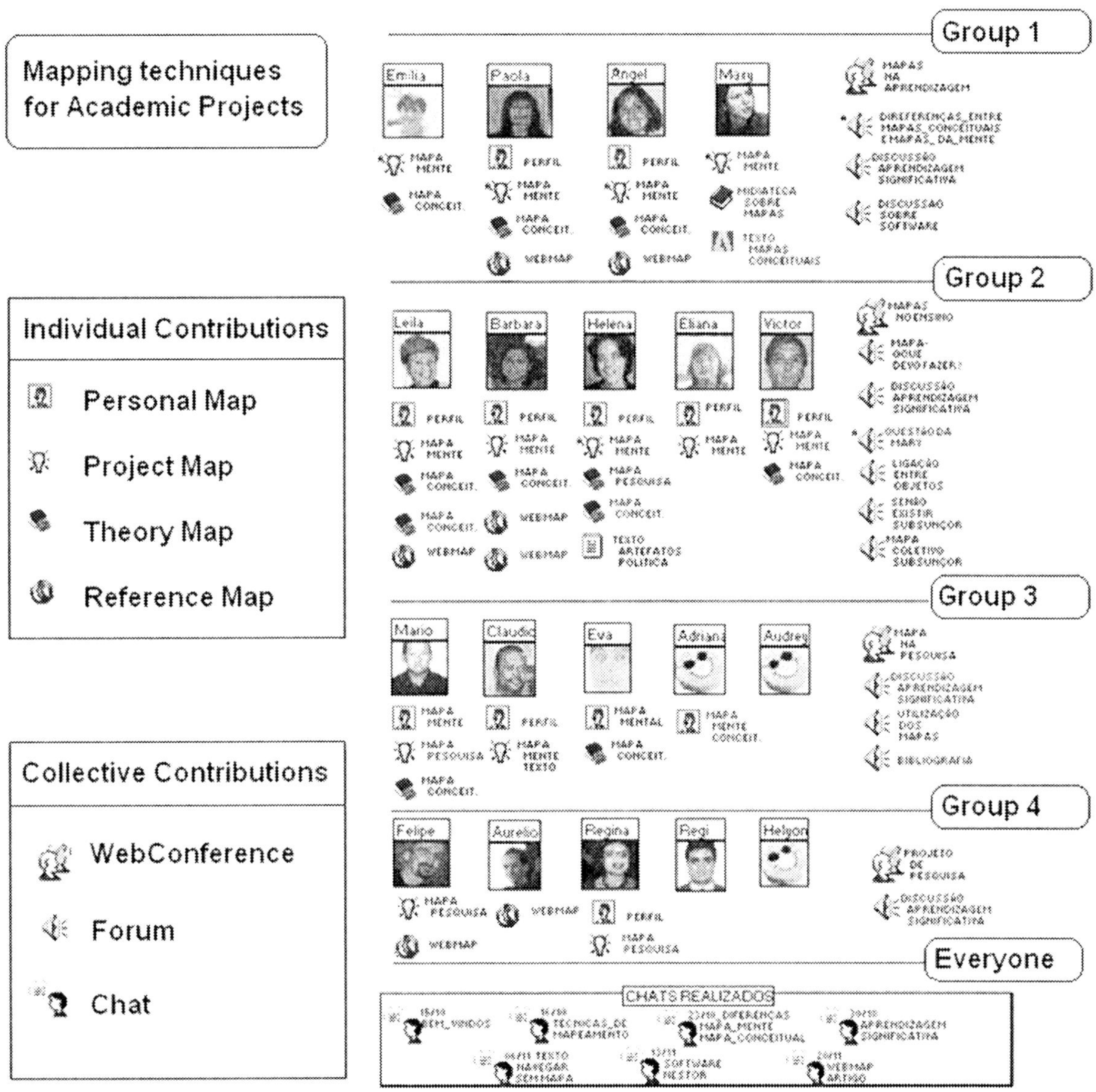

were integrated to present the participants' life history, online content and collaborative tasks.

Figure 4 shows a portfolio map, where the participants accessed interfaces to share their ideas, work samples, maps and papers. Through this map, they were able to visualize, navigate and choose different interfaces to add individual or collective contributions:

- **Forum:** asynchronous interface for exchanging ideas, where participants confronted questions and discussed meanings in order to make sense of theory and methodology.
- **Chat:** synchronous interface for exchanging ideas, where participants described their problems and contributed with solutions about technology.
- **Videoconference (FM tool):** synchronous interface for online meetings, where participants discussed their productions and feedback.
- **Wiki:** collection of web pages for constructing text collaboratively.
- **Blog:** website with regular entries of commentary, where participants shared on self-

Figure 5. E-Mapbook published in Portuguese (http://projeto.org.br/emapbook/)

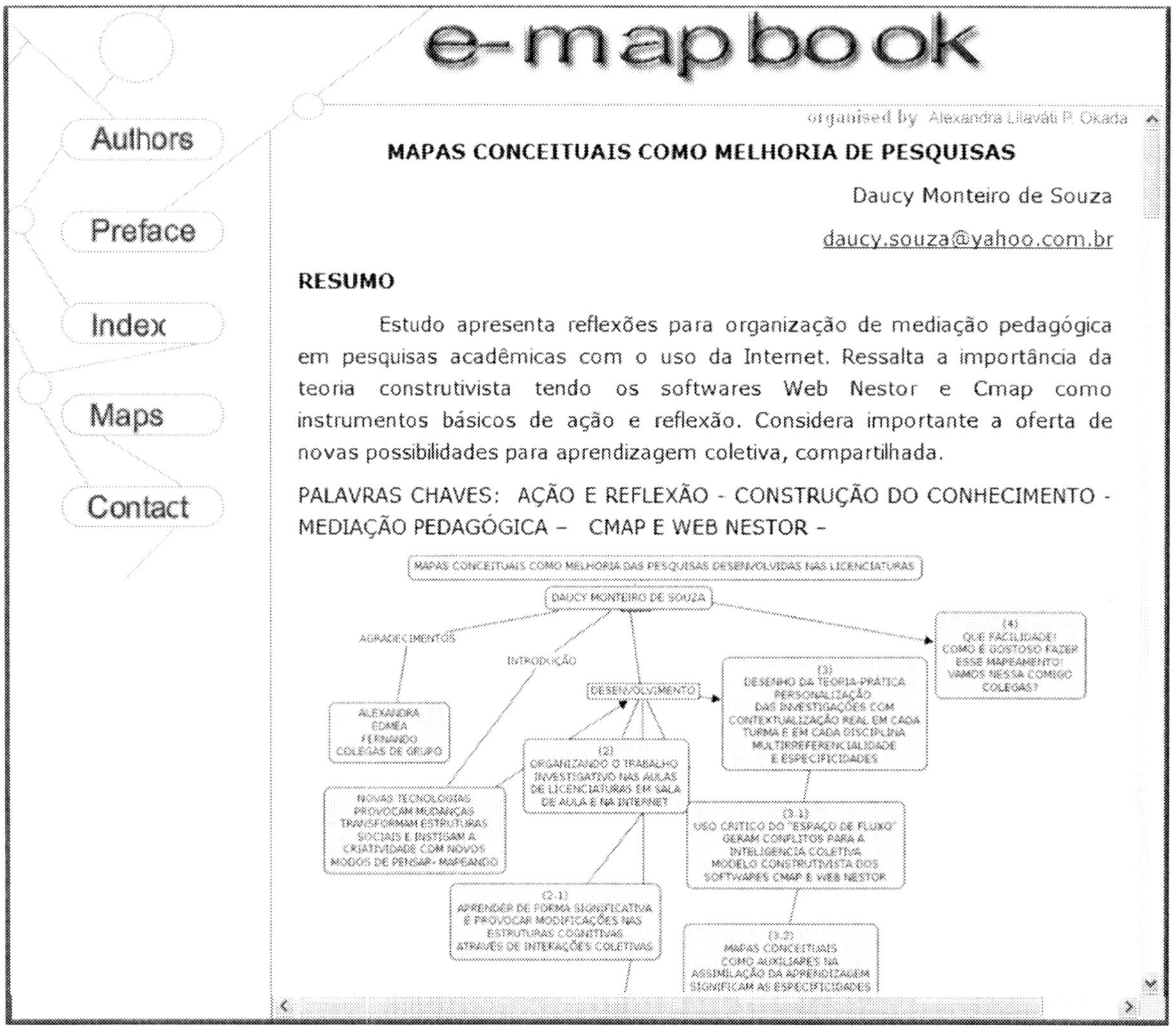

reflection about the research and learning process with feedback of all participants.
- Maps: graphical representations created in different tools.

Participants described that the portfolio map allowed them to:

- *"feel engaged to share more content by visualizing all contributions"*
- *"follow own progress and identify easily where to add or update content"*
- *"access other colleagues' contribution quickly to give them feedback"*

CLE: Collaborative Productions: E-Mapbook

The e-mapbook (*Figure 5*) is an electronic book in Portuguese, which was produced collaboratively by thirty participants (described in Table 5) of the USQR community and organized by the author.

Figure 5 shows also a map paper about "*Concept maps to improve research*" by Souza with research map, theory map and reference map. Souza commented that publishing a paper and participating in the review process helped her develop critical view of her work.

CLE: Individual Productions: Inquiry Maps for Developing Academic Research

In order to present the benefits of using inquiry maps for developing thinking skills in research projects, we selected thinking skills, which were described in Figure 1, to analyse six models of maps: research map, reference map, reading map, theory map, fieldwork map and writing map. Table 8 shows these principles and the messages written by the participants in their blogs. These messages describe how researchers analysed the contributions of mapping for developing their project and their thinking skills.

Research Map: Introducing the Research Project Through Maps

The research map in *Figure 6* shows the structure of a research project with the main key concepts to generate a brainstorm: research questions or aims (what?), relevance of research (why?), contributions in the field (for what?), methodology of investigation (how?), work field (where?) and research schedule (when?). By using this structure, the participant was able to bring forty-one key ideas that might be useful to plan the investigation.

Good inquiry projects depend on significant questions. At the beginning of a research, it is not easy to define a relevant issue. Initially, students and researchers can be lost when they have to face lots of information without questions, or when there are many questions but no significant references. Mapping the starting point of a research project can help people to find better questions (Conklin, 2006).

Learners can connect prior knowledge and initial questions in order to find an important issue to be investigated through a "research map". Mind mapping can be used to generate a brainstorm of key research issues and to organise the initial structure of a project. Finding a significant and innovative issue becomes easier when relevant information is mapped.

The research students described that mapping their project helped them find out critical topics that need to be explored better. The identification of new possibilities and difficulties facilitated a continuous process of elaborating new questions in order to the delimitating better the focus of their research.

Reference Map: Organising References

The reference map in *Figure 7* shows 82 documents related to critical thinking literature review classified by nine conceptual categories (glossary, papers, website, case study, foundation, abstract, software, thesis and books) and 16 types of icons.

Table 8. Fostering thinking skills through inquiry maps in research projects

Inquiry Map	Thinking skills	Researchers' comments
Research Map	**Problem solving**: reformulate questions, find new alternatives, build acceptance **Designing**: formulate goals, draft outcomes, revise process.	*"Through my research project I could find a focus for my research. After creating several maps of my investigation, I could visualise the main ideas and identify the key questions of my research."*
Reference Map	**Decision-making:** identify possibilities, generate alternatives, compare options.	*"Constructing a map to select references make me think what are the most relevant literature to support my research and how could I group them to facilitate future studies"*
Reading Map	**Evaluating:** define criteria, assess information, recognise fallacies	*"Structuring my reading through maps are very useful for extracting key ideas from texts."(...) "It makes me reflect more about what the meaning of each sentence an how to categorise relevant information"*
Theory Map	**Connecting:** compare and contrast, infer deductively and inductively, identify relationships	*"Using maps to connect different perspective from the same concept is very challenging. Maps can reduce the meaning of concepts and it is hard to summarise in few words complex definitions." (...) "However, they help us to compare different approaches and identify connections to reconstruct new meanings"*
Fieldwork map	**Analysing**: recognise patterns, classify main ideas, find connections **Elaborating**: reflect, widen and deepthen, update, concretise.	*"The main contributions of organizing a field work through maps, (when you have electronic data - text, image and audio), are - navigate easily in the corpus research, - classify and interpret data - making connections, - find and group relevant analysis based on different perspectives" "Although it is not easy to map lots of data, once you have your corpus well organised it is easier to recognise patterns and identify new issues to be clarified"*
Writing map	**Synthesising:** plan, hypothesise, summarise. **Imagining:** predict, speculate, visualize.	*"Maps applied to writing seems to be a great strategy because it help us visualize and integrate enough evidence to back up our claims, identify ideas to be deepened, approaches to be widened and plan a clear structure for presenting our thinking."*

Figure 6. Research map created in a FreeMind

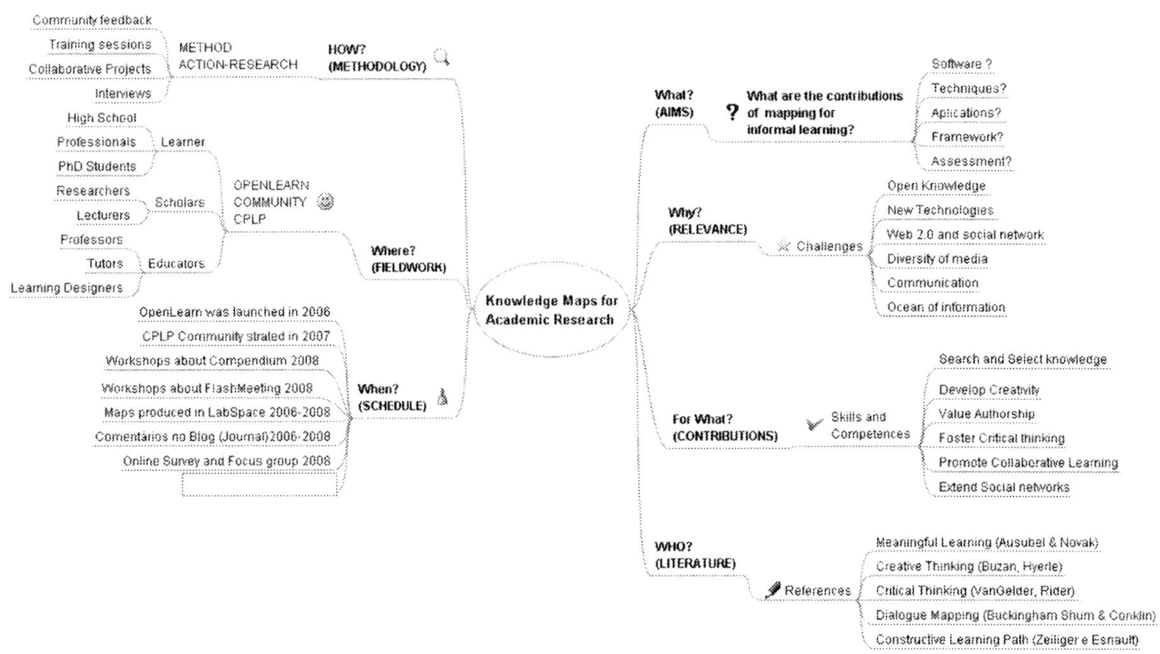

The icons can be useful to distinguish different kinds of content, or format.

During the construction of reference maps, researchers identified some benefits. Mapping papers on the web, intranet, or own computer helps to locate easily a set of documents about a specific topic. Representing many documents in a small space offers a global overview of the literature selected. Maps are also useful to answer the following key questions:

- What are the main articles, papers and other references?
- What are the key theories, foundations, concepts and origins?
- What are the main case studies?
- Is there any technique or tool involved in this process?
- What are the major debates about the topic?

Web mapping for organising references help users get an overview of relevant information from different areas of knowledge, languages, formats and media through web maps (Okada and Zeiliger, 2003).

The participants pointed out that reference maps created through web mapping were useful for selecting, organising and updating diverse references in their investigation. The graphical representation facilitates the process of storing, retrieving and sharing different kinds of documents. It was easy to add summaries, classify materials using categories, and establish connections between them. Visualising all references through maps allowed participants to compare contents and reuse them in other research projects.

Figure 7. Reference map about critical thinking created in Nestor Web Cartographer

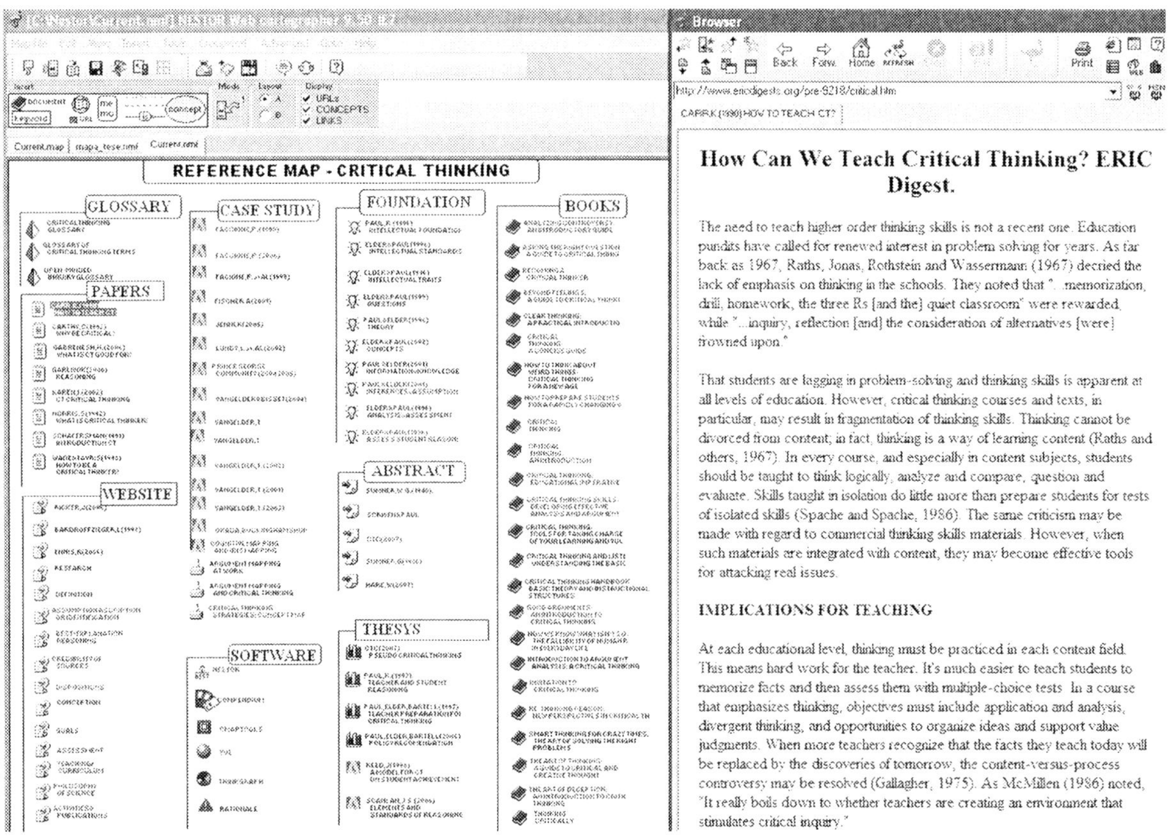

Reading Map: Interpretating and Analysing Papers

This reading map in *Figure 8* about a PhD dissertation presents initially 24 keywords related to critical thinking and the number of frequency which they appear in the text on the top of the map. Its structure is organised by 15 categories: from theme to results. A reading map helps researchers not only summarise the document, but also review, remember and reassess the content. Different categories can be chosen to analyse different kinds of documents. Identifying a significant structure helps readers to browse and study the paper. Through a set of categories, they can organise the main topics of the paper, locate easily the mains ideas and construct a memory system. Moreover, a reading map might help readers explore and make sense of big documents.

Selecting significant references entails interpreting the content. Interpretation implies analysis in order to apprehend meanings (Ricoeur, 1974). It means breaking down the complex text into simple parts. Mapping important statements of a document and their interrelationships helps readers to interpret new concepts. Through reading maps, they can visualise what is important, store and retrieve pieces of information quickly.

A deep analysis of papers allows researchers to identify a network of concepts and lines of reasoning. These abstract elements when are represented in maps turned into concrete elements able to be recombined.

The participants found reading maps useful for understanding how conclusions were drawn from a set of evidences, arguments and contra-arguments. They were able to highlight important key concepts and ideas and use the same template

Figure 8. Reading map about PhD dissertation about critical thinking created in Nestor

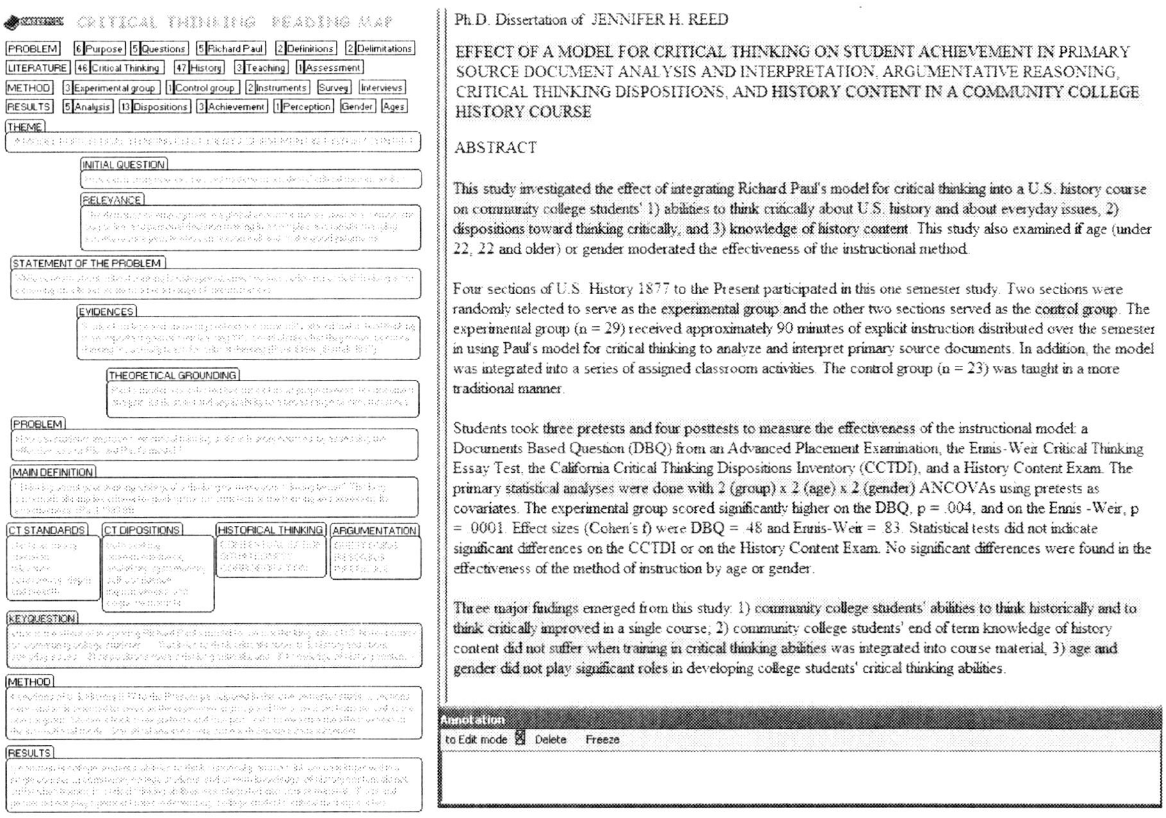

for other papers and also research projects. Mapping the reading process helped also researchers identify new references from the bibliography, which were included in their reference map.

Theory Map: Understanding Concepts

This theory map (*Figure 9*) shows three perspectives (context, key definitions and key concepts) to organise different meanings for "critical thinking". For that, 14 definitions from different authors were selected, grouped by context and ordered by date. From these definitions, 16 words were generated to capture the key ideas, which were integrated in a conceptual area. Researchers consider theory maps as a guide to help them interpret different viewpoints, compare and combine different approaches to reconstruct their own interpretations.

Clarifying concepts is an important step to understanding theories and for meaningful learning (Novak, 1998). Mapping several sources from different authors that explain the same concept helps researchers select and reconstruct maps from a wide and more significant perspective. When concepts are well mapped, learners can compare, combine and reintegrate similar groups of references.

The participants' discussions indicated that theory maps helped them visualise gaps and misunderstanding for further investigations. They identified new concepts that should be clarified. Theory maps were useful to connect concepts, definitions and the original source by organising a graphical memory system of their studies. Participants described that they were able to represent and reconstruct semantic networks from their own perspective and reuse theory maps in different inquiry projects.

Figure 9. Theoretical map about critical thinking created in CMap tools

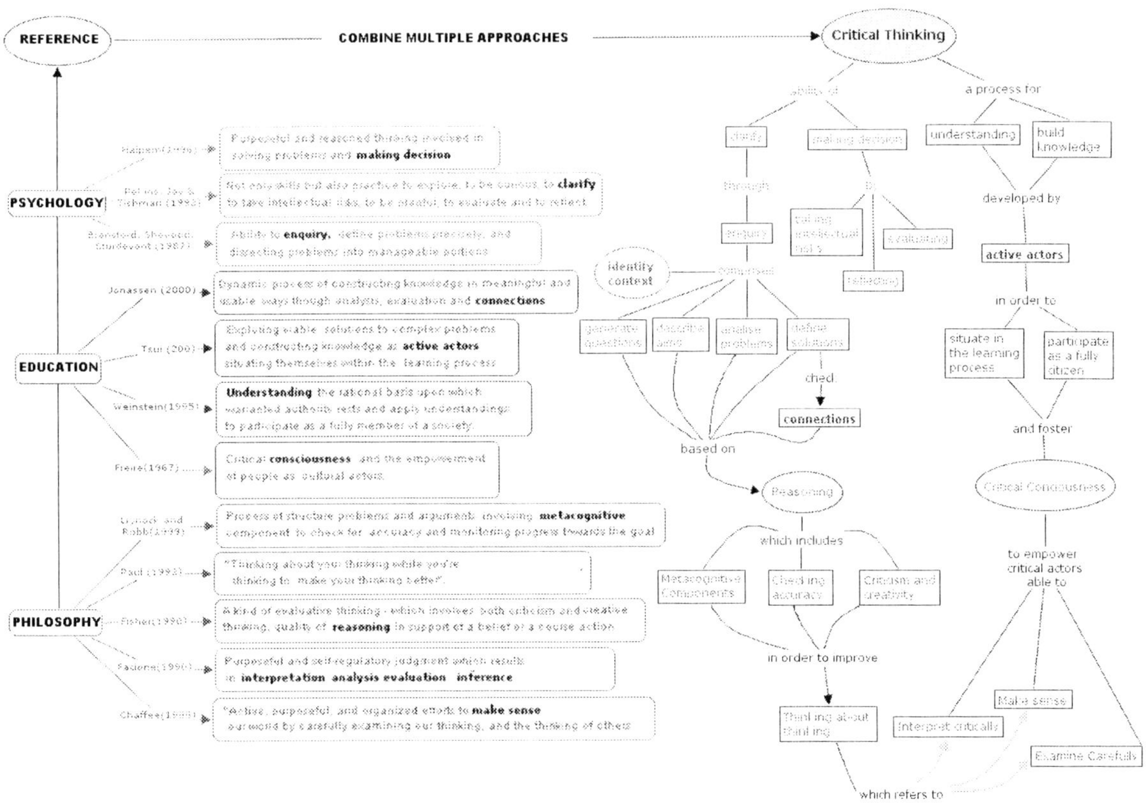

Fieldwork Map: Collecting and Analysing Data About the Fieldwork

The fieldwork map in *Figure 10* describes a case study about the online course USQR. This graphical representation created in the mapping tool Compendium shows the 30 participants and all their mapping productions and research diaries classified by time, topics and grouped by portfolios.

One of the main benefits of fieldwork maps is to classify and connect the most important data and navigate in different contexts. Visualising and analysing key data through maps is also useful for reorganising multiple views and get an overview of the most relevant findings.

Analysing lots of data during an empirical study demands deep and systematic reflection Whyte(1991). Well-designed maps can facilitate the process of analysis mainly when important components are well connected. Thus, researchers must be aware of important issues that might help them create their fieldwork map, such as:

- Criteria from the methodology which assure rigour and quality of analysis;
- Instruments to collect data, organised by categories.
- Theoretical categories from theoretical maps to guide the empirical analysis
- Reflective annotations to interpret their empirical investigation

Figure 10. Fieldwork map of a PhD research created in Compendium

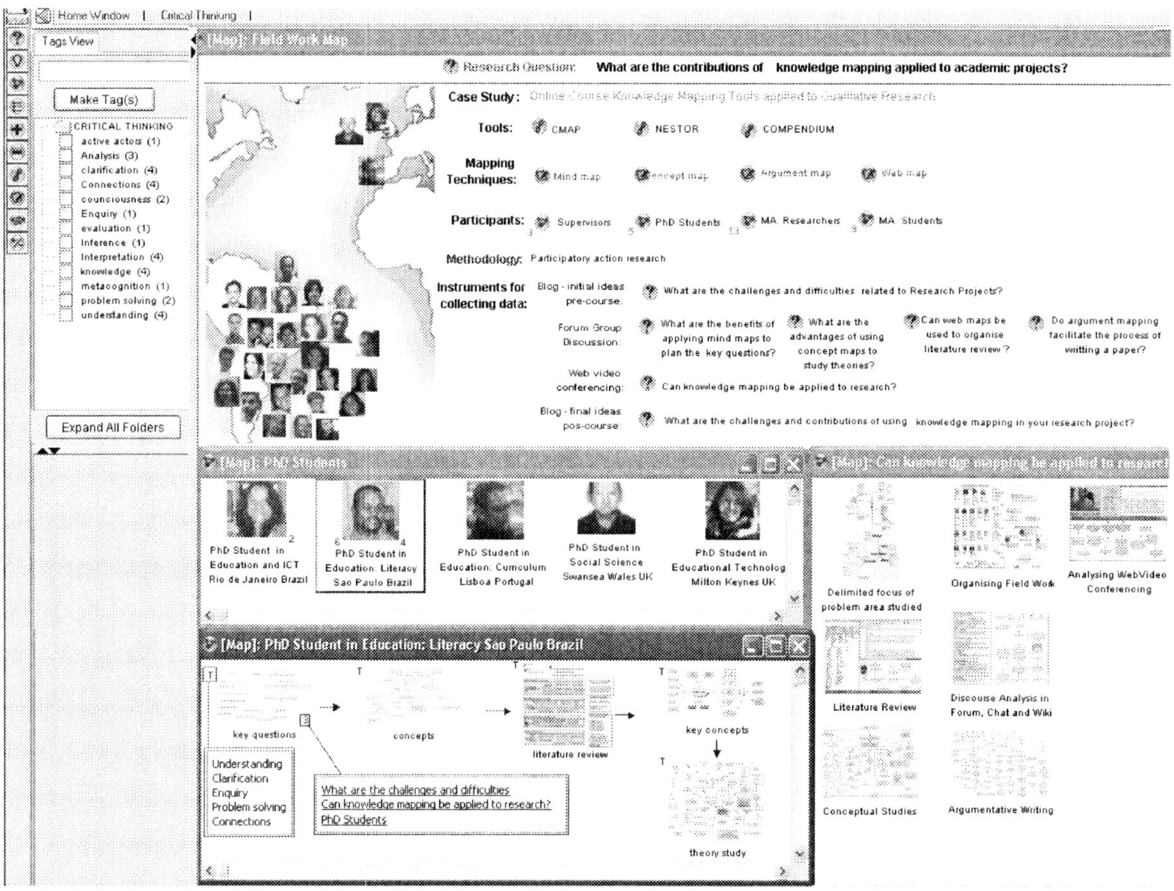

When these issues are integrated into fieldwork maps, researchers are able to reinterpret their meaning visualising parts and the whole context. They are also able to reconstruct several maps to see different perspectives. New key concepts from the field study can emerge generating new categories resulting in a reconstruction of better maps.

Some of the research students wrote that through fieldwork maps, they were able to categorise the main key concepts, include comments about the analysis and integrate relevant data in different perspectives. They also described that when lots of maps are constructed, one critical issue is to focus on the main research question. The fieldwork map can then be used as a guid-

Figure 11. Writing map of a PhD research created in Compendium

Box 2. Abstract of a PhD research developed from the writing map

ABSTRACT - "One of today's great challenges in the context of" research and learning is to find ways to deal with the overload of data. New techniques are needed to organize better the whole process of investigation to construct knowledge. When relevant and significant information is mapped, researchers are able to explore theories and practices thoroughly. Students can also immerse deeply in their learning. The key question of this thesis is "What are the contributions of knowledge mapping applied to academic projects?" The intention of this work is to offer to researchers strategies to improve their investigations. To answer this problem, I based my investigations on the course "Cartography software in the qualitative research" at PUC-Cogeae Online, from 2003 to 2006. The methodology of this research is participatory action research. Questioning, planning and reflections were developed with the research students and the course material was produced during this investigation. About limitations of this research, the experiment group was small, 30 participants: doctors, masters, PhD and M.A. students. This work focused on benefits of knowledge mapping rather than obstacles and disadvantages. In this work, I define what "Knowledge Cartography" is starting with the history of cartography and the concepts of maps and investigation. After that, I discuss epistemological principles to guide the creation of inquiry maps, with reference to systemic, hermeneutic and dialectic theories. Then, I analyse the practice, discussing the contributions of mapping in various stages of research: problematisation, literature review, conceptual studies, fieldwork plan, discourse analyses and argumentative writing. In the end, I emphasize how knowledge cartography helps researchers be more focused and engaged in their investigation, and at the same time they can expand their creative horizons and critical thinking."

ing tool, by reminding the key issues and main direction to follow.

Writing Map: Systematising the Research

After mapping theories and the fieldwork, the next step is to map the research's outcomes and synthesise a significant conclusion. It means integrating each relevant component to form a coherent whole. For that, a well-structured map is useful to organise ideas clearly and coherently. Mapping for arguing (Adriessen, Suthers, & Baker, 2003) can help researchers and learners describe and visualise their line of reasoning. Visualising argumentation (Kirschner, Buckingham Shum & Carr, 2003) in a map facilitates the process of assessing claims by checking if there is enough evidence and facts that support key ideas. Therefore, drawing a coherent conclusion is easier when claims, arguments, counter-arguments, evidence and facts are well connected in a writing map. Writing maps are useful not only for facilitating the process of writing but also for understanding how the research problem was answered.

The writing map in *Figure 11* shows a rational structure to facilitate scientific writing of a PhD research abstract (see *Box 2*). It presents 47 statements: titles, questions, answers, notes, pro and cons. The participants described that a writing map was useful for organising lines of reasoning in a logic and objective way. The writing map helped them organise an argumentative structure and communicate clearly the relevant aspects of their research. (see Figure 12)

DISCUSSION

The first purpose of this study was to identify the contributions of applying inquiry maps to academic projects.

Inquiry maps and comments written by participants indicate that they were able to:

- select and organise relevant content,
- structure and represent their thinking graphically
- integrate new concepts and their own interpretation .

The participants pointed out that research map used to represent key ideas enabled them to find their key questions. The reference map helped them organise the literature review. The reading map was useful to interpret papers. The theory map facilitated the integration of different viewpoints about the same concept. The fieldwork map provided interesting ways to analyse data. Finally, the writing map was a good strategy for summarising

Figure 12. Outline view of a writing map of a PhD research created in Compendium

key ideas with arguments and evidence. Through these inquiry maps and their discussion in the CLE, teachers could also observe that participants were very engaged in applying these mapping techniques to develop their researches.

This study indicates different uses of inquiry maps to develop thinking skills in academic research. Inquiry maps created by researchers show that graphical representations may be useful for developing *the inquiry cycle* (Llewellyn, 2005) and *the spiral of research* (Blaxter, Hughes and Tight, 2001). These two approaches used to develop a scientific investigation describe six steps that in this study were summarised as: *Problematisation, Literature Review, Interpretation, Conceptual Studies, Analysis and Synthesis* illustrated in Figure 13. These six kinds of inquiry maps applied to academic research may help researchers to implement and integrate these six steps better.

Most of the participants (80%) in this course were able to create research maps, references maps and theory maps. Few participants (10%) who had already collected electronic data from their fieldwork were able to create fieldwork maps. In addition, some researchers (30%) who were interested in improving their writing skills elaborated reading and writing maps. In this study, the participants did not apply these six kinds of maps to develop all steps in their research, because most of them were busy in different stages in their investigations and they also presented different priorities. However, the integration of these inquiry maps to develop a research project will be theme for our next investigation in the CoLearn Community.

The second purpose of this study was to identify the benefits of using CLEs with diverse mapping techniques for participants. The collected data indicate that the CLE helped students learn and apply mapping techniques and software tools in their academic research. Inquiry maps presented in the CLE such as learning path maps and portfolio maps played an important role by engaging students in exploring contents, discussions and their productions. Research students were able to visualise connections between

Figure 13. Research flow map created in Compendium

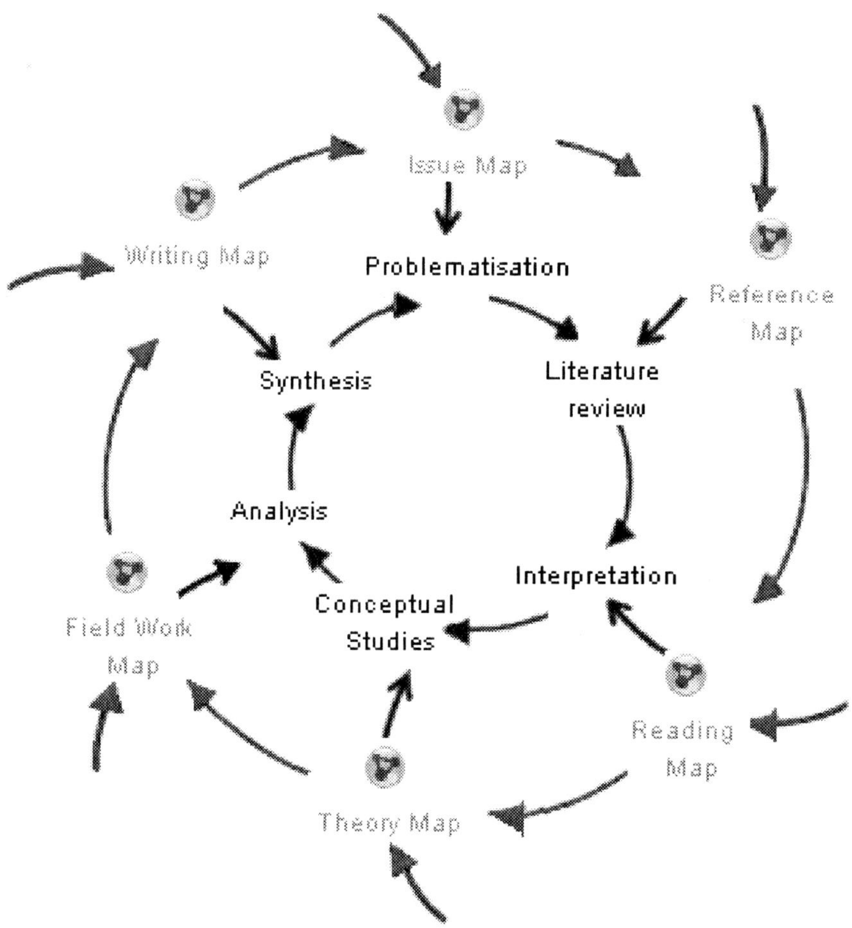

references, activities and learning goals through learning path maps, which helped them browse content and establish connections between theory and practice. Participants were able to access all contributions, discuss about their progress and identify where they could share constructive feedback through portfolio maps in order to improve their productions.

Knowledge integration environments (Bell, Davis and Linn, 1995) through inquiry maps seem to engage learners in developing and applying their thinking skills. Inquiry maps can guide them to find different spaces and groups to negotiate meanings, issues, claims and arguments with evidence and references. When CLEs stimulate learners to interact, contribute and develop productions together, they feel able to share cognitions and construct more knowledge that is significant together.

Inquiry maps can play an important role in CLE in representing collective construction of knowledge where all participants can access and connect different spaces such as questioning space, argumentative space and referential space (e.g. Figure 4) without feeling lost. Participants can negotiate meanings and add contributions connecting evidence. In this sense, these inquiry maps can help them develop and apply their thinking skills by analyzing and establishing more connections between referential space, argumentative space and questioning space in their maps and writing.

Concerning difficulties with inquiry maps in CLE, this study shows that learners (10%) who faced problems with their computers, internet or software tools, gave up learning.

The participants (14%) who mentioned be very busy in their professional activities did not

Figure 14. Research flow map created in Compendium

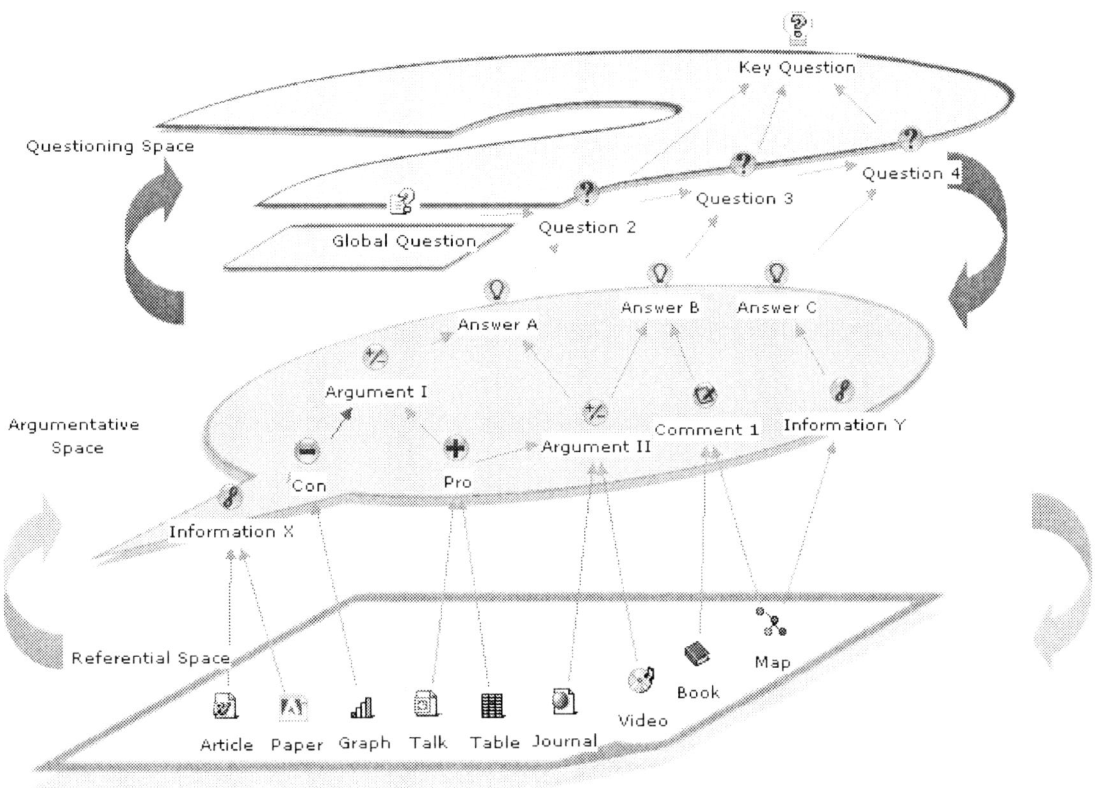

find time to interact in groups and were not able to learn and apply inquiry maps in their academic projects.

Few participants (10%) who were not familiar with graphical representations with hyperlinks found it difficult to understand the content through maps. However when they started to produce their own maps they mentioned that learning path and portfolio maps were very useful.

Regarding the challenges of using inquiry maps to elicit thinking skills, several participants (60%) described that it was hard to explore diverse methods and different technologies.

However, after getting used to mapping techniques and tools, participants could identify differences and apply different resources better. In order to develop good maps it is necessary to get used to think graphically and create several maps. Participants also described that it is hard to avoid reducing the meaning of concepts in maps and to deal with lots of data. Some of them mentioned that big maps could be confusing and maps with several levels can be difficult to navigate and get the whole picture.

CONCLUSION AND FUTURE TRENDS

Findings drawn in this study describe the use of inquiry maps in academic research for eliciting thinking skills. The outcomes of this study also highlight the importance of collaborative learning environments to support researchers in exploring tools and applying mapping techniques in their academic projects.

This work presented six kinds of inquiry maps that can help researchers implement their scientific investigation and develop thinking skills:

1. Research Map
2. Reference Map
3. Theory Map
4. Reading Map
5. Fieldwork map
6. Writing Map

This work also described three kinds of inquiry maps that may help designers to plan collaborative learning environments:

1. Personal Map
2. Learning Path Map
3. Portfolio Map

Moreover, this study has identified some difficulties of the participants in using different tools, represent thinking graphically and creating maps with lots of data. Some insights, which have emerged from this work, such as the use of inquiry maps to develop academic research and the cycle of scientific inquiry will be the focus of the next studies.

The emergence of social software and Web 2.0 which creates new scenarios for open learning and collaborative construction of knowledge also highlights the importance of the ongoing research in this field. Inquiry maps may be considered strategic and heuristic tools for representing what is important, interpreting and reconstructing meanings, recording and sharing new structures of components and connections. All this skills are essential to foster critical thinking and make better decisions in research learning communities and social networks.

ACKNOWLEDGMENT

I am thankful to Dan Suthers and Tony Sherborne for their comments on an earlier version of this chapter and to peer reviewers of this book whose comments have contributed to the improvement of this final version. I am also grateful to Simon Buckingham Shum and Fernando Jose de Almeida for their comments and discussions in support of my work. Finally, thanks to the Brazilian Education Ministry Agency (the CAPES Foundation) for supporting my PhD program and the OpenLearn team for the opportunity to continue this research.

REFERENCES

Adriessen, J., Suthers, D., & Baker, M. (2003). *Arguing to Learn: Cronfronting Cognitions in Computer-Supported Collaborative Learning Environments*. London: Kluwer Academic Publishers.

Anderson, T. (2007). *Reducing the Loneliness of Distant Learner Using Social Software*. Open and Distance Learning Conference. Retrieved on January, 17, 2007 from < http://www2.open.ac.uk/r06/conference/TerryAndersonKeynote-Cambridge2007.pdf >

Baker, M. (2003). Computer-mediated argumentative interactions for the co-elaboration of scientific notions. In J. Andriessen, M. Baker, & D. Suthers (Eds.), *Arguing to Learn: Confronting Cognitions in Computer-Supported Collaborative Learning environments* Amsterdam, the Netherlands: Kluwer Academic Publishers.

Bell, P., Davis, E. A., & Linn, M. C. (1995). *The Knowledge Integration Environment: Theory and design.* Proceedings of the Computer Supported Collaborative Learning Conference (CSCL '95: Bloomington, IN), (pp. 14-21). Mahwah, NJ: Lawrence Erlbaum Associates.

Blaxter, L., Hughes, C., & Tight, M. (2001). *How to Research*. Buckingham, UK: Open University Press.

Cañas, A., & Novak, J. (2008). Concept mapping using Cmap tools to enhance meaningful learning. In Okada, A., Buckingham Shum, S. & Sherborne, T. (Eds.) *Knowledge Cartography: software tools and mapping techniques.* London: Springer.

Chen, C., & Czerwinski, M. (1997). Spatial ability and visual navigation: An empirical study. *The New Review for Hypertext and Multimedia, 3*, 40-66. http://research.microsoft.com/~marycz/nrhm.htm

Conklin, J. (2006). *Dialogue Mapping: Building Shared Understanding of Wicked Problems*. Chicester, UK: John Wiley.

Cook, L. K., & Mayer, R. E. (1988). Teaching readers about the structure of scientific text. *Journal of Educational Psychology, 80*, 448–456. doi:10.1037/0022-0663.80.4.448

Dodge, M., & Kitchin, R. (2001). *Mapping cyberspace*. London: Routledge.

Edelson, D. C. (2001). Learning-for-use: A framework for integrating content and process learning in the design of inquiry activities. *Journal of Research in Science Teaching, 38*, 355–385. doi:10.1002/1098-2736(200103)38:3<355::AID-TEA1010>3.0.CO;2-M

Edson, E. (1997). *Mapping time and space: how medieval mapmakers viewed their world.* London: The British Library.

Freire, P. (1967). *Educação como prática da liberdade*. Rio de Janeiro, Brazil: Paz e Terra.

Harley, J. B., & Woodward, D. (1987). Cartography in prehistoric, ancient and medieval Europe and the Mediterranean. *The history of cartography* (Vol.1). Chicago: Chicago University Press.

Hmelo-Silver, C., Duncan, R., & Chinn, C. (2007). Scaffolding and Achievement in Problem-Based and Inquiry Learning. *Educational Psychologist, 42*(2), 99–107. Mahwah, NJ: Lawrence Erlbaum Associates, Inc. http://www.cogtech.usc.edu/publications/hmelo_ep07.pdf

Jonassen, D. (2000). *Computers as mindtools for schools: engaging critical thinking*. Upper Saddle River, N.J: Merrill.

Jonassen, D. H., Beissner, K., & Yacci, M. A. (1993). *Structural knowledge: Techniques for representing, assessing, and acquiring structural knowledge.* Hillsdale, NJ: Lawrence Erlbaum Associates.

Kirschner, P., Buckingham Shum, S., & Carr, C. (Orgs) (2003). *Visualizing Argumentation: Software Tools for Collaborative and Educational Sense-making.* London: Springer-Verlag.

Llewellyn, D. (2005). *Teaching High School Science Through Inquiry: A Case Study Approach.* NSTA.

Mayer, R. (2003). *Learning and Instruction.* Englewood Cliffs: NJ: Merrill Prentice Hall.

McTighe, J. (1992). Graphic Organizers: collaborative links to better thinking. In N. Davidson, & T. Worsham (Eds.)*Enhancing thinking through cooperative learning*. New York: Teachers College, Columbia University Press.

Novak, J. (1998). *Learning Creating and using Knowledge: concepts maps as facilitative tools in schools and corporations.* London: Lawrence Erlbaum Associates Mahwah.

Okada, A., & Zeiliger (2003). The building of knowledge through virtual maps in collaborative learning environments. In *Proceedings of EdMedia Hawaii USA*, (pp. 1625-1628).

Okada, A. (2004). *Software Tools and Mapping Techniques for Academic Research - Online Course*. Retrieved on January, 17, 2007 from <http://cursosonline.cogeae.pucsp.br/index2.php?wcurso=EDP>

Okada, A. (2005). The Collective Building of Knowledge in Collaborative Learning Environments. In T. Roberts, (Org.) *Computer-Supported Collaborative Learning in Higher Education.* (Vol. 1, pp. 70-99). Hershey, PA: Idea Group Publishing.

Okada, A. (2006). *Cartography for Inquiry: epistemological and communicational interfaces to map knowledge in academic projects.* Doctoral Thesis, São Paulo PUC-SP University & The Open University – Knowledge Media Institute OU-UK.

Okada, A., & Buckingham Shum, S. (2006). *Knowledge Mapping With Compendium in Academic Research and Online Education*. 22nd ICDE World Conference [www.icde22.org.br], Rio de Janeiro, Brazil.

Okada, A., Buckingham Shum, S., & Sherborne, T. (2008, in press). *Knowledge Cartography.* London: Springer.

Okada, A., & Connolly, T. (2008). Designing Open Educational Resources through Knowledge Maps to Enhance Meaningful Learning. Common Ground: The International Journal of Learning, 15(7), 209-220.

Paul, R. W. (1992). Critical Thinking: what, why and how. In C.A. Barnes (Ed.) *Critical thinking: educational imperative*. San Francisco, CA: Jossey-Bass.

Ricoeur, P. (1974). *The Conflict of Interpretations: Essays in Hermeneutics.* Chicago: Northwestern University Press.

Roschelle, J. (1994). Designing for cognitive communication: Epistemic fidelity or mediating collaborative inquiry? *The Arachnet Electronic Journal on Virtual Culture* [On-line serial], *2*(2). Retrieved from ftp://ftp.lib.ncsu.edu/pub/stacks/aejvc/aejvc-v2n02-roschelle-designing

Suthers, D. D. (2003). Representational Guidance for Collaborative Inquiry. In J. Andriessen, M. Baker & D. D. Suthers (Eds.). *Arguing to Learn: Confronting Cognitions in Computer-Supported Collaborative Learning Environments*. Dordrecht, the Netherlands: Kluwer.

Van Gelder, T. J. (2002). Argument Mapping with Reason!Able. *The American Philosophical Association Newsletter on Philosophy and Computers,* 85-90.

Veerman, A. (2003). Constructive discussions through electronic dialog. In: Andriessen, Baker & Suthers (Eds.) *Arguing to Learn Confronting Cognitions in Computer-Supported Collaborative Learning Environments*. Dordrecht, the Netherlands: Kluwer.

Whyte, W. F. (1991). *Participatory action research.* Newbury Park, CA: Sage Publications.

Willinsky, J. (2006). *The Access Principle - The Case for Open Access to Research and Scholarship.* Cambridge: MIT Press.

KEY TERMS AND DEFINITIONS

Argument and Evidence Map: was first proposed by J.H. Wigmore in the early 1900s to help in the teaching and analysis of court cases. The objective is to expose the structure of an argument, in particular how evidence is being used, in order to clarify the status of the debate. Still used in legal education today, the idea has been extended, formalised (and reinvented) in many ways (Buckingham Shum, 2003; Reed et al., 2007), but all focused on elements such as *Claims, Evidence, Premises* and supporting/challenging relations.

Concept Map: was developed by Joseph Novak around 1972, based on Ausubel's theory that meaningful learning only takes place when new concepts are connected to what is already known. Concept maps are hierarchical trees, in which concepts are connected with labelled, graphical links, most general at the top. Novak and many others have reported empirical evidence of the effectiveness of this technique, with an international conference dedicated to the approach.

Inquiry Map: is a technique for knowledge visualization in academic research, which aims to facilitate the creation and communication of knowledge in inquiry projects through graphic representation. Beyond the mere transfer of facts, inquiry maps aim to further create or transfer insights, experiences, attitudes, values, interpretations, perspectives, understanding, and predictions by using various mapping techniques.

Issue Map or Dialogue Map: derives from the "Issue-Based Information System" (IBIS) developed by Horst Rittel in the 1970s to scaffold groups tackling "wicked" socio-technical problems. IBIS structures deliberation by connecting *Issues, Positions and Arguments* in consistent ways, which can be rendered as textual outlines and graphical maps. "Dialogue Mapping" was developed by Conklin (2006) for using IBIS in meetings, extended as "Conversational Modelling" by Sierhuis and Selvin (1999) to integrate formal modelling and interoperability with other tools.

Mind Map: was developed by Tony Buzan in the early 1970s when he published his popular book "*Use Your Head.*" Mind Mapping requires the user to map keywords, sentences and pictures radiating from a central idea. The relatively low constraints on how elements can be labelled or linked makes it well suited for visual notetaking and brainstorming.

Open Learning: is a learning method for the knowledge acquisition based on open educational resources, open source technologies and online communities. Open learning aims to allow pupils self-determined, independent and interest-guided learning. It has been also focussed on collaborative study and social learning.

Social Network: refers to the acquisition of social competence that happens primarily in a social group, virtual learning environments or online communities. Social network depends on group dynamics, people with similar interests and disposition for interacting together.

Web Map: appeared relatively recently as a result of the rapid growth of the internet. Software tools provide a way for users to capture, position, iconify, link and annotate hyperlinks in a visual space as they navigate, creating a richer trail which comes to have more personal meaning than a simple bookmark list.

Chapter 5
Concept Maps as a Tool for Promoting Online Collaborative Learning in Virtual Teams with Pre-Service Teachers

Wan Ng
La Trobe University, Australia.

Ria Hanewald
The University of Melbourne, Australia.

ABSTRACT

The chapter provides an overview of concept mapping and a description of a theoretical framework that adopts concept maps as a tool to enhance collaborative learning in virtual teams in an online learning environment. It was developed by drawing on the relevant literature and on research in online collaborative learning in the pre-service teacher education programs at an Australian university. The framework is underpinned by socio-constructivist learning theories for collaborative learning with online technologies. The use of a concept map to illustrate the framework is presented and the pedagogical benefits are highlighted. The framework is applicable to all higher education courses in promoting collaborative virtual team learning.

INTRODUCTION

The primary aim of this chapter is to provide readers with an overview of concept mapping and a theoretical framework for working collaboratively with concept maps in small virtual teams in online learning environments. The framework is targeted at students in teacher education courses although it is also applicable across all higher education courses.

DOI: 10.4018/978-1-59904-992-2.ch005

The chapter begins with a clarification of the relationships between mind maps and concept maps, terms that frequently appear to be used interchangeably, before discussing concept maps and their pedagogical benefits. It is followed by an exploration of the hierarchical versus cyclical nature of concept maps and how they can be used to elicit different types of understanding (Ruiz-Primo & Shavelson, 1996; Hibberd, Jones & Morris, 2002; Safayeni, Derbentseva & Cañas, 2005). Various concept mapping tools are presented, including

Cmap Tools (http://cmap.ihmc.us/), which are online collaborative concept mapping tools used to capture staged developmental thinking and/or learning. Their practical applications are situated in socio-constructivist learning theories that underpin the theoretical framework for collaborative learning within online technologies (Hung & Nichani, 2001). Based on socio-constructivist learning theories, the use of concept maps as a tool to promote collaborative online learning with pre-service teachers in teacher education courses will be discussed. The case in point illustrates one practical application, supporting collaborative learning using online technologies to enhance virtual teamwork.

CONCEPT MAPPING

The two closely related and rather similar terms of concept maps and mind maps are often used interchangeably. It is therefore necessary to clarify the differences between them to avoid further misunderstandings.

A Definition of Concept Maps

Concept maps are graphical tools for organizing and representing knowledge. They include concepts, usually enclosed in circles or boxes of some type, and relationships between concepts indicated by a connecting line linking two concepts. Words on the line, referred to as linking words or linking phrases, specify the relationship between the two concepts. (Novak & Cañas, 2006, p. 1)

The two most distinguishing attributes of concept maps are the hierarchical structure in which concepts are represented and the use of cross-links, which indicate the relationships between the knowledge components. The hierarchical organisation of concept maps usually depicts the most general concept at the top and more specific concepts underneath. The cross-links show how concepts in the knowledge area are connected to each other. Both of these features are central in aiding creative thinking.

A Definition of Mind Maps

Mind maps use a central word, idea or other item arranged intuitively around the central word. A mind map starts with a key notion that radiates out into branches. This free flow of ideas promotes brainstorming. These tree structures generate and chart elements without the immediacy of having to establish an intrinsic conceptual framework.

Concept Maps and their Different Structures

Concept maps are visual, 'at-a-glance' representations of key concepts or ideas and the interrelationships between them. It is a technique first developed in the 1970s by Joseph Novak (born 1932) for representing understanding in multimodal forms. The concepts can be linear and hierarchial or networked consisting of nodes (concepts) and links (relations between concepts). Nodes are displayed as circles, ovals or rectangles, with the links being either uni-or bi-directional and usually labelled.

Concept maps are – at their most basic – graphic organisers of information. There are several types such as chain (or sequential), spider, hierarchical and cyclical concept maps. Each type is suitable for a particular kind of knowledge or content.

Sequential maps are also called chain maps and are typically used for time lines or for a sequence of events or a succession of steps. (see Figure 1)

Hierarchical concept maps are typically used for heredity (i.e. family trees) or food chains (i.e. the feeding relationship between species within an ecosystem), which are classical tree structures. The more general concept is at the top of the hierarchy, with more specific concepts in the lower part; hence they are read from the top to the bottom. However, most hierarchical concept

Figure 1. Basic structure of a chain/sequential map

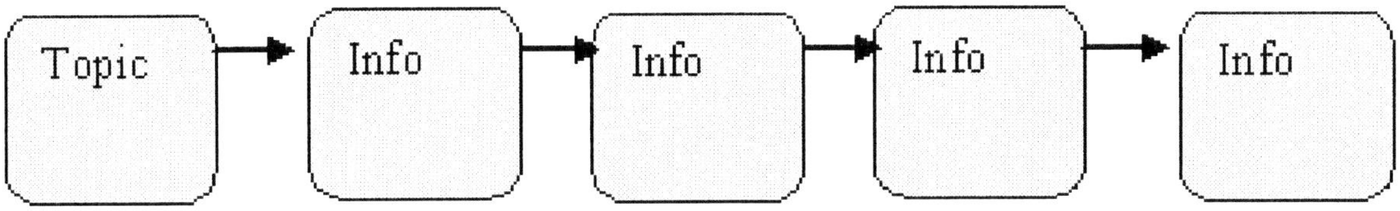

maps are not pure tree structures, but instead also have some cross-links. Both pure and cross-link variations are seen as the most appropriate tool for representing static knowledge, and relationships between concepts. (see Figure 2)

Originally, hierarchical structures for concept maps were heavily promoted (Novak, 1998) but researchers such as Ruiz-Primo & Shavelson (1996) have reservations. Their hesitation is based on the fact that the hierarchical structure of the map compels the knowledge to be shaped to fit this mode of organization. They propose instead that the structure of the knowledge should drive the structure of the map.

Cyclical concept maps are typically used for concepts that feed into one another in a closed loop. They show self-contained systems such as the water cycle, air cycle, time cycles (i.e. calendar, seasons), the nutrient cycle or ecosystems and are fundamental to biological systems; hence they are mostly found in the natural sciences. Derbentseva, Safayeni and Cañas (2004) propose that "...the structure of the map influences the type of relationships that are likely to be constructed in a proposition that links two concepts together." (p.1). They suggest that cyclical *Cmaps* are more likely to generate dynamic propositions and predict that, therefore, they will be preferred for dynamic relationships between concepts and the functional relationships amongst a cluster of concepts. (see Figure 3)

Concept maps, however, do not have to be restricted to only one of the types described above. The proposed framework in this chapter and the embedded case study does not advocate a sequential, hierarchical or cyclical concept map over the other. Rather a combination of all is outlined. This novel approach of blending and focusing on the organic growth of the concept map rather than subordinating the thinking to a given structure utilizes the advantages of each model without being constrained by their limitations and thus provides users with the most favorable environment to develop their understanding and learning.

Figure 2. Basic structure of a hierarchical concept map

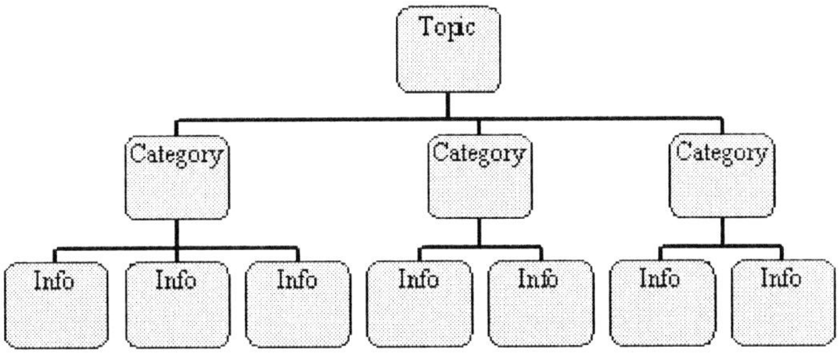

Figure 3. Basic structure of a cyclical concept map

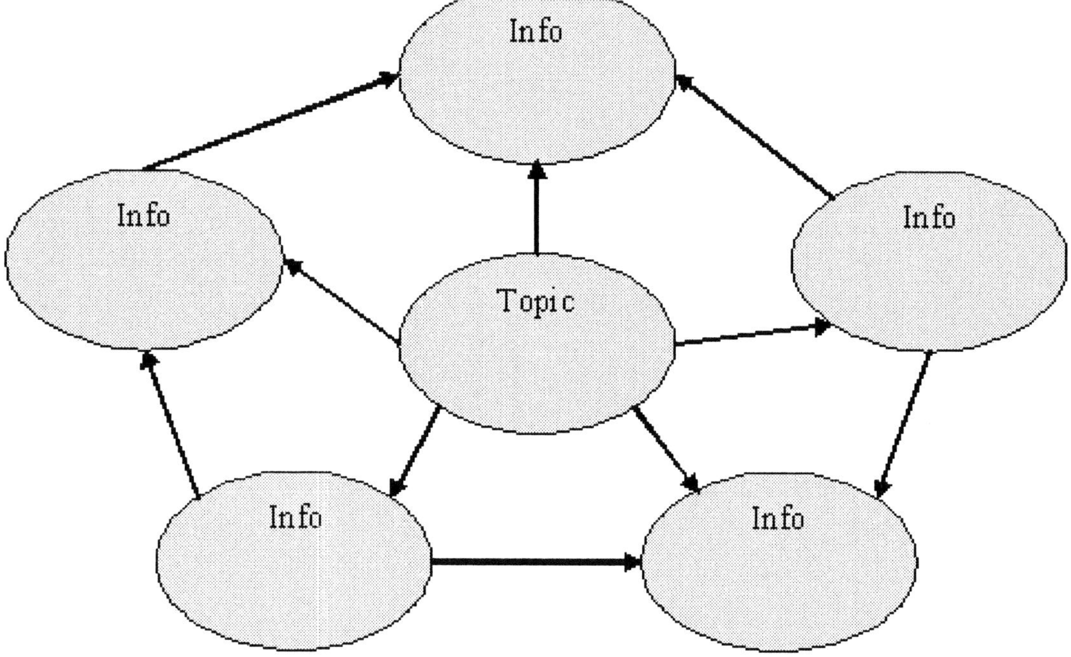

Concept mapping work is based on the work of David Ausubel (1918-2008), namely his theories on learning. Ausubel asserts that the learner's prior knowledge is the most important factor that influences learning. According to his learning theory, meaningful learning occurs only when the new understanding can relate to understanding of concepts that are already in existence in the learner's cognition. The bridging of the old and new concepts learned can be scaffolded using graphical organizers such as a concept map. Concept mapping software such as *Inspiration, Kidspiration, Gliffy, Conceptshare, Cmap and Thinkature* (see next section for more details) provides means to communicate and share thinking and understanding visually. These visual tools can help learners organize and express their ideas in different ways. Apart from using key words to represent concepts or ideas in the nodes, learners could insert images, multimedia or video files to represent the concepts. The tools also have auditory functions for learners to voice record a short statement or explanation in a node. To show relationships between these multimodal ways of representing concepts and ideas, learners create labeled links between them to further demonstrate and clarify their understanding. Hence concept mapping tools provide flexible means for organizing conceptual understanding, allowing students to organize ideas in a logical but not rigid manner, with the option to build on the maps as they progress through their learning.

Concept maps can be used for brainstorming activities to generate ideas, communicating complex ideas, presenting findings, demonstrating progressive learning as understandings evolve and assessing conceptual understanding. An advantage of concept maps over writing is that learners are compelled to *sharpen* their thinking and reflect on how much they understand about the concepts under study to make correct links between them. These processes foster the development of life-long learning skills such as critical thinking and research skills. As concepts may be constructed individually or in a group, teamwork and communicative skills are also developed as learners decide on keywords and size of the concept map, discuss what constitutes key concepts and appropriate links between them and whether to represent key concepts in text, images, sound, multimedia format or a combination of those.

Concept mapping can be used at various stages of learning. Instructors could use small concept maps with four to six key concepts or ideas as exercises for students. In working with their partners instructors assess formatively aspects of the topics under study, for example the concept of photosynthesis within the context of functioning processes in plants. Such tasks could be administered as open tasks where students decide on five keywords to show their relationships in small groups. Alternatively, the instructor could provide the five keywords and the students make the connections with linkage words. As the teaching and learning evolve, students build up understanding of respiration as another of the plant processes and demonstrate that understanding in a concept map. These two concepts of photosynthesis and respiration are then brought together in a larger concept map. The concept map could be extended at the end of a unit where a review of photosynthesis and respiratory processes in plants and respiratory processes in animals are brought together and students are required to show the connections between these concepts, including the similarities and differences between these processes in plants and animals. These sorts of activities are extremely useful to measure growth in learning and to develop critical thinking skills (i.e. analyze, compare, evaluate, research) as students organize their new understanding in relationship to existing knowledge. The maps are also useful for the instructor to quickly visualize and diagnose the students' thinking as part of a formative assessment process through recognition of misdirected links or wrong connections. This objective way of evaluating enables teachers to readjust and reinforce teaching in areas that require it before moving onto the next concept. Concept maps are useful products that could be easily displayed on overhead or data projectors for peer-review and critique by the students in the class. This process provides students with ownership of their learning in the production of the concept maps and in helping each other learn rather than being directed by the teacher.

CONCEPT MAPPING SOFTWARE

Used in the ways described above, mapping software fosters socio-constructivist and constructionist modes of learning while simultaneously helping to develop technology literacy skills. Students building concept maps are actively engaged in their own learning, demonstrating their understanding and portraying it in a visual form. The resulting artifact shows their selection of issues, content and key terms appropriate to the topic under study. Due to the open-ended nature of concept mapping activities, students have a high level of ownership and control over their learning (Safayeni, Derbentseva & Cañas, 2005; Novak & Cañas, 2006) while developing higher order thinking and life-long learning skills such as critical thinking, decision making, evaluative and creative thinking skills.

There are a number of free software packages that enable the creation of concept maps collaboratively. Examples are:

- *CmapTools* is a free web-based concept mapping tool (http://cmap.ihmc.us). It allows construction, navigation, sharing and critiquing of knowledge models in the form of a concept map by a number of web users. *Cmaps* can be created on personal computers and shared on servers anywhere on the Internet. Users can browse and edit maps produced by others while simultaneously chatting with their collaborators. They can work together by adding new discussion threads to the forum, by writing post-it notes to comment on the maps or by concurrently searching the web for relevant information, which they can add to the concept map. *Cmap Tools* was developed by the Institute for Human and Machine Cognition (IHMC), a not-for-profit research institute of the Florida University System. Their website (http://cmap.ihmc.us/Publications) offers several publications, reporting on research in the collaborative use of *Cmap Tools*. A report on *Project Quorum* proposes meaningful, collaborative learning by students within and between schools by using a feature of the *Cmap Tools* named the *Knowledge Soup*. This software system permits sharing of claims or propositions originated by students through concept maps. (Cañas, Ford, Novak, Hayes, Reichherzer, & Suri, 2001)

- *Thinkature* (http://thinkature.com/) is a web-based visual workspace for collaboration. It provides synchronous chatting, which allows individuals to engage in discussions to shape the concept map they are working on. In real-time, individuals can draw free-form shapes, add images, text and edit.

There are also several commercial concept mapping software packages available. Some are designed for the educational market, others are predominately for business professionals. Listed and described below are the most widely-known and used software packages for teaching and learning purposes:

- *Inspiration Software, Inc.* (http://www.inspiration.com/) is one package that is suitable for teaching and learning environments. It offers a variety of techniques (i.e. webs, concept maps, idea maps and plots) to encourage thinking. Inspiration 8 runs on Mac OS X and Microsoft Windows.
- *Kidspiration* (http://www.inspiration.com/productinfo/kidspiration/index.cfm) is also published by Inspiration Software, Inc. mainly for students in pre-schools to Grade 5. It allows building of concept maps, diagrams and webs by combining pictures, text and spoken words. Kidspiration has been developed for children, offering simpler and more colourful ways to represent ideas and connections visually. It assists younger students in early literacy skills development and older students in organising ideas for writing. Kidspiration is usable across all subjects (e.g. science, mathematics, social studies) and is compatible with both Macintosh and PC operating systems.
- *Gliffy* (http://www.gliffy.com/) is a Mac and PC compatible online diagramming software, which allows for the creation of flowcharts, diagrams, floor plans, etc. Its library contains shapes and interfaces but also permits importing of the user's own images (e.g. company logos and backgrounds). Read-only versions of the diagram can be created to protect documents. The image can be saved as .jpg, .png or .svg to incorporate into Word documents, Power Point presentations, wikis, blogs or web pages.
- *Conceptshare* (http://www.conceptshare.com/) is a collaborative environment to share and review designs, comment and manage participants in real-time chat. Annotation tools (i.e. pencil, line, text) facilitate mark up of the design.

ONLINE COLLABORATIVE CONCEPT MAPPING

In incorporating the above mentioned concept mapping tools, students collaborating in concept mapping activities online as a means of learning could be facilitated in one of two ways:

1. Initially as an individual effort, students use concept mapping software to construct their ideas offline. They then post the maps in the online discussion boards for other team members to view and critique. Through synchronous or asynchronous discussions, team members pick the most salient relationships between groups of concepts from each concept map before combining them into a final map. An adapted case study of this approach of collaboration is described in a later section of this chapter.
2. The team members progressively create the concept map online, including real-time collaboration on the web, for example using *Cmaps* software.

In pre-service teacher education and other higher education courses, both methods of collaboration to demonstrate understanding using concept maps are practical.

The first method requires consolidating individual understanding into concept maps and the negotiation between team members to pick the *best bits* to form the final map. Through the negotiations, individuals argue their positions and others counter-argue until a common understanding of the concepts under study is reached.

The second method involves the gradual building of a concept map that is placed in a central position (online, such as WebCT, Wikis or Google Docs) where team members visit regularly to download, add or delete concept boxes or linking phrases as their understanding develops over the course of the unit or the topic under study.

Alternatively, the construction of the concept maps on tools like *Cmaps* enables students to make changes directly on the concept maps themselves. For example, pre-service teachers' conceptions of what makes a good teacher over the course of their study will most likely evolve and change as they progress from theoretical perspectives provided during lectures (e.g. philosophers and learning theories) to practical perspectives that are influenced by school placement experiences and other educational field work.

In a study (Ng, Nicholas & Williams, in press) that tracked secondary pre-service teachers' beliefs on four occasions during their year-long Diploma in Education course, it was shown that the pre-service teachers' beliefs changed in response to their placement experiences. Despite an initial low level of focus on classroom management prior to school-experience, this factor rapidly emerged as a strong area of need throughout the rest of their preparation year.

While beliefs such as this change in response to experience, the beliefs that good teachers assist students to achieve and relate well to them remained strong and did not appear to change substantially in response to experience. Their beliefs that good teachers are kind, caring, understanding and charismatic people remained constant throughout the course. While beliefs in these social/affective characteristics remained constant, beliefs where there is emphasis on academic and communication skills in good teachers decreased as their school placements increased.

The evolving beliefs just described could be captured in concept maps as the pre-service teachers work in small teams of two or three over the year, sharing experiences and forming perceptions of good teaching based on personal experiences and observations during school placements. Instead of keeping individual reflective journals, which instructors may or may not read until the end of the semester, students in their teams construct concept maps with the central and key concept as *'good teacher'*. These are constructed at set intervals such as at the beginning of the course, after each school placement and at the end of the course with timely submission to their instructors. Teamwork in this type of activity generally requires consensus in choice of concepts and linking phrases. However, pre-service teachers could exercise innovative thinking to indicate possibilities of alternative beliefs and justify their choice within the same concept map. By communicating thinking and beliefs in this visual manner, instructors will be able to quickly view the common beliefs, differences, strengths and weaknesses of the whole cohort of pre-service teachers. Teaching could be adjusted accordingly to what the pre-service teachers needs are from the concept maps.

LEARNING IN A TECHNOLOGY-ENHANCED ENVIRONMENT

Socio-Constructivist Learning

While there are still many issues to be addressed in terms of *effective* learning using technology, at the theoretical level, learning that is mediated by technology (Semple, 2000; Nanjappa and Grant, 2003), including mobile technology (Roschelle, Patton & Tatar, 2007; Sharples, Taylor & Vavoula, 2007) and online technology (Garrison, Anderson, & Archer, 1999; Ng and Nicholas, 2007) has been discussed. Semple (2000) argued that the application of learning theories to the use of technology in the classroom is a "...matter of fitness for purpose..." (p. 27) and depending on the learning theory that is dominant at the time, appropriate learning environments can be created for meaningful learning to take place. One of the most discussed learning theories in a technology-enhanced learning environment is the socio-constructivist learning theory. Constructivism has different meanings for different people and takes many forms (Matthews, 1998) but this chapter will be concerned with socio-constructivism and the theories behind it in supporting online collaborative learning using concept maps.

Constructivism is a learning theory that has been highly influential in Western education over the last three decades and a vast amount of literature has been accumulated on it, particularly in the area of Science education (Phillips 1995, 2000). It is a theory of knowledge that offers explanations of *how we have come to know what we know*. The two notable theorists associated with socio-constructivism were Jean Piaget (1896-1980) and Lev Vygotsky (1896-1934). Piaget's constructivism is also known as personal constructivism and is based on his cognitive developmental theories, which propose that concept formation in the individual follows a clearly defined set of stages that must be experienced sequentially by that individual. In light of this, Piaget's constructivism is also called cognitive constructivism. The underlying principle in cognitive constructivism is that knowledge resides in individuals and that it cannot be given or transmitted complete to them by their teachers. Learners must construct their own knowledge in their minds and build upon the knowledge that is based on prior experiences. Learners learn only when they are actively engaged in the process, either at the operational level where learners are engaged in physical manipulations or at the cognitive level where they are mentally processing information or stimuli.

Vygotsky's social constructivism makes similar assertions to Piaget's cognitive constructivism about how learners learn in regard to knowledge as being progressively built up and continually re-interpreted. However, Vygotsky places more emphasis on the social context of learning. In his theory, the learning process involves interaction with other individuals where culture and society will influence the learning. A difference between cognitive and social constructivism is that in the former, the teacher plays a limited role, acting as a facilitator, whereas in the latter, the role of the teacher is active and involves helping students grasp concepts by guiding and encouraging group or other analytic work. Socialization within Vygotsky's theory is not confined to teacher-student interactions. The interaction between students is pivotal in aiding students to construct and build knowledge.

These cognitive and social theories of Piaget and Vygotsky form the basis of socio-cognitive constructivism. Socio-cognitive constructivism as applied to many face-to-face and online learning situations, draws on and sees learning as a dynamic and social process. In a technology-mediated learning environment, involving for example the use of concept mapping software, the interactive, open and non-linear nature of learning require learners to be actively analyzing, evaluating and making decisions while manipulating the information at hand in order to construct new knowledge or solve a problem. They will constantly have to compare their own prior knowledge of a body of information with that presented in the learning environment and seek means of either re-confirming their prior knowledge or de-constructing and re-constructing new meanings. When the learning is embedded within an online learning management system such as WebCT, Blackboard or Moodle, collaborative learning is possible at both local and global levels. The social aspect of online collaboration will be discussed further in the next section.

Blended into constructivist learning theory is the theory of constructionism. Papert (1991) linked constructivism to technology and developed the concept of construction from studies of children using the programming language Logo. According to Papert (1991), students are engaged in learning when constructing a public artefact that others will see, critique and/or use. Such artefacts could include a sand castle or a theory of the universe or a concept map. Constructionism is linked to constructivism in the students' active learning and learning by doing, all of which takes into consideration students' prior knowledge and experiences.

For example, Anderson and Witfelt (2003) discussed an online collaborative learning experience where students design a *game* to connect computer-based and experimental learning environments. The project involved 12 to 13-year old students in Danish primary and lower secondary schools studying home economics. They were working in groups to solve problems of survival by learning about food of various kinds, the principles of cooking and concepts of energy and nutrition via an interactive game prior to cooking in the kitchen. The students also engaged in *storylines* pedagogy where they used a *MOO* (MUD object oriented) environment as the software genre. The project integrated the technology to utilize the students' natural curiosity and interest in playing games by producing a learning resource, namely a prototype in the form of a MOO storyline.

A MOO is an Internet-based multi-user environment that allows multiple users to share a community of rooms, virtual spaces to explore specific concepts or engage in particular kinds of interactions and objects based on text, and to interact with each other (Holmevik & Haynes, 1998). Such an environment offers students the opportunity to undertake problem-solving tasks where they explore, investigate, and create narratives based on their own ideas and experiences. A MOO-environment encourages teamwork and promotes integration of learning areas across the curriculum. For example, the storyline could be along the theme of developing a catering business (Anderson & Witfelt, 2003). In exploring this theme, the students take up employee roles in various sections of an imaginary ecological restaurant, such as the economic department, the production planning department and the kitchen. They design a business plan, investigate the cost of production of different cooked products and physically experiment with various recipes to find out about the cheapest and best way of producing each recipe. Skills in Science, Mathematics, English and those associated with Home Economics are developed in this particular example of a storyline. As a result of this evaluative task oriented group-work, the MOO prototype promotes teamwork and higher order thinking skills in students. In this environment, where active construction of new knowledge and learning at an individual and a social level is taking place, and where students are developing both technical and cognitive skills, the capturing of the learning could be extended by using concept mapping pedagogy for each virtual department (e.g. kitchen, planning and financial departments). The group members in each department will identify key concepts and collaborate to create a concept map using the key concepts to present a visual representation of the requirements of each department and how they are connected in order to ensure the successful operation of the department. The combining of the concept maps from each virtual department showing the inter-linkages between departments' maps forms a much larger map that is visually communicated on the topic of survival.

In summary, creating concept maps collaboratively involves the cognitive, social and constructionist modes of learning to demonstrate understanding that is translated into an artifact, constructed as a result of critical selection of content and expressions through negotiation between teams members.

THEORETICAL FRAMEWORK FOR ONLINE COLLABORATIVE LEARNING WITH CONCEPT MAPS FOR PRE-SERVICE TEACHERS

The previous section discussed the theories of socio-constructivism and constructionism in supporting collaborative learning in computer enhanced environments, including online environment. This section expands on those theories, particularly the social perspective of learning online, by discussing *Computer Supported Collaborative Learning* (CSCL). CSCL has been researched since the early nineties and Stahl, Koschmann & Suthers (2006) have provided a historical perspective of CSCL. Lipponen (2002) defines CSCL as the focus on how collaboration and technology are able to facilitate the sharing and distribution of knowledge and expertise among members in a community to enhance peer interactions and group work. Through directed discourse and negotiated meanings with others, CSCL supports collaborative learning for construction of shared knowledge.

Central to knowledge construction in CSCL is a socially-immersed learning environment where the students create a virtual community with the presence of cognition, peers and teacher enabling them to construct knowledge in an interactive and iterative process across the dimensions of conceptualisation and construction (Ng & Nicholas, 2007). Through discussions that negotiate meanings, Fung (2004) asserts that by being exposed to different perspectives in collaborative learning environments, students are able to develop critical thinking and judgment skills that value, support or oppose the different views. The requirement for a social environment is based on the belief that there are pedagogical benefits in having access to discussions generated in learning dialogues, such as in developing critical reflections and constructing knowledge (Anderson, 2004; McConnell, 2000).

The Framework for Collaborative Online Learning with Concept Maps for Pre-Service Teachers

Figure 4 shows a framework for online collaborative learning undertaken by pre-service teachers working in small virtual teams where understanding is demonstrated in concept maps. For students who have not undertaken collaborative online learning, the preparation stage is necessary for fostering challenging intellectual engagement and diversity of opinions prior to the online teamwork learning experience. The main features of the framework are:

1. Two weeks preparation and introduction were scheduled for online discussions and concept mapping using stimulus material such as a

Figure 4. Framework for online collaborative learning with concept maps in small virtual teams

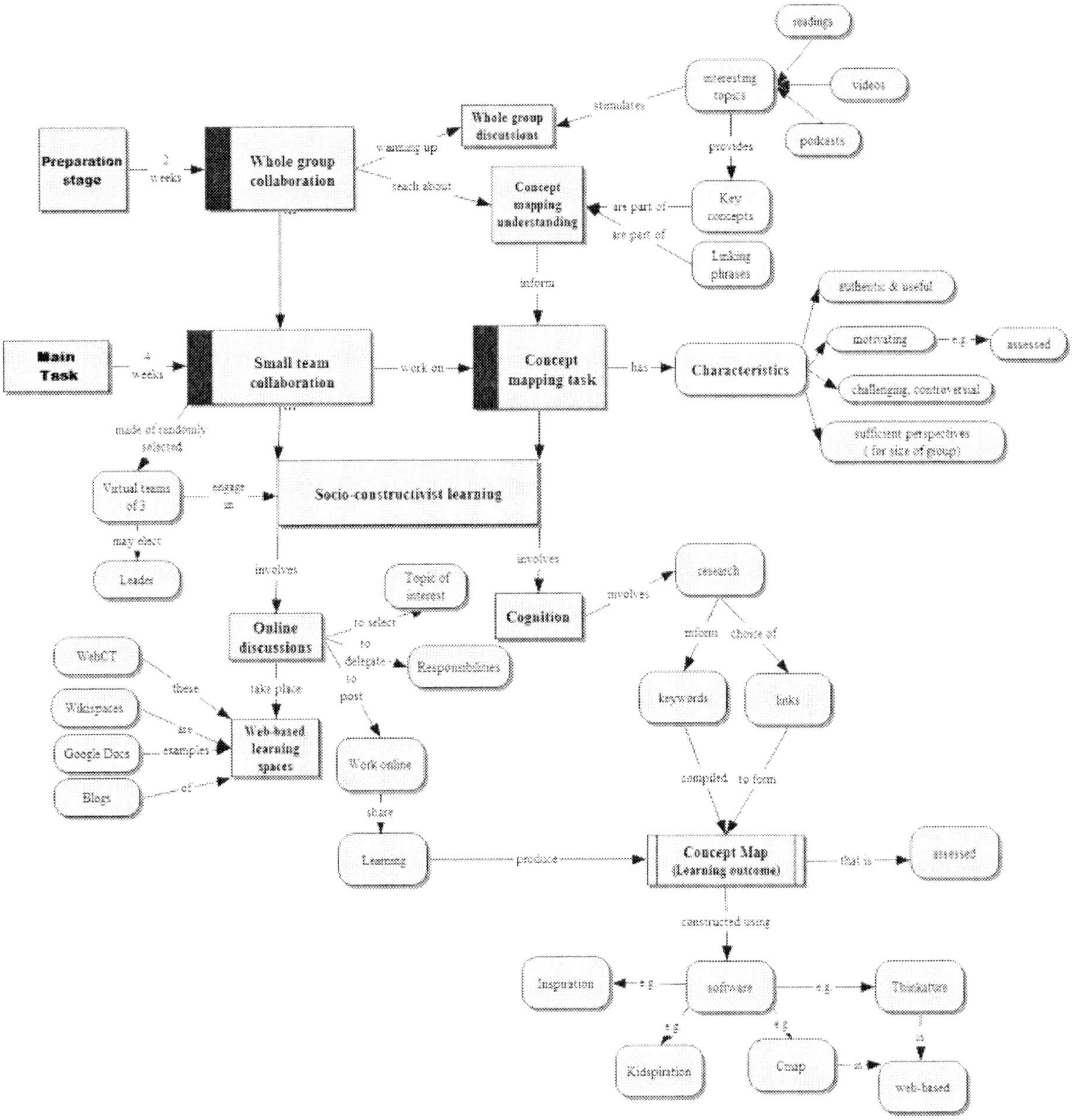

short reading, video clips or podcasts from the Internet. All students are encouraged to contribute to key ideas from the materials and linking phrases. Students volunteer to construct concept maps based on the other students' contributions. In addition, instructors could post prepared concept maps for critique, ensuring that in these maps there are concepts incorrectly placed and linking phrases that are incorrectly constructed to generate discussions that will demonstrate students' understanding of concept maps.

2. Four weeks are allowed for the *virtual team task* to produce a concept map that is assessed. Students are randomly placed in groups of no more than three to work on a

task that is authentic and useful (Basque, Dao & Contamines, 2005); challenging, possibly controversial (Bennett, Howe & Truswell, 2002), motivating, for example a reward or assessed with a group mark (Bennett, Howe & Truswell, 2002; Birch & McDonald, 2005) and contains sufficient components for every team member to work on. In undertaking the virtual team task, students work collaboratively to construct their concept maps as the learning outcome, utilizing the asynchronous or synchronous facilities present in learning management systems. The use of synchronous technology such as *Chat* has been encouraged by Levin, He and Robbins (2006) who have shown that there is more reflective thinking in synchronous online case discussions.

Virtual team collaborative learning could be an effective means of developing students' teamwork skills and fostering better learning as team members are able to cover many aspects of a topic. Pea (1993) sees this benefit of working collaboratively and sharing ideas as *sharing the cognitive burden*. Research has shown that the use of Web-based discussion tools can provide students with an appreciation for diverse viewpoints and perspectives, leading to a deeper understanding of the course content (Chickering & Ehrmann, 1996).

The Framework with a Case Study as Example

The following section provides a case study adapted from Ng (2008) to support the theoretical framework described above. The case study presents an online discussion leading to the construction of a concept map on Space Exploration and Ethics (Figure 5). The online discussion is an adapted excerpt resulting from Ng's (2008) research study which involved 31 pre-service teachers at an Australian-based university undertaking online collaborative learning in the coursework unit *Science and Technology in Contemporary Society.* The dialogue was obtained from a chat session by a virtual team of three students (Stu1, Stu2 and Stu3) discussing the research they had undertaken for an agreed topic on Space Exploration and Ethics. Their task was to create a 700-word essay using each team members' research. The dialogue has been slightly modified to turn the requirement of a short essay as a final learning product to a concept map instead. The modifications are shown in italics.

To put the case study in perspective, the class – in which the three students, Stu 1, Stu2 and Stu3 were participating - had to undertake an online task as a small virtual team. The task was worth 20% of their final mark. The aim of the online task was twofold: (i) to develop a conceptual understanding of ethics and (ii) to experience virtual collaboration and web-based communication. Ethics is an important concept in education as the fundamental concepts of ethics: value, obligation and responsibility are what many of us use in our everyday decision-making on issues that affect our lives. All students were required to participate in three tasks (Ng, 2008):

Task 1. As a warming up exercise, all students had to read articles on the topic of ethics that were posted in the *Main* strand on WebCT's discussion board. Each student had to read and express an opinion on at least one of these three articles and respond to other opinions expressed by their peers. The articles were: *Whose values, which ethics? Science education and the civil society* by Newell (2003), the *History of ethics*[1] and the *Belmont report*[2].

Task 2. Students were divided into virtual teams of three to five, randomly placed and not in friendship groups. Each virtual team selected an ethics-related topic of interest to research. They had to produce a final product in the form of a *concept map* on the topic. Learning spaces (strands) on the discussion board were created for the collaboration, one for each virtual team to

Figure 5. Concept map for space exploration & ethics resulting from online collaborating teamwork

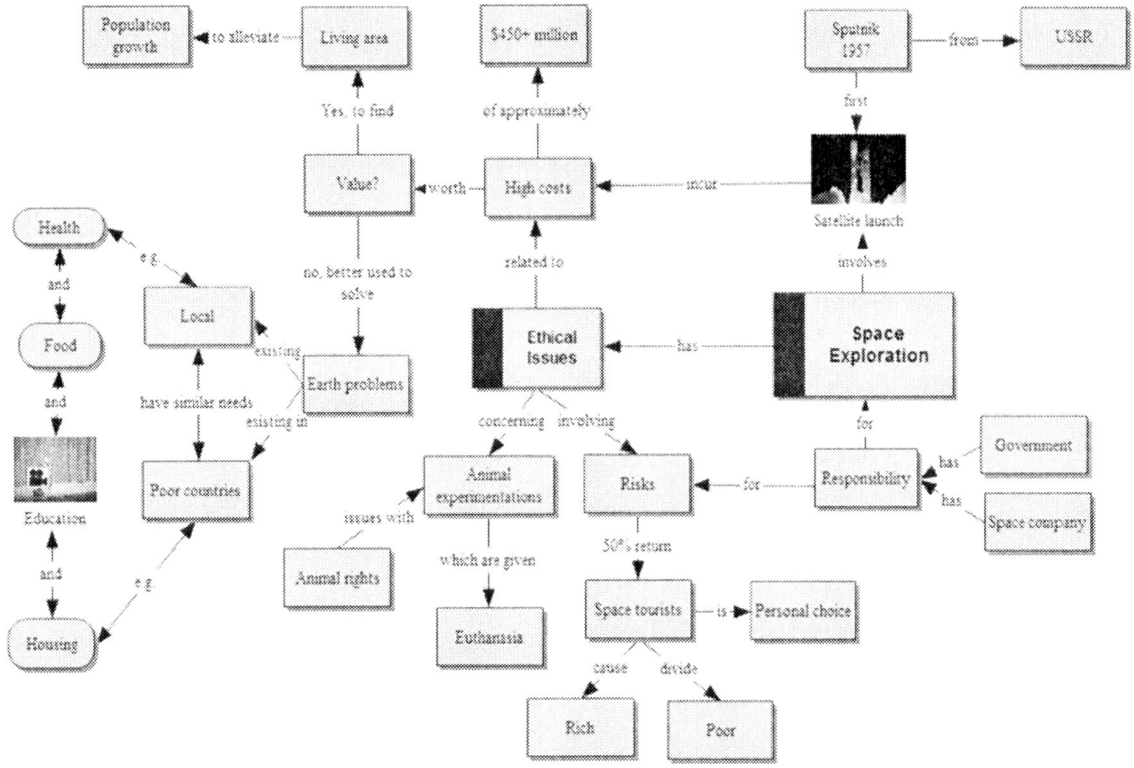

post discussions and other materials (for example, researched or written contributions). Each team's learning space was opened to the whole class so that all students could read the other teams' discussions and work.

Task 3. The completed teams' works were posted in the Main strand of the discussion board. Each team had to select another team's work to peer-review and assess. A short descriptive paragraph on the assessment was required from the review with comments on the concept maps. The teams had five weeks to complete the task.

The Case Study with Stu1, Stu2 and Stu3

In asynchronous discussions on WebCT, Stu1, Stu2 and Stu3 had chosen Space Exploration as the topic to investigate and decided that each member of the team will research space exploration and ethical issues associated with it. They agreed that each member will condense the research to about a page and post it on WebCT for other team members to read. The summary of their research did not have to be formally written and could be in point form or short paragraphs stating the main points. They continued the discussion asynchronously after reading each other's work. In order to construct a concept map on Space Exploration and Ethics as a final product, the students could undertake this in one of two ways:

1. Asynchronously. They could think independently of the key concepts based on the research work that each had carried out and post their lists on WebCT for further

discussion. They would then agree on the key concepts and elect a member to construct the concept map. Other team members would then critique the concept map and make suggestions for improvement.
2. Synchronously. They could meet in the chat room and discuss the key concepts and how these concepts relate to ethics of space exploration. They would then discuss the links between pairs of concepts and a team member would construct a map for further discussion in the next chat session.

Collaboration between the students prior to the concept map construction stage was conducted asynchronously. This is the information seeking stage where after agreement from all members of the team, each student actively researched to find information in order to put together a one-page document on space exploration and the ethical issues related to it. In searching for appropriate information, particularly on the WWW, the students were required to critically analyse and evaluate information to sort out the important points pertinent to the topic of Space Exploration. They were required to condense paragraphs of descriptions into main points in order to fit into a page that they would upload onto WebCT for others to view. The students had moved from a social decision-making process to a more personalized and active constructivist mode in this initial phase of the collaboration. In the second phase of the collaborative learning, the students moved towards socio-constructivist learning, either synchronously or asynchronously, where they interacted to build common understandings between members of the team to construct the concept map.

After considering both options, Stu1, Stu2 and Stu3 decided to meet in the chat room and arranged for a time to meet for a synchronous discussion. The dialogue as taken from the chat room script (modifications indicated in italics) is as follows:STU1>>Stu2's was really good too...i reckon start with Stu3's intro bit then move into Stu2's historySTU1>> we could construct our *concept map from key concepts* from the paragraphs and points from each others researchSTU2>>yeah that sounds like a good idea, *how big has it to be, how many nodes and links do we need to have*?STU1>>yeah its hard...there is *no size specified* but what is specified are we need to describe the topic and talk about ethical issues around itSTU3>>well do we set it up like a debate type thing - you know pros and cons *for the concept map*?STU2>>I was wondering, if we should like really explore one issue, or just slightly touch on a few?STU1>> we all researched the topic and wrote damn good stuff. Maybe we read what each has researched and pull out the main points i.e. *concepts from each and write a list and then put them together into a concept map.*STU3>>fair enoughSTU2>>ok, so we start with Stu3's bit, then have a bit of history, then go into STU1's part with rich vs poor

The script above indicates collaboration that was carefully constructed with highly sensitive dialogue to ensure that every team member was working toward the same understandings and a common direction to complete the set task. The rest of the discussion, as recorded in the chat room transcript without any modifications is reported in Table 1. Only the relevant parts that have a direct relationship to drawing out key concepts are reported. The right hand column lists the key concepts or keywords derived from the students' research and conversation. These are used to form the concept map.STU3>>how many *keywords* are we roughly up to now?STU2>>i'm writing the *key concepts* as we go alongSTU1>>cool magoolSTU3>>will the references still need to be addedSTU2>>yer i rekon

The concept map of Space Exploration and Ethics, as shown in Figure 5, is based on the keywords identified by the students during the above synchronous chat session as they compiled the relevant section of each student's research. The social dialogue leading to the collaborative

Table 1. Collaboration to identify key concepts for concept mapping

Dialogue	Key concepts & words
STU3>>agree, coz i wanna go to space STU2>>can't afford it at the moment though STU3>>so unless i win lottery............. STU3>>In terms of ethical debates and issues, the arguments underlying space exploration is something that is not often raised in the media, such as human cloning and abortion for example. However once the statement 'Should so much money be spent on space exploration with so many problems on our own planet?' is prompted, a very heated debate between opposing sides arises	Rich Poor Value of space exploration Problems on earth
STU1>>hehe..i couldnt go to space...i'm scared of being confined in small spaces	Claustrophobia
STU1>>i liked how you included some history STU2, that (cut and pasted from short essay STU2 produced) The ethical debates surrounding space exploration started at the dawn of the 'space race' in the late 1950?s when the former USSR successfully launched Sputnik, a satellite into the outer atmosphere of earth. From that instant, governments and organizations have hurled things into space where the moral and ethical concerns have been intensively debated. The initial space flights sparked debates about animal rights where numerous animals such as dogs and primates were sent into space to predict how humans would respond to this new and strange environment. These animals however never returned to earth but were euthanized near the end of their 'missions'.	History of space travel 1950 USSR Sputnik Moral and ethical concerns Animal experimentation Animal rights
STU2>>cool, so then move on with... Bottom of Form STU1 (continues cutting and pasting from her own work>>The development of space exploration has led to the possibility of tourists in space. As a result the ethical considerations of such a possibility need to be considered. The probability of not returning from a space expedition (to the moon) is 50%. Is it ethically viable to allow tourists to enter into space with such a high probability of not retuning? Would it be up to the company running the space expedition, the government or the tourist to decide whether they were going to accept this risk? Would the desire for money of the company governing the space tours outweigh their desire to make tourists fully aware of the potential hazards of such an expedition? Additionally, the cost of being a tourist on a space expedition is undoubtedly going to be rather expensive as launching a space shuttle costs approximately $450 million. Thus, space tourism will only be available to those who are quite affluent, consequently creating yet a further divide between the rich and the poor. So, is it ethically just that only the rich have access to, and the opportunity to obtain all of the benefits that might arise from, tourism in space?	Space tourists Risks 50% return chance Company greed? Government responsibility? Personal choice? Costs 450 million launch space shuttle Gap widens between rich and poor
STU1 (cut and paste continues) >>The risk to, and loss of human life is just the tip of the iceberg when it comes to the moralities surrounding space exploration. Such topics include unnecessarily excessive funds being diverted to space programs in many countries when that money could be better spent on healthcare, food and housing for the poor or education. Other people are concerned about such things as storage of nuclear waste in space, isn?t it enough that we fill our own planet to the brim with highly toxic rubbish that we no have to resort to shooting it into space?	Money better spent? Poor countries Healthcare, food, housing, education Nuclear waste in space con
STU3>>The possibilities that space exploration leads to could be enomous, however the ethics associated, and fundementally those to do with economics, could have a colossal influence on the extent to which such exploration progresses.	

eliciting of key concepts for the construction of the concept map was rich as the students reviewed and reflected on their research to reconstruct new meanings and collective understandings. In thinking about ethical issues related to Space Exploration, the students brought to the discussion their worldviews, their beliefs about social justice and the ways they perceived space exploration issues morally and ethically. The interaction was dynamic with Stu1 cutting and pasting relevant sections into the chat forum so that Stu2 could draw out the keywords or phrases that could be suitable for a concept map. They were actively engaging with the materials and constructing understandings that were meaningful to them. Stu3's contribution blended in well in this environment where both social presence and cooperation were high. The open-nature of the task provided more opportunity for interaction and to learn teamwork skills. The results are similar to those found by Kapur and Kinzer (2007) who examined the effect of problem type in a synchronous computer-supported collaborative learning (CSCL) environment. They found that the more open and 'ill-structured' problem type produced positive effect in teamwork, as demonstrated in the case study with Stu1, Stu2 and Stu3 described above.

FUTURE TRENDS

Concept mapping skills are useful at all levels of employment - from business (e.g. managing projects, developing marketing strategies) to engineering (e.g. management of manufacture processes, product life cycles) and therefore, should be taught to students in primary and secondary schools. The incorporation of concept mapping competency into the curriculum would ensure that these vital skills are taught in schools and are brought into the future workforce to ensure a competitive edge. Therefore, primary and secondary students should be encouraged to gain experience in using concept maps both individually, in partnered teams and in larger group work. For this to occur, in-service teachers and teacher educators need to undergo professional development to acquire the necessary skills and knowledge to teach concept mapping skills to their students, including pre-service student teachers, in their classes. A compulsory integration into tertiary courses and professional development for teachers would make their competency in using and teaching about concept maps a certainty. In the light of fast developing technological capabilities and increasing sophistication, concept mapping software will handle multimedia (i.e. photos, videos, audio) even more easily; with embedded web reference links providing further or deeper background information. The compatibility of concept mapping software with other devices (e.g. Interactive Whiteboards and PDAs[3]) would support their expansion. A central repository, akin to a searchable online library, would house the emerging body of concept maps, much like Google images accommodate illustrations. Relevant copyright and intellectual property rights would need to be amended to include concept maps. Already, there are dedicated concept map conferences and publications such as this; additional dissemination opportunities would further assist in raising awareness, sharing of knowledge and stimulation of research into the advancement and understanding of concept mapping.

CONCLUSION

The use of technology in education is becoming increasingly pervasive, with concept mapping software available at no cost from the Internet. The free and easy access allows the use of concept maps by individuals for their work and for collaborating with others. The advantages of concept mapping have been described in this chapter as an opportunity for meaningful learning. They facilitate organization and expression of key ideas in a visual way. Concept maps provide a way to communicate and share understanding visually and in a focused manner. They demonstrate progress through learning by the evolving of and building on maps, permit revision and summarization of material which has been learned, and promptly revealing misconceptions in learning that are not depicted in conventional assessments. Concept mapping software is an effective tool for classroom teaching and learning. For it to be utilized effectively at primary and secondary levels, pre-service teachers have to be imparted with the theoretical and practical knowledge of using these tools in order to apply them in their future practice. The proposed framework will provide a viable model for usage in teacher education programs and in promoting virtual team work.

REFERENCES

Anderson, K., & Witfelt, C. (2003). Educational game design: Bridging the gab between computer-based learning and experimental learning environments. In *Conference Proceedings of the 6th International Conference on Computer-Based Learning in Science* (Vol.1, pp. 402-413).

Anderson, T. (2004). Towards a theory of online learning. In T. Anderson & F. Elloum (Eds.), *Theory and practice of online learning* (pp. 33-60). Edmonton, Canada: Athabasca University Press.

Basque, J., Dao, K., & Contamines, J. (2005. June 27-July 2). Participating virtually in a scientific conference: A collaborative e-learning scenario for authentic learning. In P. Kommers & G. Richards (Eds.) *Higher Education, Proceedings of ED-MEDIA 2005*, Montreal, (pp. 1953-1960).

Bennett, C., Howe, C., & Truswell, E. (2002). *Small group teaching and learning in psychology.* Retrieved February 15, 2007, from http://www.psychology.heacademy.ac.uk/docs/pdf/p20030116_smallgroupsreport.pdf

Birch, D., & McDonald, J. (2005). *Students' perceptions of a virtual team assessment item (VTAI): An Australian case study*. Paper presented to The Open and Distance Learning Association of Australia (ODLAA) Breaking Down Boundaries Conference, Adelaide. Retrieved February 15, 2007 from http://www.odlaa.org/events/2005conf/ref/ODLAA2005Birch-McDonald.pdf

Cañas, A. J., Ford, K. M., Novak, J. D., Hayes, P., Reichherzer, T. R., & Suri, N. (2001). Using concept maps with technology to enhance collaborative learning in Latin America. *Science Teacher (Normal, Ill.)*, *68*, 49–51.

Chickering, A. W., & Ehrmann, S. C. (1996). Implementing the seven principles: Technology as lever. *AAHE Bulletin*, *49*(2), 3–6.

Derbentseva, N., Safayeni, F., & Cañas, A. J. (2004). *Experiments on the Effects of Map Structure and Concept Quantification during Concept Map Construction.* Proceedings of the First International Conference on Concept Mapping. Pamplona, Spain, 2004.</conf>

Fung, Y. H. (2004). Collaborative online learning: interaction patterns and limiting factors. *Open Learning*, *19*(2), 135–149. doi:. doi:10.1080/0268051042000224743

Garrison, D. R., Anderson, T., & Archer, W. (1999). Critical thinking in text-based environment: Computer conferencing in higher education. *The Internet and Higher Education*, *2*(2-3), 87–105. doi:.doi:10.1016/S1096-7516(00)00016-6

Hibberd, R., Jones, A., & Morris, E. (2002). *The use of Concept Mapping as a Means to Promote and Assess Knowledge Acquisition,* (CALRG Report No. 202).</other>

Holmevik, J. R., & Haynes, C. (1998). *High wired. On the design, use and theory of educational MOOs.* Ann Arbor, MI: University of Michigan Press.</bok>

Hung, D. W. L., & Nichani, M. (2001). Constructivism and e-learning: Balancing between the individual and social levels of cognition. *Educational Technology*, *41*(2), 40–44.

Kapur, M., & Kinzer, C. (2007). The effect of problem type on interactional activity, inequity, and group performance in a synchronous computer-supported collaborative environment. *Educational Technology Research and Development*, *55*(5), 439–459. doi:.doi:10.1007/s11423-007-9045-6

Levin, B., He, Y., & Robbins, H. (2006). Comparative analysis of preservice teachers' reflective thinking in synchronous vs asynchronous online case discussions. *Journal of Technology and Teacher Education*, *14*(3), 439–460.

Lipponen, L. (2002). *Exploring foundations for computer-supported collaborative learning.* Retrieved October 12, 2006 from http://newmedia.colorado.edu/cscl/31.html

McConnell, D. (2000). *Implementing Computer Supported Cooperative Learning* (2nd ed.) London: Kogan Page.</bok>

Nanjappa, A., & Grant, M. (2003). Constructing on constructivism: The role of technology. *Electronic Journal for the Integration of Technology in Education, 2*(1). Retrieved March 15, 2008, from http://ejite.isu.edu/Volume2No1/nanjappa.htm

Newell, K. M. (2003). Whose values, which ethics? Science education and the civil society. *Australian Science Teachers Journal*, *49*(2), 6–11.

Ng, W. (2008). Virtual teamwork: Students learning about ethics in an online environment. *Journal of Research in Science & Technological Education*, *26*(1), 13–29. doi:. doi:10.1080/02635140701847421

Ng, W., & Nicholas, H. (2007). Conceptualising the use of online technologies for gifted secondary students. *Roeper Review:Journal of Gifted Education*, *29*(3), 190–196.

Ng, W., Nicholas, H., & Williams, A. (in press). How experiences in schools influence pre-service teachers' beliefs about teaching efficacy and good teaching. *Teaching and Teacher Education Journal.*

Novak, J. D. (1998). *Learning, creating, and using knowledge: Concept maps as facilitative tools in schools and corporations*. Mahwah, NJ: Lawrence Erlbaum Associates.

Novak, J. D., & Cañas, A. J. (2006). *The Theory Underlying Concept Maps and How to Construct and Use Them.* (Technical Report IHMC Cmp Tools, January 2006). Pensacola, FL: Florida Institute for Human and machine Cognition. Retrieved March 1, 2008 from http://cmap.ihmc.us/Publications/ResearchPapers/TheoryCmaps/TheoryUnderlyingConceptMaps.htm

Pea, R. D. (1993). Practices of distributed intelligence and designs for education. In G. Salomon (Ed.), *Distributed cognitions: Psychological and educational considerations* (pp. 47-87). Cambridge, UK: Cambridge University Press.

Phillips, D. (Ed.). (2000). *Constructivism in education: Opinions and second opinions on controversial issues.* Chicago: University of Chicago Press.</edb>

Phillips, D. C. (1995). The good, the bad, and the ugly: The many faces of constructivism. *Educational Researcher*, *24*(7), 5–12.

Roschelle, J., Patton, C., & Tatar, D. (2007). Designing networked handheld devices to enhance school learning. In M. Zelkowitz (Ed.), *Advances in Computers, 70*, 1-60.

Ruiz-Primo, M. A., & Shavelson, R. J. (1996). Problems and issues in the use of concept maps in science assessment. *Journal of Research in Science Teaching*, *33*(6), 569–600. doi:.doi:10.1002/(SICI)1098-2736(199608)33:6<569:AID-TEA1>3.0.CO;2-M

Safayeni, F., Derbentseva, N., & Cañas, A. (2005). A theoretical note on concepts and the need for cyclic concept maps. *Journal of Research in Science Teaching*, *42*(7), 741–766. doi:.doi:10.1002/tea.20074

Semple, A. (2000). Learning theories and their influence on the development and use of educational technologies. *Australian Science Teachers' Journal*, *46*(3), 21–28.

Sharples, M., Taylor, J., & Vavoula, G. N. (2007). A theory of learning for the mobile age. In R. Andrews & C Haythornthwaite (Eds.), *The SAGE Handbook of E-learning Research* (pp. 221-247). London: Sage.

Stahl, G., Koschmann, T., & Suthers, D. (2006). Computer-supported collaborative learning. In R. K. Sawyer (Ed.), *Cambridge handbook of the learning sciences* (pp. 409- 426). New York: Cambridge University Press.

KEY TERMS AND DEFINITIONS

Concept map: A concept map is a diagram that depicts concepts and their hierarchical relationships.

Concept mapping: A technique to show concepts and their relationships with each other through labeled arrows.

Mind map: A mind map is a diagram uses intuition to depict words, ideas or other items in branches around a central key word or idea.

Mind mapping: A process that brainstorms ideas, words, tasks or other elements and arranges them in groups around a central notion.

Online collaboration: Also known as computer-supported collaboration (CSC). It uses technology such as email, calendaring, text chat, wiki to help people work together for a common goal.

Offline collaboration: A process by which people share knowledge and resources to support each other's activities towards a mutual outcome.

Virtual teams: A group of two or more people who work for a common goal in different geographic locations and sometimes even organizations and time zones through the use of electronic communication.

ENDNOTES

[1] http://www.cgu.edu/pages/1.asp

[2] http://www.hhs.gov/ohrp/humansubjects/guidance/belmont.htm

[3] Personal digital assistants

Chapter 6
Factors Influencing Individual Construction of Knowledge in an Online Community of Learning and Inquiry Using Concept Maps

Simone C. O. Conceição
University of Wisconsin–Milwaukee, USA

Maria Julia Baldor
University of Wisconsin–Milwaukee, USA

Carrie Ann Desnoyers
University of Wisconsin–Milwaukee, USA

ABSTRACT

This chapter describes a study that used the community of learning and inquiry and concept maps as strategies to facilitate individual construction of knowledge in an asynchronous online course. Six factors influenced the concept map creation, which in turn affected individual construction of knowledge: group characteristics, social presence, cognitive presence, facilitation style of student, discussion summary format, and teacher presence. Working in a collaborative community allowed students to explore different ideas and concepts, but it was through the individual concept map work that students refined and expanded their knowledge and constructed personal meaning. The chapter concludes with strategies to facilitate individual learning in a collaborative online environment.

INTRODUCTION

As collaborative learning becomes more prevalent in asynchronous online learning environments, the use of communities of learning and inquiry that integrate cognitive, social, and teaching presence as the main strategy for the learning experience would be explored and used by instructors in higher education. Garrison (2003) posits that the goal of the community of learning and inquiry strategy is to

DOI: 10.4018/978-1-59904-992-2.ch006

foster "independent thinking in an inter-dependent collaborative community of inquiry" (p. 49). These communities of learning and inquiry have the potential to create an environment for learners to be part of meaningful social and cognitive experiences and gain higher-order learning.

However, when students work in these communities of learning and inquiry in an asynchronous online environment, it can be challenging to foresee the individual learning outcome if the process of learning focuses primarily on the experience of the community as a whole. The community of learning and inquiry is a good foundation for designing asynchronous online learning environments. We have used this model as the strategy for online courses. We have also included as part of the online learning experience the use of concept map tools to help students construct individual knowledge after participating in collaborative learning. We believe that concept maps can effectively facilitate individual construction of knowledge.

LITERATURE REVIEW

A thorough review of the literature on concept maps and collaborative learning indicates that most articles are descriptions of the authors' experiences using concept maps in various situations within the context of the online environment. Some of these publications describe the difficulties that students have constructing concept maps. Rábago (2004) states that these difficulties lie in the students' lack of the tools necessary to build them or a lack of familiarization with the subject of the concept map. One example of this type of difficulty is found in the descriptive article by Laampere, Matsak, and Kippar (2006), which investigates the technological and pedagogical challenges due to the poor integration of concept map tools and learning management systems.

The review of 24 articles on the subject reveals also an experimental study on the use of technology such as CTOOLS Concept Connector to facilitate students' visualization of their thinking process online and immediate feedback (Luckie, Harrison, & Ebert-May, 2004). A study based on Novak and Gowin's (1984) theoretical framework of concept mapping explores methods to design didactic materials based on concept maps (Cabral, Zeve, Nicolao, & Amoretti, 2004). A descriptive article also referencing Novak's framework investigates instructional approaches to use concept maps in which education students reported that concept mapping aided them in acquiring or improving their skills in studying, thinking, and teaching (Calderón, Agüera, & Alfageme, 2004). Other studies emphasize the benefits of using concept maps to increase organized communication among participants (Tarouco, Geller, & Medina, 2006).

The majority of the literature on collaborative learning and concept maps addresses the work of small groups of learners, generally two or three, constructing concept maps together. Basque and Lavoie (2006) researched 39 published studies on collaborative concept mapping at a distance and face-to-face. The quality of concept maps, the effect of collaborative concept maps on learning, and the integration of ideas and knowledge between learners during collaborative concept mapping are some of the subjects of these studies. The articles reviewed vary in their designs from descriptive, to experimental, to analytical. They are predominantly based on Novak's theoretical framework of concept mapping. They also vary in the groups that they study. For instance, Cañas, Ford, et al., (2001) focus their experimental study on Latin American school children using asynchronous concept mapping and sharing them through a network called Knowledge Soup. Rice, Ryan, and Samson (1998) scrutinize the use of concept maps in seventh grade life science classes to assess student learning. Many studies focus on high school students; however, some studies such as the one examined by Calderón, Agüera, and Alfageme (2004) investigated the effectiveness of concept mapping as a tool for college education majors.

De Simone (2007) investigated and reviewed three commonly used applications of concept mapping in higher education. She compared physical concept mapping (hand drawing), mental concept mapping (creating the maps only in one's mind but not on paper), and electronic concept mapping with Inspiration software to share in an online environment. Fonteyn (2007) discussed the use of concept mapping in nursing education as an easy way to deepen understanding of concepts and stimulate critical analysis. MacNeil (2007) describes concept mapping as a means of course evaluation in an undergraduate wellness program. Students completed concept maps pre-lecture and post-lecture that were later compared by complexity.

Kim, Yang, and Tsai (2005) analyzed previous empirical research on collaborative concept mapping. Based on the existing literature, there seems to be little research regarding the assessment of "individual internalization" since "individual cognition is the interplay between situations where group members construct meaning together through social negotiation and where individuals actively construct their own meaning" (p. 4). Moreover, in the literature reviewed, there is no mention of individual construction of concept maps within the context of collaborative learning. We reaffirm the benefits of collaborative learning and suggest further analysis to better understand the relationship of concept maps and collaborative learning. Therefore, the purpose of this study was to investigate the use of concept maps as a tool to facilitate individual construction of knowledge after participating in asynchronous collaborative online learning.

CONCEPTUAL FRAMEWORK

The work of Garrison, Anderson, and Archer (2000) illuminate the path to the assessment of collaborative learning occurring in a community of learning and inquiry within our study. According to Garrison's model (2003), a community of inquiry, made possible by collaborative and reflective communication, is structured by the occurrence of three elements that are essential for a thriving educational experience, these elements are: cognitive, social, and teaching presence.

Cognitive presence facilitates critical reflection and discourse and it is related to the intellectual realm. Its uniqueness relates to the ability of students to initiate, construct, and confirm meaning through sustained reflection and dialogue in the online discussion. Cognitive presence reflects the acquisition and application of higher-order knowledge and it is related to critical thinking skills.

Social presence reflects personal and emotional connection to the group and for its very nature presents a challenge to online communities. It is evident when, during the online discussion, participants are able to express themselves socially and emotionally in a genuine manner. Social presence facilitates cognitive presence since in order to express their deep thoughts and ideas in the online discussion, participants need to feel comfortable relating to each other.

The adequate functioning of the educational process is facilitated by the teaching presence. Teacher presence is the voice of the moderator who models critical discourse, constructive critique, and formative feedback. It also regulates the amount of content covered, determines group size, and monitors discussions. The balance and interaction of cognitive, social, and teaching presence makes possible the educational experience (Garrison, 2003).

THE ONLINE COURSE CONTEXT

The online course addressed in this study was designed to allow students to analyze concepts, theories, and research on distance education and develop and assess distance education programs. Course content was divided into five modules dur-

Table 1. Course outline

Duration	Module & Topic	Activities/Assignments
First day of class	Course Orientation	Introduction to Media & Syllabus Course Surveys Scavenger Hunt
3 weeks	**Module 1** - Distance Education: Defining the Concept, Identifying Technologies, Historical, and Conceptual Foundations	Online Discussion
		Discussion Summary
		Module 1 Concept Map Team Project: Task 1
3 weeks	**Module 2** – Learning and Learners	Online Discussion
		Discussion Summary
		Team Project: Task 2
		Module 2 Concept Map
3 weeks	**Module 3** – Design, Instruction, Application, and Assessment of Distance Education	Online Discussion
		Team Project: Task 3
		Discussion Summary
		Module 3 Concept Map Team Project: Task 4
2 weeks	**Module 4** – Policy, Management, and Evaluation of Distance Education	Online Discussion
		Discussion Summary
		Module 4 Concept Map Team Project: Task 5
	Thanksgiving Holiday Break	
2 weeks	**Module 5** – Global Perspectives and the Future of Distance Education	Online Discussion
		Discussion Summary
		Module 5 Concept Map
		Team Project: Task 6
		Final Concept Map

ing which specific readings were assigned. Each module lasted two to three weeks. As part of the course assignments, learners were required to: participate in online discussion forums for each module, create concept maps of their understanding of the concepts addressed in the readings and online discussion forums upon completion of each module and a synthesis concept map summarizing the whole course at the end of the semester, self-reflect on the concept maps immediately after they created them, and participate in a team project to develop a distance education program. See Table 1 for detailed course outline.

Course Orientation

The course met face-to-face for an orientation in the beginning of the semester and during the last day of the class. The remainder of the course was totally online using the learning management system Desire2Learn (D2L). During the orientation portion of the course, students participated in a scavenger hunt of the course management system which involved learning about the different features of the course (i.e., content information, discussion forums, announcements, gradebook, drop box, chat, links, and survey). Also, students completed a learning style inventory and a "Getting to Know You Survey" and posted messages introducing themselves to the rest of the class. The syllabus and course timeline were provided to students with required due dates and course assignments.

The first week of the course was reserved for the course orientation; the instructor assigned individuals to groups and group members were involved in logistical tasks to prepare for the module discussions and team project. Upon completion of each module, the instructor provided prompt feedback on the individual participation in the group discussions, team project tasks, and concept map assignment. The purpose of this introductory phase of the course was to help students develop positive attitudes and perceptions about the class tasks, such as perceiving tasks as valuable and interesting, believing they have the ability and resources to complete tasks, and having a clear understanding about the tasks.

Online Group Discussions

For the online discussion forums, students were divided into groups of 3-5 students. Students were required to post a minimum of 3-5 messages per discussion module. For each discussion module, two class members were assigned the roles of facilitator and summarizer. In addition, all group members played the role of contributor. The facilitator was responsible for initiating the discussion with one or two questions from the readings. As group members responded to the facilitator's questions, the facilitator extended the discussion by posing new questions on issues that came out of the discussion. Additionally, the facilitator was supposed to refer back to the readings to initiate discussion on another aspect of the topic. The facilitator was responsible for selecting discussion topics based on the readings, setting an agenda for the length of the discussion period, providing brief summaries during the discussion, initiating new topics, and keeping an active and involved discussion going throughout the specified module dates. The summarizer was responsible for providing a brief review of the main issues discussed, the key points that participants made in their group, and any conclusions reached by the group at the end of each module. Contributors were required to respond to questions posted by facilitators and group members, as well as review and comment on the responses of others through the discussion board.

One important aspect of the online discussion was that each response posted by participants had to clearly tie back to the reading materials. Participants could post comments in a variety of different formats: introduce scholarly references from other sources to support or highlight their perspectives, discuss personal experiences, or share professional experiences related to the module topic, but each response had to refer back to a point or points in the reading materials. Participants had to make their arguments, describe experiences, or discuss alternative perspectives within the context of the reading materials. The grading of the online discussion activity was based on two aspects of the students' participation: role playing (facilitator, contributor, and/or summarizer) and critical thinking (demonstrate evidence of dynamic reorganization of knowledge in meaningful and usable ways).

Concept Maps

Concept maps were used as a teaching and evaluation tool in this online course. As a teaching tool, concept maps allowed students to gain a set of skills to process and generate information and beliefs. As an evaluation tool, concept maps assisted students to self-assess their own thinking processes. Students created concept maps that depicted their understanding of the theories and concepts in the readings and online discussion forums upon completion of each module. In addition, at the end of the semester students created a concept map of their understanding of the concepts addressed in the course throughout the semester. Students received feedback on concept maps based on the following criteria (Novak & Gowin, 1984):

- Proposition: Is the meaning relationship between two concepts indicated by the connecting line and linking word(s)? Is the relationship valid?
- Hierarchy: Does the map show hierarchy? Is each subordinate concept more specific and less general than the concept drawn above it (in the context of the material being mapped)?
- Cross links: Does the map show meaningful connections between one segment of the concept hierarchy and another segment? Is the relationship shown significant and valid?

METHODOLOGY

This study was based on an online course (N=30) offered in 2003 (n= 8), 2004 (n= 13), and 2005 (n=9) on the topic of distance education for adults at a higher education institution in the Midwest region of the United States. The following research questions guided the study:

- What factors related to social presence affect student individual construction of knowledge in an asynchronous collaborative online learning environment?
- What factors related to cognitive presence affect student individual construction of knowledge in an asynchronous collaborative online learning environment?
- What factors related to teacher presence affect student individual construction of knowledge in an asynchronous collaborative online learning environment?

Data Collection and Analysis

Data for this qualitative study were collected using: (1) transcripts from student and instructor records of online group discussions, (2) student reflections after completing the individual concept maps developed by students throughout the course, and (3) student concept maps. Students developed a total of 6 concept maps.

Data analysis was based on Garrison, Anderson, and Archer's (2000) community of learning and inquiry model to investigate the presence of cognitive, social, and teaching presence on the learners' and instructors' contributions posted in computer-mediated conferences. With coding instruments, written transcripts were analyzed. Three researchers conducted the analysis. Each researcher looked at one course year for a total of three courses. After the initial analysis, the researchers conducted an audit trail to verify that the themes and terms were consistent. Through the process of content analysis, researchers looked for indicators of cognitive, social, and teaching presence in the postings. Indicators were words or segments of the transcripts that were grouped into developed categories, such as triggering event, exploration, or resolution.

Transcript Analysis

Data from the online group discussion transcripts (only the first and the last course modules) were analyzed based on the community of inquiry model (Garrison, 2003). Investigators looked at postings or segments of postings that showed social, cognitive, or teacher presence. Message categories that elicited cognitive presence involved a triggering event, exploration, integration, or resolution. Cognitive presence indicators included sense of puzzlement, information exchange, connecting or applying ideas. Message categories that elicited social presence included emotional expression, open communication, or group cohesion. Social presence indicators included emotions, risk-free expression, or encouraging collaboration. Teacher presence was indicated by messages that provided instructional management, building understanding, and direct instruction. Postings that clarified issues, shared personal meaning, and redirected students to focus on the discussion were indicators of teacher presence (Garrison, Anderson, & Archer, 2001).

We only used transcript data from the first and last modules to compare the initial postings of the students with the contributions at the end of the semester in order to analyze developmental changes in students' learning processes. Calculations of average number of contributor messages were determined with the following equation: (total number of messages) – (number of instructor messages) – (number of facilitator messages) divided by (number of group members) – (1 for facilitator) = average number of messages per contributor.

For instance, the first category of cognitive presence (taken from Garrison's model), a triggering event, was described as the feeling of uneasiness as the result of an experience indicated by a sense of puzzlement on the part of the participants which outcome was the asking of questions in the online discussion. For example, one student felt a sense of puzzlement regarding the opinion of an article's author, expressed confusion to the class, and ended the online posting with a question to the group: "Does anyone have a thought on this?" The second category of cognitive presence, exploration, is defined by a search for information indicated by the sharing of opinions with other participants. For instance, a student expressed an opinion about a specific class reading, offering conclusions as a suggestion for the group to consider; however, still explored the subject and ended the online posting with the following question: "Anyone else offer an opinion?"

Within the social presence, three categories were used to analyze data: emotional expression, open communication, and group cohesion. In the absence of physical presence, it was difficult to show closeness or warmth and the expression of emotions was reduced. To compensate some participants used emoticons, sense of humor, and self-disclosure. A student used an emoticon to demonstrate agreement to the group: "Sound great:-)." This comment was classified as part of the emotional expression category within social presence.

The categories of teaching presence included: instructional management, building understanding, and direct instruction. It is important to note that teacher presence was not analyzed by the researcher who was also the course instructor. The other two researchers analyzed data pertinent to teacher presence. Instructional management is concerned with structural aspects of the course such as the curriculum, methods, and assessment. It addresses planning issues and guidelines before and during the online discussion. For instance, two members of a group used references to the show Star Trek as examples. The references to the show increased and another member of the group, perhaps not familiar with the television program, expressed discomfort. The instructor intervened with a posting that explained there was a section of the course, different from the discussion area, called "Sharing Almost Anything" which was more appropriate to the exchange of ideas about Star Trek. This kind of teacher presence was categorized "Instructional Management."

Concept Map Reflection Analysis

Concept map reflections were placed in the course dropbox to which only the course instructor had access. Because the reflections were viewed only by the instructor during the course, the students shared meaningful and private information about their learning process. Students were asked to respond to the following reflective questions upon completion of each concept map:

- After creating the concept map, did you see relationships among concepts that you did not see before?

- What was the easiest relationship among concepts to depict? What were the most difficult relationships to depict? Why were they easy or hard to depict?
- Look at the concept map and think back to the online discussion you participated during this module. Is there a relationship between the concepts you read and the online discussion? Were there moments in the online discussion you felt disoriented or confused? Does the concept map provide any clues about why you felt this way?

Investigators analyzed the self-reflections based on students' perceptions and attitudes toward their individual construction of knowledge.

Student Concept Map Analysis

Data analysis of concept maps was used as a way to triangulate data. The three years of data were used to compare module 1 and module 5 postings and concept maps. Concept maps served to confirm that facilitator questions and/or summary format influenced map content and structure. For example, a concept map could be organized based on the themes evident in the questions asked by the facilitator or on the format that the summary was structured.

FINDINGS

Based on the data analysis, we found that an online course that focuses on a community of learning and inquiry, the following factors influenced the concept map creation, which in turn affected individual construction of knowledge: group characteristics, social presence, cognitive presence, facilitation style of student, discussion summary format, and teacher presence.

Group Characteristics

An analysis of the educational experience, based on the transcripts of student online postings of the groups involved in this study, yielded many interesting results. One surprising result is that the size of the group did not have direct influence to the number of messages written by that group. The groups studied ranged from 3 members to 5 members. A common prediction would be that larger groups would have longer discussions due to more group members posting. In fact, the opposite seemed to be true. The groups with less members posted more messages per contributor than the larger groups did. See Table 2, Table 3, and Table 4 for a summary of the results.

For instance, in 2004 Group C had three members. The average number of postings per contributor was 29.5 messages in module 1 and

Table 2. Group A summary of results

Group Name 2003	Group A Module 1	Group A Module 5	Group B Module 1	Group B Module 5
# of Members	4	4	4	4
Messages per Mod	76	75	69	47
Instructor Messages	18	12	10	6
Facilitator Messages	20	21	24	22
Average Contributor Messages	12.7	14	11.7	6.3
Average Contributor Convergence	2.5	1.75	2.25	1.25
Average Contributor Divergence	1.25	0.75	0	0.50
Average Convergence and Divergence	3.75	2.50	2.25	1.75

Table 3. Group B summary of results

Group Name 2004	Group C Module 1	Group C Module 5	Group D Module 1	Group D Module 5	Group E Module 1	Group E Module 5
# of Members	3	3	5	5	5	5
Messages per Mod	91	55	194	117	119	71
Instructor Messages	12	9	22	7	23	8
Facilitator Messages	20	18	75	53	23	21
Average Contributor Messages	29.5	14	24.25	14.25	18.25	10.5
Average Contributor Convergence	.033	1	0.2	1.2	0	1
Average Contributor Divergence	0.33	0.66	0.2	1.2	0.2	0.
Average Convergence and Divergence	0.66	1.66	0.4	2.4	0.2	1

Table 4. Group C summary of results

Group Name 2005	Group F Module 1	Group F Module 5	Group G Module 1	Group G Module 5
# of Members	4	4	5	5
Messages per Mod	116	65	100	64
Instructor Messages	32	12	14	7
Facilitator Messages	26	23	33	15
Average Contributor Messages	19.3	10	13.25	10.5
Average Contributor Convergence	1.25	0.25	2.4	0.6
Average Contributor Divergence	0	0	0.6	0
Average Convergence and Divergence	1.25	0.25	3	0.6

14 messages in module 5. Conversely, Group E had five members and the average number of messages per contributor was 18.25 for module 1 and only 10.5 for module 5. We see this again in 2005 wherein the contributors of Group F, a 4 member group, wrote an average of 19.3 messages each in module 1, and the contributors of Group G, a five member group, averaged 13.25 messages each in the same module. Therefore, the total numbers of messages for each group discussion are very similar, regardless of group size. This may indicate that all groups are having the same level of depth and cognitive presence in their discussions. Groups with fewer members need to contribute more per person in order to fully discuss the topics. Therefore, group size influenced the amount of postings made per contributor, indirectly influenced group learning and in turn affected concept map creation.

One very interesting outcome of this analysis is that the groups that encountered some sort of conflict during their discussions engaged in more succinct discussions. Students who stated diverging and converging opinions deepened their discussions and came to conclusions more rapidly than groups who were not as forthright with opinions. For example, members of Group A stated high numbers of converging and diverging opinions and had a low average contributor response of 12.7 messages. Group C had low numbers of converging and diverging opinions in module 1 and had a high average contributor response of 29.5. The difference in the average number of responses between Group A and Group C is 16.8, which is important to note.

Social Presence

Accepting that learning occurs socially within communities of practice (Gunawardena & Zittle, 1997), students were required to participate in two group assignments: the online group discussion, composed of 3 to 5 students who would rotate on the roles of facilitator, summarizer, and contributor; and the team project (composed with the same individuals in the group discussion forums) working on the development of a program. The group discussions focused on the readings assigned to each module by the instructor. The team project focused on the completion of six tasks towards a final project.

As in any group, the issue of trust building was present. The first students' interactions in the online environment were greetings and acknowledgments. Beside the most basic and shy statements such as, "Hi group," we encountered examples of open communication such as: "I am sorry I didn't post," group cohesion statement such as "Have a great weekend," communication encouragement such as "Does anyone have a thought on this?" and emotional expression such as "I'm a bit nervous writing on [course management system]..." These signs of social presence grew deeper over time. Students used emoticons and emotional expressions such as ":)" and "Hmm" and offered suggestions for consideration such as "Have others experienced this?" and contributed positive feedback through statements such as "Interesting insights..."

For all three offerings of the course, the total number of messages posted in the online discussions decreased from module 1 to module 5. In part was due to the length of the module. See Tables 2, 3, and 4 for summary results. While the quantity of messages declined, the quality of the same increased. The amount of social presence instances also decreased dramatically. During the first module the interaction seems to indicate that students dedicated a considerable amount of time to getting to know each other and building trust. The decrease of social presence in the following modules indicates that trust was built and students could interact among themselves with more confidence. As trust increased, social presence decreased. As social presence decreased, students' concept maps contained more clearly stated ideas and connections among concepts. The evidence of increased focus on content is demonstrated by the concept maps. For example, Figure 1 and Figure 2 show concept map 1 and concept map 5 of Student A. Concept map 1 is more linear than concept map 5 and contains ideas taken dirextly from the readings. Concept map 5 is more focused, contains more clearly stated ideas, and shows more interconnections among concepts.

Cognitive Presence

Cognitive presence was evident through the participation in online group discussions that involved critical thinking. Critical thinking in this case was part of both process and product within a group dynamic (Garrison, Anderson, & Archer, 2001). As part of the online course, students were required to post a minimum of 3 to 5 quality messages per module. Quality messages were defined as postings that clearly tied back to the reading materials and/or course content. Students were free to post comments in a variety of formats. Some introduced scholarly references from other sources to support or highlight their perspectives. Others discussed personal experiences, while still others

Figure 1. Student A concept map 1 (© 2008 Conceição, Baldor, & Desnoyers. Used with permission)

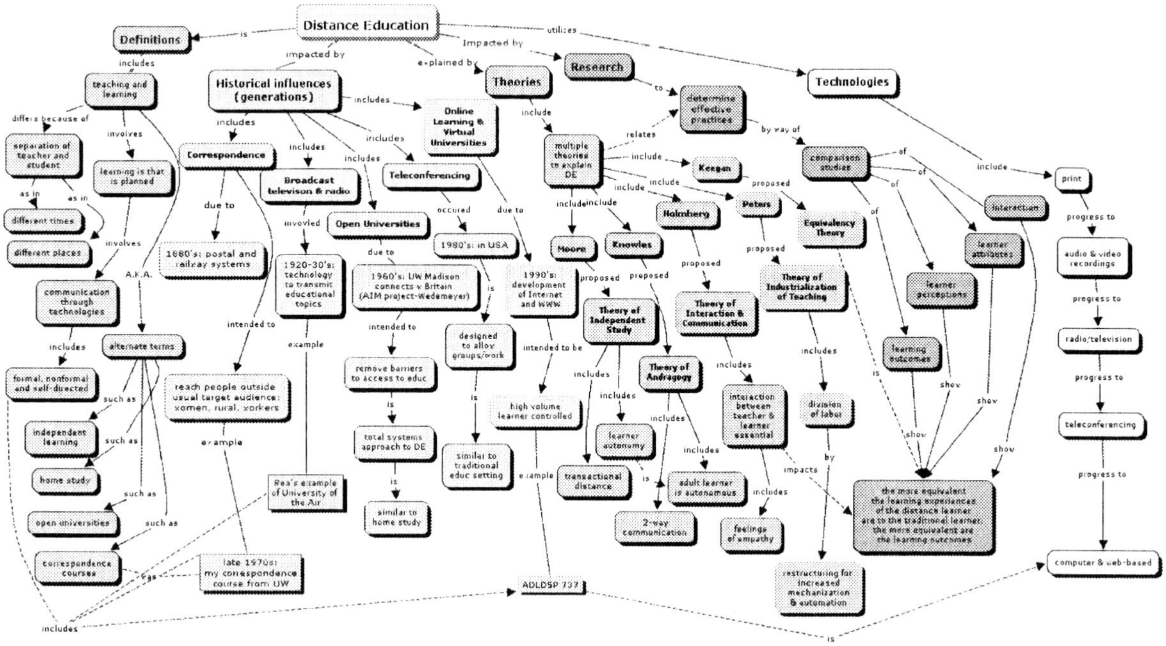

Figure 2. Student A concept map 5 (© 2008 Conceição, Baldor, & Desnoyers. Used with permission)

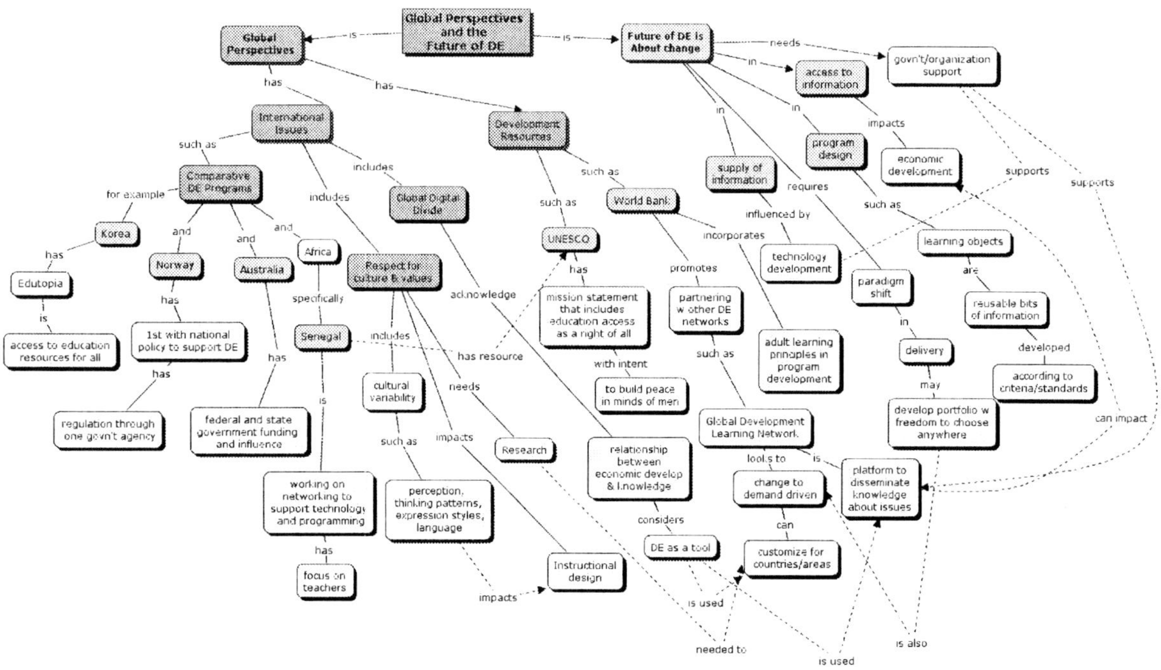

brought in professional experiences. Any of these formats were acceptable, but each response had to refer back to a point or points in the reading materials. Participants made arguments, described experiences, or discussed alternative perspectives within the context of the reading materials. The important aspect of the cognitive presence was that each comment had to explicitly connect with some aspect of the readings.

An analysis of cognitive presence postings data indicate that online discussion contributions vary from simple to complex. Students started their contributions by responding to the facilitator's questions through simple exchange of information. Then students moved the discussion to a more complex level by integrating concepts in readings with what was being discussed in the group forum. Ultimately students applied these concepts to their own personal experience. Some students were able to bring to the discussion concepts learned in other courses and integrate them to the concepts addressed in the readings and the group discussion. Some of the postings included all aspects: exchange of information, connection, and application to personal meaning. Cognitive presence increased dramatically from module 1 to module 5 especially in the areas of connecting ideas, testing solutions, and defending solutions all of which are evidenced in the subsequent concept maps.

Facilitation Style of Student

Each module discussion was facilitated by a different member of the group. The facilitation style of the student leading the discussion affected individual learning and the creation of concept maps as stated by the students in their reflections. Facilitators were responsible for selecting discussion topics, setting an agenda for the discussion, and keeping an active and involved interaction among members. When analyzing the facilitation style, we found the following: facilitators focused the discussion agenda on content, process, and on experience, or a mix of some or all of them. When the discussion focused on content, facilitator questions ranged from simple to complex. Simple questions were directive or included agree/disagree or like/dislike statements. Complex questions required more in-depth analysis of theory and research through comparisons and synthesis. Discussions that focused on process were organized on a panel discussion, scenario, or case study format and "what if...what would you have done differently" questions were posed. Discussions that focused on experience asked students to provide personal examples, defend their opinion based on personal experience, and provide practical implications.

Some students stated that being the facilitator of the discussion affected how the concept map was created based on their way of organizing information. For example, one student stated that being the facilitator, she had skimmed all the readings prior to the start of the module and felt organized and prepared for the unit with strategies she did not use in other units. Another student said that the role of a facilitator kept her "busy focusing on preparing questions and responses as opposed to focusing on the content." Once she began looking through the discussion that had taken place and looking back at the readings she was able to reflect on the specifics of the module content when creating the concept map.

Two students explained how their role as facilitators influenced the creation of their concept maps:

Student One: I was struck by how closely the completed concept map reflected my thinking as I designed the questions for facilitating (I was the facilitator for this module). As I was developing various options for questions, and deciding which to use and which to omit from the discussion group's work, I was essentially framing out the concept map. I didn't actually notice this until after the concept map was roughed-out and I saw how it matched the themes of the questions

Figure 3. Student B concept map 1 (© 2008 Conceição, Baldor, & Desnoyers. Used with permission)

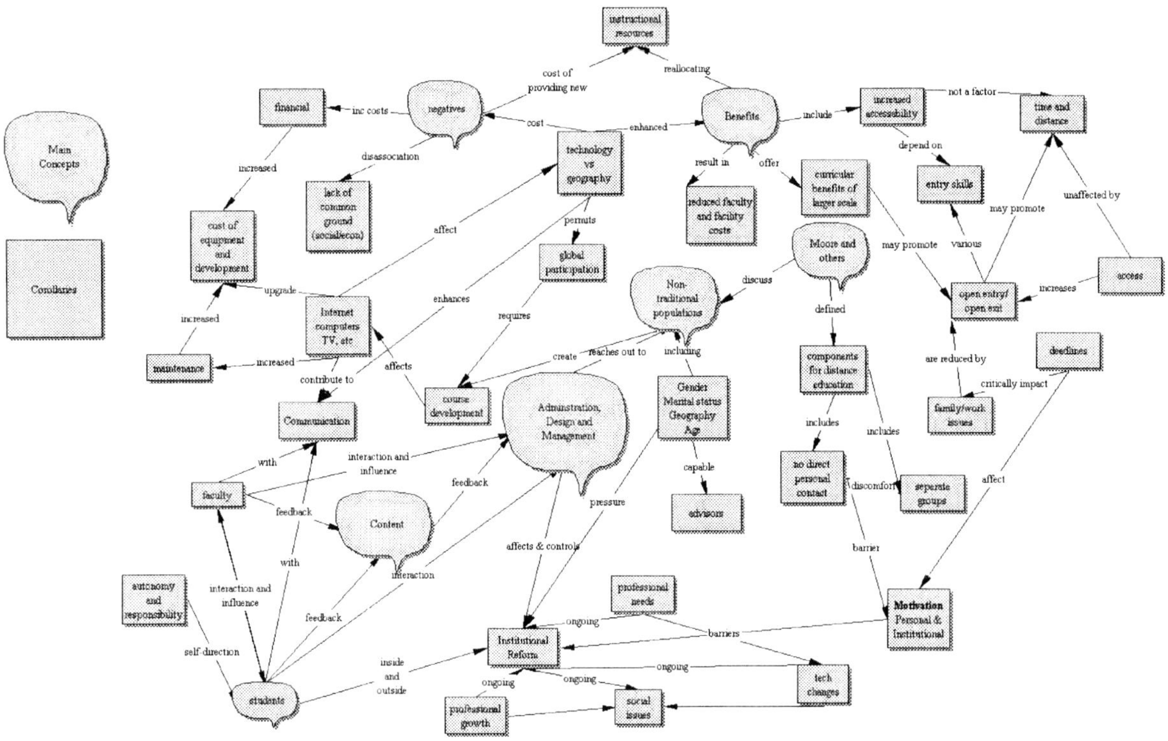

I'd prepared during the module itself. Compared to the other four maps, which were more clearly shaped by the readings and discussion, this map was first shaped by the readings and the questions I developed, and then the discussion and input from our group's thinking helped to fill it in, rather than directly shape it.

Student Two: Having the role of facilitator I was able to construct what I wanted to discuss and what I thought was most important. So, when it came to doing the concept map as a reflection I knew that I already had a summary of what stood out to me within the discussion. In previous modules I would have to do a lot of re-reading in order to pull out what was missed in the discussions that had stood out to me. Or, there were questions that I didn't feel were as pertinent but still answered to be a part of the discussion and then I simply didn't include them in my concept map.

The facilitator style affected how their concept map was created. For instance, Figure 3 shows concept map 1 of Student B organized by themes addressed in the group discussion. For some, not being a facilitator was easier as this student stated: "because I wasn't the facilitator I was able to spend more time answering and reflecting on the questions. Yes, there was a relationship between the concepts I read and the online discussion." If the facilitator provided good questions and was able to pull out key concepts that fired up the discussions, students were able to achieve higher level of comprehension as this student stated: "Our group facilitator for this module was able to formulate valuable questions from the readings that forced me to look at each article in depth. Reading the articles over the first time and then actually looking back at them to answer questions, brought out a level of comprehension I wasn't able to achieve in the previous module." One student said that she found herself attempting to link the facilitator's posting to what she was reading for the module.

When the facilitator led the discussion well, some concept maps were created based on the themes of the discussion. However, when the facilitator did not lead the discussion well, some concept maps were created based on the readings as this student commented:

...I don't feel that our facilitator did a good job in leading the conversation. There were no real questions posted at the beginning of the week because "no one seems to answer them anyway"? I hardly think that's the point. At least starter questions give you something to think about while you're reading. If other questions or comments come about, then the learning technique is working! I also didn't feel like she gave any kind of feedback or support to our group. [...] I know from personal experience in online classes, lack of personal feedback leaves me confused and lackadaisical. [...] But if nothing else, this module helped me to analyze my needs and maybe next time I feel this way, I will request more interaction from the facilitator.

Discussion Summary Format

Based on an analysis of discussion summary layout, summaries were presented in the following formats: descriptive (per person's contributions, per week contributions, per themes, or per questions), comparative (based on discussions addressing pros/cons, agreements/disagreements, and comparisons among concepts), process-based (explaining the group process during the discussion), or a combination of two or more formats.

For some students, the discussion summary was a start for the creation of their concept map. For these students, the format of the summary was evident in the structure of the concept maps:

- "the summary gave me a great outline with which to start."
- "...after reviewing my postings and [name]'s summary, I was able to portray a clear connection between the concepts with the topics discussed."
- "It was easy to identify the concepts I wanted to be sure to include, as the summary helped me identify them."

For others the summary served as a way to review the concepts addressed in the discussion as these students stated:

- "I was almost completed with the concept map by the time the Module Summary was posted, and I used it more to check my map than create it."
- "The concept map does speak to aspects of discussion, and after the first pass at the map, I reviewed the module four summary written by [name]."
- "I also used the discussion summary to aid in creating the concept map. The final summary from the online discussion was very thorough and helped me to understand many points that I may have missed."
- "I think that our group's module 2 discussion was much more focused on the assigned readings than our module 1 discussion and also contained a well-developed summary. For these reasons, there are strong connections between the assigned readings, my group's online discussion, and my personal concept map."

Teacher Presence

Teaching presence is defined by Anderson, Rourke, Garrison, and Archer (2001) as the design, facilitation, and direction of cognitive and social processes in order for personally meaningful and educational learning outcomes to occur. The strategies utilized by the instructor in the online course were fundamental in placing the students at the center of the learning process (Prestera &

Moller, 2001). Analysis of instructor's postings included the following strategies:

- Allowing students to control (or share control of) the process of meaningful learning. In this case, students not only shared the role of facilitator, summarizer, and contributor, but they also contextualized course content with their personal experiences. Students took responsibility and shared control of the learning process which involved a sense of ownership since they were able to choose problems for the discussion that were real and important to them.
- Providing positive and constructive feedback that was timely and personalized. After a student apologized to classmates online for not participating on the discussion due to family problems the instructor acknowledged online this student's circumstances and wrote: "You will catch up. In the future try to manage better your time but I understand family issues."
- Facilitating collaboration through the use of group discussions that were threaded and involved completion of group projects.
- Encouraging reflection through asynchronous communication that used a pattern of read-reflect-write.
- Facilitating the building of personal knowledge/meaning by requiring from participants the creation of personal concept maps, the rationale being that concept maps helped learners to organize and communicate their ideas.
- Guiding students to become increasingly independent thinkers by reducing interventions. As the teacher presence decreased students became more independent. The teacher presence during module one was focused on redirection and direct instruction, while in module five the focus is on positive feedback.
- Mentoring learners by modeling positive contributions to discussions.
- Coaching students by providing formative feedback during modules. One example of such feedback is in this instructor's comment to one of the groups: "Good topics, try to connect them to the readings. Look at other groups' discussions for guidance."
- Assessing students on their course outcome through reflections on their thinking process on course content upon completing concept maps.
- Making a variety of materials and resources available to students.

In this study, teacher presence influenced concept map creation which affected individual construction of knowledge. The level of instructor and facilitator presence is clearly associated with the length of the discussion. For example, in group discussion forums in which the instructor posted more messages, the number of postings of students was also higher. Similarly, discussions in which the group facilitator posted more questions, contained more postings by students in that group. It was also evident that cognitive presence affected teacher presence. As discussions became more focused, teacher presence decreased and students' messages had greater depth, which in turn helped students create more focused concept maps.

CONCLUSIONS AND IMPLICATIONS: STRATEGIES TO FACILITATE INDIVIDUAL LEARNING IN A COLLABORATIVE ONLINE ENVIRONMENT

The analysis of group characteristics; comprehensive examination of social presence, cognitive presence, and teacher presence based on the Community of Learning and Inquiry model; and the analysis of facilitation style of students and discussion summary formats show that there is no doubt that these factors influenced students' creation of concept maps, which reflects their individual knowledge construction. We suggest to instructional designers and instructors that when designing online courses that use collaborative strategies consider the following elements: strategies that build trust, allow opportunities for collaborative and individual learning, provide students facilitation guidelines, and provide students guidelines for effective discussion summaries.

Using Strategies to Build Trust

We learned that building trust is an essential component of the educational experience in an online environment. Our study shows that when students get to know each other and build trust at the beginning of the course, they are more comfortable in sharing insights and personal stories during the intellectual discussions of content. The amount of social presence decreases while cognitive presence increases overtime after students feel there is a sense of community. As a result of this increase in cognitive presence, students were able to create concept maps that had more clearly stated ideas and connections among concepts. As a result, instructors should provide activities in the beginning of the online course for students to build trust and feel comfortable with each other. For example, the instructor may distribute students to groups and require them to assign leadership roles in course tasks and communicate learning concerns and challenges regarding group experience in previous course work.

Opportunities for Collaborative and Individual Learning

In this study, as social presence decreased cognitive presence increased in quality and quantity. Contributions progressed from simple postings providing direct answers to the facilitator questions to complex analysis of concepts and theories often applying examples from personal experience. The decrease in social presence occurred because students successfully built trust with each other in a group and therefore felt more comfortable in focusing on in-depth intellectual contributions to the discussion. The concept map assignment allowed students to prioritize information in readings, integrate concepts, confirm knowledge, and construct new knowledge. The community of learning and inquiry strategy used in this online course built an environment for social learning, but it was through individual concept maps that students were able to construct their own knowledge. Hence, it is important to provide students with opportunities to work together, but at the same time it is valuable to include activities that facilitate individual construction of knowledge. Concept maps are effective tools to accomplish it.

Concept maps also provide a way for students to connect their new knowledge to knowledge that they had previously acquired. According to De Beni, Magnussen, Cornoldi, Larsson, and Rønnberg (2007), when students acquire new knowledge in a way that connects to previous knowledge already in the memory, it is much easier for the mind to retrieve. By creating a concept map, learners create many connections between new ideas and previous knowledge; therefore, permanently positioning the new knowledge into their long term memory. These memories will be made even stronger if the learners connect the new ideas to their own personal experiences.

In this study, the instructor allowed students control over their own learning, provided positive and constructive feedback, facilitated collaboration, encouraged reflection, facilitated building of personal meaning, guided students to become independent thinkers, mentored students by modeling positive contributions, and coached students throughout the process. These social and cognitive strategies helped create and maintain an active community of learning and inquiry. Teacher presence was a critical element of the online classroom in this study. Therefore, we suggest that instructors need to take into consideration

Table 5. Discussion focus

Discussion Focus	Description
Content	Facilitator questions range from simple to complex. Simple questions are directive or included agree/disagree or like/dislike statements. Complex questions require more in-depth analysis of theory and research through comparisons and synthesis.
Process	Discussion is organized on a panel discussion, scenario, or case study format and "what if...what would you have done differently" questions are posed.
Experience	Ask students to provide personal examples, defend their opinion based on personal experience, and provide practical implications.
Mixed	Combine questions focusing on content, process, and experience.

the design of the course and use interactive, collaborative, and individual strategies to facilitate student construction of knowledge in an online community.

Facilitation Guidelines

Students stated that how the discussion was framed influenced their learning process and construction of the concept map. Some students who took the role of the discussion facilitator noted how easy it was to design the concept map because they had already organized information for the discussion. Others commented that they were able to achieve higher levels of comprehension and it became more effective and efficient to create the concept map when the facilitator selected meaningful discussion topics, provided timely feedback, and organized and led the discussion on key concepts from the readings. We suggest that instructors provide facilitator guidelines in the beginning of the course on how to facilitate discussion forums, so more effective learning can take place. Table 5 provides a description of the discussion focus addressed by facilitators in this study. This list may help students in online courses to focus the discussion depending on the content of the module.

Guidelines for Effective Discussion Summaries

For some students the discussion summary provided an outline for the creation of the concept map. For others it provided a review of the concepts addressed in the discussion. It is important that students are aware of the benefits of the discussion summary format, so it can be another tool to assist them in designing their concept map and constructing knowledge in the course. Instructors can provide samples of effective summary formats and formative feedback to the summarizer of the discussion upon completion of each module. Table 6 provides a list of types of summaries encountered in this study. This may assist students in online courses to define a format of the summary before writing it.

Table 6. Types of summaries

Type of Summary	Description
Descriptive	Per person's contributions, per week contributions, per themes, or per questions.
Comparative	Based on discussions addressing pros/cons, agreements/disagreements, and comparisons among concepts.
Process-based	Explaining the group process during the discussion.
Mixed format	Combination of two or more formats.

This study shows that when students work in a community of learning and inquiry in an asynchronous online environment, individual learning occurs when cognitive strategies that involve individual construction of knowledge such as concept maps are present. In this study, six factors influenced the concept map creation, which in turn affected individual construction of knowledge. These factors were: group characteristics, social presence, cognitive presence, facilitation style of student, discussion summary format, and teacher presence. Group characteristics, social presence, cognitive presence, and teacher presence are common elements of a community of learning and inquiry. What is unique in this study is the influence of facilitator style and summary formats on the concept map creation and student individual construction of knowledge.

The community of learning and inquiry was a good base for the design of collaborative and interactive aspects of the course. The use of concept maps facilitated students' construction of individual knowledge after participating in collaborative learning. Working in a collaborative community allows students to explore different ideas and concepts, but it is through the individual work that students refine and expand their knowledge and construct personal meaning. Using a combination of both collaborative and individual strategies proves to be effective to facilitate individual construction of knowledge in an online environment.

REFERENCES

Anderson, T., Rourke, L., Garrison, R., & Archer, W. (2001). Assessing teaching presence in a computer conferencing context. *Journal of Asynchronous Learning Networks*, *5*(2), 1–17.

Basque, J., & Lavoie, M. (2006). Collaborative concept mapping in education: Major research trends. In A. J. Cañas & J. D. Novak (Eds.), *Concept maps: Theory, methodology, technology: Proceedings of the Second International Conference on Concept Mapping,* (Vol. 1, pp. 79-86). San Jose, Costa Rica.

Cabral, A. R. Y., Zeve, C. M. D., Nicolao, M., & Amoretti, M. S. M. (2004). Use of conceptual maps in distance learning courses. In A. J. Cañas, J. D. Novak, & F. M. González (Eds.), *Concept maps: Theory, methodology, technology: Proceedings of the First International Conference on Concept Mapping,* (Vol. 2, pp. 111-114). Pamplona, Spain.

Calderón, M. D., Agüera, E., & Alfageme, M. B. (2004). Los maps conceptuales, herramienta hipertextual para el trabajo colaborativo y desarrollo de habilidades comunicativas y docentes. In A. J. Cañas, J. D. Novak, & F. M. González (Eds.), *Concept maps: Theory, methodology, technology. Proceedings of the First International Conference on Concept Mapping* (Vol. 2, pp. 115-118). Pamplona, Spain.

Cañas, A. J., Ford, K. M., Novak, J. D., Hayes, P., Reichherzer, T. R., & Suri, N. (2001). Using concept maps with technology to enhance collaborative learning in Latin America. *Science Teacher (Normal, Ill.)*, *68*, 49–51.

De Beni, R., Magnussen, S., Cornoldi, C., Larsson, M., & Rønnberg, J. (2007). Memory experts. Visual learning, wine tasting, orienteering and speech reading. In S. Magnussen & T. Helstrup (Eds.) *Everyday memory* (pp. 201-227). Hove, UK: Psychology Press.

De Simone, C. (2007). Applications of concept mapping. *College Teaching*, *55*(1), 33–36. doi:10.3200/CTCH.55.1.33-36

Fonteyn, M. (2007). Concept mapping: An easy teaching strategy that contributes to understanding and may improve critical thinking. *The Journal of Nursing Education, 46*(5), 199–200.

Garrison, D. R. (2003). Cognitive presence for effective asynchronous online learning: The role of reflective inquiry, self-direction and metacognition. In J. Bourne & J. C. Moore (Eds.), *Elements of quality online education: Practice and direction*, (pp. 47-58). Needham, MA: Sloan-C, Sloan Center for Online Education.

Garrison, D. R., Anderson, T., & Archer, W. (2000). Critical inquiry in a text-based environment: computer conferencing in higher education. *The Internet and Higher Education, 2*(2-3), 87–105. doi:10.1016/S1096-7516(00)00016-6

Garrison, D. R., Anderson, T., & Archer, W. (2001). Critical thinking, cognitive presence and computer conferencing in distance education. *American Journal of Distance Education, 15*(1), 7–23.

Gunawardena, C. N., & Zittle, F. J. (1997). Social presence as a predictor of satisfaction within a computer-mediated conferencing environment. *American Journal of Distance Education, 11*(3), 8–26.

Kim, B., Yang, C., & Tsai, I. (2005). Review of computer-mediated collaborative concept mapping: implications for future research. *Proceedings of the 2005 Conference on Computer Support for Collaborative Learning: Learning 2005: The next 10 years!* (pp. 291-295).

Laampere, M., Matsak, E., & Kippar, J. (2006). Integrating a concept mapping tool into a virtual learning environment: Pedagogical and technological challenges. In A. J. Cañas & J. D. Novak (Eds.), *Concept maps: Theory, methodology, technology: Proceedings of the Second International Conference on Concept Mapping,* (Vol. 1, pp. 280-287). San Jose, Costa Rica.

Luckie, D. B., Harrison, S. H., & Ebert-May, D. (2004). Introduction to C-TOOLS: Concept mapping tools for online learning. In A. J. Cañas, J. D. Novak, & F. M. González (Eds.), *Concept maps: Theory, methodology, technology. Proceedings of the First International Conference on Concept Mapping,* (Vol. 2, pp. 261-264). Pamplona, Spain.

MacNeil, M. S. (2007). Concept mapping as a means of course evaluation. *The Journal of Nursing Education, 46*(5), 232–234.

Novak, J. D., & Gowin, D. B. (1984). *Learning how to learn*. Cambridge, UK: Cambridge University Press.

Prestera, G., & Moller, L. (2001). Facilitating Asynchronous Distance Learning: Exploiting opportunities for knowledge building in asynchronous distance learning environments. *Proceedings of the Mid-South Instructional Technology Conference, Middle Tennessee State University.* Retrieved January 24, 2008 from http://www.mtsu.edu/~itconf/proceed01/3.html

Rábago, A. R. (2004). La construction de mapas conceptuales en educacion a distancia. In A. J. Cañas, J. D. Novak, & F. M. González (Eds.), *Concept maps: Theory, methodology, technology. Proceedings of the First International Conference on Concept Mapping,* (Vol. 2, pp. 309-312). Pamplona, Spain.

Rice, D. C., Ryan, J. M., & Samson, S. M. (1998). Using concept maps to assess student learning in the science classroom: Must different methods compete? *Journal of Research in Science Teaching, 35*, 1103–1127. doi:10.1002/(SICI)1098-2736(199812)35:10<1103::AID-TEA4>3.0.CO;2-P

Tarouco, L., Geller, M., & Medina, R. (2006). Cmap as a communication tool to promote meaningful learning. In A. J. Cañas & J. D. Novak (Eds.), *Concept maps: Theory, methodology, technology: Proceedings of the Second International Conference on Concept Mapping,* (Vol. 2, pp. 44-49). San Jose, Costa Rica.

KEY TERMS AND DEFINITIONS

Community of Learning and Inquiry: Within a context of collaborative and reflective communication, the community of learning and inquiry is an educational model composed of social, cognitive, and teaching presence.

Social presence: Computer mediated interactions and communications that reflect the personal and emotional connections of the individual to the group.

Cognitive presence: Computer mediated interactions and communications related to the intellectual realm. Cognitive presence is the discourse that facilitates critical reflection.

Teaching presence: computer mediated inputs from a course instructor or discussion facilitator that facilitate the community of learning and inquiry.

Convergence: participant's stated agreement with an idea or concept posted by another participant in a previous message.

Divergence: participant's disagreement with an idea or concept posted by another participant in a previous message.

Knowledge construction: process of accessing, understanding, evaluating, connecting and refining information in order to produce personal meaning.

Chapter 7
Distance Collaboration with Shared Concept Maps

Alfredo Tifi
World Wide Maps, Italy

Antonietta Lombardi
World Wide Maps, Italy

ABSTRACT

This chapter aims to review Web-mediated practices in collaborative concept mapping which were set up in informal partnerships among teachers from different schools and countries, and their students. Starting with various examples of such practices, we will outline some models of collaboration which will be compared and criticised in the hope that they will be useful in challenging other teachers to plan suitable strategies for engaging in similar experiences. It is important firstly to examine the context in which distance collaboration can be established and then outline some theoretical background to show the reasons why this kind of collaboration should be recommended as an objective for educationalists.

INTRODUCTION

WWMaps (World Wide Maps) is a community of practitioners aimed at establishing concept-mapping collaboration teams among groups of students from different countries. Unlike other international twinning initiatives (e.g.: etwinning in the European Lifelong Learning Programme), this community is unofficial and based exclusively on concept mapping.

The topics chosen by collaborating teams are mainly drawn from their curricula and cover such issues as the environment, citizenship, history, specific science topics etc., depending on the preferences of the teachers involved. It is nevertheless possible for partners to deal with inter-cultural topics or for teacher members of the community to debate educational topics.

The choice of Collaborative Concept Mapping (CCM)is doubly valuable. Concept mapping firstly facilitates the engagement of teachers of non-language subjects (such as L1, History, Maths or Science), although support from an L2 teacher is welcome. Secondly, concept mapping is more than a medium. It can be considered as a skill in itself, being a learning tool for metacognition that

DOI: 10.4018/978-1-59904-992-2.ch007

can be applied and transferred to other fields by pupils and teachers.

If these collaborative practices and technologies are to have a positive impact on local educational communities, they should be set up with objectives and expectations that are agreed and shared by both school partners. They should also incorporate strategies to assure effective interaction and sharing among students. These are the reasons for studying in depth the educational objectives of, and strategies for, an effective CCM.

BACKGROUND 1: THE CONTEXT OF COLLABORATION

Collaborative Learning (CL) is different in *presence-based* education (for groups of students in the same classroom) from *distance-based education*, where students may collaborate from different countries. First of all, *face to face promotive interaction* is not possible in distance-based CL: the mother tongue is often different, and communication depends on different technologies and time zones (for instant messaging). There are also multiple factors which can greatly differ among partners -- their respective cultures, degrees of cosmopolitanism, educational missions, priorities, objectives and curricula and, last but not least, the ages, grades and number of students per class. All these differences, often under-estimated, strongly affect the effectiveness of collaboration.

Furthermore, teachers can also be granted very different levels of independence (or conversely restriction) by their institutions in deciding (freely or otherwise) how to manage time and plan educational projects involving international collaborations, even on curricular topics.

Finally, the relevance of such international collaborations, and the educational results they achieve, are not evenly appreciated and promoted by all institutions and their managers. These factors are somewhat related to the concept of *governance* introduced by Gowin and Novak (Novak & Gowin, 1984; Gowin & Alvarez, 2005). Governance factors affect the intrinsic meaning of the educational experience, even when the same tasks are undertaken. The environment in which the collaborating team operates may modify the development and *sustainability* of the collaboration because of the presence or absence, and quality of feedback.

Apart from governance factors, the sustainability and effectiveness of collaborative work depend strongly on efforts to attain such other well known requirements of the *Cooperative Learning* model (Johnson & Johnson et. al., 1994) as individual and group accountability, interpersonal and small group skills, and group processing.

We are aware that interpersonal and small group social skills training should be strongly implemented from the very beginning of a collaboration, while individual and group accountability need to be assured as the process develops. This is why in the collaborations we are currently undertaking daily communication takes place between student and teacher partners, helping them to construct suitable communication skills, technologies and methodologies independent of the contents to be elaborated in subsequent Collaborative Concept Mapping activity. This activity would be sterile without the vital habit of communication and feedback among partners.

Finally, *group processing*, i.e. reflection on the work of students and their interactions, focusing on achieving group goals and ensuring effective working relationships, is normally delegated to the teacher. On the contrary, however, this task should rest with team members. Given that the above conditions are rather complex and not always controllable, the working teams in *Wwmaps* cannot be considered as a flat terrain where formal research questions can be easily planned and carried out. They are, rather, fertile soil where good practices of collaborative concept mapping and friendly co-operation can be nurtured through the years.

Despite the disadvantages already outlined, there are indeed great opportunities to enhance the education of students, to stimulate positively the governance of institutions and to challenge teachers engaged in new methodologies and technologies. We strongly believe that collaboration in an international team group helps to improve command of a second language, acquisition of an inter-cultural sense, widening of students' self-perspective and interest in others and overcoming of the sense of closure arising from repetition of curricula and crystallisation of roles. Beyond the opportunities to give our students a leading role, we wish them to share new communicative experiences - collaboration tools and resources through the Web. In this sense, differences become a resource.

Finally, we believe we are achieving our first objective as founders of this community (promoting concept mapping in education) since there are some students creating concept maps for their own study purposes, outside of their working team.

BACKGROUND 2: CONCEPT MAPPING AS A MEDIUM AND A LEARNING OBJECTIVE

Concept mapping can be mainly viewed from two non-exclusive perspectives. From a representational perspective, a concept map serves as an effective way to show or "inform" how a network of concepts are related and organized to form a knowledge domain (Cañas, Hill, Carff, Suri, Lott, Eskridge, et al., 2004). From this perspective, a concept map is viewed as a substantially static object, or as a "finished" product. From a constructivist perspective, a concept map is viewed as a *process* that reflects the steps that contribute to changing the organization and definition of individual or group cognition of the same knowledge domain. This kind of concept mapping (Henao Cálad, 2004) is dynamic in all its aspects, decreasingly useful to learner, to educator-mediator and to external readers. These two perspectives can obviously be superimposed on one another (Novak & Cañas, 2004) and are inter-changeable, but we believe it is important to stress that the second is most important from pedagogical and collaborative points of view, given that learning and collaboration are dynamic processes.

In detail (Novak & Cañas, 2008) a concept map includes concepts, usually inside closed shapes, and relationships between concepts made explicit by linking phrases. A concept is defined by Novak as "a perceived regularity in events or objects, or records of events or objects, designated by a label". The label is generally one or more words, although sometimes symbols or images can be used. A proposition is a statement that contains two or more concepts, connected by a linking word or phrase to form a meaningful statement, or unit of meaning. Nevertheless, a concept map is more than just a network of propositions. It has a context that can be identified in a text, a laboratory activity or a particular problem that one is trying to understand. The awareness of the context and of the domain framework will help to determine the hierarchical structure of the concept map. A good way to define the context for a concept map is to construct a *Focus Question*, i.e. a question that clearly specifies the problem or issue the concept map should help to resolve. It is important to stress that a concept map that describes a domain of knowledge aimed at answering a focus question is very different from a concept map that purely describes that domain without a precise goal, the latter being characterized by a higher level of arbitrariness.

From its dynamic perspective, a concept map can also be viewed as an open-ended platform where a learner structures his/her attempts to accommodate previous knowledge and to subsume new knowledge. The integrative process that takes place has been described as electively facilitated by concept mapping (Novak & Cañas, 2006). In our opinion, this is due to three fundamental and unique properties of concept mapping: *a)* its "immediacy", i.e. the opportunity offered by the two-dimensional plane of making co-existing and clearly visible declarations separated in time and in logical flow; *b)* its "flexibility", i.e. the opportunity of freely changing concatenations and ranks of concepts and of adjusting the type of relationships; and *c)* its "disciplining", i.e. the requirement to make every proposition a meaningful, explicit and unambiguous statement, or semantic unit, inserted in a precise and explicit definition of the framework. We maintain that these properties of concept mapping also facilitate integration between the cognition of different people as they interact, as can be required in educative mediation and in collaborative concept mapping. In this way we believe we can justify the contention in the title of this section. We see concept mapping as potentially both a privileged medium of interaction, comparison and negotiation of beliefs, meanings and opinions on the same subject, and also as an educational objective as a learning tool that can make a difference in the education of individual learners.

In a collaborational framework, which is this chapter's subject matter, it is evident that partners collaborating on the same concept map should jointly decide a knowledge domain and a focus question, should agree upon criteria to alternate and differentiate each one's contributions and should keep a chronological log of changes. (see Figure 1)

In the remaining part of this chapter we will use the generalized term "knowledge claims" to speak of propositions, statements, or even small concept maps, that answer a single specific focus question, and can be proposed as contributions by the members of a team. In a collaborative team we can describe a concept map as being constructed of *knowledge claims*, i.e. of statements that are in some way integrated or processed. In a *knowledge model,* i.e. a group of related concept maps and linked resources, answering different focus questions about the same knowledge domain, a *generalized knowledge claim* is made up of one of these cmaps.

An important additional aspect of concept mapping concerns multilingual collaboration. As we have illustrated elsewhere (Tifi & Lombardi,

Figure 1. WWMaps goals and distance collaboration using concept mapping

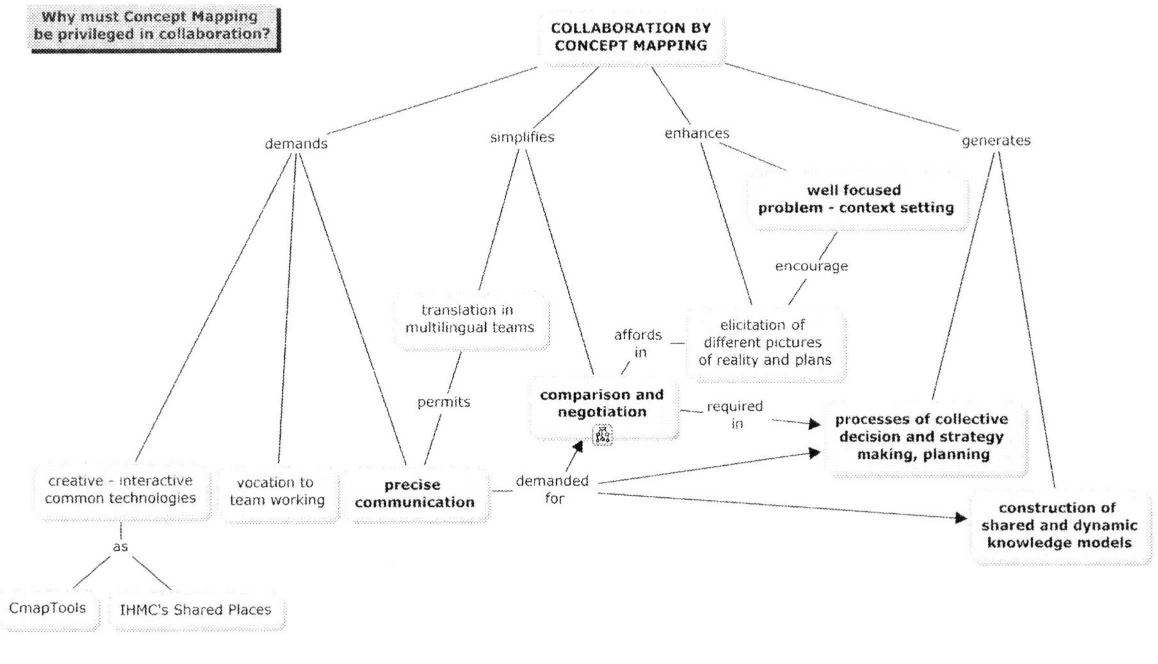

2006), there is a great advantage in sharing ideas and knowledge claims in L2 if these are coded in concept map language, rather that in plain text, for pupils with very basic skills in L2. Due to the elementary propositional structure of concept maps, collaborative concept mapping facilitates elicitation and representation of knowledge, reducing speech to its essentials and simplifying negotiation and convergence of views for effective communication and sharing, even if collaborators speak different languages. We have proposed the use of bilingual concept maps in two colours to facilitate the passage from L1 to L2 and the achievement of reciprocity and mutual understanding of contents and contexts. The decoding of an L2 concept map to L1 is analogous to but easier than translating a linear text.

PRACTICES IN WEB-BASED COLLABORATIVE CONCEPT MAPPING

Our first experiments in collaborative concept mapping dealt with the basics of the process: twinning of teachers; creating a community; sharing of tools (IHMC CmapTools ®) and technological know-how; finding a common model for concept mapping, i.e. the Novak, Cañas (2008) model; adapting its rules for students facing concept maps for the first time (see Appendix); searching for relevant and interesting topics to develop in the teams; and thinking about how to coordinate and alternate contributions of students in each team.

We have tested, and are testing different strategies for Web-based collaborative learning concept mapping which are aimed at enhancing positive interdependence among the members/students. Our experiences of these tests will be discussed in depth in this section from a theoretical point of view, and general conclusions will be drawn.

The Concept of "Active Collaborative Process"

The objective of Collaborative Concept Mapping via the Web is to realise, as much as possible, the five basic criteria of Cooperative Learning, i.e. e: Positive Interdependence, Individual Accountability, Face-to-Face Promotive Interactions, Use of Collaborative Skills, and Team Self-Assessment (Johnson et al. 1994). In an absolute sense, F2F promotive interactions are impossible in distance collaborative learning but, as we have suggested in 3.2., a sort of promotive interaction is potentially enabled through the coexistence of feedback and contributions, as knowledge claims, from different members of the team. These "ingredients" have the capacity to generate what we define as an "*active collaborative process*". (see Figure 2)

A concept map or knowledge model can include knowledge claims from different members of the learning community that are conflicting, faulty, wrong, superfluous, duplicated or unclear or incomprehensible to some as well as coherence and connectivity flaws or other faults. If this is the case, we can observe what develops over time. If these faults provoke proper feedback, or interactions between team members, we can affirm that an *active collaborative process* is taking place. The presence of feedback is more important than subsequent editing or adjustments that can take place under the supervision of teachers but unknown to some members. Communication of feedback is the first sign that integration between individual views or cognitions is possible. It is the condition for the construction of awareness that a learning community exists. We therefore attempt to make an analogy between differing integration processes, both facilitated by concept mapping -- the one that takes place in progressive differentiation and integrative reconciliation of Ausubel's-Novak's principles and the one (Novak & Cañas, 2006) which concerns *collective knowledge* (Reigeluth, 1999).

Figure 2. Cmap illustrating how Collaborative Learning differs from Cooperative Learning

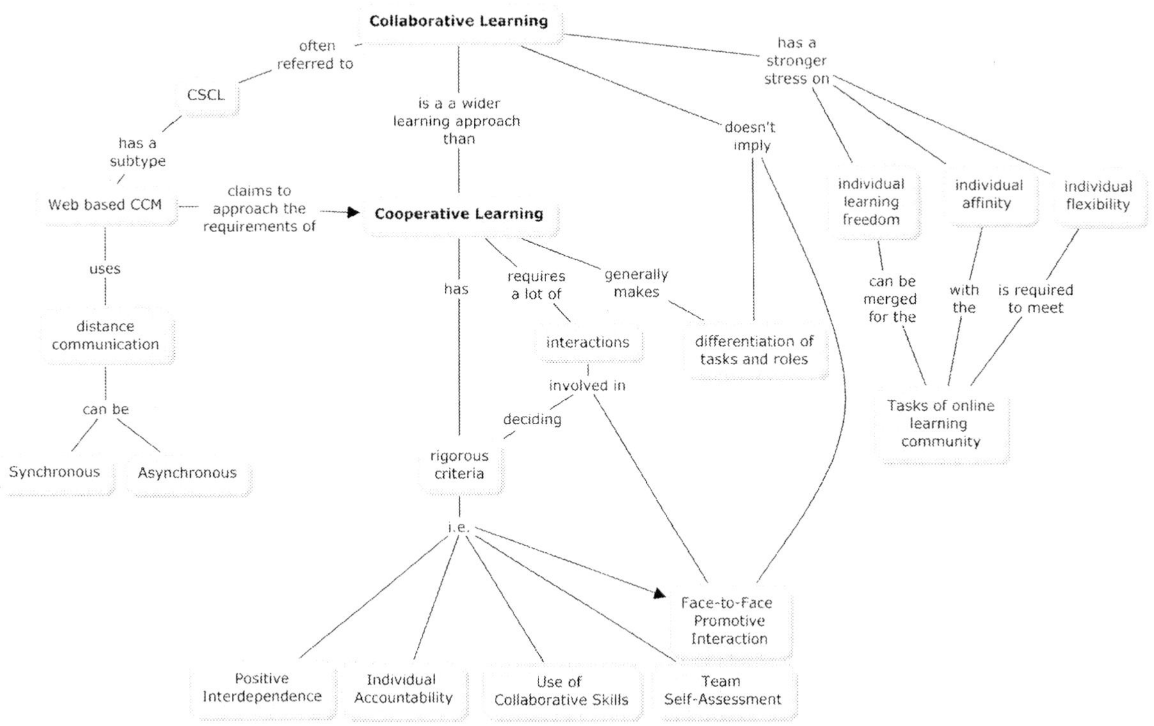

The time lag between the generation of a problem in a dynamic knowledge concept map and the generation of feedback, or the time interval between feedback and a reply from other members, can be considered as the "clock frequency" that measures the rate of activity of the active collaborating team. An online learning community will obviously have a reduced activity rate in comparison with face-to-face interacting teams, but we believe that a minimal threshold exists under which the process can be declared collaboratively inactive. Such inactive, or minimally interactive, collaborating processes constitute the most probable outcome of attempts to establish full interactive experiences because the sustainability of the latter is highly demanding in terms of psychic energy, motivation, time, collaborative skills and boundary conditions (governance). Nevertheless such processes can be classified as examples of collaborative learning experiences because some outcomes have been achieved as the result of reciprocal help and contributions. However, we hope that a stronger awareness of the difference between distance collaboration and face-to-face interaction will help us and future partners to bring about more effective collaboration teams, where proper criteria to attain positive interdependence, individual accountability, use of collaborative skills and self-assessment of the process could be planned

Motivation

In this paragraph we are *not* going to talk about the motivation of students as it is often perceived, i.e. as an independent variable that is a prerequisite at the beginning of a learning process. We want instead to deal with *motivation* as a good outcome of the process, generated by clever choice of strategies (at the preparatory stage), detailed planning setting and complete initial agreement on sharing of objectives.

First of all, we should ask ourselves why as teachers we want to engage in a concept mapping Web-based collaborative learning project, taking into consideration that the conditions for learning, in such an unusual method of education, are hard enough to meet even within the boundaries of a single class. We take the answer to this question from the children themselves. On several occasions, we have perceived their positive emotive feedback as they recognize they are going to establish contact with children "on the other side of the computer" and that those children overseas have the same expectations, the same enthusiasm and the same longing to collaborate as they themselves do. The first rule, therefore, is to amplify the emotive impact by complying with their eagerness to exchange personal information with their partners. One or more chat meetings or videoconferences can be a suitable way to achieve this objective, which is also valuable in exploiting children's instinctive cosmopolitanism. In our view such effort is well worthwhile.

Secondly, we must keep the flame of curiosity alive. The *choice of topic* is a key step. We should support and guide the children's interest. We often choose the topic of collaboration, but a different way of proceeding is more convincing: within a given general theme or subject, dictated by the class curriculum or by the general purposes of the partnership, we should leave the children free to choose the specific areas of interest that they want to develop. Children are often unexpectedly autonomous in defining a field in which they have a burning interest, either because of previous experiences or something they partially know. It may also be that they want to engage themselves in something new for the sake of surprising their partners.

The next key stage for motivation is the choice of *documentation resources*. For younger children, the documents are best chosen by the teachers, to allow autonomous and easy research and reading by the pupils. We know of cases where children decide for themselves to meet out of school time to study documents. For older students, it is advisable to let them search for resources or browse a pre-determined list of links. Moreover, to foster interest and engagement, it is possible to assign a topic that is pertinent to the other partner (for example, Italian students may study alterations to the environment of the Nile valley while Egyptian students in the same team might study alterations to the Po valley, about which Italians are more informed.) This exchange of competences could enhance reciprocity and encourage the subsequent stage of peer reviewing.

Furthermore, another factor that might encourage collaboration is the availability of the same sources of documentation in both languages of the different speaking partners. For example, if the theme is about the European Community, the portal http://europa.eu is a viable resource as it is translated into twenty three languages. Some famous books in the mother tongue of each partner are also easily tracked down. Availability of multilingual sources -- in addition to facilitating ease of translation of knowledge claims to share and integrate in concept maps or knowledge models -- helps motivate children by making sure that the other partner shares exactly the same resources, in the same words. Such sources, when available, combined if possible with the learners' autonomous free choice of an area of interest, should be preferred in planning educational collaborations.

The third element in our survey of motivation factors is the *quality and timing of inte-ractions* among partners. In this context we can suggest a few de-motivating factors. First of all, if a partner has to wait a long time to read a contribution from other members of the team, this will not only slow down the process but also discourage the team. This negative factor can be avoided if members of the team agree to a timetable which they themselves monitor. For primary school children the observance of scheduled times is entrusted to the teacher-coordinators.

The quality of collaborative interactions also depends on the availability of technological tools.

The software CmapTools (available from the Institute of Human and Machine Cognition – IHMC – at http://cmap.ihmc.us/) includes features that allow students, as well as teachers, to interact and question each other. The easiest way is to use the "Annotation tool", a 'post-it note'-type of digital tool which enables students to post short notes or comments on each others' concept maps or on the same shared concept map. These annotations have the advantage of being easily inserted or changed, but they also have the drawback of being unlabelled for data, colour, status (read/not read, answered/not answered) etc. If a single knowledge claim has more than one of these annotation icons, therefore, it becomes confusing. This drawback can be overcome, in a few more clicks, by attaching a "Discussion Thread" instead to every knowledge claim or concept map element (concept node, linking phrase or folder). A discussion thread needs to be given a title, sends an email notification when a post is answered or added and permits attachment of documents and images. Through such threads students can get involved in more in-depth discussions than with annotations.

At a higher level of complexity and potential, there is another collaboration tool, called "Knowledge Soup". This is a repository of shared and independent knowledge claims (linear text statements) that are subjected to questioning and negotiation before being picked out and adapted in the concept maps. A knowledge soup can be considered as a preliminary stage that permits knowledge to become meaningful to all team members, facilitating the next stage of integration of heterogeneous knowledge in the concept maps. All these powerful collaboration tools, if correctly managed by the students, may stimulate positive interdependence and promote interactions, thus enhancing motivation.

The *concept mapping skill* of every individual member can affect individual motivation for the collaborative task of coding knowledge as propositions in concept maps. The syntax of concept mapping is often hard for older students, no matter how much metacognitive aptitude they have and how meaningful their learning style is, because the cognitive structure of adults can be characterized by high levels of "connectiveness" or a prevalence of complex sequential structures that aren't easily reducible to minimal elementary propositions, characterized by relative independence from one another. Some people do not immediately perceive the added value of concept mapping language in its capability to elicit knowledge and reduce ambiguities that is so useful in collaborative learning and integration of heterogeneous knowledge. Similarly, not everybody accepts the worth of deconstructing in depth and re-coding their knowledge in such a "format" that follows a set of "strange" rules. This is especially true if adult students have never had any experience in concept mapping before. The same problem is attenuated with children, as their natural language is closer to the basic articulation of concept maps. To prevent a possible "discouragement effect" in the "predisposed" cases, we suggest some precautions: a mini-training session, including a video or presentation of the several steps in the construction of a concept map (where the application of each criterion is well highlighted) and an exercise in constructing a concept map from a short and familiar text, followed by proper feedback from the teacher-trainer. In CmapTools there are two useful features for preparing presentations. They are "Presentation Builder" and "Cmap Recorder". The former produces a slideshow of the main stages in the building of a concept map and the latter provides a continuous record of every editing step. Similar training can also be useful to prepare team members to use the collaboration tools and other basic features of the software. Another aid to overcoming the first impact with the conventions of concept mapping can be the preparation of "Skeleton Concept Maps" (Novak & Cañas, 2004). These are a sort of "embryo" concept maps, in which the focus question has already been stated, the root concept (that is, the top concept in the pyramidal structure) and some of the first level concepts have been already placed in the lay-out, together with some other nodes that need to be labelled, suggesting a possible structure to be continued. The "Octopus Self-Introduction Maps" have pre-defined nodes and descriptors and a member is invited to customise it by introducing herself or himself to the team community. (see Figure 3)

This is a very simple and humorous way to introduce the basic elements and creative editing of concept maps to novices. Moreover, these maps encourage learners' curiosity to know each other and motivate them towards collaboration.

The last stage of our survey on *motivation* concerns the *final products* of a collaboration team. We know, as educators, that a good process is more important than the final product and for us the best result is when the online learning community wants to keep collaborating in the future. On the other hand, the real product of a collaboration is open-ended and continuous, namely the creation of new collective knowledge within the team. Nevertheless students in the learning community need a precise and definite target to attain. This could consist of a reviewed concept map or knowledge model made up of a certain number of pre-assigned focus questions and concept maps

Figure 3. A preset "octopus self-introduction map" (upper, yellow map) and how it was edited by Eleonora in her first approach to concept mapping (lower pink map.)

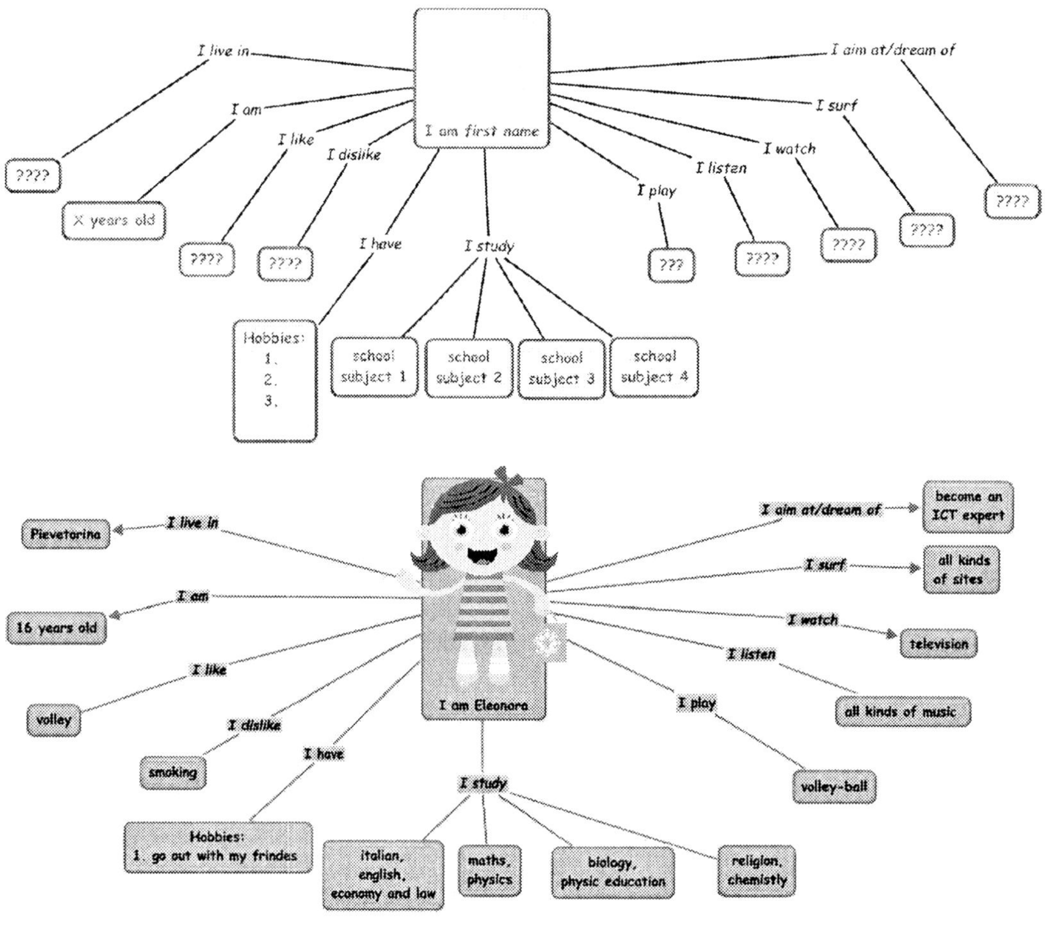

that answer these questions. Sometimes the lack of such a well defined target (because of the "open-endedness" of sources and of all knowledge) is interpreted as a sign of arbitrariness that may be used to "justify" a sort of permission to ignore individual responsibility and the need for inter-dependent team processing.

To avoid such a de-motivating factor, the role of the educator as a "manager" of the collaborating team is to assist it to focus on a clear general objective from the beginning and to frequently re-design a realistic form (qualitatively and quantitatively speaking) of the concept maps or knowledge models that are expected to be constructed on the basis of the observed trend in the collaborative process. That target shouldn't be continuously changed, but just progressively focused, as the time remaining for active work reduces. Behind this role of the educator is the idea that having a clear objective to pursue, which is perceived as not too far from current attainment levels, can give a motivating push. This kind of motivation is strictly connected with the level of autonomy within the team. The level of autonomy in determining the team's working process also reflects its capacity to plan the timetable necessary for carrying out the task, to recognize/define the target and to assess its own work. These objectives of high levels of autonomy and motivation should enable progress towards

Figure 4. Key influences in Web based CCM, as elicited in the analysis of motivation related factors. Double arrows indicate a synergy effect.

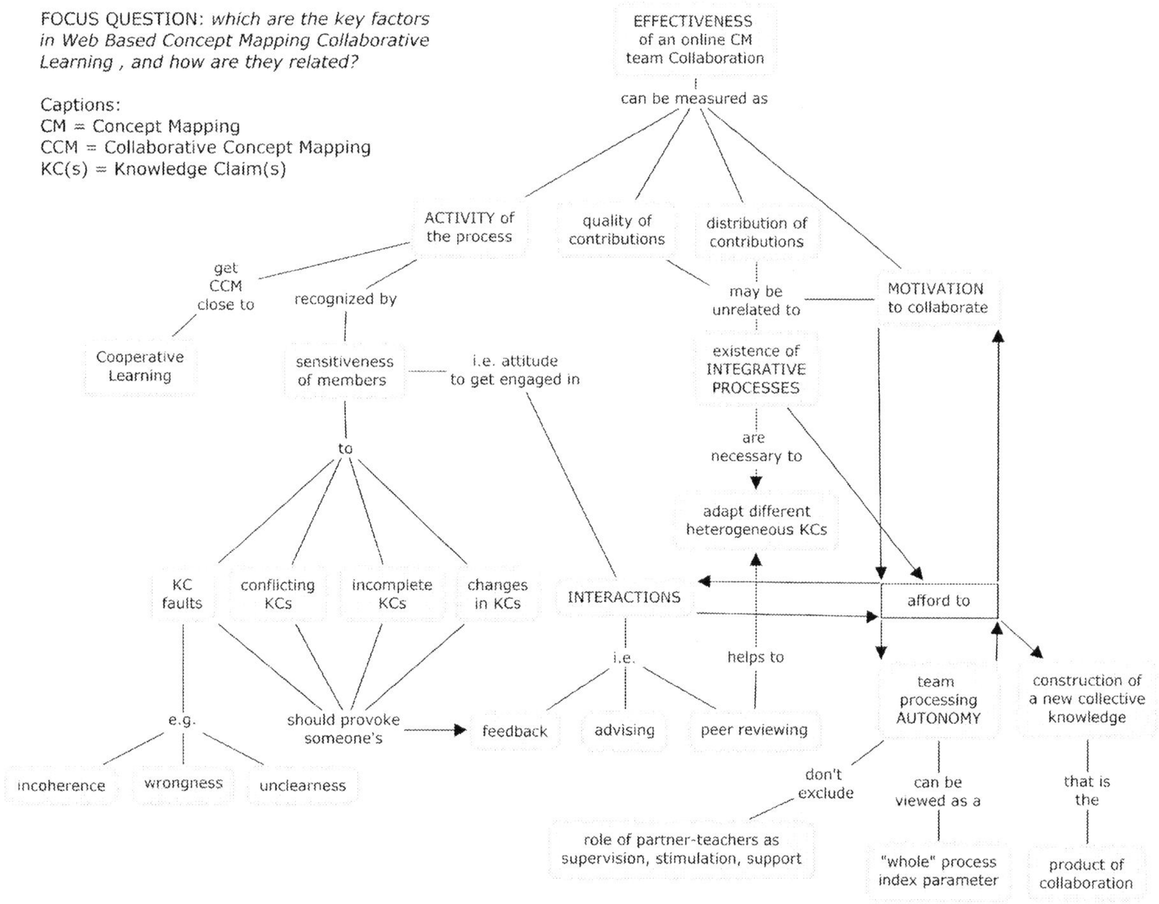

favourable official evaluations of the products, if the partnership has been formed around an official competition.

On the other hand, self-evaluation of the overall process and its products is demanded of teacher-organisers and team coordinators as an obvious condition of continuing collaboration. (see Figure 4)

Analysis of Past and Current Cases and Generalization of Types of Collaborative Processes

The key aspect of a concept mapping collaborative learning process is the method used to integrate knowledge claims and sharing of productions. Nevertheless we will start describing some examples of how projects can be planned collaboratively among teacher-partners, through the use of concept maps.

Figure 5 and Figure 6 show the result of agreement between three partners and two partners respectively. Yellow "post-it"-type icons are annotations added by the three teacher-partners as the map was processed asynchronously. Figure 5 has the structure of a concept map while the map in Figure 6a has a hybrid-free scheme (flowchart + concept map). It also indicates when the most important changes have been applied and by whom, saving each version separately (Fig 6b).

Figure 7 refers on the other hand to "one shoot" planning about "Atom" and peer tutoring among Italian and Romanian students.

Once a plan has been agreed, if the partnership is dealing with a wide-ranging topic or a variety of topics, it is necessary to parcel the task into sub-topics and entrust them to the students. The following maps in Figure 8, Figure 9 and Figure 10 show several examples of domain mapping and assignation of sub-tasks.

Figure 5. A three-partner bilingual project map for a partnership about materials and waste recycling

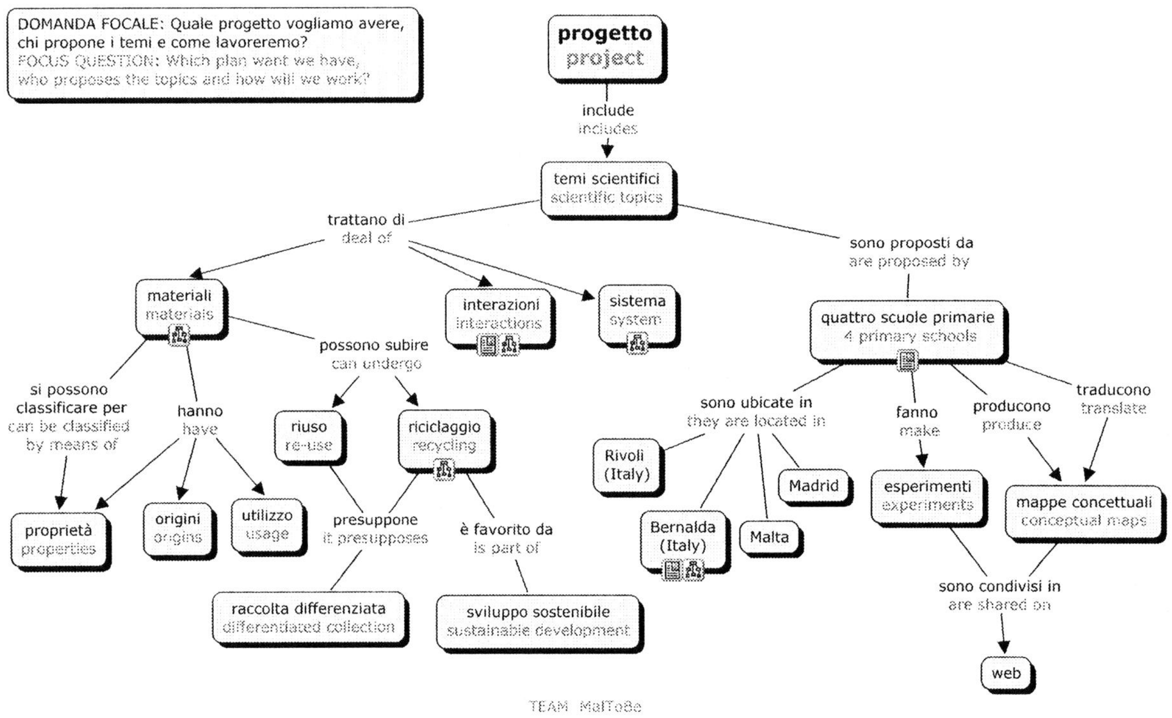

Figure 6. a) Planning-free scheme realized between two partners (translated from the original in Spanish) and b) history of previous versions inside its CmapServer folder

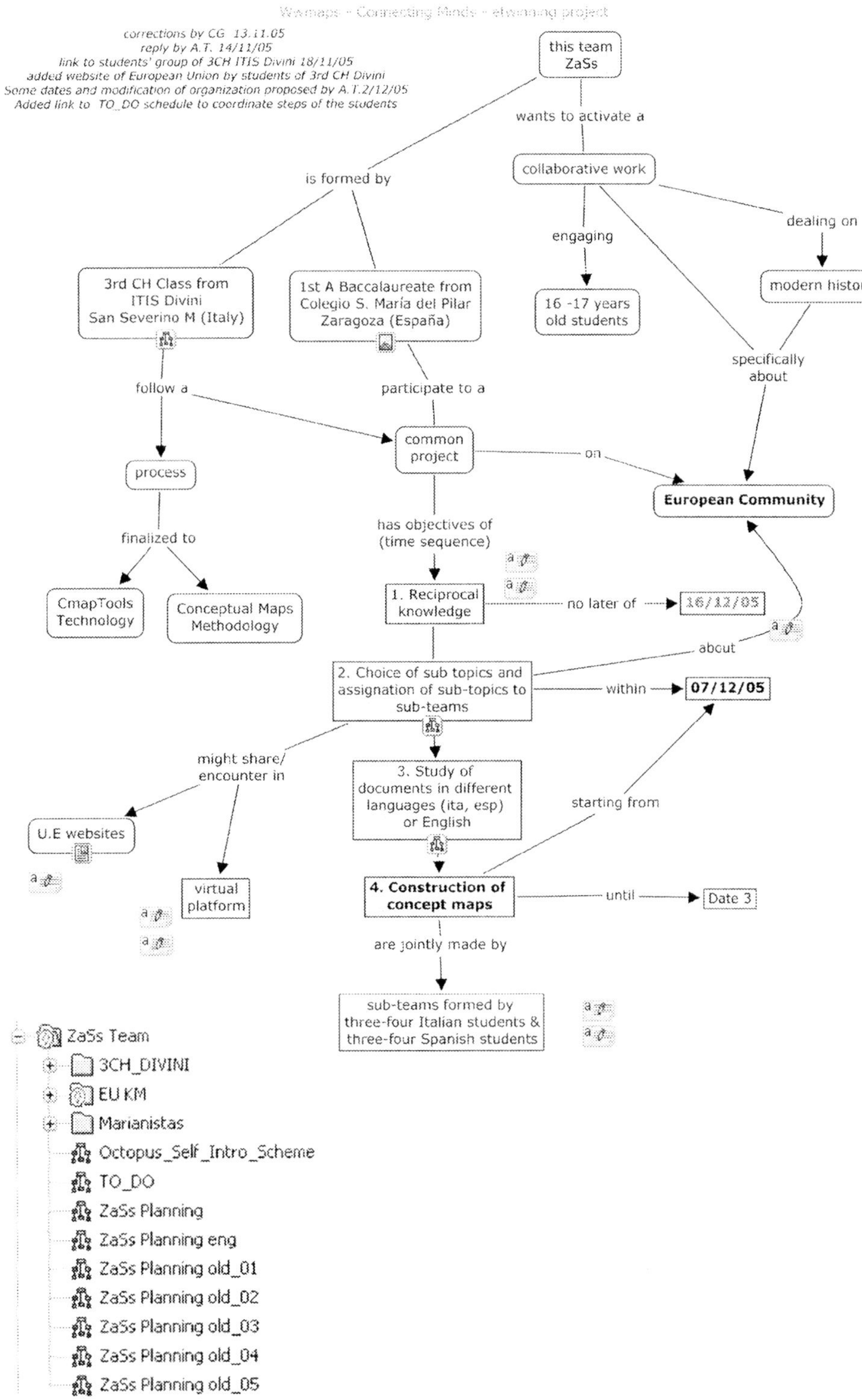

Figure 7. Model of collaboration about atom with peer reviewing. It was accepted as a whole, without discussion (except for subtopic selection

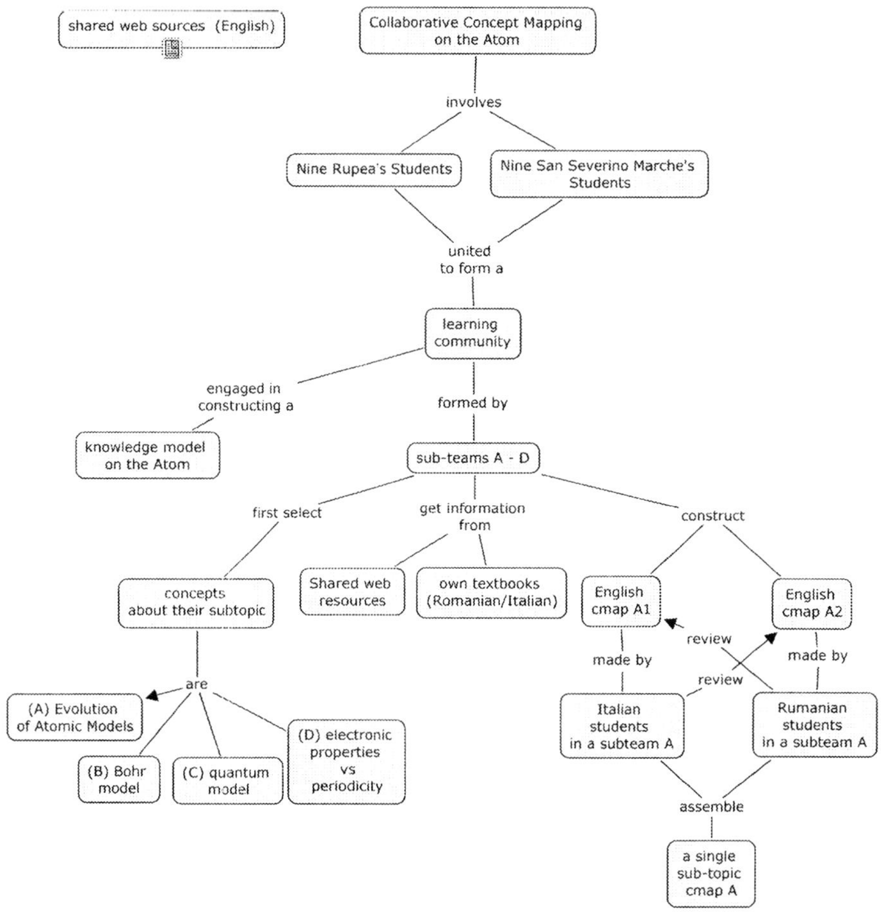

The following project map was discussed and planned by adult students in grades 11-13, starting with an analysis of different aspects of the main theme.

Students from Italy, the U.S. and Japan mapped the theme of *Teenager and Drugs* in five main areas that were explicitly interconnected (in contrast to the plan in Figure 8)

Figure 10 shows a current plan (March 2008), taking in student contributions in choosing and defining unrelated inter-cultural topics. The top node has a discussion thread icon added in a second phase to replace annotations in subsequent discussions among teacher- coordinators. This scheme serves as an index to all the resources in the project and access platform. The variety of colours helps to identify each independent team in the partnership. It is noteworthy that there is a set of instructions to create annotations at the bottom of the plan.

Most of the project maps or schemes we have illustrated so far do not help us to see how each partner's contributions have been integrated with each other in the shared content-maps (except for Figure 7, where this is detailed). There is an increasing complexity of cases.

The easiest form of collaboration is when partners work on separate multilingual (for the sake of mutual comparison) concept maps on the same topic (maybe with the same focus question,

Figure 8. Sub-topics of a complex theme and assignation of students to every team

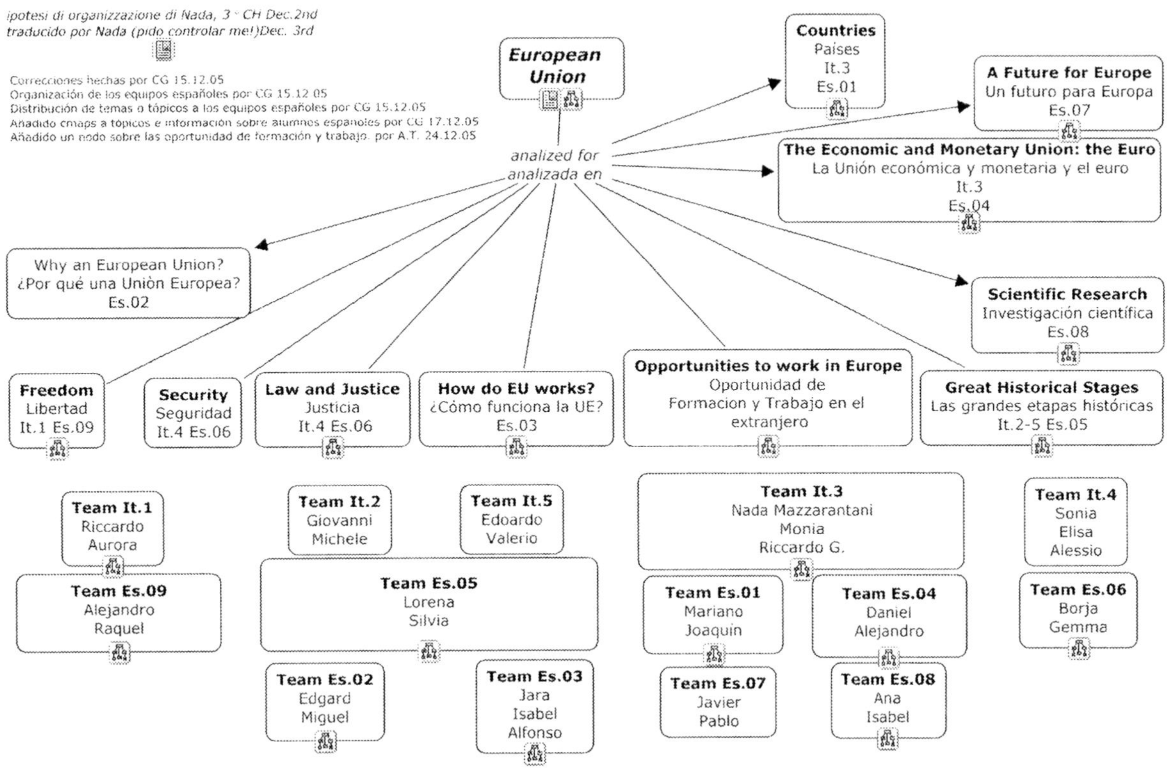

Figure 9. Theme analysis by an international team formed by adult students from grades 11 to 13. One of the annotations has been opened to give an idea of the flavour of the discussion

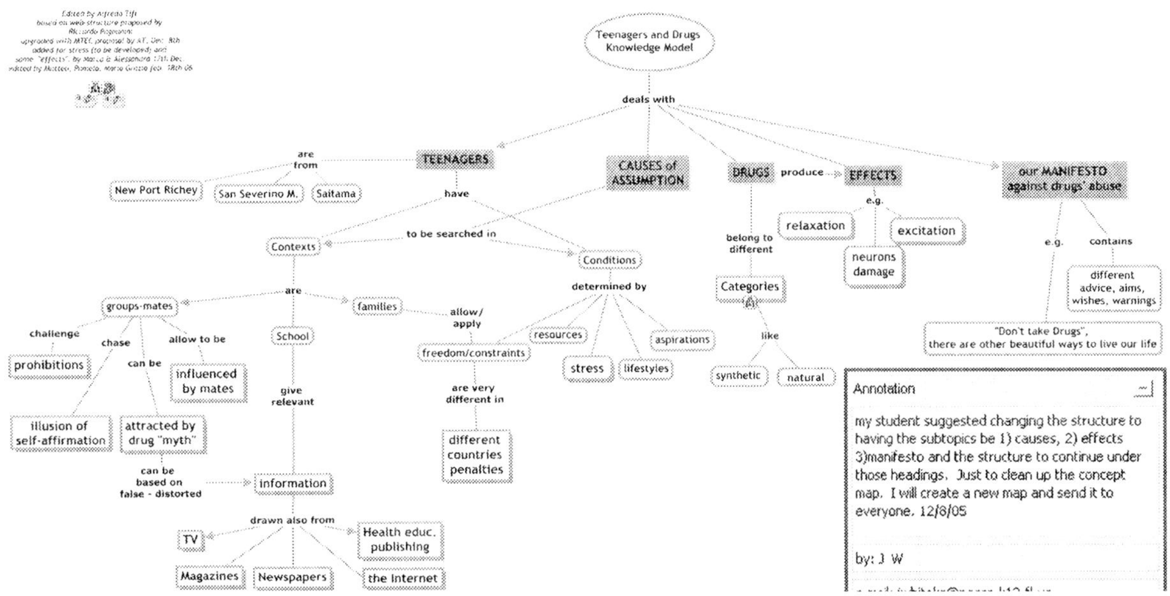

Figure 10. Map representing the chosen topics in an intercultural project between Italian and Egyptian partners. Each student has selected the preferred intercultural topic and has been connected to it by a line

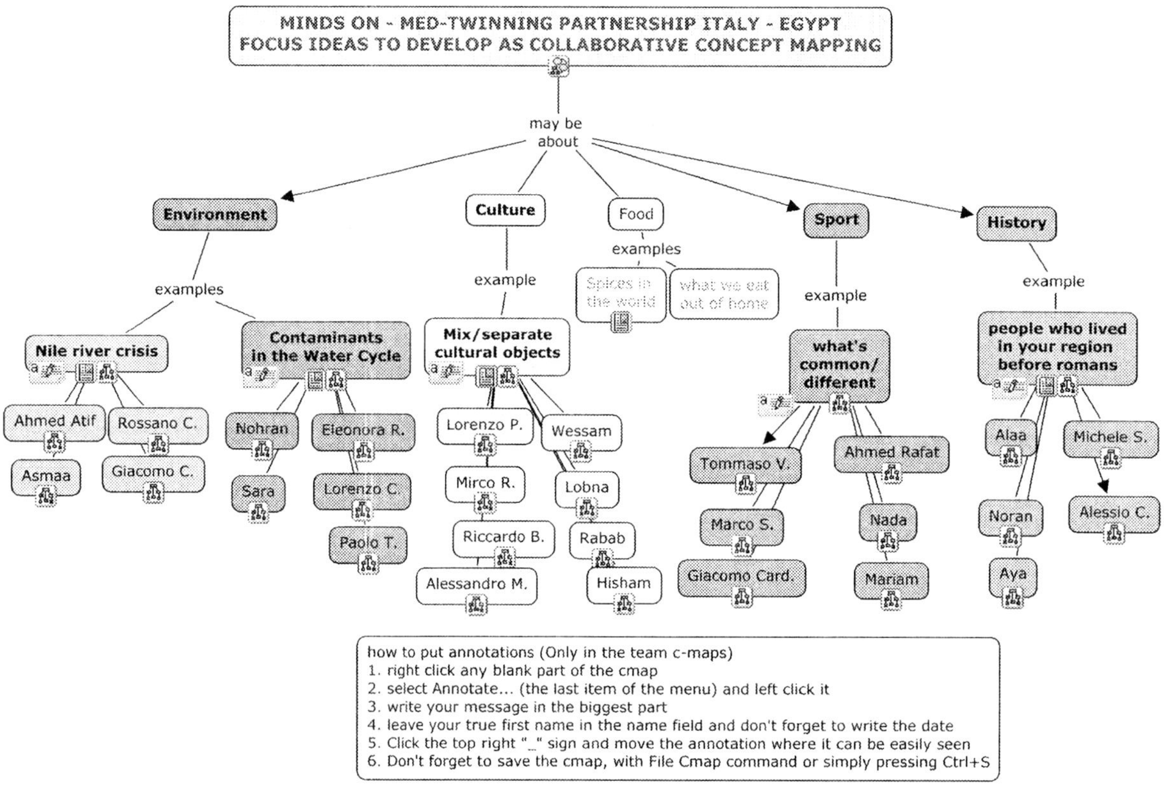

but generally applied to the local context, environment or culture). Partners are linked by sharing the same task applied to their local context. The motivation to see contributions, and interact with the partner comes from children's curiosity and the teacher's drive to learn from peers instead of textbooks, and is not a prerequisite of true interdependence. Collaboration among younger children was often set up in such a way. Two independent multilingual maps from three partners are shown in Figure 11.

A slightly more complex type of collaboration alternates contributions from several partners on the same shared concept maps or uses a mixed strategy of independent development followed by the addition of contributions and revisions from a different partner.

The first method of "alternation" seems more practicable than the mixed one for younger children because the latter requires greater mastery in comparing and criticising concept maps and each other's knowledge claims in a *non-additive-integrative* process that is more feasible with high school students and adults. Children are more likely to be concerned with exploring and taking note of the contributions of the other partner and with finding other evidence and knowledge claims to answer the focus question in a substantially *additive* process. The extent of interdependence is evidently higher in both these forms of collaboration than in the comparative strategy, as can be argued from the following examples (Alternation method: Figure 12, Figure 13, & Figure 14; mixed method: Figure 15).

Figure 11. Examples of multilingual cmaps constructed in an instance of collaboration through independent realization of resources biased by local context, but referring to the same theme of endangered animal species

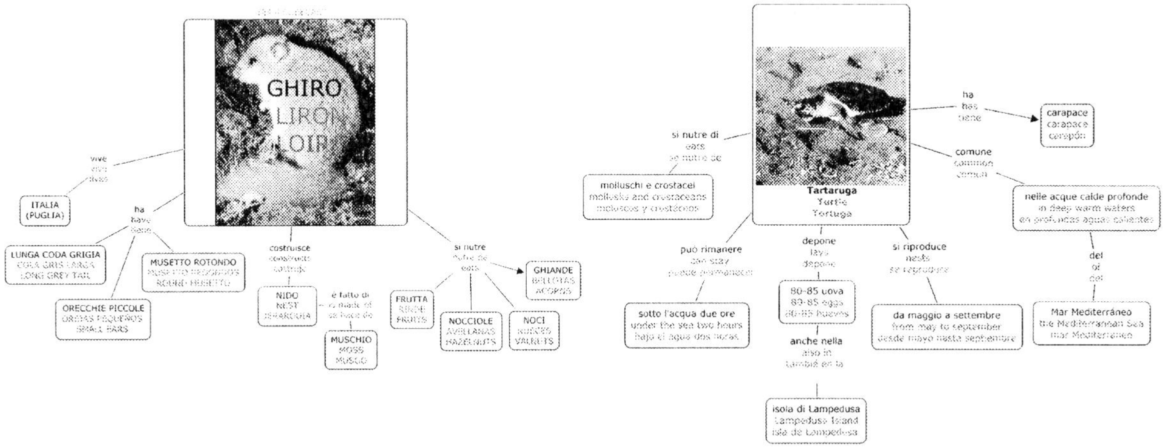

Figure 12. A true collaborative Concept Map created by three partners, after two stages of alternating contributions

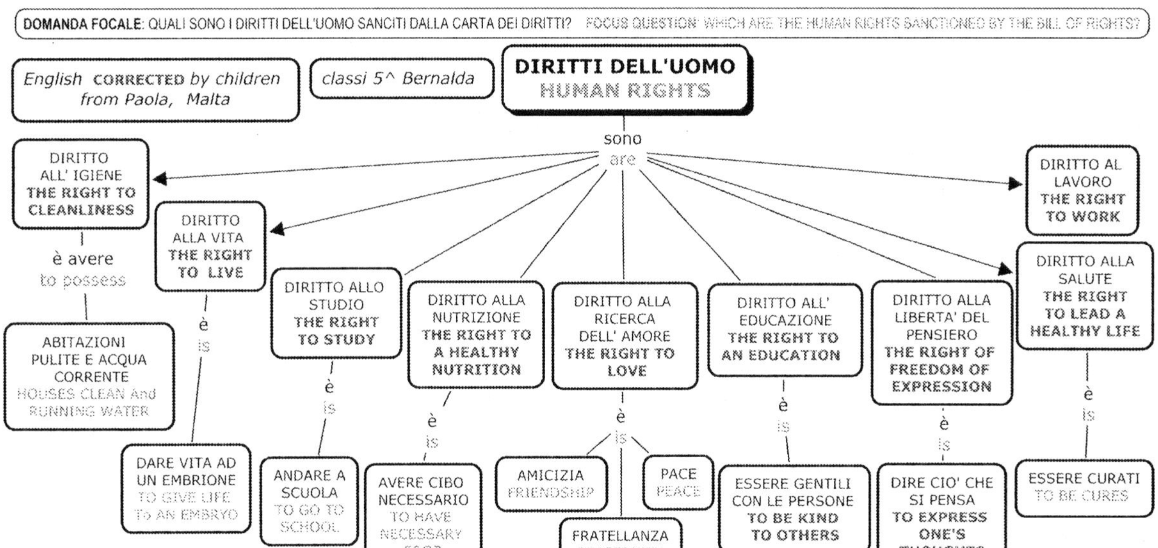

In Figure 15, initial work by Daniele & Loris has undergone several revisions (distance peer reviewing) by the other online team-mate. This can therefore be considered as a *mixed type* of team process because it brings together independent concept mapping and the alternation method, characterized by a high level of interdependence, autonomy and integration of knowledge claims.

It is important to make it clear that both processes (pure alternation-addition CCM and the mixed strategy) can be managed in very different ways, depending on the method of decision mak-

Figure 13. Some further knowledge claims added by a third partner

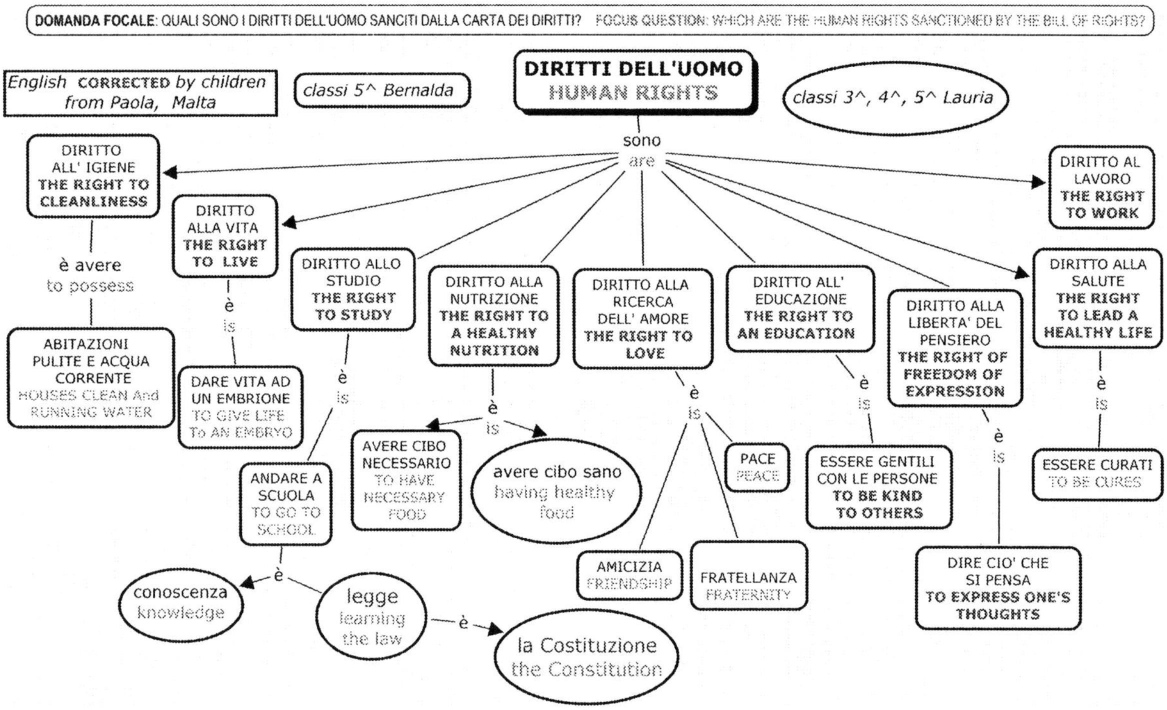

Figure 14. The same concept map after several additions of knowledge claims from all partners. This process wasn't exclusively additive, because some little adjustments were made and linking phrases were changed

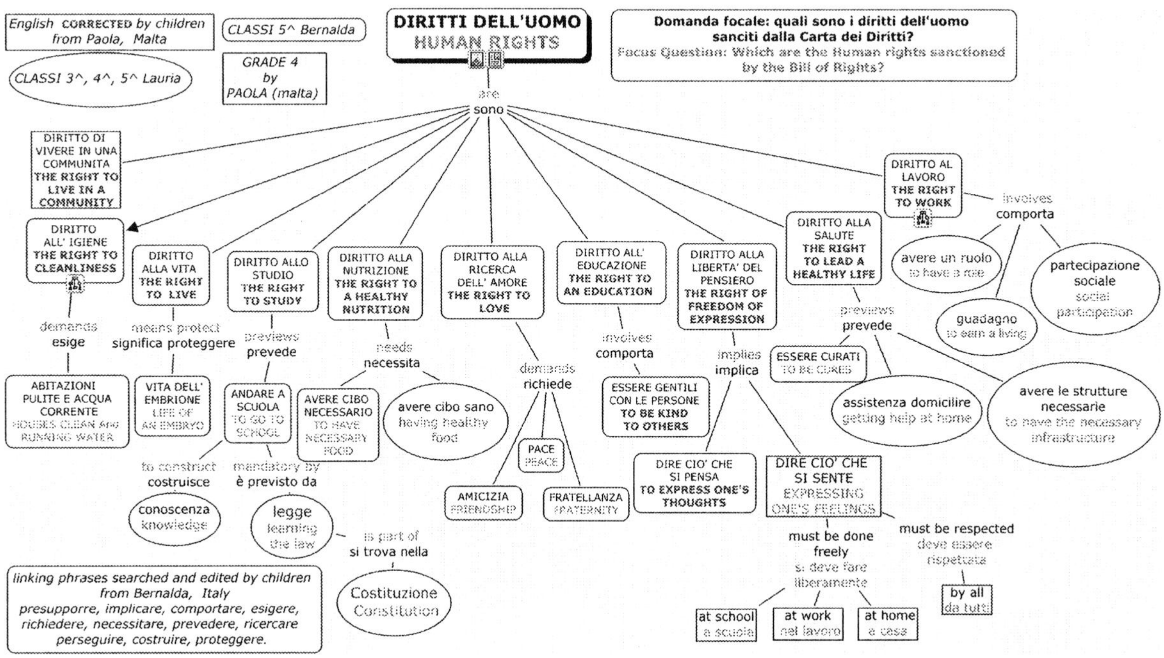

Figure 15. Resources history in a sub-team formed by three people working at a sub-topic of photo-electric effect

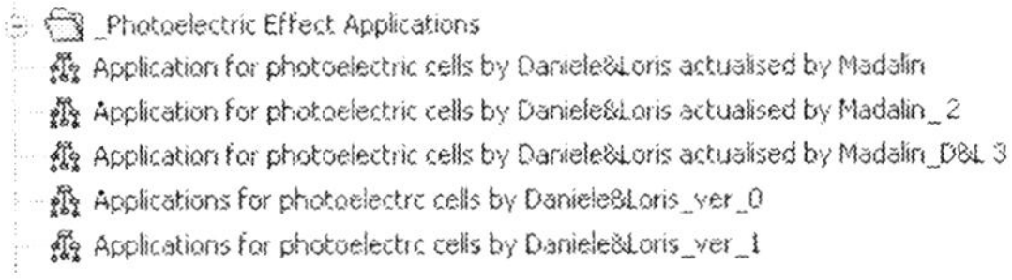

ing chosen locally. If a single resource is shared within the team, it may be that there are many students in every partnership group engaged on reading the map modified by the other partner, translating it and discussing and negotiating how to integrate the resource with knowledge claims, all under the supervision of a teacher carefully seeking to make sure all students are involved. Such construction of the resource is a collective task.. In this case involvement is easily controllable and achieved, but at a lower rate and at the expense of individual students' decision making autonomy. Alternatively, it is possible to assign the same task or similar tasks to smaller groups of 2-4 pupils, twinned with the same number of pupils from the other partner country. Within these small teams it is easier for students to have a creative approach and it is more likely that knowledge claims will be elaborated through peer discussion. Mediation by the teacher is still possible and desirable in this case, but would not best benefit children if it were a constant in the social and learning processes. This solution obviously implies greater complexity in organising and managing the processes, but it is worth the effort if students have the necessary skills to support their autonomous work.

To complete this analysis we will go through some CCM processes (more sophisticated from the technological and methodological points of view) that make use of shared concepts or knowledge soups, and peer reviewing.

"Atom:" An Online CCM Experience Based on the Sharing of Concepts

The strategy based on the *sharing of concepts* has been used successfully to maximize the interdependence in a CCM collaborative learning team created with the purpose of tackling a *new curricular topic* (atom, Figure 7) previously chosen by the two teacher-partners. They divided the theme into four sub-themes, described by as many focus questions as formed a continuum in the curriculum. Each sub-theme was assigned to a sub-team. The first stage was the *search for the key concepts* in every context. As Figure 14 shows, this task was accomplished collaboratively. Sixteen-year-old Italian and Romanian students picked out concepts through the examination of resources in their mother tongue that were selected in advance by the teachers, mainly from the course textbooks belonging to the students, but also from shared resources on the Web in English, i.e. the common "bridge" language in the learning community. It is noticeable that *a)* the work of concept eliciting has also been continued in the subsequent work of concept map construction, as students became more involved with the topic; *b)* they were informed of the focus question for each sub-topic when they started the task (that was the reason for the expanded annotation in Figure 16); *c)* contributions and exchanges of relevant concepts from other sub-teams in the same community were welcomed at this stage; *d)* the stages in the gathering of concepts have been recorded (by means of the Recorder Tool of CmapTools) for the sake of the supervising teachers (this means

Figure 16. One of the four gathering schemes in atom CCM experience, as it appeared when the collaboration ended. It is related to the focus question "How were the electronic models of the atom born and how did they evolve through experimental evidence leading to the rejection of the classical physics principles?". An exploded annotation (blue bordered rectangle) shows a feedback-process query by the Italian student Monia. Two concepts ("plum pudding model" and "Gold foil experiment") have also icon links to Web resources (applets), to allow a better understanding of their meaning by the team-mates. One of the many relationships that are implicit, among concepts, is made explicit by the linking phrase: "are solved by". Concepts aren't placed randomly. Instead their position reveals semantic contiguity and inclusion relationships that are a prelude to a sort of "proto concept mapping"

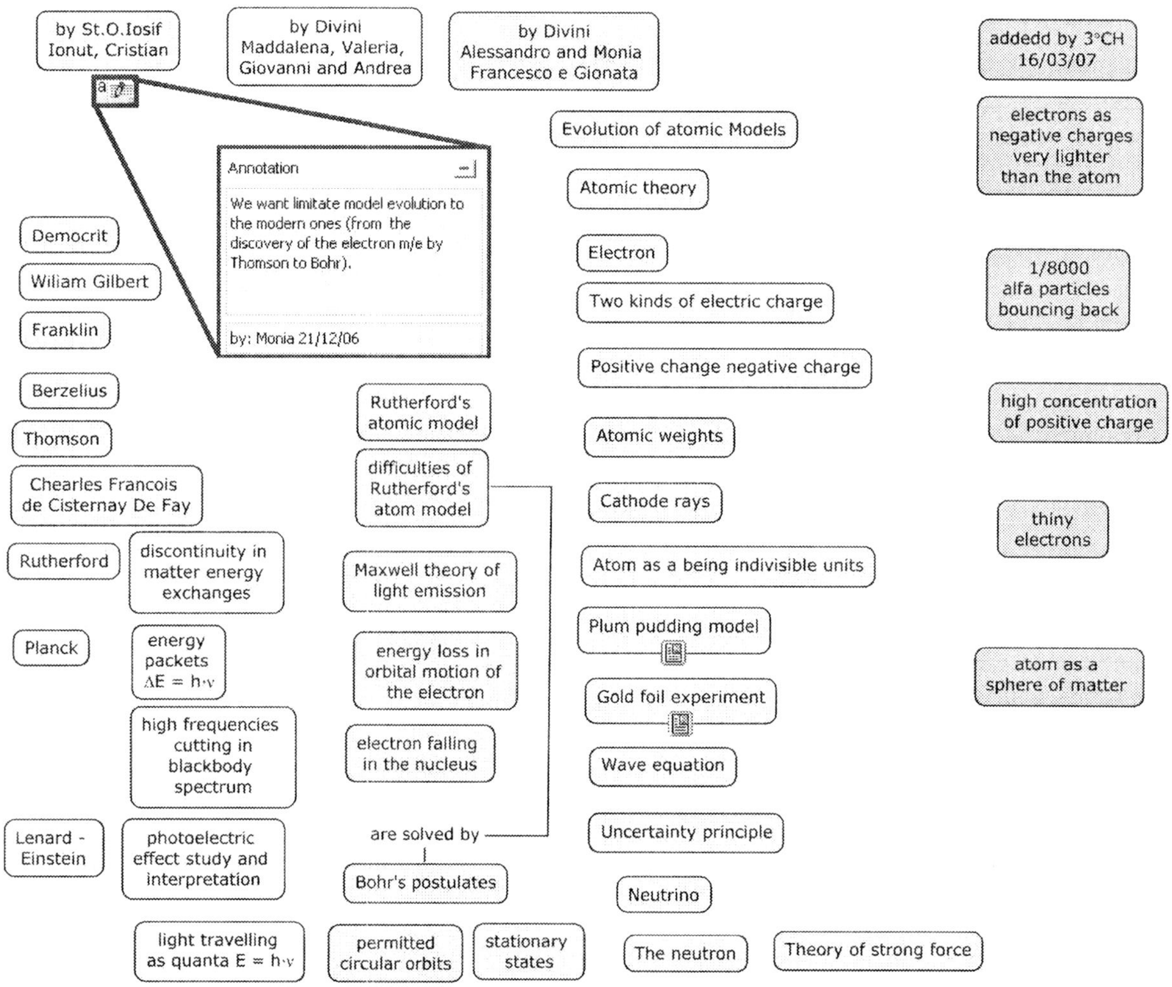

the teacher can see the order in which concepts are attached to the resource as it develops).

As a second stage, the Italian and Romanian teams separately copied the concepts from the developing page and pasted them into a new, empty resource file, where they started to construct new concept maps, while bearing in mind the sub-topic focus question. This was the most difficult stage because the topic was new for both partners, and the Italian students had a poor mastery of English. They therefore needed some help to translate the focus questions into Italian and to prepare

Figure 17. Structure of the Atom project of collaborative learning in CmapTools Places. From it an idea of the organization of the whole project can be grasped. The fourth team sub-folders have been all opened, to see from the names of the resource files of each group how peer reviewing has progressed

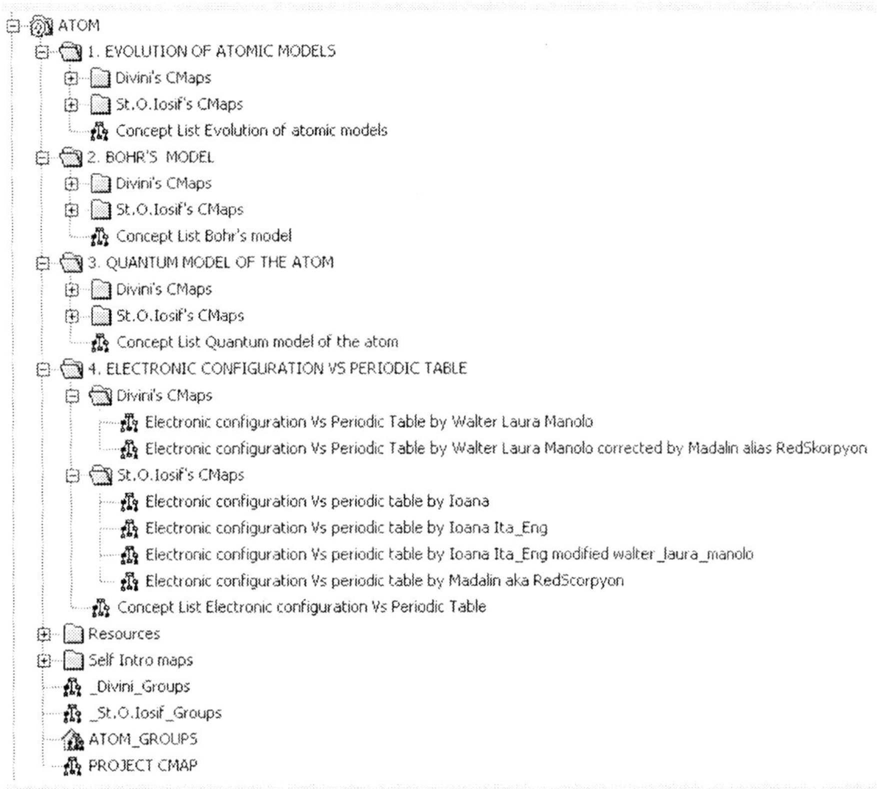

skeleton Cmaps (concept maps initiated with the root concept which had some first level concepts added). During this phase the students frequently asked their teachers for support in understanding the points of the subject they found difficult to grasp and for suggestions about the organization of knowledge in their maps.

The final stage involved a peer reviewing task, in which the Italian students reviewed the Romanian concept map in the same sub-topic and vice versa. Figure 17 shows the structure of the folders of CMapTools Places. From it, the organization of the peer exchanges and of the whole project can be easily grasped even without analysing the nature of the reviewed concept maps (these are available on "IHMC Public CMaps(2)" Cmap Server, under "2WMAPS" folder, "ATOM" sub-folder). However, in some cases reviewing was made interactive through the addition of annotation queries addressed to the partner. Unfortunately, the stage of peer reviewing was only initiated in the last month of the school year, so that only a single reciprocal reviewing exchange was possible. The experience was successful, however, and laid the basis for friendship and future collaboration.

The critical innovation in this project has been the shared gathering of concepts. This task was appreciated for its simplicity and usefulness (fundamental to trigger a peer collaboration), but we learned from it that a preliminary idea of the relationships between concepts, a sort of "proto-concept map", is needed to evaluate the relevance of each concept and its responsiveness to the focus question, where a simple list of words would be

Figure 18. How to transform linear text into knowledge claims and fragments of concept maps

These three sentences from the text:	These excerpts from the text:
Louis XI warranted reliefs from silk tax; Louis XI warranted silk orders for the Royal Court;	*Arab traders introduced pepper to Europe...*
	led through Damascus and across the Red Sea
Louis XI decreed planting of mulberry trees	*Pepper was known in Greece by the fifth century B.C ...its use was medicinal rather than culinary*
Can be represented in a fragment of map:	Can be represented as:

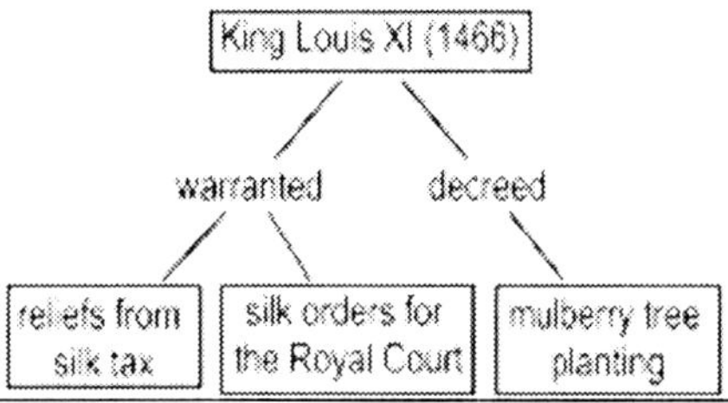

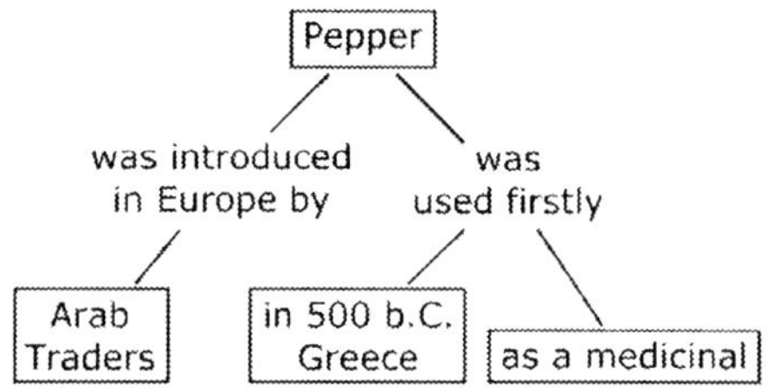

meaningless. Eventually, concepts can be refined later on in the true C-map.

"History Maker Molecules:" An Online CCM Experience Based on Knowledge Soups

The three partners in this project, which we will call HMM (History Maker Molecules) from now on, came together in an experiment about *collaborative reading of a shared book,* facilitated by concept mapping and Knowledge Soups. The two Italian partners and the Spanish one shared the same book, *'Napoleon's Buttons: 17 Molecules that Changed History',* which has seventeen relatively independent chapters, each one of which deals with how a different kind of chemical has changed, or provoked great events in, the history of the world.

Both teachers and students were highly motivated by this book, which they found very intriguing. Six students, aged from 14 to 16 from each partner school, chose three chapters of the book via a poll based on chapter titles and a paragraphs index. Two students from each class were assigned to each chapter-team according to the priorities they had shown in the poll, so forming three teams of six students. Working on the English (L2 for all) edition of the book, students started to gather relevant knowledge claims from *portions* of the chapter by highlighting paper copies of the text. The students were then instructed to transform such elemental knowledge claims in "mini maps", as in the examples shown in Figure 18

These fragments were copied from the highlighted text and shared as CmapTools resources, as in the examples in Figure 19, gathered by the student Aurora (chapter 5, Nitro Compounds).

This step is not as easy as it seems because this kind of analytical reading implies *a)* eliciting concepts and relationships, *b)* deciding which concepts are subordinate to which and *c)* choosing

Figure 19. Fragmented maps as gathered by Aurora. The yellow non-expanded annotation suggests to her (in Italian) to cancel the lower fragment, as it is irrelevant to the topic of explosives

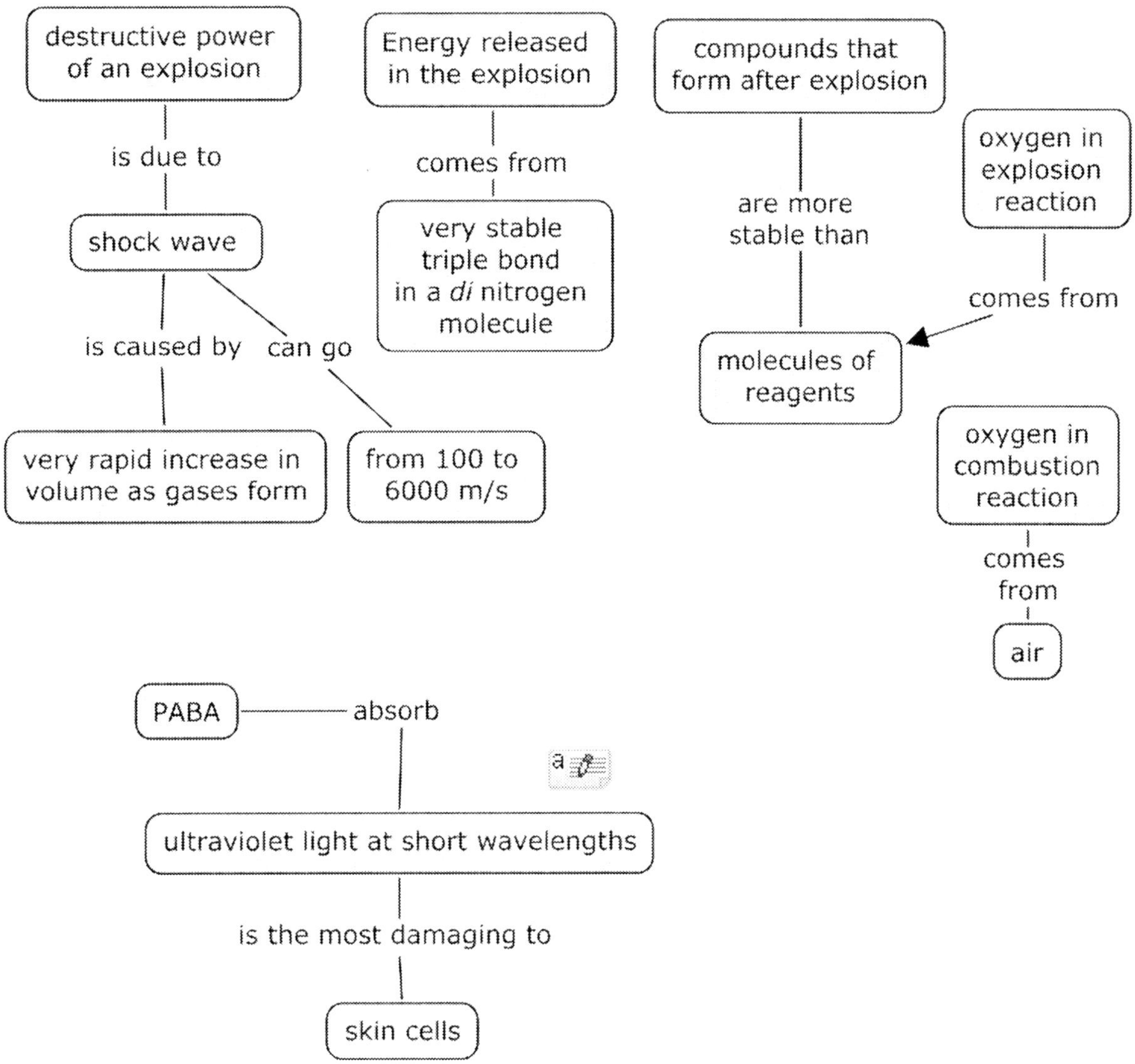

the way to "quantify" the top-concepts, i.e. how to cluster attributes creating articulated concepts adaptable to the context. The effort involved in such analytical reading was the main reason why all the teachers in the project found it so interesting. On the other hand, the task was harder for the Spanish students, who had the option of reading both the Italian and English editions of the book (a Spanish edition has not been published yet).

After gathering Cmaps, the next stage was to collect Knowledge Claims in a knowledge soup, a task automatically accomplished by associating the gathering resource-map to a *Knowledge Soup*. Knowledge claims in all the Cmaps from the same chapter were easily published in that chapter's Knowledge Soup. As the next figures show, this project was constructed in such a complex way in order to maximize the interdependence of the teams. It was therefore decided in initial planning to start concept mapping from pre-set skeleton

Figure 20. Initial planning of History Maker project. The "Skeleton Cmap" wasn't applied and this was a serious error for the success of the collaboration

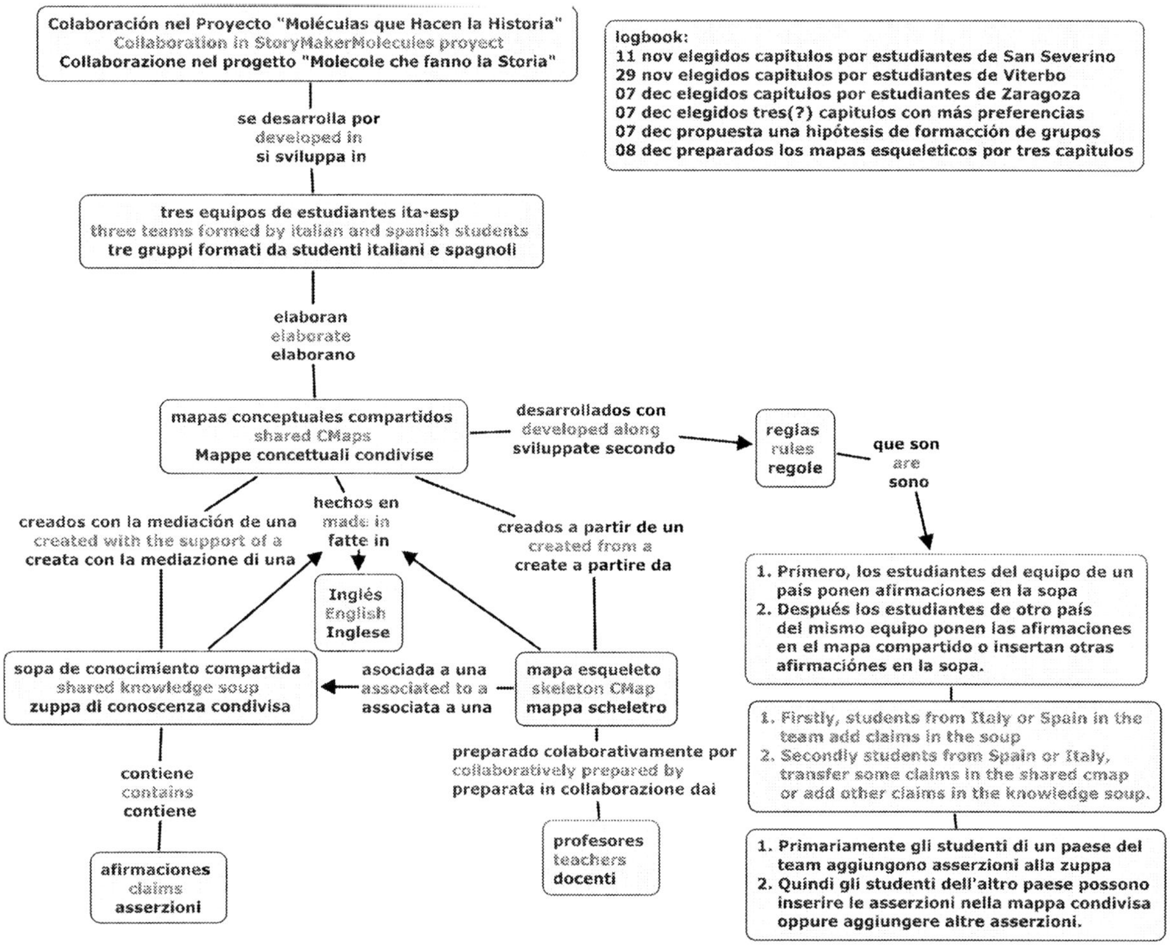

concept maps (Figure 20) in three languages.

The teams were left free to co-ordinate the timing and method of assembling a "Knowledge Model" (an assembly of concept maps) for each chapter. One suggestion of waiting to complete the collection of knowledge claims before starting to assemble a Knowledge Model for the chapter, and another of starting from the skeleton Cmap to construct the knowledge model, were both rejected by the teams. They started to construct concept maps from the fragmented maps and through a selection of suitable propositions from the knowledge soups, which were steadily becoming longer. They were obliged to insert at most four propositions at a time in the same map and then await a partner's contribution before adding further knowledge claims. This rule was brought in to prevent asymmetrical development of the shared concept maps by one partner, and it was effective in promoting interdependence. (see Figure 21)

The process of "bottom up" concept mapping, i.e. the construction of a concept map from analytical knowledge *without having* a detailed vision of the whole or a shared organizing preliminary choice, made this task very hard. In fact it caused initially erratic and arbitrary mapping and, by the end, resulted in teams being virtually unable to

Figure 21. A Skeleton Cmap that was prepared as key "milestones" for Chapter 5, "Nitro compounds"

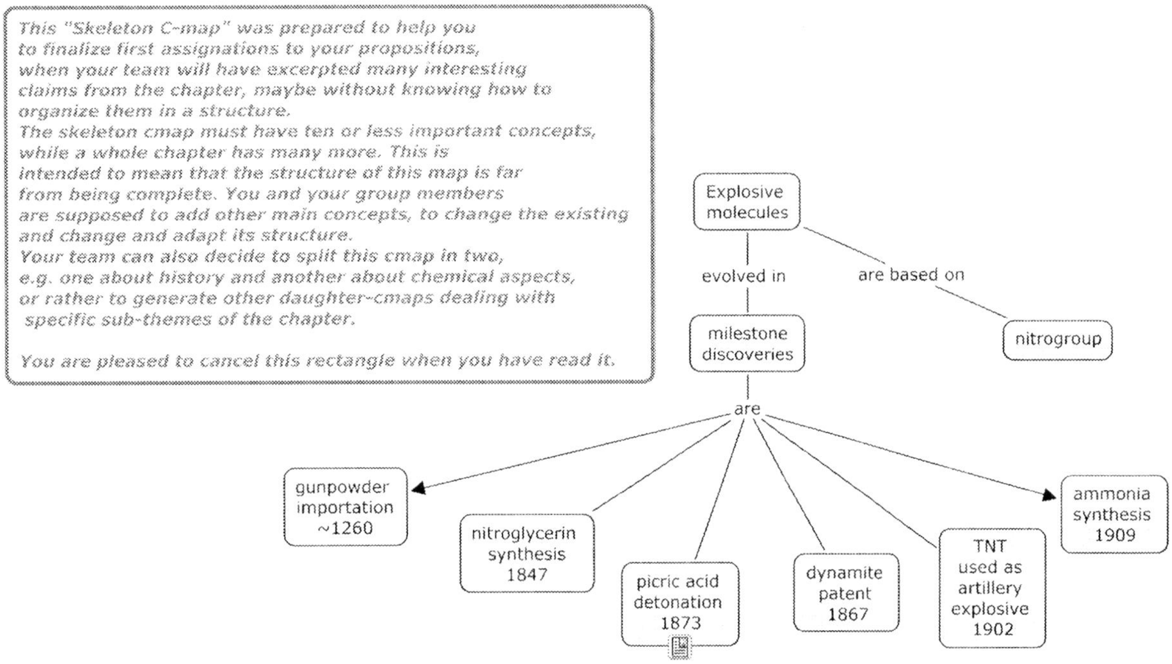

cover their entire chapter in a single shared concept map. Nevertheless the students appreciated the challenge of the collaboration, and the resulting resources they created together were highly interesting, as demonstrated in Figure 22, Figure 23 and Figure 24. Chapter 5 was completed in a single collaboratively constructed Cmap, which gives an unsatisfactory answer to the focus question. Chapters 12 and 13 were completely covered by a set of concept maps realized by different partners. The product was therefore quantitatively acceptable, but showed only limited integration and interdependence.

The knowledge claims turned out to be much too detailed to be fitted into a general concept map. The resultant knowledge soup was much too long to be easily interrogated for useful claims. With hindsight, it is possible to argue that if it had been suggested to the teams that the task would be helped by Skeleton Cmaps, they would have been encouraged to quickly divide the chapter into several topic-concept maps and transform the skeleton cmap into a sort of useful index-cmap. The knowledge soup claims would not have been too detailed for a single-topic concept map, which could have been split into a shorter list of knowledge claims. Finally, the Skeleton Cmap would have been a better guide for sharing parts of the chapter to be analytically read by each team partner.

Recalling Section 4.2's description of the effect of a definite and feasible target on motivation, it is worth noting that in the HMM collaboration the unattainability of a single concept map for the whole chapter acted as a discouraging factor.

In conclusion, the lessons for future experiences of collaborative reading in L2 (perhaps also in L1) can be synthesized as follows:

1. units of reading from the book should be narrowed to single specific contexts, each containing no more than 30 to 40 concepts, before being tackled collaboratively;

Figure 22. Collaborative Concept Map of Chapter 5 "Nitro compounds". Notwithstanding its complexity, it doesn't properly answer the focus question "How did Nitro Molecules Change History?"

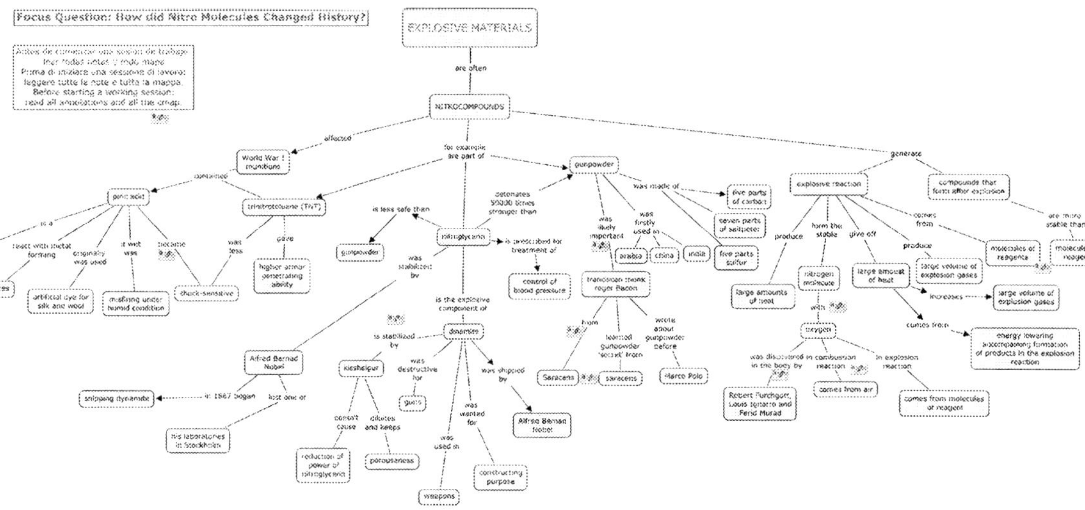

Figure 23. Collaborative Concept Map of Chapter 12 "Molecules of Witchcrafts". Only that part of the chapter was developed collaboratively. Other sections of the chapter that are relevant for the answer to the focus question are distributed in other concept maps constructed separately by single partners, while assembling the knowledge soup

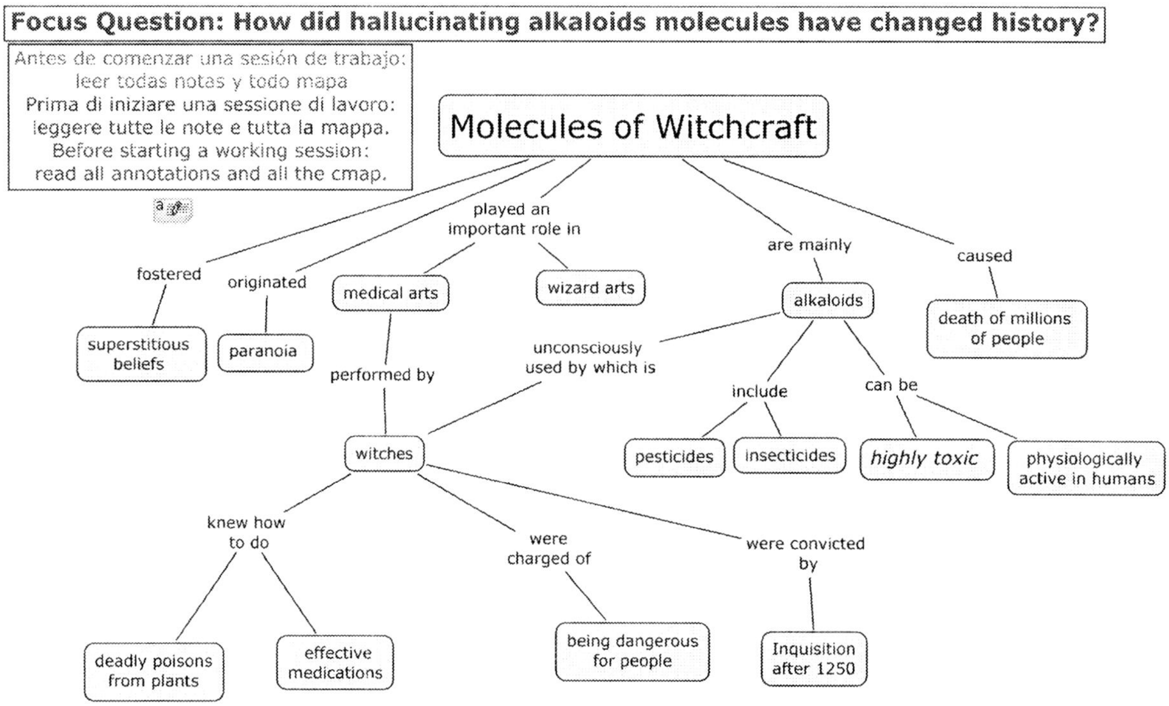

Figure 24 .Collaborative Concept Map of Chapter 13 "Morphine, Nicotine and Caffeine". As with the other collaborative concept maps, this one doesn't properly answer the focus question, but the answer could be given by a knowledge model composed by all the concept maps that were constructed by the team partners while realizing the knowledge soup of this chapter

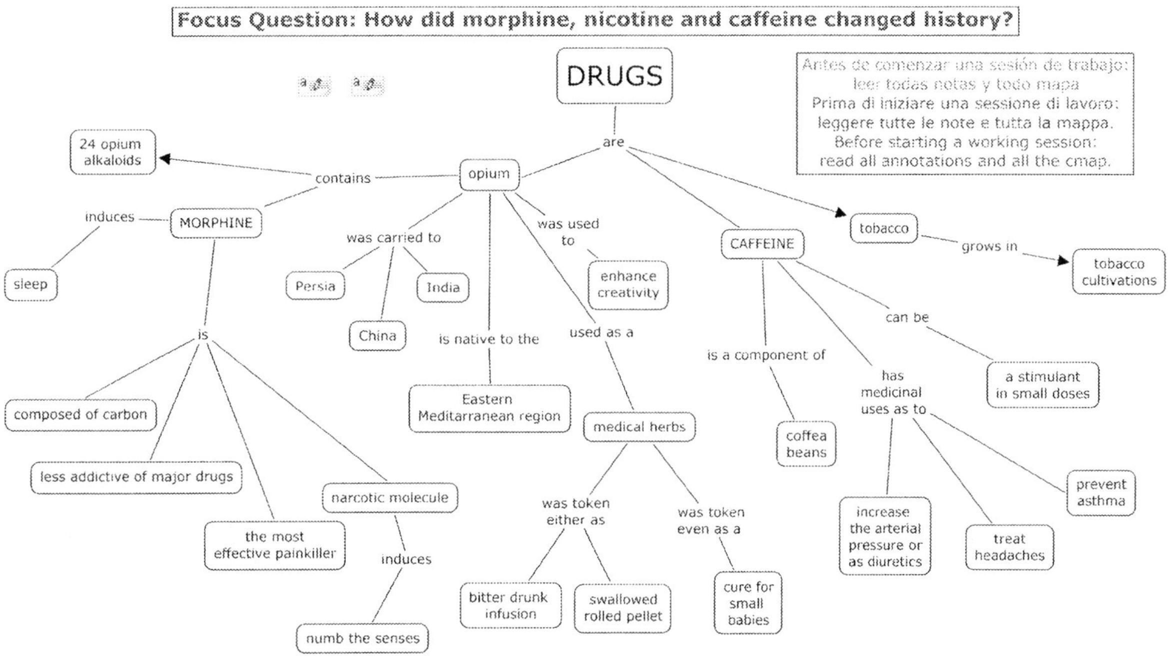

2. analytical reading of each reduced unit should be accompanied by direct concept mapping by every member and direct publication of knowledge claims in a single shared knowledge soup for that unit;
3. the whole team should deal with one such unit at a time, progressing from each unit to the next, until the complete chapter is read and transformed into customised, individual concept maps. Knowledge Soups should be used for sharing claims and interacting with the partners through discussion threads, questions and advice, and propositions in the concept maps;
4. a skeleton cmap could be used to create a hyperlink-index guide to browsing the knowledge model about the book, and, at the same time, it can be further developed top-down to give a deeper vision of the chapter as a whole.

FUTURE TRENDS

2008 is the European year of Intercultural Dialogue, which serves to remind us of the multitude of initiatives and projects aimed at increasing awareness of a European conscience and intercultural principles. These elements of global citizenship could be introduced into the education system through such exchanges and activities as the EU Lifelong Learning Programme. Alternatively, inter-cultural themes can be tackled like every other subject in the curriculum, with their own epistemology, provided we have, without being highly skilled in foreign languages, a methodology for the elicitation of concepts and, above all, *ethical values*, to highlight the commonness and universality of their meanings as well as to distinguish slight differences in their contexts. We need a platform, which could be a shared concept map, where related concepts may be systematically compared; correlated with the experience of each culture through connections and enrichments with qualifying attributes; and displayed in a hierarchical-inclusive-type structure. If this organizational task takes place so that educational curricula are planned to be made up of collections of scientific Concept Maps for every foundation subject, characterized by an inter-cultural bias, we will get a platform for educating citizens in the universal ideals of interdependence and communality of resources and faiths for all people and cultures. In such a scenario the educational community, if it is to have an inter-cultural future, must be open to "compromises" with other cultures and should intensify opportunities for true collaborative learning much more than is currently done. Teachers should realize the need for a wider model of education, where the opportunities for genuine collaborative learning at a distance, based on free peer exchange and discussions, should be ranked above informal experiences of twinning.

We have used many arguments that make concept mapping an almost obligatory choice for monitoring processes of knowledge integration at every level, from that of individuals, interacting with the teacher, to that of learning communities. We must, therefore, ask ourselves why this methodology struggles to be acknowledged by the community of e-learning research and by associations dedicated to the founding of international collaborative initiatives.

Despite the lack of acknowledgement of the role of CCM in e-learning and official e-twinning, we have become more and more convinced – partially because of the analysis of current and past experiences of collaboration set out in this chapter - of its worth. This is why we seek to strengthen our links with colleagues in Italy and all over the world so that they are as persuaded as we ourselves are that a conscious use of customized CCM methodology is the right choice.

Table 1. Collaborative concept mapping best-practice methodologies

Methods	*Complexity-age*	*Strengths*	*Weaknesses*	*Interdependence*
Comparison of individual concept maps	Primary school	Simplicity, freedom in timing	No integration of knowledge demanded	Low
Alternated contributions to shared concept maps	Primary school	High integration demanded to create knowledge claims	Risk of additive-type only contributions; Reciprocal limitation of freedom in concept mapping	Medium
Mixed individual/inter-dependent concept mapping and peer reviewing	Middle-Secondary school	Relative simplicity; Flexibility	Interdependence not required in the first stage	Medium-high
Preliminary sharing of concepts + peer reviewing	Secondary school	Relative simplicity; High interdependence	Difficulty in elicitating good concepts before concept mapping Strict control of timing required, Interdependence not required in the middle stage	High
Collaborative reading of texts + knowledge soup sharing	Secondary school	Facilitated construction of concept maps High interdependence	Complexity; availability of the same source; Integration not required in shared knowledge soups = risk of fragmentation of knowledge	High

CONCLUSION

We hope that future collaborative learning opportunities undertaken in our community, will be more successful as a result of the awareness of the basic principles of cooperative learning we have applied – in a very elementary way – to the description of our own experiences. It may be useful to pull together the soundest Collaborative Concept Mapping methodologies in a table (Table 1), while bearing in mind that that flaws can only ever be partially compensated for and that the effectiveness of a strategy also depends on a multiplicity of factors that have been only partially discussed in this chapter.

Other factors could be invoked in this survey, such as the demand for teacher support or the autonomy permitted to teams and the kind of feedback permitted/demanded. We would finally add that we expect other combinations of strategies to be devised in future which will optimise the quality of collaboration in each learning task for specific grades of learners.

REFERENCES

Cañas, A. J., Hill, G., Carff, R., Suri, N., Lott, J., Eskridge, T., et al. IHMC (2004). CmapTools: A knowledge modelling and sharing environment. In A. J. Cañas, J. D. Novak & F. M. Gonzáles (Eds.), *Concept Maps: Theory, Methodology, Technology, Proceedings of the First International Conference on Concept Mapping* (Vol. I, pp. 125-133). Pamplona, Spain: Universidad Pública de Navarra.

Gowin, D. B., & Alvarez, M. (2005). *The Art of Educating with V Diagrams* (pp. 19-21). New York: Cambridge University Press.

Henao Cálad, M. (2004). Experiencia con el Uso de Mapas Conceptuales como Estrategia de Enseñanza en un Curso de Ingeniería del Conocimiento. In A. J. Cañas, J. D. Novak & F. M. Gonzáles (Eds.), *Concept Maps: Theory, Methodology, Technology, Proceedings of the First International Conference on Concept Mapping* (Vol. I, pp. 325-332). Pamplona, Spain: Universidad Pública de Navarra.

Johnson, D. W., Johnson, R. T., & Holubec, E. J. (1994). *The Nuts and Bolts of Cooperative Learning*. Edina, MN: Interaction Book Company.

Novak, J. D., & Cañas, A. J. (2004). Building on New Constructivist Ideas and CmapTools to Create a New Model for Education. In A. J. Cañas, J. D. Novak & F. M. Gonzáles (Eds.), *Concept Maps: Theory, Methodology, Technology, Proceedings of the First International Conference on Concept Mapping* (Vol. I, pp. 469-476). Pamplona, Spain: Universidad Pública de Navarra.

Novak, J. D., & Cañas, A. J. (2006). The Origins of the Concept Mapping Tool and the Continuing Evolution of the Tool. *Information Visualization Journal, 5*(3), 175-184. Retrieved April 2, 2008, from http://cmap.ihmc.us/Publications/ResearchPapers/OriginsOfConceptMappingTool.pdf

Novak, J. D., & Cañas, A. J. (2008). *The Theory Underlying Concept Maps and How to Construct Them* (Technical Report IHMC CmapTools, 2006-01 Rev 01-2008). Retrieved April 2, 2008, from http://cmap.ihmc.us/Publications/ResearchPapers/TheoryCmaps/TheoryUnderlyingConceptMaps.htm

Novak, J. D., & Gowin, D. (1984). *Learning How to Learn*. New York: Cambridge University Press.

Reigeluth, C. M. (1999). *Instructional-design Theories and Models*. (pp. 284-286). Mahwah, NJ: Lawrence Erlbaum Associates.

Tifi, A., & Lombardi, A. (2006). WWMPAS: A Community on Education through Collaborative Concept Mapping. In A. J. Cañas & J. D. Novak (Eds.), *Concept Maps: Theory, Methodology, Technology. Proceedings of the Second Int. Conf. on Concept Mapping*. San José, Costa Rica: Univ. de Costa Rica. Retrieved April 2, 2008, from http://cmc.ihmc.us/cmc2006Papers/cmc2006-p92.pdf

KEY TERMS AND DEFINITIONS

Active Collaborative Process: is operatively defined as a collaborative process where any incoherence or incompleteness in the shared resources soon provokes feedback and discussion. A strong interdependence characterizes this process as it is very similar to the processes that accompany cooperative learning.

Autonomy or Team Independence: is a second feature and a goal of CCM, consisting of a self-sustained leading process within the team, characterized by maximum productivity with minimum direct contribution or inputs from the teacher(s). This parameter is highly related to the age of collaborator-learners and their motivation, and obviously requires internal communication, interactions and decision making.

Collaborative Concept Mapping (CCM): can be viewed as an extension of teacher-mediated learning assisted by concept mapping, in a socio-constructivist perspective, where students construct concept maps in small groups by questioning, discussing, sharing or alternating their contributions and by peer reviewing, under the supervision of the teacher(s), in order to integrate multiple views and create shared understanding.

Collaborative Learning (CL): is a general term indicating a joint intellectual effort by groups of students working together when searching for understanding and/or constructing shared artefacts. CL is usually characterized by high levels of individual learning freedom, by space and time independence and by absence of specialized roles in the goal oriented process.

Concept Map: According to Novak & Cañas (2008) concept maps are graphical tools for organizing and representing knowledge characterized by: labelled nodes (closed shapes) to represent concepts, linking phrases and connecting lines as a way to make explicit the relationships between concepts and to form oriented *propositions*, i.e. meaningful statements with two or more concepts that are hierarchically ordered. The main use of a C-map in education is to provide a shareable document for teacher and students to negotiate meaning.

Cooperative Learning: is an instructional environment in which teams of students work on structured tasks under conditions that meet five criteria: positive interdependence, individual accountability, face-to-face interaction, appropriate use of collaborative skills, and self-assessment of team processing. Students usually have well defined and differentiated roles and high interdependence in the attainment of the shared goal.

Governance: (Gowin & Alvarez, 2005) is the combined effect of all external factors that control the meaning of learning contents, i.e. social construction and feeling of instances perceived as counting or not counting, the efforts put into teaching, into the curriculum, and into learning.

Interdependence.: At its best (with the '*positive*' attribute in cooperative learning acceptation), Interdependence is the fundamental and structured characteristic of a cooperative team. It means that each group member's efforts are required and indispensable for group success and that each group member exerts a unique role in the joint effort because of his/her specific resources and task responsibilities. From the general perspective of this chapter, *interdependence* is a highly valuable goal of CCM, where team partners stimulate each other by questioning, reviewing and regulating, by depending to some extent on contributions from their peers, but without the prerequisite of indispensability as inherent to the process.

Metacognition: is the active and conscious control over the process of thinking that is used in learning situations. It is based on the intentional use of stored experiences of differentiation, comparison and organisation among different pieces of knowledge and from different external sources, and on the use of strategies for challenging and rehearsing knowledge. Metacognition can be supported by mediated concept mapping and even more by CCM.

APPENDIX: BASIC RULES FOR GOOD CONCEPT MAPPING, DRAWN FROM A TEAM THAT WAS QUICKLY TRAINED TO COLLABORATE ON PHOTOELECTRIC EFFECT

1. Every node should contain only a single concept (even though it can be formed by two or more words, as "main Light Theories");
2. Concepts should not be repeated (exceptions: most important concepts, as "Light" in a c-map about light physics, or concepts that are out of context (such as "Physics") or auxiliary (e.g. Scientist));
3. Linking phrases should have verbs (predicates);
4. Linking phrases should not include relevant concepts (with the same exceptions as in point 2);
5. Concepts having higher rank must be placed above the subordinated concepts in the layout;
6. Every proposition (concept - link phrase - concept) should be self- consistent, i.e. every node should be a starting point to read the c-map meaningfully;
7. Every concept should not have more than four sub-concepts linked;
8. Every good c-map should proceed to satisfy a pre-established "Focus Question" (such as: "How were theories about light developed");
9. After a preliminary map is constructed, cross-links should be sought to show relationships between the sub-domains in the map;
10. Fixing relationships between different domains in the map can require some extensive re-arrangement of part of the already constructed c-map.

A set of less demanding and less detailed rules, such as the following, might be preferred:

1. A concept map is a network of nodes labelled with concept words connected (related) by linking phrases;
2. Each node has a single concept which (if possible) should not be repeated in the c-map, and:
    2.1. it may include attributes and words of qualification (to describe a single object or event);
    2.2. it should be a starting point for reading valid propositions;
3. The linking phrases between every two concepts:
    3.1. should not be used to divide both sides of the same concept;
    3.2. should not include other concepts having a relevant role in the context of the c-map;
    3.3. should not refer to other concepts besides those being connected;
4. Each connection should go from top to bottom, i.e., from a more inclusive or general concept to a more specific one. A connection with reverse direction (bottom-up) requires an arrowed connection line;
5. The context of a Cmap should be explicitly pre-defined by means of a Focus Question.

Section 2

The Use of Concept Mapping and Collaborative Learning in Face-to-Face Situations

Chapter 8
Collaborative Learning:
Leveraging Concept Mapping and Cognitive Flexibility Theory

Chaka Chaka
Walter Sisulu University, South Africa

ABSTRACT

This chapter explores the interface between collaborative learning (CL), concept mapping (CMing) and cognitive flexibility theory (CFT). The major argument of the chapter is that concept maps (CMs) are versatile and multi-purpose tools with cross-disciplinary applications. In view of this, the chapter reports on 15 research studies to serve a dual purpose: to support its argument and to demonstrate the link between CL and CMing on the one hand, and between CMing/CL and CFT on the other hand. The focal points of the chapter are: CMs as tools for supporting and facilitating CL; CMs as assessment tools in CL environments; CMs as drivers for collaborative curricular initiatives; CMs as higher-order thinking and problem solving tools; and CL/CMs and CFT.

INTRODUCTION

Collaborative learning (CL) lies at the heart of collective learner-directed learning approaches. It represents a major shift away from teacher-dominated learning approaches to a more egalitarian view of learning. It is a methodological innovation which encourages a co-production of knowledge, a co-determination of meaning, collective problem solving, and multiple perspectives among learners and between learners and teachers. It also enhances cognitive skills and harnesses different learning styles. Above all, CL serves as an antidote to individualistic and competitive *winner-takes-all* learning tendencies. As such, it fosters a communitarian spirit and a sense of togetherness among learners. Most importantly, it acts as a vehicle for navigating and bridging cultural diversity and linguistic pluralism.

It is against this background that both concept mapping (CMing) and cognitive flexibility theory (CFT) come into the picture. For example, CMing is used to generate ideas, design complex structures, communicate complex ideas and evaluate under-

DOI: 10.4018/978-1-59904-992-2.ch008

Figure 1. Chapter outline represented through a concept map

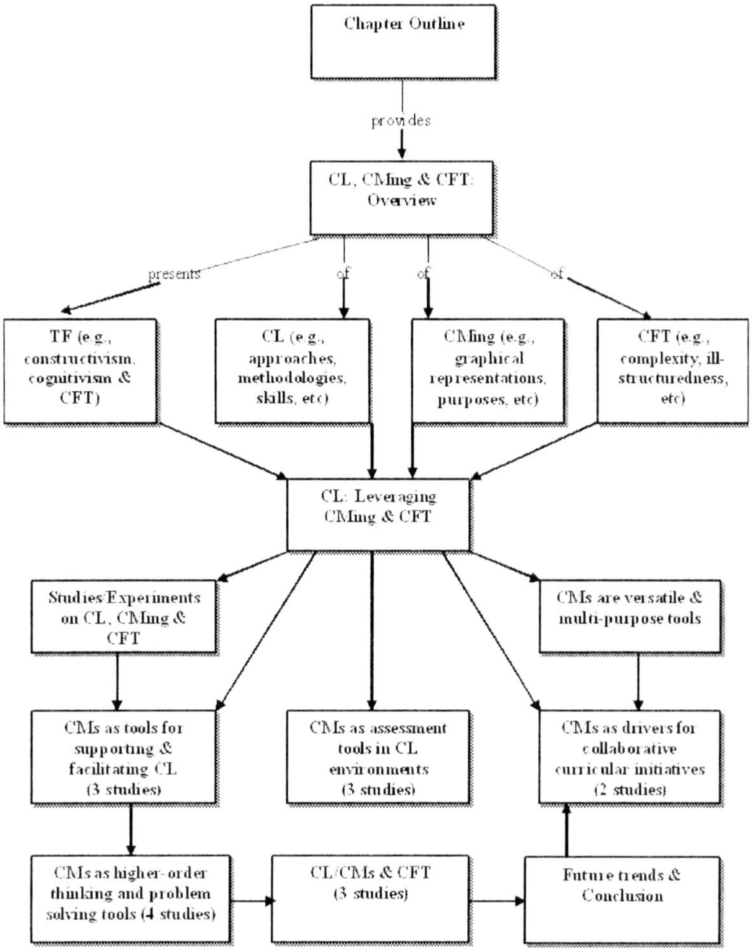

standing or identify misunderstanding. Additionally, it promotes relational and critical thinking. It also helps abstract important information; organise and structure knowledge; integrate new knowledge with old knowledge; retain knowledge for longer spells; and apply knowledge to new contexts (Milam, Santo & Heaton, n.d.).

In this context, CFT espouses the view that learning is better leveraged through a variety of problems and cases. It argues that learners tend to acquire complex and ill-structured knowledge through multiple perspectives or multiple representations. It further maintains that the proof of the newly acquired knowledge lies in its successful application by learners to new and different knowledge domains. Thus, the multiple perspectives or representations of knowledge and the successful application of any knowledge to new contexts can be better facilitated by both CL and CMing. This is particularly so as CL leverages the collective views of several learners and since CMing encourages representing knowledge from different angles. On this basis, *Figure 1* provides a CM outline of this chapter.

COLLABORATIVE LEARNING, CONCEPT MAPPING AND COGNITIVE FLEXIBILITY THEORY: OVERVIEW

The theoretical framework (TF) informing this chapter draws on constructivism, cognitivism and CFT. The first incorporates social construction while the second subsumes assimilation theory. In this context, both constructivism and cognitivism serve as a theoretical base for CL and CMing. For example, CL is rooted in social construction. Most importantly, through leveraging CL - in a social constructionist environment - learners are able to develop complex conceptual frames based on their collective prior knowledge. Similarly, CMing is grounded in assimilation theory that posits new knowledge can be learnt effectively by relating it to prior knowledge. Thus, assimilation theory displays subsumption, integrative reconciliation and progressive differentiation (Maas & Leauby, 2005). On this basis, both constructivist epistemology and CFT are underpinned by the common view that knowledge domains are complex and ill-structured.

Collaborative learning (CL) is an umbrella term for various approaches in education encompassing collective learning endeavours by learners or by learners and teachers. It refers to both learning methodologies and learning environments whereby learners engage in common activities with all of them depending on and accountable to each other. One of its core objectives is to enable learners to work together as groups so as to search for content understanding, meaning or solutions or so as to undertake a learning task. So, it is dialogic and multimodal in nature.

As an approach, CL is more learner-centred and highly open-ended. For instance, it fosters the following collaborative virtues: social and interpersonal skills; leadership skills and team spirit; self-esteem; equal participation; complementary interdependence; cooperative interaction; and inter-ethnic/cross-cultural relationships. Other attributes it nurtures among learners are understanding; caring; kindness; courtesy; empathy; tolerance; respect for diverse abilities; and appreciation of multiple viewpoints. In the latter case, CL enables learners to accommodate others' divergent perspectives and to be amenable to opposing views (Joritz-Nakagawa, 2004; Van Boxtel, 2000). Additionally, it facilitates collective information sharing, joint mastery of content material and collaborative achievement. Above all, CL allows learners to jointly develop communication skills, and problem solving and critical thinking skills. It also helps them acquire community norms and values, and develop joint decision making and conflict resolution skills. In so doing, it eliminates the need for competition and individualism among learners as supported by other approaches to learning (Johnson & Johnson, 2007; Joritz-Nakagawa, 2004).

Furthermore, CL promotes deep learning among learners. Deep learning allows learners to apply knowledge to other contexts. In addition, CL enables learners to engage in in-depth analysis and creativity. Most importantly, as Joritz-Nakagawa (2004) points out, CL - through peer collaborative tasks - has the potential to leverage learners' multiple intelligences (MI) (see also Chaka, 2008). Here CL serves as a basis for harnessing learners' collective abilities and collective intelligences. For example, collaborative group or team tasks involving interpersonal skills and social/community skills (e.g., interaction, dialogues, or group activities) are thought to be linked to interpersonal intelligence. CL tasks encouraging learners to be introspective in their approach (e.g., following one's own goals, interests, wisdom, and intuition) are believed to have intrapersonal intelligence as their basis (see also Gardner, 2003; Lane, 2000).

Similarly, collaborative language activities (e.g., oral, written, reading, using words, and auditory skills) are likely to activate learners' verbal/linguistic intelligence. Logical/mathematical intelligence is harnessed by collaborative tasks focusing on reasoning, calculating, experimenting, solving puzzles, asking cosmic questions, logic games, and abstract and conceptual thinking. Likewise, CL activities encouraging sensitivity to sounds and rhythms, speaking rhythmically, and appreciating aesthetics are more likely to trigger musical intelligence. And CL tasks enabling learners to draw, read maps, do jigsaw puzzles and learn through drawings, and physical and verbal imagery have the prospect of enhancing visual/spatial intelligence. In the same way, bodily/kinaesthetic intelligence is likely to be tapped by CL activities allowing learners to engage in body movement, touching, body language communication, hands-on learning, and role playing. Lastly, CL tasks concentrating on recognising and classifying several species (e.g., the flora and fauna) populating given habitats may lead to the application of naturalistic intelligence (Joritz-Nakagawa, 2004).

For their part, CMs are graphical representations of knowledge. They are diagrams displaying inter-relationships among concepts and exhibiting conceptual frameworks for specific domains of knowledge (Van Boxtel, 2000; see also Milam et al., n.d.). They represent concepts hierarchically with the most general concepts placed at the top of the map and the more specific concepts located below. The hierarchical typology of a CM is determined by a given knowledge domain and by the context in which that knowledge is applied (Novak & Canãs, 2008). Concepts are, here, perceived regularities in objects or events designated by labels. There are concrete and abstract concepts. The former are ideas of common objects (e.g., stone or aeroplane) or object qualities (e.g., green or round). The latter are rules classifying events that require verbal definitions.

Often concepts are enclosed in circles or boxes. Such circles or boxes have conceptual relationships signalled by linking lines connecting various circles or boxes. Linking lines can be uni-, bi-, or non-directional while concepts may have associative, specified, causal or temporal relations. Accordingly, words on the linking lines spell out the relationships existing between concepts. Concept-link-concept units comprise propositions (meaningful statements) that are regarded as semantic units. Moreover, CMs have cross-links that depict relationships between or among concepts stemming from different domains or regions within a given CM (Novak & Canãs, 2008).

CMs can serve several purposes. These include:

- Generating ideas (e.g., brainstorming, organising and structuring ideas)
- Representing knowledge and information graphically or visually
- Analysing content and tasks
- Solving problems, stimulating thinking and facilitating creativity (e.g., creative writing, or hypertext design)
- Communicating knowledge or ideas
- Designing complex structures (e.g., hypertext, hypermedia, websites, or innovations
- Learning by integrating new and old knowledge
- Organising and planning curriculum
- Assessing understanding or diagnosing misunderstanding (Lanzing, 1997; Milam et al., n.d.)

CMs have diverse uses as well. These range from planning, scaffolding, consolidating, organising and summarising instructional content to being advance/graphic organisers, teaching/learning tools and evaluation tools.

Overall, CMs tend to promote and enhance meaningful non-linear learning - learning for understanding multi-dimensional information as opposed to the rote learning of linearly sequenced items. They facilitate knowledge discovery, knowledge externalisation and knowledge visualisation. Above all, they allow new concept learning to take place through subsumption, progressive differentiation and integrative reconciliation. That is, through CMs, learners are able to identify concepts, hierarchically represent them, differentiate between them, and display more complex hierarchical relations (Novak & Canãs, 2008).

Furthermore, CMs help stimulate metacognition and concept understanding. The latter is the ability to apply knowledge across a spectrum of contexts or circumstances. It is different from declarative knowledge learning in that it does not require memorising any given association between two or more concepts (Darmofal, Soderholm & Brodeur, 2002). Most significantly, CMs help develop learners' higher-order thinking skills (critical thinking, divergent and multi-lateral thinking, and logical reasoning skills). They are, therefore, worthwhile CL and instructional tools.

The foregoing characterisation of CL and CMs leads to a short overview of cognitive flexibility theory (CFT). CFT is a conceptual model of instruction based on both flexibility theory and cognitive learning theory. It deals with factors contributing to the failure to learn complex and ill-structured knowledge as encountered at advanced instructional levels. Even though it is a theory, CFT is also an approach. Some of its contentions are: linear and simplified material results in the inability to handle complex and ill-structured knowledge; oversimplified material makes it impossible to transfer knowledge to new and varied contexts; and as the difficulty of the material increases so does its complexity and ill-structuredness (Mendes, Mosley & Counsell, 2001; Spiro, Feltovich, Jacobson & Coulson, 1992).

In general, CFT encompasses the following principles and assumptions:

- Instruction needs to reflect the complexity and ill-structuredness of the problems rather than treating them as simple, linear processes. As such, CFT stresses interconnectedness and conceptual patterns and avoids oversimplification of material.
- Emphasis has to be given to case-based learning rather than to grounding learning on a single case or example. Thus, cases must be used as the central point of learning.
- The same concept should be used and applied differently across varied sets of cases.
- Context-dependent knowledge is best suited for learners.
- Flexible knowledge lends itself to being acquired from and applied to many different contexts.
- Multiple representations using multiple analogies and multiple metaphors must be provided (Mendes et al., 2001).

Finally, CFT draws on and enriches lateral thinking theory as it requires learners to look at a conceptual structure of a learning problem from several fronts with a view to fitting such a structure into their existing conceptual frameworks. Hence, its view that the use of multiple and flexible concepts, schemata and thematic perspectives better represents the multi-pronged nature of the content domain (Mendes et al., 2001).

COLLABORATIVE LEARNING: LEVERAGING CONCEPT MAPPING AND COGNITIVE FLEXIBILITY THEORY

Several studies and experiments have been conducted on the uses of CL and CMing in various disciplines. The focal points of such studies and experiments range from the effectiveness of CL and CMing to the use of CMs as assessment tools in CL (Basque & Lavoie, 2006). Likewise, there are studies that have investigated the interface between CMing and CFT. In the light of this scenario, this section of the chapter reports on the research studies exploring CL and CMing on the one hand, and on research efforts examining either CL and CFT or CMing and CFT. It contends that CMs are versatile and multi-purpose tools with cross-disciplinary applications. The section has the following focal points: CMs as tools for supporting and facilitating CL; CMs as assessment tools in CL environments; CMs as drivers for collaborative curricular initiatives; CMs as higher-order thinking and problem solving tools; and CL/CMs and CFT.

CMs as Tools for Supporting and Facilitating CL

There are studies that have investigated the link between CL and CMs. One such study is Van Boxtel's (2000) study which reports on three related experiments. All the three experimental studies examined whether collaboration on given tasks resulted in the improvement of the individual use of physics concepts (as measured by comparing scores on the pre- and post-tests). Since these three experimental studies are sequels highlighting the potential CMs have in CL environments, I will only concentrate on the first one. The first experiment - which was piloted in 1997 - involved 40 students from 2 physics classes at a Dutch intermediate general secondary school in the Netherlands. It compared a CMing task with a poster task and focused on how the nature of the group task done in a CM affected the quality of peer talk as opposed to a poster task. The main questions of the research experiment were: How do the characteristics of the collaborative learning task affect the features of peer talk and which of these features contribute to the learning of physics concepts? The collaborative tasks used in the experiment were expected to elicit elaborative talk and the verbalisation of scientifically incorrect conceptions about electricity concepts (e.g., current consumption, non-recursive current and the lack of differentiation between current strength and voltage).

Students were randomly assigned to same-sex dyads - 4 girl-girl dyads, 15 boy-boy dyads and 1 mixed-gender dyad. Within each class the same number of dyads was also randomly assigned to one of the four conditions: 1) CM with individual preparation; 2) CM without individual preparation; 3) poster with individual preparation; and 4) poster without individual preparation. A pre-test and a post-test consisting of three units - the concept definition, problem-solving and essay units - were administered to the students. The students' attitudes towards physics were measured by a 5-point Likert scale questionnaire. Their experiences with the CM and poster tasks, and with the individual and collaborative preparations were also tested using a Likert scale questionnaire. And their verbal interactions were video-recorded. The CM and poster tasks were compared and so were the individual and collaborative tasks.

In this experiment, overall, the collaborative tasks were found to have improved students' use of electricity concepts through verbalising, elaborating, negotiating and co-constructing conceptual meanings. That is, both collaborative CM and poster tasks - as opposed to the individual tasks - yielded much elaborative talk and an improved verbalisation of conceptual understanding. But of the collaborative tasks, the CM task showed to be a better instrument to elicit elaborative talk about concepts than the poster task. Also noteworthy in the experiment is that, student interaction not only led to elaborative talk, but it resulted in collaborative elaboration in relation to argumentation and co-constructed reasoning as well. This underlines the versatility of CMs and the multiple purposes they can serve. However, the talk about the electricity concepts reflected misconceptions concerning voltage and current strength (Van Boxtel, 2000).

Another study reporting on the use of CMs as tools for supporting and facilitating CL is Hill's (2004) study. The study was conducted in the fall of 2002 at the Virginia Commonwealth University (VCU) School of Pharmacy. It had 109 *Communications in Pharmacy Practice* students as its participants. The former is a 2-credit core course offered in the first semester of the doctor of pharmacy programme at VCU. Its content focuses, inter alia, on the following: communication skills employed in the pharmacy practice with reference to patient counselling and education; communicating with such diverse populations as geriatric patients, mentally ill patients and disabled patients; and the instruction on low health literacy and cultural competence.

A CMing assignment was added to the course since the existing course structure did encourage engagement with the material and students could not retain information they needed to successfully function in the advanced pharmacy practice experiential programme. In the course, students received handouts explaining the assignment. This entailed: a CM detailing the general structure; providing an example of a relevant CM; and specifying the criteria for grading. In this regard, students were expected to include 3 concepts: the pharmacist; the patient; and pharmacy care. They were encouraged to add material from other courses as long as they were able to demonstrate the connection of this material to pharmacy communications. Students were then allowed to work in self-selected groups of 3 to 5 to collaboratively create a common CM and a narrative explaining the decisions they made in its creation. In all, students were encouraged to cooperate and negotiate meaning within their group discussions. CMs were to be produced by hand or by using computer software programmes such as *Inspiration* (Version 7).

In drawing their CMs, students had to take the following 5 features into account: concepts selected; choice of linking words explaining the relationship between 2 concepts; a hierarchical ordering of concepts within the map; cross-links established between sections of the map; and examples illustrating the map's concepts. The quality of the CMs was satisfactory, and most students indicated that they had found the CMing technique very useful. The technique helped them organise their ideas, retain information and relate the communications course material to other courses in the curriculum. It also helped them better understand the role communications played in pharmacy care. Some indicated their eagerness to apply CMing to other courses. Moreover, the students reported that they had experienced both disequilibrium and equilibrium (Hill, 2004).

A further study underscoring the positive link between CL and CMs is the one undertaken by Tifi and Lombardi (2006). Reporting on its first results, the project - which was coordinated in Italy - involved the use of *WWMaps* (*World Wide Maps*) for sharing knowledge by a multilingual online community using CMs. *WWMaps* facilitates the dissemination of CMs and the sharing of knowledge and teaching and learning practices among global participants. It also promotes the mutual understanding of diverse cultures. In short, it encourages collaborative efforts among diverse distant global audiences (Tifi & Lombardi, 2006).

The project monitored several distance online collaborative teaching and learning teams in primary and secondary schools in Italy. It was made up of teachers and students. Some of the data the participants had to provide when joining the *WWMaps* initiative were students' personal details such as age, town, preferred topics of collaboration, and language within a map called *MeToo*. This then allowed the potential member to download and install the software programme, *CmapTools v4*. Collaborative teams were formed according to topics (covering human rights, psychology, physics and mathematics) that enabled students to engage in collaboratively constructing CMs. Each team developed a project map in collaboration with a member of the *WWMaps* staff. Collaborative teams in the project consisted of 10 pairs and 2 triads with the following racial composition: 2 Italian-Spaniard primary school (P.S.) pairs; 2 Italian-Costa Rican (P.S.) pairs; 2 Italian-Romanian secondary school (S.S.) pairs; 1 Italian-Maltese (P.S.) pair; 1 Italian-Panamanian (P.S.) pair; 1 Italian-Spaniard (S.S.) pair; 1 Romanian-Mauritian (S.S.) pair; 1 Italian-Maltese-Spaniard (P.S.) triad; and 1 Italian-Mauritian-Spaniard (S.S.) triad.

In the design phase of the project, schools, students and teachers were required to get to know each other and to be familiar with the *WWMaps* and its tools (e.g., a tutorial on how to use *CmapTools* and on how to build good CMs). Students were encouraged to actively participate in creating CMs. They also practised a given foreign language when writing personal descriptions through using the "octopus self-intro" map which was easy to customise and read. On this basis, the teachers had the students work in collaborative teams focusing on the complexity of the tasks, the mutuality among the students, and the autonomy of the students.

Some of the results from the project include the following. The teams were able to work both autonomously and collaboratively and kept contact with each other through e-mails. Thus, the *WW-Maps* afforded the participants the opportunity to cooperate and share knowledge. The students, in particular, were able to translate multilingual CMs into three different languages. Above all, CMs were used as effective tools of communication and for engaging the students in complex forms of knowledge (Tifi & Lombardi, 2006). All this highlights the versatile and multiple uses CMs can be put to.

CMs as Assessment Tools in CL Environments

There are different forms of assessment that can be employed in CL environments using CMs. To this effect, this section of the chapter discusses some of the studies that have used CMs as tools for assessing learning content or students' understanding and knowledge of the learning content.

One such study is the one reported by Gouli, Gogoulou and Grigoriadou (2003). This was an experimental study which was conducted in the spring semester of 2001/2002 on the postgraduate course, *Distance Education and Learning*, offered by the Department of Informatics and Telecommunications at the University of Athens (Greece). The course had the following content areas: Open and Distance Learning (ODL) and Open and Distance Systems; the role of the tutor in ODL; the design and implementation of educational material suitable for ODL; and the exploitation of ICT in ODL. There were 51 postgraduate students who participated in the study. They ranged from secondary school teachers to systems managers and software engineers. The course expected the students to gain a deep understanding of given concepts and of their inter-relationships.

A set of assessment activities was developed, and was experimentally evaluated following a three-step process outlined in the *AssessToLearn* framework. The latter is a coherent and integrated framework consisting of a three-step process incorporating principles from contemporary theories of learning. Each step addressed specific assessment functions and served specific assessment purposes. The three steps entailed: ascertaining the students' prior knowledge - activating knowledge; promoting knowledge construction and identifying conceptual changes - constructing and enriching knowledge; and assessing knowledge construction - refining knowledge. The assessment tools used were various CMing tasks such as a free construction task, a concept list task and a partial recognition framework task. The questions asked involved free-response questions; questions based on short cases; multiple choice questions; true/false questions; ordering questions; and fill-in-the-blank questions. And the forms of assessment employed were collaborative assessment, self-assessment, peer assessment and summative assessment. Most importantly, the participants were supported by a Web-based course management system, called *E-Classroom*. In addition, there was a two-hour training session on CMing at the beginning of the course to familiarise the students with the use of CMs.

Some of the significant results of this experimental study are noted here. The students' CMs on prior knowledge revealed the unknown concepts, their incomplete understanding and their false beliefs. The results concerning "The process" showed that the combination of the three assessment tools (e.g., the free-response questions, the questions based on short cases and the CMing tasks) were effective and tended to support each other. Thus, the three tools managed to supplement each other in activating and eliciting the students' prior knowledge. The free-response questions and the questions based on short cases acted as a scaffolding framework and helped the students to think of the concepts and the relationships they had in the CMs. In addition, the researchers point out that the assessment activities and the use of CMs enabled the students' progressive conceptual changes regarding the fundamental concepts of the content matter that had to be identified and evaluated accordingly (Gouli et al., 2003). The findings of this experimental study further highlight the versatile and multi-purpose nature of CMs in CL.

Another study focusing on CMs and assessment is Rodriguez's (2006) project. This was a three-year intervention project involving teachers and students from culturally diverse schools. It was called *Integrating Instructional Technology with Science Education* (*I^2TechSciE*). It had a dual objective: to help science teachers plan and implement an inquiry-based and culturally relevant curriculum using CMs; and to assess students' cognitive growth through CMs.

The teachers who participated in this *I^2TechSciE* project were all grade 4, 5 and 6 teachers from one elementary urban school located in the Pacific Southwest. There were 10 of them: 2 Latinas, 3 Latinos, 1 Anglo male, and 4 Anglo females. The student racial composition was 56.5% Latino, 18.8% Anglo, 5.6% African, 2.6% Asian, 0.6% First Nations, and 16% Other. 37% of the students were classified as English Language Learners. In the first year of the project, a representative cohort of 10 students was selected from each of the grade 4, 5 and 6 classrooms. Each year the *I^2TechSciE* students in the cohort groups were placed in classrooms with project teachers to maintain continuity and to assess the effect of the project on students' learning process. Semi-structured, fill-in-the blank CMs were made available to the participants due to their inexperience with CMing. Activities modelling how to make the science curriculum units multicultural, inquiry-based and hands-on were conducted.

The project yielded some of the following key findings. The analysis of the data gathered in the first two years of the project demonstrated that all the teachers had gained new science knowledge and skills, and increased proficiency in CMing. Additionally, all of them had found the use of CMs as curriculum planning tools very useful for organising and structuring the science units. Finally, the analysis of the students' unit CMs indicated significant knowledge growth across grade levels for all the units tested (Rodriguez, 2006). Thus, in all this, lie elements of CMs as flexible and multi-purpose devices.

One of the recent studies to have explored the application of CMs as assessment devices is the one reported by Moni and Moni (2008). The study was undertaken in 2005 and involved 61 students enrolled in a Foundation Biological Sciences for Dentistry (DENT2012) at the University of Queensland in Brisbane (Australia). DENT2012 serves as an academic background for second-year dentistry. The current study reports on the second phase of the project. The aims of this second phase were: to explore the opinions of students about the task and the rubric; and to investigate the processes by which students understood and used the criteria of the rubric to construct a group CM. The key research questions were: 1) What are the opinions of students about the CMing task in physiology?; 2) What are the opinions of students about the assessment rubric for the CMing task?; and 3) How do groups use the criteria to complete the CMing task? The study was conducted in two stages. The first stage sampled students' opinions concerning both the assessment task and the rubric at the end of the assessment period. The second phase entailed observations of two student groups as they completed their CMs. Surveys were designed to record the opinions of the 61 second-year dentistry students concerning: 1) CMing assessment task (5 items); and 2) the assessment rubric (6 items). For both surveys, the participants were asked to complete 5-point Likert scales whose items ranged from *very strongly disagree* on the one end to *very strongly agree* on the other. Furthermore, two groups volunteered to have their group work videotaped. Both groups (*Groups A* and *B*) comprised 4 males and 4 females respectively.

The results of the study were as follows. The teaching faculty staff involved in the course evaluated the CMing task on students' opinions to be potentially valuable to the students in fostering high-order reasoning skills and team work needed for professional dental practice. Also, the students who favoured the CMing task achieved higher grades for their CMs. This underlined the correlation between favouring the assessment task and the final grades. Again, most students reported that working in groups had helped them learn more. Moreover, the assessment rubric was strongly favoured by most students. Lastly, the two volunteer groups favoured the CMing task and achieved higher grades (Moni & Moni, 2008). This study once more serves to emphasise the versatile and multi-faceted uses of CMs.

CMs as Drivers for Collaborative Curricular Initiatives

As evidenced by Moni and Moni's (2008) study, CMs can be used for planning, designing or implementing curricular efforts. Two more instances of initiatives exploring CMs as instruments for curricular intervention are provided here. The first initiative relates to Vilela, Austrilino and Costa's (2004) project. The project, focusing on a curricular reform and implementation of a medical course, took place in 2002/2003 at the Universidade Federal de Alagoas (the Federal University of Alagoas) or (UFAL) in Brazil. A curricular reform commission was established to drive the project. The participants were medical faculty staff, student unions, teaching monitors, and technicians from both the General Health Office and a municipal district. The commission had to produce a new curricular reform proposal for the medical course.

Some of the questions the project had to deal with were: 1) What are the objective of the course?; 2) What is the manager's role in the running of the curriculum?; and 3) What administrative structures are planned? The project used the technique of conceptual mapping to guide it in producing a new curricular proposal. The key concepts guiding this conceptual mapping effort were: 1) changing propeller factors (social, institutional, and individual); 2) the objectives of the medical course at UFAL; 3) the steering principles of the curricular reform; 4) the necessary changes in the curriculum to follow the new curricular guidelines; and 5) the presentation structure of the new general pedagogic proposal.

Overall, one of the important results of this project is that a CM was constructed which reflected a new collaborative curriculum. During the construction phase, all the participants were actively involved in identifying the steering principles and in influencing the changing curricular needs of the medical course. Therefore, for the whole reform team, the construction of the curricular CM was a powerful tool for visualising and identifying curricular demands (Vilela et al., 2004).

The second initiative concerns McDaniel, Roth and Miller's (2005) study. The project - a faculty project - was about the application of CMing as a tool for understanding, collaborating in, validating and integrating curriculum content designed to develop specific competencies in graduate level academic programmes. The programmes are offered at the National Defense University's (NDU's) Information Resources Management (IRM) College at Ft McNair in Washington. These programmes (offered by 4 academic departments) cater for a diverse student population: military students; senior leaders from government agencies; foreign military officers and civilians; and private sector students working with the government.

The project's major focus was: the specific application of CMing to curriculum; and how *curriculum mapping* was useful to faculty and academic administrators charged with maintaining curriculum quality at the IRM College. Its purpose was: using CMing at the NDU's IRM College as a tool for designing curriculum and for maintaining academic programmes that meet articulated competencies leading to certification, accreditation and stake-holder interests. The first use of CMing involved mapping competencies to clarify and validate the Information Assurance certificate programme. The software programme, *CmapTools*, was employed for this purpose. From here, the project created and produced CMs of the courses (together with articulating competencies) offered at the college. These were then submitted to the National Security Agency (NSA) for accreditation purposes.

The project shed the following insight. The use of CMing to represent the academic programme curriculum proved beneficial to the faculty, the administrators and the accreditation agency. The curricular CMs as employed in this project were a critical driver for curriculum design. They also served as a valuable tool for creating a cross-disciplinary curriculum. In addition, they acted as a framework which helped articulate cross-curricular needs and expectations of the programmes offered at the college. The use of CMing facilitated the accreditation and certification process (McDaniel et al., 2005).

In all, the two projects reported here further stress the flexibility of CMs and the multiple functions they serve.

CMs as Higher-Order Thinking and Problem Solving Tools

There are several studies that have investigated the use of CMs as both higher-order thinking and problem solving tools. Maas and Leauby (2005) report on one such study exploring students' development of higher-order thinking skills through the use of CMing. The study sample comprised 64 Accounting 202 students during the fall and spring semesters at a comprehensive state university. The fall semester class of 29 students served as a control group while the spring semester class of 35 students acted as an experimental group. Traditional instructional methods were used in teaching the control group whereas two end-of-unit CMing activities were administered to the experimental group. The major result of this study is that the experimental group not only performed better than the control group, but also showed significantly better thinking skills than the control group (Maas & Leauby, 2005).

In a related but different context, Nelson's (2007) study highlights the role played by CMing in improving students' critical thinking skills. The study targeted first semester freshman students entering Louisiana State University (LSU) in the fall of 2006. The purpose of the study was to determine the effects on the academic performance of freshman students participating in activities involving the Academic Concept Mapping (ACM) instrument. The latter is an academic advising instrument. One of the objectives of the study relevant to this chapter was: To compare, for the purpose of providing preliminary results, the propensity for incorporating critical thinking skills, as measured by the California Critical Thinking Disposition Inventory (*CCTDI®*), and the ability to apply critical thinking skills, as measured by the ACM quiz, among entering freshman students. The corresponding research question for this objective was: Will freshman students at a four-year comprehensive university in the South, who participate in the ACM instrument, develop and more effectively implement critical thinking skills into their academic planning, while enrolled in a freshman seminar course? (Nelson, 2007).

The platform for conducting the study was a freshman seminar course in which students volunteered to enrol after the orientation advising process. A total of 258 students, with ages ranging from 17 to 24, participated in the study. The ACM programme replaced class activities for the treatment group for 5 class sessions needed to complete the instrument. All the students completed a persistence factors questionnaire, an ACM quiz and the *CCTDI*® post-test. Half of the treatment group completed the *CCTDI*® as a pre-test; and so did half of the control group.

The conclusions relating to both the objective and the question of the study are briefly summarised here. The activities and critical thinking skills (e.g., analytical reasoning, rational problem solving, and projection) employed in the ACM resulted in the application of critical thinking skills as measured on the ACM quiz. The treatment group's mean score was significantly higher than the control group's mean score. This indicated that the ACM was effective in teaching the application of critical thinking skills in the designated academic areas. So, this trend implied that the ACM activities did have a positive effect on the propensity for incorporating critical thinking skills (Nelson, 2007).

Two studies attest to the critical role played by CMing in facilitating collaborative problem solving. The first study (Khamesan & Hammond, 2004) reports on both learning effectiveness and personal and interpersonal awareness in synchronous collaborative problem solving CMing via ICT (information and communications technology) with two computer communication channels. It addressed two questions: 1) Is collaborative, synchronous, Web-based, CMing as effective as collaborative CMing using FtF (face-to-face)?; and 2) What CMC (computer-mediated communication) channels are needed to improve collaboration in synchronous CMing via ICT? It consisted of 30 pairs (12 males and 48 females aged between 17 and 23) from York University and employed a between-subject design. Its between-group variable was the CMC channels in a collaborative Web-based problem solving CMing task with three levels: online with written chat (text) versus online with audio/text connection versus FtF.

The 30 pairs (60 participants) drew their shared CMs in the three conditions of the study: FtF, ICT with text, and ICT with audio and text connection. 10 pairs (20 participants) served in each condition. Initially, the participants were given a short training session on the use of a CM through the software programme, *CmapTools*. They were, then, allowed to draw their individual CMs on planning a new university campus. Thereafter, they were randomly assigned to the three conditions of the experiment. In the first condition, the participants used the *CmapTools* software to draw a CM with FtF collaboration. They used the online collaborative part of *CmapTools* with different CMC channels for collaboration. In the second and third conditions, the participants were given the same task but in separate rooms. In the second condition, the participants employed *Yahoo! Messenger*'s written chat room while those in the third condition used *Yahoo! Messenger*'s written and audio connection. At the end of the experiment, the participants filled out technology and awareness questionnaires.

The significant results emanating from the study are as follows. All the FtF and ICT participants concurred about the positive performance of *CmapTools* both in collaborative and individual CMing. The two ICT groups assessed the use of *Yahoo! Messenger* positively. The participants in the ICT group (with audio and text connection) had slightly higher personal awareness than the two other groups. The text group generated more CMs than the two other groups. In general, the study showed that computer-based CMing could be used in collaborative learning with remote communication as effectively as with FtF communication (Khamesan & Hammond, 2004).

The second study in this regard is the one by Keller, Tergan and Coffey (2006). This study investigated whether a specific kind of awareness tool, the *KIA-Tool*, led to more efficient collaboration and to more efficient computer-supported collaborative problem solving in the context of resource-based learning (RBL). It took place at the University of Tuebingen, Germany. The participants were 90 students (58 females and 32 males). They were randomly assigned to both the experimental and control groups. There were 15 control and experimental groups apiece, each consisting of 3 participants.

The participants were made to work in a spatially distributed, synchronous form with net-based, shared and unshared tools. They were given a task requiring them to care for a fictitious spruce forest. The task had information elements consisting of 13 concepts, 30 relations between the concepts, and 13 background resources. The experimental group used two different software programmes. The first was *Bebop*, a *Zope3*-based groupware containing different information units the group members needed for solving problems. The second was *CmapTools*. In addition, the experimental group was provided with a *KIA-Tool* which enabled them to see the individual CMs of their collaboration partners.

Prior to the experiment, a diagnostic test was administered to measure control variables including the participants' experience with computers, mapping techniques, and group working skills. The participants also received instruction in using *CmapTools*. There were 3 phases to the experiment. At the beginning of phase 1, all the participants were told that they were experts who had to protect a spruce forest. However, they were requested to refresh their domain expertise before starting collaborating and finding a common solution to the problems. This phase lasted 23 minutes and had group members working separately. In phase 2, each control group member had 5 minutes to examine his/her own map. By contrast, each participant in the experimental group had 5 minutes to view his/her own map as well as his/her collaborators'. After this activity, the participants filled out a questionnaire used as a manipulation check. In phase 3, three group members had to collaborate to solve two problems (i.e. which pesticide and which fertiliser they would use to protect and to cultivate the spruce forest). At the end of the study, the participants filled out a questionnaire asking for the difficulties concerning communication and collaboration, the use of *CmapTools*, and how helpful the *KIA-Tool* was.

This study yielded the following preliminary results that are relevant to this chapter. The experimental groups indicated that having an overview of the collaborators' CMs was helpful. The participants in the experimental groups reported that the collaboration with each other had led to a better overview regarding the relations of the domain knowledge as compared to the control groups. With respect to the domain knowledge performance, the experimental groups had a higher performance on domain relations that were shared by a participant-collaborator dyad, in comparison to the control groups. This supported the helpfulness of the *KIA-Tool*. Concerning the problem solving tasks, the experimental groups solved the two tasks correctly, as compared to the control groups. This demonstrated that using the *KIA-Tool* was useful for problem solving performance. All this highlights the value of Web-based collaborative learning and problem solving as a way of enhancing knowledge and information awareness (Keller et al., 2006). Overall, the four studies explored in this section, like those analysed earlier, attest to the versatile and multiple uses CMs can be put to in CL contexts.

CL/CMs and CFT

Cognitive flexibility theory (CFT) seems to have the potential to be harnessed better by either CL or CMing or by both, especially regarding complex and ill-structured problem solving situations. To this end, there are certain initiatives that have explored the interface between either CL and CFT or CMing and CFT. As such, this section examines three such initiatives, thereby contributing to the debate in this regard. One of these initiatives does not employ any CMing, but has a CL element built into it.

The first initiative is Stoyanov and Kommers' (2006) experiment. It was conducted in one university institution in the Netherlands on 22 fourth-year undergraduate students. It attempted to answer two main questions: 1) What is the effect of the type of CMing instruction on solving ill-structured problems?; and 2) What is the effect of individual differences on the construction of CMs given an ill-structured problem solving situation? The two hypotheses it set out to test were: 1) The experimental group using the new method for CMing instruction would score significantly higher on mapping production than the control group, which applied the classical CMing method; and 2) Individual differences in learning styles would predict the differences in mapping production and would generate an interactive effect with the type of instruction.

All the participants were randomly selected and then equally assigned to the control and experimental groups on the basis of their learning styles (i.e. thinkers and doers). The experiment was conducted in a one-day session with the two groups remaining constant throughout the duration of the study. The experimental design included two independent variables: type of instruction on CMing and learning style. The type of instruction had two levels: the classical CMing instructional method and the new CMing instructional method. The learning style also had two levels: doers and thinkers (Stoyanov & Kommers, 2006).

The dependent variable was CMing production. This variable was operationalised through two criteria - broad perception and divergence - each having several indicators. Broad perception was defined as the extent to which a problem solver represented a problem situation comprehensively. Its indicators were: fluency (number of nodes and links); and flexibility (variety of nodes, labels and links). On the other hand, divergence was defined as the extent to which the problem solver produced alternative solutions. This criterion also had two indicators: fluency (number of ideas); and flexibility (variety of ideas). A learning style questionnaire (LSQ) was used to measure the students' learning styles. It consisted of two combined learning styles: doers (activists and pragmatists); and thinkers (theorists and reflectors). The LSQ was distributed to the participants for them to fill it in. The control group was introduced to the classical CMing method while the experimental group had to use the new CMing method. A case, *George's Career Dilemma*, was presented to the two groups to solve. It represented a situation in which a previous year university student was confronted with a problem to take a decision about his future. Each group had to use the procedure introduced to it in solving this problem.

The study produced important results. For example, the experimental group generated considerably more items than the control group. It also demonstrated significantly higher results on the flexibility of nodes. This group was even superior to the control group in relation to the divergence criterion. Moreover, the new CMing method supported and facilitated problem solving performance better than the classical one. With reference to the learning style variable, the thinkers tended to use significantly more structural types of links than the doers. They also formulated substantially more assumptions than the doers. They classified information into clusters and generated more hypotheses. However, the doers expressed more feelings in the perception of the problem solving space than the thinkers.

Furthermore, the results supported the hypothesis that the new CMing instructional method was significantly better than the classical one in solving ill-structured problems. It proved to be more effective in analysing the problem situation and in generating ideas. Additionally, it fostered a broad perception with more and diverse information items and more complex labels on the links. Most significantly, it promoted a broader and more complex cognitive structure with a pronounced effect on the structural, interrogative, causal and remote associative types of links. By contrast, the classical CMing method used more simple and descriptive links (Stoyanov & Kommers, 2006). Needless to say, that this study, furthermore, highlights the versatility and the multi-purpose nature of CMs.

The second initiative is Shuyska's (2007) case study. The purpose of this study was to assess the potential the *CONTEXT* tool had in supporting students' analytical thinking and coherent understanding as regards complex material. To this effect, *CONTEXT* is a didactic technological tool designed to facilitate deep and coherent understanding of complex material. It seeks to engage the learner in actively constructing knowledge and in deeply exploring the knowledge domain. The tool itself is a combination of computerised CMing and hypertext writing in a wiki. It is mainly used for creating an interlinked Web-based knowledge repository in a wiki and a corresponding CM.

The study was conducted to 12 history A-level students with a view to helping them revise their Russian Revolution history material. Convenience sampling served as a sampling strategy for the study. The study comprised: 1) two working sessions with the class teacher; 2) two classroom sessions; 3) a short planning session with the teacher (between the two sessions); 4) short follow-up interviews (with the students at the end of the second class session); and 5) a closing interview with the teacher. At the beginning of the first class session, the students were introduced to *CONTEXT* as a tool which could aid their learning and revision. They were then required to use the tool to work on the task (topics) assigned to them, and were also allowed to use their history textbooks. For the task, they had to prepare CMs and write articles in groups for the wiki.

Two aspects emerged as part of the tentative results of this study: motivation and cognitive flexibility. Concerning the first aspect, all the interviewees reported that CMing and wiki writing had been an enriching learning experience for them. The students, in particular, indicated that *CONTEXT* had motivated them to do their work. As regards the second aspect, the students pointed out that both the CM and the wiki were idea-ordering tools that had helped them see things in context and as part of the bigger picture.

The third initiative - even though not employing any CMing - is Lima, Koehler and Spiro's (2004) study which evaluated *Panteon*, a Web-based system for diagnosing complex business cases. *Panteon* is delivered through the Web. In it, authoring and learning are closely related. It enables students to, first, do individual cases (e.g., to diagnose cases), and then, mount a collaborative group synthesis (Lima et al., 2004).

The study's focus was to design CL environments that fostered critical thinking, creativity and reasoning from multiple perspectives. The study involved 29 students (19 men and 10 women whose mean age was 23.6) from two 4th year classes in the School of Business Administration in a private university in Brazil. It was conducted in a computer laboratory with one computer per student. In addition, 4 business and education experts (with the mean age of 35.75) participated in the study as observers. The case study given to the participants comprised nine Conduit Technology (a pseudonym) employees' perceptions about the three most urgent problems faced by their firm: a) internal and external communication problems; b) uneven cash flow; and c) lack of commitment. These problems had to be viewed from the perspective of a five-category model of analysis which made employees reflect upon the following organisational dimensions: production (tasks); structure (hierarchical relations); strategy (market- and environment-related); technology (including available infrastructure); and psychology (human resources motivational policies and leadership). 256 perceptions were loaded into the *Panteon* interface.

A questionnaire was administered for the participants to assess the advantages and disadvantages of the *Panteon* approach as compared to the *Harvard Method*. A 4-point Likert scale was also used to evaluate the two approaches in terms of the four dimensions of ***Collaborative***

Interactivity: *Interactivity*; *Authenticity*; *Reflectivity*; and *Collaboration*. The same was done to rate the two approaches concerning the eight dimensions of ***Integrated Thinking***: *Establishing Connections*; *Analyzing*; *Designing*; *Evaluating*; *Problem Solving*; *Synthesizing*; *Decision Making*; and *Imagining/Elaborating*. The participants were then assigned roles in their groups either as a *Generalist*, a *Marketing Specialist*, a *Human Resources Specialist*, or an *Organizational Systems Specialist* (Lima et al., 2004).

Here follows a brief overview of the results of this study. On every dimension of ***Collaborative Interactivity***, the mean rating of the 29 participants for the *Panteon* approach was higher than for the *Harvard Method*. The 4 experts also rated *Panteon* higher than the *Harvard Method* on all measures of ***Collaborative Interactivity***. Similarly, the 29 participants rated *Panteon* significantly higher than the *Harvard Method* on each dimension of ***Integrated Thinking*** with *Establishing Connections* topping the ratings. In this context, problem solving featured as an inherent part of interacting, reflecting, and collaborating during the diagnosis stage of the *Panteon* approach. The latter also tended to engage the participants in a complex decision making activity. Overall, the experiment suggested that *Panteon* had acted as a valuable cognitive trigger which had allowed the participants to engage in multiple perspectives characteristic of the cognitive flexibility principle. It had afforded participants the opportunity to apply both their basic and their higher-order complex thinking skills (e.g., critical and creative thinking skills and complex reasoning skills).

FUTURE TRENDS

One of the long-term future trends in relation to the points raised in this chapter is mainstreaming CMing not only in CL environments but also in all teaching and learning domains at all levels of education (schools and higher education institutions). This entails popularising CMing as a global instructional practice and not employing it only as a once-off or an ad hoc approach as some of the studies explored in this chapter demonstrate. One key to realising this move will be to integrate CMing into professional teacher training and curricular programmes.

Related to the preceding trend is the move to use CMing in CL contexts as a formal and accredited means of assessment and curricular design and implementation. This trend requires the same mainstreaming approach spelt out in the previous trend. Another trend likely to be a determining factor for CL and CMing is using the latter to simultaneously foster a set of integrated multiple higher-order skills in learners in CL environments. The skills in question are critical, creative, analytical, conceptual, lateral, convergent, divergent, logical, reflective thinking skills and reasoning and decision making skills. Most studies seem to examine and treat these skills discretely or in isolation.

A further trend in this regard is mounting more sustained initiatives to closely study the interface between CL, CMing and CFT. This is necessary as CFT lends itself well to studying complex, ill-structured and non-linear knowledge domains and problem solving situations from multiple case perspectives. These two factors also lend themselves well to both CL and CMing. So, CL, CMing and CFT have the potential to leverage each other. Lastly, the potential future value of CL, CMing and CFT lies in the Web 2.0 and Mobile Web 2.0 learning environments. This is more so as applications such as blogs, wikis, social networks and social bookmarks are ideal spaces for collaborative learning, collaborative writing, collaborative problem solving, and co-production of knowledge. In fact, blog- or wiki-based learning does not only stand to add collective intelligence to CL, CMing and CFT, but also stands to bring multiple perspectives and multiple intelligences (sourced from the crowd) to these three variables.

CONCLUSION

From the studies and experiments explored in this chapter, there appears to be a symbiotic relationship between CL, CMing and CFT. For instance, both CMing and CFT tend to facilitate CL. Similarly, both CL and CMing have the potential to foster cognitive flexibility. In particular, CMing serves as a valuable tool for assessment, curricular design/implementation, and higher-order thinking and problem solving skills. Likewise, CFT promotes complex and ill-structured problem solving and higher-order thinking skills through multiple case perspectives. Moreover, CMing develops coherent understanding in students and helps students revise learning material. Thus, CMs are versatile and multi-purpose cross-disciplinary tools.

REFERENCES

Basque, J., & Lavoie, M.-C. (2006). *Collaborative concept mapping in education: Major trends.* Retrieved January 15, 2008, from http://cmc.ihmc.us/cmc2006Papers/cmc2006-p192.pdf

Chaka, C. (2008). Revisiting knowledge management and learning organisation: Factoring in multiple intelligences and cognitive flexibility theory. In A. Koohang, K, Harman & J. Britz (Eds.), *Knowledge management: Theoretical foundations* (pp. 191-222). Santa Rosa, CA: Informing Science Press.

Darmofal, D. L., Soderholm, D. H., & Brodeur, D. R. (2002). *Using concept maps and concept questions to enhance conceptual understanding.* Retrieved March 11, 2008, from http://www.cdio.org/papers/concept_maps_fie.pdf

Gardner, H. (2003). *Multiple intelligences after twenty years.* Retrieved November 12, 2006, from http://pzweb.harvard.edu/Pls/HG_MI_after_20_years.pdf

Gouli, E., Gogoulou, A., & Grigoriadou, M. (2003). *A coherent and integrated framework using concept maps for various educational assessment functions.* Retrieved March 11, 2008, from http://jite.org/documents/Vol2/v2p215-240-23.pdf

Hill, L. H. (2004). *Concept mapping in a pharmacy communications course to encourage meaningful student learning.* Retrieved March 19, 2008, from http://www.ajpe.org/ aj6805/aj6805109.pdf

Johnson, D. W., & Johnson, R. T. (2007). *Cooperative learning and moral education.* Retrieved March 10, 2008, from http://www.co-operation.org/pages/Volume% 2022(1)%20Internet.doc

Joritz-Nakagawa, J. (2004). *Spencer Kagan's cooperative learning structures.* Retrieved March 10, 2008, from http://www.jalt.org/pansig/PGL2/HTML/Nakagawa.htm

Keller, T., Tergan, S.-O., & Coffey, J. (2006). *Concept maps as a "knowledge and information awareness" tool for supporting collaborative problem solving in distributed groups.* Retrieved March 18, 2008, from http://cmc.ihmc.us/cmc-2006Papers/cmc2006-p.125.pdf

Khamesan, A., & Hammond, N. (2004). *Synchronous collaborative concept mapping via ICT: Learning effectiveness and personal and interpersonal awareness.* Retrieved March 18, 2008, from http://cmc.ihmc.us/papers/cmc2004-212.pdf

Lane, C. (2000). *Implementing multiple intelligences and learning styles in distributed learning/IMS projects.* Retrieved May 04, 2007, from http://www.tecweb.org/styles/ imslsindl.pdf

Lanzing, J. (1997). *What is concept mapping?* Retrieved January 17, 2008, from http://users.edte.utwente.nl/lanzing/cm_home.htm

Lima, M., Koehler, M. J., & Spiro, R. J. (2004). Collaborative interactivity and integrated thinking in Brazilian business schools using cognitive flexibility hypertexts: The Panteon project. *J. Educational Computing Research, 31*(4), 371-406. Retrieved March 30, 2007, from http://mkoehler.educ.msu.edu/OtherPages/Koehler_Pubs/Lima_Koehler_Spiro_ 2004.pdf

Maas, J. D., & Leauby, B. A. (2005). *Concept mapping: Exploring its value as a meaningful learning tool in accounting education*. Retrieved March 03, 2008, from http://gpae.bryant.edu/~gpae/vol2/04-038%20Concept%20Mapping.pdf

McDaniel, E., Roth, B., & Miller, M. (2005). *Concept mapping as a tool for curriculum design.* Retrieved March 19, 2008 from http://proceedings.informingscience.org/ InSITE2005/142f49McDa.pdf

Mendes, E., Mosley, N., & Counsell, S. (2001). *The cognitive flexibility theory: An approach for teaching hypermedia engineering.* Retrieved March 30, 2007, from http://www.cs.auckland.ac.nz/~emilia/publications/ITICSE2001.pdf

Milam, J. H., Santo, S. A., & Heaton, L. A. (n.d.). *Concept maps for web-based applications: ERIC technical report.* Retrieved February 07, 2008, from http://www.highered.org/docs/milam-conceptmaps.PDF

Moni, R. W., & Moni, K. B. (2008). *Student perceptions and use of an assessment rubric for a group concept map in physiology*. Retrieved March 19, 2008, from http://advan. physiology.org/cgi/reprint/32/1/47

Nelson, D. B. (2007). *Academic concept mapping (ACM): A critical thinking tool in academic advising for improving academic performance in college freshmen*. Retrieved February 07, 2008, from http://etd.lsu.edu/docs/available/etd-06262007-190651/ unrestricted/DorothyNelsonDissertation.pdf

Novak, J. D., & Canãs, A. J. (2008). *The theory underlying concept maps and how to construct and use them.* Retrieved March 17, 2008, from http://cmap.ihmc.us/ publications/ResearchPapers/TheoryUnderlyingConceptMaps.pdf

Rodriguez, A. J. (2006). *Using concept maps as a tool for curriculum planning and student assessment in culturally diverse schools*. Retrieved March 19, 2008, from http://cmc.ihmc.us/cmc2006Papers/cmc2006-p166.pdf

Shuyska, J. A. (2007). *The CONTEXT tool – Going beyond the narrative*. Retrieved March 25, 2008, from http://www.reflectingeducation.net/index.php/reflecting/ article/viewFile/44/47

Spiro, R. J., Feltovich, P. L., Jacobson, M. L., & Coulson, R. L. (1992). Cognitive flexibility, constructivism, and hypertext: Random access instruction for advanced knowledge acquisition in ill-structured domains. In T. M. Duffy & D. H. Jonassen (Eds.), *Constructivism and the technology of instruction: A conversation* (pp. 57-76). Hillsdale, NJ: Lawrence Erlbaum Associates

Stoyanov, S., & Kommers, P. (2006). *WWW-intensive concept mapping for metacognition in solving ill-structured problems.* Retrieved January 17, 2008, from http://www.ou.nl/Docs/Expertise/OTE/Publicaties/slavi%20stoyanov/10%20Stoyanov1. pdf

Tifi, A., & Lombardi, A. (2006). *WWMaps, a community on education through collaborative concept mapping.* Retrieved March 07, 2008, from http://cmc.ihmc.us/ cmc2006Papers/cmc2006-p.92.pdf

Van Boxtel, C. (2000). *Collaborative concept learning - Collaborative learning tasks, student interaction and the learning of physics concepts.* Retrieved March 17, 2008, from http://edu.fss.uu.nl/medewerkers.gk/files/Boxtel_dissertation.pdf

Vilela, R. B., Austrilino, L., & Costa, A. C. (2004). *Using concept maps for collaborative curriculum development*. Retrieved March 19, 2008, from http://cmc.ihmc.us/papers/ cmc2004-239.pdf

KEY TERMS AND DEFINITIONS

Cognitivism: This epistemology treats conceptual understandings as a pattern of connections between similar elements (e.g., concepts) and learning as strengthening or weakening those connections (e.g., conceptual relations).

Coherent understanding: This is a term drawing strongly on the concept of cognitive flexibility. It puts emphasis on a bird's eye overview of the material and suggests a multifaceted picture of the learning process.

Constructivism: Even though there is no single constructivist theory, constructivism mainly contends that learners, through interacting with their social world, actively construct, test and refine knowledge.

Disequilibrium: Disequilibrium signifies that students experience uncertainty and search for information to clarify their understanding regarding material during the CMing process.

Equilibrium: This refers to regaining comfort with material by achieving understanding and the restructuring of mental schema held about the material.

Harvard Method: This is the Harvard Business School's case method meant to stimulate critical and creative thinking among students so as to help them solve problems and make decisions for organisational purposes.

Integrative Reconciliation: This is relational cross-links representing new relationships between new concepts and existing conceptual structure. Thus, integrating new concepts entails reconciling new information with the old information.

Panteon: It is a Portuguese acronym for *Applied Project of New Technologies for On-Line Case Studies*. It is a Web-based databank allowing business students to create and diagnose hypertextual case studies about complex organisational environments.

Progressive Differentiation: This is when new concepts are introduced first in one's conceptual framework and are then progressively differentiated in terms of detail and specificity.

Subsumption: This refers to when new information is linked with existing concepts and conceptual relations in a student's conceptual structure during conceptual change.

Chapter 9
Teaching Critical Thinking and Team Based Concept Mapping

Dawndra Meers-Scott
Texas Tech University, USA

LesLee Taylor
Texas Tech University, USA

John Pelley
Texas Tech University, USA

ABSTRACT

Critical thinking cannot be fully developed without involvement in collaborative learning activities that elicit problem solving dialogue. Concept maps are effective tools for dialogue because they require decisions about the organization of and the relationships between facts and concepts. This active decision making process develops both long term memory and the ability to apply that knowledge. The authors describe a new method for incorporating scored concept maps into an established collaborative learning method, Team-Based Learning, as a way to improve the effectiveness of individual preparation and for enhancing the problem solving dialogue during group activities. Their new method, Team-Based Concept Mapping, has advantages for students with different personality types and with different backgrounds because it provides greater clarity and precision in the group dialogue. The effect of concept mapping on the interaction between different personality types is discussed and suggestions for future studies to develop this method are offered.

INTRODUCTION

Collaborative learning is necessary to help students move beyond assimilative learning to become, as described by Mezirow (1991, p. 167), "...critically aware of how and why our assumptions have come to constrain the way we perceive, understand, and feel about our world..." This "critical awareness" by students is revealed to us when we observe the dialogue that occurs during team problem solving. Dialogue allows team members to hear alternative ways of perceiving a situation and to reflect on their own approach to solving a problem. During the collaborative learning process, the individual team members use the critical thinking skills of analysis, interpretation, inference, evaluation, and

DOI: 10.4018/978-1-59904-992-2.ch009

explanation (Facione & Facione, 1997) to reach decisions that produce a new understanding for all members. New understanding by the students, in turn, converts their "working memory" into long-term memory. Thus, students participating in the process of team problem solving avoid the pitfalls of assimilative learning which simply layers new information onto old understanding. The process of using current knowledge to create new knowledge has been termed transformative learning (Boyd & Myers, 1988), and this functional transformation is accompanied by a parallel anatomical transformation occurring in the brain itself (Zull, 2002). Collaborative learning physically transforms the brain by establishing a greater number of long lasting synaptic connections through the growth of nerve cell dendrites. This physical transformation occurs in two interactive areas of the brain: 1) the temporal area and 2) the prefrontal area. The temporal area accesses existing memory and adds to that memory when new learning takes place. The prefrontal area uses knowledge from the temporal area to establish new possibilities and to make logical decisions about them. The active use of both of these areas of the brain is necessary to develop critical thinking skills thus indicating that the growth of dendrites occurs both in the area of the brain that stores memory and in the area of the brain that uses that memory for decision making (Zull, 2002; Bransford, Brown, & Cocking, 2000).

Prior to our research on team problem solving, we discovered that individual preparation for participation in team problem solving exercises is enhanced by concept mapping. This is because the construction of a concept map requires analytic reading through the constant formulation of focused questions (Cañas & Novak, 2006). The back-and-forth process of asking a question ("Where does this go in my map?") and then answering it ("It is connected here…and here…") helps the student discover how new knowledge can be organized. This facilitates, in turn, the retrieval of this knowledge during the team dialog when each student must defend their decisions to the rest of the team members. The formulation of a rationale for suggesting new possibilities, or for choosing among optional solutions, requires more than recall knowledge of factual content. Such a rationale also requires an understanding of the meaning of factual content and the construction of a concept map reveals that meaning through patterns and organization.

Concept maps are effective tools for helping students understand their individual learning style and how their preference for processing information affects their individual learning strategy, both for remembering information and for using it in problem solving. Students who are either Sensing or Intuitive personality types, as determined by the Myers-Briggs Type Indicator (MBTI), process information very differently with dramatically different outcomes in learning and test achievement (Pelley & Dalley, 1997). If the Sensing type students follow their preferences, their learning is focused on facts and details that are committed to recall memory (a temporal area brain function). They learn in linear order and consequently do not spontaneously look for patterns and relationships. This is in contrast to their opposite, the Intuitive type student, that spontaneously seeks out new patterns and relationships (a prefrontal area brain function). Because they tend to emphasize the use of one area of the brain over others, both types can have learning "blind spots" that are corrected by the use of concept maps. The concept mapping process motivates the Sensing type to seek out relationships in order to construct a concept map complete with cross-links and it motivates the Intuitive type to seek out details that are overlooked when they focus on the "big picture." The use of concept maps has led to improvement in learning skills by both types of students in health care professional education (Pelley & Steele, 2002)

We have found an exciting new way to extend the use of concept maps in individual learning to group learning by incorporating them in a highly effective collaborative learning method, Team-

Based Learning (Michaelsen, Knight, & Fink, 2004). This learning system was developed to permit small group learning in large classrooms and utilizes multiple choice questions to not only score learning achievement, but to serve as the vehicle for dialogue in team problem solving. Although the process is highly engaging and students are able to debate and decide on their choice of answers, the discussions frequently suffer from difficulties in communication of knowledge. It occurred to us that the substitution of scored concept maps would provide a method for assessing learning while communicating knowledge more accurately and reliably. As a result, we describe here a modification of Team-Based Learning (TBL) that incorporates concept maps. We have given our method a similar name, Team-Based Concept Mapping (TBCM).

Upon completion of this chapter you will be able to:

1. Describe the different ways that students prefer to process information and how collaborative learning makes the most efficient use of those differences in the development of critical thinking skills.
2. Implement Team-Based Concept Mapping in your learning environment.
3. Develop improved methods for teaching the process of concept mapping and for evaluation of critical thinking using concept maps produced by groups.

BACKGROUND: PERSONALITY TYPE, CONCEPT MAPPING, AND TEAM-BASED LEARNING

Personality Type Influences on Teaching Critical Thinking Skills

It is logical that a learning tool such as concept mapping, designed to develop critical thinking skills (Novak and Gowin, 1984), is used differently by students who process information differently. These differences can be identified with the Myers-Briggs Type Indicator (MBTI), a personality instrument developed to sort preferences within several dimensions of thinking (Myers, McCaulley, Quenk, & Hammer, 1998). The Myers-Briggs personality types represent different mental habits used for information processing and, as such, they represent predictable aptitudes and attitudes with respect to the construction of concept maps. Additionally, they represent predictable aptitudes and attitudes with respect to the critical thinking process. Since the Myers-Briggs preferences do not represent limitations in thinking, they become opportunities for using concept maps in the development of critical thinking skills.

The Myers-Briggs Type Indicator is a psychological instrument that was developed to reliably determine the personality preferences first described by Jung (Myers, McCaulley, Quenk, & Hammer, 1998). When taken together, the mental functions observed by Jung describe mental habits that lead to a consistent thinking style in information processing. These mental functions involve: 1) information input, 2) generation of alternative possibilities, 3) deduction of the most logical alternative, and 4) assessment of human outcomes/values. Although all types use all of the functions in their learning, they are differentiated from each other by the emphasis of one of these functions over the others. If left unbalanced, the student's critical thinking skills will have strengths in the most used functions, but weaknesses in the less frequently used functions. Identification of the emphasized function allows the adoption of strategies to develop the lesser used functions.

The MBTI only measures normal differences in behavior that are established as mental habits, and thus does not measure abnormal behavior. Since all of the preferences that are measured are found in normal thinking, personality type is nothing more than a comfort zone where thinking occurs with less effort and with the greatest amount of trust. When a student of a given personality type uses a non-preferred mental function, for example at work, they use more mental energy and tire more easily. But, like any motor skill, use of the non-preferred side can be developed and mastered. Thus, no one is limited, or compartmented, by their type; the limitation is usually a lack of awareness of type preferences and, therefore, a lack of awareness of why some aspects of thinking are more difficult than others. The available data do not support a relationship between personality type and intelligence nor a connection with any psychopathology. Each dimension of type, as described below, is exercised to different degrees by individuals of the same type preference due to the influence of other personality traits and life circumstances (Myers, McCaulley, Quenk, & Hammer, 1998).

Extraversion and Introversion Preferences

Students differ in their preference for whether their best thinking is achieved through "talking it out" or through "thinking it through." Those students who have extraversion preferences will have an easier and more effective learning experience if they can verbalize their learning as it is happening. Although verbalizing learning is impractical in some learning settings, such as lecture presentations or seminar settings where verbalization would be disruptive, it has an energizing effect on study groups where verbalization is expected.

The opposite preference for introversion is seen in students who have a more effective learning experience when they are able to process new information quietly before it is discussed. While this type of learning can lead to isolation from others, if used regularly in a study group it will bring depth of thinking to the group process. Thus, extraverts talk-to-think while introverts think-to-talk.

Sensing and Intuitive Preferences

Students differ in their preference for the way they give their attention to new information. Those students who have sensing preferences tend to trust information that is perceived directly by the senses, i.e. vision, hearing, touch (manipulation), taste, and smell. This information exists in the present as facts and details and carries a high degree of certainty. If a pattern or relationship exists, it is also perceived as a fact...but is only "discovered" as a relationship with great effort. When studying in a group, the sensing student is always alert to completeness of the facts.

The opposite preference for intuition leads students to trust their ability to find patterns and relationships. This information exists in the future as a possibility and several ways of organizing these relationships might be perceived. While a minimal set of facts is needed to form a pattern, once the pattern is decided the remaining facts are ignored by the intuitive type student. When studying in a group, the intuitive students help with the discovery of patterns and bring the alternative points of view that are essential to critical thinking. Thus, sensing types think about "what is" while intuitive types think about "what if."

Thinking and Feeling Preferences

Students differ in their preference for the way they react to or make a decision with new information. Those students who have thinking preferences tend to trust their logic to evaluate the facts and possibilities. They are impersonal and objective in their analysis since they seek to obey the laws of deductive and inductive reasoning. While they have feelings, they do not trust their feelings in reaching a decision or making a judgment. When studying in a group, the thinking types draw attention to the "correctness" of relationships and the clarity of thinking.

The opposite preference for feeling judgment leads students to trust learning decisions based on personal and subjective analysis. This analysis is still a rational process, but it references against human outcomes rather than laws of logical reasoning. When confronted with a conflict between a logical alternative or a personally valued alternative, the feeling type will choose the latter, even while realizing the former is also valid. When studying in a group, the feeling type students bring harmony and enhance communications. Thus, thinking types need to include feelings and values as facts in their thinking while feeling types need to realize that thinking types have difficulty trusting feelings and values.

Judging and Perceiving Preferences

Students differ in their preference for the way they conduct their learning activities. Those students who have judging preferences tend to organize their time around a plan. They are motivated to obtain closure by completing their plan or checking off items on their task list. They will tend to sacrifice learning additional information if that learning prevents them from completing their schedule. When studying in a group, they keep the group on task and help it to be more efficient.

The opposite preference for perception leads students to conduct their learning in a flexible and adaptable manner. They are motivated to discover new information that makes a more complete set of facts or a more complete pattern. They may attempt to follow a schedule, but they will value the acquisition of additional information over meeting a deadline. When studying in a group, they keep the group open to new information in resolving learning issues. Thus, judging types seek "the joy of closure" while perceiving types seek "the joy of discovery."

Concept Mapping as a Tool for Dialogue

Individually constructed concept maps have been used to understand the development of team cognition (O'Connor, Johnson, & Khalil, 2004). Aggregate team concept maps were synthesized from separate individual maps that were constructed by team members at predetermined stages of a team task. The aggregate maps were considered to represent a shared mental model that represented team understanding. While this study showed that aggregate concept maps are one way to visualize group knowledge and understanding, it was not designed to assess the effect of the process of constructing a concept map on the group dialogue. We have studied concept mapping during the team task as a method to help focus dialogue while producing a shared mental model.

Our earlier work involved group study sessions where individually constructed concept maps were discussed. We found that the visual construct provided an effective vehicle for students to explain their own thinking and to share it with others. However, it was noted that the dialogue rarely ended with the maps in their original condition. Instead, the groups questioned existing structure and proposed alternative structures without prompting from faculty mentors (Pelley, 2006). The repeated spontaneity of this process caused us to organize a more structured process to facilitate the dialogue between team members by assigning construction of group maps from previously constructed individual maps as a team task (Scott, Taylor, & Pelley, 2006). The dialogue during construction of the group map involved discussion not only of the informational content to be learned but the rationale for the organization of this content, i.e. branching and cross-linking. We found that the contribution from Sensing type students in the group, at first, centers on the content of facts with less attention given to cross-link relationships between branches. In contrast, the contribution from Intuitive type students, at first, centers on the relationships and novel approaches to organization of facts and concepts. If the same group meets for several sessions, both the Sensing type students and the Intuitive type students display more balance in their thinking and demonstrate the skills of their opposite type. The reason for this is that, while personality type preferences are relatively constant, the development of thinking skills in the non-preferred mental functions is not restricted (Pelley & Dalley, 1997). Sensing types gain a "big picture" perspective by discovering and constructing organizational and conceptual relationships and Intuitive types gain a greater grasp of the facts by attaching them to their "big picture" concept map. This helps to avoid the problem of construction of primarily descriptive maps in place of the desired explanatory maps as cited by Cañas and Novak (2006). They point out that the choice of focus questions can prevent a map from serving as a classification of facts but, rather, can influence the student to construct a dynamic map that shows explanations of cause and effect.

Concept maps can be effective in facilitating the maturation of a team from a collection to a community. While a collection of students can be directed to work together, they do not achieve synergy until they trust each other as do members of a community. Team maturation has been reported to progress through four stages before students develop the trust needed for effective critical thinking skills (Tuckman, 1965). During the first two stages of team maturation, "forming" and "storming," the dialogue that is critical to effective team learning is less efficient and thus more difficult to bring to a focus. These early stages are characterized by the uncertainty of the students about their role and concern for their compatibility with other team members, thus diverting attention away from the team task. Even the more productive latter stages, "norming" and "performing," are vulnerable to inefficiency if team members are unclear in explaining their rationale for a particular decision. Our research has been aimed at addressing these issues with team dialogue by using concept mapping not only as a team task but as a vehicle for clarity in communication. We want to facilitate the maturation of teams as healthy, trusting communities that can quickly develop a focus and effectively reach a decision. We have found that team concept mapping indeed helps teams develop a more immediate focus for discussion and produces a more effective and efficient learning process.

Team-Based Learning as a Collaborative Learning Process

Team-Based Learning (TBL) is a unique collaborative learning approach that has been adopted around the world in business and science courses (Michaelsen, Knight, & Fink, 2004) and recently in health care education (Michaelsen, Parmelee, McMahon, & Levine, 2007). TBL brings together theoretically-based and empirically-grounded strategies for ensuring the effectiveness of small-groups working independently in classes with high student-faculty ratios (e.g., up to 200:1) without losing the benefits of faculty-led small groups with lower ratios (e.g., 7:1). The opportunity for whole-class inter-group dialogue confers a distinct learning advantage over methods where the groups work in isolation from each other. TBL can be used in conjunction with other teaching modalities to provide an efficient system for developing critical thinking skills in a team problem solving setting.

One of the primary characteristics of TBL (Michaelsen, Knight, & Fink, 2004) is that all of the activities involve teacher constructed multiple choice tests. In the TBL method, all teams work on the same multiple choice problems at the same time. This maximizes the opportunity for two valuable comparisons: 1) team members can compare the knowledge that each has brought to the session and 2) each team can compare their own performance with how other teams used their knowledge to develop a rationale for solving the problems. Thus, the teacher-directed whole group discussion establishes a common understanding of the basic concepts and facts and how they are applied in more challenging case-based multiple-choice questions.

This collaborative learning method balances individual and group accountability by including both individual test performance and group test performance in the final score. The requirement for individual test performance prior to the beginning of group activities maintains the learning responsibilities that are expected of each student. This assures that group morale and team motivation don't suffer from one or more students arriving for the group activities without making an effort to prepare. However, it is not a safe assumption that students who responsibly attempt to prepare for a team exercise have done so competently. Many students study in vain, unable to comprehend critical facts or organization. Thus, this otherwise highly effective collaborative learning process is vulnerable to the effectiveness of the individual learning skills of each student. We addressed this vulnerability by merging an effective method for individual preparation, concept mapping, with a proven method for collaborative learning. Because the student uses concept maps for both individual preparation and for participation in group dialogue, we have named this method "Team-Based Concept Mapping." Since the assigned material has already been organized into a concept map, the degree of mastery of the material can be assessed by deriving a score using a grading system weighted so that links representing higher order thinking, e.g. cross-links, receive a greater number of points. The substitution of scored concept maps for the multiple choice test assessment maintains the individual accountability for each student and provides a motivation for the deep analytical reading needed to find the more complex links to include in their maps. Thus, by substituting scored concept maps for multiple choice tests, we assure that individual preparation will be maximized while maintaining all of the important collaborative learning features of Team-Based Learning.

MAIN FOCUS OF CHAPTER: TEAM-BASED CONCEPT MAPPING

In order to describe the process of Team-Based Concept Mapping (TBCM) we will compare it to the established process of Team-Based Learning. This will further detail the advantages of TBL as a unique method of collaborative learning and will permit us to emphasize the advantages of incorporating concept mapping as part of the process. A section on Fundamentals will be followed by:

1. Teaching the process of concept mapping
2. Scoring concept maps
3. Composing teams
4. Conducting a TBCM session

Fundamentals

The format of Team-Based Learning as originally developed by Michaelsen, Knight, & Fink, (2004) is a structured sequence consisting of 3 phases. In Phase 1, students study independently outside of class to master learning objectives designed for the session. Phase 2 begins the classroom activities as individual learners complete a multiple-choice test to assess their readiness to apply knowledge from Phase 1. This "readiness assurance test" contributes points toward their final score. Then, groups of 5-7 students re-take the same test and turn in their group consensus answers for immediate scoring and posting toward their final score. As with typical examinations, outside reference materials are not used during the readiness assurance exams. At this point, the teacher leads a whole class discussion that enables clarification of concepts where needed. Phase 2 takes approximately one hour. In Phase 3, groups complete in-class case oriented assignments that require collaboration to answer multiple choice questions concerning the case. Answer choices for these questions are constructed with some ambiguity so that groups must select the most correct choice, thereby stimulating vigorous discussion and debate. At designated times, all groups are led by the teacher to simultaneously share their groups' answers with the entire class for easy comparison and immediate feedback. This stimulates an energetic total-class discussion with groups defending their answers and the teacher helping to consolidate learning as needed. Phase 3 takes from 1 – 2 hours. A complete TBL session addresses all three of the recommendations of the National Research Council that produce effective learning: 1) addressing pre-existing understandings, 2) teaching some subject matter in depth, and 3) teaching metacognitive skills that aid reflection on learning (Bransford, Brown, & Cocking, 2000). For a more thorough description of the TBL process, the reader should consult either Michaelsen, Knight, & Fink (2004) or Michaelsen, Parmelee, McMahon, & Levine (2007).

Our adaptation of TBL substitutes scored concept maps for the multiple-choice readiness assurance tests. This has shown promise for facilitating group maturation and more effective use of higher order thinking skills by all students. The student first develops concept maps from the individual preparation in Phase 1 (see Figure 1). The content of the maps is guided by the learning objectives which serve in the role of focus questions. As mentioned previously, effective focus questions are one of the essentials in effective map construction as discussed by Cañas & Novak (2006). Phase 2 of TBCM begins with the exchange of maps between classmates for scoring. This step has some advantages over a multiple choice test. First, each student has a motivation during Phase 1 to construct a map that is readable and readily interpreted by a randomly assigned classmate. This can encourage students to study together ahead of time to refine their maps for scoring during Phase 2. When more thought goes into the concept map, reading is deeper and more analytical. A second advantage of this scoring method is that each student is using the first part of the session to compare another student's thinking, as represented by their concept maps, with their own. This helps to mentally prepare them to enter the discussion during the group activities. Students have the opportunity to appeal the scoring of their maps with review by the teacher at a later date. This process takes longer to complete than scoring a multiple choice test and we have found that, rather than extend the time of the session, it is more effective to simplify the concept maps by narrowing the focus and reducing the cognitive load for the session. For example, instead of using an entire chapter from a text as the source material for the TBCM session, limit the exercise to a section within the chapter. It is important to emphasize that the goal of both TBL and TBCM is analytical problem solving, not comprehensive review.

Phase 2 continues after the score for the individual maps has been recorded with each team collaborating to complete a group concept map (See Figure 2). Just as in the original TBL method, the groups receive a score on their "team map." The maps are turned in for scoring at a later date by the teacher, but are posted in the classroom for reference and discussion of any concepts that remain unclear. The teacher has a similar opportunity in this step for review and clarification of the topic by referring to each of the group maps to point out strengths and opportunities for improvement. This is comparable to the systematic whole-class review in TBL by the teacher of each question on the "readiness assurance" multiple choice test at the end of Phase 2.

Phase 3 involves team analysis of one or two cases by the teams. The team score for each case is based on two factors: 1) answering one or more multiple choice questions concerning the case and 2) concept maps that illustrate analysis of the case. Teams should be encouraged to construct two separate maps, one that shows the relevance of each important element in the case and one that shows the relationship for each answer choice with the elements of the case. The first map directs discussion of the group to the material they have just reviewed in class and begins the process of

Figure 1. Reproduction of an original individual concept map of a reading assignment in "ethics." The student was only able to identify three cross-links with the remaining elements of the map in a branched hierarchy. This student has sensing preferences in her learning and thus does not seek out these relationships. The scoring is discussed later in the chapter below

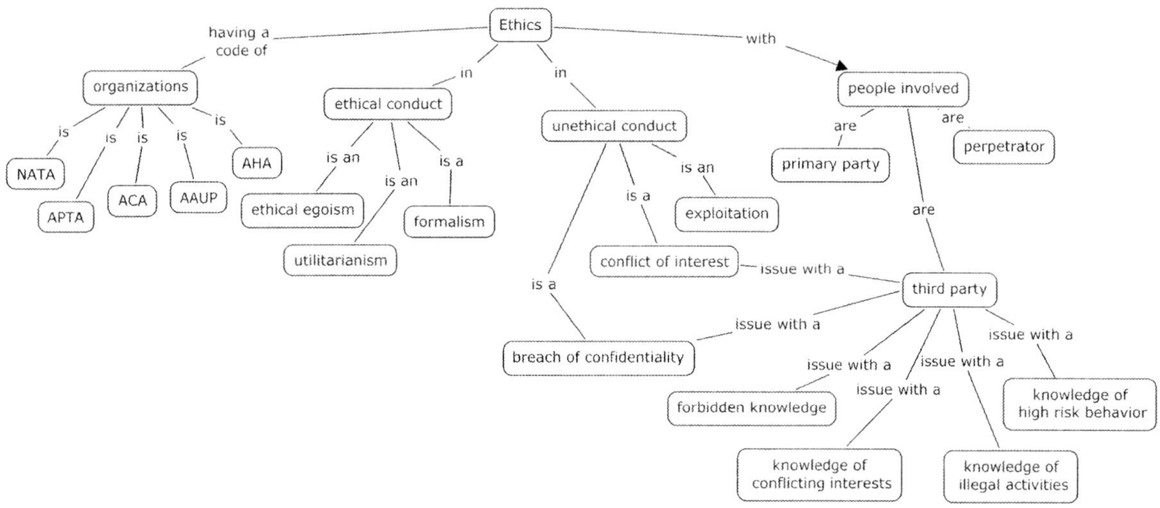

Figure 2. Reproduction of an original group concept map of a reading assignment in "ethics." Many more cross-links are present indicating a stronger input by intuitive type members of the team who tend to see relationships more readily. The scoring is discussed later in the chapter

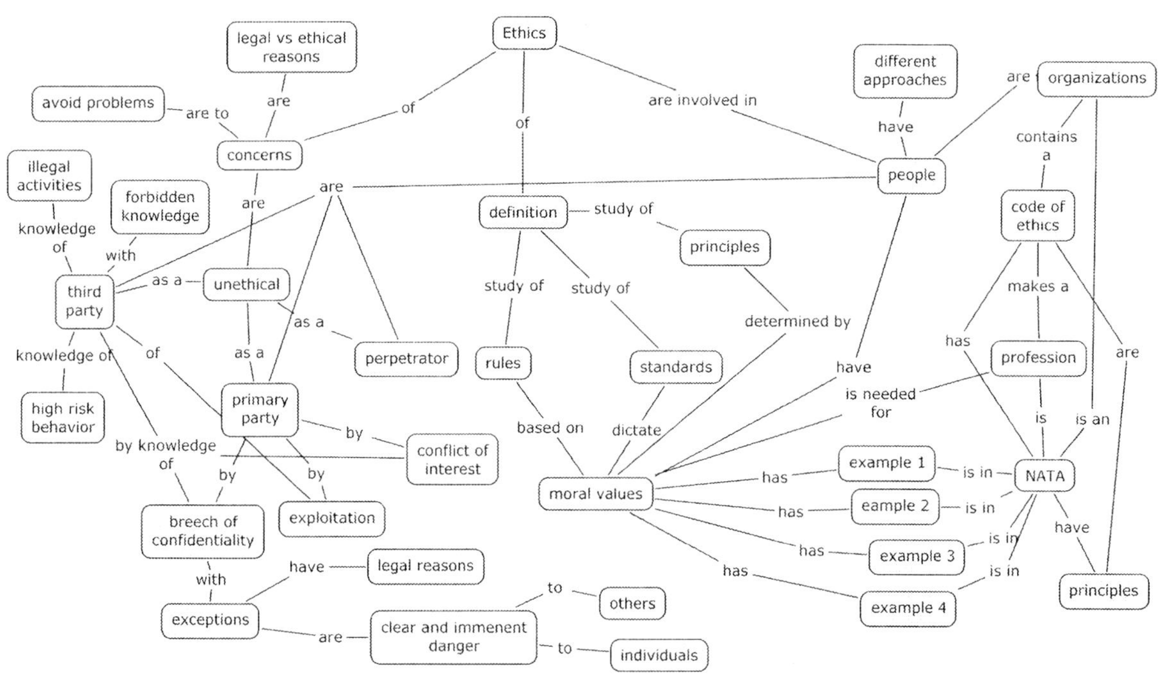

prioritizing importance and relevance. The second map uses deductive reasoning to associate the relevant elements of the case to each answer choice. This second step is the most powerful in eliciting higher order thinking because it is a visual demonstration that, while the greatest number of links will be associated with the correct answer, other incorrect answers also have some links associated. This is of great importance to the Sensing type student that expects all answers to be either totally correct or totally incorrect.

By the end of the TBCM session, all students have had an opportunity to discuss facts and concepts with a common visual focus and to develop the greatest number of relationships within the topic area. The discovery of relationships is a thinking skill that will carry over into all future learning for each of the team members. The analysis of the relevance of different elements of a case is aided by a common visual focus to illustrate that most real world problems are not "black-or-white."

Teaching the Process of Concept Mapping

Because Intuitive types and Sensing types react differently to the process of concept mapping, it is important to include training sessions prior to conducting TBCM. This is accomplished effectively with a mock TBCM session so that the training occurs in a setting that begins training in team problem solving skills.

We teach concept mapping as a communication tool. When the student is studying alone, concept maps help authors or teachers to communicate with them through the written word. Similarly, when the student is learning in a group, concept maps help them communicate with team members through the spoken word. Communication can become very frustrating when Intuitive type students are talking about patterns and relationships that the Sensing type students missed and when Sensing type students talk about details and facts that the Intuitive type students missed. The following approach will help in training all students to become proficient with concept mapping so that each type can see what the other type is trying to say.

"List-Group-Compare" – Students can commit this simple phrase to memory for use during individual study. It reminds them to follow a sequence that guides reading in a way that facilitates construction of a concept map. The first step in the sequence is a reminder to survey the material and pull out terms that are more general and inclusive. Even though this appears to be a simple and obvious step, the Sensing type student will in many cases overlook that there are more general terms that are different in nature than other more specific terms. This type of student can be so focused on details and facts that all are given equal importance. This serves as an initial frustration in construction of a proper concept map. It serves as an equally powerful motivation when this type of student "sees" the reading in a new light through concept mapping.

The grouping step involves extracting the first terms to be included in the map and beginning the organization of the map itself. This should be illustrated first by the teacher with constant reference to the list. Reassurance is important at this stage for the students to understand that the initial list will become more accurate and useful over time. Additionally, it is important for Sensing type students that the map is constructed from the top-down and not center-out. It is of no consequence to Intuitive type students where the map begins, so it gives a common ground for ease of understanding and dialogue if the top-down model is used. If necessary, a left-to-right map will also be easier for Sensing type students to construct and interpret.

Sensing type students have a need for certainty that affects their acceptance of and proficiency with concept mapping. This need can be satisfied readily if the teacher reviews their maps with them to help support the overall process. It is always important that the teacher does not show the student how to map the material, but instead asks the student about their thinking. Students can converse with each other about their maps and acquire this sense of certainty through the dialogue. Whenever a teacher shows all or part of a map, they are showing their thinking and missing what the student is thinking.

The comparing step overlaps with the grouping step as the map is completed in greater detail. At this stage students are asked to complete the map by including all subtopics within their main topic branches and to seek out comparisons, i.e. relationships, between major branches that are represented as cross-links. This is a natural step for the Intuitive type student and an extra step for the Sensing type student. Many Sensing type students never seek out relationships and rely on the teacher to point them out as additional facts to memorize. The process of seeking out relationships is taken for granted by Intuitive types, but it is a skill that can be learned by Sensing types.

Scoring Concept Maps

A simple scoring system will drive the desired behavior of seeking out integrative relationships in the learning assignment. We have used a modification of the scoring system reported by West, Pomeroy, Park, Gerstenberger, & Sandoval (2000).

1. Links (1 point). Any two concepts or facts, enclosed to form nodes of a variety of shapes, correctly connected receive one point to reward the acquisition of factual knowledge. It has a special value for the Intuitive type students who tend to overlook facts that do not contribute a "big picture."
2. Branch points, excluding convergences (5 points). Nodes that branch into two or more links receive 5 points to recognize the discovery and documentation of an indexing hierarchy. This is a higher order thinking skill that is needed during problem solving and this score emphasizes the importance of organization of knowledge over simple memorization.
3. Cross-links, including convergences (10 points). Links between branches, including links that converge on a common node receive 10 points, instead of 1 point, to recognize the discovery of valid comparisons between branches. This develops the ability to identify cause-and-effect relationships, similarities, and differences. This is most valuable to the Sensing type students who tend to overlook patterns and relationships.

The scoring for the map in Figure 1, above, produced: 22 links, 5 branches, and 3 cross-links yielding a final score of 77. The scoring for the map in Figure 2, above, produced: 42 links, 9 branches, and 9 cross-links yielding a final score of 177. The additional 100 points represents the additional learning produced when students of different learning styles discuss the same reading assignment.

We have noticed from subjective feedback by our students that the identification of cross-links to achieve a higher score is a function of the number of branch points. Likewise, the establishment of valid branch points which are more valuable than simple facts are a function of the number of facts identified. Thus, the higher scoring components of the maps drive the learning of the lower scoring components.

Composing Teams

It is very important to the development of team maturation, that care be given to the composition of teams. It has been shown by Michaelsen, Knight, & Fink (2004) that when students self select into teams, maturation is delayed. The reason for this is that self selection tends to compose teams of students who think alike. This deprives a team of the varied intellectual resources needed to solve problems. Instead, the recommendation is that teachers direct the composition of teams so that resources are distributed as evenly as possible. For example, a science course would want to distribute students who are science majors evenly among the groups followed by students with some science background and finally with those remaining students with minimal science background. This type of heterogeneity has been shown to produce the fastest and strongest team maturation.

One caveat regarding insights gained from the Myers-Briggs research is that type only indicates a preference for a mental function and does not measure the degree to which the preference is developed as a thinking skill. Just as students of the same intelligence use their intelligence differently and students with the same life or work experience have used that experience differently, so do students of the same type use their type skills differently. Thus, it is better not to use personality type to compose teams but rather, after the teams are composed, to discuss how each member can use their type best. Placing the focus on best use of type encourages personal (and thus professional) development instead of creating a stereotypical expectation.

Research shows that group size for this type of collaborative learning has a range of 5-7 students (Michaelsen, Knight, & Fink, 2004). This is smaller than the generally accepted size of 8-10 members for small group work. Groups that are too small generally don't have enough different contributions to effectively apply their knowledge in solving problems and groups that are too large break down into subgroups so that collaboration is reduced.

Conducting a TBCM Session

After the students have received training in the fundamentals of preparing concept maps, they can start their individual preparation for the first TBCM session. Since the construction of concept maps is a learned skill, students will demonstrate increased proficiency and higher scores after several TBCM sessions are conducted. Much of their skill development will occur during the TBCM session itself as the students compare and defend their maps.

The TBCM session starts with each member handing copies of their concept maps to another teacher-assigned student in the group. This allows rotation among different members of the group from session to session. After the maps are scored, the maps are then scored by one more member of the group to check for discrepancies. If a student wishes to appeal a score, time does not have to be taken from the session. Copies of the scored maps can be retained by the student who can then prepare a written appeal to be turned in for a decision by the teacher at a later date. The appeals process is an important part of the development of mapping skills and should be encouraged.

After the individual maps have been scored and turned in, each team starts the construction of their group map. In our limited experience with this step, we have found that the usual strategy is for team members to agree on one map as a "core" to which other elements from the remaining maps are added. Subjective data indicate that students derive a great deal of satisfaction from their learning at this stage since they can see the thinking of others who studied the same material. When the teacher has determined that the group maps have been completed, they are turned in for scoring by the teacher or an assistant at a later date. The teacher uses the group maps to diagnose any concepts that may need further clarification.

Groups are now permitted to use any resources they wish in analyzing a case. The teams are given a case and one or two multiple choice questions concerning the case to answer. The answer choices are chosen by the teacher to correlate with the case information to different degrees, thus requiring teams to discriminate the "best-fit" answer choice. The teams first create a new map of the case information to establish what is known about the problem and to evaluate its relevance. These maps will be used by the teacher for a subjective assessment of the team rationale in evaluating the information in the case. The teams then create a second map that links the greatest amount of case information to each answer choice. The goal is to determine the answer choice with the greatest number of links with the case information, thus forming the strongest rationale. When all maps are completed and turned in, the teacher conducts a discussion session where groups compare and defend their answers. This process is very comparable to that of TBL with the exception that the team problem solving dialogue has been facilitated by the use of concept maps. The teacher led discussion brings out the rationales used by each team so that inter-group dialogue can occur, further enhancing the development of higher order thinking skills.

FUTURE TRENDS: TEACHING SKILLS AND EVALUATING RESULTS

Both the economy of faculty resources and the orientation toward collaborative problem solving provide a large return on investment in the college classroom. This will drive a greater adoption of this teaching method in a greater variety of educational settings. In our experience, there will be several factors that may serve as barriers and, therefore, need to be better understood.

1. Personality Type Influences on Concept Mapping Behavior

The motivation to learn and use concept mapping skills is dramatically different for sensing and intuitive type students. Our experience has shown that sensing type students resist concept mapping because the construction of a map requires them to abandon their usual linear reading style. This resistance is overcome by metacognitive training that teaches them not only about their own learning but about how their learning compares with other learning styles. When coupled with step-by-step training in map construction, these students become devoted converts. Their personalities do not change, but instead they become more balanced in their learning skills.

The research challenge suggested by these observations is to understand more completely the most efficient approach in teaching both intuitive and sensing types the necessary metacognitive insights into concept mapping as a learning tool.

2. Mapping as a Dialogue Facilitator

Collaborative learning requires more than simple dialogue. It requires clear and unambiguous dialogue. Our experience has shown that students acquire both an increased clarity in their dialogue and an increased enthusiasm for the collaborative process (Pelley, 2005). When the collaborative effort involves students with both sensing and intuitive type preferences, it helps both preferences understand the thinking of their opposite. It is important to take research design into account regarding the impact of personality type on learning as evidenced by the study reported by Laight (2004). This study showed no significant influence of learning style as self-reported by pharmacy students on the utility of concept maps, but the study supplied the students with maps prepared in advance. This study also showed the student's reaction when the faculty revealed their own thinking to the students, but it did not reveal how receptive the students were to constructing their own maps or whether maps helped them discuss their learning.

The research challenge is to document more objectively the effects of concept mapping as a tool for organizing group dialogue and to avoid the use of maps as an additional representation of the instructors' knowledge. Our data are only from subjective evaluations and direct observation at this point. The development of instruments that document time on task for each team member as well as other behaviors related to collaborative problem solving are needed.

3. Scoring and Evaluation of Maps

The method for scoring the content of a map appears to be simple and straightforward when provided to the students. The reality, however, is that scoring requires judgment and thus improves with experience. Students are at different stages of understanding when they construct maps and their maps reflect that thinking. The inspection of one student's map by another student is a moment of truth where learning and knowledge are exposed. In the end, scoring drives analytic reading.

The scoring process as we have implemented it in Team-Based Concept Mapping needs further study to improve accuracy and objectivity. We have found that members of the same team are naturally supportive and this eases the tensions. It would be important to know what steps can be taken to hasten the scoring skills and maintain accuracy. We want to compare the scoring of individual maps when conducted by members of other teams to determine if there is a scoring bias due to team loyalty or to peer pressure from members of the same team. We also see a need to assess the effect of awarding points for finding mistakes. While this sets up a competitive attitude, it could be a stronger incentive for analytic reading since all students will use the same learning objectives to prepare for the session where they will be scoring concept maps.

4. Peer Evaluation Within Teams

Peer evaluation is an integral part of the original derivation of Team-Based Learning. It is intended to provide an additional control on individual preparation so that no single team member can rely on others to make up for their lack of preparation. If this is detected by team members, peer evaluation provides a mechanism for removing some credit toward their final score. We think that training in peer evaluation can be incorporated into the early stages of team training.

We have not yet employed peer evaluation in our TBCM sessions. Future studies will help to show the most effective format for training and implementation of peer evaluation in this setting.

CONCLUSION

Concept maps are a tool for enhancing communication. When used during individual preparation for collaborative problem solving, they facilitate communication between the author of the text or the teacher of a lecture and the student. When used during the group process in collaborative problem solving, they facilitate communication between the team members. Our system of Team-Based Concept Mapping uses scored individual and group concept maps to achieve a more complete, accurate and analytical dialogue. The construction of concept maps during individual preparation will help students regardless of different learning styles acquire a better grasp of the information and thus it will help them participate more effectively in the team dialogue. The consolidation of individual maps during the early steps in group tasks will allow team members to synthesize and strengthen their grasp of the material before applying it to the case-based problems. Concept maps further enhance the problem solving dialogue by providing a tool for collaborative discussion of the relative importance of information given in the case and for prioritizing the answer choices for questions concerning the cases. Since case problems contain real-world ambiguity that can suggest more than one possible solution, each team is forced to prioritize their choice. Collaborative agreement on cross-links between answer choices and information mapped from the case will elicit and develop the ability of students to prioritize the importance of information to support their conclusions. It is this prioritization process that develops higher order thinking skills and communication skills.

REFERENCES

Boyd, R. D., & Myers, J. G. (1988). Transformative education. *International Journal of Lifelong Education, 7*, 261–284. doi:10.1080/0260137880070403

Bransford, J. D., Brown, A. L., & Cocking, R. R. (Eds.). (2000). *How People Learn: Brain, Mind, Experience, and School*. Washington, DC: National Academy Press.

Cañas, A. J., & Novak, J. D. (2006, September). *Re-examining The Foundations For Effective Use Of Concept Maps.* Presented at the Second International Conference on Concept Mapping, San Jose, Costa Rica.

Facione, N., & Facione, P. (1997). *Critical thinking assessment in nursing education: An aggregate data analysis*. Millbrae, CA: California Academic Press.

Laight, D. W. (2004). Attitudes to concept maps as a teaching/learning activity in undergraduate health professional education: influence of preferred learning style. *Medical Teacher, 26*(3), 229–233. doi:10.1080/0142159042000192064

Mezirow, J. (1991). *Transformative dimensions of adult learning.* San Francisco, CA: Jossey-Bass.

Michaelsen, L. K., Knight, A. B., & Fink, L. D. (Eds.). (2004). *Team-based learning: A transformative use of small groups in college teaching.* Sterling, VA: Stylus Publishing, LLC.

Michaelsen, L. K., Parmelee, D. X., McMahon, K. K., & Levine, R. E. (2007). *Team-Based Learning for Health Professions Education.* Herndon, VA: Stylus Press.

Myers, I. B., McCaulley, M. H., Quenk, N. L., & Hammer, A. L. (Eds.). (1998). *MBTI manual: A guide to the development and use of the Myers-Briggs Type Indicator* (3rd Ed.). Palo Alto, CA: Consulting Psychologists Press, Inc.

Novak, J. D., & Gowin, D. B. (1984). *Learning how to learn.* New York: Cambridge University Press.

O'Connor, D. L., Johnson, T. E., & Khalil, M. K. (2004, September). *Measuring team cognition: Concept mapping elicitation as a means of constructing team shared mental models in an applied setting.* Presented at the First International Conference on Concept Mapping, Pamplona, Spain.

Pelley, J. W. (2005). *Concept Mapping: A Tool For Teaching Integrative Thinking.* Presented in the International Association of Medical Science Educators webcast series: Learner Centered Education. Retrieved from http://www.iamse.org/development/2005/was_2005_spring.htm.

Pelley, J. W. (2006, September). *Effect of concept mapping on Myers-Briggs personality types*. Presented at the Second International Conference on Concept Mapping, San Jose, Costa Rica.

Pelley, J. W., & Dalley, B. K. (1997). *SuccessTypes for medical students: A program for improving academic performance*. Lubbock, TX: Texas Tech University Division of Extended Learning.

Pelley, J.W., & Steele, M. (2002, April). *Teaching tools for medical knowledge competency: Concept mapping.* Presented at meeting of the Society for Academic Continuing Medical Education, Charleston, SC.

Scott, D. M., Taylor, L. L., & Pelley, J. W. (2006, September). *The use of concept mapping in integrative learning with allied health profession students.* Presented at the Second International Conference on Concept Mapping, San Jose, Costa Rica.

Tuckman, B. W. (1965). Developmental sequence in small groups. *Psychological Bulletin, 63*, 384–399. doi:10.1037/h0022100

West, D. C., Pomeroy, J. R., Park, J. K., Gerstenberger, E. A., & Sandoval, J. (2000). Critical thinking in graduate education: A role for concept mapping assessment? *Journal of the American Medical Association, 284*(9), 1105–1110. doi:10.1001/jama.284.9.1105

Zull, J. (2002). *The art of changing the brain.* Sterling, VA: Stylus Publishing, LLC.

KEY TERMS AND DEFINITIONS

Concept Map: A visual construct composed of encircled concepts (nodes) that are meaningfully inter-connected by descriptive concept links either directly, by branch-points (hierarchies), or indirectly by cross-links (comparisons). The construction of a concept map can serve as a tool for enhancing communication, either between an author and a student for a reading task, or between two or more students engaged in problem solving.

Critical Thinking Skills: The ability to solve problems by generating alternatives from existing facts and to prioritize these alternatives with respect to their logical justification and/or human outcomes. Both recall skills and higher order thinking skills are utilized in this process.

Group Concept Maps: Concept maps constructed through active collaboration among group members. More content and cross-links result due to contribution by different learning styles that bring different knowledge to the task.

Intuitive Type: A preference for perceiving new information by discovering patterns and relationships in an integrative sense. This type tends to construct and interpret concept maps easily and to include more cross-links in their maps.

Myers-Briggs Type Indicator: A psychological instrument that determines preferences in normal thinking processes. Provides a self-assessment that indicate constructive strategies for personal development.

Sensing Type: A preference for perceiving new information by observing details and facts in a literal sense. This type tends to be uncomfortable when either constructing or interpreting concept maps.

Team-Based Concept Mapping (TBCM): Team-Based Learning with scored concept maps substituted for multiple choice exams. Concept maps become a tool for more effective dialogue.

Team-Based Learning (TBL): A three step process that progresses from individual learning to collaborative learning by teams in large classroom settings. Requires specific decisions on the same problem by all teams in order to share rationales.

Team Maturation: A change in the nature of the interactions and dialogue between members of a team over time. Dialogue is more meaningful and effective as members develop trust in each other.

Transformative Learning: The process of using information received from a teacher and creating new knowledge by using higher order thinking skills.

Chapter 10
Intersubjective Meaning-Making in Dyads using Object-Typed Concept Mapping

Josianne Basque
LICEF Research Center, Télé-université, Canada

Béatrice Pudelko
LICEF Research Center, Télé-université, Canada

ABSTRACT

In this chapter, we investigate, with an intersubjective epistemology approach, how a concept mapping software tool that integrates a typology of knowledge objects (nodes) and a typology of links mediates the process of meaning-making and of meaning-negotiation of a dyad of adult learners engaged in a collaborative concept mapping activity, more specifically in the context of a text comprehension task. This case study shows that the tool and its object-typed concept mapping language induce certain types of epistemic actions as well as the formation of diverse representational rules by participants, which were jointly and progressively elaborated by them in an intensive effort to share meaning.

INTRODUCTION

Combining the advantages of the learning strategy of concept mapping[1] (CM) with those of collaborative learning, collaborative concept mapping (CCM) has become a topic of interest for an increasing number of researchers in the field of education (Basque & Lavoie, 2006; Gao, Shen, Losh, & Turner, 2007; Kim, Yang, & I-Chun, 2005; Nesbit & Adesope, 2006).

A close examination of the methodologies of 39 studies reported in our own review of research in this field (Basque & Lavoie, 2006), along with over 20 additional studies reviewed since then, made it possible to pinpoint many differences in the structure of the CCM tasks proposed to learners by researchers. For instance, a list of concepts and/or links may be provided to subjects; links may be labelled or not; links may be arrowed or not; roles may be given by researchers to each member of the CCM group, communication constraints may be imposed, etc. Also, CM software tools, such as *Inspiration*, *CMapTools*, or others (some of them still being R&D products), are becoming increasingly popular. Actually, a total of 43 of the 65 studies that we investigated so far provided students with a

DOI: 10.4018/978-1-59904-992-2.ch010

CM software tool, either in a face-to-face context (21 studies) or at a distance (24 studies[2]). In this chapter, we argue that the CM tool and the CM method used in CCM activities can significantly affect the processes of meaning-making and that of meaning-negotiation amongst learners and, consequently, upon learning that may result from such activities.

The idea that CM software are "cognitive tools" (Kommers, Jonassen, & Mayes, 1992; Lajoie & Derry, 1993) or "mindtools" (Jonassen, 2000) to the same extent as databases, microworlds or visualization tools was put forth by Jonassen in the beginnings of the nineties (Jonassen, 1992). Such tools facilitate external representations of information and enhance cognitive functioning (Kommers, Jonassen, & Mayes, 1992; Olson, 1985). This notion of cognitive tool is somewhat similar to the notion of "cognitive artefact" proposed in the field of Human-Machine Interaction by Norman (1991) and by other authors involved in Computer-Supported Collaborative Learning (CSCL) (Suthers, 2006) or working within the Activity Theory framework (Engeström, Miettinen, & Punamäki, 1999). Such a notion acts as a kind of "boundary object" (Star & Griesemer, 1989) for researchers from different fields sharing the idea that external knowledge representation tools guide and influence the learner's activity and, thus, must be considered when investigating potential learning benefits. In the field of CSCL, Suthers (2003) suggested the expression "representational guidance" to refer to the fact that the properties of cognitive tools constrain which knowledge can be expressed in a shared context, and, in making some characteristics of that knowledge more salient, promote certain types of "epistemic actions" to the detriment of others.

In this chapter, we investigate how a CM tool that integrates a typology of knowledge objects and a typology of links mediates the process of meaning-making and of meaning-negotiation of learners engaged in a CCM activity, more specifically in the context of a text comprehension task.

BACKGROUND

We view the CCM activity as a tool-mediated *intersubjective meaning-making activity* (Suthers, 2006). Our approach is based then on what Suthers (2006) calls an "intersubjective epistemology", which differs from an "individual epistemology". In the latter, the individual is the unit and the agent of learning, and collaboration simply provides learning conditions and support. Although it is stimulated by social interactions, the cognitive process remains predominantly individual. In the former, the group is the unit of learning, within which "interpretations can be *jointly created* through interaction in addition to being formed by individuals before they are offered to the group" (Suthers, 2006, p. 317). Intersubjectivity also includes a participatory component: "it is a simultaneous process of mutual constitution that may involve disagreement as well as agreement about shared information" (Suthers, 2006, p. 317) and is comparable to a "polyphonic nonharmonious concert characterized by synchronic movements, as well as by distinct, conflicting and dissonant voices" (Smolka, De Goes, & Pina, 1995, in Suthers, 2006, p. 317).

This intersubjective meaning-making activity is a *tool-mediated activity*, which means that it is situated in a socio-cultural environment where tools and signs are imbricated with actions and thinking that provide them with meaning (Vygotsky, 1978). According to Vygotsky, qualitative transformations induced in the cognitive activity through "psychological tools" or "cultural tools" constitute the main factor of cognitive development and learning in a given socio-historical context.

Our theoretical position thus leads us to suggest that in order to define how a CCM activity can prompt or hinder learning, we must (1) study the communication and collaboration processes which take place among the partners involved in the activity, such processes being closely linked to the joint actions undertaken and (2) consider the representational properties of the tools used in the CCM activity.

As for Suthers (2006), intersubjective learning has not yet been sufficiently explored in work conducted in the field of CSCL. We think that this is the case in the specific field of computer-based CCM in educational contexts. Studies in this field seldom go beyond categorization of social interactions occurring during the learning activity, without associating them with the participants' joint actions and without taking into account the dynamics of the interactional process. The influence of the representational properties of tools on the process of co-construction of knowledge is also neglected in most of these studies.

In the following paragraphs, we first delve into the representational properties of CM software tools and techniques. Then, we present results of some studies on CCM, more particularly those conducted in face-to-face learning contexts, since such modality is addressed in this chapter.

Representational Properties of Concept Mapping Techniques and Software Tools

The CM technique developed by Novak (1990; Novak & Gowin, 1984) based on Ausubel's (1968) theory of "meaningful learning" became a reference for most CM research and applications carried out in the field of education. A "standard format for concept maps" according to this technique was presented in 1992 at the National Convention of the National Science Teachers Association in the U.S.A. (Wandersee, 1992), and some of the most popular CM tools (e.g. *CMapTools*[3], *Inspiration, SemNet*[4]) have been designed or are used according to this technique.

Basically, this technique uses a knowledge representation language composed of two main types of "primitives": nodes and links. When combined, these two elements make it possible to create a graphic representation of a field of knowledge organized into a hierarchical network. Each node of the network represents a concept defined as "perceived regularities or patterns in events or objects, or records of events or objects" (Novak & Canas, 2006). Each concept is designated with one or a few words. Links among knowledge objects are represented with arrows, on which concept mappers add their own "linking words" (Novak, 1990). Two concepts connected through a labelled link constitute a "proposition".

In this technique, minimal constraints are placed on the process of naming concepts and links, except that the concepts must be designated with nouns and the links usually with verbs, and that a special link (*e.g.*) is used to designate "examples" of some concepts. However, the map structure is constrained. The most general concepts must appear at the top of the map, and the more specific concepts must be placed at the bottom. Hence, the spatial layout and the directions of the arrowed links aim to express the idea of a hierarchy of concepts, going from the most general to the most specific ones. According to Novak & Gowin (1984), representing a knowledge domain in such a way enhances meaningful learning (Ausubel, 1968).

However, numerous researchers slightly modify this technique and pre-structure the activity by providing participants with a predefined list of concepts, nodes or both. Such is the case for almost half of the 65 studies we examined. Most researchers fail to justify this task organisation in their publications, probably most of the time for practical reasons rather than theoretically founded ones. Indeed, since language ambiguity can be reduced with the use of pre-labelled nodes and links, it becomes easier for researchers to evaluate the maps produced by learners. In other cases, such a strategy could have been used to decrease the time span of the experiment or to alleviate the level of difficulty of the task (for instance, with children or with learners creating maps at a distance).

Nonetheless, a few researchers clearly opt for an explicit theoretically-driven position on this issue, arguing that, when the goal is learning, it can be beneficial to constrain the CM activity in some way to guide learners through this cognitively challenging task.

Many authors agree that the greatest difficulty for learners in creating concept maps resides in structuring concepts into a coherent collection, that is, in representing relations among concepts through labelled links (Basque & Pudelko, 2003; Canas, Valerio, Lalinde-Pulido, Carvalho, & Arguedas, 2003; Faletti & Fisher, 1996; Fisher, 1990; Novak & Gowin, 1984). Jo (2001) reports that students tend to use very general terminology to label links, which may suggest rather superficial thinking. According to Fisher (1990; Faletti & Fisher, 1996), one of the reasons is that school settings focus on teaching isolated concepts rather than structuring them into complex conceptual representations.

For those who consider learning as a process of building knowledge structures of increasing complexity, creating concept maps with a limited set of field-independent links would be a good strategy to enhance learning (Holley & Dansereau, 1984a; Kharatmal & Nagarjuna, 2006). For instance, the networking strategy proposed by Holley & Dansereau (1984a)[5] to support comprehension of scientific texts includes six types of links to represent not only hierarchical knowledge (*type of/example of*; *part of*), but also chains (*leads to*) and clusters (*analogy*, *characteristic*, *evidence*) of knowledge. The limit imposed on the link representation process would help in disambiguating natural language used to designate links, in making students more aware of the existence of various types of relations between knowledge objects (for instance hierarchical, mereological, temporal or causal) and in making different types of knowledge structures more salient for them.

Other researchers disagree with such a strategy arguing that, although certain relations can be considered as ubiquitous across domains, others remain domain-dependent (Faletti & Fisher, 1996). As Fisher (2000) puts it, "specialized knowledge requires specialized relations" (p. 155). Consequently, these authors believe that a constrained set of relations would severely limit the possibilities of expressing meaning.

In order to favour learning, other researchers propose guiding the CM activities with other knowledge representation constraints, which are actually some kind of "micro-structures" made up of a combination of typed links and nodes. For instance, Reader & Hammond (1994) led a study to compare constrained and non-constrained approaches to CM in an educational context. The participants (university students) had to create concept maps that represented the content and the arguments of a text on eating disorders. Students using the constrained approach were provided with a limited set of nodes (*statements, proofs*, *critique*, etc.), links (*supports*, *contradict*, etc.) and different types of predetermined propositional structures (for example, *proof-support-statement* is permitted, while *proof-contradict-proof* is not). The non-constrained approach imposed no restriction. The concept maps produced by participants using the constrained approach were more complete, more efficient in communicating the main ideas of the text and more compatible with the goal of the activity. More recently, Komis, Ergazaki, & Zogza (2007) compared the cognitive processes of a dyad of high-school students in two knowledge modeling conditions: a Novakian paper-and-pencil CM condition and a more structured computer-supported condition based on the use of the software *ModelsCreator*. In both situations, students were provided with a set of objects and properties pertaining to the topic of photosynthesis. However, within the *ModelsCreator* environment, they had to use, in addition, a limited set of semi-quantitative relationships (*increases-increases*; *increases-decreases* or *increases-increases less*). Researchers analyzed the students' cognitive processes using a combination of the OCAF (Object-oriented Collaborative Analysis Framework) coding scheme to model basic operations (Avouris, Dimitracopoulou, &

Komis, 2003) and the Stratford *et al.*'s coding scheme (1998, in Komis *et al.*, 2007) for cognitive strategies. They found that *ModelsCreator* "mediated a discernable and more challenging learning environment than 'paper-and-pencil', especially in terms of self-assessment and monitoring, as well as moving towards a more mathematically informed understanding of photosynthesis" (Komis, Ergazaki, & Zogza, 2007, p. 1013).

Other studies showed the potential learning benefits of a constrained approach to creating different types of graphical representations of knowledge in small groups in an educational context. For example, Fischer, Bruhn, Gräsel, & Mandl (2002) investigated the social interactions between university students in Educational Psychology who created representations of three lesson plans in dyads, with either a graphic editor or a structured automated tool (*CoStructure*). The *CoStructure* tool provides a list of concepts pertaining to each lesson and it proposes two types of relations among concepts (positive and negative relations). Results show that the task structure provided by the tool encouraged dialogues on conceptual aspects of the task as well "conflict-oriented consensus building" discussions. The structure of the tool "represents a semantic co-ordinating element, which helps learners posing constraints in working on the case" (Fischer *et al.*, 2002, p. 229).

Suthers and his colleagues (Suthers, 1999; 2001; Suthers, Girardeau, & Hundhausen, 2002; Suthers, Toth, & Weiner, 1997) led a set of studies on the use of a software tool called *Belvedere*. It makes it possible to construct "inquiry diagrams" or "evidence maps" (a special kind of concept maps) which relate data (empirical statements) and hypotheses (theoretical statements) with two types of evidential relations: for (consistency) and against (inconsistency). A data conjunction link is also proposed. Such investigations show that the representational or primitive properties of *Belvedere* constitute an ontology of categories and structures to organize the task domain and can have a significant effect on the learners' knowledge-building discourse, on learning outcomes as well as on the content of students' collaboration. Activities carried out with Belvedere thus permit to enhance students' ability "to address scientific hypothesis testing in an organized and analytical way" (Suthers, Toth & Weiner, 1997, p. 6).

Social Interactions During Face-to-Face CCM Learning Activities

Roth & Roychoudhury (1992; 1993) are among the first researchers who addressed the issue of CCM in an educational context. Using qualitative research methods, they observed sustained science discourse during this activity in junior and senior physics classes who created paper-and-pencil concept maps in small groups. They noted that such discourse replicates typical interactions in scientific communities, that is, "co-construction interactions", "adversarial interactions" and "formation of alliances". However, they also observed that dialogues are often reduced to short sentences or even one-word utterances, which led them to wonder whether this may prevent the development of more complex concepts, as well as more elaborated arguments. Sizmur & Osborne (1997) analyzed the degree of elaborated exchanges between 9- to 11-year-old children creating paper-and pencil concept maps collaboratively in science classes. They observed "the phenomenon of children's continuing each other's contributions" during CCM, which allowed participants to make more scientifically valid propositions than when the exchanges were not elaborated. Nevertheless, they also noticed that many ideas introduced in the conversations were not retained by the group, perhaps since they generally had been verbalised with non elaborated utterances.

Van Boxtel, van der Linden, & Kanselaar (2000) observed a larger quantity of elaborated cognitive conflicts and constructed reasoning episodes between pairs of children creating concept maps compared to those creating posters. However, in the Chang, Sung, & Lee (2003) study, adult participants are not as prone to negotiate their ideas and consensus is rare, although it should be mentioned that in that case, one member of each group of three or four was elaborating the map, while others were providing comments and suggestions.

Other studies conducted with adults show that social interactions during CCM were quite cognitively engaging. Based on observations of four triads of university students who co-elaborated concept maps during a semester, Steketee (2006) noted that "structural discourse" had a strong presence in each recorded session: "groups reflected on their combined prior knowledge, made inferences about it, challenged each other, determined the implications of interrelationships and made attempts to fit ideas into a coherent explanation" (p. 11). Ryve (2004) showed that communication among university students who created concept maps in triads in the domain of Linear Algebra contains the elements that are characteristic of a "mathematically productive interaction" (p. 157). This researcher found several examples of "explicit interpersonal elaborations of the intended foci" (p. 172) in the students' discourse.

We found very few studies examining correlations between interactions during CCM and the quality of the maps produced by the group (e.g. Chung, O'Neil, & Herl, 1999; Sizmur & Osborne, 1997). In general, such studies conducted with children show that more interactions and more elaborated, high-level and complex interaction, lead to improved performance.

A single study exploring correlations between post-test learning measures and social interactions during CCM was found. This study was also conducted with children (Van Boxtel, van der Linden, & Kanselaar, 1997, 2000). Results show that the frequency of "elaborative episodes" during CCM correlates with comprehension measures.

This overview of research reveals that (1) most studies that investigate social interactions during face-to-face CCM in an educational context have been conducted with children and relatively few have involved adult participants; (2) such studies show that, in general, CM can trigger cognitively productive interactions; (3) the very few studies that investigate the effect of constraints imposed on activities of knowledge modeling tend to show that they have a positive effect on the quality of the external representation produced and on the interactive dynamics among learners, and (4) using an intersubjective perspective to observe both actions and communications between partners remains to be explored in CCM studies.

THE MAIN FOCUS OF THE CHAPTER

Our literature review in the field of CCM in education, as well as informal conversations with researchers, lead us to think that many investigators hesitate to restrict the "flexibility of expressiveness" (Alpert, 2004) of CM tools by imposing some kind of typologies of knowledge objects and/or of links, although, as we mentioned earlier, many of them self-contradictorily provide the learners with a list of concepts and/or links, which is an even more constraining modality.

Some authors argue that the "obligation of freedom" in link labelling is an essential condition to engage learners in active elaboration of knowledge relations and in more meaningful learning. This practice would help learners in the identification and progressive differentiation of domain-specific relations. Indeed, some studies show that learning increases link diversity: experts establish more links than novices, and the links of the former are more information-rich (Herl, 1996; Markham, Mintzes, & Jones, 1994). When creating maps, novices tend to use essentially the verbs "to be" and "to have" (Fisher, 1990), or "includes" and "are related" (Baroody & Bartels, 2000) as linking terms. In that sense, maps created by experts are more domain-specific than those built by novices.

On the other hand, expert language and scientific language tend to be parsimonious. As stated by Kharatmal & Nagarjuna (2006), "though we may think that there can be innumerable number of linking words, if we look closely into any expert domain in any science, we realize that the number of *kinds* of linking words is limited, and does not increase as the knowledge advances". These authors thus suggest some refinement to the Novakian CM technique in order to make it more effective for science education by inculcating some "discipline in choosing the right kind of relation types". That explains why some researchers (e.g. Holley & Dansereau, 1984b) introduced a typology of links in their CM technique. Such position is similar to the one claimed by Medland (2007) who proposes that young children be trained in the use of ontologically- and epistemologically-based language to help them analyze, synthesize and share knowledge. The main idea is that if we want students to develop thoughts that meet the requirements of scientific reasoning, they should be provided with language that matches scientific thoughts, such as the ones used by expert scientists.

We propose a CM approach that, we think, offers an original compromise. Given that links are usually represented with verbs (which express *actions*) in concept maps, one may envision representing actions in nodes rather than in links. These nodes would then be a special kind of nodes, representing *procedural* knowledge (albeit in a declarative format)[6]. Such a solution makes it possible to preserve both the diversity and the specificity of knowledge from a given field. Moreover, the links are then used to represent *only* generic relations, resulting in a more economical and more parsimonious representational language. In addition, the expressiveness of the CM language is enhanced by making the domain-specific procedural knowledge more salient. In other words, the representation of "actions" into the nodes would help in focussing attention of the mapper not only on the meaning of *concepts* but also on the meaning of *actions*, given that the mapper is now able to work on linking these actions between them and on linking these same actions to concept nodes.

This solution was selected for MOT[7], an object-typed CM software tool that integrates a typology of links *and* a typology of knowledge objects (nodes). We have been working with this tool for over a dozen years as a teacher and as a researcher. We used it as (1) an instructional design tool (Doré & Basque, 2002), (2) a learning tool proposed to students enrolled in online courses (Basque & Pudelko, 2002, 2003), (3) a support tool to conceptualize theoretical constructs, either individually (Basque, 2004) or in groups (Basque *et al.*, 2002; Basque, Rocheleau, Paquette, & Paquin, 1998), and (4) a tool to elicit expert knowledge and to transfer expertise in organizations (Basque, Imbeault, Pudelko, & Léonard, 2004; Basque, Paquette, Pudelko, & Léonard, 2008). For us, such a wide experience with object-typed CM in various individual and collective activities convinced us of the potential of this tool for meaning-making and meaning-negotiation for our own purpose.

A few years ago, we launch a research program in the field designed to address, among others, the mediations of such a tool in collaborative learning situations where adults acted as participants. Our main research question was the following: does this object-typed CM tool favour or hinder the participants' learning and meaning-making and meaning-negotiation processes?

Pudelko (2006) explored this question through a microgenetic study of epistemic mediations of this tool in an individual CM activity designed to enhance text comprehension. Data shows the transformations of the external and internal activity structures induced by the use of the tool. Based on this work, we examine, in this chapter, a CCM activity which also involves text comprehension and which was performed with MOT by one dyad of adult learners working face-to-face. The data come from a larger study designed to compare

Table 1. The typology of knowledge objects (KO) in MOT

Types of KO	Descriptions	Examples
Concept	Class of objects from a given field (what?) which share common properties. Property "values" are used to differentiate objects from one another.	• Square • Book • Vertebrate animal.
Procedure	Set of operations that permit actions on objects (*how?*).	• Multiply two-digit numbers. • Search information on the Internet. • Manage a project.
Principle	Statement to describe object properties, establish cause-effect links (*why?*) or determine in which conditions a procedure applies (*when?*). Principles often take the shape of "if Condition X, then Condition Y or Action Z".	• When the soufflé is ready, it must be served immediately. • Road safety rules. • Metal dilatation laws pertaining to the effect of heat. • Instructional design principles.
Fact	Instantiation of knowledge of the type concept, procedure or principle. When a fact instantiates: • a concept, it becomes an *example*. • a procedure, it becomes a *trace*. • a principle, it becomes a *statement*.	• Example: A specific book. • Trace: The procedure I used to manage a specific project. • Statement: If I heat my silver bracelet to a temperature superior to 200oF, it becomes longer.

three CCM conditions: a face-to-face setting and two distance conditions, one being synchronous and the other asynchronous (Basque & Pudelko, 2004).

In the remainder of this section, we will first describe the MOT software tool and its object-typed CM language. Then, the CCM situation under study and the data analysis methodology used for this case study are briefly presented. Finally, results are revealed through the presentation of a collection of vignettes taken from the data protocol.

MOT: An Object-Typed Concept Mapping Tool

The MOT software tool has been developed at the LICEF Research Center[8] in the early nineties by Paquette and his team (Paquette, 2002). This knowledge modeling tool includes three types of *abstract* knowledge objects (nodes) that are classified with different graphic shapes: *concepts* (rectangles), *procedures* (ovals) and *principles* (hexagons). MOT also differentiates *concrete* knowledge objects (KO), called *facts* (rectangles with indented corners), which refer to instances of abstract knowledge objects (see Table 1).

This classification of knowledge objects reaches a consensus in the educational literature, despite certain divergent opinions relative to terminology and associated definitions (e.g., Merrill, 1994; Romizowski, 1999; Tennyson & Rasch, 1988; West, Farmer, & Wolff, 1991).

To represent a certain type of knowledge object in a map, concept mappers must *first* select a type of knowledge object from a menu before they drag-and-drop the specific graphic shape assigned to such knowledge object in the MOT windows and adjust it to the desired size. They can then add its label.

Knowledge objects are connected with each other through arrowed links. When links are selected from the menu and drawn between two knowledge objects, the first letter of the link label is automatically displayed on the link (see Figure 1). The link typology comprises six types of links: *Composition*, *Regulation*, *Specialisation*, *Precedence*, *Input or Product (I/P)* and *Instance*.

The representation of links must conform to "grammar rules" established in the software. For example, a "specialisation link" (S; equivalent to '*sort of*") can only be used between two objects

Figure 1. An example of a knowledge model created with the MOT software

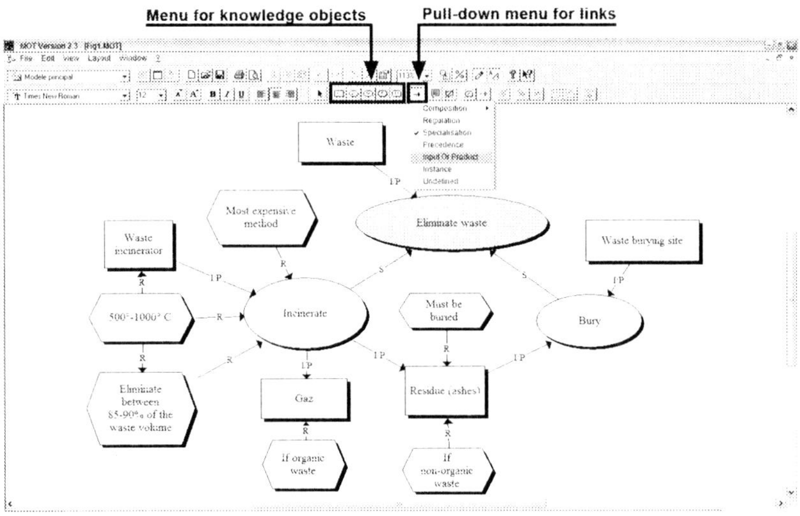

of the same type. Consequently, if the user relates two knowledge entities of different types with the S-link, the software will automatically display the default link, that is, the best-suited and the most probable according to the knowledge modeling grammar implemented in the software. If users disagree with the suggested link, they can right-click on the link and choose another one from the pool of "permitted" links (the invalid links not being clickable). Table 2 provides a summary of valid links that can connect different types of knowledge objects according to the MOT grammar. This grammar can be viewed as a collection of representational "micro-structures" that guides the representational activity. It is based on a sort of "natural" and pragmatic semantic, which aims at enhancing coherence and reducing ambiguity in the knowledge models produced, which thus facilitates their interpretation (Paquette, 2002).

However, users can put their own label on an "untyped" (or undefined) link. A specific shape is also provided for "untyped" knowledge objects. The tool can thus be also used in accordance with an unconstrained concept mapping approach.

Figure 1 shows an example of a map in the domain of waste elimination. The map describes two main types (*sort of*) of procedures (*incinerate* and *bury*) to *eliminate waste*, which is also defined

Table 2. Valid links between different types of knowledge objects according to the MOT grammar (Adapted from Paquette, 2002)

Destination Origin	Abstract Knowledge Objects			Concrete Knowledge Objects (Facts)		
	Concept	Procedure	Principle	Example	Trace	Statement
Concept	C, S	I/P		I, C		
Procedure	I/P	C, S, P	C, P		I, C	
Principle	R	C, R, P	C, S, P, R			I, C
Example				C	I/P	
Trace				I/P	C, P	C, P
Statement				R	C, R, P	C, R, P

as a procedure. Concepts are defined as *input to* or *products of* (I/P) such procedures. Principles are linked to these procedures and concepts with *regulation* links. No facts are represented in the map. The perspective adopted here to describe the domain is thus primarily procedural. Concepts are used here to indicate the inputs and the outputs of the procedures represented, and principles are used to specify some conditions or constraints regulating these procedures.

Other types of knowledge structures, from simple to complex ones, can be represented with MOT. Paquette (2002) identifies thirteen different types of knowledge structures that can be represented, using the set of primitives of the MOT language, such as taxonomies or typologies of concepts, procedures, principles or facts (all links are of the *specialisation* type), component structures (part-whole; all links are of the *composition* type), flowcharts (including iterative procedures), laws and theories, decision trees, etc.

Among other functionalities of the MOT software tool, we find the possibility of creating sub-maps attached to each knowledge object represented in the higher-level map, as well as the feasibility to attach documents of different formats (with OLE or URL links) to each knowledge object. It is also possible to attach a "comment" to knowledge objects or links.

The Collaborative Concept Mapping Situation

As mentioned above, the issue discussed in this chapter is addressed with data taken from a larger research project conducted with dyads of voluntary adult participants who elaborated a concept map with the MOT tool, either in face-to-face or remote conditions. The analysis of the whole data collected is on-going. Here, we focus on the CCM activity of one of the eight dyads who worked in the face-to-face condition. This dyad obtained the best concept map score, as determined by an evaluation method based on a comparison with an "expert map", although the learners' score falls far beyond the maximum score[9]. Our goal in choosing this dyad was to study how the MOT tool and language contributed or not to the relative success of this dyad in creating their map.

The experimental research methodology used to generate such data is detailed in Basque & Pudelko (2004); only the gist of the investigation is reported here.

The experiment took place at the LORIT, a distance learning engineering research laboratory based at Tele-université, Canada[10]. Participants volunteered to participate to this study by responding to an invitation sent to different discussion lists at this university. Only those corresponding to the following criteria have been retained: (1) they were postsecondary students or had post-secondary instruction; (2) they had no or low familiarity with MOT or with other graphical node-link representations; (3) they had no or low familiarity with the domain described in the texts used in the experimentation.

Three dyads participated in each experimental session, which proceeds as follows. First, participants completed a short comprehension pre-test. Second, they were trained on the MOT software and technique (75 minutes). Third, they practiced concept mapping by using MOT to create a map individually on the topic of waste elimination (20 minutes). Fourth, after a 15-minute break, participants were paired arbitrarily and asked to perform the CCM task. This task consisted of elaborating a concept map with MOT, representing the domain described in a one-page text. This text has been written by the first author as part of course material used in a distance education course in Cognitive Psychology. It describes the main components of the Human Information Processing System (Sensory Memory, Short-Term Memory and Long-Term Memory) and the Cognitive Information Process (CIP). After having read the text individually for 5 minutes, pairs were allotted 45 minutes to construct their CM using the MOT tool. They had access to a printed version of the text during the CCM activity. At the beginning of the session, one member of each pair was arbitrarily identified as the "editor" of the map (the one who manipulates the mouse), yet participants were told that they could freely change roles during the session. Finally, after a second 20-minute break, participants filled out the post-test (identical to the pre-test).

Each dyad session was audio- and screen-captured with the *Windows Media Encoder* (WME) software and the dyad's final map was collected.

Data Analysis Method

All dialogues of the chosen dyad were transcribed verbatim, and the verbatim protocol was cut up into *episodes*. Each episode represents an *intentional act* expressed explicitly either by the participants' actions, by their utterances or both. Here are some examples of intentional acts: selecting a knowledge object (KO) from the text, indicating that a link should connect two specific KO (verbally only and/or by using the cursor). Some higher-level intentional acts imply a series of other intentional acts expressed altogether. For example, an utterance such as "*Here, we should put a procedure called 'select information'* " expresses the intentional act of creating a KO, which includes a series of implied operations of identifying a KO from the text, labelling it and categorizing it. The intentional act can thus concern different level of the activity (*operations*, *actions*, *activity*), as suggested in the Activity Theory (Leontiev, 1974).

Aside from identifying the intentional act expressed in each episode, the interactional style of the dyad is also analyzed concurrently, using an adapted version of Gilly's coding scheme (Gilly, 1988), based on the socio-cognitive conflict theory (Doise & Mugny, 1984). To do so, the propositions expressed by participants in each episode are investigated in order to specify (1) if the proposition initiating the episode is explained or not, (2) the reaction of the partner, which can take the form of an implicit or explicit statement of agreement and disagreement, or no reaction at all, (3) whether statement of agreement and disagreement are explained or not by the partner. We also note disagreements expressed by the software tool, that is, when it does not display the link chosen by the learners.

Description of the Dyad's Meaning-Making and Meaning-Negociation Process

Our analysis aims to illustrate meaning-making and meaning-negotiation actions and utterances which seem to have been induced by the properties of MOT and of its language during the session. After a brief presentation of the dyad under study in this chapter, we will first describe the actions and utterances related to the representation of each of the three main types of abstract KO (concepts, procedures, principles[11]). Then, we will present how partners of the dyad acted and negotiated the representation of links.

Presentation of the Dyad

The dyad was composed of a woman of 24 years-old who teaches French as a Second Language and a male student of 21 years-old. They will be designated in this chapter by their gender: F (female) and M (male). Both of them declared having low prior knowledge in Cognitive Psychology before taking part in the experiment. Only M had used the MOT tool before the experiment, for a two-hour period.

At the beginning of the session, F was arbitrarily designated as the "editor" of the map, and the viewing of the WME file shows no indication that such a role changed over the session. This dyad used paper and pencil during a segment of the session, along with MOT. Participants lacked sufficient time to finish their map, even though throughout the session, they remained task-focused and produce scarce irrelevant utterances,

Table 3. Vignette 1

Line Number	*Subject*	*Utterance*
1	F	Okay, cognitive system. Hum... Would you agree that it is composed of three main memories ?
2	M	Indeed, indeed.
3	F	Do you think it is split up into other elements than memory or only memory?
4	M	Hum... well... let's say that we could split the cognitive system in three memories, and then... make another one... something that describes the infomation processing process.

most being short humorous statements.

The dyad's final map includes five "submaps" (or sublevels), in addition to the first-level map. The whole map includes a total of 37 nodes, some of them being repeated in sub-maps.

Concepts

Actions of representation of concepts essentially consist of (1) identifying an object in the text and (2) deciding whether to represent them as concepts. Overall, 22 concepts appear in the map (including 5 concepts reused in sub-maps), which represents 61% of the total quantity of nodes. From the perspective of the MOT language, all of these concepts are valid, that is, they have been correctly defined as concepts. A comparison between concepts found in participants' map and those in the "expert map" shows that participants added 5 supplementary concepts not found on the experts' map and that 13 concepts defined by experts are absent in the dyad's map.

Partners' actions related to the representation of concepts triggered little discussions: they were performed without or with very few utterances. Hence, it seems that participants implicitly agreed on the categorisation of KO as concepts and on their labels, as shown in Vignette 1 (Table 3), an excerpt from the first two episodes that generated the resulting map shown in Figure 2[12]. The suggestion of F to represent "cognitive system" and its three "memories" (SM, STM, LTM) in the map is accepted at once by M. Subject M responded to F's question ("*Do you think it is split up into other elements than memory or only memory?*") by saying that if there is something else to add here, such an entity would be the "*information processing process*". As will be seen in Vignette 2, this last proposition will be re-enunciated and applied by F as soon as the three "memories" have been added as concepts in the map.

The distinction made in the MOT language between concepts and procedures seems cognitively productive for this dyad, as it allows participants to recognize, right from the start, the distinctive nature of a procedural representation (the information processing process) compared to a conceptual representation (the three memories),

Figure 2. Map related to the Vignette 1

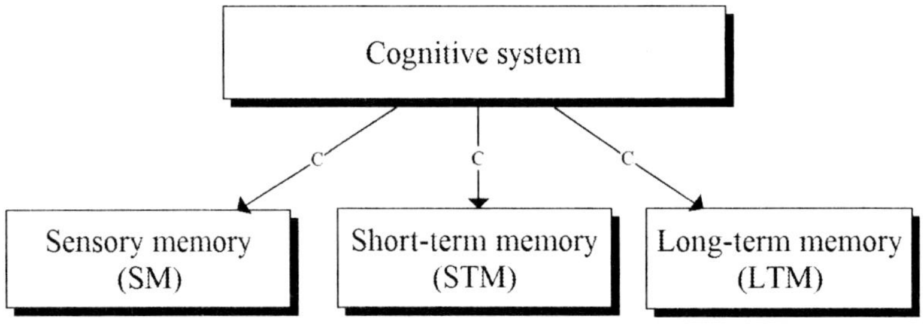

Table 4. Vignette 2

Line Number	*Subject*	*Utterance*
23	F	And then, information processing, do you think it should be something... is it a process?
24	M	Yes, I think so. Indeed a process. But I'm not sure where to put it.
25	F	Hum... a cognitive system process. Hum, let's put it as a process that is subordinate to this, here.
26	M	Indeed. Yes, indeed. I think that's a good idea. A process that is linked to the cognitive system...
27	F	Syst... hum, information processing?
28	M	Well,... I dunno.
29	F	Would you call it that way?
30	M	Hum, information processing. Yes, that's it. Information processing.

thus guiding the gist of the intentional acts that will follow. Indeed, the dyad will then begin to describe the information processing procedure and its sub-procedures, to which the three memory entities will be linked as inputs at certain points. This perspective is exactly the one adopted in the text.

As we will see later, the procedural viewpoint adopted by the subjects all along the rest of the CCM activity also leads the participants to elaborate the following shared implicit complex rules related to the creation of concepts: "A procedure has inputs and outputs" and "If a KO is the input of a procedure, then it is a concept", which were expressed in the map by the representational micro-structure *Concept→ I/P→ Procedure→ I/P→ Concept*. These rules are not implemented as such in the software. Stemming from the basic properties of the MOT language, they have been inferred by participants. Each time a procedure was added to the map, the subjects were trying to specify the input and output concepts of this procedure.

Procedures

Compared to the representation of concepts, depicting procedures generated more discussion. Such dialogues generally started with a participant's proposition (acted out or uttered) of the type "X is a procedure", justified by the CM rule that states that actions or processes are represented as procedures (e.g., "X is a procedure since it is an action"). For example, as already mentioned, right from the start, M suggested representing the information processing process as a procedure. Note that although he stated that this procedure is linked in some way to the concept "cognitive system" (see Vignette 2 [Table 4]), neither participant specified the nature of such a link (see Figure 3), which remained unspecified until the end of the session.

All other procedures appearing in the final map have been created following the inferred rule: "actions are represented as procedures composed of successive steps (or sub-procedures)." Participant M clearly asserted that when he said (Vignette 3, line 53) that it is necessary to *"split up* [the information processing procedure] *into steps"* and then *"split up those steps over again"*. When F agreed, M enumerated these steps (Vignette 3, line 55) and repeated such a comment in response to F's objection (Vignette 4, line 60): *"They're still steps, which are processes, which are connected through Precedence links"* (Vignette 4, line 61). Participant F concurred with such reasoning. In the remainder of the CCM activity, no other disagreements or questions surfaced when representing procedures, which amount to 12 in the final map, all of them correctly represented. (see Vignettes 3 and 4 [Table 5 & Table 6])

Figure 3. Map related to the Vignette 2

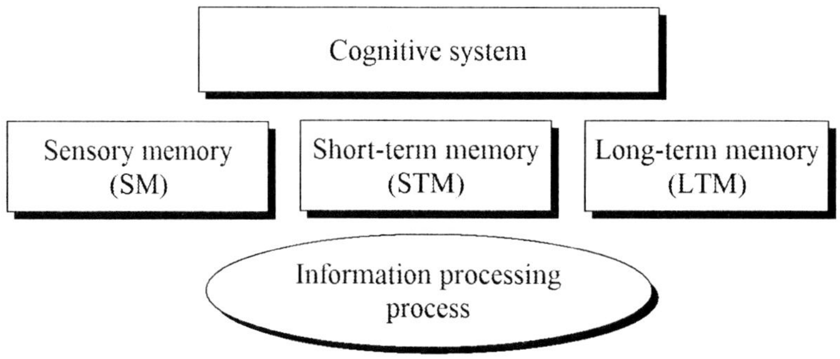

Principles

Actions related to the representation of principles have been scarce: only three principles have been represented in the final CM. Such a discovery is even noticed by participants themselves, who seemed to adhere to an implicit (and inaccurate) shared rule: "the map must include various types of knowledge". Hence, on Line 338, more than 25 minutes after the session started, F worried that they "*haven't put in a single principle*". Participant M replied that they already had inserted "*one or two*", which is actually a single one at that point.

It seems that actions related to the representation of principles were based essentially on two inferred representation rules which seemed equivalent to participants: "X regulates, then X is a principle" and "X is a principle, then X regulates". For participants, a symmetrical implication seems to imply a *Regulation* link (R). In other words, principles require a *Regulation* link and, conversely, a *Regulation* link requires a principle. Note that the former solution is incorrect in the MOT language. Such a deduction probably stemmed from the short training provided to participants, which addressed only *Regulation* links between principles and other types of knowledge, whereas other links, such as *Composition* and *Precedence*, were not covered in relation with principles.

Participants' limited interpretation of possible ways to illustrate relations between principles

Table 5. Vignette 3

Line Number	*Subject*	*Utterance*
53	M	I think it should be split up, see, kind of into seven steps, and then, those seven steps should be split up over because here, also, I'm afraid it will be too loaded. You see what I mean?
54	F	Okay.
55	M	See, for example, the first one would be perceive, then, hum, reconize... select...hum store.... encode.

Table 6. Vignette 4

Line Number	*Subject*	*Utterance*
60	F	Hey, do you find that here, we just have to specify the seven steps, very schematically? Or if we have to represent them more in a quasi-linear fashion, huh!?
61	M	Well, hum, they're still steps, which are processes...which are connected through *Precedence* links.

Figure 4. Sub-map that includes the principle "Time constraint" which regulates the procedure "Storing"

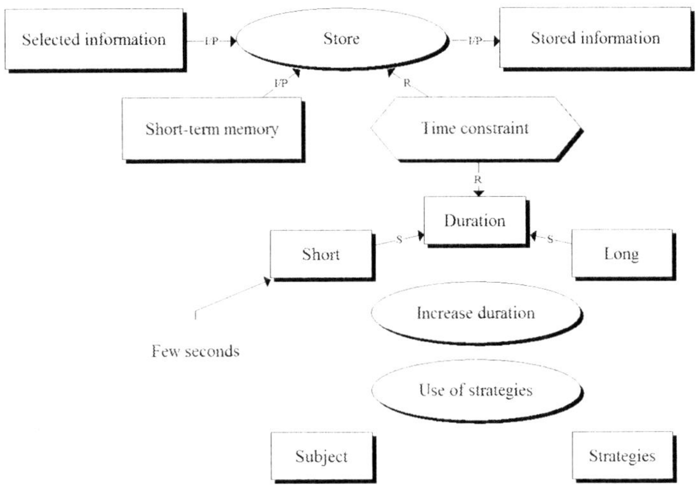

and other types of KOs created a major difficulty at the end of the session. Attempts were made to represent the idea that temporal constraints regulate the process of storage of information in Short-Term Memory, which is described in the text in the following sentence: "*the selected information is stored in the Short-Term Memory (or Working Memory) for a few seconds; it can be stored longer if certain cognitive strategies are used*". Although participants spent the last twelve minutes of the session attempting to solve this challenge, they failed to represent their comprehension of this idea in a satisfactory way. We think that this is partly due to the fact that they were not sufficiently aware of other possible relations that can be established between principles and other types of KOs. Furthermore, we observe that they self-imposed an additional constraint in the CM method, which complicated their meaning-making effort: they labelled nodes with the fewest words possible. Indeed, if this implicit rule generally applies to concepts and procedures, it is ill-suited for principles, as they are often formulated with short sentences, of the type "If…then". For example, in the "expert map", temporal constraints linked to the process "Store in Short-Term Memory" are translated through two principles linked to that process with a *Regulation* link and formulated as such: "A few seconds" and "If cognitive strategies are used, stored longer". Thus, it would seem necessary that CM trainers emphasize explicitly that principles can be labelled with more than one word, if need be. This would prevent participants from over-dissecting principles beyond the point where integrity and meaning is lost, as was the case in this dyad. Indeed, as shown in Figure 4, in their attempt to illustrate the sentence mentioned above, the dyad added a proliferation of KOs in trying to specify the principle "time constraint", which leads M to generate convoluted propositions which F can hardly understand (i.e., "*Here, I'd say that we could add a small 'Zero Plus' meter, so your time constraints, then they count, plus the temporal bonus….*"). Such dialogues, seemingly based on a misinterpretation of the MOT constraints, would have hindered the construction of shared meaning.

We also found that participants' actions and utterances regarding the representation of principles are based on another MOT rule they were taught: principles can represent "constraints" or "conditions", such as time and space, which

Table 7. Vignette 5

Line Number	Subject	Utterance
105	F	"In the environment"... do you think it's a regulation?
106	M	Well, hum, gess so, no? Well, comes from the environment. Well, it's a space unit, right...
107	F	Yes, it's a space unit. Okay. Yes.
108	M	That's another thing we must indicate.
109	F	Okay.

regulate other KOs (see Vignette 5 [Table 7] and Figure 5).

However, participants disagreed on the level of importance of this rule, which, as we will see, lead them to discuss, later on (see Vignette 9), whether "sensory memory" falls into the category principle or concept. For F, who perceives sensory memory as a "space" where information "storage" occurs, it should appear as a principle (which is not necessarily a valid inference in the MOT language). However, for M, who ended up convincing F, the sensory memory (as all sensory receptors, the STM and LTM) does not consider it a "space" but rather an "instrument" for storing processes and should consequently be presented as a concept (according to a rule presented at the training session, where inputs to procedures constitute concepts). Hence, in this case, the representational properties of the MOT language lead participants to identify and discuss in detail issues for representing structural components of the cognitive system as a "space" or a "location" where information "travels" (according to the typical metaphor found in cognitive psychology scientific literature) or as active "registers" that are defined as functions of the human cognitive system.

Links

Links Between Knowledge Objects of the Same Type

Links between KOs of the same type are of three types: composition (*part-whole*), class inclusions (*sort of*) and chains (*temporal precedence*). The third micro-structure can only be used between procedures, while the other two apply to all types of knowledge.

Participants discussed the composition micro-structure between *concepts* few times during the CCM activity. One occurred when identifying the link between the concepts "cognitive system" and the "three memories", where the dialogue triggered by F's initial proposition to use a composition link (see Vignette 1) is very brief and mutual agreement is quickly established. Moreover, the

Figure 5. Map related to the Vignette 5

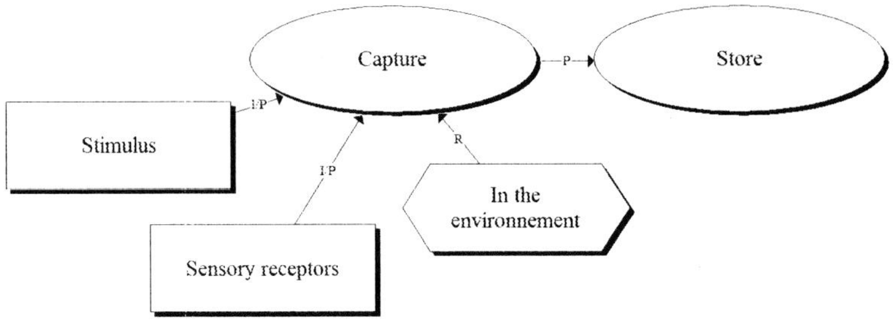

Table 8. Vignette 6

Line Number	Subject	Utterance
247	M	I'd say that… I'd say that we could put… long term memory… we could add another concept that would be… hum, you see, I'd say that the long term memory… I would not link it right away to attributing meaning to stimulus, but I'd say that long term memory is composed of information… you see the long term memory.
248	F	Yeah.
249	M	Inside, you have a composition of information, with this information… it's a composition link.
250	F	Yes.
251	M	So it's this information that go into the… into the process.

action was immediately applied, along with F's utterance: "*Okay, so this is decomposed, right ? So we'll add* [C] *links on it*".

A second instance where the composition micro-structure is used by participants occurred when time comes to illustrate M's perception that the Short-Term Memory is composed of information "that makes it possible to attribute meaning to the sensory trace" (see Vignette 6 [Table 8]).

Hence, this composition link allowed M to elaborate the idea that long-term memory does not directly provide meaning to the sensory trace but rather, *the information it contains*, hence the information stored in it (see Figure 6).

As to composition links between *procedures*, we observed that they are not represented in the map through the *Composition* (C) link, but rather through sub-maps. As mentioned above, at the onset of the activity, M proposes "splitting" the "cognitive processing" method into main steps (perceive, recognize, select, store), before conducting additional decompositions of each of these procedures into sub-procedures. He also proposed representing each of these actions by "decomposing" them into sub-maps in order not to "overload" the first-level map. His partner accepted such a strategy. This approach is actually accurate and often useful when the first-level is replete with a large number of knowledge units. However, it seems that, in that case, it prevented the participants from seeing certain relations between the "steps", such as the presence of "inputs" and "products" between the steps of superior levels[14].

It is interesting to note that participants specified, in a sub-map decomposing the "information processing process" (defined as a procedure in the

Figure 6. Map related to Vignette 6[13]

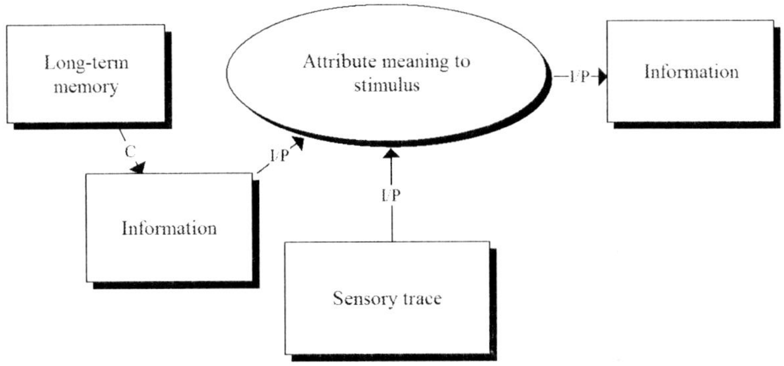

Figure 7. Submap of the "information processing" procedure

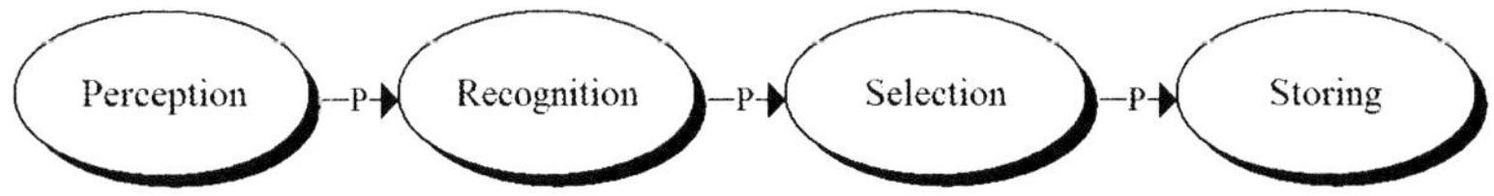

first-level map), a relation of temporal precedence between the various processes that pertain to this procedure (see Figure 7). This was done before decomposing successively each of them. It seems that the participants viewed the *Precedence* link as an intrinsic element of the "decomposition" rule. In other words, they adopted the following reasoning: "if an action is decomposed in steps, then such steps follow one another in a temporal manner". Although this inference is usually valid, it can sometimes prevent the participants from seeing that certain actions can be simultaneous. However, in the CCM activity of this dyad, such a shared interpretation allowed them to quickly and exhaustively identify and represent the different information processing phases described in the text.

Links Between Knowledge Objects of Different Types

As mentioned above, most concepts have been created by subjects through the application of a micro-structure rule that is used to illustrate relations between (1) an input represented as a concept connected to a procedure through an I/P link and (2) a product or output also represented as a concept connected with a procedure through an I/P link (going from the procedure to the concept).

When the two links (input and product) are connected to the procedure, they express a "transformation" process, which is an essential micro-structure to comprehend a processing system (a functional system). We noted that this micro-structure rule has been progressively elaborated and implemented by the dyad during the CCM activity. Participant M is the initial bearer of this idea. Looking at the map reproduced in Figure 7, he declared: *"there is something intriguing here ... there are links between these processes. Kinda like... you have processes that generate products"*. This proposition, which revealed a certain "cognitive discomfort" in M, is at that time ignored by F. However, M did not abandon such an idea and brought it up again once they agreed on the set of steps for information processing. This time, however, his partner considered the issue and formulated questions in trying to grasp the meaning of M's proposition and inviting him to be more specific, which M was not able to do then (see Vignette 7 [Table 9]).

Then, the dyad will be engaged repeatedly in an active joint elaboration around this rule of "transformation of an input into an output" in the following episodes. For example, when elaborating the sub-map of the "perception" procedure (see the resulting map in Figure 8), F proposed that the

Table 9. Vignette 7

Line Number	*Subject*	*Utterance*
86	M	Hum, indeed, these steps, they each produce, in fact, each time they produce something that is the common denominator of the following step, it seems.
87	F	The common denominator and what? Pardon?
88	M	At the following step. See, for example, the stimulus, it produces… wait, it produces… ah, never mind, it doesn't produce anything.

Figure 8. Map related to Vignette 9

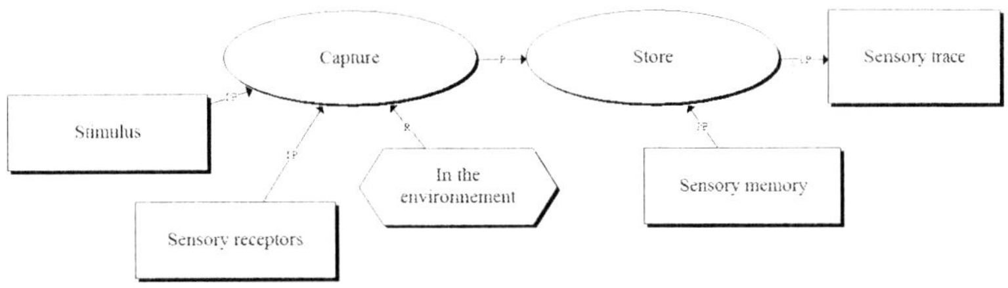

sub-procedure "capture" process *"takes stimuli and takes sensory receptors"* (line 103), while M indicated that *"capture produces something, but we don't know its name"* (line 110), and that such a unit is both a product of the "capture" procedure and an "input" of the following procedure, that is, the procedure labelled "storing" (line 112). While discussing this issue, each partner completed the action or utterance of the other (see Vignette 8 [Table 10]).

In the map presented in Figure 8, however, the two "inputs" linked to the "capture" procedure ("stimulus" and "sensory receptors") do not play the same role: the "stimulus" constitutes the object of the "capture" process, while the "sensory receptors" is the instrument that makes such processing possible. Such a distinction is not clear to F, as shown in the Vignette 9 (Table 11).

Later, M reapplied these two same inferred rule ("procedures transform an input into a product" and "there may be two types of inputs") when the dyad specified the procedure "attribute meaning to stimulus", which, again, required explanations, as F continued to express doubts (see Figure 6). In the end, the first of these rules had become a main constraint guiding the structuring of the map, which makes the dyad asking systematically: "what is the input of this procedure?" and "what is the product of this procedure?".

Finally, the participants' reasoning regarding the actions of linking the knowledge objects seems to have been partly influenced by the fact that they sought to respect an explicit instruction they were given: avoid the use of "untyped

Table 10. Vignette 8

Line Number	*Subject*	*Utterance*
162	F	Yes, okay. And then we said that... inputs.
163	M	Inputs are stimuli and the others....
164	F	Which are concepts.
165	M	There! Exactly.
166	F	How is the first one called? Stimulus?
167	M	Stimulus or stimuli.
168	F	Stimulus, hum, and the other is...?
169	M	Sensory receptors.
170	F	Okay. So these ones, they were hum... inputs.
171	M	Inputs-outputs, yes.
172	F	So the relation goes that way.
173	M	And the other this way.

Table 11. Vignette 9

Line Number	Subject	Utterance
208	F	Well, I don't get it. Explain the link between this and that.
209	M	Here, you have two components arriving. Here, you have an information, hum a captured stimulus. So there! And you store it into the sensory memory.
210	F	Yes.
211	M	The sensory memory is the storage instrument for the captured stimulus.
212	F	Okay.
213	M	So, see, they are two inputs.
214	F	Okay.
215	M	Because we took both the stimulus input and the sensory receptors...
216	F	Yes, yes, right, okay, I get it...

links" as much as possible. For example, when discussing the following sentence of the text "the selected information is stored in the short term memory *for a few seconds*; it can remain *longer* if the participants use certain strategies" (see Section 3.3.3 and Figure 4), M proposed to split the "Time" principle (which was then already represented in the map and which later became "Time constraint") into two units in order to be able to *"represent it with a composition [link] rather than an untyped link*".

When attempting to express this temporal constraint which included two conditions ("if no use of cognitive strategies, then duration lasts a few seconds" and "if use of strategies, then lasts longer"), the participants tried firstly to add specialisation links between the "Time" principle and the "Short" and "Long" concepts, which is not allowed by the software. Thus, just a few minutes before the session is over, MOT got involved for the first time as an "active arguer" into the conversation, causing a three-way dialogue. When MOT refused to accept the specialisation link that the participants tried to add between the "Short" concept and the "Time" principle, M realized that "*the problem is that apparently, we cannot put just about anything here. It doesn't want us to do the the... the software prevents us from putting the the....*" This caused the introduction of a new concept labelled "Duration" so that the "Short" and "Long" concepts could be connected to it through specialisation links. The label of the principle "Time" was then changed for "Time constraint", which was linked to the "Duration" concept through a regulation link (see Figure 4).

Such a compromise, the product of a three-way negotiation, conforms to the constraints of the MOT language. However, it still did not satisfy the participants, whose intentional act expresses an alternative for two conditions (whether strategies are used or not). They then pursued their meaning-making effort, adding numerous KOs: two procedures ("Increase duration" and "Use strategies") as well as two concepts ("Subject" and "Strategies"), which they failed to link due to the session time limit.

All the discussion around the representation of the principle "Time constraint" illustrates how the software (which applies the object-typed grammar rules) can become an active participant, joining the dyad in their meaning-making and meaning-negotiation efforts throughout the CCM activity. The interpretation given by the participants to the representational properties of the CM language and to the CM method also guides strongly the activity, usually in a productive way but not always.

CONCLUSION

Our qualitative analysis of joint actions and utterances by a dyad of adult participants involved in a CCM activity designed to enhance text comprehension and mediated with an object-typed CM tool, shows that co-learners are actively involved in intense meaning-making and meaning-negotiation processes. The text, an excerpt from typical instructional material in the field of Cognitive Psychology, includes numerous propositions which are more or less salient or easily grasped than they initially appear. In this sense, this may consist of the main advantage of CCM activities for text comprehension: they provide an opportunity to work collaboratively on the meaning of words and propositions in order to co-construct shared meaning.

In such a space of shared external representation, questions asked, arguments stated and rules inferred are strongly biased by the representational properties of the CM software tool and language, as well as by the CM technique proposed to learners. In this particular investigation, where a CM tool integrates category constraints for both nodes and links, as well as a grammar that determines valid links between different types of nodes, it is clear that the participants used such constraints to guide their meaning-making and meaning-negotiation actions.

The users' interpretation of representational properties of the selected CM tool and language is based on an active construction of meaning, the stabilisation of which depends on both the software and the participants' agreement. A preliminary analysis of the dyad's interactional style reveals that instances of disagreement were infrequent in that dyad, who adopted instead an interaction mode that Gilly (1988) calls "co-construction", where A initiates an action or utters a proposition, which is accepted by B, who performs the action or further refines A's initial proposition. Almost four times less disagreement occurs in this dyad. The software also rarely disagreed with the dyad, thus participating in setting up an interaction mode of the "co-construction" type with the participants. Furthermore, disagreements were discussed, as it was also the case for over half of the instances where participants agreed.

Representational properties of an object-typed concept mapping tool language and method can have an authentic epistemic dimension. Most of the time, it seems that they helped learners building knowledge which is valid from a scientific perspective, and, to a certain extent independent from the field. For example, the procedural micro-structure "transform an object with a tool" built by the participants, expresses knowledge and relational structures which characterize various knowledge domains such as functional systems in biology, ecology, mechanics, etc.

This being said, it should be noted that the goal of the CM activity proposed to learners must be considered when addressing the issue of selecting a constrained or an unconstrained concept mapping language in educational contexts. When the main goal is to allow teachers or researchers to track misconceptions in students' cognitive structures, it may be best to constrain the CM activity the least possible and eventually use computerized "disambiguation" tools to analyze and assess maps (da Costa, da Rocha, & Faveo, 2004). However, it seems to us that this practice is not optimal for the learners' active participation in joint and dynamic processes designed to structure and negotiate knowledge. On the contrary, in this case, the tool becomes an epistemic representational guide for co-learners. To this end, we hypothesized and hopefully demonstrated that a more disciplined language of concept mapping, such as an object-typed one, would be a promising avenue.

FUTURE TRENDS

Obviously, our conclusion, based on an analysis of a single case, must be validated and further investigated in future studies. We believe this issue contains much research potential in the field of CCM, and that intersubjective epistemology is a rich framework to help in shedding some light on the meaning-making and meaning-negotiation processes of concept mapping groups.

To better grasp the contribution of concept mapping tools, languages and methods in collaborative learning situations, and in order to verify how more or less constrained concept mapping approaches can support or hinder knowledge co-construction, more research is needed. Here are some research issues that we find particularly relevant for future investigations in the field:

- How do learners who produce less well structured maps use the representational properties of the CM tool?
- Do the processes of meaning-making and meaning-negotiation, as well as learning results, differ when learners use an object-typed versus an untyped concept mapping tool ?
- What are the correlations between the dyads' interactive and argumentation styles and (1) the quality of the map produced and (2) learning results?
- How should participants be trained to object-typed concept mapping in order to optimize the meaning negotiation process and the potential learning benefits?

REFERENCES

Alpert, S. (2004). Flexibility of Expressiveness: State of the Practice. In P. Kommers (Ed.), *Cognitive Support for Learning: Imagining the Unknown* (pp. 253-268). Amsterdam: IOS Press.

Ausubel, D. (1968). *Educational Psychology: A Cognitive View*. New York: Rhinehart & Winston.

Avouris, N. M., Dimitracopoulou, A., & Komis, V. (2003). On analysis of collaborative problem solving: An object-oriented approach. *Computers in Human Behavior, 19*(2), 147–167. doi:10.1016/S0747-5632(02)00056-0

Baroody, A. J., & Bartels, B. H. (2000). Using concept maps to link mathematical ideas. *Mathematics Teaching in the Middle School, 5*(9), 604–609.

Basque, J. (2004). Le transfert d'apprentissage: Qu'en disent les contextualistes? In M. Frenay & A. Presseau (Eds.), *Le transfert des apprentissages: Comprendre pour mieux intervenir.* Québec, Canada: Presses de l'Université Laval.

Basque, J., Dufresne, A., Léonard, M., Lundgren-Cayrol, K., Paquette, G., & Prom Tep, S. (2002). *Vers un modèle générique de système d'assistance pour le téléapprentissage.* Paper presented at the ACFAS conference, May 13-17, 2002, Université Laval, Québec, Canada.

Basque, J., Imbeault, C., Pudelko, B., & Léonard, M. (2004). Collaborative knowledge modeling between experts and novices: A strategy to support transfer of expertise in an organization. In A. J. Canas, J. D. Novak & F. M. Gonzalez (Eds.), *Proceedings of the First International Conference on Concept Mapping (CMC 2004), Pamplona, September 14-17, vol. 1* (pp. 75-81). Pamplona, Spain: Universidad Publica de Navarra.

Basque, J., & Lavoie, M.-C. (2006). Collaborative Concept Mapping in Education: Major Research Trends. In A. J. Canas & J. D. Novak (Eds.), *Concept Maps: Theory, Methodology, Technology - Proceedings of the Second International Conference on Concept Mapping* (Vol. 1, pp. 79-86). San Jose, Costa Rica: Universidad de Costa Rica.

Basque, J., Paquette, G., Pudelko, B., & Léonard, M. (2008). Collaborative Knowledge Modelling with a Graphical Knowledge Representation Tool: A Strategy to Support the Transfer of Expertise in Organisations. In A. Okada, S. Buckingham Shum & T. Sherborne (Eds.), *Knowledge Cartography. Software Tools and Mapping Techniques* (pp. 357-382). London: Springer-Verlag.

Basque, J., & Pudelko, B. (2002). *La (co)construction des cartes de connaissances dans le contexte du téléapprentissage: quelques pistes de recherche.* Paper presented at the Atelier CIRTA (Centre Interuniversitaire de Recherche sur le Télé-apprentissage), Télé-université, Montréal, Canada.

Basque, J., & Pudelko, B. (2003). Using a concept mapping software as a knowledge construction tool in a graduate online course. In D. Lassner & C. McNaught (Eds.), *Proceedings of ED-MEDIA 2003, World Conference on Educational Multimedia, Hypermedia & Telecommunications, Honolulu, June 23-28, 2003* (pp. 2268-2264). Norfolk, VA: AACE.

Basque, J., & Pudelko, B. (2004). The effect of collaborative knowledge modeling at a distance on performance and on learning. In A. J. Canas, J. D. Novak & F. M. Gonzalez (Eds.), *Proceedings of the First International Conference on Concept Mapping (CMC 2004), Pamplona, September 14-17,* (Vol. 1, pp. 67-74). Pamplona, Spain: Universidad Publica de Navarra.

Basque, J., Rocheleau, J., Paquette, G., & Paquin, C. (1998). An Object-Oriented Model of a Computer-Enriched High School. In T. Ottmann & I. Tomek (Eds.), *Proceedings of ED-MEDIA/ED-TELECOM 98.* Charlottesville, VA: Association for the Advancement of Computing in Education.

Canas, A., Valerio, A., Lalinde-Pulido, J., Carvalho, M., & Arguedas, M. (2003). *Using WordNet for word sense disambiguation to support concept map construction.* Paper presented at the SPIRE 2003 - 10th international symposium on string processing and information retrieval, Manaus, Brazil.

Chang, K.-E., Sung, Y.-T., & Lee, C.-L. (2003). Web-based collaborative inquiry learning. *Journal of Computer Assisted Learning, 19,* 56–59. doi:10.1046/j.0266-4909.2003.00006.x

Chung, G. K., O'Neil, H. F. J., & Herl, H. E. (1999). The use of computer-based collaborative knowledge mapping to measure team processes and team outcomes. *Computers in Human Behavior, 15*(3-4), 463–494. doi:10.1016/S0747-5632(99)00032-1

da Costa, J. V. J., da Rocha, F. E. L., & Faveo, E. L. (2004). Linking phrases in concept maps: A study of the nature of inclusivity. In A. J. Canas, J. D. Novak & F. M. Gonzalez (Eds.), *Concept maps: Theory, Methodology, Technology - Proceedings of the First International Conference on Concept Mapping* (Vol. 1, pp. 2M-249). Pamplona, Spain: Universidad Publica de Navarra.

Doise, W., & Mugny, G. (1984). *The social development of the intellect.* Oxford, UK: Pergamon Press.

Doré, S., & Basque, J. (2002). *Why not apply an engineering methodology when creating courses?* Paper presented at the American Society for Engineering Education Annual Conference & Exposition (ASEE), Montréal, June 2002.

Engeström, Y., Miettinen, R., & Punamäki, R.-L. (1999). *Perspectives on Activity Theory.* Cambridge, UK: Cambridge University Press.

Faletti, J., & Fisher, K. M. (1996). The information in relations in biology, or the unexamined relation is not worth having. In K. M. Fisher & M. R. Kibby (Eds.), *Knowledge acquisition, organization, and use in biology* (pp. 182-205). Berlin, Germany: Springer.

Fischer, F., Bruhn, J., Gräsel, C., & Mandl, H. (2002). Fostering collaborative knowledge construction with visualization tools. *Learning and Instruction, 12*(2), 213–232. doi:10.1016/S0959-4752(01)00005-6

Fisher, K. (1990). Semantic networking: The new kid on the block. *Journal of Research in Science Teaching, 27*(10), 1001–1018.

Fisher, K. M. (2000). SemNet® Semantic Networking. In K. M. Fisher, J. H. Wandersee & D. E. Moody (Eds.), *Mapping biology knowledge* (pp. 143-165). Dordrecht, the Netherlands: Kluwer Academic Publishers.

Gao, H., Shen, E., Losh, S., & Turner, J. (2007). A review of studies on collaborative concept mapping: what have we learned about the technique and what is next? *Journal of Interactive Learning Research, 18*(4), 479–492.

Gilly, M. (1988). *La fonction sociale des outils cognitifs*. Montréal, Canada: CIRADE.

Herl, H. E. (1996). Construct validation of an approach to modeling cognitive structure of U.S. history knowledge. *The Journal of Educational Research, 89*, 206–218.

Holley, C. D., & Dansereau, D. F. (1984a). Networking: The technique and the empirical evidence. In C. D. Holley & D. F. Dansereau (Eds.), *Spatial learning strategies: Techniques, applications and related issues* (pp. 81-108). Orlando, FL: Academic Press.

Holley, C. D., & Dansereau, D. F. (1984b). *Spatial Learning Strategies. Techniques, applications, and related issues.* New York: Academic Press.

Jo, I.-H. (2001). The effects of different concept mapping techniques on college-level writing. *Educational Technology International, 3*(1).

Jonassen, D. H. (1992). Semantic networking as cognitive tools. In P. Kommers, D. H. Jonassen & J. T. Mayes (Eds.), *Cognitive Tools for Learning* (Vol. F81, pp. 19-21). Berlin: Springer Verlag.

Jonassen, D. H. (2000). *Computers as mindtools for schools: Engaging critical thinking* (2nd Ed.). Upper Saddle River, NJ: Prentice Hall.

Kharatmal, M., & Nagarjuna, G. (2006). A proposal to refine concept mapping for effective science learning. In A. J. Canas & J. D. Novak (Eds.), *Concept Maps: Theory, Methodology, Technology - Proceedings of the Second International Conference on Concept Mapping* (Vol. 1, pp. 151-158). San José, Costa Rica: Universidad.

Kim, B., Yang, C.-c., & Tsai, I.-C. (2005). *Review of computer-mediated collaborative concept mapping: Implication for future research.* Paper presented at the 10th international conference on Computer-Supported Collaborative Learning (CSCL), Tapei, Taïwan, May 30 to June 4, 2005.

Komis, V., Ergazaki, M., & Zogza, V. (2007). Comparing computer-supported dynamic modeling and 'paper & pencil' concept mapping tehnique in students' collaborative activity. *Computers & Education, 49*, 991–1017. doi:10.1016/j.compedu.2005.12.007

Kommers, P. A. M., Jonassen, D. H., & Mayes, J. T. (Eds.). (1992). *Cognitive tools for learning.* Berlin: Springer-Verlag.

Lajoie, S. P., & Derry, S. J. (Eds.). (1993). *Computers as cognitive tools.* Hillsdale NJ: Lawrence Erlbaum.

Leontiev, A. (1974). The problem of activity in psychology. *Social Psychology, 13*(2), 4–33.

Markham, K. M., Mintzes, J. J., & Jones, M. G. (1994). The concept map as a research and evaluation tool: Further evidence of validity. *Journal of Research in Science Teaching, 31*(1), 91–101. doi:10.1002/tea.3660310109

Medland, M. D. (2007). Tools for knowledge analysis, synthesis, and sharing. *Journal of Science Education and Technology, 16*(2), 119–153. doi:10.1007/s10956-006-9019-1

Merrill, M. D. (1994). *Principles of instructional design*. Englewood Cliffs, NJ: Educational Technology Publications.

Nesbit, J. C., & Adesope, O. O. (2006). Learning with concept and knowledge maps: A meta-analysis. *Review of Educational Research, 76*(3), 413–448. doi:10.3102/00346543076003413

Norman, D. A. (1991). Cognitive artefacts. In J. M. Carroll (Ed.), *Designing interaction: Psychology of the human-computer interface* (pp. 17-38). Cambridge, UK: Cambridge University Press.

Novak, J. D. (1990). Concept mapping: a useful tool for science education. *Journal of Research in Science Teaching, 27*(10), 937–949.

Novak, J. D., & Canas, A. J. (2006). The origins of the concept mapping tool and the continuing evolution of the tool. *Information Visualization, 5*, 175–184. doi:10.1057/palgrave.ivs.9500126

Novak, J. D., & Gowin, D. B. (1984). *Learning how to learn*. Cambridge, UK: Cambridge University Press.

O'Donnell, A. M., Dansereau, D. F., & Hall, R. H. (2002). Knowledge maps as scaffolds for cognitive processing. *Educational Psychology Review, 14*(1), 71–86. doi:10.1023/A:1013132527007

Olson, D. R. (1985). Computers as tools of the intellect. *Educational Researcher, 14*(5), 5–8. doi:10.2307/1174200

Paquette, G. (2002). *Modélisation des connaissances et des compétences*. Sainte-Foy, Canada: Presses de l'Université du Québec.

Pudelko, B. (2006). *Étude microgénétique des médiations épistémiques d'un outil informatisé de représentation graphique des connaissances au cours d'une activité de compréhension de texte: Propositions pour une approche instrumentale étendue des médiations des outils cognitifs dans l'apprentissage*. Doctoral thesis, Université Paris 8. Paris: Bibliothèque de Paris.

Reader, W. R., & Hammond, N. (1994). A comparison of structured and unstructured knowledge mapping tools in psychology teaching. In N. Hammond & A Trapp (Eds.), *Proceedings of CiP 94*. University of York.

Romizowski, A. J. (1999). Designing instructional systems: Decision making in course planning and curriculum design. London: Kogan Page.

Roth, W., & Roychoudhury, A. (1993). The concept map as a tool for the collaborative construction of knowledge: A microanalysis of high school physics students. *Journal of Research in Science Teaching, 30*(5), 503–554.doi:doi:10.1002/tea.3660300508

Roth, W.-M., & Roychoudhury, A. (1992). The social construction of scientific concepts or the concept map as conscription device and tool for social thinking in high school science. *Science Education, 76*(5), 531–557. doi:10.1002/sce.3730760507

Ryve, A. (2004). Can Collaborative Concept Mapping Create Mathematically Productive Discourses? *Educational Studies in Mathematics, 56*(3), 157–177. doi:10.1023/B:EDUC.0000040395.17555.c2

Sizmur, S., & Osborne, J. (1997). Learning processes and collaborative concept mapping. *International Journal of Science Education, 19*(10), 1117–1135. doi:10.1080/0950069970191002

Star, S. L., & Griesemer, J. R. (1989). Institutional ecology, 'translations' and boundary objects: Amateurs and professionals in Berkeley's museum of vertebrate zoology, 1907-39. *Social Studies of Science*, *19*, 387–M0.

Steketee, C. (2006). Modelling ICT integration in teacher education courses using distributed cognition as a framework. *Australasian Journal of Educational Technology*, *22*(1), 126–144.

Suthers, D. (1999). *The Effects of Representational Bias on the Collaborative Inquiry*. Paper presented at the HCI.

Suthers, D. (2006). Technology affordances for intersubjective meaning making: A research agenda for CSCL. *Computer-Supported Collaborative Learning*, *1*, 315–337. doi:10.1007/s11412-006-9660-y

Suthers, D. D. (2001). Toward a systematic study of representional guidance for collaborative discourse . *Journal of Universal Computer Science*, *7*(3).

Suthers, D. D. (2003). Representational guidance for collaborative inquiry. In J. Andriessen, M. Baker & D. Suthers (Eds.), *Arguing to Learn* (pp. 27-46). London: Kluwer.

Suthers, D. D., Girardeau, L. E., & Hundhausen, C. D. (2002). *The roles of representation in online collaborations*. Paper presented at the AERA 2002, New Orleans, LA.

Suthers, D. D., Toth, E. E., & Weiner, A. (1997, 14 octobre). *An integrated approach to implementing collaborative inquiry in the classroom*. Paper presented at the Computer support of collaborative learning, Toronto, Canada.

Tennyson, R. D., & Rasch, M. (1988). Linking cognitive learning theory to instructional prescriptions. *Instructional Science*, *17*(4), 369–385. doi:10.1007/BF00056222

van Boxtel, C., van der Linden, J., & Kanselaar, G. (1997). Collaborative Construction of Conceptual Understanding: Interaction Processes and Learning Outcomes Emerging From a Concept Mapping and a Poster Task. *Journal of Interactive Learning Research*, *8*(3/4), 3F–361.

Van Boxtel, C., van der Linden, J., & Kanselaar, G. (2000). Collaborative learning tasks and the elaboration of conceptual knowledge. *Learning and Instruction*, *10*, 311–330. doi:10.1016/S0959-4752(00)00002-5

Vygotsky, L. S. (1978). Mind in society: The development of higher psychological process. Cambridge, MA: Harvard University Press.

Wandersee, J. (1992). *A standard format for concept maps*. Paper presented at the National Convention of the National Science Teachers Association, Boston.

West, C. K., Farmer, J. A., & Wolff, P. M. (1991). *Instructional design: Implications from cognitive science*. Englewood Cliffs, NJ: Prentice Hall.

KEY TERMS AND DEFINITIONS

Activity Theory: A psychological meta-theory initiated by a group of revolutionary Russian psychologists in the 1920s and 1930s, mainly L.S. Vygotsky and his colleagues A. R. Luria and A. N. Leont'ev. Basic principles of Activity Theory include object-orientedness, the dual concepts of internalization/externalization, tool mediation, hierarchical structure of activity, and continuous development.

Boundary Object: Term proposed by Star & Griesemer (1989) to designate an entity (artifact, document, vocabulary) that can help people from different communities build a shared understanding. Boundary objects will be interpreted differently by the different communities, and it is an acknowledgement and discussion of these differences that enables a shared understanding to be formed.

Cognitive Artefact: An artificial device designed to maintain, display, or operate upon information in order to serve a representational function.

Cognitive Tools (or Mind tools): Computer-based tools and learning environments that have been adapted or developed to function as intellectual partners with the learner in order to engage and facilitate critical thinking and higher-order learning.

Meaningful Learning: According to Ausubel (1968), it is a non-arbitrary, non-verbatim, substantive incorporation of new knowledge into cognitive structure requiring deliberate effort to link new knowledge with higher order concepts in cognitive structure.

Object-typed concept mapping: A technique of concept mapping based on the use of typologies of knowledge objects and links.

Representational Guidance: According to Suthers (2003), it designates a set of constraints and facilities offered by the representational notation that provides a set of primitive elements out of which external representations can be constructed. A cognitive tool is a software implementation of a representational notation.

ENDNOTES

1. Various terminologies have been used to designate more or less similar graphical knowledge representations (*knowledge maps, semantic networks, mind maps, knowledge graphs, cognitive maps, visual thinking networks*, etc.). Some characteristics of graphical representations are occasionally used to differentiate them (hierarchical or non-hierarchical map structures, labelled links or not, use of a link typology or not, etc.). The terminology has yet to be standardized. In this chapter, the expression "concept map" is employed, as it is most commonly found in educational science literature. However, it is used in a very generic sense to designate all graphical knowledge representations based on nodes and links.
2. Some of these studies compared face-to-face and distant conditions.
3. Created by Alberto Cañas' team, Institute for Human and Machine Cognition, University of West Florida (USA): http://cmap.ihmc.us/
4. Created by Kathleen Fisher & John Faletti's team, SemNet Research Group (San Diego, USA): http://trumpet.sdsu.edu/SemNet_About_SemNet.html
5. The term "network map" used by Dansereau and colleagues at that time has been replaced later by "knowledge map" (O'Donnell, Dansereau, & Hall, 2002).
6. When juxtaposed to the term "knowledge", the adjective "declarative" comprises two different meanings which are often misinterpreted. On one hand, all overtly "verbalised" knowledge (i.e., expressed with words) is said to have a declarative format. Procedural knowledge can thus be represented in a declarative format. On the other hand, "declarative knowledge" can refer to knowledge pertaining to objects and object properties (the *know-what*), as opposed to "procedural knowledge" or knowledge pertaining to actions (the *know-how*).
7. MOT stands, in French, for "*Modélisation par Objets Typés*", which means "*Object-typed modeling*".
8. Based at Télé-université in Quebec, Canada, the LICEF Research Center is a laboratory dedicated to cognitive informatics and training environments. For further details on MOT and to download an English version of the software freely, visit the LICEF Website at http://www.licef.ca.
9. The dyad obtained a "Knowledge Object Score" (KO Score) of 34/82 and a "proposi-

tion score"(P Score) of 50/192, for a total of 84/260. These results show that the quality of the maps created by all participants was quite low (KO Score: M = 25,65 ; SD = 6,16; P Score: M = 26,50; SD = 10,04), as measured by comparing them with an expert map created consensually by two MOT experts and two content experts (also knowledgeable in knowledge modeling with MOT). See Basque & Pudelko (2004) for more details on the scoring method and on group results.

10 LORIT is a French acronym that stands for "*Laboratoire Observatoire de Recherche sur l'Ingénierie du Télé-apprentissage*", basically, a laboratory and an observatory for research pertaining to telelearning engineering. Télé-université is a distance education university. For more information about the LORIT, please visit www.licef.ca

11 Participants have not been introduced to concrete knowledge objects (facts) in this study.

12 Vignettes and maps have been translated from French.

13 In French, the term "information" can have a plural or a singular form. In figure 6 reproducing a part of the dyad's map, the term "information" appearing at the left side has the plural form and the one appearing at the right side has the singular form.

14 This could be due to the fact that the participants were not introduced to the MOT feature that allows to "copy with a reference" a KO that comes from another level of the map. This feature facilitates finding KOs which are replicated in sub-maps. Moreover, when a modification is made to the label of a referenced KO, it will be done in all sub-maps where the KO appears.

Chapter 11
Collaborative Learning by Developing (LbD) Using Concept Maps and Vee Diagrams

Päivi Immonen-Orpana
Laurea University of Applied Sciences, Finland

Mauri Åhlberg
University of Helsinki, Finland

ABSTRACT

Collaborative Learning by Developing (LbD) was researched in a University course. The focus was reflective metacognitive competence development (Appendix 1.) of Physiotherapy students. The authors used both individual and collaborative concept mapping and improved Vee heuristics in learning process evaluation. The content of the design experiment was 'Coping at Home'. As educational research it was a design experiment, a multi-case, multi-method study. The core concept of the study unit and development project was 'successful aging'. Both Cmap Recorder and videotaping of discussions during group concept mapping were used. The main result was that plenty of face-to-face dialogue was needed before the shared understanding and group concept maps were created. First the main concepts were fixed and then other concepts and their relationships were elaborated. Differences between individuals and two groups are analysed. In the collaborative learning process, the feeling in both groups was as if they had a unified and shared thinking process. Students continued each others talking and thinking very fluently like they had had "common brains".

DOI: 10.4018/978-1-59904-992-2.ch011

INTRODUCTION

The purpose of the study is to research on Collaborative Learning by Developing in university course of physiotherapy students. We focused on students' ability to evaluate their own learning and competence and formulate a personal development plan; to identify development challenges and problems; to develop their competence independently and with others, and to share what they have learnt in a work community; to assume a role in a group/team and acquire and analyse knowledge systematically. Learning by Developing (LbD) is developed in Laurea University of Applied Sciences, Finland. It is a teaching and learning method. We focus on meaningful conceptual change. Physiotherapy students' learning is monitored and evaluated by the concept maps and improved Vee heuristics. The following aspects are researched: (1) individual conceptual changes and (2) collaborative learning of two student groups.

BACKGROUND

Learning as Conceptual Change and Collaborative Learning

According to Pintrich & Sinatra (2003, p. 6) intentional conceptual learning is goal-directed. The goal is to change conceptual understanding. Intentional conceptual change is characterized by conscious initiation and regulation of cognitive, metacognitive and motivational processes to bring about a change in knowledge. According to Vosniadou & Kollias (2003, 2) the conceptual change is the outcome of a complex cognitive as well social process. Studies of conceptual change have shown that this is a slow and gradual process.

Collaborative learning is used in this report in the sense of Adey & al. (2007, 93): Collaborative learning means that learners learn to listen to another, to argue and to justify, and become accustomed to change their positions. Laurillard (2002) calls for learning technologies and discourse to promote collaborative learning.

Concept Mapping as a Method to Monitor and Promote Learning, Thinking and Acting

In this study concept mapping is used to represent the conceptual understanding and change of it. According to Novak & Cañas (2008) concept maps are graphical tools for organizing and representing knowledge. They include concepts, usually enclosed in circles or boxes of some type, and relationships between concepts indicated by a connecting line linking two concepts. Words on the line referred to as linking words or linking phrases, specify the relationship between the two concepts. They define concept as a perceived regularity in events or objects, or records of events or objects, designated by a label. The label for most concepts is a word, although sometimes they use symbols such as + or %, and sometimes more than one word is used. Propositions are statements about some object or event in the universe, either naturally occurring or constructed. Propositions contain two or more concepts connected using linking words or phrases to form a meaningful statement. Sometimes these are called semantic units, or units of meaning. (Novak & Cañas 2008, 1)

Åhlberg (1993 – 2005) regards concept mapping as a research method to monitor and promote high quality learning, thinking and acting. Åhlberg (1993) defines concept as a basic element or unit of thinking. Concept becomes accurate only in relation to other concepts, as a part of theory. Importance or centrality of a concept can be estimated from a concept map, by counting the number of links connecting a concept to other concepts. The more links a concept has with other concepts the more important it is in that concept map. If you would take the highest linked concept away from

Table 1. Comparing the main elements of Gowin's Vee diagram and improved Vee heuristic

The main elements of the Gowin's Vee diagram (Novak & Gowin 1984; Novak 1990, 1998)	The main elements of the improved Vee heuristic (Åhlberg 1993 – 2006; Åhlberg and Ahoranta 2002; Åhlberg, Äänismaa & Dillon 2005)
1. Focus questions	1. Focus question(s)
2. World view	2. Value basis: Why do you want to spend your life, time and resources to answer the focus question(s)?
3. Philosophy/epistemology	3. Theoretical basis: What is your tentative theory in the beginning of your inquiry? What do you know in the beginning of your inquiry?
4. Theory	
5. Principles	
6. Constructs	
7. Concepts	4. Conceptual basis: What are the main concepts of your theoretical basis? They act like lenses or a net by which you try to answer the focus question(s).
-	5. Methodological basis: What methods do you plan to use to answer your focus question(s)?
8. Events and/or objects	6. Description of what has been really done in order to answer the focus question(s).
9. Records	7. Records
10. Transformations	8. Transformations
11. Knowledge claims	9. Knowledge claims
12. Value claims	10. Value claims

a concept map, it would cause the greatest damage to that conceptual and propositional network.

Promoting High-Quality Learning

According to Åhlberg (1993 - 2005) high quality learning has at least 20 aspects. It is divided into individual and social learning, learning from personal level to organizational and the humankind levels. The first four aspects of high quality learning are as follows: 1) Meaningful in the sense that it corresponds real needs of individual, society and humankind. Through the concept of 'real needs' high quality learning is connected to the general definition of 'quality'. It is meaningful also in the Ausubel's sense that new learnt knowledge is connected to earlier knowledge. 2) Deep in the sense that grounds and justifications for knowledge are actively sought after, and consequences of knowledge is actively tested both theoretically and empirically. 3) Proactive, creative, expanding, transformative, surpassing earlier knowledge and expertise in the sense that real human needs are better and better met, real problems are solved, or at least alleviated, and better future is created as a result of creative learning. It often means reframing problems, seeing the world, its problems and real human needs in different perspectives. 4) Metacognitive, in the sense that ways to monitor and promote own learning are learnt. According to Biggs & Tang (2007, 115 – 117) concept maps promote quality learning.

Novak and Gowin (1984) have published an influential book about concept maps and Gowin's Vee diagram or Vee heuristic to promote meaningful learning. Åhlberg (1993) created and presented his improved versions of both concept maps and Vee heuristic, which have been used and improved later on (e.g. Åhlberg 2005; Åhlberg, Äänismaa & Dillon 2005). The problem with original Vee

Figure 1. Competence development in Laurea University of applied sciences (Adapted from Raij, 2007)

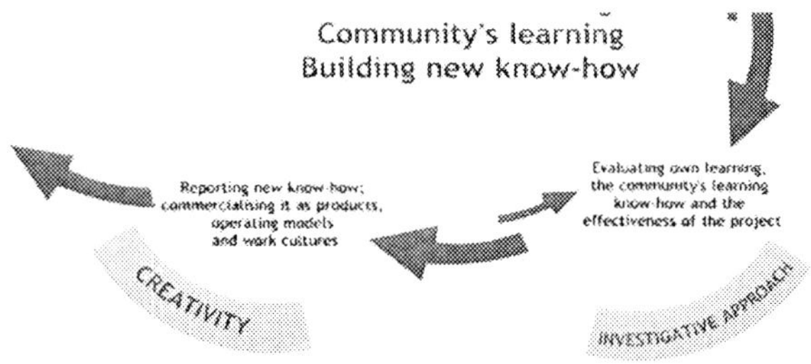

heuristic was that university students in Finland did not know what their 'World View' is. Åhlberg changed 'World View' to 'Value Basis'. It was described by a question: 'Why do you want to spend your life, time and resources to answer the focus question'? Åhlberg integrated the basic ideas of action research and Deming's cycle from Continual Quality Improvement movement into the design of Improved Vee Heuristic. This tool provides for teachers, pupils and researchers valuable data (e.g. Åhlberg & Ahoranta 2002, Åhlberg, Äänismaa & Dillon 2005). Gowin's original Vee diagram is compared to improved Vee heuristic in the Table 1.

Learning by Developing

In the late 1990s, Laurea University of Applied Sciences chose as its strategic approach the integration of learning, education, research and development, and regional development. This strategy was revised in spring 2007 (Raij 2007). According to the strategy, learning at Laurea takes place through instruction, research and development. While implementing the pedagogical strategy, Laurea's developers refined this principle into the Learning by Developing (LbD) model. Competence development in the LbD- model includes values of authenticity, partnership, experiential nature, investigative approach and creativity. (Figure 1.) Learning by Developing combines professional education (learning) and research-oriented higher education (developing). With the new competence-based core curriculum, Laurea continues to strongly define itself as a university of applied sciences specialising in service innovations, whose specific task is to foster the competitiveness and regional development of the Helsinki metropolitan area. Two of the most evident outcomes of this long-term development effort are Laurea's appointment as a centre of excellence in regional development for 2006-2007 evaluated by Higher Education Evaluation Council of Finland and the fact that according to the latest statistics (graduates from 2000-2004), Laurea graduates are employed better than students from any other university of applied sciences in Finland. (Rauhala 2007a)

The development objectives of the European Higher Education Area and research on curricula carried out by Finnish higher education institutions led to the adoption of a competence-based curriculum. The model's focus is not on contents but on broader competences (Appendix 1) needed in the workplace of the future. In Laurea's theory, subject-specific competences are also generic competences (knowledge-based and skill-based professional competences). (Rauhala 2007b)

The curriculum of physiotherapy education is based on Laurea's development work and European Physiotherapy benchmark work (WCPT 2003). Laurea's Physiotherapy curriculum examines physiotherapy with the help of its four main phenomena (movement, mobility, functional ability and activity). The investigation of the phenomena of physiotherapy is customer-oriented and takes place in various contexts and situations in accordance with the principles of sustainable development and welfare. It focuses particularly on operational challenges and problems.

Professional growth is based on recognising and reflecting on the students' own activities and their desire to become investigative and developmental experts in their field. The LbD model includes meta-level generic competences, which are also essential to expertise in physiotherapy. Professional growth starts with the basic level of knowledge-based and skill-based competence, and progresses through the intermediate stage to the advanced level. At the basic level, the student's competence is personal. Student can identify and describe the principles, knowledge base and conception of humanity used in physiotherapy, and work appropriately, safely and reliably in various physiotherapy-related learning environments. At the intermediate level, competence expands to encompass the community; at the advanced level, it influences society. Advanced-level students understand and can explain the knowledge base of physiotherapy from various perspectives and evaluate existing research data, applying them to work in the field. They can produce new analyses and develop new practices for physiotherapy. The expertise and skill-based professional competence descriptions included in Laurea's Physiotherapy curriculum have been compiled from different sources. In the curriculum the professional Therapeutic skills are divided into five areas: synthetic skills; pedagogical skills; manual and technical skills; encountering skills; and research and development skills. Expertise also includes network, ethical, reflective, globalisation and innovative competence, which are described in the generic competences. These areas of expertise together form the skill-based professional competence needed by a physiotherapist. Curriculum process in physiotherapy in Laurea is described by Piirainen, Julin and Immonen-Orpana (2007).

SUCCESSFUL AGEING AS A CONCEPT FOR WELLBEING PROFESSIONALS OF THE FUTURE

The Purpose of the Study and Research Questions

The purpose of the study is to make research on quality of learning and the development of the professional and generic competence of the physiotherapy students (Laurea 2006). The teaching and learning method was learning by developing. We focus on meaningful conceptual change. Physiotherapy students' learning is monitored and evaluated by concept maps both at individual and group levels and improved Vee heuristics The core concept of the course and developmental project is 'successful aging':

1. What kind of individual conceptions do the students build on the concept of successful ageing? What kind of concepts and propositions they have a) before the project, b) after the project and c) after they have built together a summarising concept map?
2. What happens in collaborative learning? What happens when a concept map of the same theme is built by group of students? What kinds of differences (if any) are in collaborative learning processes and products when the two groups are compared?
3. What kind of collaboratively created concepts and propositions do the students have after the project?

4. What did the students learn self-reflected by Vee diagrams?

Method and Subjects

As educational research this is a design experiment, a multi-case, multi-method study (applying Brewer & Hunter 2006 and Yin 2003). The subjects are Bachelor of Health Care students in Laurea University of Applied Sciences in Finland. They are third year Physiotherapy students, and all are female. They are studying their professional study unit Physiotherapy of the elderly in the Coping at Home developmental project and in its Caring TV- project. In the two study groups there are 22 students. Out of these 22 students, a purposive sample of nine students was selected: three high achieving, three average achieving and three low achieving students based on their academic achievements before this course. The nine subjects differ in the matriculation examination of Finnish language test (one of the qualifications of entrance to university).

Developmental Project as a Research and Learning Environment

The learning and research environment - the developmental project - is based on real needs of elderly people (over 75 years) who still live at home and wish to stay home and even after having been hospitalized for a while, wish to return to their homes. Research and development targets:

1. The processes leading to maintain and promote wellbeing in elderly people living at home
2. New models of wellbeing and technology based products which support living at home
3. The two way interactive channel to offer guidance and support services for elderly people and their significant others at homes
4. Quality measurement tools of welfare services which support living at home

The project partners are Espoo City, Laurea University of Applied Sciences and IT–specialist TDC Song. The physiotherapy students are going to design programs for an interactive TV, Caring TV-project to support the functional ability, activity, social and mental welfare of the elderly and also to give possibilities to the elderly to discuss with other elderly people in similar positions with the help of different counseling, teaching and instruction methods. The aim of the project is to support successful ageing and sustainable welfare. The needs and expectations of the elderly (Piirainen & Sarekoski 2008, 43) exist in four themes: 1. Promotion of family caregivers work, (welfare and well being of care receivers) 2. Security, (safety) receiving help, taking care of the relatives when absent. 3. Action competency, functional activity, to inform, to help and to see the instructor. 4. Possibility to participate, to take part in activities - doing together, participating in their own groups. (Figure 2.) Students try to find out these themes in their professional basis and to plan evidence based functional sessions as programs of that interactive television. To plan the programs they need to understand and learn more about the main concept of project: successful ageing.

Data Collection as a Part of Design Experiment

In September 2006, before the study unit of Physiotherapy of elderly starts, the students built their first individual concept maps of the concept of successful ageing based on their own common sense knowledge. After that they discussed the themes and tried to find out their own professional possibilities to support the aims of the project. They wrote evidence based reports of the theme and planned programs to television. They had their 7 week sessions; one to two programs

Figure 2. Successful ageing according the elderly in the Coping at Home Project (Piirainen & Sarekoski 2008, 43). (© 2007, 2008 Laurea University of Applied Sciences. Used with permission)

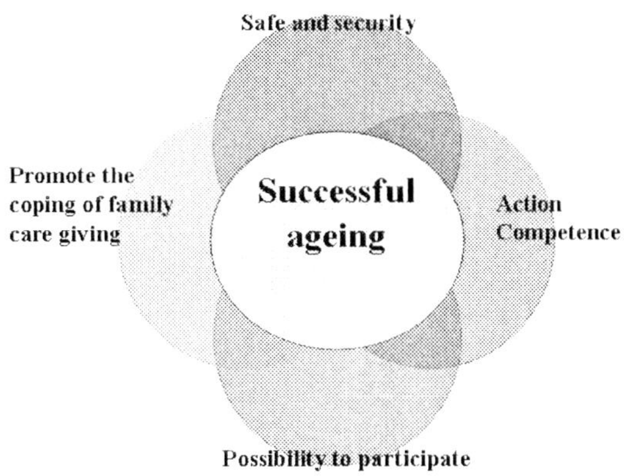

Table 2. The categories in dialogue analysis applying Aarnio (1999, 77) and Aarnio & Enqvist (2001)

1. synthesizing concepts (CS)
2. synthesizing propositions (PS)
3. expressing own opinion (O),
4. asking a clarifying question, (CQ)
5. asking a direct question (DQ)
6. supporting other's ideas by agreeing or linking own ideas to other's (S)
7. answering a question (AQ)
8. continuing other's idea with a new idea (C),
9. expressing own attitudes (A)
10. expressing an unnecessary idea, that is not linked to aims (UN)

every weekday and 3-4 programs per a student pair. The students had a possibility to discuss in a computer based learning environment of their own experiences and evaluate each others work and reports. They visit in homes of the elderly and there to observe and evaluate the functional ability and activity, quality of life, stress and experiences of the elderly by World Health Organisation's (WHO) measures. After that they built their second individual concept maps of successful ageing based on experiences with the elderly, evidence based literature and reports and work in television. In January 2007, the students wrote their self-reflection by improved Vee heuristics of the process of planning their TV programs. Because of practical scheduling reasons the students were divided into two groups: group A had 15 members and group B had 7 members. In January both group A and B created collaboratively a group level concept map of the theme' successful ageing' These face- to – face interaction sessions were both videotaped and recorded by Cmap Recorder of CmapTools. In February 2007 the students built third individual concept maps, individual professional concept map based on shared collaborative learning.

Figure 3. The total number of concepts and propositions in improved concept maps

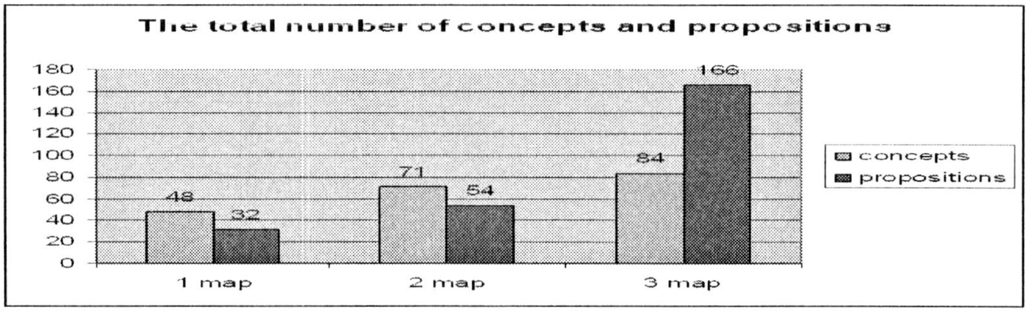

Data Analysis

Both individually and socially built concept maps and improved Vee heuristics were evaluated and analysed by content analysis and by simple statistical analysis. Collaborative group level concept mapping processes, videotaped sessions, were analyzed by dialogue analysis. Before process analysis, the dialogue was divided into dialogue turns. A turn is an interval of expression by a single participant. The units were either turns or parts of turns, such as sentences or single words. Dialogue analysis of Aarnio (1999, 77) and Aarnio & Enqvist (2001) was applied (Table 2).

Results

Individual Conceptual Change

What kind of individual conceptions do the students build on the concept of successful ageing? What kind of concepts and propositions they have a) before the project, b) after the project and c) after they have built together a summarising concept map?

There were differences in process but no qualitative differences between students based on their academic achievements before this course. The number of concepts and propositions increased during the process. In Figure 3 is described the chances of total number of concepts and propositions.

Before the study unit, in first concept maps of a concept of "successful ageing" the students created 48 concepts and 32 propositions. The following kinds of abstract concepts were used: 'life' (good life, quality of life, life satisfaction) 'health' (physical and mental health), 'welfare', 'functional ability', 'human relationships', 'economical situation', 'home' and 'environment'. In Figure 4, 5 and 6 is an example of concept maps of high achieving student.

Immediately after the study unit, in the second individual concept maps the total number of concepts increased from 48 to 71 and the total number of propositions from 32 to 54. The content categories remained similar as earlier. Professional, active functional aspect was very prominent and concepts of 'independency' and 'activity' were mentioned many times. In propositions there was found five themes: "supporting the health", "physical -", "social -", "environmental -" and "psychological existent".

After the social concept mapping, in the third individual concept maps, the total number of concept increased from 71 to 84 and the total number of propositions increased from 54 to 166. In the third individual concept maps, there were lot of new propositions and concepts like "optimistic attitude", "sexuality", "good physical condition" "social support" and "pets" which did not exists in first or second individual concept maps. In propositions there

Figure 4. "Successful ageing" according a high achieving student before study unit

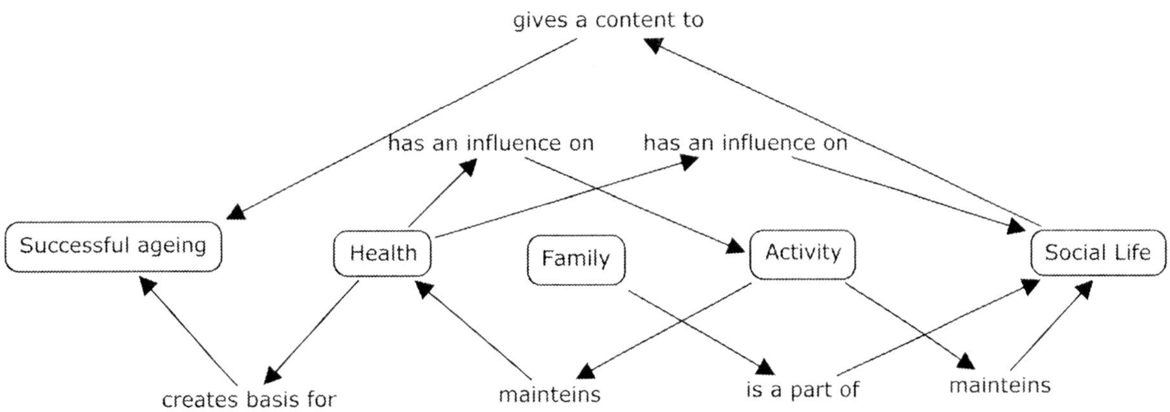

Figure 5. "Successful ageing" according a high achieving student immediately after study unit

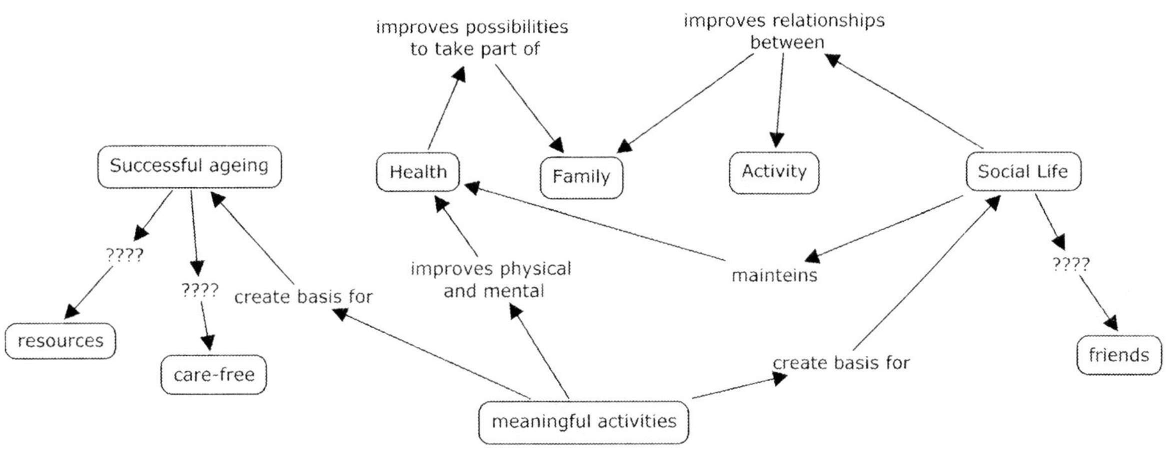

were found very similar themes like earlier and a new theme "safety". Themes in third concept maps had changed to become more concrete compared with second maps like "environmental existent" which had changed to "stimulating environment". Even those students who didn't participate orally in social concept mapping used socially build concepts in their third concept maps.

There were differences in process between students in different groups (Figure 7 and 8). In the second concept mapping the number of concepts increased in the "high achieving" group from 18 to 26; in the "average achieving" group from 12 to 23 and in the "low achieving" group from 18 to 22. The number of propositions increased in the "high achieving" students from 12 to 28, in "average achieving" students from 12 to 22 and decreased in "low achieving" students from 8 to 4. The increase of relevant propositions indicates increase in meaningful learning.

After the social concept mapping, in the third individual concept maps the "high achieving" students increased their total number of propositions

Figure 6. "Successful ageing" according a high achieving student after social concept mapping

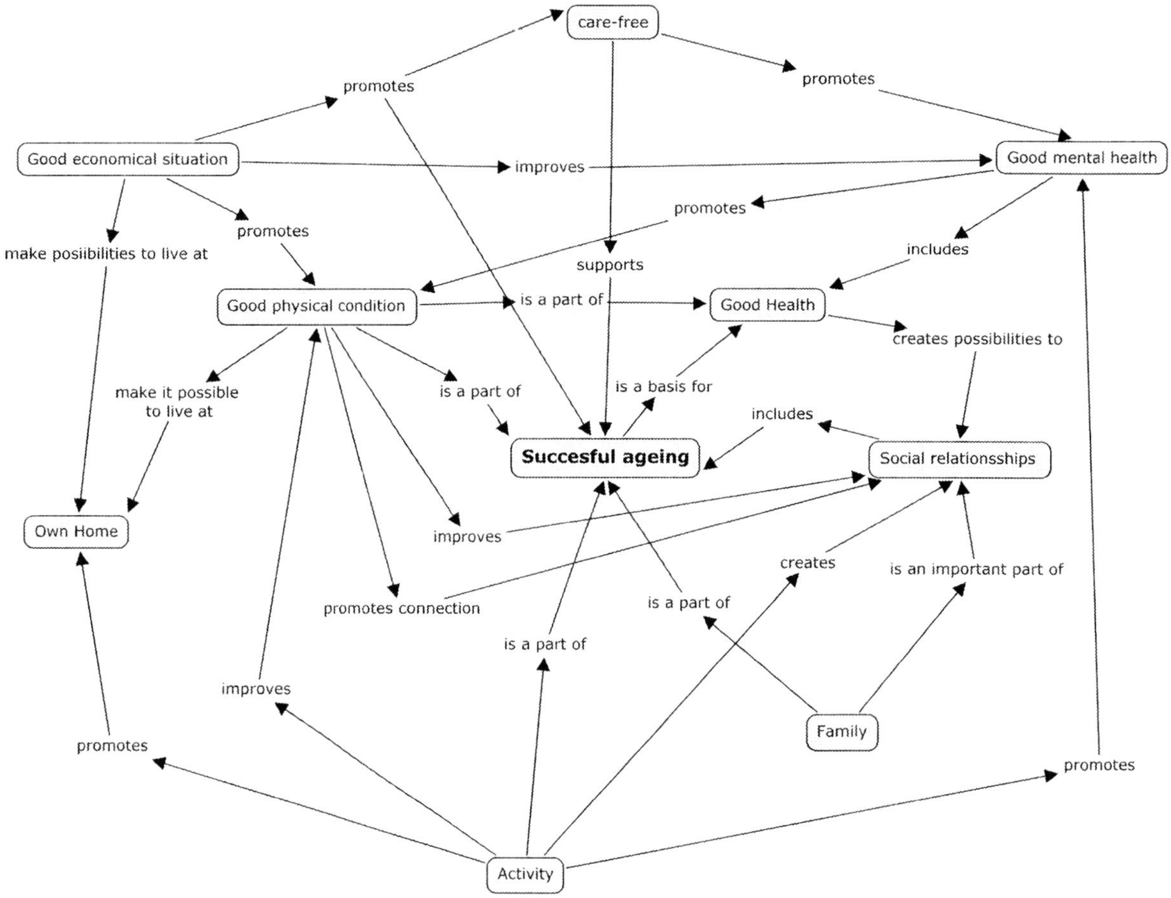

Figure 7. The number of concepts in different groups

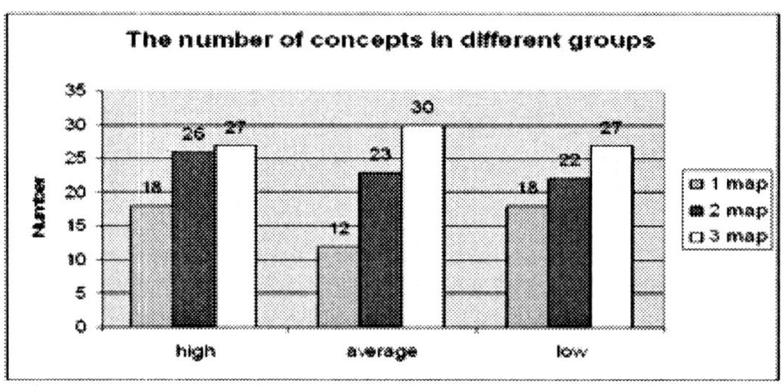

only from 28 to 49 and the "average achieving" students from 22 to 61. The "low achieving" students' group increased the number of propositions from 4 to 56. The number of relevant concepts increased only a little. The increase of relevant propositions indicates increase in meaningful learning. In the third individual concept maps, multidisciplinary, stage, concrete links are made

Figure 8. The number of propositions in different groups

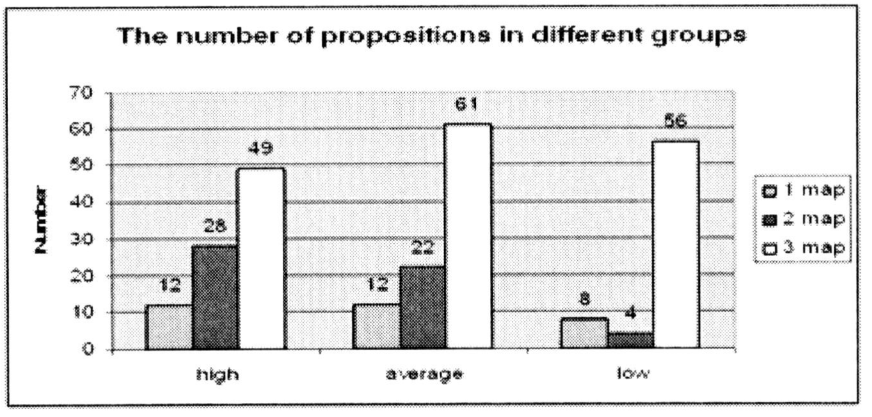

Figure 9. Number of speech acts in the dialogue of the first group

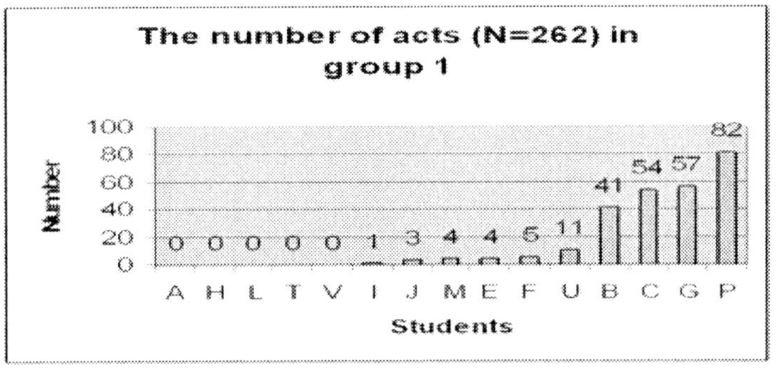

between all concepts, interconnectedness is understood and the wider context dependency of the concepts is recognized.

Collaborative Learning

What happens in collaborative learning? What happens when a concept map of the same theme is built by group of students? What kinds of differences (if any) are in collaborative learning processes and products when the two groups are compared?

Between the second and third individual concept maps a group level collaborative concept map was created in both groups. The first group created its collaborative concept map in January 2007. The whole process took one 45 minutes session. Students of this bigger group (15 members) created 262 speech acts in their dialogue. (Figure 9) The variance of speech acts in their dialogue was from 0 to 82 speech acts. In that bigger group five of the 15 students did not say anything during the process. The most active speaker was the student, who acted as a secretary in social concept mapping.

The second group created its collaborative concept map also in January 2007. The whole

Figure 10. Number of speech acts in the dialogue of the second group

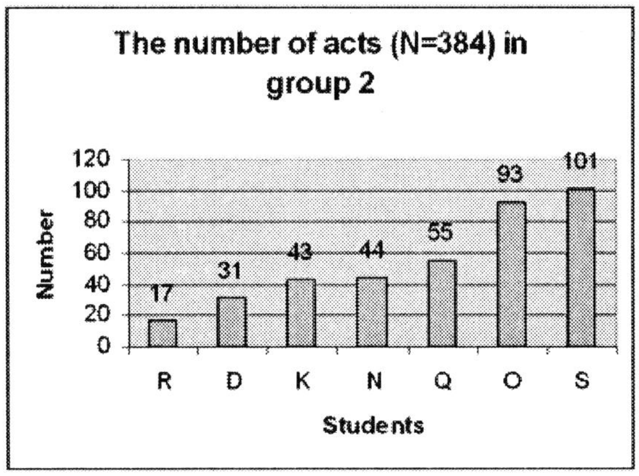

Table 3. The categories of dialogues in collaborative concept mapping of the first group

Category	Number
1. synthesizing concepts (CS)	15
2. synthesizing propositions (PS)	40
3. expressing own opinion (O),	42
4. asking a clarifying question, (CQ)	34
5. asking a direct question (DQ)	32
6. supporting other's ideas by agreeing or linking own ideas to other's (S)	133
7. answering a question (AQ)	22
8. continuing other's idea with a new idea (C),	51
9. expressing own attitudes (A)	5
10. expressing an unnecessary idea, that is not linked to aims (UN)	5
SUM	379

process was over in one 45 minutes session. Students of that smaller group (7 members) created 384 speech acts and the variance was from 17 to 101, so in that smaller group (Figure 10) all seven students participated also orally. The most active student was, like in first group, the secretary of that social concept mapping. The discussion during the concept mapping process was like a shared thinking process. Students continued each others talking and thinking very fluently like they had had "common brains".

In the first group, the videotaped conversations 262 speech acts were divided into dialogue turns as a part of building group level concept maps. All dialogue turns were analyzed applying the category system of Aarnio & Enqvist (2001) (Table 2). In the Table 3, there are the results of dialogue analysis of the first group conversation.

The sum of all dialogues was 379 speech acts. In that first group the category used mostly was to support others opinion to connect or to agree. Almost a half of all dialogical turns were classified in that category. There were 66 questions (CQ+DQ) but only 22 answers. There were very few unnecessary opinions. It took many speech acts of dialogue before the synthesis of concepts or propositions was finished. In their concept map there were 15 concepts and 40 propositions. (Figure 7)

In the second group, the videotaped conversations 384 speech acts of dialogue were analyzed as a part of building group level concept maps. All dialogue turns were analyzed by the category system applying Aarnio & Enqvist (2001). In the Table 4 there is the result of dialogue analysis of the second group.

In the second group the category used mostly was also to support others opinion or to agree. And also almost a half of all dialogical turns were classified in that category. There produced quite many question (123 questions) and only 36 answers. Unnecessary opinions were 9, also only a few like in the first group.

In the collaborative improved concept mapping of the second group, they produced 11 concept synthesis and 22 proposition synthesis. So it means lot of discussion and conversation before the synthesis had existed. (Figure 11.)

What kind of socially built conceptions (concepts and propositions) do the students have after

Table 4. The categories of dialogues in collaborative concept mapping of the second group

Category	Number
1. synthesizing concepts (CS)	11
2. synthesizing propositions (PS)	22
3. expressing own opinion (O),	5
4. asking a clarifying question, (CQ)	87
5. asking a direct question (DQ)	36
6. supporting other's ideas by agreeing or linking own ideas to other's (S)	190
7. answering a question (AQ)	48
8. continuing other's idea with a new idea (C),	35
9. expressing own attitudes (A)	21
10. expressing an unnecessary idea, that is not linked to aims (UN)	9
SUM	464

Figure 11. The number of concepts and propositions in socially built concept maps

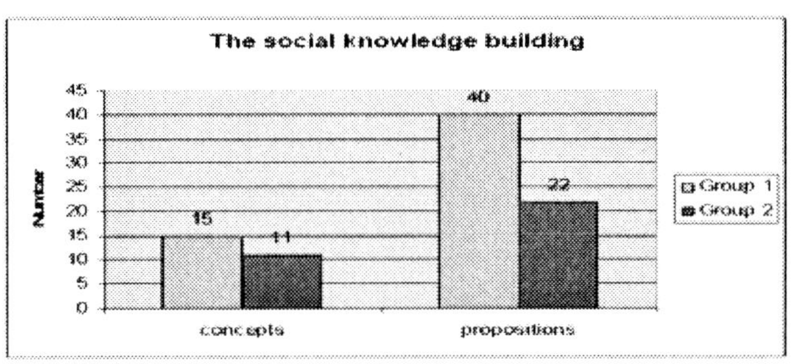

the project?

In the socially built concept map, there were few concepts, which did not exist in first or second individually built concept maps, like "supporting network", "accepting the death", "free environment" and few concepts, which was so meaningful, that students had them in their third individual concept maps created after social concept mapping, like "safety", "social network" and "good economical resources". The safety – theme was also mentioned in many propositions.

The process and content of collaborative concept mapping of two groups was quite similar according to data from Cmap recorder. Both groups started by the concept of "successful ageing" and continued by themes like health (good health, comprehensive health), home (home of own), independency (to make own decisions), environment, services (good services) and social network (supportive social network). In the first group the important concepts were connected to mental, social and environmental area like "accepting to become old and to die" or "good economical situation". There were no mentions of physical aspect (Table 5.) In the second group the free living environment was important, but there were no mentions of physical or mental aspects. (Table 6)

Table 5. Concepts in the collaborative concept mapping of the first group

	INNER RESOURCES			ENVIRONMENTAL RESOURCES
Common level	**Physical aspect**	**Mental aspect**	**Social aspect**	
Good health	0	Good mental health	Social network	Active environment
Independency		Accepting becoming old	Good economical balance	Own environment
Activity of own		Accepting dying		Home of own
Functional ability		Safety		Good services

Table 6. Concepts in the collaborative concept mapping of the second group (using CmapTools)

	INNER RESOURCES			ENVIRONMENTAL RESOURCES
Common level	**Physical aspect**	**Mental aspect**	**Social aspect**	
Comprehensive health	0	0	Supportive social network	Living environment
Possibility to make own decisions Active life style			Family, Society	Free environment Home Services

Self-Reflection by Improved Vee Heuristics

What did the students learn self-reflected by improved Vee heuristics?

Students reflected their learning themselves and created improved Vee heuristics by pairs of 10 small groups. Students described three of four themes: functional activity, welfare and safety. The possibility to participate and take part in activities was the way to implement the CaringTV programs and that's why it was not a theme.

1. **Focus question(s):** The focus questions of the improved Vee heuristics were professional and connected to the needs and expectations of the elderly. (Table 7.) The most meaningful issues were "welfare" (4 groups / 9 students) and "functional activity" (5 groups / 11 students). One group had chosen both the theme "safety" and the other theme "activities of daily living" (ADL) and one group only the theme "safety". All focus questions were connected to themes, for example: does the music and dance have any effects to the welfare of the elderly; does the balance exercise promote ADL? All questions were also connected to the expertise areas of physiotherapy like functional ability and human movements. For example balance training, physical fitness and strength training was involved in programs of six small groups and relaxation techniques in programs of two small groups.
2. **Value basis:** Why do you want to spend your life, time and resources to answer the focus question(s)? According to two of ten small groups it had to take part to study unit because it is a part of curriculum. They described the value basis:

"For credits, this is a part of study"

"Part of curriculum"

One of the small groups wrote that the study unit is the part of curriculum but it also gives intensive knowledge of physiotherapy of the elderly people. Seven small groups agreed that the issue is very important because of progress in society. There will be so many elderly in future in Finland.

Table 7. Themes of elderly and TV programs of the student small groups

Expectations of the elderly and themes	Content for the TV programs	Students' group number:
Action competence: **Functional activity**	Functional capacity, ADL (Activities of Daily Living)	3, 7
	Power Muscle strenght	6, 9
	Endurance	4
Promotion of family caregivers work: **Welfare**	Nutrition	8
	Relaxation	1, 2
	Creativity: music, dance	5
Possibility to participate	Participative way implement	All groups
Security, **Safety**	Balance	7, 10

They also wanted to increase their professional knowledge in physiotherapy of elderly.

"Balance is very important part in prevention of falling risk of elderly and supports possibility to stay home"

"We wanted to create an opportunity for recreation and give family care givers some time for themselves"

"We were very interested in this issue"

3. **Theoretical basis:** What is your tentative theory in the beginning of your inquiry? What do you know in the beginning of your inquiry? Students described their experimental basis of knowledge and theoretical basis they have learned earlier. They wrote that the theme was very important but that they had not enough knowledge how to apply that to elderly. Student described their theoretical basis:

" I had only some own experiences"

"Activities of daily living (ADL) are very important for the psychological, social and physical welfare"

"I know something about becoming old and something about training but not these issues together"

4. **Conceptual basis:** What are the main concepts of your theoretical basis? They act like lenses or a net by which you try to answer the focus question(s).

The conceptual basis was partly involved with the concept of successful ageing and partly with the themes and contents of TV programs.

5. **Methodological basis:** What methods do you plan to use to answer your focus question(s)?

All students found knowledge in databases of internet (for example Health Gate, Cochrane, EBM, Medline, PEDRO).

6. **Description of what has been really done in order to answer the focus question(s):** Students described the process of gathering knowledge, reading the material and references. They also described the process of evaluation the reliability of material. Description of what students had really done in order to get answers:

"We try to find out evidence-based reports"

"We find the articles of "health-gate" (=database of evidence based health care research reports and articles)

"We try to find reliable material, we red them and processed the knowledge and wrote the report"

7. **Records** (What kind of material is gathered and got?): The real material included books, research reports and articles. Students tried to find out evidence-based literature, but it was possible to find it out only to some themes. Description of records:

"We found out some material and literature of relaxation, but relaxation and its meaning to elderly people was not at all in research reports"

"We found some evidence-based articles of muscle training for the elderly"

8. **Transformations (Decision making):** Students described their process of transformation and decision making – how to pick up the right information from the material. One of the groups wrote that their transformation was based on experience and other group took all material they had got. Two groups used the results of the collected research reports when they planned the programs. Two groups wrote that they picked up articles by using core concepts. Two groups created hypothesis and tried to find out explanations for limitations. Only three groups got directly material of their chosen theme connected to elderly or physical education of elderly.
9. **Knowledge claims:** Seven student groups wrote that producing the TV program their knowledge of evidence based knowledge of elderly increased and they could support the welfare of the elderly. Three of the ten groups did not learn new or intensive knowledge. One group felt that simply searching for information supported learning process.
10. **Value claims:** Students wrote that their attitude against elderly changed by the process because of that evidence based material. They could respect more the elderly and understood better the problems of ageing. They saw the situation to learn as a new possibility to learn and the process was very interesting way to work. The method – the interactive TV – was new and innovative way for physiotherapists.

"We got new evidence based knowledge and our learning process was different compared with the earlier, because we worked in television"

The evidence based biomechanical and physiological knowledge got a new dimension – they learned values and ethical aspects in promotion of elderly people's functional activity.

"Endurance training is not so important with elderly people – it is more important to improve muscle strength for getting a good balance"

FUTURE TRENDS

Because of good experiences of CmapTools, in future it will be applied as a teaching and learning environment for discussions, instead of the earlier used local digital environment for individual and collaborative learning. The Cmap Recorder data gave a lot of new information of the process. For scientific purposes it is important to replicate these kinds of design experiments. New similar design experiments are planned for the becoming academic years. For us this has been an optimal way to make research, and to monitor and promote quality of learning in this kind of course.

CONCLUSION

As far as we know, this is the most many-sided research project, done at university level, on student collaborative learning, using concept maps. Results are very encouraging. The change between first and second individual concept map was quite small (Figure 3) This was very ordinary way to evaluate students' learning: before and after the study unit. In that case we continued the process with collaborative learning. That was very meaningful to all students because in the third individual concept maps all students improved their conception of the core concept "successful ageing" very highly. Differences between high, average and low achieving students almost disappeared. Especially the average achieving students increased their learning.

In the collaborative learning process, the feeling in both groups was like a unified and shared thinking process. Students were continuing fluently with each others' talking and thinking. The social atmosphere was excellent in both groups and there were plenty of informal qualitative positive evidence for collaborative learning and thinking. Concept mapping centred teaching, studying and learning was a success in these university level groups. Students discussed much more than in traditional teacher centred teaching of the earlier years.

The improved Vee heuristics were integrated parts of the total process and provided plenty of important knowledge of the overall quality of the course and learning experiences. All students learned something important, and they learned to describe and evaluate their learning by improved Vee heuristics. When students evaluated the value of study unit some of students really wanted to know more and some only did the task, because it was a part of the course. The theoretical basis was connected partly to the professional basis concepts of physiotherapy like functional activity, activities of daily living, movement, balance, relaxation and wellbeing and partly to the core concept of study unit and project: successful ageing. It was easy to the students to collect data in internet and library. They could easily use databases in internet and they knew quite good evidence-based databases. Decision making was very difficult. Actually few of student small group did not even describe how their chose the content to CaringTV programs. Maybe it was difficult to understand the question. Most of student did learn something new for their professional competence. One very valuable result was the change in attitudes. Both the new way to work and elderly people as a client in television programs was something special.

For the supervisor the answers to ten questions of the improved Vee heuristic provided plenty of new knowledge of student thinking. For students the answers were worthwhile for continual improvement of their metacognition. The improved Vee heuristic was a window to student values thinking. The researchers got impression that the students were very honest in their answers to the questions used in the improved Vee heuristic.

REFERENCES

Aarnio, H. (1999). *Dialogia etsimässä* [Searching for dialogue]. Doctoral dissertation, Acta Universitatis Tamperensis, (p. 676).

Aarnio, H., & Enqvist, J. (2001, June 25-30). Dialogic knowledge construction as the crucial issue in network-based learning in vocational education. In C. Montgomerie, & J. Viteli, (Eds.), *Proceedings of ED-MEDIA 2001: World Conference on Educational Multimedia, Hypermedia & Telecommunications* (pp. 1-6). Tampere, Finland: AACE.

Adey, P., Csapo, B., Demitriou, A., Hautamäki, J., & Shayer, M. (2007). Can we be intelligent about intelligence? Why education needs the concept of plastic general ability? *Educational Research Review, 2*(2), 75–97. doi:10.1016/j.edurev.2007.05.001

Åhlberg, M. (1993). *Concept maps, Vee diagrams and rhetorical argumentation (RA) analysis: Three educational theory-based tools to facilitate meaningful learning.* Paper presented at The Third International Seminar on Misconceptions in Science and Mathematics, August 1-5, 1993, Cornell University. Retrieved June 20, 2008 from http://www.scribd.com/doc/2868664/Ahlberg-Cornell-University-August-1-5-1993

Åhlberg, M. (1998). Education for sustainability, good environment and good life. In Åhlberg, M. & Leal Filho, W. (Eds.) *Environmental education for sustainability: good environment, good life.* (pp. 25-43). Frankfurt, Germany: Peter Lang.

Åhlberg, M. (2004a). Varieties of concept mapping. In A. Cañas, J. Novak, & F. Gonzales, (Eds.), *Concept Maps: Theory, Methodology, Technology. Proceedings of the First International Conference on Concept mapping (CMC 2004),* Pamplona, Spain, Sept 14 – 17, (Vol. 2, pp. 25-28). Retrieved January 26, 2008 from http://cmc.ihmc.us/papers/cmc2004-206.pdf

Åhlberg, M. (2004b). Concept mapping for sustainable development. In Canas, A., Novak, J., Gonzales, F. (Eds.) *Concept Maps: Theory, Methodology, Technology. Proceedings of the First International Conference on Concept mapping (CMC 2004)*, Pamplona, Spain, September 14-17, (Vol. 1, pp. 39 – 44). Retrieved January 26, 2008 from http://cmc.ihmc.us/papers/cmc2004-233.pdf

Åhlberg, M. (2005). Integrating education for sustainable development. In W. Leal Filho (Ed.), *Handbook of Sustainability Research*, (pp. 477-504). Frankfurt, Germany: Peter Lang.

Åhlberg, M., Äänismaa, P., & Dillon, P. (2005). Education for sustainable living: Integrating theory, practice, design and development. *Scandinavian Journal of Educational Research*, *49*(2), 167–186. doi:10.1080/00313830500048923

Åhlberg, M., & Ahoranta, V. (2002). Two improved educational theory based tools to monitor and promote quality of geographical education and learning. *International Research in Geographical and Environmental Education, 11*(2), 119 – 137. Retrieved June 20, 2008 from http://www.multilingual-matters.net/irgee/011/0119/irgee0110119.pdf

Åhlberg, M., Kaasinen, A., Kaivola, T., & Houtsonen, L. (2002). Collaborative kowledge bilding to pomote in-service teacher training on environmental education: First year results. *Journal of Information Technology for Teacher Education, 10*(3), 227–238.

Åhlberg, M., & Kaivola, T. (2006). Approaches to learning and teaching environmental and geographical education for sustainability. In Lee, J. & Williams, M. (Eds.) *Environmental and Geographical Education for Sustainability* (pp. 79 – 93). New York: Nova.

Åhlberg, M., Turja, L., & Robinson, J. (2003). *Educational research and development to promote sustainable development in the city of Helsinki*: Helping the accessible Helsinki Programme 2001 - 2011 to achieve its goals. *International Journal of Environment and Sustainable Development*, *2*(2), 197–209. doi:10.1504/IJESD.2003.003321

Biggs, J., & Tang, C. (2007). *Teaching for quality learning at university*, (3rd ed.). Maidenhead, UK: Open University Press.

Brewer, J., & Hunter, A. (2006). *Foundations of multimethod research.* Thousand Oaks, CA: SAGE.

Laurea. (2006). *Physiotherapy Education Curriculum.* Espoo, Finland: Laurea University of Applied Sciences.

Laurillard, D. (2002). *Rethinking University Teaching: A Framework for the Effective Use of Learning Technologies*. London: Routledge-Falmer.

Novak, J. D. (1990). Concept maps and Vee diagrams: two metacognitive tools to facilitate meaningful learning. *Instructional Science*, *19*, 29–52. doi:10.1007/BF00377984

Novak, J. D. (1998). *Learning, creating and using knowledge: Concept maps as facilitative tools in schools and corporations*. Mahwah, NJ: Erlbaum.

Novak, J. D., & Cañas, J. D. (2008). *The Theory Underlying Concept Maps and How to Construct and Use Them* (Tech. Rep. IHMC CmapTools 2006-01 Rev 01-2008). Florida Institute for Human and Machine Cognition, 2008. Retrieved June 20, 2008 from http://cmap.ihmc.us/Publications/ResearchPapers/TheoryUnderlyingConcept-Maps.pdf

Novak, J. D., & Gowin, D. B. (1984). *Learning how to learn*. New York: Cambridge University Press.

Piirainen, A., Julin, M., & Immonen-Orpana, P. (2007). Curriculum process in Physiotherapy. In Kallioinen, O. (Ed.) The Competence-Based curriculum at Laurea. *Laurea Publications B 25*, 147- 155.

Piirainen, A., & Sarekoski, I. (2008). Family care giver's CaringTv concept. In Arja Piirainen & Irma Sarekoski (Eds), Customer-driven CaringTV® concept for elderly family care givers living at home. Final report of the Coping at Home research. *Laurea Publication A 62.* 32- 67

Pintrich, P. R., & Sinatra, G. M. (2003). Future Directions for Theory and Research on Intentional Conceptual Change. In G. M. Sinatra & P. R. Pintrich (Eds.) *Intentional Conceptual Change* (pp. 429-443). Mahwah, NJ: Lawrence Erlbaum Associates.

Raij, K. (2007). Learning by Developing. *Laurea Publications A, 58.*

Rauhala, P. (2007a). Foreword. In O. Kallioinen, (Ed.) The Competence-Based curriculum at Laurea. *Laurea Publications B* 25, 5.

Rauhala, P. (2007b). The Competence –Based Core Curriculum as Laurea's Choice. In O. Kallioinen, (Ed.) The Competence-Based curriculum at Laurea. *Laurea Publications B* 25, 11-12.

Vosniadou, S., & Kollias, V. (2003). Using Collaborative, Computer-Supported, Model Building to Promote Conceptual Change in Science. In E. De Corte, L. Verschaffel, N. Entwistel and J. Van Merrienboer (Eds.) *Powerful learning environments: Unravelling basic components and dimensions* (pp. 181–196). Oxford, UK: Elsevier Press.

WCPT. (2003). *European Physiotherapy Benchmark Statement*, Barcelona, Spain. Retrieved from http://www.fysiot.ee/dok/01.pdf

Yin, R. (2003). *Case study research. Design and methods* (3rd Ed.). London: SAGE.

KEY TERMS AND DEFINITIONS

Collaborative Learning by Developing (LbD) means integration of learning, education, research and development, and regional development

Collaborative Learning: means that learners learn to listen to another, to argue and to justify, and become accustomed to change their positions. (Adey, Csapo, Demitriou, Hautamäki, & Shayer 2007, p. 93)

Concept Maps: are graphical tools for organizing and representing knowledge. At the same time concept mapping is also an educational research method to monitor and promote high quality learning, thinking and acting.

Dialogue: means discussion for promoting high quality learning.

Dialogue Analysis: is a method to analyze and describe basic elements of dialogue.

High Quality Learning: in this report means learning that is meaningful, deep, creative and metacognitive.

Improved Vee Heuristic: is an action research framework for planning, implementing and evaluating a course of action with the aim of improving a situation.

Learning as Conceptual Change: means conscious initiation and regulation of cognitive, metacognitive and motivational processes to bring about a change in knowledge.

APPENDIX A: LAUREA'S GENERIC COMPETENCES 2007

LAUREA	Ethical competence	Reflective competence	Network competence	Innovative competence	Globalisation competence
Personal Competence Level Students are able to:	-reflect on their own values and on the consequences of their actions for other persons' welfare -work on the basis of the sector's value base following ethical principles -apply the field's ethical norms to problem-solving -examine situations from different parties' perspectives -work in a tolerant and understanding way in multicultural networks -apply ethical principles to R&D projects	-evaluate their own learning and competence and formulate a personal development plan -identify development challenges and problems -develop their competence independently and with others, and share what they have learnt in a work community -assume a role in a group/team -acquire and analyse information systematically	-present things in writing, verbally and visually using various communication styles -act in situations requiring professional communication and interaction -identify various team work models and work in teams -participate collaboratively and actively in a network -take into account cultural differences and work with people with different cultural backgrounds	-analyse their own thought processes, knowledge and actions -plan, organise and develop their own work -set targets and turn them into appropriate, concrete actions -come up with new ideas, working creatively, expressing their thoughts and convincing others to participate in developing them -use various ideation methods -manage themselves, work during change and in diverse projects -conduct intermediate and final evaluations of a project	-make use of international sources of information in the field -take into account the effects and opportunities of multiculturalism and internationalisation on their fields -recognise the significance of their professional actions in relation to global interactions and sustainable development -evaluate their own value bases from the point of view of sustainable development and globalisation, and apply them to their work
Organisational Development Level Students are able to:	take into account the major customer groups' special characteristics, interests and rights -build customer-oriented interactions and work dialogically -assume responsibility for the welfare of the community and the ethics of the work done -examine ethical issues from the point of view of the organisation and the community -develop fair methods of proceeding in their work	-work dialogically, collaboratively and with self-evaluation -evaluate their self-management and professional competence as part of a community -evaluate their own work as part of a group, community or professional field -work during change, identifying and making use of various opportunities for learning and action -critically argue the principles of their work -apply the concepts of development challenges appropriately	-participate in and manage negotiations -participate in group communications -communicate on topics related to their specialist fields to a lay audience -develop the networked operations of a professional organisation -take initiatives in developing a network	-monitor and analyse organisation's work processes and results -use benchmarking in developing their work -use diverse teamwork methods -lead projects, processes and networks or participate in leading them in ways that inspire commitment in others -link innovation to organisation's strategic targets -take into account the nature and demands of innovation work	-take into account the effects of their own sector and its organisations on sustainable development, responsibility and international connections -promote the values of environmental and social responsibility in organisations in the sector, taking into account global interactions -apply responsibility for sustainable development to organisations in the sector -work in international networks

LAUREA	Ethical competence	Reflective competence	Network competence	Innovative competence	Globalisation competence
Social Reform Level Students are able to:	-critically reflect on their own attitudes and prejudices -actively participate in social debates on values -assume responsibility at societal and global levels -make arguments in favour of fair societal and global changes -solve conflicts and influence issues	-analyse their own actions in relation to societal and historical backgrounds -evaluate and develop the functionality of regional operating methods and models -evaluate systematic information that has been structured, produced and written, and develop societal activities on the basis of it	-work as experts in communication situations -plan communications from a strategic point of view, and work with the media -initiate networked operations and lead a strategic network -adopt a developmental and predictive approach to work, and influence the work environment -monitor and analyse networks' operating models	-gather ideas from a network, convince others of them and link them to the societal framework -ensure that participants in innovation work are sufficiently committed (guidance, encouragement) -create innovations for productization -participate productively in innovation networks, and identify and investigate innovation mechanisms	-recognise the global nature of societal issues and the characteristics of sustainable development -critically evaluate societal values and practices from the point of view of sustainable development, taking into account global interactions -make arguments at societal level, applying the principles of sustainable development and taking into account global effects

Chapter 12
A Systematic Review of Research on Collaborative Learning with Concept Maps

Olusola O. Adesope
Simon Fraser University, Canada
John C. Nesbit
Simon Fraser University, Canada

ABSTRACT

This chapter reviews research about the collaborative use of concept maps for learning. Although decades of research have produced some understanding of the cognitive effects of using concept maps in individual learning, theories about their effects in collaborative learning are less firmly established. The review incorporated a systematic literature search, analysis of dependent variables as effect sizes, and discussion of representative studies. Students who learned collaboratively by constructing concept maps outperformed those who learned from other activities such as studying texts, outlines, lists and lectures. However, no effect of studying pre-constructed concept maps in collaborative settings was statistically detected. There was homogeneity across the subsets of studies investigated. The review concludes with suggestions for future research in learning with concept maps in collaborative environments.

INTRODUCTION

Concept maps represent knowledge as node-link networks that show concepts as nodes and the relationships among the concepts as arrows or links. Researchers and teachers have used concept maps as media for constructive learning activities, instructional presentations, and as shared materials for co-construction of knowledge (Cañas et al., 2003). Researchers have suggested that concept maps may also serve as an effective learning medium when implemented as interactive software (Cañas et al., 2003; Novak, 2002). An example of a concept map is shown in Figure 1. The map in Figure 1 has links that are directional and labeled with a relational term, but the maps used in educational research sometimes have non-directional, unlabeled links.

Dansereau and his colleagues (e.g., Hall, Dansereau, & Skaggs, 1992; Lambiotte & Dansereau, 1992; McCagg & Dansereau, 1991) have used the term 'knowledge map' to refer to a type of concept map with directional links labeled

DOI: 10.4018/978-1-59904-992-2.ch012

Figure 1.

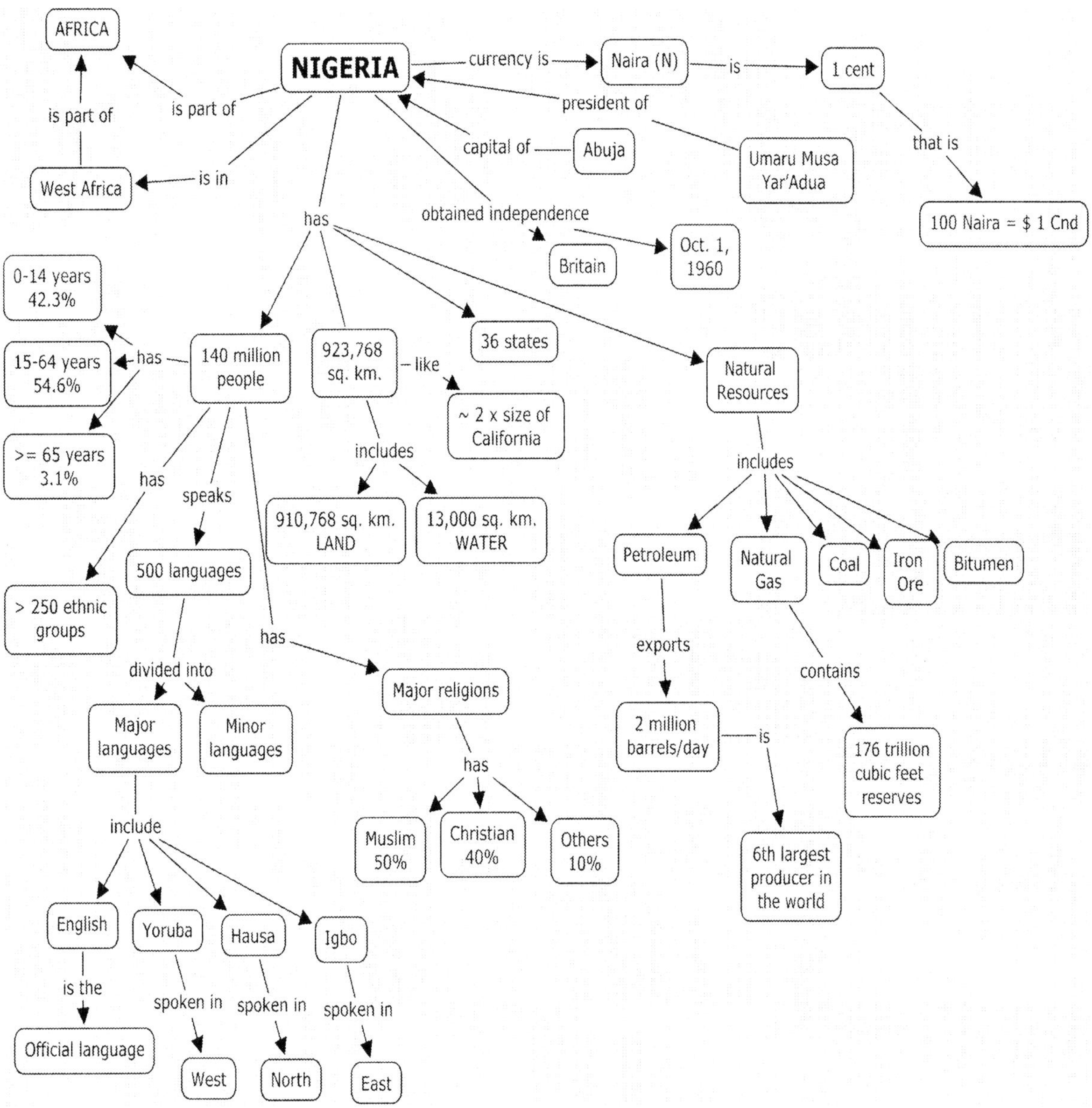

by symbols such as P (part) or C (characteristic), drawn from a fixed set of relations. Figure 2 shows the same information in figure 1 displayed using a knowledge map.

Research on the instructional applications of concept maps has proliferated since they were first used as educational tools in the early 1970s. Figure 3 shows the growth of research literature on concept maps over the past three decades. A period of rapid growth from 1985 to 1990 followed the publication of Novak & Gowin's (1984) *Learning How to Learn,* which presented a comprehensive argument for the educational use of concept maps.

When used for learning, concept maps can be (a) created by instructors and presented to students (Cliburn, 1986; Lambiotte & Dansereau, 1992), (b) constructed by students (McCagg &

Figure 2.

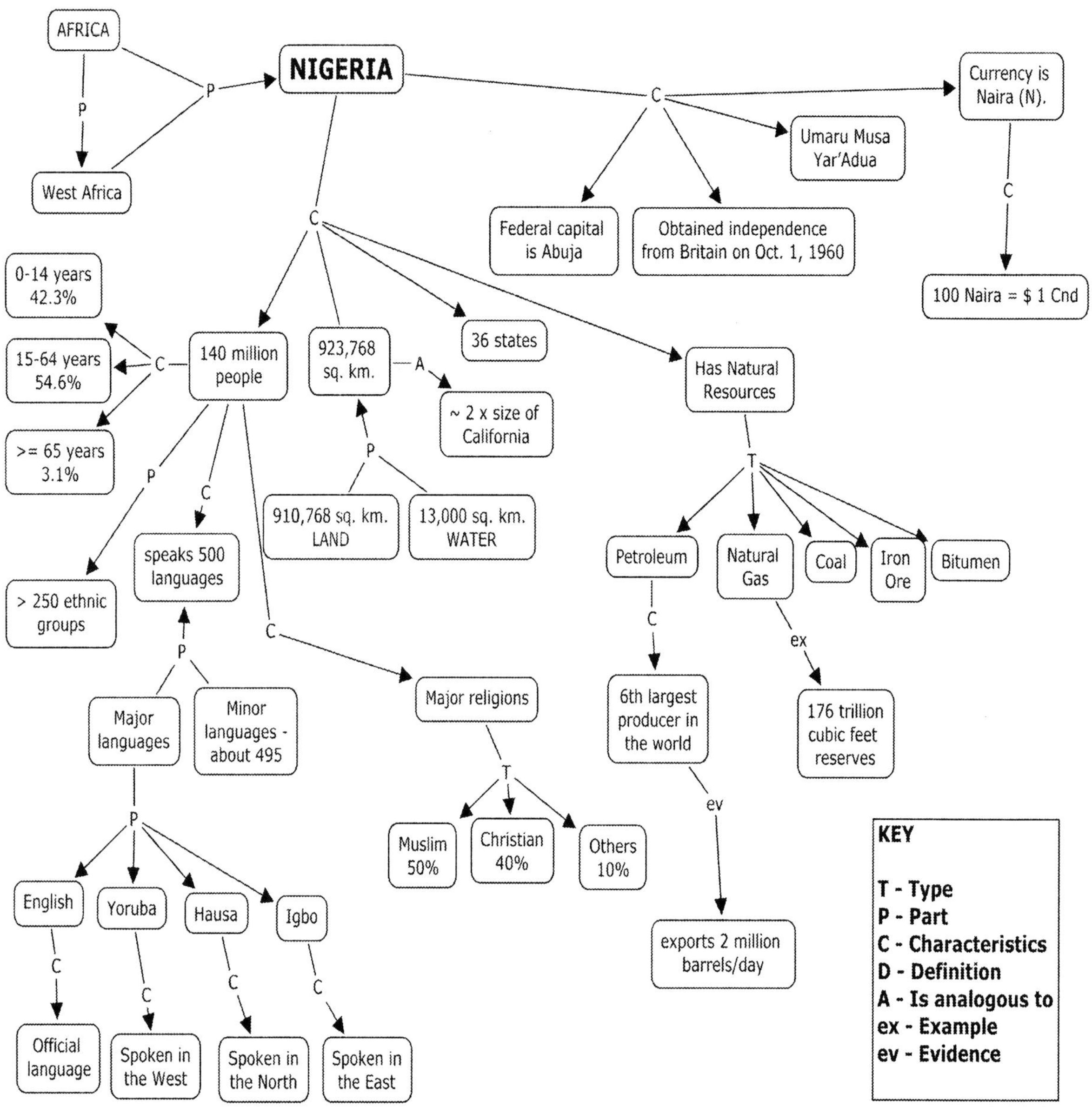

Dansereau, 1991), or (c) received and modified by students (Chang, Sung, & Chen, 2002). Recent reviews on the instructional use of concept maps show their growing prominence as a tool for supporting learning (Nesbit & Adesope, 2006; O'Donnell, Dansereau & Hall, 2002). In a recent meta-analysis, Nesbit and Adesope (2006) found that, in different instructional conditions, settings and experimental features, the use of concept and knowledge maps produced increased retention and transfer of knowledge when compared with control conditions where students studied with texts, outlines, lists or listened to lectures. The meta-analysis also found that studying with instructor- or researcher-generated concept maps rather than texts yielded an overall effect size of $d = .4$ standard deviations. The effect was higher ($d = .82$) when students constructed their own concept

Figure 3.

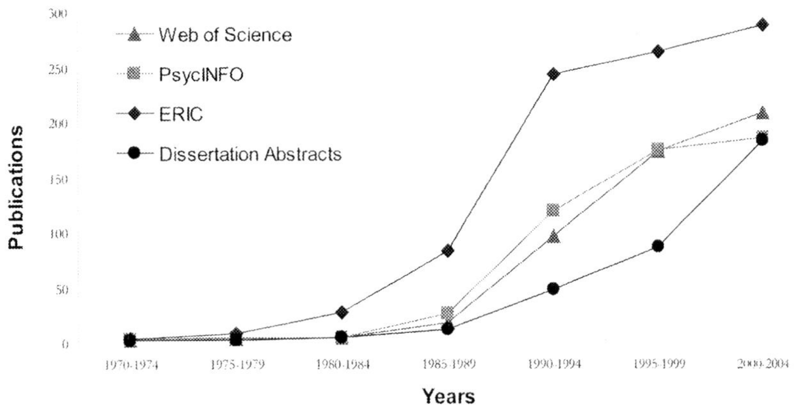

maps or modified maps constructed by teachers. In his influential book on statistical power analysis, Cohen (1988) suggested that an effect size of d = 0.2 could be regarded as "small", an effect size d = .50 could be regarded as "medium", and an effect size d = .8 could be regarded as "large".

In addition to evidence about the effects of concept maps on achievement, studies have found that students often report positive attitudes toward learning with concept maps (Nesbit & Adesope, 2006). There is further research evidence that the use of concept mapping strategy can lower anxiety and frustration while increasing motivation during learning (Bahr & Dansereau, 2001; Czerniak & Haney, 1998; Okebukola & Jegede, 1988).

The effectiveness of concept maps as communication tools has been attributed to their minimal use of text, which may help students to scan for information (Plotnick, 1997). Indeed, there is evidence that concept maps communicate effectively because they are less wordy than typical instructional texts (O'Donnell, Dansereau, & Hall, 2002; Blankenship & Dansereau, 2000). Compared with the language typically used in textbooks, concept maps offer a simpler node-link-node syntax that learners with lower reading ability or studying in a second language may comprehend more easily. Unlike conventional texts that have only one dominant processing order that progresses from left to right and top to bottom; concept maps usually have many different processing routes (Lambiotte et al., 1989). This flexibility may allow learners to choose a processing route that suits their prior knowledge, thus affording deeper cognitive processing and more meaningful learning.

While research in the instructional use of concept maps has expanded, Figure 4 shows that only a small fraction of the studies examined the use of concept maps in collaborative learning environments. In addition to the limited number of studies, research on the collaborative use of concept maps has produced inconsistent results. One of the goals of this review was to investigate these inconsistencies and suggest different ways by which future research in this area might address this issue.

Objectives of the Chapter

Although there are many reviews of collaborative learning research (Johnson & Johnson, 1989; 1991; O'Donnell, Hmelo-Silver, & Erkens, 2006; Slavin,

Figure 4.

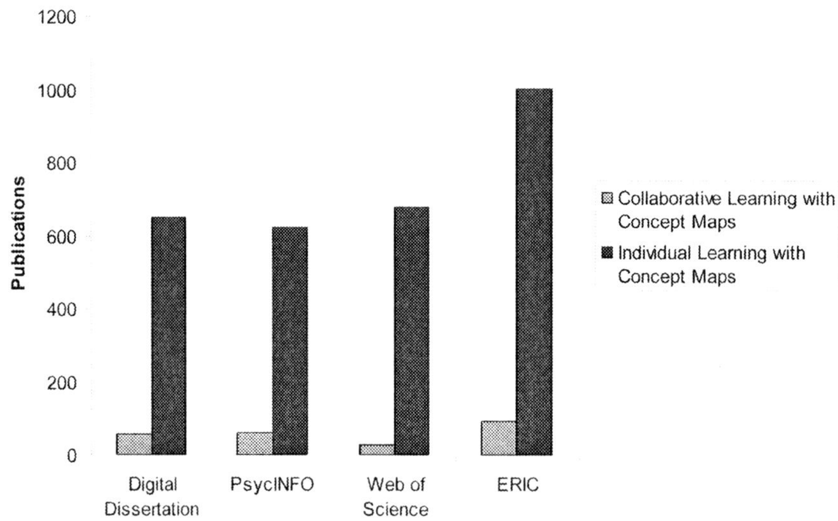

1989; 1990), we could find no meta-analytic reviews of research specifically on the use of concept maps in collaborative and cooperative learning environments. Cooperative learning is often described as a type of collaborative learning in which students work together in small groups on a structured activity that promotes individual accountability. However, in the method adopted for this review, collaborative and cooperative learning are not distinguished because the level of individual accountability could not be determined in many of the included studies. The main goal of this chapter was to synthesize research on the collaborative use of concept maps for learning. Specifically, we investigated the effect of using concept maps in collaborative learning in comparison with other information formats, such as text, outlines, or lists.

THEORETICAL BACKGROUND

Novak and Gowin (1984) developed their ideas for using concept maps in education as an extension of Ausubel's assimilation theory in science education. Ausubel's theory describes cognitive learning as a process in which new information is linked to relevant, preexisting cognitive structures such that both the newly acquired information and the preexisting structures are modified to form a more highly differentiated cognitive structure (Ausubel, 1963; 1968). Ausubel claimed that the most important factor influencing learning is what the learner already knows. To investigate Ausubel's theory, Novak and his research team attempted to determine how children's knowledge of science changed over a period of 12 years (Novak & Musonda, 1991). They introduced concept mapping as a method for tracking children's knowledge over time.

Schema theory also emphasizes the importance of prior knowledge in learning. A schema is a cognitive structure that organizes knowledge in memory. Schema theorists (Rumelhart, 1980; Norman & Rumelhart, 1981) have posited that schemas are dynamic, and existing schemas can be modified or expanded as new schemas are constructed or added to the existing ones. Learners use schemas constructed from previously experienced situations to interpret new, related experiences. Integration of new information with prior knowledge allows learners to develop more generalized cognitive structures and meaningfully interpret the current information environment.

Paivio's dual coding theory (Paivio, 1986; 1991) and Baddeley's three-component model of memory (Baddeley, 2000) are also relevant to understanding how learners process concept maps. Paivio theorized that there are different memory representations for verbal and visual information and connections between these representations allow easier retrieval of information and lead to improved learning and retention. Because working memory is limited in capacity, information processing can fail when dealing with too much information. Baddeley theorized that working memory deals separately with visuospatial and verbal information, with the implication that cognitive operations are most efficient when both types of information are used and the two modalities share the processing load. Concept maps, which represent information both verbally and spatially, may enable more efficient cognitive processing than unimodal information formats. Research has found that visual information helps learners to process and remember verbal information and vice versa (Mayer & Anderson, 1991). It may be that interpreting concept maps requires translation between visual and verbal information and thus promotes germane cognitive processing (Paas, Renkl, & Sweller, 2003; Sweller, Van Merrienboer, & Paas, 1998).

COLLABORATIVE USE OF CONCEPT MAPS

Collaborative learning is often understood as a pedagogical approach in which learners socially interact in small groups to facilitate construction of knowledge. Collaborative learning accesses scaffolds co-produced by the learners or provided by an instructor (O'Donnell, Hmelo-Silver, & Erkens, 2006). Cañas et al. (2003) speculated that since the separate use of concept maps and collaborative learning has been shown to produce educational benefits, their combined use may synergistically amplify these benefits by promoting more active involvement in learning and further knowledge construction.

Collaborative Concept Mapping

When concept maps are drawn on large paper sheets or white boards, they may be concurrently edited, extended or corrected by group members without reorganizing the entire map (Nesbit & Adesope, 2006). Proponents of collaborative concept mapping have argued that the social interactions it entails can successfully sustain critical thinking and meaningful co-construction of knowledge (Okebukola, 1990; Okebukola & Jegede, 1988; Roth & Roychoudhury, 1993, 1994; Stoyanova & Kommers, 2002; van Boxtel, van der Linden, Roelofs, & Erkens, 2002). For example, Stoyanova and Kommers (2002) found that collaborative generation of a concept map vividly revealed the conceptual structure of a given topic, facilitated group negotiation of meaning and promoted deeper mutual understanding among group members.

However, research on collaborative concept mapping has produced mixed results which seem to vary as a function of the nature and extent of interaction among group members. For example, van Boxtel et al. (2002) noted that collaborative concept mapping may inhibit learning because of an inability of participants in collaborative groups to sufficiently explain reasons for inclusion or exclusion of certain propositions. Similarly, Roth and Roychoudhury (1994) observed instances where incorrect notions were not challenged by group members. Other researchers have found that off-task interaction took substantial time away from discussion of important content-oriented concepts (Chiu, 2003). In another study, Chung, O'Neil, and Herl (1999) inferred that the quality of concept maps constructed in a group setting was not a result of group work processes because correlations were not statistically detected between the team processes and team outcomes (constructed concept maps).

Collaborative Learning with Pre-Constructed Concept Maps

The use of pre-constructed concept maps has been studied in small group learning activities such as scripted cooperation (Rewey, Dansereau, Dees, Skaggs, & Pitre, 1992) and peer teaching (Patterson, Dansereau, & Wiegmann, 1993). In their meta-analysis, Nesbit and Adesope (2006) found that the use of pre-constructed maps did not produce benefits when students learned through dyadic and scripted cooperation.

Differences across studies in the abilities of participants as well as differing contextual factors may be responsible for the mixed findings on the collaborative use of pre-constructed maps. Ability factors include learners' familiarity with concept map studying strategies. Contextual factors include the type and composition of groups, the type of learning environment (e.g., face-to-face or networked collaboration), or the quality of artifacts and scaffolding (e.g., link labels) provided to the learners. For example, in one experiment, Patterson, Dansereau, and Wiegmann (1993) found that low verbal ability learners benefited from peer-teaching with concept maps, compared to peer-teaching with text. In another experiment, Patterson et al. (1993) found that high verbal ability learners did not benefit from peer-teaching using concept maps. It appears that the mixed results in research on the use of pre-constructed maps may result not only from differences in the process of collaboration but also ability differences of participants involved in these studies.

Software Technologies that Support Collaborative Use of Concept Maps

Although concept mapping research predates the advent of the World Wide Web, internet technology has made collaborative work with concept maps feasible in a wider range of instructional settings. *CmapTools*, developed by the Institute of Human and Machine Cognition, was designed to enable users, even those from different geographical locations, to collaborate while constructing concept maps (Cañas, Hill, Carff & Suri, 2003). With this tool, users can concurrently edit, share and publish the product of collaboration. *CmapTools* uses a client-server architecture that allows sharing of concept maps stored on their server through a global network. An apparent advantage of *CmapTools* is that it supports synchronous collaboration involving two or more people concurrently constructing a concept map. The tool also allows asynchronous collaboration, search, annotation of nodes for signaling, as well as the addition of discussion threads to concept maps. Cañas, Hill and Lott (2003) reported that *CmapTools* supports aggregation of concept maps to represent a knowledge domain.

Winne et al. (2006) developed a tool for collaborative construction of concept maps within an educational annotation system called *gStudy*. In *gStudy*, information objects such as notes and information tags created by a learner while studying are automatically included in a concept map and are graphically sortable and filterable within the map. Aside from an auto-generated concept map, students can manually create a concept map. *gStudy's* concept maps serve as both navigational interfaces and media for information representation. *gStudy's* chat tool allows learners to adopt a role while chatting (e.g., critic) and will automatically-generate a concept map showing the participation structure and the roles learners adopt (Perry & Winne, 2006). *gStudy* also records logs that learners and researchers can use to analyze the collaborative process.

Recent technological advances have ushered in the development of sophisticated tools for assessing concept maps constructed by learners. For example, multiagent systems (MAS) sensitive to arrangement of concepts have been used to compare teachers' and learners' concept maps (Anohina, Graudina, & Grundspenkis, 2007).

Effect of Collaborative Use of Concept Maps on Learning

Search Procedures and Inclusion Criteria

We conducted a systematic review of research to assess the achievement effects of using concept maps compared to other information formats in collaborative learning. The meta-analysis adhered to specified search and inclusion criteria and computed effect sizes (Cooper & Hedges, 1994; Hedges & Olkin, 1985; Slavin & Cheung, 2005).

Six different online databases were searched to locate relevant articles: Academic Search Elite, Digital Dissertation, ERIC, PsycARTICLES, PsycINFO and Web of Science, using keywords "concept map*"OR "knowledge map*" OR "node-link map*" and "collaborat*". Table 1 shows the number of papers returned for each database.

Further literature search was conducted by examining review articles and books on concept maps and collaborative learning (Gao et al., 2007). Articles were selected if they met the following criteria:

(a) Investigated the effect of using concept maps in collaborative learning by comparing the outcomes of a group that used concept maps collaboratively to the outcomes of a group that used information formats such a text passages, lists, outlines, or lectures
(b) Measured cognitive outcomes, such as knowledge retention or problem-solving transfer
(c) Reported enough statistical information to calculate effect size. Articles were selected if they reported statistics such as means and standard deviations that allowed extraction of an effect size.
(d) Published in English

Table 1. Results of searches conducted March 2008

Database	Number of Studies
ERIC	93
Web of Science	30
PsycInfo	61
PsycARTICLES	2
Academic Search Elite	27
Digital Dissertations	58

When a study was reported in a peer-reviewed journal and another outlet (e.g., conference proceedings), the journal version of the study was selected. Out of 271 studies returned by the database searches, only 21 studies met the specified inclusion criteria and were selected for coding and analysis. Three of the 21 studies reported two effect sizes each, resulting in a total of 24 effect sizes. Out of the 24 effect sizes that were included for review, 8 reported the effect of using pre-constructed concept maps while 16 examined concept maps constructed by students as they learned. We extracted all effect sizes and other statistics with particular attention paid to statistical independence of the effect sizes. Specifically, we obtained a single effect size from Amer (1994) article that has two statistically dependent effect sizes by calculating a weighted average over the two different control groups that were compared with one experimental group in the article. Therefore, all subsequent analyses show a grand total of 23 effect sizes. In all analyses, guidelines from Lipsey and Wilson (2001) were followed.

Computation of Effect Sizes

Effect sizes were computed for each study by calculating the difference between the experimental group mean (concept map group) and the control mean (non-concept map group), with the result of this difference divided by the pooled standard deviation of both groups to obtain Cohen's *d*, a measure of effect size (Hinkle, Wiersma & Jurs, 2003; Lipsey & Wilson, 2001). When other statistics such as *F* or *t* were provided, these were also used to derive effect sizes or in some cases, to verify the Cohen's *d* effect sizes obtained from the formula expressed above (Cooper & Hedges, 1994). All studies that met the inclusion criteria provided information sufficient to allow extraction of effect sizes.

Results

Characteristics and effect sizes obtained from the included studies are summarized in Table 2. The overall sample size was 1,984. The table shows significant characteristics of each study, including the grade level of participants, the total number of participants involved in each study (*N*), the setting of the study, whether concept maps were constructed by the students (constructed) or preconstructed and given to students as a learning aid (studied), and Hedge's *g*, the unbiased effect size,.

Figure 5 shows a distribution of statistically independent effect sizes. The distribution is presented only as a characterization of the effect sizes found in the analysis and should not be used as an estimate of the benefits of using concept and knowledge maps because the effect sizes it shows are not weighted in accordance with their individual sample sizes.

Table 3 shows mean effect sizes for learning with collaborative concept maps. The table includes the number of participants (*N*) in each subset, the number of studies (*k*), the weighted mean effect size (*M*) and its standard error (*SE*), the 95% confidence interval around the mean, the test of homogeneity (*Q*) with the corresponding degrees of freedom (*df*) and the magnitude of heterogeneity (I^2). Due to the sensitivity of the Q statistic with respect to large sample sizes, researchers have recommended the use of I^2 statistic as a complement to the Q statistic for proper interpretation of homogeneity (Higgins & Thompson, 2002; Huedo-Medina, Sánchez-Meca, Marin-Martinez, & Botella, 2006). I^2 represents heterogeneity in proportional terms as the percentage of variability in point estimates that is due to heterogeneity rather than sampling error. I^2 was calculated as: $I^2 = 100\% \times [(Q - df)/Q]$, where *Q* is the heterogeneity statistic and *df* is the degrees of freedom. Negative values of I^2 are assigned a value of zero so that I^2 lies between 0% and 100%. A value of 0% indicates no observed heterogeneity, and larger values show increasing heterogeneity. Higgins and Thompson (2002, p. 1553) also recommended the following interpretations of I^2: percentages of around 25% ($I^2 = 25$), 50% ($I^2 = 50$), and 75% ($I^2 = 75$) would mean low, medium, and high heterogeneity, respectively.

When compared with other learning activities, the effect of constructing concept maps in a collaborative learning environment was statistically detected. The mean effect size was large, suggesting that collaborative concept mapping is an effective learning activity. Statistically detectable benefits were observed for constructing concept maps compared with attending lectures or participating in class discussions, using independent or self-selected study strategies (McCagg & Dansereau, 1991), and performing an additional experiment requiring problem solving (Pankratius, 1987). Interestingly, students presented with outlines outperformed students who constructed concept maps collaboratively. The large standardized mean effect size found for collaborative concept mapping compared with lecture or class discussion may result from the use of non-interactive, whole class settings for these comparison activities. That is, the effect size may be mainly due

Table 2. Characteristics of coded studies and associated effect sizes

Study	Grade	*N*	Setting	Use	Effect Size (g)
Amer (1994)	PS	9	Classroom	Constructed	
Effect of collaborative mapping vs. underlining on recall					0.49
Effect of collaborative mapping vs. text reading on recall					0.95
Brown (2003)	11	64	Classroom	Constructed	
Effect of collaborative concept mapping vs. text reading on biology achievement					2.30
Chang (1994)d	7	140	Classroom	Constructed	
Effect of collaborative mapping vs. outlining on biology achievement					-0.23
Chularut & DeBacker (2004)	PS	79	Classroom	Constructed	
Effect of concept mapping vs. study and discussion on achievement					2.16
Esiobu & Soyibo (1995)	10	538	Classroom	Constructed	
Effect of collaborative concept and vee mapping vs lectures on biology achievement					1.34
Guastello, Beasley & Sinatra (2000)	7	124	Varied	Constructed	
Effect of group mapping vs. teacher-led discussion in low achieving students					5.94
Jolly (1998)d	6	36	Classroom	Constructed	
Effect of collaborative mapping on science problem solving					0.44
Lambiotte, Skaggs & Dansereau (1993)	PS	47	Classroom	Studied	
Effect of collaborative studying of maps vs. lists on recall and problem solving					0.03
McCagg & Dansereau (1991)	PS	81	Varied	Constructed	
Effect of group mapping vs. self-selected study strategies on psychology recall					0.33
Nicoll, Francisco & Nakhleh (2001)	PS	20	Classroom	Constructed	
Effect of group mapping on understanding of relationships among chemistry concepts					1.03
Novak (1994)d	PS	30	Classroom	Constructed	
Effect of collaborative mapping vs lecture and demo on AV ed-tech achievement					0.23
Okebukola (1990)	PS	138	Classroom	Constructed	
Effect of collaborative mapping vs. lectures on ecology achievement					1.53
Okebukola (1992)	11	116	Classroom	Constructed	
Effect of collaborative mapping vs lectures and discussions on biology achievement					1.54
Pankratius (1987)d	12	87	Classroom	Constructed	
Effect of group mapping vs. additional experiment on physics problem solving					0.45
Patterson, Dansereau & Wiegmann (1993)	PS	38	Laboratory	Studied	
Effect of map vs. text in peer-teaching low verbal ability learners					0.52
Effects of map vs. text in peer-teaching high verbal ability learners					-0.22
Patterson, Dansereau & Newbern (1992)	PS	101	Laboratory	Studied	
Effect of collaborative learning with maps vs. text on recall					0.39
Prater & Terry (1988)	5	30	Classroom	Constructed	
Effect of teacher-led mapping vs. teacher-led discussion on story comprehension					0.76
Reynolds et al. (1991)	PS	38	Laboratory	Studied	
Effect of studying with hypermaps vs. hypertexts on recall					0.03
Rewey et al. (1992)	PS	82	Laboratory	Studied	

Study	Grade	*N*	Setting	Use	Effect Size (g)
Effect on low-ability learners of collaborative study with map vs. text on recall					0.38
Effect on high-ability learners of collaborative study with map vs. text on recall					0.07
Rewey et al. (1989)	PS	54	Laboratory	Studied	
Effect of dyadic study of map vs. text on recall					0.08
Smith (1992)	PS	42	Classroom		Constructed
Effect of reviewing readings with maps and vee heuristics vs. discussion on recall					0.91

d = Dissertation, PS = Post-secondary, N = Number of subjects, Use = Whether concept maps were constructed or studied by participants.

Figure 5.

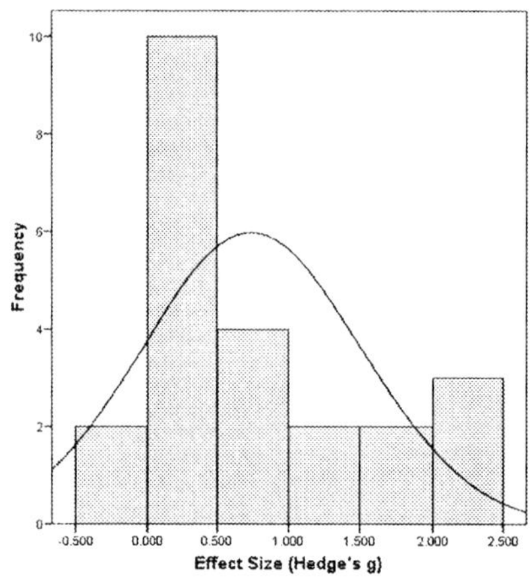

Table 3. Weighted mean effect sizes for collaborative use of concept maps vs. comparison treatments

			Effect size (g)		95% confidence interval		Homogeneity of effect size		
	N	*k*	*M*	*SE*	Lower	Upper	*Q*	*df*	I^2
Learner-Constructed Maps	1,624	15	1.05*	.20	.66	1.44	14.67	14	4.54
Outlining (Chang, 1994)	140	1	-.23						
Text reading (Amer, 1994)	99	1	.67						
Writing summaries (Brown, 2003)	64	1	2.40						
Self-selected study strategy and problem solving	168	2	.38*	.06	.25	.51	.94	1	.00
Lecture/Group discussion	1,153	10	1.28*	.17	.94	1.62	12.85	9	29.97
Pre-Constructed Maps	360	8	.19	.11	-.02	.40	3.51	7	.00
List (Lambiotte et al.,1993)	47	1	.03						
Text reading	313	7	.22	.11	-.01	.44	3.14	6	.00

* $p < .05$.

to small group collaboration in the experimental treatment rather than to concept mapping. In many of the studies we reviewed, the extent of collaboration especially in the comparison group was not explicitly stated.

In studies of dyadic, scripted cooperative learning, no effect was detected for studying with pre-constructed concept maps rather than text. It may be that participants who do not know strategies for collaborative learning with maps are not able to take advantage of concept maps for learning. For example, one of the studies reviewed here (Patterson, Dansereau, & Newbern, 1992), examined the effects of communication aids and strategies on collaborative learning. They found that students exposed to dyadic learning about the interrelated physiological effects of alcohol performed better with concepts maps when they were given a strategy for using the maps in collaborative learning environment ($d = .66$) than when no strategy was provided ($d = .30$). Furthermore, students presented with a map to study in a group may not have as much incentive to process the meaning of the map as much as students who are tasked with constructing a map. All the subsets of collaborative concept map studies were homogeneous, indicating that the variability among effect sizes was not greater than that expected from sampling error. The I^2 statistic confirms the homogeneity results obtained by the Q statistic, and shows that most of the subsets of collaborative construction of concept maps have low heterogeneity. The I^2 statistic shows no observed heterogeneity for studies in which participants collaboratively studied pre-constructed concept maps.

FUTURE RESEARCH DIRECTIONS

This review helps elucidate the gaps in research on collaborative use of concept maps. There is clearly a need to investigate the effects of collaborative concept maps on higher level cognitive processes such as problem solving transfer, application and analysis (Anderson & Krathwohl, 2001). As noted by O'Donnell, Dansereau and Hall (2002), research in concept maps might obtain more informative results by pre-training participants in collaboration strategies. There is a need for more studies implementing training not only in concept mapping but also in productive collaboration among participants.

There is also a need to conduct methodologically-sound research that properly investigates the nature of collaboration in the concept mapping and comparison conditions. In many of the studies we reviewed, the nature and extent of collaboration among participants in the comparison group could not be determined. More rigorous research design and reporting would alleviate this problem.

CONCLUSION

In this chapter, we have reviewed research on the collaborative use of concept maps for learning. We summarized the cognitive theories most commonly cited to describe learners' cognitive processes as they use concept maps for learning and discussed the features of concept maps that have been hypothesized to enhance collaborative learning. We concluded with a systematic review of literature on the use of concept maps in collaborative learning.

The advantage of concept mapping was especially apparent in small group activities. Collaborative work with concept maps was more effective than activities with non-map materials. A large effect was obtained in research that had learners work in groups to construct concept maps. However, no effect was statistically detected in research that had learners collaboratively study pre-constructed concept maps. There was no evidence that the well-planned use of concept maps is in any way deleterious. Finally, we believe that to more deeply understand the educational uses of concept maps, researchers should investigate new ways of using them in collaborative settings.

REFERENCES

Amer, A. A. (1994). The effect of knowledge-map and underlining training on the reading comprehension of scientific texts. *English for Specific Purposes, 13*, 35–45.* doi:10.1016/0889-4906(94)90023-X

Anderson, L. W., & Krathwohl, D. R. (Eds.). (2001). *A taxonomy for learning, teaching and assessing: A revision of Bloom's taxonomy of educational objectives.* New York: Addison, Wesley, Longman.

Anohina, A., Graudina, V., & Grundspenkis, J. (2007). Using concept maps in adaptive knowledge assessment. In G. Magyar, G. Knapp, W. Wojtkowski, W. G. Wojtkowski, & J. Zupancic (Eds.), *Advances in Information Systems Development New Methods and Practice for the Networked Society.* New York: Springer.

Ausubel, D. P. (1963). *The psychology of meaningful verbal learning.* New York: Grune & Stratton.

Ausubel, D. P. (1968). *Educational psychology: A cognitive view.* New York: Holt, Rinehart & Winston.

Baddeley, A. D. (2000). Short term and working memory. In E. Tulving & F. I. M. Craik (Eds.). *The Oxford handbook of memory* (pp. 77-92). New York: Oxford University Press.

Bahr, G. S., & Dansereau, D. F. (2001). Bilingual knowledge maps (BiK-Maps) in second-language vocabulary learning. *Journal of Experimental Education, 70*, 5–24.

Blankenship, J., & Dansereau, D. F. (2000). The effect of animated node-link displays on information recall. *Journal of Experimental Education, 68*, 293–308.

Brown, D. H. (1994). *Teaching by principles: An interactive approach to language pedagogy.* Englewood Cliff, NJ: Prentice Hall, Regents.

Brown, D. S. (2003). High school biology: A group approach to concept mapping. *The American Biology Teacher, 65*, 192–197.* doi:10.1662/0002-7685(2003)065[0192:HSBAGA]2.0.CO;2

Cañas, A. J., Coffey, J. W., Carnot, M. J., Feltovich, P., Hoffman, R. R., Feltovich, J., & Novak, J. D. (2003). *A summary of literature pertaining to the use of concept mapping techniques and technologies for education and performance support.* Report from The Institute for Human and Machine Cognition.

Cañas, A. J., Hill, G., Carff, R., & Suri, N. (2003). *CmapTools: A knowledge modeling and sharing toolkit.* (Technical Report IHMC CmapTools 93-01). Institute for Human and Machine Cognition.

Cañas, A. J., Hill, G., & Lott, J. (2003). *Support for Constructing Knowledge Models in CmapTools,* (Technical Report CmapTools 93-03). Institute for Human and Machine Cognition.

Chang, K., Sung, Y., & Chen, I. (2002). The effect of concept mapping to enhance text comprehension and summarization. *Journal of Experimental Education, 71*, 5–23.

Chang, W. (1994). The effects of using concept mapping to supplement class notes on the biology test scores of seventh-grade students in Taiwan, R.O.C. (Doctoral dissertation, The University of Iowa). *Dissertation Abstracts International, 55,* 3800.*

Chiu, C. H. (2003). Exploring how primary school students function in computer supported collaborative learning. *International Journal of Continuing Engineering Education and Lifelong Learning, 13*, 258–267. doi:10.1504/IJCEELL.2003.003276

Chularut, P., & DeBacker, T. K. (2004). The influence of concept mapping on achievement, self-regulation, and self-efficacy in students of English as a second language. *Contemporary Educational Psychology, 29*, 248–263.* doi:10.1016/j.cedpsych.2003.09.001

Chung, G. K. W. K., O'Neil, H. F. Jr, & Herl, H. E. (1999). The use of computer-based collaborative knowledge mapping to measure team processes and team outcomes. *Computers in Human Behavior, 15*, 463–493. doi:10.1016/S0747-5632(99)00032-1

Cliburn, J. W., Jr. (1986). An Ausubelian approach to instruction: The use of concept maps as advance organizers in a junior college anatomy and physiology class. (Doctoral Dissertation, University of Southern Mississippi). *Dissertation Abstracts International, 47*, 852A.

Cohen, J. (1988). *Statistical power analysis for the behavioral sciences* (2nd Ed.). Hillsdale, NJ: Lawrence Erlbaum Associates.

Cooper, H. M., & Hedges, L. V. (1994). *The Handbook of Research Synthesis*. New York: The Russell Sage Foundation.

Czerniak, C. M., & Haney, J. J. (1998). The effect of collaborative concept mapping on elementary preservice teachers' anxiety, efficacy, and achievement in physical science. *Journal of Science Teacher Education, 9*, 303–320. doi:10.1023/A:1009431400397

Dansereau, D. F. (2005). Node-link mapping principles for visualizing knowledge and information. In S. O. Tergan & T. Keller (Eds.). *Knowledge and information visualization: Searching for synergies*. (LNCS 3426, pp. 61-81). Heidelberg, Germany: Springer-Verlag.

Esiobu, G. O., & Soyibo, K. (1995). Effects of concept and vee mappings under three learning modes on students, cognitive achievement in ecology and genetics. *Journal of Research in Science Teaching, 32*, 971–995.* doi:10.1002/tea.3660320908

Gao, H., Shen, E., Losh, S. C., & Turner, J. (2007). A review of studies on collaborative concept mapping: What have we learned about the technique and what is next? *Journal of Interactive Learning Research, 18*, 479–492.

Guastello, E. F., Beasley, M., & Sinatra, R. C. (2000). Concept mapping effects on science content comprehension of low-achieving inner-city seventh graders. *Remedial and Special Education, 21*, 356–365.* doi:10.1177/074193250002100605

Hall, R. H., Dansereau, D. F., & Skaggs, L. P. (1992). Knowledge maps and the presentation of related information domains. *Journal of Experimental Education, 61*, 5–18.

Hedges, L. V., & Olkin, I. (1985). *Statistical methods for meta-analysis*. San Diego, CA: Academic Press.

Hinkle, D. E., Wiersma, W., & Jurs, S. G. (2003). *Applied statistics for the behavioral sciences* (5th Ed.). New York: Houghton-Mifflen.

Hmelo-Silver, C. E. (2004). Problem-based learning: What and how do students learn? *Educational Psychology Review, 16*, 235–266. doi:10.1023/B:EDPR.0000034022.16470.f3

Johnson, D. W., & Johnson, R. T. (1989). *Leading the cooperative school.* Edina, MN: Interaction.

Johnson, D. W., & Johnson, R. T. (1991). *Learning together and alone. Cooperative, Competitive, and Individualistic Learning,* (4th Ed.). Englewood Cliffs, NJ: Allyn & Bacon.

Johnson, D. W., & Johnson, R. T. (1994). *Learning together and alone: Cooperation, competition, and individualization,* (4th ed). Englewood Cliffs, NJ: Prentice Hall.

Jolly, A. B. (1998). The effectiveness of learning with concept mapping on the science problem solving of sixth grade children (Doctoral dissertation, Indiana University of Pennsylvania, 1998). *Dissertation Abstracts International, 59,* 3356. *

Lambiotte, J. G., & Dansereau, D. F. (1992). Effects of knowledge maps and prior knowledge on recall of science lecture content. *Journal of Experimental Education, 60,* 189–201.

Lambiotte, J. G., Dansereau, D. F., Cross, D. R., & Reynolds, S. B. (1989). Multirelational semantic maps. *Educational Psychology Review, 1,* 331–367. doi:10.1007/BF01320098

Lambiotte, J. G., Skaggs, L. P., & Dansereau, D. F. (1993). Learning from lectures: Effects of knowledge maps and cooperative review strategies. *Applied Cognitive Psychology, 7,* 483–497.* doi:10.1002/acp.2350070604

Lipsey, M. W., & Wilson, D. B. (2001). *Practical meta-analysis.* Applied Social Research Methods Series (Vol. 49). Thousand Oaks, CA: SAGE Publications.

Mayer, R. E., & Anderson, R. B. (1991). Animations need narrations: An experimental test of a dual-coding hypothesis. *Journal of Educational Psychology, 83,* 484–490. doi:10.1037/0022-0663.83.4.484

McCagg, E. C:, & Dansereau, D. F. (1991). A convergent paradigm for examining knowledge mapping as a learning strategy. *The Journal of Educational Research, 84,* 317–324.

Nesbit, J. C., & Adesope, O. O. (2006). Learning with concept and knowledge maps: A meta-analysis. *Review of Educational Research, 76,* 413–448. doi:10.3102/00346543076003413

Nicoll, G., Francisco, J., & Nakhleh, M. (2001). An investigation of the value of using concept maps in general chemistry. *Journal of Chemical Education, 78,* 1111–1117.*

Norman, D. A., & Rumelhart, D. E. (1981). The LNR approach to human information processing. *Cognition, 10,* 235–240. doi:10.1016/0010-0277(81)90051-2

Novak, A. D. (1994). The effects of cooperative learning with concept mapping in a preservice teacher education course, (Doctoral dissertation, Indiana University). *Dissertation Abstracts International, 56,* 895.*

Novak, J. D. (2002). Meaningful learning: The essential factor for conceptual change in limited or inappropriate propositional hierarchies leading to empowerment of learners. *Science Education, 86,* 548–571. doi:10.1002/sce.10032

Novak, J. D., & Gowin, D. B. (1984). *Learning how to learn.* New York: Cambridge University Press.

Novak, J. D., & Musonda, D. (1991). A twelve-year longitudinal study of science concept learning. *American Educational Research Journal, 28,* 117–153.

O'Donnell, A. M., Dansereau, D. F., & Hall, R. H. (2002). Knowledge maps as scaffolds for cognitive processing. *Educational Psychology Review, 14,* 71–86. doi:10.1023/A:1013132527007

O'Donnell, A. M., Hmelo-Silver, C., & Erkens, G. (2006). *Collaborative learning, reasoning, and technology.* Mahwah, NJ: Lawrence Erlbaum.

Okebukola, P. A. (1990). Attaining meaningful learning of concepts in genetics and ecology: An examination of the potency of the concept-mapping technique. *Journal of Research in Science Teaching, 27*, 493–504.* doi:10.1002/tea.3660270508

Okebukola, P. A. (1992). Concept Mapping with a cooperative learning flavor. *The American Biology Teacher, 54*, 218–221.*

Okebukola, P. A., & Jegede, O. J. (1988). Cognitive preference and learning-mode as determinants of meaningful learning through concept mapping. *Science Education, 72*, 489–500. doi:10.1002/sce.3730720408

Paas, F., Renkl, A., & Sweller, J. (2003). Cognitive load theory and instructional design: Recent developments. *Educational Psychologist, 38*, 1–4. doi:10.1207/S15326985EP3801_1

Paivio, A. (1986). *Mental representations: A dual coding approach*. Oxford, UK: Oxford University Press.

Paivio, A. (1991). Dual coding theory: Retrospect and current status. *Canadian Journal of Psychology, 45*, 255–287. doi:10.1037/h0084295

Pankratius, W. J. (1987). Building an organized knowledge base: Concept mapping and achievement in secondary school physics (Doctoral dissertation, Georgia State University College of Education). *Dissertation Abstracts International, 49*, 474.*

Patterson, M. E., Dansereau, D. F., & Newbern, D. (1992). Effects of communication aids and strategies on cooperative teaching. *Journal of Educational Psychology, 84*, 453–461.* doi:10.1037/0022-0663.84.4.453

Patterson, M. E., Dansereau, D. F., & Wiegmann, D. A. (1993). Receiving information during a cooperative episode: Effects of communication aids and verbal ability. *Learning and Individual Differences, 5*, 1–11.* doi:10.1016/1041-6080(93)90022-K

Perry, N. E., & Winne, P. H. (2006). Learning from learning kits: gStudy traces of students' self-regulated engagements with computerized content. *Educational Psychology Review, 18*, 211–228. doi:10.1007/s10648-006-9014-3

Plotnick, E. (1997). *Concept Mapping: A graphical system for understanding the relationship between concepts*. (ERIC document, EDO-IR-97-05).

Prater, D. L., & Terry, C. A. (1988). Effects of mapping strategies on reading comprehension and writing performance. *Reading Psychology, 9*, 101–120.* doi:10.1080/0270271880090203

Quillian, M. R. (1967). Word concepts: A theory and simulation of some basic semantic capabilities. *Behavioral Science, 12*, 410–430. doi:10.1002/bs.3830120511

Rewey, K., Dansereau, D., Dees, S., Skaggs, L., & Pitre, U. (1992). Scripted cooperation and knowledge map supplements: Effects on the recall of biological and statistical information. *Journal of Experimental Education, 60*, 93–107.*

Rewey, K. L., Dansereau, D. F., Skaggs, L. P., Hall, R. H., & Pitre, U. (1989). Effects of scripted cooperation and knowledge maps on the processing of technical material. *Journal of Educational Psychology, 81*, 604–609.* doi:10.1037/0022-0663.81.4.604

Reynolds, S. B., Patterson, M. E., Skaggs, L. P., & Dansereau, D. F. (1991). Knowledge hypermap and cooperative learning. *Computers & Education, 16*, 167–173.* doi:10.1016/0360-1315(91)90023-K

Roth, W. M., & Roychoudhury, A. (1993). The concept map as a tool for the collaborative construction of knowledge: A microanalysis of high school physics students. *Journal of Research in Science Teaching*, *30*, 503–534. doi:10.1002/tea.3660300508

Roth, W. M., & Roychoudhury, A. (1994). Science discourse through collaborative concept-mapping: New perspectives for the teacher. *International Journal of Science Education*, *16*(4), 437–455. doi:10.1080/0950069940160405

Rumelhart, D. E. (1980). Schemata: The building blocks of cognition. In R. J. Spiro, B. C. Bruce & W. F. Brewer (Eds.), *Theoretical issues in reading comprehension*. Hillsdale, NJ: Erlbaum.

Salomon, G. (1993). No distribution without individuals' cognition: a dynamic interactional view. In Salomon, G. (Ed.). *Distributed Cognition: Psychological and educational perspectives*. Cambridge, UK: Cambridge University Press.

Slavin, R., & Cheung, A. (2005). A synthesis of research on language of reading instruction for English language learners. *Review of Educational Research*, *75*, 247–284. doi:10.3102/00346543075002247

Slavin, R. E. (1989). Cooperative learning and student achievement. In R. E. Slavin (Ed.), *School and classroom organization* (pp. 129-151). Hillsdale, NJ: Erlbaum.

Slavin, R. E. (1990). *Cooperative learning: Theory, research and practice*. Englewood Cliffs, NJ: Prentice Hall.

Smith, B. E. (1992). Linking theory and practice in teaching basic nursing skills. *The Journal of Nursing Education*, *31*, 16–23.*

Stoyanova, N., & Kommers, P. (2002). Concept mapping as a medium of shared cognition in computer supported collaborative problem solving. *Journal of Interactive Learning Research*, *13*, 111–133.

Sweller, J., Van Merrienboer, J., & Paas, F. (1998). Cognitive architecture and instructional design. *Educational Psychology Review*, *10*, 251–296. doi:10.1023/A:1022193728205

van Boxtel, C., van der Linden, J., Roelofs, E., & Erkens, G. (2002). Collaborative concept mapping: Provoking and supporting meaningful discourse. *Theory into Practice*, *41*, 40–46. doi:10.1207/s15430421tip4101_7

Winne, P. H., Nesbit, J. C., Kumar, V., Hadwin, A. F., Lajoie, S. P., Azevedo, R. A., & Perry, N. E. (2006). Supporting self-regulated learning with gStudy software: The Learning Kit Project. *Technology, Instruction . Cognition and Learning*, *3*, 105–113.

KEY TERMS AND DEFINITIONS

Best-evidence synthesis: An approach for synthesizing research. Similar to meta-analysis but may use representative sampling of studies to arrive at a given conclusion.

Collaborative learning: A pedagogical approach in which learners socially interact in small groups to facilitate construction of knowledge.

Concept map: A diagram that facilitates organization, presentation, processing and acquisition of knowledge by showing relationships among concepts as node-link networks. Ideas in a concept map are represented as nodes and connected to other ideas/nodes through link labels.

Cooperative learning: A form of collaborative learning structured to promote individual accountability towards group performance.

Information map: A type of knowledge map used by counselors to convey information to and improve clients' recall on preventing relapse.

Knowledge map: A node-link diagram that has ideas located in nodes and connected to other related ideas through labeled pre-fixed links.

Meta-analysis: A quantitative research review that applies statistical techniques to examine, standardize and combine the results of different empirical studies that investigate a set of related research hypotheses.

Node-link map: A diagram that has ideas in nodes connected with other related ideas through links. Thus, concept and knowledge maps are sometimes referred to as node-link maps.

Chapter 13
Exploring Semiotic Approaches to Analysing Multidimensional Concept Maps Using Methods that Value Collaboration

Christina J. Preston
University of London, UK

ABSTRACT

This chapter focuses on teachers' multidimensional concept mapping data collected at the beginning and end of a one-year Masters level course about e-learning. A multidimensional concept map (MDCM) defines any concept map that is multimodal, multimedia, multilayered and/or multi-authored. The teachers' personal and professional learning priorities are analysed using two semiotic methods: the first is a traditional analysis of the words used to label the nodes; the second is an innovative analysis method that treats the whole map as a semiotic artefact, in which all the elements, including the words, have equal importance. The findings suggest that these tools offer deep insights into the learning priorities of individuals and groups, especially the affective and motivational factors. The teachers, as co-researchers, also adopted MDCM to underpin collaborative thinking. These research tools can be used in the assessment process to value multimodal literacy and collaborative engagement in new knowledge construction.

INTRODUCTION

This chapter reports on the research, development and modification of tools for the semiotic analysis of hand-drawn, desktop-published and digital concept maps called multidimensional concept maps (MDCM) that will be defined in the first section. These MDCM were collected at the beginning and end of a Continuing Professional Development (CPD) programme for teachers. The Masters level module was about e-learning, a term that refers to any learning that is electronically mediated using digital technologies. Examples would include a learning episode when information is derived from digital resources online or where the learning discussion is hosted in a virtual learning environment (VLE).

The objective of the chapter is to provide innovative socio-cultural tools for the analysis of MDCMs that provide insights into learners' priorities. These insights are expected both to improve understand-

DOI: 10.4018/978-1-59904-992-2.ch013

ing of learning processes for the researcher, the tutor and the map-maker, and to provide emerging multidimensional tools for assessing learning priorities. In particular, the benefits are considered for teachers who are conducting their own research projects with colleagues and pupils.

The chapter covers the meaning of the key terms, concept maps, mind maps and multidimensional concept maps as well as offering a definition of semiotics. The findings from the maps are described and the potential roles for researchers and co-researchers in the map-making process are also explored. The findings consider the value of the tools in identifying collaborative learning and future trends.

DEFINING A MULTIDIMENSIONAL MAP

The terms concept map and mind map are often used interchangeably. In fact, they refer to different methods of mapping and different map-shapes. The term concept map is most often associated with Novak who has worked in this field with other colleagues since the 1980s (Novak and Godwin, 1984; Cañas and Novak, 2007). The Novakian system requires the careful teaching of agreed topics followed by the construction of a prescribed hierarchical map shape by single learners or groups (Figure 1). Ålhberg (2007) is critical of the high levels of prescription and points out some inconsistencies. Nevertheless, in his own work he only reduces the number of rules and clarifies them rather than questioning the prescriptive method.

A different map-shape is promoted by Buzan (2002), called a *mind map*, to refer to maps that radiate from the centre. For Buzan *mind maps* mirror how the mind works. In contrast to Novak he promotes a free mapping exercise where the map-maker provides the content from professional or personal thinking exercises. However, Buzan's belief that these mind maps reflect the ways in which the brain works is not supported by the limited research into the topic (Anderson-Inman and Ditson 1999).(see Figure 2)

So far, much of the research into mapping has continued to follow this pattern of prescriptive teaching leading towards drawing shapes that are agreed in advance. This study, on the other hand, concentrates on the map-makers' creativity in exploring their own concepts, rather than following instructions from the teacher. This willingness to listen to the learner is a key principle in this alternative approach to mapping associated with the socio-cultural school of semiotics.

Saussure (1916), a founding theorist in this area, defined semiotics as the science of the life of signs in society. Since then semiotics has developed as an all-encompassing term for the study of any kind of sign that is used in a culture to communicate meaning. These signs can be in many different modes including sound, animation, graphics, gaze and gesture. Signs are often, therefore, described as multimodal and the capacity to read them is regarded as multimodal literacy (Jewitt and Kress 2003). Another important aspect of the sociocultural semioticians approach is their explanation of the four communicative strata: discourse; design; production, and dissemination (Kress and Van Leeuwen 2001). These strata emphasis the dynamic set of collaborative processes that result in effective multimodal communication. These strata are not valued in many traditional learning and assessment situations.

The hypothesis in this study was that the teachers might not be as multiliterate as their students. As a result it was not practical, when the study began, to request the teachers to use digital mapping software. The resourcing and training challenges were too great. Pen and paper was the default option. However, during the one-year data collection period the teachers' multimodal literacy improved. Some of the map-makers elected to use desktop publishing. A few used digital mapping packages and one used a sophisticated mapping package linked to the internet that allowed authors

Figure 1. A Novak-style hierarchical map

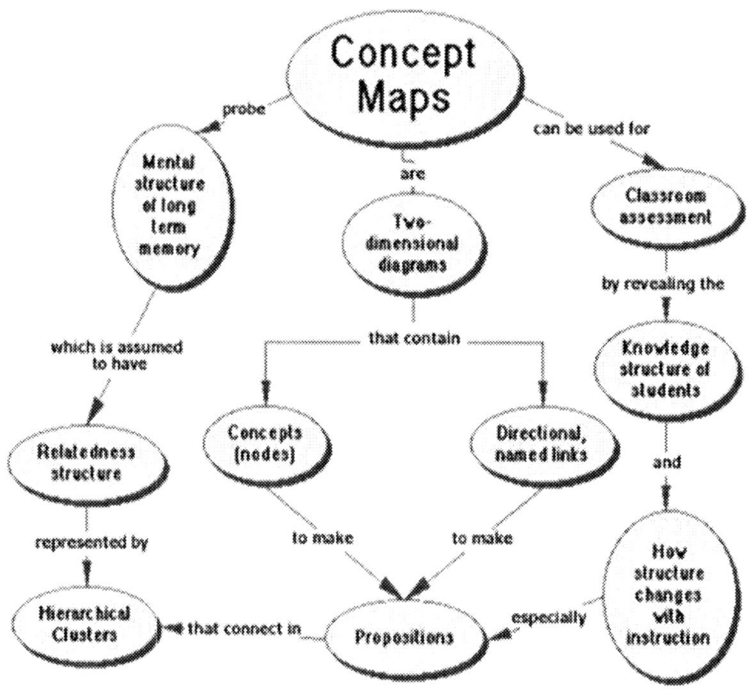

to work together remotely on a single map. This growing list of affordances gave rise to a new term used in this study, the multidimensional concept map (MDCM). The elements might be multimodal, multimedia, multilayered and multi-authored in various combinations. The analytical tools defined in this paper can be applied to any kind of MDCM.

Nevertheless, in this study the focus is not on the range of multidimensional semiotic resources that the teachers choose to use, but on how the analytical tools helped to reveal the learners' priorities at two fixed points. The aim was to investigate what MDCMs can reveal about learning priorities that the researchers, tutors and map-makers themselves cannot easily discover from more traditional methods of assessment like essays or multi-choice questionnaires that privilege writing.

The focus on the unique nature of each map and map-maker was inspired by the approach of the UK ImpaCT2 team who analysed two thousand hand-drawn maps by children from 10-14 years old entitled Computers in My World (Mavers, Somekh and Rosterick 2002). This team comments that learners' maps are 'fascinating, cultural artefacts, full of unique detail and often aesthetically pleasing' (ibid p. 191). This statement indicates their respect for the map-makers' ingenuity and creativity.

THE BACKGROUND

This section provides some background to the underlying principles of the CPD programme in e-learning that is the subject of this research. The topics covered are: three key theories of learning; the national CPD landscape in England; and, the meaning of both action research and communities of practice (CoPs) in relationship to collaborative learning.

The module under discussion, called e-learning, concentrated on two aspects of electronically mediated learning, online learning in a VLE and the

Figure 2. A Buzan-style radiant map

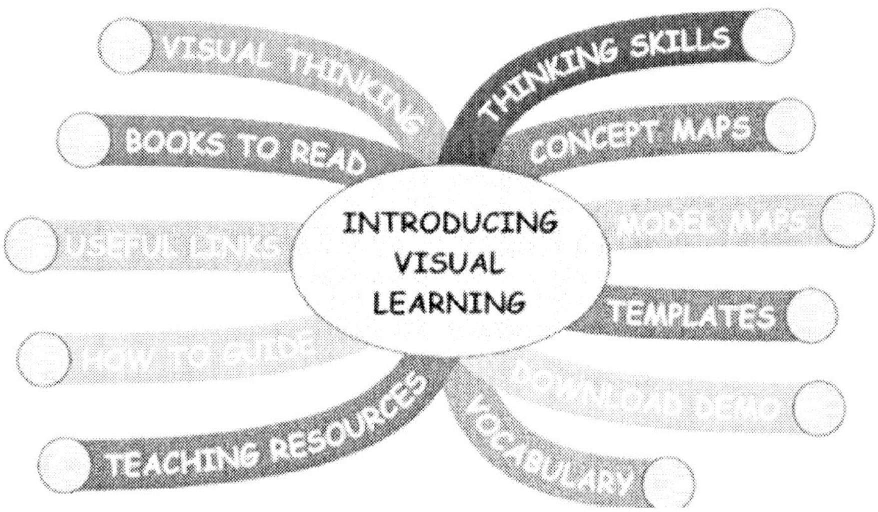

e-facilitation of collaborative knowledge building. The module was designed by tutors belonging to an international professional organisation called the MirandaNet Fellowship. This organisation is committed to the use of digital technologies as a catalyst in systemic change in pedagogy (www.mirandanet.ac.uk).

The MirandaNet Fellowship evaluation of the UK National Opportunities Fund (NOF) training for one thousand teachers indicated that the programme had not been entirely successful in transmitting an understanding of Information and Communications Technology (ICT) pedagogy in classrooms (Preston 2004). The NOF evaluation highlighted the challenge of using computers well when teachers' skills were poor. However, further MirandaNet research suggested that computer skills courses for teachers did not inspire them to be innovative in the classroom (Preston and Danby 2004). Fellows used the findings from these reports to develop innovative ICT CPD programmes for the twenty-first century. To begin with the Fellows looked back at the history of theories of learning in order to apply them to contexts where e-learning technologies were being used. In his summary of the history of learning theories, Pachler (2005), a teacher educator, provides three useful terms about learning: *information transmission, constructivism and social interaction*. On the one hand, Pachler warns of a general tendency to perceive the value of the new technologies in terms of *information transmission*, a means of delivering facts. On the other hand, he explains that *constructivists*, influenced by Piaget, argue that humans learn most effectively if they reconstruct their own meaning from what they have been taught rather than simply learning by rote. The third theory, the *social interaction* approach to learning, associated with Vygotsky, emphasises the wider value of collaborative learning in social and cultural contexts. These three cognitive approaches, *information transmission, constructivism* and *social interaction*, provide useful terminology for describing the different pedagogical approaches adopted by the designers of the CPD programmes.

The pilot module at Masters level in this study, 2005-2006, designed by MirandaNet Fellows, concentrated on two aspects of e-learning, online learning and e-facilitation. The participant teachers negotiated projects with the MirandaNet tutors that focused on e-learning challenges they had identified in their own classroom or school. This project based learning is often called *action research* - the process where teachers develop an understanding of constructivist learning techniques by applying their in classroom-based research projects (Elliott 1991; Somekh 2005; Somekh 2007). In this paper the term, *practice-based research,* is used rather than *action research.* Practice-based research refers to a more rigorous, scholarly approach to reflection on practice at Masters level (Lawes 2003). This practice-based design for learning represents a major shift away from tutor-dominated learning approaches in teacher education and from the notion that only academics should publish research. A more egalitarian view of teachers' own expertise is adopted in this design that links with Third Space theory (Zeichner and Liston 2006). The value of this theory is the conceptualisation of a space where the worlds of the academic and the practitioner join. In Zeichner's view (2008) this productive relationship can result in learning on both sides. He would no longer refer to CPD in these circumstances, but to *knowledge creation programmes.* The MirandaNet e-journal designed for this module like a VLE provided an online Third Space in which tutors and teachers could meet as equals. In fact, this relationship continued because twenty-five percent of the participant teachers and all the tutors in this CPD pilot were already MirandaNet members. A further fifty per cent of the teachers became members during or after the programme that meant that they also had a commitment to the MirandaNet vision that was beyond the normal expectation of course participants.

The MirandaNet Fellowship has been cited as community of practice (CoP) by UNESCO because it displays many of the indicators required to be identified as a CoP (Resta 2002). The term was first developed by Wenger and Lave (1998) who traced a modified version of the medieval guilds' approach to sharing expertise in business practice. The CoP is the focus of learning, meaning and identity exemplifying a greater will to collaborate in learning underpinned by a social interaction tradition (Wenger, McDermott and Synder 2002). By 2004, Wenger is emphasising the ways in which digital technologies can enhance collaboration in business. In education similar developments of CoPs have emerged that have been assisted by the development of electronic networking and, more recently, familiarity with social networking (Preston 2008). MirandaNet Fellows, in their development of several professional networks have developed two related theories about CoP practice: communual constructivism and braided learning. Communual constructivism emphasises the knowledge building role when teachers are working together often across national boundaries (Holmes, Tangney and Fitzgibbon 2001; Leask and Younie 2001; Cuthell 2002; Leask and Younie 2002; Cuthell 2005; Younie, Constable, Leask, Preston, 2008). Braided learning is a theory based on the observation of online social interaction practices and the ways in which professionals use the braided texts that are created (Preston and Holmes; Preston 2007a; Preston 2007b). Another set of studies by Fellows focuses on the building of CoPs for teachers, on evidence that shows that CPD is more effective if teachers are given greater professional freedom to decide their own transformational agenda within their local community (Preston 2004; Davis, Preston and Sahin, 2008 in press; Davis, Preston and Sahin 2008 in press; Davis 2008 in press). Sachs (2003) calls this kind of teacher who is engaging with others to set the professional agenda, *an activist professional.*

METHODOLOGY

The aim of the semiotic analysis of the MDCM in this study is to compare the map-makers' priorities at two different points in time: the beginning and end of a CPD programme, with a particular emphasis on the collaborative features of their learning. There were twenty-five teachers in the cohort. In this section the key principles of the research methodology are explained: the research questions outlined; the selection of the teacher-participants explained and the data collection procedures.

The Research Questions and Themes

The main question of the study was:

How can semiotic theory provide insights into teachers' learning priorities in the context of CPD that focuses on e-learning i.e. use of digital technologies in the mediation of learning?

How can analytical tools provide insights into the priorities of a whole group and of individual learners?

How does the role of the researcher impact on the development of collaboration and collegiality?

The theme that runs through the chapter is the relationship of the researcher in this study to the map-makers and their CoP. In semiotics research observation is on the relationship between *the viewer* and *the map-maker*. This is akin to an information transmission relationship. The *viewer* is just absorbing information and may have no relationship with the map-maker. A *reader* of the MDCM is a preferable term in this study because this implies more interactivity and engagement with the map-maker's meaning (Jewitt 2003). A further complexity of relationships is introduced in this study between those who create the MDCMs and those who read the images. The mediating role of the researcher, in particular, becomes the key to unlocking the meaning of the MDCM for the creator. The researcher becomes a particular kind of objective viewer who has a dual role: setting up the mapping process according to agreed conditions and judging the design. But in this study the creator of the map is also a co-researcher, a further complexity that is to be discussed later in this section.

As the main researcher, myself in this study, I undertook a complex, mixed role that is best explained through the lens of Adler and Adler (1987), who pioneered 'insider' and 'outsider' research. These researchers define three different perspectives on research activity. The first is the *peripheral member researcher* (PMR) who merely observes and has no impact on the situation. The second role is as an *active member researcher* (AMR): the situation when the researcher impacts on the learning situation for the duration of the project. For example, a researcher might take on a role as a teacher in a school for a year in order to report on what is happening with deeper understanding than timed and expected visits would afford. This is in the spirit of constructivist methods of learning. Thirdly, Adler and Adler describe the *complete member researcher* (CMR) characterised by complete full-time immersion before and after the research intervention. Adler and Adler favour this last approach arguing that, while CMRs may sacrifice some detachment, the depth of the data they collect is a more truthful representation of the realities of the situation being researched. In their opinion, CMRs also have the opportunity to acquire *understanding in use* rather than *reconstituted understanding*. From their point of view, data gathering should not only occur through a detached observational role, but also through the subjectively immersed role as well. In this role the researcher is learning through social interaction. This role proved to be important in the design of this research project and in the unexpected elements of the findings as they relate to collaboration.

In fact, in this study, all three research roles are fulfilling in different ways: PMR, AMR and CMR. Firstly, the conventional objective researcher, the PMR, intermingled with my role as the AMR because I also designed the programme, trained the tutors and became a tutor myself for the duration of the course. I was also engaged with the teachers in reflecting on the findings of this practice-based research project. However, as the founder and the chair of the MirandaNet Fellowship who were running this course I also had a CMR role characterised by complete full-time immersion within and beyond the research period.

Selecting the Participating Teachers

Twenty-five participants were selected from seventy applicants in a national competition advertised in the educational press in England. These participants were all Information and Communications Technology teachers or managing the deployment of computers in schools and regions. They had chosen to take this module on e-learning because this was a major thrust of UK education policy for schools. The education network, the Learning Grid, had been established as early as 1995. E-learning quickly became a key aspect of the Information and Communications Technology (ICT) UK National Curriculum. However, this CPD programme extended further than presenting ICT as a classroom subject. The module concentrated on cross-curricula ICT as well as the general use of digital technologies underpinning the learning process. The key focus was the use of virtual learning environments for collaborative learning including the development of the participants' e-facilitation skills. These skills were considered to be essential in the promotion of social interaction and knowledge building.

This group of classroom teachers, senior managers, staff trainers and regional advisers who were selected from all over England will be referred to simply as 'teachers'. In order to avoid dominance in the group by teachers with a technical approach to computers, a high level of technical expertise was not a key criterion for selection for a funded place on this course. Instead, for a fully-funded place, evidence of good social interaction was required. The selection process was conducted by a combination of interviews, reference letters from employers and questionnaires. These processes were designed to identify teachers who were willing to learn and to collaborate with their peers. Other selection criteria included a keenness to reflect on practice and expert knowledge in a relevant curriculum area as well as a willingness to collaborate with colleagues and to contribute to an active community of practice by web publication and forum discussion. To ensure that there would be an impact on the professional ecosystem locally and nationally, the applicants had to provide proof of their potential as staff tutors and of their established links into subject associations and regional professional groups.

Data Collection Procedures

At the start of the first workshop the CPD programme director explained the underlying design principles of change and the expectation that the participants would be building a community of practice during and after the CPD intervention. Discussions were held about the value of this collaborative approach in the context of e-learning, e-facilitation, practice-based research, communities of practice and the creation of knowledge.

The teachers, who were the map-makers in this study, were shown examples of both kinds of MDCM; 'concept maps' and 'mind maps'. The researcher emphasised that these were just two examples of many possible shapes. The map-makers were assured that any interpretation of the term 'mind map' and 'concept map' would be acceptable and that there were no right answers.

The heading for the MDCM used at the beginning and at the end of the programme was negotiated with the teachers using the ImpaCT2 heading as a model: Computers in My World. (Harrison, C., Comber C., Fisher, T. Haw, K. Lewin, C.Lunzer, E., McFarlane, A. Mavers, D., Scrimshaw, P., Somekh, B., Watling, R. ; Mavers, Somekh et al. 2002; Mavers 2004b; Mavers 2007). The participants decided on the following heading:

The impact of computers on my personal and professional life with particular relevance to e-learning.

This heading was a more complex interpretation of the heading by the ImpaCT2 team for children's MDCM which was 'Computers in my World'. Like the Impact 2 project, the participants had 20 minutes to work on their maps.

Although the teachers were not expected to follow any particular mapping model some were reluctant to engage in this exercise as they were doubtful about their ability to draw. At this point, one of the teachers stepped into the role of a supportive tutor by calling on his expertise in mapping. This teacher, in this study called by the pseudonym, Neil, injected an element of enthusiasm for MDCM into the proceedings because he had already published about his use of maps with other academics as an action researcher. He explained his own practice in using MDCM in his classroom as a vehicle for assessing children's learning. His intervention was a significant catalyst in motivating the group to tackle the mapping exercise.

Of the nineteen advisers who made full practice-based research submissions only sixteen were present at the beginning of the workshop session. There were, therefore, sixteen double maps in this sample available for analysis. In addition, there were four thirty-minute transcripts from the four group discussions at the end of the course. Extra motivation for the teachers to work collaboratively on their evidence from their practice-based research was provided because they were invited to present their studies to the Head of the Department for Education and Skills (DFES) Schools ICT division. The way in which these teachers chose to present their evidence to the public consultation with the UK E-learning Strategy are explained in the findings (DfES 2003).

TOOLS FOR ANALYSIS

The tools described in this section are designed to identify what the teachers' priorities were before the course began and afterwards. The MDCM and the transcripts are also investigated for evidence of an awareness of the value of collaborative learning, in particular, as the CPD was underpinned by constructivist and social interaction principles.

Tools for Content Analysis

This section focuses on two kinds of content analysis: the content analysis of the word labels in the nodes and the content analysis of transcribed discussions between the map-makers.

In the simple content analysis exercise the frequency with which concept words are repeated in the maps was recorded on a spreadsheet for the group as a whole. This method was adopted to investigate whether new concepts about digital technologies and the quality of these concepts were growing amongst the cohort of teachers as the year progressed. The agreed heading was so close to Computers in My World that initially the indicators were adapted from the scoring sheet in ImpaCT2 (2002). These were: computer systems and networks; information and communication; publishing and correspondence; games; sound, and learning.

However, other than the *learning* category, the ImpaCT2 indicators are factual. For this reason a different set of indicators are used identified in, *Learning with others in mind,* outlined in Table 1. These are: knowledge construction, community, meta-learning, autobiography and cognition. These indicators offer a qualitative method designed to investigate the social interaction and collaborative aspects of learning (Daly and Pachler 2007). The headings were established in the analysis of online discussions between teachers on an MTeach course designed in a similar way to the e-learning CPD programme under discussion.

Table 1. Categories and indicators (modified from Daly and Pachler 2007)

Categories and indicators of professional learning through participation in CMC	
Category	**Indicators**
Knowledge construction	Reassessments New ideas/proposals Questions/enquiries Endorsements/verified ideas Modified ideas
Community	Shared values/goals Seeking/giving support Statements expressing mutuality Practice-based exchanges
Metalearning	Verbalising the learning process Verbalising understanding Verbalising difficulties
Autobiography	Critical incidents Personal reflection Teacher identity Learner identity
Cognition	Statements of understanding: theoretical critical practical references to personal learning

The headings provide a means of investigating into the participants' awareness of the quality of the learning process.

Tools for Analysis by Sign

The framework for the analysis of these MDCMs has been pieced together from a number of studies that include concept mapping analysis (Table 2)

The core of the framework has been extrapolated from the list of topics covered in *Reading Images: the Grammar of Visual Design* (Kress and Van Leeuwen 2007): concepts; modalities; compositional elements and their interrelations; materiality of meaning; third dimension; dimensionality; narrative and ludic qualities; dynamics; representations and interactions.

More detail information about the analysis of maps has been adapted from the UK government ImpaCT2 research that has already been mentioned (Mavers, Somekh and Rosterick 2002). These two thousand MDCMs, titled Computers in My World, offer a particular emphasis on pictures representing concepts about computers, computer networks and online environments. Although they are children's maps the analytical features relate as well to adults. Mavers, the main researcher in the ImpaCT2 mapping analysis, has looked more closely at children's maps in three later studies. In these she explains how individual children use the resources around them to make meaning (Mavers 2004a; Mavers 2004b; Mavers 2007). She examines children's individual modes of drawing, writing, layout and linkage separately, as well as defining how they are combined in multimodal ensembles, and, provides a robust analysis of the differences between the meanings conveyed by word labels and pictures that can be used together to communicate concepts to an unknown audience. Comparing different representations of the internet by pupils, she shows how neither the writing nor the drawing can function without each other. Each mode takes on related but complementary functions that she calls *co-fixing*. From this work

Table 2. The framework for semiotic analysis

MCM features	Evidence	Key Theorists
Information and Transmission		
Concepts	Grouping of ideas and themes towards a key summarizing node; symbols used for ideas and how they are juxtaposed and connected in clusters; a classificational or an analytical design with some political implications: a hierarchical shape or network style map perhaps denoting authoritarian or liberal knowledge patterns.	Kress and Van Leeuwen 2007 (second edition)
Modalities	The features of the map that promote veracity from the point of view of the map-maker – shading, colour, brightness, etc.	Kress and Van Leeuwen 2007 (second edition) Mavers, Somekh & Restorick 2002
Compositional elements and their interrelations	Framing, Positions on the page, sizes, foregrounding and marginalising etc.	Kress and Van Leeuwen Mavers, Somekh & Restorick 2002
Materiality of Meaning	Surface textures, inscriptions and additions	Kress and Van Leeuwen 2007 (second edition) Mavers 2004
The Third Dimension	3-D concept maps	Kress and Van Leeuwen 2007 (second edition)
Dimensionality	Multilayering and hyperlinking	Kress and Van Leeuwen 2007 (second edition) Mavers 2004
Constructive Learning		
Narrative	Trajectories that tell a story	Kress and Van Leeuwen 2007
Affectual factors, ludic qualities	Indications that the learning was not only cognitive but affective	Kress and Van Leeuwen 2007 (second edition)
Dynamics	Direction of links and arrows: animation of images and lettering	Mavers 2004
Social Interaction		
Representations and interactions	Indications of the relationship between the map-maker and the viewer or audience	Kress and Van Leeuwen 2007 (second edition) Jewitt 2003

by Mavers' another analytical category has been identified called *dynamics*. This is a reference to her analysis of the direction of links and arrows in hand-drawn maps and to the digital animation of images and lettering that she discovers on a young learners' web page. Although she has not analysed a digital map where these functions are used, the column has been added in Table 2. in anticipation of this possibility in the data collection. Practice-based researchers may want to investigate her work more fully if they are dealing with digital maps.

The last category in the framework, *representations and interactions* is listed under evidence of social interaction. This is an adaptation of Jewitt's study (2003) that is referring to audience engagement rather than co-construction of knowledge. Her work is not about concept maps. Nevertheless her attendance to the relationships between the creator of a resource and the viewer are explored more fully in a long study of the development of multimodal theory, based on studies of school children learning from digital resource. In this study she suggests a major shift in the focus of analysis from identifying the roles of modes, like image and sound effect. Jewitt's suggestion is that the researcher will learn from looking at how the role of the learner is acted out. This emphasis on activity becomes very significant in looking at how digital technologies are employed to make meaning in ways that are not rooted in the conventions of writing and reading.

The framework, illustrated in Table 2, is divided into three columns: MDCM features, evidence and key theorists. The descriptors are written in accessible language particularly for practice-based researchers. The topics have also been divided into three categories that relate to the three cognitive theories: *information transmission, constructivism and social interaction* (Pachler 2005). Under *information transmission* are the analytical units that inform the researcher about which new concepts have been learnt. In the *constructive learning* rows are the units that reveal elements of the teachers' learning journey as they apply, reconstruct and make new constructs from the concepts they have assimilated. What is also important here is the focus on the affectual in learning; the emotional response that impacts on engagement and motivation. The final heading, *social interaction*, has only one unit so far which is used to record any evidence of an understanding of the relationships between the map-maker and the viewers, readers or audience for the map.

A key inference that emerges from the consideration of the *representations and interactions* category is that communication is not just a two way process between the map-maker and the viewer or audience. The findings explore the potential of social interactions in greater depth. This is also a way of entering a discussion about the children's learning role. Are they passive imbibers of information or are they becoming exploratory researchers actively engaged in understanding how meaning is made? In the young learners' case the resources have been created by a film maker but, these questions are also relevant to the teachers are map-makers when they share their concept maps with others, and even more when they engage in collaborative construction of a map.

THE FINDINGS

In this section the data is analysed from the semiotic point of view. The aim is not only to seek the value of analysis for the researcher, but also for the tutor and the map-maker. This raises again questions about who is the researcher and who is the subject.

Content Analysis

In this content analysis, the words used by the participants were extracted from the pairs of MDCMs and placed under score sheet headings from ImpaCT2 that indicate the teachers' awareness of the applications of computers and networks. The tables are too large to publish and so a summary of the results is offered in Table 3. In the group content analysis of the MDCM at the beginning and end of the course, the scores are represented for each subject under the headings Map One and Map Two.

The seven headings listed on the left are taken from the ImpaCT2 study. The headings indicate the kind of computer applications that the children had prioritised: computer systems and networks; information input; communication devices; publishing and correspondence; games; sound software as well as learning strategies. These provided an appropriate starting point for the analysis of the teachers' MDCM because the heading of their map was so similar.

This overview of the words the teachers use to label the nodes indicates that their focus is measurably more sophisticated by the time they map their understanding of the impact of computers in their personal and professional lives for the second time. Over a year the most

Table 3. Gains and losses in computer concepts

The ImpaCT2 content categories	Word counts	Differences
Computer systems and networks		
Map One	12	
Map Two	19	+17
Information systems		
Map One	31	
Map Two	32	+1
Communications systems		
Map One	30	
Map Two	32	+2
Publishing and correspondence		
Map One	13	
Map Two	12	–1
Games		
Map One	4	
Map Two	8	+4
Sound		
Map One	2	
Map Two	8	+6
Learning		
Map One	9	
Map Two	17	+8

significant gain is in the row *computer systems and networks*. Relevant word references more than double in the second map from 12 -29. The increase of 17 new words also illustrates an increased sophistication in the references to digital technologies. These references seem to be the result of growing awareness of the role of computers in e-learning during the year. For example, Interactive Whiteboards were not mentioned in the first map, but are often mentioned in the second map. References to *educational games software* also double. This may be due to the group projects.

At first inspection, the incidents of increased knowledge in other fields look minimal from the numerical point of view. For example, the fourth row, *publishing and correspondence,* shows a reduction in numbers of references but the second map references are more sophisticated. They change from general comment on word processing in the first map to a more specific mention of school newspapers and the collaborative course e-journal as new means of publication. This awareness of the publishing potential is another new idea that has emerged during the year.

In this analysis of content, it becomes clear that many of the more sophisticated references in the second map are due as much to home use as training in school or the CPD programme. A closer examination of the detail illustrates a new emphasis on interactive network tools like chat, text and the Messenger software that are replacing less specific references to online learning in general. This suggests the teachers are becoming more aware of interactive communication opportunities. Digital technologies like mobile phones that are also used in the home show significant increases and references to the use of 'music' in a digital context like an MP3 player quadruple. These scores suggest that the multimodal and interactive affordances of computers are increasingly appreciated as the year progresses. There is also a tentative indication in the maps that this is not just because of the presence of digital equipment at home, but also because of the influence of their own children who are frequently depicted in the maps as a source of teachers' information.

Whereas the most outstanding numerical gain is in the factual area of computer and network applications, the second most outstanding gain over the year is found in the references to learning with computers. The scores nearly double and the words in the second map indicate a growing understanding of the subject and the potential impact on professional learning. The participants are more aware of their learning. They discuss more long-term approaches to CPD overall than they do in the first map. Some examples that relate to collaboration are: facilitating learning, e-learning and self-development. In [INSERT FIGURE 001] the second MDCMs the teachers also focus more on their learning needs and achievements rather than on the learning needs of the pupils. This highlights a greater awareness of their own needs as teachers and of their achievements as a CoP.

Transcript Analysis

Two Individual Learners' Priorities

The transcripts were limited to the conversations of sixteen teachers talking in groups of four for half-an-hour (Table 4). As a result only exemplar comments from the transcripts have been organised under the five headings provided by Daly and Pachler(2007): knowledge construction, community, meta-learning, autobiography and cognition. These extracted comments provide latent evidence that the strategies developed in the CPD programme design are being absorbed and articulated even though the scale is small. In these exchanges it is possible to trace elements of knowledge building in a social and collaborative context that are based on group analysis of the meaning of the MDCM. These learning themes are developed in more detail in the semiotic analysis of the individual maps in the next section

Analysis by Sign

Two pairs of maps are selected to illustrate how the different aspects of semiotic analysis in the Table 2 framework can be used to analyse individual learning priorities. The semiotic framework provides the means to explore firstly what the teacher's priorities at the start of the programme are and, secondly, whether this view of e-learning has altered by the end of the programme. The Daly and Pachler indicators (2007) are also relevant in investigating both manifest and latent evidence of explicit understanding of the learning model that underpins the development of this CoP of teachers as learners. All the names are pseudonyms.The findings from the two individual map-makers MDCM are triangulated in this section with the biographical details provided in the participants' initial questionnaire to help to focus on each individual's professional context.

Tim: The Professional E-Facilitator

At twenty-eight, Tim was six years younger than the next youngest participant in the cohort. Yet, the questionnaire data indicated that Tim was an experienced professional e-learner in this group at the UK National College for School Leadership. He was moderating discussion between head teachers using a sophisticated learning platform called Talking Heads. He needed to work collaboratively with other e-facilitators to improve his technique.

An extract from the transcript offers the PMR

Table 4. Exemplar comments under the Daly and Pachler headings

The analytical headings: Daly and Pachler (2007)	Examples of comments made by members of the PBR cohort in comparing their MDCM pairs
Knowledge construction	Because as a child you then throw things out the window you don't care, so you can draw anything you like, I mean I wouldn't draw smiles! but then it makes you think about what you've drawn ...you can say well that's good, so there has been a good aspect to that because it frees your mind a bit!
Community	Here's a little picture of me looking at other people telling me things cos it allows me to listen to other people and learn from other people, which obviously didn't get a chance before because we've got all that information highway, dig things up on the Internet that you never thought were there, broadens everything you know...
Metalearning	I thought it was all about, the teachers I was working with online. Now I've realized that actually working through the process my own process has deepened! It seems to me a lot of talk in education circles out there is about taking learners from where they are to where you want them to be and often doesn't reflect on the learners own position and make connections between those two.
Autobiography	Now I'm very lucky, I spend large box of time at my desk or I go out to schools and I work with schools and teachers, so the separation is easy for me, if I was back working in school again it would be much more fragmented and difficult and I would find myself in your position, at 10 pm at night, thinking well I can't be bothered reading that and switching off and watching newsnight or something for a bit of light relief.
Cognition	Yeah things have moved on for me its developed but I think the links have become much more complex.

insights into Tim's cognitive processes as he undertook the actual drawing of the TDCMs that are analysed below.

I did one train of thought and then I kind of abandoned that because it got a bit cluttered and I started another one... ...then I looked at the concepts more in the second map

His first map (Figure 3) already indicates a conceptual understanding, both in the labels and the connecting links, of how e-learning works. His specific professional knowledge is expressed autobiographically in his references to all the organisations that are linked online into the National College for School Leadership (NCSL). The phrase *web-crossing* seems to refer to the activity going on between key groups of professionals that are referred to by acronyms like GTC-General Teaching Council. Tim's technical knowledge is exemplified by the sophisticated package he names (LPSM means Linux Persistent Memory, a storage method that is open source).

The composition of the MDCM elements highlights collaboration between himself and other learners and tutors in the online CoPs where he already has membership. His experience is already about spanning more than one community. The composition does not reflect the division in the heading between personal and professional uses of e-learning in the first map. This lack of division gives the impression that e-learning is all pervasive in his life. All the elements in this image technical and human, are given the same weighting in size. The swirling dynamic connectors represent a nest of cables that also serve as mental connections between the participants in the network and between their institutions. His drawing also seems confident and the sketch of the person rather humorous which suggests that he feels in charge of his learning.

It is possible that personal and professional differences may seem irrelevant to Tim. This impression is reinforced by his depiction of his

Figure 3. Tim's first map

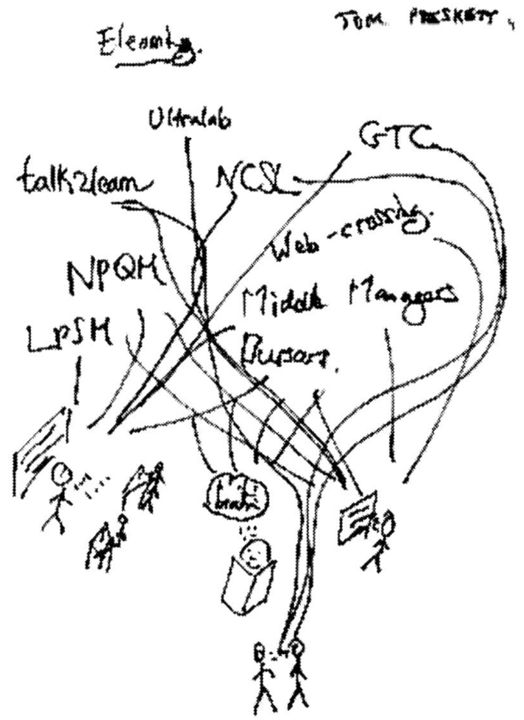

e-communities as a unifying concept overarching the learning platform. The integration of digital power and human thinking is shown as an electronic brain linked to his own brain. Learning inputs are derived from human agents using interactive whiteboards and traditional books, a mode he does not reject.

In terms of composition Tim's second map, Figure 4. is another individualistic interpretation of an MDCM design that is even less dependent on words than the first map. This time, a central line provides a clear division between the personal and the professional. In the professional top-half he repeated the links between all the e-communities he works with and repeated some of technical terms from his first map. The new personal section is liberal with graphics particularly light bulbs that indicate the creation of ideas. The light bulbs indicate his knowledge about computer icons in general use. The curving, iterative connecting links provide a strong sense of a collaborative e-community. He promotes personal research as well giving a strong visual focus to the e-learning benefits of access to information sources, time (saving) and convenience.

A second passage from the transcript confirms Tim's commitment to the processes of human communication within the network in addition to his sophisticated understanding of the technicalities of e-facilitation.

Basically what e-learning is all about is getting people to learn just using a different context and with the environments that we have... there's community forums, lifethink.com, where group discussions with tutors er... imparting knowledge and then allowing them to reflect on their learning and then there's solo learning environments where they find things out for themselves... it's at their own convenience, in their own time and they have control of what they do.

Figure 4. Tim's second map

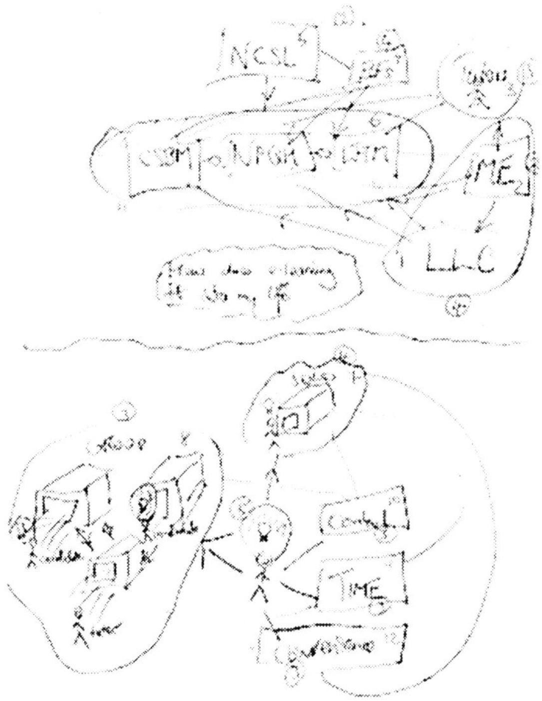

Neil: The Expert Map-Maker

Neil, his pseudonym, is the primary teacher who was keen to share his four-year experience of MDCMs with his colleagues at the beginning of the CPD programme. He also uses MDCMs in his own practice-based research project for the programme. What he learns over the year becomes clear in the design of his two maps and the transcripts of his discussions about the maps with the group. For example, in the first map he labels many of the links with adverbs like *by* and *through* indicating that he is familiar with Novak's methods of designing concept maps to illustrate learning in a pre-agreed way. This clear vision of design was based on the projects he had developed for first practice-based publications. Sharing his expertise in the transcripts proved revealing for the teachers and for the researcher without more diligent delving. For example, his design decisions are well considered:

> *Talking about mess... no drawings, there's absolutely no drawings on this because I've been doing mapping for a very long time and teaching the children how to map so it becomes almost second nature...*

This extract indicates that he had developed this particular approach to building a map with this class in which he followed clear rules that would not be explicit to a PMR. Within his own context, this first MDCM (Figure 5) is the most sophisticated effort in combining words with other semiotic features. Conceptually, Neil uses a full range of terms to describe his *personal* and *professional* priorities for e-learning. To the *personal* and the *professional* he also adds *home* and *family* with an emphasis on his collaboration with his own children. Professionally he makes a distinction between CPD activities like collaborating online with colleagues and the different ways in which he uses the computer in school including some of

Figure 5. Neil's first map

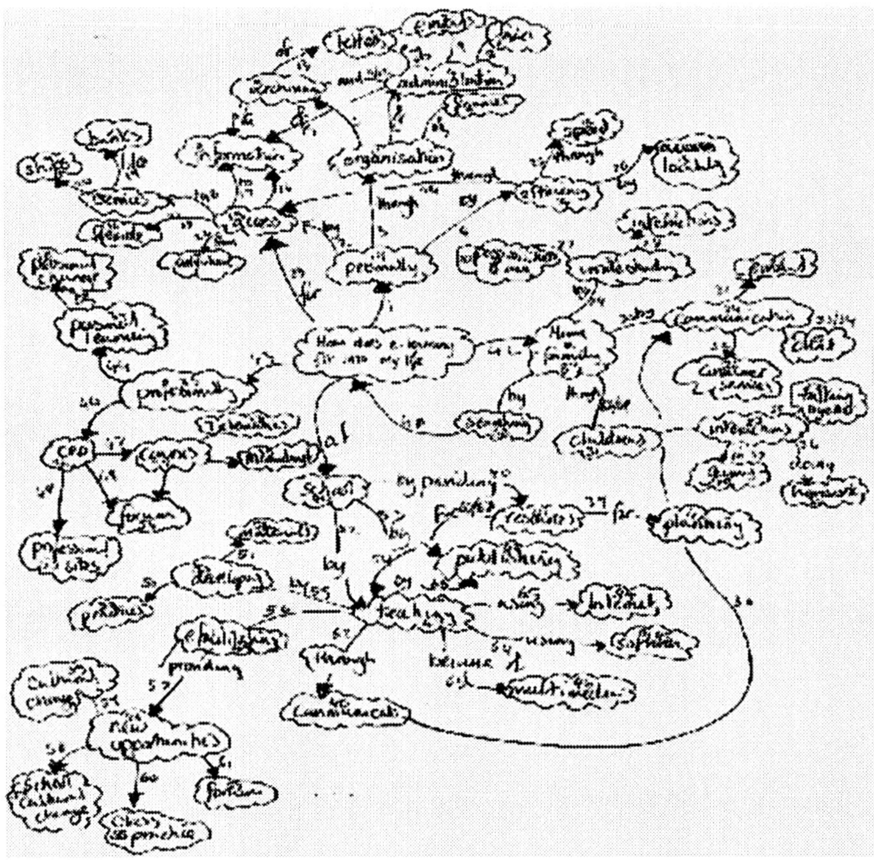

the applications. He is also aware of forums for cultural exchange suggesting a keenness to engage in collaboration beyond his local horizon in the citizenship sphere as well as in the CPD sphere. Neil's enthusiasm for e-learning, the affectual content, is underlined by the *new opportunities* node he introduces and his references to the potential for personal learning and personal change.

This first MDCM immediately gives a visual impression of semiotic fluency in composition. The myriad links and nodes indicate business with a wide range of knowledge about computer applications. The arrows with their emphatic heads create an impression of dynamic iterative activity. The curling edges to the nodes also have significant aesthetic appeal. However, the map is so detailed it is not easy to read. In addition, Neil has not made the relationship of each concept clear by positioning and size or the connections between the categories and the category components. The composition does not distinguish by levels either between higher order concepts and subordinate ideas. Technicalities and abstractions are given equal semiotic weight that makes it quite difficult for the PMR to interpret the relative values of each. The busyness of the MDCM also interferes with the communication of meaning although the drawing skill is impressive.

In his transcript, Neil explains that he has four year's experience of using MDCM in the classroom. During this year-long episode of formal learning Neil has been experimenting with digital maps in his practice-based project with his class. This experience explains why he was the only person in the cohort who offered a digital map at the end of the CPD programme since by then he had the software on his laptop.

The immediate impression is that this second map has far less content. However, in terms of composition, it is easier to see what the main concepts and the subsidiaries are by position and size. A creative limitation of this digital software is that arrows cannot be curved thus losing the dynamic characteristics of a hand-drawn TDCM. Another limitation is the time required to learn all the features. For example, Neil is not yet ready to employ the affordances of colour provided by the software developers. Only the default colour, green, had been utilised for the nodes although he has experimented with different coloured arrows.

The immediate impression is that this second map has far less content. However, in terms of composition, it is easier to see what the main concepts and the subsidiaries are by position and size. A creative limitation of this digital software is that arrows cannot be curved thus losing the dynamic characteristics of a hand-drawn TDCM. Another limitation is the time required to learn all the features. For example, Neil is not yet ready to employ the affordances of colour provided by the software developers. Only the default colour, green, had been utilised for the nodes although he has experimented with different coloured arrows.

In the final presentation of his case study findings to his peers, Neil explained that his negotiated practice-based research project involved experimenting with MDCM for language analysis with his pupils. For this he had used the taxonomy of higher order thinking developed by Bloom and his colleagues (Bloom, Englehart, Furst, Hill, & Krathwohl, 1956). Since the fifties there have been several new version of this cognitive domain. Neil cites the Officeport version(2006) in his explanation of how he had shared this theory with his pupils in order to help them build more complex maps and analyse them. These views about higher-level thinking were well expressed in the transcript where he gave his own verdict on his maps. He told his peers that he judged his first map to be overcomplicated, difficult to read and inadequate in categorising the concepts to prove higher order thinking. As a result, he explained that in his second map, he had consciously drawn on Bloom's theory in using the affordances of digital mapping to refine and categorise his ideas to illustrate higher order thinking. The outcomes of his study were that studying Bloom's theoretical approach during the year had led him to change his practice with his pupils. Secondly the affordances of digital mapping allowed him to experiment more effectively with hierarchies of ideas. However, the layers underneath the top surface that he had created could not be seen on the paper version. On his laptop screen he demonstrated how clicking on the nodes of first order concepts revealed the layers of detail below. This dimensionality had not been recognised in the initial PMR analysis. However, in Neil's discussions with this colleagues and the transcript the value of a layered digital map was revealed.

For researchers this episode raises the issues about collecting multidimensional digital data. Although this is now a serious consideration for a researcher there are many operational difficulties that need to be overcome. (see Table 5)

THE FINDINGS FROM A PMR PERSPECTIVE

The PMR is able to use the evidence to make some statements about the learners' priorities from a detached standpoint. In this mode, the PMR can infer from the word analysis that the growing frequency and quality of their computer concepts suggests that the teachers' knowledge about e-learning has increased during the year. The reference to the quality of learning in the first

Table 5. The impact of the role of the researcher on teachers as activists

The role of the researcher	The role of the teacher as learner	The impact on policy and practice
Peripheral member researcher (PMR)	The teacher as the subject of the research is not engaged in the process and is unlikely to read the subsequent report.	No impact on policy or practice.
Active member researcher (AMR)	The teachers are invited to comment on the results.	The teachers achieve individual and group insights into their own learning and into the value of reflection on practice.
Complete member researcher (CMR)	The teachers negotiate their own topics for practice-based research that address their current challenges and are planned to impact on their policy and practice.	Impact on the teacher's practice and possibly on their institutional development plan depending on the topic and the status of the teacher.
Collaborative complete member researcher (CCMR)	Teachers collaborate in designing joint research projects and in reviewing their own evidence and creating new knowledge and theory about practice.	Publication of the teachers' collaborative knowledge based on evidence has impact on national policy and practice.

content analysis exercise indicates that the teachers are prioritising their learning and thinking more widely about learning models, about knowledge creation and about communities of practice than they were at the start of the course.

The tutors cannot take all the credit for the development of e-learning concepts. The variables are too great to prove this direct relationship. Social interaction was not a feature of the first content analysis indicators modified from ImpaCT2. However, the second qualitative method of analysing the transcripts, adapted from Daly and Pachler (2007), offers a productive way forward for analysts interested in the impact of community membership on learning because these indicators focus in more detail on the social interaction aspects of the learning processes. More research is required, however, as the data was too limited to provide conclusive proof of the success of these methods of content analysis. The PMR will be able to refine methods to analyse social interaction more deeply when another opportunity appears for similar research. However, the challenge for the PMR is that even if a new research report is published some time later this is unlikely to have an impact on the collaborative CoP processes themselves. Therefore, the report will not influence the kind of collaborative learning advocated by Sachs (2003) in her activist professional thesis. Indeed the CoP may not have access to the report.

This division between the publication of PMR activity and the absence of action by the subjects of the research is a challenge. Often when a PMR research report is published the subjects of the research are unaffected by the results as they have no access to the information. This is particularly true of teachers who have little access to academic publications. If teachers do have access they often find the style inaccessible. Another problem with a rigid research design is that there is no opportunity to include data that is unexpected and unplanned for. In fact, in most observation routines the PMR may not be in the right place to see that their intervention has stimulated a change in behaviour.

Findings From the Perspective of the AMR

The PMR cannot be sure that new priorities are a result of the course alone. In fact, the nature of practice-based research suggests that the teachers will have learnt from their practice with students and other colleagues. On the other hand the AMR who is promoting collaboration will not see this as a problem. The AMR role as the tutor in this

CPD design was to develop habits of social interaction as a process for learning rather than to teach a set of agreed e-learning concepts. A confident AMR who is also a tutor will welcome new knowledge. This is an important approach for tutors in a subject like e-learning where so little is yet known about the digital technologies and the pedagogical applications.

The ImpaCT2 team established the value of interviewing map-makers. The interviewer also showed respect for the map-makers' opinion which is an important standpoint in the AMR role. Joining in the group conversations and asking for my judgments to be considered from their point of view was a pivotal method of converting from a PMR to an AMR. Without access to the map-makers, the PMR cannot be sure that the truth has been exposed because the map-makers reveal information in the transcripts that is not revealed in a solitary study of the MDCM. In this study collaborative discussion reveals new truths. It also indicates how the map-makers can mentor each other to extend the group's understanding of different kinds of learning that are valid. Overall this approach has the potential for individuals and groups to pinpoint opportunities for change. For this reason the analysis of the exemplar pairs of maps included evidence from the transcripts about the teachers' own interpretation of meaning. Each map-maker emerges as a unique individual. It must be appreciated, therefore, that if other maps had been selected for this analysis the discussion of the data would be different. The transcripts provided vital triangulation and also provided new directions for interpretation when they were included as an element of the CoP discourse.

In the two following sections from the AMR perspective the priorities expressed in the Tim and Neil's pairs of MDCMs are considered in detail. Each commentary shows how the maps have provided significant opportunities for the tutors, researchers and map-makers to think about the learning priorities from multiple perspectives that cannot easily be developed from an essay.

Tim's Learning Priorities

There are two factors in Tim's professional background that an AMR/tutor might wish to discuss with him in view of his fluency in drawing: his youth and his professional role as an e-facilitator rather than a conventional teacher educator. His two maps indicate that he has a good understanding of e-learning as it relates to online learning (Figure 3 and Figure 4). The use of the established computer icon, the light bulb, to represent creative thinking suggests that he expects others to understand these new elements of communication. This evidence of his acculturisation into sophisticated semiotic features of the digital world is also underpinned by his use of technical terms. This provides an insight into the sophisticated concepts that precede the CPD intervention.

Just as the notion of community runs through Tim's discussion with his peers, his drawings of *representations and interactions* also indicate again the role of the community in e-learning. The two maps comment on the activities of other players in the networked community. The frequency with which he depicts people in animated poses reinforces his concept of the network as a dynamic tool for human engagement in a collaborative community. He shows awareness of this audience in using graphics to promote an autobiographical slant as well.

The ways in which he combines words, pictures and lines in a unique combination using well-known icons and fresh images suggests a talent for expression in this medium that is worthy of celebration whatever advantages he might be perceived to enjoy. For example, his drawings of stick people are an effective graphic means of indicating the priority he gives to the human being within the electronic network. The use of the brain image also draws attention to his enthusiasm for e-learning opportunities. In the second map the connections between all the learners represented by the light bulbs suggest knowledge construction as well as community connections. The light bulbs also seem to suggest cognition and meta-learning processes. He seems to have covered all the Daly and Pachler (2007) indicators either manifestly or by inference. His maps, therefore, show a high level of understanding and practice in collaborative learning from the beginning of the course that improves further in the second map.

Neil's Learning Priorities

In a first impression a PMR might conclude that Neil knows less at the end of the course than he did at the beginning because visually there are fewer connections in his second map. To the uninitiated his complex first map, drawn in twenty minutes, seemed to be an outstanding example of how an MDCM should look, except that there are no pictures. The complexity is outstanding. Up to this point I had tended to rate MDCM with pictures more highly than those without assuming that a capacity to use graphics showed more familiarity with multimodal communication. However, in the transcript Neil said that he tutored his pupils not to use graphics in a MDCM. I realised, as an AMR, that this kind of design decision should be respected. However, this kind of decision can only be respected if the map-makers' intention is revealed through interview of group conversations.

Visually, the second digital map does not seem to be as aesthetically pleasing as the first hand-drawn maps to me. This was another subjective judgment that was overturned by Neil's comments in the transcript. I began to question how much latitude a researcher should have in allowing subjective ideas of aesthetics to count in judging a map. This consideration of Neil's expertise in mapping led me to observe that in this context a PMR might not realise, unless they have contact with the subject, that their observations are based on their own lack of knowledge about MDCM. These weaknesses might be lack of experience in using the package for example.

Neil's judgments about the higher level thinking in his maps were particularly illuminating because they had been developed from the evidence of his practice-based research. Drawing on Bloom's taxonomy of higher order thinking (1956), he had consciously used the affordances of digital mapping to refine and categorise his ideas to illustrate higher order thinking by using layers. This semiotic framework allows for this specifically in the multidimensional category. Perhaps multilayering should be included specifically in this section to help the researcher.

FINDINGS FROM THE PERSPECTIVE OF THE CMR

The Compete Member Researcher can have more impact on the outcomes of research than a Peripheral Member Research. CMR takes more account of the impact of research intervention. The collaborative MDCM, for example, was an unplanned student-generated outcome that had not been designed into the CPD programme by the researcher or the tutors. However, as a CMR I had access both to the data and the opportunity to understand how this opportunity had been generated. If I had only been operating as a PMR I might not have seen this data collection opportunity.

In this case my personal knowledge of these map-makers over a longer period than the year of the course was an important factor in interpreting the MDCM as a CMR. As a member of the CoP that ran the course, and as a tutor myself, I was able to incorporate the significance of the collaborative map into this research about the value about CoPs and knowledge creation in learning. This required a collegiate perspective on my research subjects as professional equals.

The collaborative map and Neil's expertise also led me to recognise my own limitations. For example, these days MDCM expertise is bound to be present in varying degrees amongst the participants in a CPD programme about e-learning. In this context, questions should be asked about what kinds of objective judgments a PMR is qualified to make about MDCMs and what is outside the researchers' remit. A key question if the MDCM were used as a summative test is the advantage to map-makers who already have significant experience in mapping. Another question is what kinds of allowances should be made for mappers who adhere to the methods of one or other mapping school. MDCM appear to have advantages in multimodal testing over essays. Nevertheless some sophisticated procedures would have to be devised to ensure validity. In this case, active involvement in the community meant that not only was I better informed as a CMR, but through collaborative discussion, Neil's colleagues also had more insight into the potential of MDCM in learning. His experience was invaluable in the collaborative learning that took place in this CPD intervention.

Collaboration of Complete Member Co-Researchers in a CoP

The collaborative map and the three e-journal outcomes from this CPD intervention indicate the potential of collaborative learning when it is valued by the tutors. In many traditional academic courses there is no time for collaborative outcomes after the solitary work is marked. There may also be some suspicion on the part of the tutors that work researched with others is cheating the exam system. This notion of cheating, however, assumes that the tutors can be the arbiters of what knowledge is important and what is not. This approach to learning in the digital technologies also fails to acknowledge innovation that takes place after the syllabus is set and future trends that are not yet fully formed. In these modern learning circumstances Jewitt and Kress (2003) question the value of assessment in any form because it is always biased towards the limited vision and culture of the tutor/tester. In contrast, by developing ideas about theory in this MDCM, the group have themselves created new knowledge about e-learning theory rather than reproducing what they had been taught (Preston and Cuthell 2005).

The participants in this concept mapping study have been engaged in all four of communication strata presented by Kress and Van Leeuwen (2001). They have been initiated into the appropriate *discourses* for learning. They reconstruct their understanding of that discourse by *designing* and *producing* single authored and multiauthored MDCMs. Furthermore they take the opportunity to *disseminate* their growing knowledge professionally. However, *dissemination* suggests a one-way flow of information. The nature of the MirandaNet CoP's submission to the e-learning consultation had some impact on the policy makers. Therefore, this submission fulfilled Sachs' aim to encourage teachers to become *activist professionals* by providing evidence that impacts on policy and practice. A case might be made, in this light of this event, to add another communication strata called *interactive action* (Preston 2006). This extra category acknowledges the complexity of multimodal communication that aims to achieve a result. This can only happen once learners or readers are no longer seen as empty vessels to be filled - the stance of experts who subscribe to the information transmission mode of teaching.

FUTURE TRENDS

Future trends in collaborative mapping are already being explored. When the collaborative map was drawn at the end of this CPD programme, multi-authorship for MDCM was virtually unknown the MirandaNet CoP. Three years later several Fellows have followed Neil's lead in developing new practice-based projects that focus on using digital maps collaboratively with young people in order to develop social interaction as a means of knowledge construction. Collaborative digital mapping are now frequently used in MirandaNet events to scaffold discussions, identify priorities and to explore collaborative solutions to problems. An MDCM has been designed as an innovative index to highlight the incorporation of different multimodal genres in a volume of an academic e-journal on mapping (Howell-Richardson and Preston 2007).

Further MDCM research is planned to investigate whether further collaboration between the researcher, the tutors and the teachers in analysing priorities deepens a CoPs understanding of the nature of teachers' learning and of knowledge creation techniques. This knowledge will add to the body of information about methods of building CoPs for learning and about knowledge creation. Most importantly this research might provide methods of multidimensional assessment that resonate with the challenges of learning in the future.

CONCLUSION

In conclusion the core questions driving this research are revisited. Overall semiotic theory has provided new insights into teachers' e-learning priorities as individuals and as a group. The research was not designed to establish right and wrong answers about the teachers' priorities, but to develop semiotic multidimensional analysis tools that would help researchers, tutors and map-makers to gain insights into learning from their different perspectives. Semiotic theory has also proved useful in stimulating thinking about the ways in which learners might now present their work in multimodal ways, why this is important in the twenty first century and how multimodal communication might be empowering for an activist professional. The discussion has also raised questions about the relationships between researchers, tutors and map-makers. This raises questions about where the assessment power should lie. A radical conclusion to draw is that assessment by a tutor or a PMR can be reliable because they must assess from personal and cultural bias as well as from their limitations in the field. Unless the researcher is prepared to acknowledge these limitations, there is a danger that the originality and creativity of the learner will not be understood or encouraged.

The particular merits of the different analytical tools that have been piloted have been identified. There were two methods: analysis by word content and by applying a semiotic framework. The words on the maps were helpful in gaining insights into the priorities of a whole group and of individual learners. These techniques could be used in a needs analysis as well as providing data for tutors about those who appeared to have sophisticated conceptual understandings of already those who did not. In addition, it was possible to trace patterns in the groups of increasing knowledge that had accumulated in the period between the two maps. MDCMs are likely to be useful in situations where the researcher is aiming at a wider and deeper needs analysis than written tasks or verbal interviews can provide, particularly at the beginning of a course when essay writing is inappropriate.

These discussions that used the maps as scaffolds suggest that MDCMs can be an important stimulus in learning collaboratively in two ways: firstly in stimulating the map-makers to think about their own learning and the learning of others and, secondly, to help the group to focus on individual differences and misconceptions and to celebrate new insights together. Evidence of spontaneous peer-tutoring emerged in the transcripts because the group was given ownership of the research process. However, there are differences in the results from the maps and the transcripts depending on whether the researcher was looking for the facts to build an increasing body of knowledge or an awareness of meta-cognitive and collaborative learning processes.

The greatest difficulty in presenting the findings has been in unravelling the impact of the researcher in different roles. In this project I had the role of the researcher, the designer of the CPD programme, a tutor of this group of advisers, and eventually a CoP e-journal editor. It seems important, at this point, to evaluate who has benefited from this exercise in MDCM; myself as researcher and tutor, or the map-makers and their institutions. My view changed substantially as the project unfolded. The more I engaged the map-makers in the research the more I learnt about the possibilities of collaboration.

I realised that by acting as a CCMR there were three important outcomes. The first was the improvement in the depth of my analysis of the teacher's priorities. The second was the value for the teachers of being seen as co-researchers. The third was the confidence this gave the CoP in their own collective judgment and capacity for theory creation.

The drawing of the collaborative MDMC was the turning point in my understanding of the role of co-researchers in a CoP who feel empowered to share knowledge and build theory. This group of teachers represented a new perspective on professional ownership: the collaborative complete member co-researcher (CCMCR). What emerges from this study is the potential of the MDCM as a tool for developing a new learning theory called the *collaborative construction of new knowledge*: a learning theory for the future. In this way the mapping task becomes a focus for a Third Space (Zeichner and Liston 2006; Zeichner 2008) blurring the roles of the researchers, tutors and map-makers still further in the cause of learning.

REFERENCES

Anderson-Inman, L., & Ditson, L. (1999). Computer-based concept mapping: A tool for negotiating meaning. *Learning and Leading with Technology*, *26*(8), 7–13.

Bloom, B., Englehart, M., Furst, E., Hill, W., & Krathwohl, D. (1956). *Taxonomy of educational objectives: The classification of educational goals. Handbook I: Cognitive domain*. New York, Toronto: Longmans, Green.

Clark, W. (2007). Mapping the syllabus: multimodal digital mapping tools. In *Reflecting education: Fascinating cultural artefacts: multimodal (concept) mapping in teaching and learning* (pp. 43-60). Retrieved November 3, from www.reflectingeducation.net/index.php/reflecting

Cuthell, J. P. (2002). A learning community: A community of learners. *Journal of Interactive Learning Research*, *13*(1/2), 169–188.

Cuthell, J. P. (2005). *What does it take to be active? Teacher participation in online communities*. Geneva, Switzerland: Inderscience.

Daly, C., & Pachler, N. (2007). Learning with others in mind. In J. Pickering, C. Daly & N. Pachler (Eds.) *New Designs for Teachers' Professional Learning*. London: Bedford Way Papers.

Davis, N., Preston, C., & Sahin, I. (2008 in press). ICT teacher training impacts multiple ecologies: evidence from a national initiative. *British Journal of Education Technology (BJET)*.

Davis, N. E. (2008 in press). How may teacher learning be promoted for educational renewal with IT? In J. Voogt & G. Knezek (Eds.), *International handbook of information technology in education*. Amsterdam: Kluwer Press.

Davis, N. E., Preston, C., & Sahinl, I. (2008 in press). Theoretical and evaluation frameworks to inform technology-related professional development for teachers, tested with evidence from a national study of ICT professional development for teachers. *British Educational Research Journal*.

DfES. (2003). *Towards a Unified e-learning strategy: consultation document*. Retrieved from http://www.dfes.gov.uk/elearningstrategy/strategy.stm.

Elliott, J. (1991). *Action Research for Educational Change*. Buckingham, UK: Open University Press.

Harrison, C., Comber, C., et al. (2002). *ImpaCT2 The Impact of Information and Communication Technologies on Pupil Learning and Attainment.* London: Becta for DfES.

Holmes, B., Tangney, B., & Fitzgibbon, A. (2001). Communal Constructivism: Students constructing learning for as well as with others. In *Proceedings of the 12th International Conference of the Society for Information Technology & Teacher Education* (SITE 2001), Charlottesville, VA, Association for the Advancement of Computing in Education.

Howell Richardson, C., & Preston, C. (Eds.). (2005). *Special Inaugural Edition: E-learning.* London: Reflecting Education.

Howell Richardson, C., & Preston, C. (2007). Multimodal concept mapping in teaching and learning: editorial: Fascinating cultural artefacts: multimodal concept mapping in teaching and learning. *Reflecting Education.* Retrieved from www.reflectingeducation.net/index.php/reflecting

Jewitt, C. (2003). *A multimodal framework for computer mediated learning: The ReShaping of curriculum knowledge and learning culture, language and communication.* London: Institute of Education, University of London.

Kress, G., & Van Leeuwen, T. (2001). Multimodal Discourse: The modes and media of contemporary communication. Arnold, UK: Hodder Headline Group.

Kress, G., & Van Leeuwen, T. (2007). *Reading Images: The Grammar of Visual Design* (2nd Ed.). London: Routledge.

Lawes, S. (2003). What, when, how and why? Theory and foreign language teaching. *Language Learning Journal, 28*(1): 22–29. doi:10.1080/09571730385200171

Leask, M., & Younie, S. (2001). Building On-Line Communities for Teachers: Ideas Emerging from Research. In M. Leask (Ed.), *Issues in Teaching Using ICT.* London, Routledge.

Leask, M., & Younie, S. (2002). Communal constructivist theory: ICT pedagogy & internationalisation of the curriculum. *Journal for IT for Teacher Education, 10*(1&2), 117–134.

Mavers, D. (2004a, June). *Observations on ImpaCT2 concept mapping.* MirandaNet seminar, Institute of Education, University of London, London.

Mavers, D. (2004b). *Multimodal design: the semiotic resources of children's graphic representation.* Doctoral dissertation, Culture, Language and Communications, Institute of Education, University of London, London.

Mavers, D. (2007). Investigating how children make meaning in multimodal maps. Reflecting Education: Fascinating cultural artefacts: multimodal (concept) mapping in teaching and learning. *Reflecting Education* (pp. 24-28). Retrieved November 3, from www.reflectingeducation.net/index.php/reflecting

Mavers, M., Somekh, B., & Rosterick, . (2002). Interpreting the externalised images of pupils' conceptions of ICT: methods for the analysis of concept maps. *Computers & Education, 38*, 187–207. doi:10.1016/S0360-1315(01)00074-4

Novak, J., & Godwin, D. (1984). *Learning how to learn.* Cambridge, UK: Cambridge University Press.

Officeport (2006). *A version of Bloom's Taxonomy.* Retrieved from http://www.officeport.com/edu/blooms.htm

Preston, C. (2004). *Learning to use ICT in Classrooms: teachers' and trainers' perspectives: An evaluation of the English NOF ICT teacher training programme 1999-2003.* London: MirandaNet and the Teacher Training Agency. Retrieved from www.mirandanet.ac.uk/tta.

Preston, C. (2006). Etopia: Mapping interactive learning in an e-community context. *MirandaNet Braided Learning e-journal.* Retrieved from www.mirandanet.ac.uk

Preston, C. (2007a). Braided Learning: Promoting active professionals in education. In C. Haythornthwaite (Ed.), *New International Theories and Models Of and For Online Learning.* Chicago: First Monday.

Preston, C. (2007b). Social networking between professionals: What is the point? In *Self-regulated Learning in Technology Enhanced Learning Environments: Individual Learning Individual Learning and Communities of Learners.* Amsterdam: Springer Verlag.

Preston, C. (2008). Braided Learning: an emerging practice observed in e-communities of practice. *International Journal of Web Based Communities, 4*(2), 220-243. doi:10.1504/IJWBC.2008.017674

Preston, C., & Cuthell, J. (2005). Teaching in ICT-rich environments: Using e-learning to create a knowledge base for 21st century teachers. In M. Leask & N. Pachler, (Eds.), *Learning to teach using ICT in the Secondary School,* (2nd Ed.). London: Routledge.

Preston, C., & Cuthell, J. (2007). Mapping Inspiration. *Braided Learning e-Journal.* Retrieved from www.mirandanet.ac.uk/ejournal.

Preston, C., & Danby, M. (2004). *An evaluation of ICT skills training for educators*: An internal report, Business Link Hereford.

Preston, C., & Holmes, B. (2002). Capturing the Online Knowledge, Building of Educator: ICTS, Authorship and Living Design. ITTE Conference, Dublin, Ireland, ITTE

Ralston, J., & Cook, D. (2007). Collaboration, ICT and mind mapping. *Reflecting Education,* (pp. 61-73). Retrieved November 3 from www.reflectingeducation.net/index.php/reflecting

Resta, P. (2002). *Information and communications technologies in teacher education: A planning guide.* Paris: UNESCO. Retrieved from http://unescobkk.org/education/ICT or http://unesdoc.unesco.org/images/0012/001295/129533e.pdf

Riley, N. (2005). Methods for evaluating critical learning using online discussion forums. *Technology, Pedagogy and Education, 15*(1), 61–76.

Riley, N. (2007). Investigating the impact of ICT-based multimodal mapping in developing effective learning dialogues. *Reflecting Education,* 74-97. Retrieved November 3 from www.reflectingeducation.net/index.php/reflecting

Riley, N., & Alhberg, M. (2004). Investigating the use of ICT (information and communication technologies)-based concept mapping techniques on creativity in literacy tasks. *Journal of Computer Assisted Learning, 20,* 244–256. doi:10.1111/j.1365-2729.2004.00090.x

Rojas-Drummond, S., & Anzures-Tapia, A. (2007). Oracy, Literacy, Conceptual maps and ICT as mediators of the social construction of knowledge amongst peers. *Reflecting Education.* Retrieved from www.reflectingeducation.net/index.php/reflecting

Saussure, F. (1916). *Course in general linguistics.* W. Baskin (Trans.). New York: Philosophical Library.

Somekh, B. (2005). *Action Research: A methodology for change and development.* Milton Keynes, UK: Open University Press.

Somekh, B. (2007). *Pedagogy and Learning with ICT: Researching the art of innovation.* New York: Routledge.

Wenger, E. (2004). *Learning for a small planet: A research agenda.* Retrieved from www.ewenger.com/research.

Wenger, E., McDermott, R., & Snyder. (2002). *Cultivating Communities of Practice: A Guide to Managing Knowledge.* Boston: Harvard Business School Press.

Wenger, R., & Lave. (1998). *Communities of Practice: Learning, Meaning and Identity.* Cambridge, UK: Cambridge University Press.

Zeichner, K. (2008). *Keynote: The Third Space: where teachers and academics meet.* British Education Research Association (BERA) Annual Conference, Edinburgh, BERA.

Zeichner, K., & Liston, D. P. (2006). Teaching Students to Reflect. In D. Hartley & M. Whitehead, (Eds.), *Teacher Education: Major Themes in Education.* London: Routledge.

KEY TERMS AND DEFINITIONS

Community of Practice: A term associated with Wenger that refers to the ways in which people learn in groups with a common purpose. These groups were first noticed in business. Studies have now extended to education or leisure.

Concept map: The term is used generically in this paper. It term is often associated with Novak. Novak's style maps are hierarchical, although they also radiate from a central idea. Buzan 'mind maps' are designed to radiate from concept placed at the centre of the map.

Content analysis: Content analysis is a general term for a number of approaches to analysing written or spoken language. In this paper it refers to the analysis of vocabulary.

Multimodality: This term refers to meanings made in many representational and communicational modes that include speech, writing, gesture, music, drawing and animation.

Social semiotics: The study of the signs that people make in order to communicate and of the sign makers and their motivations.

Practice-based research: A variety of action research that emphasises scholarly approaches to work-based learning

Chapter 14
Expanded Collaborative Learning and Concept Mapping: A Road to Empowering Students in Classrooms

Paulo Rogério Miranda Correia
Universidade de São Paulo, Brazil

Maria Elena Infante-Malachias
Universidade de São Paulo, Brazil

ABSTRACT

The synergic effects of knowledge accumulation, information technology development, and globalization have produced a new set of social paradigms. The transition towards a post-industrial society became evident at the beginning of the 21st century. New challenges are presented to the educational system, and a revision of methodological procedures is imperative to prepare citizens capable of dealing with complex contemporary issues. This chapter proposes 'expanded collaborative learning' for didactic activities involving peer review of any material produced by groups of students. We present a particular experience in which higher education students produced peer reviewed collaborative concept maps during an introductory course about natural sciences. Besides posing an unusual task for students, peer review changes the evaluation paradigm in traditional classrooms, reduces the power asymmetry between teacher and students, and promotes a truly collaborative atmosphere. This learning environment presents favorable conditions for empowering students, fostering them to act as autonomous citizens capable of transforming society.

NEW CHALLENGES FOR EDUCATION IN THE 21ST CENTURY

The beginning of the 3rd millennium was anticipated with some anxiety by humankind. This new era has brought a promise of changes based on the revision of our practices. The year 2000 software problem (Y2K bug), for instance, is a distant memory of the early 21st century. The millennium bug was a consequence of using two-digit dates in order to conserve computer memory, which was expensive and scarce in the 1960s. Media speculation as well as corporate and government reports at the end of

DOI: 10.4018/978-1-59904-992-2.ch014

the 1990s caused widespread concern that critical industries and government would cease operating at the stroke of midnight between December 31, 1999 and January 1, 2000 (Murray & Murray, 1996). Despite these catastrophic expectations involving the millennium roll-over, the Y2K consequences were minimal because the majority of problems had been fixed correctly through a worldwide collaborative effort. Indian engineers were responsible for correcting a huge number of US computers, illustrating an example of international collaborative work through the Internet (Friedman, 2007).

This computational crisis embodied the changes that have impacted our society since the second half of the 20th century. The knowledge explosion, information technology development and globalization have dramatically affected our society (Friedman, 2007; Hobsbawn, 1996). New social paradigms have emerged since then, making the end of industrial society a certainty. Knowledge, post-modern and post-industrial are labels used to identify this new society that has shaped our contemporary way of life (Bell, 1999; De Masi, 2000; Touraine, 2007; UNESCO, 2005). While industrial society was based on work and goods manufacturing, post-industrial society is centered in free time, creative idleness, and service production in the form of symbols, information, values and esthetics (De Masi, 2000). The power in industrial society depended on the possession of manufacturing resources (e.g., factories). On the other hand, the power in post-industrial society depends on the possession of ideation resources, such as research labs, and information (e.g., mass media).

The new challenges posed by the post-industrial society are consequences of scientific-technological development and globalization. Machines and automated systems can carry out repetitive, routine and/or brute tasks more efficiently than humans. Therefore, the desired attributes of the 21st century workforce are quite different than those used by industrial society. A revision of the educational system is desirable in order to meet the new demands of the labor market and to prepare citizens capable of dealing with the complex scenarios posed by post-industrial society.

In contrast with the pronounced social changes involved in the transition to a post-industrial society, the majority of schools have not changed and the education designed for industrial society is still prevalent. They resemble an industrial factory in that their classrooms are equal, their teachers have a standardized discourse, and there is an expectation that all students answer the same questions in the same way (de Menezes, 2000). Such standardization, one of industrial society's main features, affected the educational system by allowing only one model to satisfy teachers' and students' diverse expectations.

Traditional schools were built up under industrial paradigms, and their methodological procedures must be revised to respond to the new demands of post-industrial society. Beyond transmitting disciplinary knowledge, education for the 21st century requires the development of skills related to lifelong learning, teamwork, creative thinking, and collaborative knowledge construction (Burnard, 2006; Fischer, Bruhn, Gräsel, & Mandl, 2002; Pintrich, 2004; Sawyer, 2006). The powerful combination of these cognitive and communicative skills with confidence, which is related to emotional behavior, can foster students' empowerment in classrooms. Any methodological change in classroom activities in order to fit the needs of post-industrial society must pursue a truly collaborative and empowering environment involving both teachers and students (Mintzes, Wandersee, & Novak, 1998). The ultimate lesson to be taught in post-industrial classrooms is to learn how to learn (Georghiades, 2004; Novak & Gowin, 1984).

A schematic representation of the key features of the new post-industrial classroom is presented in Figure 1. Self-evaluation, motivation, creativity, and metacognition are the pedagogical building blocks that compose the key features to pursue

Figure 1. Key features of post-industrial classrooms (upper case) and their contributing pedagogical building blocks (lower case)

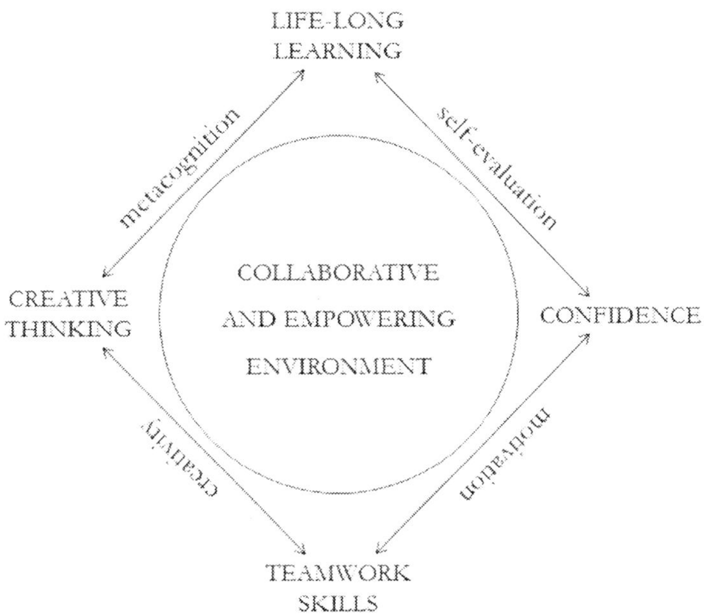

lifelong learning (metacognition + self-evaluation), confidence (self-evaluation + motivation), teamwork skills (motivation + creativity), and creative thinking (creativity + metacognition). From these key features, methodological strategies can be designed to foster collaborative learning and students' empowerment through teacher-student and student-student interactions.

The focus of the projects carried out by our research group is on changing instructional and methodological procedures used in schools to teach natural sciences in junior, high school and college classrooms. The need for a response to the complex post-industrial issues involving science, technology, society, environment, life and ethics justifies our efforts. An activity involving concept mapping and peer review was devised to include the key features of the post-industrial classroom in higher education. An *'expanded collaborative learning'* experience involving concept mapping is presented and discussed after some theoretical considerations.

CREATIVITY, CREATIVE GROUPS AND COLLABORATIVE KNOWLEDGE CONSTRUCTION

Creativity is a monopoly of mankind and is highly appraised nowadays. One interesting view of the creative process is described by Domenico De Masi. In his opinion, creativity is a synthesis between fantasy and accomplishment (De Masi, 2003). People who present both features well-developed can be considered individual geniuses. Michelangelo was a great artist not only for being appointed architect of St. Peter's Basilica in the Vatican with more than 70 years. Beyond designing its marvelous dome, he was also responsible for getting the Pope's approval and funding for his project, as well as for recruiting and managing a large team of workers (masons, carpenters, craftsmen) over a period of 20 years until his death (Condivi, 1999). Michelangelo had both features, fantasy and accomplishment, required to be a creative individual genius.

Table 1. Main steps for describing the collaborative knowledge construction of creative groups in the classroom

Step	Itemized description
1	Externalization of task-relevant knowledge
	• *Students bring individual knowledge into the situation*
	• *Different points of view can be clarified*
	• *Exchange of different individual concepts is the starting point for negotiating common meaning*
	• *Diagnosis and resolution of misconceptions can take place*
2	Elicitation of task-relevant knowledge
	• *Learning partners express their knowledge related to the task*
	• *Elicitation occurs frequently in the form of questions, which lead to externalizations in the form of explanations*
	• *Elicitation could be partly responsible for successful learning*
3(a)	Conflict-oriented consensus building
	• *Students seek a common solution or assessment of the given facts*
	• *Conflict plays an important role in reaching a consensus*
	• *Different interpretations made by learning partners can lead to a modification of knowledge structure*
3(b)	Integration-oriented consensus building
	• *Consensus can be reached through the integration of various individual perspectives into a common interpretation or solution for the given task*
	• *Superficial conflict-avoiding cooperation style may be verified in this attempt to incorporate individual views in a common perspective*
	• *There is a tendency on the part of the learners to reach an illusionary consensus*

Creation is a central value for the post-industrial society. The increasing demand for innovation and creative thinking can not be addressed by few individual geniuses. On the other hand, collaboration among people who complement one another can foster collective geniuses. In other words, creative groups can be intentionally composed to create a balanced team in terms of fantasy and accomplishment (De Masi, 2003). As collaboration and creation require a lot of practice, students should work together and collaborate in creative groups, which can be assigned by considering students' profiles in the continuum between fantasy and accomplishment. Opportunities for practicing creativity and collaboration in school must be devised by teachers using innovative methodological strategies.

The creative process that takes place in collaborative groups can be described in three steps. The externalization and elicitation of task-relevant knowledge precede consensus building, which can be conflict- or integration-oriented (Fischer, Bruhn, Gräsel, & Mandl, 2002). The itemized information presented in Table 1 can be useful for understanding how collaborative knowledge construction can take place in classroom.

Visualization tools can foster the collaborative knowledge construction process (Fischer, Bruhn, Gräsel, & Mandl, 2002). Among other options, concept maps appear as powerful visualization tools for representing knowledge. They can be defined as a set of concepts embedded into a propositional framework (Novak, 1998). Concept mapping can be useful for making idiosyncratic mental models explicit for revising (intrapersonal activity) and sharing ideas (interpersonal activity). Both purposes are important during collaborative knowledge construction because all participants can visualize, interpret and organize their own ideas (intrapersonal), before starting conflict-oriented and/or integration-oriented consensus building (interpersonal). Moreover, collaborative concept mapping can also support discursive meaning mediation and conflict negotiation when on-line tools are used (Bennett, 2003; Chiu, 2004; Jones & Issroff, 2005). CMapTools, for example, is the software developed by the Institute of Human and Machine Cognition (IHMC) for drawing concept maps, and its features allow synchronous and asynchronous collaboration among different authors. Concept mapping is based on the Assimilation Theory of Meaningful Learning proposed by David Ausubel, and it can support students' option for meaningful rather than rote learning (Ausubel, 2000; Mayer, 2002; Novak, 2002).

Collaborative concept mapping is an interesting methodological strategy that responds to some of the educational demands posed by post-industrial society. Recent findings presented in the literature confirm the effectiveness of this strategy when appropriate training is provided to students in order to avoid the naive use of concept maps in the classroom (Basque & Lavoie, 2006; Cañas & Novak, 2006; Correia, Infante-Malachias, & Godoy, 2008). In our contemporary context, collaborative concept mapping allows the development of synthesizing and creating minds (Gardner, 2006), as well as teamwork skills, which are formative requisites for 21st-century citizens (De Masi, 2003; Sawyer, 2006).

Peer review can be coupled to collaborative concept mapping to expand collaborative interactions among students in the classroom. After preparing and presenting concept maps in small groups, students from different groups can also evaluate the concept maps made by their counterparts. This approach has the potential to change traditional classroom dynamics towards a renovated post-industrial education.

THE IMPACT OF EXPANDED COLLABORATIVE LEARNING IN THE CLASSROOM

Collaborative works should be more than intentions stated in educational projects and plans - they should be present in teachers' practice. Classrooms should allow time for discussing how the disciplinary knowledge acquired during formal education can be used to understand the complex issues to be solved during the 21st century. Students must have the ability to put all this knowledge into a meaningful context in order to make independent judgments about global problems such as inequality, democracy and environmental crisis. This condition is compulsory for autonomous citizenship in the post-industrial society. Therefore, teachers must be aware of the role of collaborative works as a methodological strategy for developing students' abilities and attitudes to prepare them for transforming our society.

Peer review is rarely explored as a means of changing traditional assessment procedures; the challenge of students evaluating themselves breaks a paradigm in the classroom and reduces the power asymmetry between teacher and students. Furthermore, students are in about the same zone of proximal development as one another. Peer review offers a different opportunity for them to share knowledge compared to interactions with the teacher, who is not in the same zone of proximal development (Novak, 2002; Vygotsky, 1978).

Peer review goes beyond the collaborative activities developed by small groups of students and, for this reason, the term *'expanded collaborative learning'* is proposed to make a distinction between them. Moreover, the peer review of any students' material produced collaboratively merges *'conventional'* and *'expanded'* collaborative learning. Expanded collaborative learning through peer review changes the roles in the classroom, and students are faced with the new responsibility to judge the quality of the material produced by their peers. This activity can increase

Figure 2. Collaborative interactions that predominate in traditional classrooms (a) and in post-industrial classrooms (b+c)

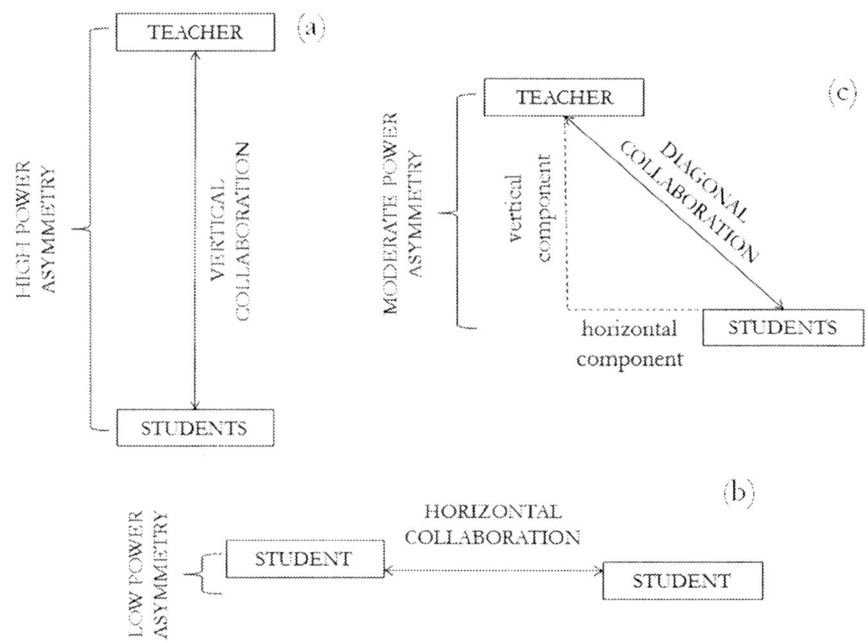

their awareness about their own achievements and failures during the learning process. As a consequence, peer review as an assessment exercise can be a safe road towards self-evaluation, which is an important pedagogical building block of post-industrial classrooms (Figure 1).

The repercussion of using peer review to intensify collaboration in the classroom is represented in Figure 2. Power asymmetry and the zone of proximal development are the two variables selected for differentiating three types of collaborative interactions. Vertical collaboration between teacher and students (Figure 2a) is prevalent in traditional classrooms, where the former's task is to transmit disciplinary knowledge, and the latter's task is to receive the transmitted knowledge. High power asymmetry and the difference between the teacher's and students' zones of proximal development characterize this vertical collaboration. On the other hand, a horizontal collaboration is observed when students interact with their counterparts (as happens during peer review, for example). In this case, power asymmetry is minimized and the students' zones of proximal development are similar (Figure 2b). The use of visualization tools to foster collaborative knowledge construction has its maximum potential in the context of horizontal interactions. An intermediate condition can be created when teacher and students interact in a constructivist learning environment (Figure 2c). Power asymmetry in this situation is moderate, because roles in the classroom are not related to transmission and reception of disciplinary information; rather, knowledge is under a continuous process of construction and revision. The teacher's role is to scaffold and stimulate students to overcome their own conceptual difficulties (intrapersonal level) and to improve their teamwork skills (interpersonal level). The difference between teacher's and students' zones of proximal development remains, and this collaborative interaction has a diagonal resultant due to the simultaneous presence of vertical and horizontal collaboration components.

The constructivist nature of the post-industrial classroom matches with expanded collaborative learning possibilities. Beyond the new roles assigned to teachers and students, it increases the horizontal collaborative component, which is not present in traditional classrooms. Power asymmetry reduction in collaborative interactions is common nowadays, and horizontal collaboration is becoming more and more important in the 21st century (Friedman, 2007). The changes catalyzed by expanded collaborative learning can address the key features of a post-industrial classroom (Figure 1), and they also create favorable conditions for empowering students and fostering them to learn how to learn.

THE SOCIAL IMPACT OF EXPANDED COLLABORATIVE LEARNING

Formal and informal education are always socially contextualized. The teacher's pedagogical task goes beyond the mere teaching and learning process because it has social meaning and significance. At any of its levels, scholar education can be considered as a means to promote students' social formation, because ideas, opinions and ideologies can be socially framed in school (Freire, 2005).

The social environment exerts a great influence on people. The relationships to be socially established depend on the experiences, beliefs and customs assimilated throughout our lives. Such relationships are important for implementing transforming actions in our globalized world. For this purpose, the evaluation of dominant ideas, opinions and ideologies must occur from a critical and questioning perspective. Teachers, who must be intentionally concerned with students' learning, also act as mediators of the influence of the social environment on their students. As a consequence, the teacher/student/knowledge relationship can neither be arbitrary nor authoritarian; on the contrary, it should be dialogical and democratic in order to value the teachers' pedagogical task (Freire, 2000; Freire, 2001). Teachers have the duty to organize content, to select methodologies and to design assessment strategies intentionally to promote students' learning in its broad sense, *i.e.*, considering conceptual and attitudinal changes in order to encourage the development of skills and values. As a social institution, schools have the responsibility to disseminate knowledge to ensure access to culture, work, social justice and citizenship.

Teachers' leadership in the classroom requires professional and moral qualities to mediate adequately between students and society. Communication and dialogue among people is a fundamental condition for supporting the formation process of the individual. As stated by Freire (2001) "the dialogue is the moment when men and women meet to reflect about their reality, about what they know and don't know, to build new knowledge as conscious and communicative subjects". Freire's educational perspective highlights the role of each person in this collective building process to promote profound social changes leading to a more egalitarian, democratic, solidary and fair society (Freire, 2005). His Freedom Pedagogy states that all individuals can venture to unveil the deterministic ideologies and concepts hidden in their consciences and, from them, promote social changes (Freire, 2001).

The understanding of social system behavior from a biological perspective is proposed by Maturana and Varela (1998). In their opinion, the sum of the individual conduct of each person defines the characteristics of a social system. Profound changes in society depend on changing the individual conduct of its members. As a whole society is manifested in the conduct of the individuals that constitute it, genuine social changes only occur when a real change in the conduct of its members takes place. The human social phenomenon occurs when individual changes assume a contagious state, spreading out and affecting the surrounding members in the society. The cumulative effects of the human social phenomenon can result in

Figure 3. Concept map for merging Freire's and Maturana's perspectives to highlight the social impact of expanded collaborative learning. Focal question: How can expanded collaborative learning foster social transformations? Concepts from Paulo Freire's ideas are in boxes, concepts from Humberto Maturana's ideas are in shadowed boxes, and the concepts related to expanded collaborative learning are in circles

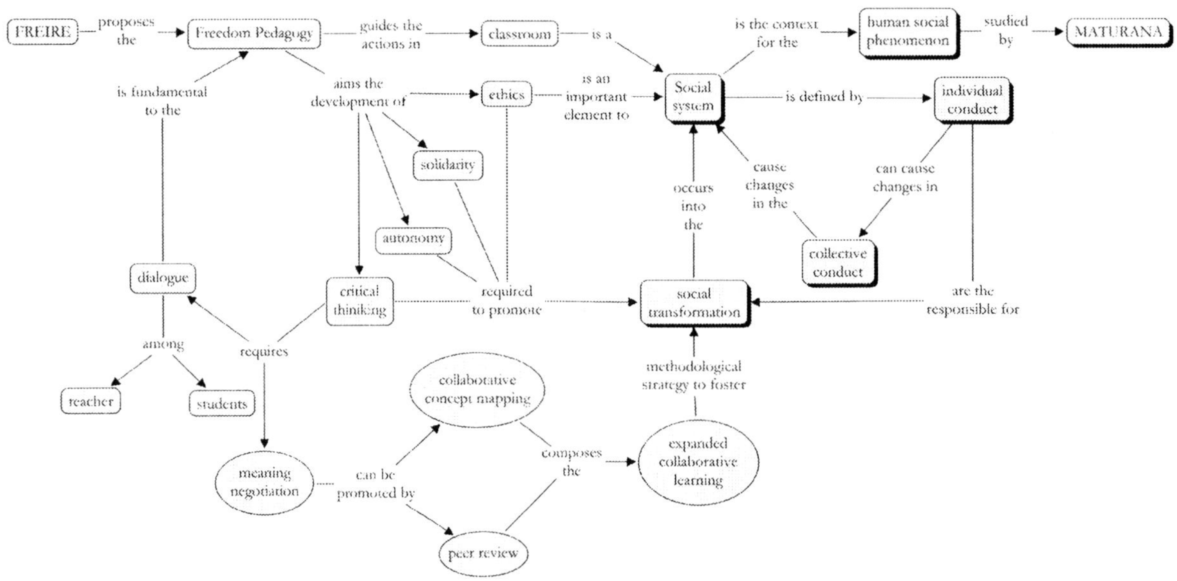

genuine social changes.

The classroom can be considered as a complex system where the human social phenomenon can take place. The merging of Freire's and Maturana's perspectives allows us to understand the classroom as the place to foster social transformations through the educational formation of citizens capable of analyzing and changing their own conduct (Figure 3). From our methodological point of view, teachers must promote collaborative learning experiences to challenge students to revise their individual attitudes in a critical and autonomous way. Changing the individuals' conduct from an individualistic (industrial) to a more collective (post-industrial) perspective may promote the required changes in the social system in order to deal with the global problems that have affected us since the second half of the 20th century (Infante-Malachias & Correia, 2009).

The classroom is a privileged social place to foster students' empowerment to allow them to become active participants in the task of promoting profound social transformations. Intentional choices in terms of content selection and methodological approaches must be carefully made by teachers in order to produce small cumulative changes in their students throughout their formal education (Infante-Malachias & Correia, 2007). The use of well-designed methodological approaches such as expanded collaborative learning should be part of any licentiate course which aims to prepare teachers ready to face the challenges posed by the education in the post-industrial society.

Table 2. Students' interest in ACH 0011 Natural Science discipline at the beginning of the semester

Undergraduate course	Low	Moderate	High
Environmental Management			XXX
Gerontology		XXX	
Information System		XXX	
Leisure and Tourism	XXX		
Licentiate in Natural Sciences*			XXX
Marketing	XXX		
Obstetrics		XXX	
Physical Activity Science		XXX	
Public Policy Management	XXX		
Textile Technology and Fashion		XXX	
*4-year degree that qualifies students to teach in primary education.			

EXPANDED COLLABORATIVE LEARNING IN HIGHER EDUCATION: AN EXPERIENCE TO EMPOWER 1ST-YEAR STUDENTS

The challenge of facing the reality of the classroom occurred during the discipline ACH 0011 Natural Sciences. It is offered for all 1st-year students at Escola de Artes, Ciências e Humanidades (School of Arts, Science and Humanities at São Paulo University). The main goal of this discipline is to provide a broad view of the impact caused by scientific and technological developments in our society. Scientific literacy, a new post-industrial demand, is a requisite for an autonomous citizenship. A new contract involving society and science is under negotiation and all citizens must have the right to make their own judgments about ethical aspects of scientific and technological issues (DeHaan, 2005; Donnelly, 2004; Holbrook & Rannikmae, 2007; dos Santos, 2009). Therefore, scientific literacy needs to be nurtured throughout formal education (Fourez, 1997; UNESCO, 2005).

Sixty students from five different undergraduate courses were grouped in the same classroom to attend two-hour weekly classes over a period of 15 weeks. Considering the discipline's introductory scope and heterogeneous audience, the challenge to satisfy students' expectations required innovative methodological strategies. Table 2 presents a qualitative estimate of the students' interest level at the beginning of the discipline considering their undergraduate courses. High, moderate and low interested students can be found in any ACH 0011 Natural Science classroom, according to instructors' testimonies.

Some instructors chose to use concept maps throughout the discipline. Students are introduced to them and trained to elaborate both manuscript and digital concept maps using CMapTools. The aim is to ensure an adequate training period for the students in order to avoid the inappropriate use of this tool, as highlighted in the literature (Cañas & Novak, 2006, Correia, Infante-Malachias and Godoy, 2008; Novak & Cañas, 2008). Table 3 summarizes the discipline schedule and the activities involving concept mapping.

The universe is the central theme during classes 1 to 5. Beyond the discussion involving natural science topics, the students are introduced to concept mapping and they make an individual manuscript concept map. As preparation for test 1, the students revise the initial concept map about the universe and prepare an individual digital version using CMapTools. This final concept map can be

Table 3. Didactic organization of the ACH 0011 Natural Science discipline

Class #	Discipline schedule	Concept mapping activities
Theme #1: Universe Mysteries and the Modern Scientific Revolution		
1	*Historical development of scientific thinking*	*Introduction to concept mapping and general guidelines for making manuscript concept maps*
2	*Galileo and the scientific revolution*	*Individual concept map*
3	*Knowledge explosion and astronomy discoveries in the 20th century*	*Revision and discussion of produced concept maps and introduction to the CMapTools software*
4	*Hubble, the galaxies and cosmology*	*Individual concept maps for synthesizing information up to the first test*
5	*Test 1: Preparing a text in pairs*	*-x-*
Theme #2: Climate Change and the Environmental Challenges of the 21st Century		
6	*Search for information, selection and set up of a conceptual framework*	*Collaborative concept maps for externalizing and eliciting knowledge (3-student groups)*
7	*Film session: An Inconvenient Truth*	*-x-*
8	*Discussion about the film and revision of the conceptual framework*	*Conflict- and integration-oriented consensus building (3-student groups)*
9	*Test 2: Preparing a short communication in pairs for publishing in a periodical*	*Revision and discussion of collaborative concept maps (3-student groups) as preparation for test 2*
10	*Double blinded peer review of counterparts' short communications*	*-x-*
Theme #3: Molecular Biology, Modern Medical Procedures and Bioethics		
11	*Search for information, selection and set up of a conceptual framework*	*Collaborative concept maps for externalizing and eliciting knowledge (3-student groups)*
12	*Film session: DNA – The Promise and the Price*	*-x-*
13	*Discussion about the film and revision of the conceptual framework*	*Conflict- and integration-oriented consensus building (3-student groups)*
14	*Debate about bioethics and final revision of the conceptual framework*	*Revision and discussion of collaborative concept maps to be presented in test 3*
15	*Test 3: Presentation of collaborative concept maps and single blinded peer review*	*Peer review of counterparts' collaborative concept maps and final revision from the received comments*

consulted during test 1.

Climate change is the following subject, and discussion occurs during classes 6 to 10. Collaborative concept maps are prepared by 3-student groups for fostering the collaborative knowledge construction steps of knowledge externalization, knowledge elicitation and consensus building (Fischer, Bruhn, Gräsel, & Mandl, 2002), as shown in Table 1. The final version of this collaborative concept map can be used by the students to prepare a short communication about the environmental challenges of the 21st century. These texts are peer reviewed during class 10 to introduce the students to this practice. All 1st-year students experience peer review, which is a procedure widely adopted in science. This strategy makes even more sense considering the scope of the discipline. From the methodological point of view, expanded collaborative learning begins in the classroom.

Bioethics is the last subject to be discussed during classes 11 to 15. Collaborative concept maps are explored again, but test 3 consists of their presentation and peer review. Students present their own maps and pose comments to assigned concept maps prepared by their counterparts. They fill in a form to help the authors to correct, improve and/or include new conceptual relationships. At the end of class 15, each group receives suggestions made by four other students. Moreover, the authors can also take advantage of the good ideas appearing during all presentation sessions. The last adjustments can be incorporated into the concept maps before students send their final versions by e-mail, with a due date of 2 days after class 15.

It should be stressed that collaborative concept mapping is preferred over individual concept mapping to explore the diversity of opinions and interests from a heterogeneous audience and to foster collaborative knowledge construction (Basque & Lavoie, 2006; Fischer, Bruhn, Gräsel, & Mandl, 2002). It is also important to ensure enough time for the continuous process of revision and improvement of collaborative concept maps. The students have at least 1 month (4 classes) to prepare the final version of the concept maps about climate change (classes 6 to 9) and bioethics (classes 11 to 15).

Three concept maps from different students are presented in Figures 4-6. An individual concept map about the universe and the historical evolution of modern science is shown in Figure 4. It is possible to identify 4 different related issues that respond to the focal question *(What are the relations between the observation of celestial events and historical scientific development?)* considering the topics discussed during the classes: historical view of scientific development (concepts in circles), religion and science conflict (concepts in shadowed circles), modern astronomy discoveries (concepts in shadowed boxes) and a description of our location in the universe (concepts in boxes). The student's efforts to organize and revise the conceptual relationships can also be noted by the creative spatial organization of the concepts. He added a comment for the readers: "I made the concept map following the shape of a spiral galaxy in order to match the visualization and the subject under study. The universe is intentionally located at the center of the concept map because it is the focus of our studies". On the other hand, unclear propositions such as 'power involves paradigms' and 'beliefs are paradigms' suggest possibilities for further improvements. As stated by Novak and Cañas (2008) "it is important to recognize that a concept map is never finished. After a preliminary map is constructed, it is always necessary to revise this map. Other concepts can be added. Good maps usually result from three to many revisions". Despite some variations, all individual concept maps made by the students after the 4-week training period were acceptable and collaborative concept mapping could be started at class 6 (Table 3).

A collaborative concept map about climate change is presented in Figure 5. The result of the efforts of three students shows the following related issues to address the focal question *(What are the problems and the possible solutions to preserving the environment?)*: the problems associated with fossil fuel burning (concepts in boxes), glacier melting as a consequence of global warming (concepts in circles), possible solutions to mitigate the climate change (concepts in shadowed boxes) and two central concepts related to social issues (concepts in shadowed circles). The interconnection among environmental problem concepts and possible solution concepts was made by using cross-links from the non-scientific concepts 'progress' and 'cities'. This fact highlights the typical awareness promoted by scientific literacy, which is the main goal of the ACH 0011 Natural Science discipline. Moreover, the cyclic nature of this concept map should be emphasized, as it attempts to establish cause-effect relationships between an environmental problem (climate change) and possible mitigation actions (use of renewable energy and reforestation). This external cyclic structure may indicate the students' attempts toward dynamic thinking, which can confirm that they are skilled map-makers (Derbentseva, Safayeni, & Cañas, 2007; Hilbert & Renkl, 2007). Interesting research questions involving the importance of collaborative knowledge construction to stimu-

Figure 4. Individual concept map done by a student for organizing information (concepts) for test 1. Focal question: What are the relations between the observation of celestial events and historical scientific development? The form and shadows of the boxes were changed by the authors

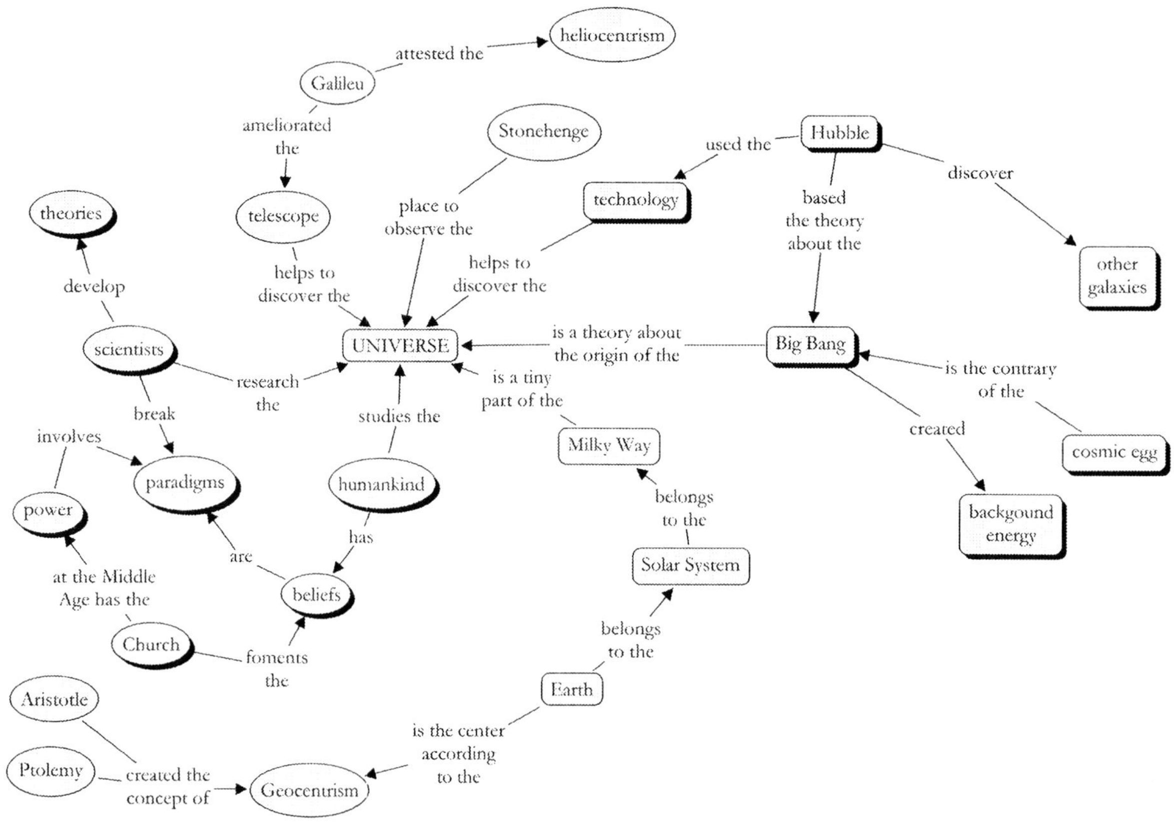

late scientific literacy as well as the role of the conceptual map as a visualization tool to support this process demand further investigation.

Figure 6 presents a collaborative peer reviewed concept map about bioethics. The authors used three different related issues to answer the focal question *(What are the relations among religion, bioethics and science?)*: religion (concepts in circles), bioethics (concepts in shadowed circles) and science (concepts in shadowed boxes). The cyclic structure appears again and the bioethics concepts are used to mediate the differences between religious and scientific points of view. The students also add three examples (abortion, cloning and stem cells) for differentiating the 'controversial issues' concept. Non-scientific concepts such as "moral", "controversial issues" and "conflicts" confirmed the students' awareness involving the impact of science on our society, as well as the need for an ethical discussion to "set parameters for research" (Figure 6). Considering life as the main subject, students presented bioethics to mediate scientific research involving reproduction. However, a tendency can be seen for approving genetic work and for disapproving restrictive actions from the Church. As any open question, this debate does not have a right answer, and continuous discussion must take place. Finally, the evolution concept was used by the authors in an elegant and sophisticated way: it has different meanings, considering the propositions "conflicts are important for evolution", "Darwin published

Figure 5. Collaborative concept map done by a three-student group for externalizing knowledge, eliciting knowledge and building a consensus through conflict mediation. Focal question: What are the problems and the possible solutions to preserving the environment? The form and shadows of the boxes were changed by the authors

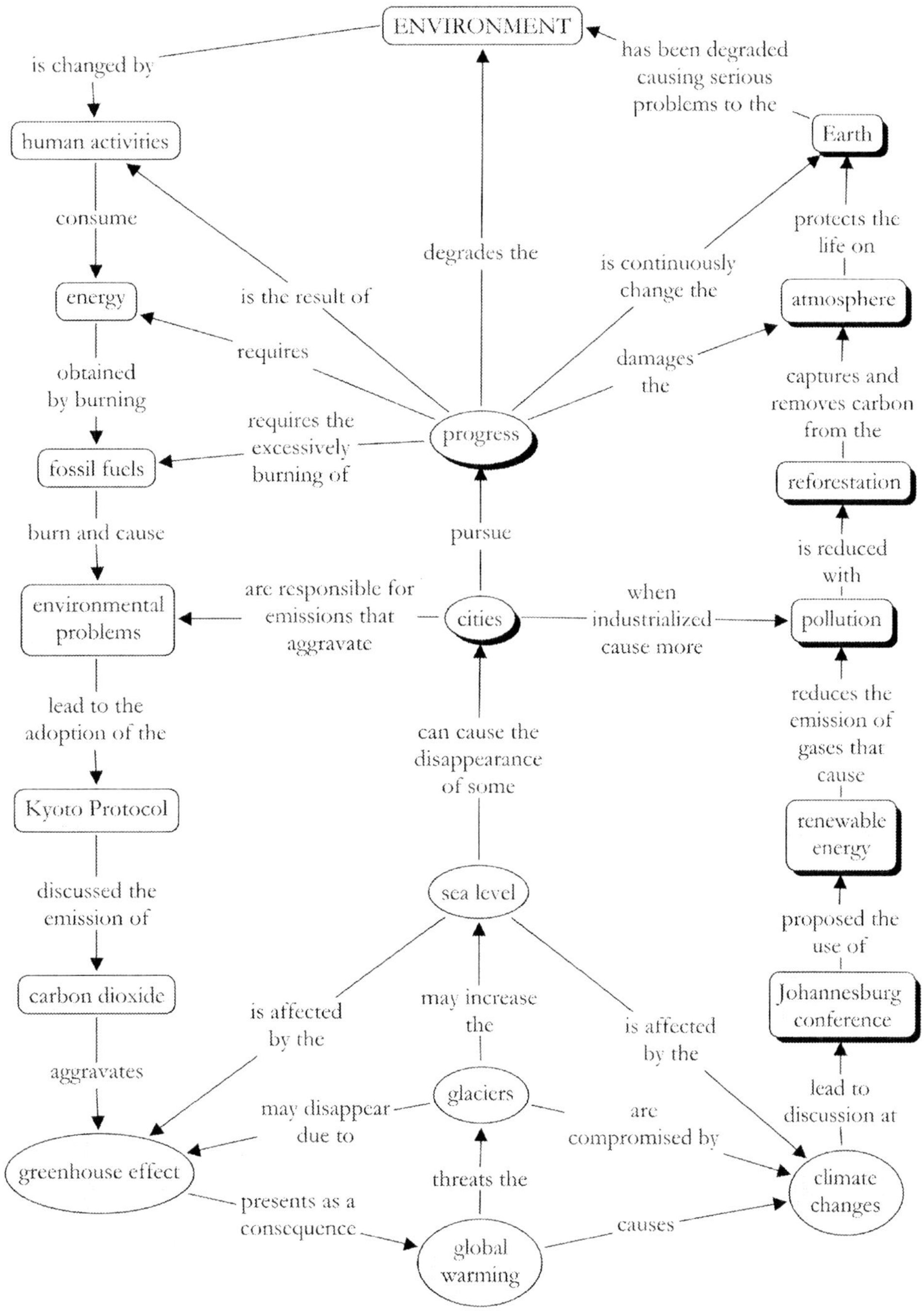

the Theory of Evolution" and "genetics can change the pathway of evolution".

A comparison among the students' concept maps is presented in Table 4. The qualitative fea-

Figure 6. Peer reviewed-collaborative concept map done by a three-student group for externalizing knowledge, eliciting knowledge and building a consensus through conflict mediation. Focal question: What are the relations among religion, bioethics and science? The form and shadows of the boxes were changed by the authors

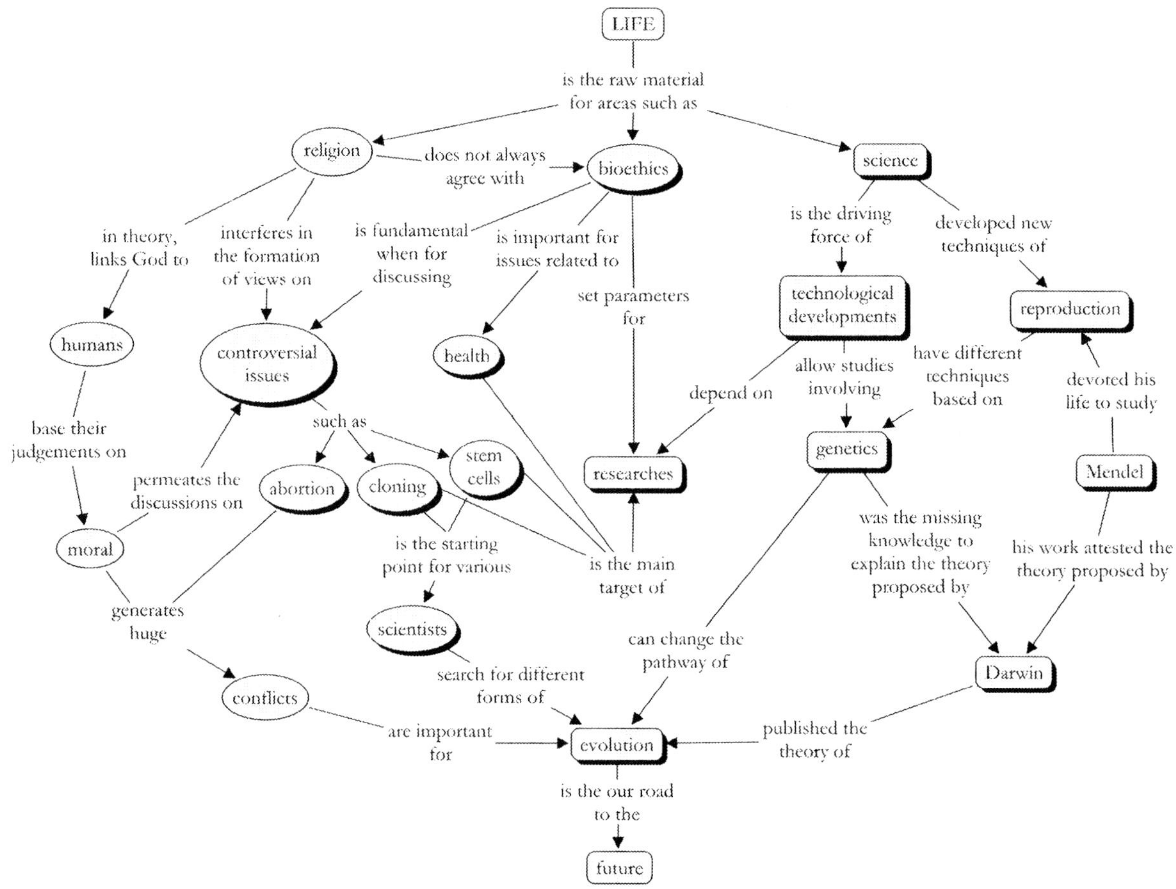

tures suggest that collaboration and peer review can support the development of better concept maps in comparison to individual production. Cyclic structures and the increase in the number of ramification points and cross-links may be attributed to the collaborative knowledge construction that only happened when three students were co-responsible for creating a concept map. Peer review further improved the quality of the revised concept maps because the authors took advantage of the good ideas presented by other students. All concept maps could be viewed by all students during the presentation sessions, and the good concept maps (in their opinions) were used as model for those students who had difficulties creating a good map. Examples that indicate a refined concept map were mentioned only in the collaborative peer reviewed concept map about bioethics (Figure 6).

FINAL CONSIDERATIONS

A new society asks for a new education. Despite not solving all educational problems, innovative methodological strategies can change the traditional dynamics which still prevail in the majority of contemporary classrooms. The new

Table 4. Comparison of the concept maps produced by students throughout the ACH 0011 Natural Science discipline.

Features	Concept map in Figure 4	Concept map in Figure 5	Concept map in Figure 6
Qualitative			
Individual	Yes	No	No
Collaborative	No	Yes	Yes
Peer reviewed	No	Yes	Yes
Cyclic nature	No	Yes	Yes
Quantitative			
Related issues	4	3	3
Concepts	24	20	21
Cross-links	1	7	6
Ramification points	3	6	7
Propositions	26	32	31
Unclear propositions	2	2	0
Examples	0	0	3

formative demands posed by post-industrial society require more than disciplinary knowledge transmission, and some skills – such as lifelong learning, creative thinking, and teamwork - must be simultaneously developed throughout formal education. Emotional aspects must also receive attention, and confidence is necessary to allow the synergic merge of cognitive and communicative skills to empower students in classroom.

Students' empowerment has a profound social impact considering the new formative demands for 21st-century citizenship. All individuals should be aware of the roles that they need to assume in order to transform society. As a complex social system, the classroom is an important place to prepare youth to take part in the construction of a more egalitarian, democratic, solidary and fair society. Even though it is a challenging process, teachers must be aware that changing the classroom to foster a collaborative and empowering environment is essential for a high-quality post-industrial education.

Peer review and concept mapping can be combined to create an expanded collaborative learning experience, as presented in this chapter. More than a theoretical discussion, this chapter also describes the implementation of expanded collaborative learning in a higher education classroom during an introductory course about natural sciences. The obtained results were promising, and they call for more research on the formation of creative groups. This research should explicitly consider students' idiosyncratic characteristics, the role of concept maps for fostering collaborative knowledge construction, and the potential of expanded collaborative learning as an empowering agent supporting the ultimate lesson to be taught in post-industrial classrooms: learning how to learn. This is the ultimate condition for preparing students to act as autonomous citizens capable of transforming society.

ACKNOWLEDGMENT

The authors thank the Conselho Nacional de Desenvolvimento Científico e Tecnológico (CNPq) and Fundação de Amparo à Pesquisa do Estado de São Paulo (FAPESP) for funding our research projects. We are also in debt to the students responsible for the concept maps presented in this chapter (Beatriz Adorno, Beatriz M. F. Dutra, Glauce C. F. Soares, Jefferson L. Cescon, Mariana S. Raya, Nayara R. Cintra and Zheng Xuee).

REFERENCES

Ausubel, D. P. (2000). *The acquisition and retention of knowledge: a cognitive view.* Dordrecht, the Netherlands: Kluwer Academic Publishers.

Basque, J., & Lavoie, M.-C. (2006). Collaborative concept mapping in education: major research trends. In A. J. Cañas & J. D. Novak (Eds.), *Proceedings of the Second International Conference on Concept Mapping,* (Vol. 1, pp. 79-86). San Jose, Costa Rica: Universidad de Costa Rica.

Bell, D. (1999). *The Coming of Post-Industrial Society: A Venture in Social Forecasting.* Jackson, TN: Basic Books.

Bennett, S. (2003). Supporting collaborative project teams using computer-based technologies. In T. S. Roberts (Ed.), *Online collaborative learning: theory and practice* (pp. 1-26). Hershey, PA, USA: Information Science Publishing.

Burnard, P. (2006). Reflecting on the creativity agenda in education. *Cambridge Journal of Education, 36*(3), 313–318. doi:10.1080/03057640600865801

Cañas, A. J., & Novak, J. D. (2006). Re-examining the foundations for effective use of concept maps. In A. J. Cañas & J. D. Novak (Eds.), *Proceedings of the Second International Conference on Concept Mapping,* (Vol. 1, pp. 247-255). San Jose, Costa Rica: Universidad de Costa Rica.

Chiu, C.-H. (2004). Evaluating system-based strategies for managing conflict in collaborative concept mapping. *Journal of Computer Assisted Learning, 20,* 124–132. doi:10.1111/j.1365-2729.2004.00072.x

Condivi, A. (1999). *The life of Michelangelo.* University Park, PA: Pennsylvania State University Press.

Correia, P. R. M., Infante-Malachias, M. E., & Godoy, C. E. C. (2008). From theory to practice: the foundations for training students to make collaborative concept maps. In A. J. Cañas; J. D. Novak; Reiska, P.; Ahlberg, M. K. (Eds.), Proceedings of the Third International Conference on Concept Mapping, (Vol. 2, pp. 414-421).Tallin, Estonia; Helsinki, Finlândia: OÜ Vali Press.

De Masi, D. (2000). *Ozio creativo. Conversazione con Maria Serena Palieri.* Italy: Rizzoli.

De Masi, D. (2003). *Creatività individuale e di gruppo.* Italy: Rizzoli.

de Menezes, L. C. (2000). Ensinar ciências no próximo século. In E. W. Hamburguer & C. Matos (Eds.), *O desafio de ensinar ciências no século XXI* (pp. 48-54). São Paulo, Brazil: Edusp.

DeHaan, R. L. (2005). The impending revolution in undergraduate science education. *Journal of Science Education and Technology, 14*(2), 253–269. doi:10.1007/s10956-005-4425-3

Derbentseva, N., Safayeni, F., & Cañas, A. J. (2007). Experiments on dynamic thinking. *Journal of Research in Science Teaching, 44*(3), 448–465. doi:10.1002/tea.20153

Donnelly, J. F. (2004). Humanizing science education. *Science Education, 88*(5), 762–784. doi:10.1002/sce.20004

dos Santos, W. L. P. (2009). Scientific literacy: A Freirean perspective as a radical view of humanistic science education. *Science Education, 93,* 361–382. doi:.doi:10.1002/sce.20301

Fischer, F., Bruhn, J., Gräsel, C., & Mandl, H. (2002). Fostering collaborative knowledge construction with visualization tools. *Learning and Instruction, 12,* 213–232. doi:10.1016/S0959-4752(01)00005-6

Fourez, G. (1997). Scientific and technological literacy as a social practice. *Social Studies of Science, 27*, 903–936. doi:10.1177/030631297027006003

Freire, P. (2000). *Pedagogy of the Oppressed* (30th Ed.). New York: Continuum International Publishing Group.

Freire, P. (2001). *Pedagogy of Freedom: Ethics, Democracy and Civic Courage* (2nd Ed.). Lanham, MA: Rowman & Littlefield.

Freire, P. (2005). *Education for Critical Consciousness*. New York: Continuum International Publishing Group.

Friedman, T. L. (2007). *The World Is Flat [Updated and Expanded]: A Brief History of the Twenty-first Century*. New York: Picador.

Gardner, H. (2006). *Five Minds for the Future*. Boston: Harvard Business School Publishing.

Georghiades, P. (2004). From the general to the situated: three decades of metacognition. *International Journal of Science Education, 26*(3), 365–383. doi:10.1080/0950069032000119401

Hilbert, T. S., & Renkl, A. (2007). Concept mapping as a follow-up strategy to learning from texts: what characterizes good and poor mappers? *Learning and Instruction, 12*, 213–232.

Hobsbawn, E. (1996). *The Age of Extremes: A History of the World, 1914-1991*. New York: Vintage.

Holbrook, J., & Rannikmae, M. (2007). The nature of science education for enhancing scientific literacy. *International Journal of Science Education, 28*, 1–16.

Infante-Malachias, M. E., & Correia, P. R. M. (2007). Problemas complejos en el mundo post-industrial. *Novedades Educativas, 203*, 29–33.

Infante-Malachias, M. E., & Correia, P. R. M. (2009). Elaboración colaborativa de mapas de conceptos. *Novedades Educativas, 219*, 72–76.

Jones, A., & Issroff, K. (2005). Learning technologies: affective and social issues in computer-supported collaborative learning. *Computers & Education, 44*(4), 395–408. doi:10.1016/j.compedu.2004.04.004

Maturana, H. M., & Varela, F. (1998). *Tree of knowledge: the biological roots of human understanding*. Boston: Shambhala.

Mayer, R. E. (2002). Rote versus meaningful learning. *Theory into Practice, 41*(4), 226–232. doi:10.1207/s15430421tip4104_4

Mintzes, J. J., Wandersee, J. H., & Novak, J. D. (1998). *Teaching science for understanding: a human constructivist view*. San Diego, CA: Academic Press.

Murray, J. T., & Murray, M. J. (1996). *Computing Crisis: A Millennium Date Conversion Plan*. New York: McGraw-Hill.

Novak, J. D. (1998). *Learning, creating, and using knowledge: concept maps as facilitative tools in schools and corporations*. Mahwah, NJ: Lawrence Erlbaum Associates.

Novak, J. D. (2002). Meaningful learning: the essential factor for conceptual change in limited or inappropriate propositional hierarchies leading to empowerment of learners. *Science Education, 86*, 548–571. doi:10.1002/sce.10032

Novak, J. D., & Cañas, A. J. (2008). *The Theory Underlying Concept Maps and How to Construct and Use Them*. (Technical Report IHMC 2006-01 Rev 01-2008). Pensacola, FL: Institute for Human and Machine Cognition. Retrieved February 1, 2008, from http://cmap.ihmc.us/Publications/ResearchPapers/TheoryUnderlyingConceptMaps.pdf

Novak, J. D., & Gowin, D. B. (1984). *Learning how to learn*. Cambridge, UK: Cambridge University Press.

Pintrich, P. R. (2004). A conceptual framework for assessing motivation and self-regulated learning in college students. *Journal of Educational Psychology Review*, *16*(4), 385–407.

Sawyer, R. K. (2006). Educating for innovation. *Thinking Skills and Creativity*, *1*, 41–48. doi:10.1016/j.tsc.2005.08.001

Touraine, A. (2007). *New Paradigm for Understanding Today's World*. Cambridge, UK: Polity Press.

UNESCO. (2005). *Towards Knowledge Societies: UNESCO World Report*. Paris, France: UNESCO Publishing.

Vygotsky, L. (1978). *Mind in society: the development of higher psychological processes*. Cambridge MA: Harvard University Press.

KEY TERMS AND DEFINITIONS

Conventional collaborative learning: learning experience based upon consensus building through cooperation by group members.

Creative idleness: a new condition imposed by the post-industrial society. Work time, leisure time and study time are no longer separated as occurred during the industrial society. They are simultaneously merged during the creative idleness, which has changed the relationship between man and work. Post-industrial citizens work, have fun and study at the same time.

Empowerment: brings students into a state of belief about their ability to act effectively.

Expanded collaborative learning: learning experience based upon consensus building through cooperation by group members, when most interactions occur among pairs (horizontal collaboration).

Globalization: interweaving of markets, technology, information systems and telecommunications systems shortening time and distance.

Innovation: the process of making improvements by introducing something new. It begins with creative ideas from individuals or teams.

Lifelong learning: the process of acquiring knowledge or skills throughout life via education, training, work and general life experiences. It also describes the modern global phenomenon where it is has become the norm for people to return to formal education and training at periods during their working life.

Metacognition: the awareness of one's cognitive processes and the efficient use of this self-awareness to self-regulate these cognitive processes. This active control over the process of thinking is necessary to regulate one's the lifelong learning.

Peer review: is a process of subjecting an author's scholarly work, research or ideas to the scrutiny of others who are experts in the same field. In unblind peer review the names are not shielded. In single-blind peer review the reviewers know the name of the author. In double-blind peer review neither the author nor the reviewers know each others' identities.

Post-industrial society: is the society that has moved past the stage of heavy industry to an economy that is mainly centered on knowledge-based and service professions.

Scientific literacy: capacity to use scientific knowledge to identify questions and to draw evidence-based conclusions in order to understand and help make decisions about the natural world and the changes made to it through human activity.

Chapter 15
Mapping Concepts with Fisherfolk

Denis Hellebrandt
School of International Development, University of East Anglia, UK

ABSTRACT

This chapter aims to show how concept mapping is a technique which is capable of representing complex systems in an accessible format and offers excellent opportunities for collaboration and meaningful learning. Effective communication is at the foundation of collaborative learning and concept mapping is expressly used in this research to facilitate the dialogue between participants and researcher. The chapter starts out by situating the reader by way of a conceptual background about complex systems, followed by the basis for the application of concept mapping in this project and the specific research context - a case study of small-scale fisheries in southern Brazil. Then, an account of the use of concept mapping during the fieldwork is given, with an assessment of the technique. The chapter ends with a reflection on the experience gained so far and comments on the application of collaborative learning in similar research projects.

INTRODUCTION

Conceptual Background

Complex Systems

We live in a world where science has largely replaced religion in explaining how things work and how people behave and relate to each other. For instance, instead of having to invoke a superior being to justify the existence of the sun and how it affects us, we can rely on scientific explanations about how stars are formed and emit radiation. This was made possible because about 500 years ago some people chose to believe that humans can understand reality without recourse to deities, but using instead our capability to observe, record and organize all that we experience. In other words, we became able to justify knowledge using our reason – these were the first days of a rationalist

DOI: 10.4018/978-1-59904-992-2.ch015

epistemology that would dominate our view of the world until Thomas Khun and Paul Feyerabend offered alternatives, as recently as 40 years ago (Lazar, 2004).

Although choosing reason represented liberation from a world view dictated by tradition and which was almost impossible to challenge, it also presented us with an immense task: once the myths had lost their power to explain reality, we were faced with the need to create a new way of making sense of the world. How could we approach reality, with all its diversity and everchanging nature?

The solution to this question came through a "new" science, which was now based on "clear and distinct ideas" and had man as its centre - a detached, objective and rational observer. A scientific method was devised which aimed precisely at eliminating uncertainty and simplifying reality in order to understand it (Morin, 2005). Descartes was one of the most influential characters in this movement and the following quote is typical of rationalist epistemology: "my method imitates that of the architect. When an architect wants to build a house [...], he begins by digging out a set of trenches from which he removes the sand, [...] so that he can lay his foundations on firm soil. In the same way, I began by taking everything that was doubtful and throwing it out, like sand ... (Replies 7, AT 7:537)" (Newman, 2005).

This search for orderly explanations of the world would lead to the development of the so-called reductionist approach, which seeks to explain complex phenomena – those which consist of many parts whose relationships cannot be easily explained – by dividing them into smaller component units in which information can be more easily dealt with, and then combining the partial explanations (Cilliers, 2000; Ramalingan and Jones, 2008).

If we take relatively simple systems such as a ball running down a slope, it is possible to reduce it to its basic components (ball, slope, the material they are made of, the inclination of the slope) and to identify the most important relationship between these parts (how the speed of the ball varies in time) to achieve a meaningful understanding of its functioning that will hold true in other similar systems. Given that the basic conditions do not vary too much, it is possible to accurately predict how fast any ball will run down any slope.

However, not everything can be adequately explained by this approach. There are systems - groups of interconnected components - which cannot be explained by means of analysis and subsequent integration, even if these are carried out in the most sophisticated manner. Contrasting the example above, if instead of a ball running down the slope we take gas bubbles forming in a glass of soda, it is unlikely that we will succeed in explaining such a system by the reductionist approach. There are several bubbles forming simultaneously, each with different shapes and sizes, moreover, some will attract others and merge, while others will not. Still, this example is a far cry from social systems in terms of complexity. The challenge in this case is to understand the nature of the relationships between elements that make up complex systems - the interactions are not the result of stable, predictable and separable chains of relations, but emerge from a large number of intricate and dynamic sets of connections, which are meaningless when analyzed out of their context and almost impossible to approach through the examination of separate units (Cilliers, 2000).

In order to highlight the most conspicuous characteristics of complex systems, Table 1 shows a comprehensive list of its features, with a detailed description of each and the corresponding example in the case of a hypothetical fishery. It is worth noting that any social system could have been used as an example, and readers are invited to try to find similar examples in systems with which they are familiar while exploring this table. The list in Table 1 is adapted from Cilliers (2000), which also inspired the comparison of each feature in conceptual and empirical terms.

Table 1. Features of complex systems, followed by a detailed description and the corresponding example considering a hypothetical fishery

Feature	Detailed description	Example
Number of elements	**Large:** a large number of elements is necessary for the richness of connection.	A fishing community comprising of several hundred people.
Relation between elements	**Dynamic:** interactions between elements change with time, having different results or re-arranging themselves.	Several interactions take place: work partnerships, fish sales, lending and borrowing of money and materials, for example.
	Rich interaction: elements influence each other through diverse connections. The specific result of a given set of interactions can be reproduced by a completely different set. Moreover, some interactions can be totally redundant.	The system has several interacting elements, including "natural" - fish, crabs, vegetation, etc. - and "social" components - fishers, middlemen, enforcement agent, politicians, etc.
	Non-linear behavior: even small changes in a restricted set of the interactions can cause a great shift in the system.	The continued absence of an influential middleman due to illness might trigger the formation of new buyer-seller relationships. A small investment in new freezers can result in huge gains when selling fish out of season.
	Local interactions: elements interact mainly with others in their vicinity (which can refer directly to spatial terms, but not necessarily). Nonetheless, the influence of interactions is wide-ranging, as rich connections mean that information or energy can flow in few steps.	The majority of fishers only sell their catch to buyers located near the landing sites. Buyers, even if physically capable of reaching distant markets (e.g. owning a truck fleet), have a limited amount of partners and intermediaries, and are likely to use them to access consumers.
	Feedback: the result of an interaction can have an effect on itself, either stimulating or constraining it.	Profits made in one season make it possible to upgrade fishing gear, resulting in better catches in the next period. Or in failure, if the fish stocks are not capable of sustaining the new level of exploitation.
Relation with other systems	**Open systems:** There is usually interaction with other systems and the surrounding environment. Boundaries are difficult to identify, they are usually only meaningful as a way to describe the system.	Fish stocks migrate, and so do fishers. Both new ecosystems and markets become accessible as a result of these movements, and systems might overlap or even fuse.
Behavior throughout time	**Non-equilibrium:** Complex systems do not achieve an state of equilibrium. Instead there is constant flow of energy and reorganization of the system	The fishery never achieves stability – it is everchanging and adapting to each season's characteristics.
	History: Each stage of its development has an influence on the subsequent stages – to understand these systems it is therefore essential to know their history.	Fishing techniques and technology result from the knowledge created, passed on, and transformed through several generations of fishers.
Emergent properties	**The elements in a complex system only carry information about a restricted set of interactions:** no single element, or number of elements, can represent the whole system. Complexity emerges from the interactions between elements.	Each actor only has access to a limited amount of information, even those most well connected in the system. Decisions, from household expenditure to fishery regulations, are always carried out with this sort of restricted knowledge, they can, however, potentially affect the whole system
	Self organization: the system is capable of developing structure from seemingly chaotic and unrelated elements. This capability enhances the possibilities of adaptation to unpredictable changes within and outside the system	If environmental conditions favor fishing of previously inaccessible and valuable species, it is likely that techniques and market networks previously never used will be either activated or created based on existing ones. This configuration may, or may not become the norm for the system.

Although these remarks on complex systems may sound like common sense to many, and in fact they have been raised in several fields of science for at least 30 years (Ramalingan and Jones, 2008), the reductionist approach is firmly positioned as the best explanation of reality available and is still the basis for understanding and decision-making in our societies. Simplicity and predictability are desirable qualities, which when not naturally occurring in social systems are usually forced into them. For instance, in modern agricultural systems, the relationship between the basic components of the system – water, soil, plants and humans – is mediated by stable technological, economic and institutional interfaces through which the prediction and control of outcomes is made possible – irrigation and fertilizers control the input of water and nutrients, while market incentive (e.g. high commodity prices) provides the funds for the development of crops which are highly resistant to diseases and variations in climate. Finally, if policies favoring export crops are maintained by the government, such as access to credit and protection of land ownership of vast areas, the system will function adequately, allowing for precise control of each component and prediction of its outcome. Agriculture does not necessarily follows this pattern as its organic and familiar roots attest, nonetheless intensive monoculture is the predominant configuration of this system in current societies.

In complex systems, on the other hand, there are several possible ways in which components can interact, depending to a great extent on what conditions are in place at the onset of the interaction. Furthermore, even these relations might change as a result of feedback, when the components or the nature of the relationship between them is changed by the result of their interaction. In systems such as small-scale fisheries, formal mechanisms of control are either lacking or too weak, being replaced instead by informal economic activities and institutions. These, coupled with the high dependency on external factors such as climate and access to fish populations, make these systems more prone to unexpected changes. For example, a new buyer might appear and change financing and market opportunities for fishers, then a sudden change of weather might make a whole fish stock divert its migration route. As a result, fishing in forbidden areas might become possible, profitable or the only option left – a collection of seemingly unrelated changes can result in a complete reconfiguration of the system. These systems do not lend themselves to the same analysis of input-output and nor do they allow for easy control.

As Edgard Morin points out, "complexity does not only involve elements and interactions whose quantities defy our computing capabilities[,] it also encompasses uncertainties, indeterminations, random phenomena" (Morin, 2005:35). These features require a totally new approach, based not on division, detachment, certainty and objectivity at all costs, but on the integration of different views, the acceptance of uncertainty and the understanding that the observer cannot totally dissociate himself from the observed (Morin, 2005).

Uncertainty and change are inherent to fisheries and also to other systems based on the exploitation of natural resources (Wilson et al., 1994; Berkes et al., 2003). It has been shown how strategies are developed in the case of small-scale scale fisheries in response to uncertain conditions at different levels of social organization and on different spatial scales (Allison and Ellis, 2001). Diverse and richly connected social networks (e.g. kinship networks or mutual support among work partners or neighbors) and flexible institutions (forums for discussion and improvement of policies; regulations which include mechanisms for assessment and revision by stakeholders) are fundamental in building the system's resilience, as the lessons learned by resource users are more readily shared and incorporated and management structures can be modified on the basis of this learning process (Allison and Ellis, 2001; Tompkins and Adger, 2004).

Thus, summing up, social systems are highly complex ones, and as such are likely to change drastically due to unpredictable forces. Although this can have potentially negative effects on people's livelihoods, these systems also show a capacity for self-organization and increasing resilience. As a result, an adequate understanding of how people develop livelihood strategies is only possible if complexity is fully taken into account and uncertainty and change are incorporated into the analysis.

Concept Mapping and Collaboration

I have adopted concept mapping in my research project as a technique to, firstly, describe people's perception of their livelihoods and secondly, to explore the possibilities of collaboration with the research participants. The sections below which look at the development of the technique in this case study present more details of how it was applied, in particular how researcher and participants interacted. There were two main reasons for choosing this technique: on the one hand, it is able to describe complex relationships between concepts in a meaningful and concise way, which allows the detail and complexity of livelihoods to be examined without losing focus. On the other hand, constant revision is possible, resulting in a dynamic description of processes. In the words of Novak and Cañas (2006), "[it] is important to recognize that a concept map is never finished", thus uncertainty and change are not considered as a hindrance but a feature of the technique.

It can be argued that graphic representation makes it possible to challenge preconceived ideas about hierarchy and connections between concepts in a more straightforward way, inviting the observer/reader to manipulate and test new arrangements (even if this is done mentally and not by actually transforming the graphics). This is in contrast to their description in a textual form, where the text structure and style impose a much more static view of the theme - although it does not necessarily mean total closure, as an interested reader with an inquisitive mind will nevertheless be able to "play" with the concepts presented in a text.

However, a method which is able to deal successfully with the description of complexity is not a complete solution, as it still yields explanations which are determined by people's perceptions and interpretations of patterns and events – and it is worth highlighting that this is applicable to both researchers and participants. If this subjective element is not properly considered, particularly if the effect of power in the negotiation of different views is neglected, it can often lead to a distorted understanding of the problem (Haugaard, 2003). Therefore, in order to take this concern into consideration, I have avoided the view of people as informants or passive sources of data, and have opted instead to make it possible for them to participate in the analysis of information, playing an active role in making sense of concepts and their linkages.

The choice of a collaborative approach has two interrelated objectives. Firstly, I seek a description and explanation of social process which is as close as possible to the actors' experience, acknowledging and openly dealing with theoretical or personal assumptions. Secondly, and most importantly, by promoting participation in my methods I hope to provide fishers with the means to have a better understanding of their actions and the social structures of which they are part. Consequently, this could lead to greater chances of them gaining control over the management of the resources which are fundamental to sustaining their livelihoods (Ashby, 2003; Vernooy and McDougall, 2003).

RESEARCH CONTEXT

The small-scale fishery in the estuary of the Patos Lagoon, southern Brazil, is a typical example of the complex systems discussed in the previous sections. It is the setting of a research project in which I have applied the collaborative concept mapping technique, and the specific findings of which I will discuss in this paper. In this research project I sought to better understand how fisherfolk construct their livelihood strategies, by exploring the role of informal institutions in creating the means for these people to learn and adapt to change. In order to achieve this, I designed a case-study and applied a qualitative analysis approach which included participant observation, in-depth interviews and focus groups. The findings presented below are the result of two three-month periods of fieldwork – from mid October 2006 to mid January 2007, and from October 2007 to mid January 2008 (see also Box 1 for details on the research epistemology and methodology). The following paragraphs provide a summary of recent developments and the current status of the Patos Lagoon estuarine fishery, and their purpose is to allow readers to contextualize the application of the concept mapping technique and the collaboration between researcher and participants.

The fisheries in the estuary[2] reached a peak of productivity during the early 1970s, mainly due to the development of a heavily-subsidized fishing industry sector. Overcapitalization and too great a -reliance on subsidies led to the over-exploitation of stocks and economic collapse of the fishery, both small and large-scale (D'Incao & Reis, 2002). While the large-scale fishing fleet was able to compensate minimally by seeking different targets and keeping some of its status, the social condition of small-scale fishers deteriorated as a result of their lack of any alternative means of making a living. This process became more intense in the 1980s and early 1990s, when the situation became critical for small-scale fishers and a radical alternative was sought. In 1996 a totally new management system was implemented in the estuary through the combined action of a Christian grass-roots non-governmental organization, fishers' representatives and government officials (D'Incao and Reis, 2002). The new institution, called *Forum da Lagoa dos Patos,* replaced the previous centralized, top-down system, which was perceived as a total failure, with a decentralized system, with representatives from fisherfolk communities playing a major role in the policy making process (D'Incao and Reis, 2002; Kalikoski and Satterfield, 2004).

The change in the decision-making process became a reality in 1998 when a *Portaria*[3] was issued with the direct participation of fishers. This piece of legislation dealt with several aspects of fisheries management in the Lagoon, such as limiting the fishing effort and defining fishing seasons. Most importantly, it tackled the issue of access to the fishery by "outsiders" - fishers who migrated every year from the neighboring state of Santa Catarina for the lucrative pink-shrimp season. This was perceived by fishers in the Patos Lagoon as a major cause of the collapse of the estuarine fishery and just as detrimental as, or for some worse than, the consequences of mismanagement by the centralized system. The new regulation strictly limited access to the estuarine fisheries to people who could prove that they resided in the towns delimiting the Patos Lagoon estuary areas. Despite protests and political pressure by "outsiders" (D'Incao and Reis, 2002), the fishery has been restricted to "locals" since the 1999/2000 fishing season .

However, this new institutional context also has severe limitations. D'Incao and Reis (2002) present a detailed account of the interaction between fishers and researchers in the participatory management process. The authors show empirical evidence on the conflict between fishers, who seek immediate results and returns from the fishery; and researchers and managers, who favor a precautionary approach and struggle to strike a balance between attending fishers' needs and guaranteeing the long term sustainability of the fishery (D'Incao and Reis, 2002). The relationship between these "opposing" sides has yet to be resolved, as several aspects of the fishery regulations and policies are still disputed by several actors. Despite the efforts of several initiatives in participatory research (D'Incao and Reis, 2002), the fishers' concerns are still not fully considered in the management process (Kalikoski and Satterfield, 2004). Furthermore, the lack of legitimacy of the fishers' representatives and institutions, along with their poor participation in the management process further complicate the achievement of a system in which decisions are adequately shared by all participants, and fishers are able to both contribute to, and benefit from the management of fishery resources. As Kalikoski and Satterfield (2004) point out, the Patos Lagoon Forum currently only manages to partially empower fishers as "elite representatives", who do not effectively represent fishers, have control of the decision-making process.

Mapping Concepts with Fishers: Experiences from the Field

The qualitative analysis approach I follow in this research draws on the framework proposed by Miles and Huberman (Punch, 2005), in which data display plays a prominent role in the course of analysis. Thus, the use of different mapping techniques as a means of experimenting with the graphical representation of the information gathered through observation, interviews and focus groups is a fundamental part of the data analysis in my project.

Concept mapping is particularly helpful in the analysis of qualitative data using the approach mentioned above, by providing the means to reduce data without losing meaning, making explicit connections between concepts and integrating apparently unrelated frameworks in comprehensive models. Novak and Cañas (2006) provide guidelines for the creation of concept maps which can fulfill these objectives adequately. I have followed these recommendations and adapted them according to the conditions of my case study. The main requirements and a discussion of their application follow below.

Working with Fisherfolk

The technique was applied in both stages of fieldwork. As far as collaboration is concerned, in the first stage the technique was introduced in informal sessions, which consisted of discussing issues noted during observation or raised by the researcher in interviews with fisherfolk. On these occasions, the basic elements of the technique were used by the researcher during the conversation, using a pencil and notepad to draw simple diagrams using the concept map structure. The main issues in each specific conversation were represented as concepts, with propositions linking them where appropriate. In the meantime, the researcher used the technique with the help of specific software (see below for details) as a means of organising the preliminary findings from the field, a task which did not have the participation of the fisherfolk.

Then, in the second stage of fieldwork, the collaborative work was intensified. The technique was used in two different settings. Firstly, it was applied in a focus group, which dealt with a specific question on the configurations of the fishing market. This was carried out in two sessions of approximately 3 hours each, with 12 and 20 participants respectively. The participants were all male, professional fishers, aged from 19 to 56 . The focus group sessions took place in a room used by a local fishers' association for their meetings (itself an improvised space in the back of a bar owned by one of the members of the association). It is worth noting that concept mapping was one of several tools used in these sessions. It was also used in informal meetings in which three to seven fishers participated (all of whom also took part in the focus group sessions mentioned above), in a total of seven meetings happening approximately once a week. In these cases the technique was applied more thoroughly when the fishers were literate and also willing to experiment with the tool, which turned out to be only three participants, with the work done in three individual meetings. Again, in all cases, only pen and paper were used.

Defining the Map Context

The basis for a meaningful map is the careful definition of the main question which the participants wish to find answers to. The choice of focus question in this case study shares many of the characteristics of its use in education, in spite of the widely different context in which it is applied. The common application of the concept mapping technique in education is likely to involve topics in a certain subject or course which need to be explored by students, usually determined by their teachers or external school boards responsible for general curriculum guidelines. Thus, the foundation of the map is given to the map makers, who have little influence over this choice. This limits the outcome to some extent, although it does not necessarily imply that freedom is denied, as there is plenty of room for exercising creativity in the definition of concepts and propositions.

In the same way as the students, the fisherfolk participating in this case study played a limited role in the definition of the focus questions – and, as suggested above, this is balanced with opportunities for a more active role in the other steps of map creation. Therefore, the process started in advance of the fieldwork, when general research questions were defined based on previous experience of the researcher, input from colleagues and a review of the relevant literature. These general questions generated sets of more specific ones, which could be either directly posed to participants in interviews, or used as guide for observation. Thus, the general research questions provided the starting point for the map and determined its general nature depending on their characteristics, ranging from the descriptive ("what" types of question) to the exploratory ("why" and "how" types of questions), while specific questions originated units of concepts and propositions which ultimately make up the concept map.

The understanding of the implications of this first step in the creation of a concept map is fundamental in order to make the distinction between graphical techniques as aids to representing observations and thoughts, and as an integral part of the research process – the difference between a complementary element used to support an argument and a necessary element employed to develop and critically appraise a narrative. It is worth highlighting that the difference is not denoted in the visible, graphic result of the map, which can be exactly the same in either case. Instead, the difference is in the critical and reflexive approach which underlies the definition of the focus question.

With this in mind, I considered the focus questions themselves to be open to revision at any moment, therefore allowing the feedback from the fishers and unforeseen changes in the system to be fully considered in the research process – even if that meant challenging the basic questions guiding the research. The account from the field is necessarily dynamic, as it represents a non-linear complex system. It cannot therefore be expected to follow a fixed, predictable structure typical of a simpler linear system. This concern is reflected in the development of procedures which allow for these unavoidable changes to be accommodated during the course of the research, while also maintaining a meaningful structure.

However, concept maps are certainly not a pre-condition for this to happen, they are simply tools. Researchers must be reflexive and willing to take action when the need for new directions is found - the means they will employ to carry out the changes are secondary. Nonetheless, concept mapping offers the possibility of promptly identifying the need for revision - for example, concepts which consistently appear in the course of interviews and cannot be clearly related to a question might indicate the necessity for new questions or the reformulating of existing ones. Furthermore, maps are helpful in keeping track of the various stages necessary in adapting the research questions, as each step can be recorded in a clear way as subsequent map versions. The latter allows for a particularly straightforward assessment of possible paths, especially when compared to a collection of reflections kept in the form of separate notes, for instance.

Defining Concepts

A concept is defined by Novak and Cañas (2006) as "a perceived regularity in events or objects, or records of events or objects". In this case study, concepts first arose from information gathered from interviews, focus groups or participant observation, which I recorded as notes in a field diary. Therefore, the initial sets of concepts originated from my perception of the events I experienced, the people I met and the conversations I had, transcribed in my field notes. It was only when these concepts were presented to participants that the technique really started to fully include the fisherfolk in the process. We could verify and refine the concepts together, with the objective of obtaining a revised list in which concepts were selected according to two criteria ; relevance to the focus question and simplicity of the term. This list is equivalent to what Novak and Cañas (2006) call "parking lot" (see definitions of key terms).

Even though some collaboration already occured between participants and researcher in the definition of the map context, it was somewhat limited. The definition of concepts was the first stage in which all participants were fully engaged in the task of giving shape to the concept map. In order to open up the opportunities for collaboration, the list of concepts identified by the researcher was used as a stepping-stone towards a more elaborate list, with no special significance being given to any of its concepts - all of them were considered as equally likely to be modified or even eliminated. Then, the relevance and simplicity criteria were gradually applied to the assessment of each concept.

As far as simplicity of concept formulation was concerned, a key role played by this initial list of concepts was to serve as an example of the form in which concepts should be expressed, and most importantly, why complex statements should be avoided. It was common for the participants to come up with full statements which contained more than one concept and its relations. An example which occurred frequently when fishing gear was the subject of interviews or discussions was to combine the gear and its most common effects or species targeted, as in "destructive shrimp bottom trawl"- which then seemed to be a natural unit. Although statements such as this are meaningful, the combination of concepts hides subtle but important differences, as while the shrimp is the species most commonly targeted with bottom trawl, the same gear is also used to fish blue crab. Both uses of the gear have a negative, but different impact, as they are employed in different areas and seasons. By considering each concept individually, it is possible to identify the different relations between them and explore further connections with other concepts, resulting in a map with greater explanatory power.

An important contribution of this case study to the original method proposed by Novak and Cañas (2006) is the categorization of concepts. Taking into account that perception is at the heart of the definition of concept in this approach, it seemed to me that it might be useful to make a distinction between events and objects according to the degree to which people's perceptions of them could be disputed. This is in line with an epistemological position which considers that although the process of knowledge construction starts with an actor's perception of events and objects, which occurs at the level of each individual, it is only "finalized" when claims over the "truthfulness" of these perceptions are negotiated with other actors via social relations. The quotation marks in the last sentence highlight that the construction of knowledge must be thought of as an open-ended process, and that perceptions do not have a conclusive nature: consequently, neither does the knowledge they originate. Moreover, by openly considering the role of social relations in the construction of knowledge, it is possible to identify the effects of social structure in determining whose accounts dominate the narratives often taken for granted as "reality" (Long, 2001).

Agency, which is here broadly understood as the capability to act, change reality and be changed by it (Giddens, 1984; Long, 2001), is a fundamental concept of my research. Therefore, first and foremost, the agents in this system needed to be identified and highlighted. In order to do this, social actors were differentiated from other concepts[4]. Furthermore, providing fishers with the means to enhance their capability to improve their livelihoods is a general objective underlying all the participatory methods I have applied in this case study. This requires, among other things, an understanding of which actors make up the system and how their actions affect each other and its functioning.

Thus, special attention was given to clarifying the difference between broad conceptual categories and concepts representing true actors. For instance, if we look at "enforcement" and "enforcement agent", it is possible to see that the former is a broad concept including regulations, personnel, etc., to which actions cannot be adequately related, while the latter is a specific concept, representing a category of civil servants with specific duties which can be clearly related to actions. This difference has a direct effect on the proposition that will be related to each concept: while we cannot say that law enforcement is negligent without sounding vague, it is possible to affirm with clarity that agents fail to be sufficiently vigilant in ensuring that laws are enforced, including how and why they act in this way.

Secondly, concepts were classified according to their degree of subjectivity. Only two broad categories were created: on the one hand, concepts which could be unequivocally classified as objects and facts independent of personal opinion, such as gear and fish species, or more loosely as constructs relatively free of subjectivity, such as regulations and income, were labelled *objective*. On the other hand, concepts which were totally dependent on people's ideas and beliefs were classed as *subjective*, "risky" and "destructive" are two such examples. This distinction was used as an attempt to clarify the diverse nature of concepts, but is certainly not without caveats. In particular, what I call here "human constructs", as their name implies, are subjective by nature and are only differentiated from more subjective concepts because the former have a concrete existence, as technical measurements or documents[5]. Although I encouraged the initial reflection on the different nature of each concept, this categorization was the result of discussions with the participants, who highlighted the need to treat concepts differently when connections between them were explored.

Defining Propositions

Once the list of concepts was considered sufficiently succinct and inclusive, which was usually achieved after two or three consecutive sessions[6], the next step was to work on the relationship between concepts, or "propositions" in the terminology of Novak and Cañas. Some more closely related concepts were selected, two at a time, by the participants and I, and we then worked together to find words or phrases to link them. A balance was sought in order to achieve the simplest representation without loss of meaning. Once all were satisfied with the result, the units of concepts and propositions were represented on sheets of paper.

This is perhaps the most time-consuming of all stages in the creation of the map, and in this case study it also proved to be potentially frustrating. Great effort must be put into phrasing the propositions adequately, and as occurred in the previous stage, it is common to have a group of separate relations considered as a unit. It can be a complicated process and in these situations, frustration had to be combatted with breaks to refocus the participants and provide mutual reassurance that it was a worthwhile task. Nevertheless, the fishers proved resilient and eventually we managed to work out the statements in all cases.

Figure 1. Screenshot of a concept map created with CmapTools during fieldwork in Southern Brazil

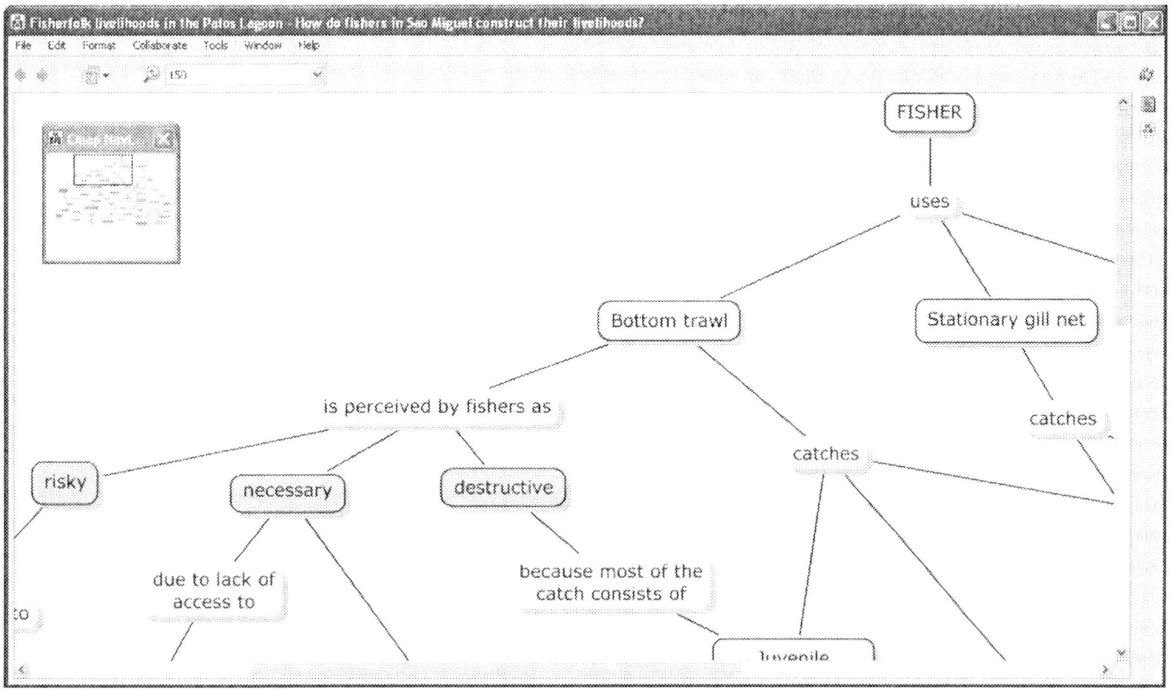

Creating the Map

The units of concepts and propositions created in the previous step were combined into larger structures, by linking units with new propositions and working out connections between separate frameworks. It is when these connections, or *cross-links*, are made that the leap is taken from linear thinking to complex arrangements and participants are able to learn in a creative and meaningful way.

However, this step also poses a major challenge as it requires those involved to perform tasks which are unusual for the majority of people, who are trained in rote learning and linear thinking. Given this, this task was only carried out with a smaller number of participants – in individual sessions with three fishers (see also the section "Working with fisherfolk" above). However, in the majority of cases, the map was finalized by the researcher alone, with the resulting preliminary structure being submitted afterwards for assessment, again by only a limited number of participants – the maps were presented in five different meetings with a varying number of fishers (from three to seven people), all from the same group who participated in the focus groups (see also the section "Working with fisherfolk" above). Nevertheless, feedback on, and new input to the map were obtained, mainly in terms of refining propositions and cross-links. It is worth noting, however, that the structure of the map was not seriously challenged in any case, despite participants being encouraged to propose rearrangements of concepts.

The software CMapTools version 4.7 (http://cmap.ihmc.us/download/, latest version: 4.18 - at time of writing, September 2008) was applied throughout the fieldwork for all tasks involved in concept mapping, although it was used only by the researcher. Participants in this case study often have basic literacy, and almost without exception, have no training in the use of computers. Although relative illiteracy was an obstacle for the use of the tool to its full capacity, it did not interfere with the ability of participants to communicate concepts to each other and the researcher. It is worth noting that it was always possible to have groups of participants where most people had at least the necessary basic reading and writing skills so as to have autonomy in the task. Nevertheless, most of the exchange of information was in the form of dialogue, with the spoken word clarifying and reinforcing what the written word showed. The writing served as reference to those able to read and write, but was never the only medium for manipulating the concepts and propositions. Thus, although the software is incredibly helpful in creating the maps, it would not be useful in this research context. Fortunately its main functions can be replicated with materials and techniques accessible to any audience. Basically, pen and paper are sufficient and if one follows the premises of the technique, the whole process can be transferred to the software with no information being lost, as was done successfully in this case. A screenshot of one of the concept maps created in CmapTools is shown below. (see Figure 1)

CONCLUSION AND FUTURE TRENDS

Concept mapping proved to be an invaluable technique, enabling the understanding of complex situations with a level of clarity and openness which would be difficult to achieve with other tools. I consider the technique to have been successfully adapted to the specific research context of my case study, and this was confirmed by the feedback obtained from the participants.

The experience obtained in this case study indicates that the most challenging of all tasks in the creation of a concept map is the definition of concepts. The discussion of all possible meanings and interpretations of concepts was sometimes daunting, and both researcher and participants had to show great determination in order to avoid simply opting for a collection of words which was more or less acceptable. Maintaining this determination was a rewarding task, as was confirmed in the feedback from participants, who mentioned how it had changed their views of things and ideas which they had previously taken for granted. There is a quote by Wenger (1998) on the process of *negotiation of meaning* which is fitting for this discussion: "[the negotiation of meaning] constantly changes the situations to which it gives meaning, and affects all participants. In this process, negotiating meaning entails both interpretation and action." (p. 54).

Nonetheless, there are difficulties in applying concept mapping. These are mainly related to its innovative nature and, consequently, the way it challenges the assumptions of those who are used to rote learning and working individually. Moreover, in this case-study, the process of generating cross-links and structuring larger models was limited as far as collaboration is concerned. Central to these caveats is the fact that the technique is highly dependent on written text.

As pointed out by Barton and Hamilton (2005), "ordinary everyday spoken interaction – which is usually referred to as face-to-face and somehow viewed as "natural" and unmediated – is in fact highly mediated, most often by texts, but also by other artifacts." (p. 18). The use of texts in the whole process of crafting a concept map is therefore nothing alien to the lives of fisherfolk. Consequently, I would argue that the successful application of this technique seems to lie more with understanding the implications of this fact and looking at ways of dealing with the resulting problems than seeking alternative ways of communicating. One such method is the use of drawings instead of words, as is commonly used in participatory research methods (Maine et al., 1996). It certainly has its merits, in particular the capacity of images to convey complex messages and leave room for subjective interpretation. When texts are accompanied by interaction and dialogue, however, they can also communicate meaning similarly, even to an audience which is supposedly incapable of understanding them.

In this case study, there was a clearly unequal "distribution of power" (Barton and Hamilton, 2005) involved in the creation of the concept maps, as the process involved fishers with very basic levels of literacy and no knowledge of the technique and myself, an academic who had mastered it previously. Being aware of this, I made a great effort to act as if I was merely another participant. Although I played a special role, providing guidance on the process, whenever possible I steered clear of unilaterally setting the direction it would take. Special attention must be given to reaching a balance of power between participants in contexts similar to the case study I presented here. This must be done in order to guarantee a *de facto* collaborative process, so that the full potential of concept mapping may be explored.

REFERENCES

Allison, E. H., & Ellis, F. (2001). The livelihoods approach and management of small-scale fisheries. *Marine Policy*, 5(5): 377–388. doi:10.1016/S0308-597X(01)00023-9

Ashby, J. (2003). Uniting Science and Participation in the Process of Innovation – Research *for* Development. In B. Pound, S. Snapp, C. McDougall, and A. Braun (Eds.) *Managing Natural Resources for Sustainable Livelihoods: Uniting Science and Participation*. London: Earthscan.

Barton, D., & Hamilton, M. (2005). Literacy, reification and the dynamics of social interaction. In D. Barton, & K. Tusting, (Eds.) *Beyond communities of practice: language, power and social context*. Cambridge, UK: Cambridge University Press

Berkes, F., Colding, J., & Folke, C. (2003). Introduction. In F. Berkes, J. Colding, and C. Folke, (eds.) *Navigating Social-Ecological Systems: building resilience for complexity and change*. Cambridge, UK: Cambridge University Press.

Cilliers, P. (2000). *Complexity and postmodernism – understanding complex systems*. London: Routledge.

D'Incao, F., & Reis, E. G. (2002). Community-based management and technical advice in Patos Lagoon estuary (Brazil). *Ocean and Coastal Management*, 45(8): 531–539. doi:10.1016/S0964-5691(02)00084-4

Giddens, A. (1984). *The constitution of society: an outline of the theory of structuration*. Cambridge, UK: Polity Press.

Haugaard, M. (2003). Reflections on seven ways of creating power. *European Journal of Social Theory*, *6*(1), 87–113. doi:10.1177/1368431003006001562

Jones, S. (2002). Social constructionism and the environment: through the quagmire. *Global Environmental Change*, *12*, 247–251. doi:10.1016/S0959-3780(02)00062-6

Kalikoski, D. C., & Satterfield, T. (2004). On crafting a fisheries co-management arrangement in the estuary of Patos Lagoon (Brazil): opportunities and challenges faced through implementation. *Marine Policy*, 28(6): 503–522. doi:10.1016/j.marpol.2003.12.001

Klein, J. (2002). Deforestation in the Madagascar Highlands – Established 'truth' and scientific uncertainty. *GeoJournal*, *56*, 191–199. doi:10.1023/A:1025187422687

Lazar, D. (2004). Selected issues in the philosophy of social sciences. In C. Seale, (ed.) *Researching society and culture*. London: Sage.

Long, N. (2001). *Development sociology: actor perspectives*. London: Routledge.

Maine, R. A., Cam, B., & Davis-Case, D. (1996). *Participatory Analysis, Monitoring and Evaluation for Fishing Communities: a manual* (FAO Fisheries Technical Paper. No. 364). Rome: FAO.

Marshal, C., & Rossman, G. B. (1999). *Designing qualitative research*. London: Sage.

Mason, J. (1996). *Qualitative researching*. London: Sage.

Morin, E. (2005). *Introducao ao pensamento complexo*. Porto Alegre, Brazil: Sulina.

Newman, L. (2005). Descartes' Epistemology. *The Stanford Encyclopedia of Philosophy (Summer 2005 Edition)*, E. N. Zalta (Ed.). Retrieved September 23, 2008, from http://plato.stanford.edu/archives/sum2005/entries/descartes-epistemology/

Novak, J. D., & Cañas, A. J. (2006). *The Theory Underlying Concept Maps and How to Construct Them*, (Technical Report IHMC CmapTools 2006-01). Pensacola, FL: Institute for Human and Machine Cognition. Retrieved September 23, 2008, from http://cmap.ihmc.us/Publications/ResearchPapers/TheoryCmaps/TheoryUnderlyingConceptMaps.htm

Punch, K. (2005). *Introduction to social research: quantitative and qualitative approaches*. London: Sage.

Ramalingan, B., & Jones, H. (2008). *Exploring the science of complexity: Ideas and implications for development and humanitarian efforts* (Working Paper 285). London: Overseas Development Institute.

Seale, C. (2004). *Researching society and culture*. London: Sage.

Tompkins, E. L. & Adger, W. N. (2004). Does adaptive management of natural resources enhance resilience to climate change? *Ecology and Society*, (9).

Vernooy, R., & McDougall, S. (2003). Principles for good Practice in Participatory Research: Reflection on Lessons from the Field. In B. Pound, S. Snapp, C. McDougall, & A. Braun (Eds.) *Managing Natural Resources for Sustainable Livelihoods: Uniting Science and Participation*. London: Earthscan.

Wenger, E. (1998). *Communities of practice: learning, meaning, and identity*. Cambridge, UK: Cambridge University Press.

Wilson, J. A., Acheson, J. M., Metcalfe, M., & Kleban, P. (1994). Chaos, complexity and community management of fisheries. *Marine Policy*, 184(4), 291–305. doi:10.1016/0308-597X(94)90044-2

Yin, R. B. (2003). *Case study research - Design and methods* (3rd Ed.). London: Sage.

KEY TERMS AND DEFINITIONS

Actor: individual persons with *agency*, i.e. *knowledge* and *capability* to assess situations and devise responses. Adapted from Giddens (1984) and Long (2001).

Concept: "a perceived regularity in events or objects, or records of events or objects, designated by a label". (Novak and Cañas, 2006: p.10).

Fisherfolk: fishers themselves, their families and other people who make up a larger group which is dedicated to fishing as a primary source of income.

Focus Question: "a question that clearly specifies the problem or issue the concept map should have to resolve.". (Novak and Cañas, 2006: p.1)

Livelihood: "the assets[,] the activities, and the access to these (mediated by institutions and social relations) that together determine the living gained by the individual or household." (Allison & Ellis, 2001)

Parking Lot: during the first stage of creation of the concept map, a list of concepts related to the focus question is generated, in no particular order, - using brainstorm, for example. These concepts are left in one margin of the working area (thus the analogy with a "parking lot"), with the most relevant being used in the map.

Participant: people who have an active role in the research process, taking part in the gathering of information, as either sources or contacts for other potential participants, as well as contributing directly to the decisions made during the research through their feedback.

Proposition: "statements about some object or event in the universe, either naturally occurring or constructed. Propositions contain two or more concepts connected using linking words or phrases to form a meaningful statement.". (Novak and Cañas, 2006: p.1).

ENDNOTES

1 Supported by the Programme AlBan, the European Union Programme of High Level Scholarships for Latin America, scholarship no. E05D059985BR.

2 The small-scale fishery dealt with in this case study takes place at the Patos Lagoon, which is located at the southernmost tip of Brazil (*Rio Grande do Sul* state). It is one of the largest coastal lagoons in the world, with an area of approximately 11 000 km^2 (300km long and 40km wide), of which 5 to 9% is estuary (area where salt water enters the lagoon and mixes with fresh water, which varies according to coastal currents and winds). It is estimated that between 3500 and 5000 fishers participate in the small-scale fishery in this region.

3 A category of Brazilian legislation which regulates the activities of the civil society, such as the use of natural resources. It is issued by the national government, but might have an effect at national, state or municipal level.

4 Important to note here that I do not assume this to be the only possible way to approach the concepts of actor and agency. Actor-Network Theory, for instance, do not distinguish between human and non-human entities when exploring agency, offering a very interesting and different perspective.

5 The process of *reification* (Wenger, 1998; Barton & Hamilton, 2005), which explores the "realization" of ideas in social relations, might also shed some light on this discussion.

6 In the case of the focus group practice, one meeting was sufficient to prepare an adequate list of concepts.

Chapter 16
Using Concept Mapping to Improve the Quality of Learning

Maria Luisa Pérez Cabaní
University of Girona, Spain

Josep Juandó Bosch
University of Girona, Spain

ABSTRACT

The European Higher Education Area, created to contribute to the development of quality education and encourage cooperation between Member States of the European Union, has created an opportunity for reflection on teaching methods in universities. These circumstances provide momentum for the use of learning strategies. Along these lines, this chapter presents work done on concept mapping, one of the methods that could significantly help prepare both students and teachers for the new roles expected of them. The results of two lines of study are presented: on the one hand, research that analyses the influence of the differential use of concept maps on the quality of learning; and on the other, an innovation in the use of concept maps in a course developed in the area of initial teacher training. Both initiatives serve to highlight increased interest in concept mapping in the field of learning and especially in collaborative learning.

INTRODUCTION

The European Higher Education Area was created to contribute to the development of quality education by encouraging cooperation between Member States through a wide range of actions: promoting the mobility of teachers and students, designing joint study programmes, facilitating recognition of studies abroad, establishing networks, exchanging information and promoting lifelong learning for all citizens of the Union. A specific, quality-related goal within this framework is incorporating in-depth reflection on university teaching practices.

Such a task is not as simple as it sounds. Even though it is widely acknowledged that reflection on teaching methods is a good way to assure continuous improvement, it is no less true that every professional creates his or her own working methods which, once they have reached a stable and sustainable balance and are perceived as comfortable and satisfactory

DOI: 10.4018/978-1-59904-992-2.ch016

Figure 1. Vicious circle representing the consolidation of procedures

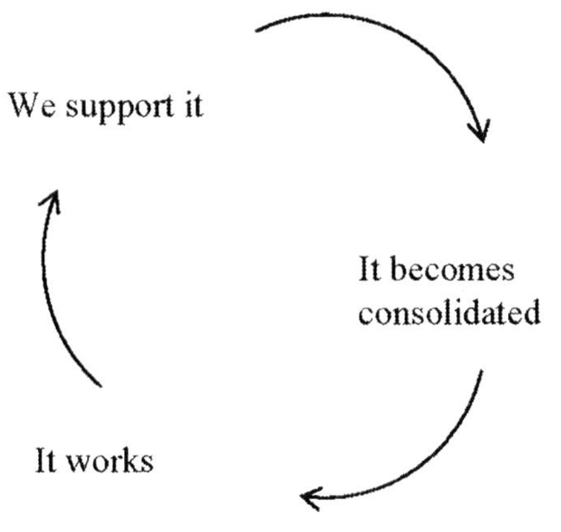

by the subject, acquire an inertia and solidity that make them particularly resistant to change.

University teaching staff are not immune to this phenomenon. When somebody finds themselves, with greater or lesser motivation, teaching classes at a university, they create their own teaching methods, often based on intuition or on their own prior experience as a student, and, once these methods have been consolidated, it is not easy to place them under scrutiny. Only a broad perspective on the part of the teaching staff, accompanied by an expressed willingness to consider ways of improvement, can break the vicious circle represented in Figure 1.

The proposal to think about teaching, from the planning process to the evaluation of learning (including its development), from the point of view of competence, obliges university teachers to enter into a process of reflection. It is not necessary to raise the issue of the quality of the teaching methods that each university, each faculty and each teacher consider to be their own; it is a matter of seeing to what extent these fit in with the new teaching needs arising out of a focus on skills.

One way of defining the teaching function of a university in this new context is by focusing on student aims: the acquisition of particular skills through the performance of learning activities related to certain contents. We consider this to be a correct interpretation of the spirit of the so-called Bologna Process, which aims to create a European Higher Education Area by 2010, when students will be able to choose from a wide and transparent range of high quality courses. Certain elements of teaching, although not new, take on greater significance: self-monitoring of learning on the part of the student and collaboration between students to facilitate collaborative learning as well as between teachers (the work of teaching skills necessarily supposes sharing aims and therefore teaching strategies).

At this moment it is especially opportune to take a deeper look at everything related to learning processes and strategies, collaborative work, regulation and self-regulation mechanisms and reflective processes for knowledge construction. It is necessary to invest in courses of action that will develop these processes and mechanisms, through research and innovation, in order to provide the teaching community with instruments that will facilitate a better adaptation of teaching methods to emerging needs.

Reflection on practice, which would be very useful for teachers and knowledge of new teaching methods, must be the motors driving this professional renewal.

We adhere to the current consensus among researchers in the field of learning strategies regarding some of the questions guiding research.

Firstly, it has become clear that, in order to learn the contents to which the construction activity is applied, the teaching-learning process consists of the student's activity together with the guidance and advice offered by the teacher during this activity. Intervention focused on improvement and innovation will have to take these parameters into account.

There is also a clear need to teach students to act consciously and intentionally to achieve certain learning goals, considering the nature of the task they have to perform, the environment in which they have to carry it out and the personal resources available to do it. Teaching "universal" learning methods, then, without teaching how to appraise the aims the performance of the task is meant to achieve or the conditions in which these methods should be applied, results in methods with little flexibility or transferability.

This consideration regarding the need to act consciously and intentionally is closely linked to the possibilities that students have to make proper use of the learning methods, in such a way that they can decide which is the best action at each moment and determine when and why it is better to use one learning procedure or another. Therefore, it is not enough to teach students learning procedures, it is also necessary to teach them how to make best use of them.

A final consideration with respect to our work is the reason that led us to choose concept mapping as a working procedure. Concept mapping, presented by Novak and his collaborators to aid learning comprehension and based on Ausubel's Theory of Meaningful Learning, has been used successfully in various studies with university students (Novak and Symington, 1982; Novak, Gowin and Johansen, 1983; Novak, 1985; Monereo & Pérez Cabaní, 1996) and has been shown to be an effective method, as will also be demonstrated throughout this chapter, to improve the quality of learning.

Although it is true that concept mapping is only one of many methods that can be used to facilitate the construction of conceptual knowledge, and that not all the contents that are taught and learnt in a formal educational context refer to concepts (since work is also done on procedural and attitudinal contents), concept maps do constitute an especially useful instrument in the process of knowledge construction for various reasons:

- They form a structure that graphically reflects the knowledge of a person or group and that can facilitate meaningful learning through the hierarchical organisation of the knowledge that this group or person is in the process of constructing.
- They allow both teacher and students to observe how different structures can be equally correct and they provide a schematic summary of what the students are constructing and learning or of what the teacher is constructing and teaching, which can be amplified and modified throughout the process.
- They offer the possibility of sharing prior knowledge and of establishing relationships, as well as exchanging points of view and negotiating meanings, both between the students themselves and between the teacher and the students. In this way students reflect on their learning process and actively participate, through the negotiation of meaning, in the construction of their knowledge, and the teacher can appreciate the differences between the students' conceptions and take decisions on how to

make his/her intervention more effective.

- They constitute an ongoing instrument of evaluation and self-evaluation of conceptual knowledge: at the beginning of a teaching/learning unit, in order to assess prior knowledge and misconceptions; during the process, in order to analyse the changes produced in the meaning of experiences and to appraise the new relationships established on the basis of new information; and at its conclusion, to assess the changes that have been produced in knowledge and how that knowledge has been constructed.
- They facilitate learning and comprehension in the performance of different teaching-learning activities: acting as a guide when extracting meaning from text books, articles or any written information that has to be studied; helping to give meaning to laboratory or field work, demonstrating the basic aspects that have to be observed or analysed and their relationships; organising ideas that have to be presented orally or through written texts; and providing an efficient way of taking notes or making a schematic summary of specific information.

These are the reasons that have led us to consider concept maps as a method that facilitates profound, meaningful learning in a classroom context, since their use requires the active participation of the student as well as the help of the teacher in the construction of knowledge. In this respect, it brings to mind the analogy made some years ago by Malone and Dekkers (1984, pg. 221) when they said that concept maps could be called "the windows of knowledge", as they allow the teacher and other students "to see inside" (to know a student's own conceptions), they enable the student to "see out" (to know how others conceive things) and they reflect everybody's own perception of a subject or question.

On the other hand, within the teaching-learning model we have as a reference, constructive activity requires dialogue and interaction between the professor and students and symmetrical interaction between students. Thus collaborative learning becomes essential.

Taking as a point of departure the considerations mentioned above, we present the first project we carried out to analyse how the teaching of the differential use of concept maps (use of concept maps versus regulative use of concept maps) impacts on the quality of the learning outcome.

The second project reported in this chapter, called an innovation project, was developed in the context of teacher-training courses at the University of Girona, and specifically in the course "Collaborative Working Strategies: Professional Teams and the School Environment". In this course, concept mapping is put into practice on a continual basis in one of its roles, specifically that of facilitating the sharing and joint construction of knowledge. An added value of this application is its connection with the ideas of collaboration, interpersonal relationships and teamwork, of clear importance in the field of teacher training.

THEORETICAL-CONCEPTUAL FRAMEWORK

As the basis for our work, we have taken into consideration the contributions and conclusions of the most representative researchers in this field of study. With the aim of bringing together in summarised form the aspects of these works that are essential for the study and the teaching innovation project we present here, we raise, also in summarised form, some questions of special importance, grouped into three sections. The first section, teaching and learning at the university, refers to the university teaching-learning model and learning as a process of construction in which the learner has to take an active role and the teacher the role of guide and 'scaffolder' during this construction process. The second section, collaborative learning and knowledge construction, is dedicated to showing the added value of collaborative work in knowledge construction. In the third section, progressive autonomy and quality of learning, we will talk about the need to teach and to learn how to learn making regulative use of learning procedures that will lead progressively to the autonomy needed to assess skills and achieve a quality of learning that responds to the needs of present-day society.

Teaching and Learning at the University

Concerning the first of these sections, the construction of a European Higher Education Area (where our work is located), an important transformation of the dominant teaching and learning model in university education, traditionally very centred on the transition of contents and on face-to-face classes, is proposed. Far from being merely a formal process, European convergence – as it is reflected in current legislative developments – makes clear the need to introduce profound changes aimed at modifying a large part of the current qualifications structure, as well as the model that serves as the basis for training. This process generates new demands on the universities that are already working to plan and initiate processes of change in response to the transformations involved in adapting to the new system.

The introduction of innovations and changes capable of producing improvements in university studies is not a linear, mechanical process but a complex and uncertain one, requiring the clear articulation of different measures in order to achieve them. But at the same time, it also represents an excellent opportunity for improvement and innovation in the processes of teaching and learning, giving greater learner autonomy and responsibility to students and modifying the role of the teacher, understood more as a guide or facilitator in the process of the students' knowledge construction than as a transmitter of scientific knowledge. These changes (which go beyond formal, structural or semantic changes) come close to what Guban (1992, pg. 226) called *second-order changes*, in other words, changes of a fundamental nature that modify the procedures, norms and rules of those who have to carry them out.

In this way, European convergence becomes an opportunity for an in-depth review of the university – its function, its organisation and the meaning of the teaching offered in it – as universities are expected to define an academic offering capable of responding to a new educational paradigm that is no longer based on teaching and the teacher but is focused on learning and the student.

However, it has to be said that universities have traditionally shown themselves to be highly resistant to this kind of second-order change. Additionally, the changes capable of producing improvements in education are costly and do not come about because they have been imposed hierarchically, nor because they have been thoroughly evaluated on a theoretical level or because the teachers, who are the true agents of change, have sufficient information. Studies of the issue show that these processes are complex have a multidimensional character and depend on the social, cultural and personal practices existing in specific contexts (Escudero 2001; Fullan, 2002; Gairin, 2000; Michavila, 2003; Ribas, 2000).

Especially interesting as an indication of the advances already being made in this direction is the work done by Monereo and Pozo (2003) in which they consider the new challenges universities must meet, the question of how to promote more strategic training for students and teachers, and a curriculum that facilitates autonomous learning.

We briefly mention here some points of these studies that have provided us a basis for our own work:

- Professionals capable of facing the complex problems of our world with multidisciplinary vision are increasingly appreciated, but in our universities there is still a radical separation of knowledge relating to the different disciplines. It will therefore be necessary to work in this direction, and collaboration is one possible way of achieving this.

- The false dilemma of teaching contents or teaching how to learn. Knowledge of learning procedures and strategies must be based on the contents proper to each discipline. Firstly, because what is being talked about is operating on something, and secondly, because what is being operated on, the contents, have a decisive influence on the way of operating itself.
- It has traditionally been considered that the first goal is to acquire basic knowledge, and that thinking about that knowledge takes place later, so that one day it can be used in a reflective and active manner. However, it seems more appropriate to teach students to think about the contents they are learning *while they are learning them*, so that learning is a consequence of thinking.
- University classrooms continue to be, for the most part, places designed for the reception of knowledge from a single source or direction: the teacher. More research must be done in this respect to demonstrate how teaching and learning, to cooperate and to work collaboratively, facilitate joint knowledge construction.

The constructivist concept of learning and teaching from which the above ideas are derived is itself based on three fundamental ideas (Coll, 1990). The first refers to the student as an active learner and the person most responsible for his/her own learning process. The second fundamental idea is that constructive activity is applied to certain highly developed contents: knowledge and cultural forms that both teachers and students find, for the most part, already developed and defined. The third idea concerns the teacher's role, which cannot be limited to creating the optimal conditions for a rich and diverse constructive mental activity; the teacher must also try to orient and guide this activity so that the knowledge construction progressively approaches what the contents signify and represent as cultural knowledge.

Collaborative Learning and Knowledge Construction

These ideas lead us to the second section of ideas that we consider basic for our work: the added value represented by collaborative work. As we said earlier, we are immersed in a culture of individual work. Each of us forms part of one particular culture, and when we realise this, we can take actions to alter the fact. This means working on the context to modify it, probably through our own relationship with it. It will also suppose a previous modification of our habits, because habits are linked to culture. In a culture of individual work, people's habits are of an individual nature; in a team-working culture those habits would necessarily be different.

To give an example of the question of habits: if somebody gives us a task and the first thing we do is consider how to carry it out, this is indicative of the habit of working individually. An indication of team-working habits would be if our first reaction were to think who we could count on to help carry it out. A basic difference: when faced with a task, do we ask ourselves *how* to carry it out or *with whom* to carry it out?

Teamwork is not easy. To a great extent because of our lack of knowledge about it, but above all, because of our lack of practice in it. A possible definition of teamwork is presented in Figure 2.

In other words, working in a team brings together certain characteristics, requires certain conditions and its essence lies in constant negotiation.

Obviously, the context also imprints its own characteristics on the teamwork. In the field of education, for example – where the staff (the teachers) have certain specific characteristics,

Figure 2. Diagram of a definition of teamwork (Pérez Cabaní & Juandó, 2001)

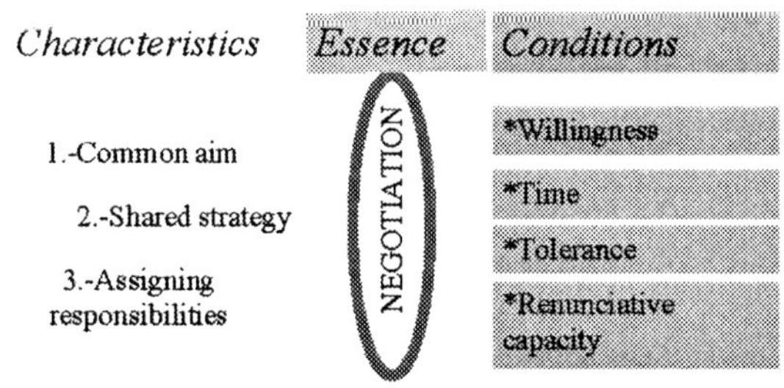

enjoy a broad range of freedom of action and enter their job (especially in the public sector in certain countries) with more or less random conditions – teamwork is not the same as teamwork in a multinational firm, where the staff's working conditions are probably much stricter and their functions more highly specified.

Nonetheless, there are certain, probably universal, consequences of teamwork whose relationship to learning is relevant.

The following extract describes the experiences of Biott and Nias (2000) and directly introduces the question of collaborative learning:

"... we learned to clarify our own thinking, as thoughts became tentative words and thus accessible to the scrutiny of the speaker, as well as to the listeners, and as we questioned each other's use of words. We saw each other as rational people and we wanted to explore the other person's meanings as well as our own." (page 84)

Being in the same context, sharing the same aim and having a willingness to collaborate make people listen to each other in a way that is different from the way they would outside this context. This is a major discovery. In the words of Scott and Flanigan (1996):

"Reaching consensus is about the journey that begins with "I" and ends in "We". (page 4)

And this is not only a process of joint knowledge construction but also of personal growth. In the area of collaborative working – teamwork – people learn to find the other, they learn to search for anchoring points for consensus in the discourse of the other, to move towards the new idea which, being shared and negotiated, takes on a value and a force that differentiates it substantially from a product of individual working. It is a path towards a new culture that has to do with the world of values and that brings people to mutual recognition that goes beyond an interest in the resolution of a specific problem or in the finalisation of a specific process.

Parrilla and Gallego (1999) illustrate, in a certain way, the idea that we have just put forward:

"When we choose to support and learn among people we know, a journey is begun that means, in the first instance, recognising the knowledge of others, of our colleagues, and secondly, assuming our own involvement and participation..." (page 56)

Another contribution of great interest along these lines is that of Järvelä & Järvenoja (2007), in which they study socially constructed self-regulation, specifically the regulation of motivation in collaborative learning. The authors propose a process of collective regulation that allows groups to develop a shared consciousness of aims, progress and tasks.

Figure 3. Level of responsibility and autonomy in the process of teaching and learning the strategic use of learning procedures (Pérez Cabaní & Juandó Bosch, 2001)

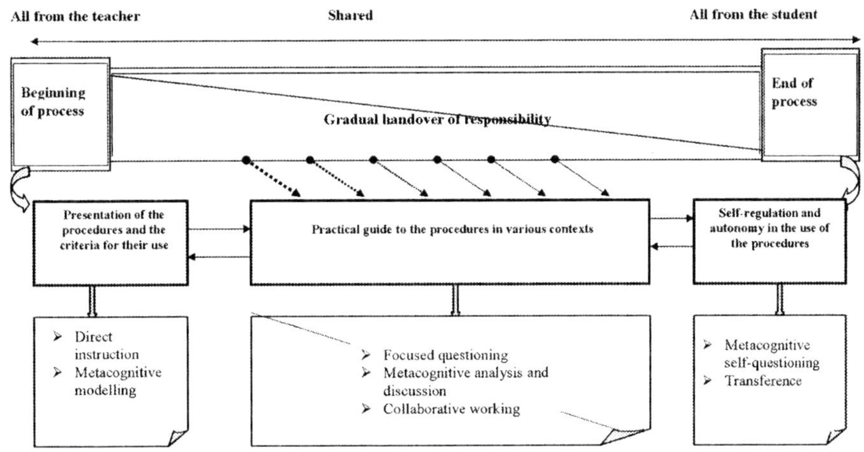

Progressive Autonomy and Quality of Learning

The third theoretical focus we have based our work on (and strictly linked with the previous points) is teaching and learning to learn, that is, making regulative use of learning procedures that lead progressively towards autonomy in order to attain certain skills and a quality of learning that responds to the needs of today's society.

During our research, we propose the regulative use of learning procedures as a basis for autonomous learning. We believe that, apart from knowing the characteristics and specifications of a procedure (declarative knowledge) and how to use it (procedural knowledge), students also have to know in what situations and under what conditions it is appropriate to use it (conditional knowledge). This difference in the use of procedures is what enables a student to arrive at autonomous learning (Monereo, Clariana, Castelló, Palma & Pérez Cabaní, 1994; Pérez Cabaní, 1995; Monereo & Pérez Cabaní, 1996; Pérez Cabaní, 1997; Pérez Cabaní and Juandó, 2001).

The assistance offered by the teacher in this process can take many forms, as no specific methodology is prescribed. It is only necessary to follow two basic principles of the constructivist perspective: to offer the student the most suitable help at each moment and to pass control of the learning process progressively from the teacher to the student.

As is shown in Figure 3 and Table 1 this handing over of responsibility until the students reach a state of autonomy can be achieved by using teaching methods that require them to take more and more decisions.

In our research this progression towards autonomy is of special interest in the two cases we present. In the first we worked only with direct instruction in one group and the progression model in the other group. In the second case, we applied the progression model directly.

Analysing a little more thoroughly the role of teachers in the process of handing control over learning to students, the figures above represent a possible progression towards autonomy. In the first instance the decisions fall basically to the teacher. It will be very important for the teacher to reflect and assess whether the decision-making process followed is sufficiently explicit and whether the students can use it as a model. The role of the teacher in the intermediate phases, in which the methodology used is focused questioning, col-

Table 1. Example of progression towards autonomy (Pérez Cabaní, 2007)

	Direct instruction	Metacognitive modelling	Focussed questioning	Collaborative working	Analysis & discussion	Self-questioning
Activity 1	■	■				
Activity 2		■	■	■		
Activity 3	■		■	■		
Activity 4		■		■	■	
Activity 5			■		■	
Activity 6				■		■
Activity 7			■		■	■
Activity...				■	■	■

laborative working analysis and metacognitive discussion, consists of encouraging students to describe their decision-making process explicitly through different kinds of intervention, such as formulating questions, contrasting students' decisions, recalling the aim of the activity and the characteristics of the procedures chosen, etc. At the end of the process, the teacher's function will be to assess whether the students' autonomous decision-making is leading them to the proposed learning outcome. Thus, the teacher's role at each moment acquires a different nuance, and handing control over to the students does not mean his or her role is any less relevant (Pérez Cabaní and Juandó Bosch, 2001; Pérez Cabaní, 2007).

In this conceptual framework a final element of the third section to take into consideration is the concept of quality of learning. The consideration of qualitative differences in learning is necessarily linked to the concept of learning that is taken as a reference. Thus, if learning outcomes are considered, quality can be defined as the achievement of goals and objectives; if the focus is more on the process, quality can be considered as coherence, or as the availability of resources. More up to date and in tune with the educational needs of the 21st century is the consideration of quality of learning in terms of transformation, change and improvement, on the basis of which it is considered that the essence of education is to teach and train students to think and act autonomously, independently and articulately. In other words, to be critical and decisive about the scientific content they have to learn, about ways of analysing and resolving different kinds of problems and situations, and to be able to eventually transfer the knowledge thus acquired to other academic and professional situations (Coll & Edwards, 1996; Salomon, 1998; Entwistle, 2000).

This new consideration implicitly embraces a constructivist conception of the teaching and learning process, but within the broad framework of this conception, different nuances can also be appreciated, depending on how the *interactive process* required by construction is emphasised, or how emphasis is placed on the study of *the student's activity*, or, in the words of Salomon (1998), according to whether attention is focused on the construction process or on the object being constructed.

From the first of these viewpoints – which focuses on the object of study in the interactive process required by construction – analysing the quality of learning will require studying the development of the instructional process of an activity in a specific scenario. From this perspective, which has been widely studied and compiled in various publications (see Mercer and Coll, 1994), the focus would be on the activity of teaching and learning in which people take part. Thus, the quality

Figure 4. Approaches to learning and quality of learning (Pérez Cabaní, 2001)

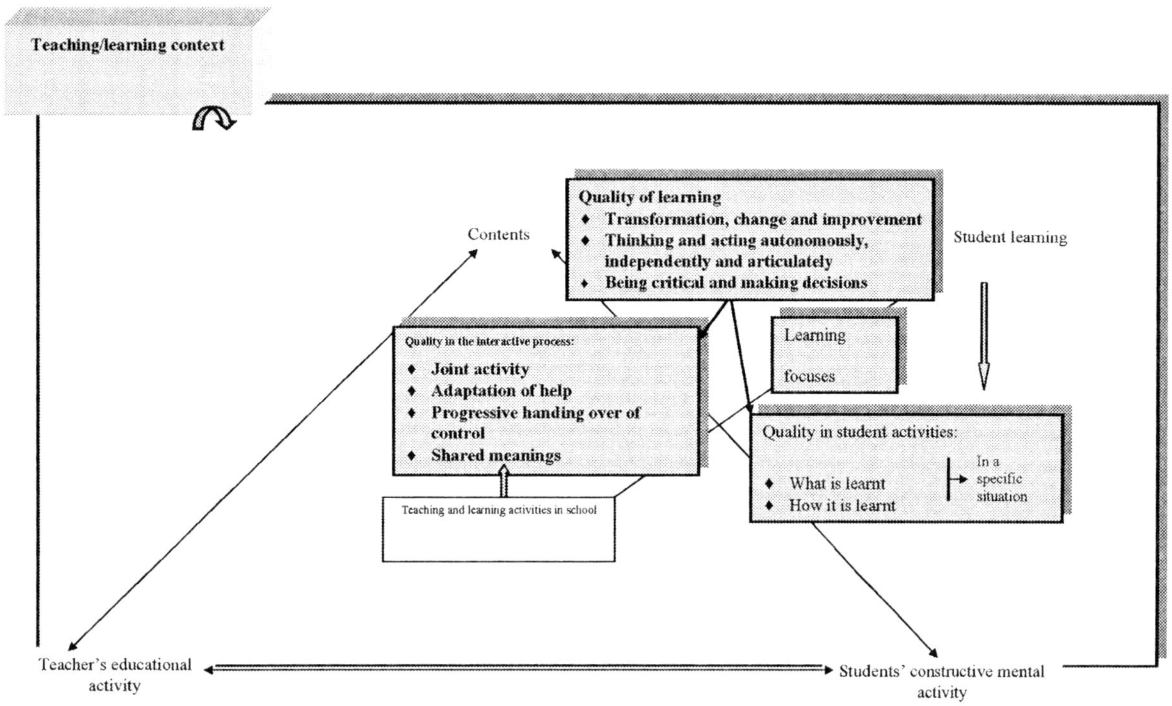

of learning would depend on the characteristics of the interactive process, in other words, on the performance of the joint activity, on the type of help offered in the construction process, on the adaptation of the help to the construction process, on the progressive handing over of control to the student, and on how, through the interaction of all the intervening variables, shared meaning is constructed in the classroom. From this point of view, the quality and focus of learning would be established jointly in the performance of the activities (Pérez Cabaní, 2001).

On the other hand, the second perspective mentioned above advocates studying the quality of learning from the standpoint of the *student's activity*, which will be conditioned by the adopted focus of learning. This is the perspective adopted by research on learning from the student's perspective (Entwistle, 2000). Quality of learning, from this viewpoint, requires attention be centred on *what the students learn* (to relate concepts, to solve problems, to apply knowledge in a specific situation, to remember facts…), *how they learn it* (what learning procedures they know, and how and when to use them), and *why* they learn something (what their intentions and aims are) in a *specific teaching and learning situation*. From this second position, the focus of learning is considered as a variable of the process that will affect the quality of the outcomes obtained at the end of the process. (see Figure 4)

Placing oneself in one position or the other can, on some occasions, respond to different constructivist epistemologies and, in other cases, to the need to focus the analysis, as we mentioned earlier, on a specific aspect of the construction process. In this respect, we find it productive to consider Salomon's proposal (1998) that it is necessary to emphasise a conception of knowledge as a social construction in a twofold sense, including both the construction process and the object the student comes to know. To put it another way,

Table 2. Difference in the use of concept maps (Pérez Cabaní, 1996)

Non-regulative use of concept maps	Regulative use of concept maps
Before starting the activity	
Initiating the concept map on the basis of the materials available on the content and of an external demand.	Analysing what is known and not known about the contents on which the concept map will be created, the quality of this prior knowledge, the difficulties or limitations that its study will entail, the personal interests regarding the study and the theme, and the goals sought. Assessing the personal resources available and the suitability, in comparison with other procedures, of creating a concept map to carry out the activity correctly.
Initiating the concept map on the basis of the external demand without analysing the characteristics of the task to be carried out.	Prior analysis of the goals of the task and the characteristics of the content on which the concept map will be created. Studying the level of difficulty of the content: structure, terminology, complexity, etc.
Initiating the concept map without assessing the influence of the context on the performance of the task and the possibility of modifying it.	Assessing the influence that the variables of the context will have on the performance of the activity. Determining actions to be undertaken when the influence can be negative.

he says that it would be just as unsatisfactory to study a learning context taking into account only the results that emerge from shared understanding and the performance of the joint work as it would be to study the process of social participation without taking into account the characteristics of the learning to which this participation leads. This would be a good area for further research and a perspective from which to deepen our knowledge of approaches to learning in the near future.

RESEARCH CARRIED OUT: THE REGULATIVE USE OF CONCEPT MAPPING

As is demonstrated in the theoretical frame of reference, the interest of research in this field is turning more and more towards the analysis of the construction of knowledge, and concept mapping is an effective educational instrument for developing the skill of learning to learn, particularly in interactive situations in which students can exchange, discuss and share their knowledge on a certain topic, among themselves and with the teacher.

In accordance with the approach adopted by Novak and his collaborators at Cornell University, concept mapping constitutes in itself a procedure that facilitates meaningful learning. However, taking into consideration that research carried out in the field of learning strategies demonstrates that the regulative use of learning procedures facilitates in-depth, meaningful learning to a greater measure than their non-regulative use, our approach argues that students who learn to use concept maps in a regulative manner will attain more in-depth, meaningful learning than students who use them without this condition.

The difference of this use of concept maps is represented in the Table 2.

On the basis of the approach set out in Table 2, the objective of this research is to study how the quality of university students' learning is affected by receiving instruction in concept mapping. The research studies the different ways in which the students can use concept mapping and its influence on their learning processes and the quality of their learning.

The empirical research studies four hypotheses: (1) the teaching and use of concept mapping will increase in-depth, meaningful learning more than not using it; (2) the teaching of the regulative use of concept mapping (planning, monitoring and evaluating) will increase in-depth, meaningful learning more than not teaching it; (3) the regulative use of concept mapping will increase the regulative use of procedures for learning in other situations more than not using it in a regulative manner; and (4) the regulative use of concept mapping will increase the transfer of this procedure to other activities more than not using it in a regulative manner.

A quasi-experimental method is designed with university students (educational psychology, teaching students, in the second year the course 1993) in three groups – control, experimental 1 (teaching in the use of concept mapping) and experimental 2 (teaching in the regulative use of concept mapping) – and in four stages – initial evaluation, intervention, final evaluation and deferred evaluation (two months after the end of the intervention).

In the first phase, an assessment was made of the initial situation of the groups and their homogeneity in relationship to their knowledge of concept mapping, the results of learning and regulation in different teaching-learning situations, the performance of class tasks, autonomous learning activities and evaluation activities.

In the second phase, the goal of the action performed was to assess the influence of the differential teaching of concept mapping in the learning of the subject content. This action was based on a training plan carried out on a weekly basis during an eight-week period. The differential characteristics of the training plan in each group were as follows: one group was trained in the use of concept maps, while the second group was also trained in the regulative use of those maps. The third group served as control: concept maps were not used in the teaching-learning process, but all the other methods were the same.

During the third phase, the variations produced by the actions taken within each group and between them were evaluated. The level of regulation was also evaluated in the three situations established: the performance of class tasks, autonomous study and evaluation activities.

The fourth and final phase was conducted two months after the actions were taken, in order to evaluate the transfer of the use of concept mapping to other teaching-learning activities of the same subject or to other subjects of the course.

In relation to the instruments, as our interest was centred on assessing teaching-learning of the differential use of concept mapping in class groups of more than thirty students, instruments were selected to make it possible to obtain quantifiable data on the evolution of each group at the different moments of the evaluation and on the differences between the groups at the same moment.

In addition, some variables refer to the learning process and the students had to be able to report on their own decisions. Others refer to the learning outcome and could be measured via the performance of a task or the replies to questions referring to the content learned. Those that refer to the transfer of some procedures to other situations were measured via the analysis of the resulting materials.

More specifically, the instruments used were:

- Self-assessments before, during and after the creation of a concept map in different learning situations (class tasks in small groups, autonomous study and evaluation activities). This tool was constructed with an open format so the students could report on all the questions they considered relevant concerning the process they were following in performing the activity and taking decisions. To analyse the content of the self-assessments, the information was grouped into categories and units of analysis that were previously defined on the basis of indicators of monitoring and regulation of learning (Zimmerman, 1990).

- SOLO taxonomy: an instrument for measuring the quality of the learning outcome (Biggs and Collis, 1982). The highest levels of the SOLO taxonomy correspond to students with more profound learning, who give a personal interpretation to the content, who establish relationships with other relevant areas of knowledge and with materials from other sources. In addition, it is an instrument that has been used in different areas of knowledge and has shown its utility in evaluating the quality of learning through the performance of different tasks (Entwistle, 1987; Biggs, 1987; Pérez Cabaní, 2001). For the use of the SOLO taxonomy in our research, the following considerations were taken into account: formulating evaluation questions in an open format, enabling students to structure the response with the desired breadth, making the request explicit and not formulating questions that could be answered with a simple "yes" or "no." The questions posed always referred to conceptual content, although some responses could manifest the student's own attitude to an issue. The correction process was conducted by two expert teachers, who previously agreed on the evaluation criteria and subsequently reached a consensus on the evaluations made.
- Questionnaires were also used to gather information on transfer in the use of concept maps.

Results Achieved and Discussion

We would like to emphasise the following results obtained.

Concerning the teaching of learning procedures, the following points are confirmed:

- The importance of guidance and mediation by the teacher in the construction of knowledge about learning procedures.
- The need to teach the regulative use of learning procedures in the curriculum in order to increase in-depth meaningful learning.

Concerning the regulative use of learning procedures, the following points are confirmed:

- The relationship between learning processes and the quality of the learning outcome.
- The regulative use of concept mapping increases the regulative use of other learning procedures (including schemes, synoptic graphics and abstracts).

Concerning the use and transfer of concept mapping, the following points are confirmed:

- The utility of concept mapping to increase students' learning means that it is useful to teach it.
- Finally, this research observed, in the two experimental groups, the maintenance and transfer of the use of concept mapping in different situations, but there is no significant data between them on the level and nature of this transfer.

Making a more detailed analysis of the results obtained, we would highlight, in relation to the differential use of concept mapping, that the teaching of concept mapping is a necessary but insufficient condition for learning its regulative use. This is learned through a training process in which emphasis is placed on conditional knowledge (Monereo, Clariana, Castelló, Palma, and Pérez Cabaní, 1994).

As for the quality of the learning outcome, although the total of the population studied shows a progressive improvement during the four phases under evaluation, there is variation in each of the groups involved. The higher level achieved by the group that has been taught regulative use, in both the final and deferred evaluations, indicates that the effect of the regulative use of concept maps on the quality of learning is not limited, in this case, to the period of the action but continues to progress thereafter.

It has also been observed that the regulative use of concept mapping favours to a greater degree the regulation of learning procedures in other learning situations (in the same subject, other activities, and in different subjects, other procedures and contents). The verification of this hypothesis lends support to the idea that a training plan in which the direct action of the teacher is gradually reduced, progressively increasing the students' own internal monitoring of their performance, has given them the possibility of taking decisions on when and where it is appropriate to use concept mapping, how to adapt its use to new situations and how to take decisions on the use of other learning procedures in different situations (Pérez Cabaní and Juandó, 2001).

The data also confirm the relationship between the regulative use of learning procedures and the quality of the learning outcome. This relationship between process and learning outcome underscores the need to plan training that allows students to learn the regulative use of learning procedures and enables teachers to carry out continuous evaluation and observe the evolution of the students' learning and thereby foster their improvement with a view to achieving the best possible result on completion of the process.

One final point to be highlighted is the observation that a large number of students who have learned to use concept maps will continue to use them in similar situations. Various authors (among them, Pressley, Wolosshyn, Lysynchuk, Martin, Wood, & Willoughby, 1990) agree in saying that the teaching and learning of learning procedures is meaningful as long as the learners know how to use them in similar situations or even in different situations in which the procedure facilitates the students' learning. Although we agree with this statement, we also believe that it is necessary to determine the factors that come into play for this transfer to occur. We need to do research that delves more deeply into these factors.

THE INNOVATION PROJECT

The possible uses of concept maps demonstrated by this research have been applied in an innovation project that was conducted as part of a university subject for three courses carried out in 2002-03, 2003-04 and 2004-05.

In the next pages this innovation project will be explained to show the knowledge building process added to the collaborative learning approach and the fundamental role that concept maps play in one of the phases of the teaching-learning process in the chosen subject.

The subject "Collaborative Working Strategies: Professional Teams and the School Environment," taught in teacher training courses at the Faculty of Education and Psychology since the academic year 2003-2004, is one of the subjects in which concept mapping has been applied in its most instrumental facet. This is a subject whose contents focus on the idea of teamwork to such an extent that the teacher and students have, in practice, changed its name to "Treball en equip" ("Teamwork"), and have even abbreviated this name to "TE" ("TW"). Consequently, in this text we will refer to the subject "Collaborative Working Strategies: Professional Teams and the School Environment" with the initials "TW."

In order to better understand the role of concept maps in TW, it is advisable to have an idea of what its aims are. As in all other subjects of the University of Girona, TW is oriented towards students' acquisition of certain skills.

The skills proposed by TW are the following:

- Effective participation in open debates.
- Effective participation in small-team tasks, boosting the potentialities of the group.

- Self-analysis of the evolution of students' own procedures and attitudes and evaluation of improvement.

A study of the design of the subject TW (programme) indicates that the working contents are:

- Definition, characteristics and conditions of teamwork.
- Evolution of the concept of working in teams. Its implantation in the different professional fields and the effects sought. Different models of teamwork.
- The characteristics of the present educational system that make teamwork necessary. What kind of teamwork? Teamwork for students and teamwork for teachers.
- Characteristics of the teamwork needed in schools today. The attitudinal work this signifies.
- Leadership and its styles and effects.
- Reflection and self-analysis as a source of knowledge and improvement.
- Participation as a tool of joint construction.

In addition, the learning activities proposed to students are:

- Reflection on the learning acquired, individually and as part of a team, using scripts supplied by the teacher.
- Attendance at explanatory sessions by external professionals with experience in teamwork.
- Participation in talks following the teacher's presentation.
- Collective tutorial sessions for monitoring and evaluation.
- Participation in tasks by teams of 4 or 5 members, debating conceptual contents based on lectures.
- Participation in tasks by teams of 4 or 5 members to carry out a common project related to school education.
- Individual tutorial.
- Participation in tasks by teams of 4 or 5 members, debating current affairs issues.
- Team-work in role-play situations presented by the teacher.

What we have here, then, is a very procedural subject. The teacher proposes that, through participation in 25 two and a half hour classes, the students identify their teamwork skills and handicaps and work to improve those skills and overcome those handicaps. This is achieved by means of a tutored process of reflection which culminates in the formulation, in the context of TW, of a series of personal goals for each student based on the facilities and difficulties identified.

What Role Does Concept Mapping Play in All This?

The goal of TW is to foster teamwork skills. In the belief that teamwork is learned by working in a team, teamwork objectives must be sought. We cannot work on nothing, just as we cannot multiply without numbers or swim without water. One of the working objectives of TW is the concept of teamwork itself. Consequently, we dedicate each of the sessions to working, in a team, on the concept of teamwork. We form teams of between 3 and 5 students. The sequence of work is as follows:

1. Work is done on selected texts on teamwork by various authors from different fields of knowledge: education, social psychology, business management, etc.
2. Each person of each team reads the text of the day in search of key ideas on teamwork. As the text is always short, the reading is quick.
3. The teamwork begins: the group must agree on the most significant ideas in the

Figure 5. Portion of a concept map constructed in 2007-2008 edition of TW

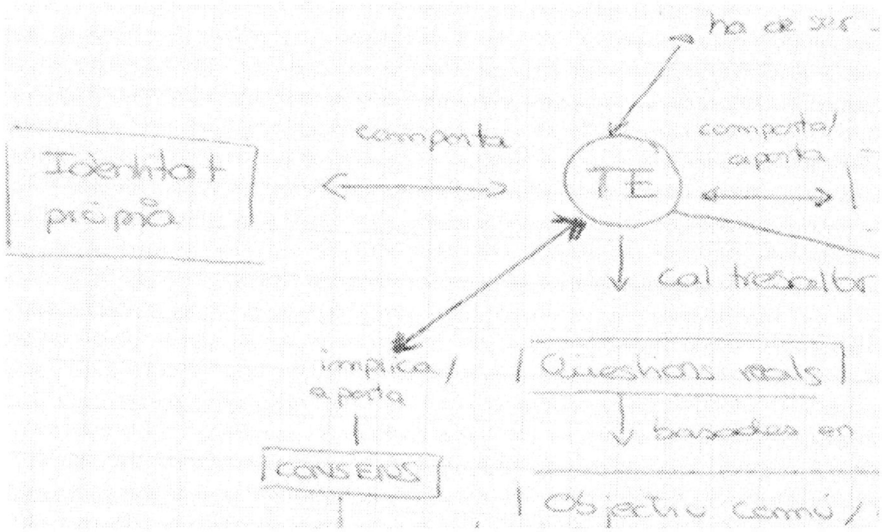

text in relation to the central concept of teamwork.

4. Once the list of ideas has been compiled with a common criterion, a concept map is constructed with these ideas.

Each working session deals with a new text. New text -> more key ideas (sometimes only variations on previous ideas) -> growth of the concept map. It is not a question, then, of constructing a concept map of each text studied, but of creating a map that grows and is sometimes modified as new readings are made. A newly-emerged concept can easily suggest a change of structure of the map. The debate continually established around the construction of the map is a rich discussion that questions preconceptions and sometimes the interpretations made of the texts, allowing conceptions to be shared, contrasted and agreed upon. What it certainly does is lay the foundation for a continuous sharing of knowledge and of points of view while contributing to collaborative work.

Periodically throughout the semester, the members of each team explain to the rest of the group the state of their map and the rationale behind the structure they are giving to it, describing the problems they have encountered and the discussions of different concepts, and expounding on their discrepancies and the way they have reached the final consensus.

Naturally, the wealth of the debate depends on many factors, as does everything in the field of education, and therefore each of the TW teams establishes its own level of depth and draws its own particular benefit from the working process.

Another of the elements that define the TW sessions is the fact that one member of each group acts as an observer of the dynamics of the session. Each session is followed by a meeting of observers, whose aim is to draw conclusions from the working process.

The process of creating a concept map in the context of the TW subject is, therefore, a process that generates discussion and joint construction of knowledge. In this way, when the various teams show their maps to the other teams there is a very interesting and appreciably deep exchange of ideas.

As an example, we present in Figure 5 a section of one of the concept maps constructed in the 2007-2008 edition of TW.

The authors of the map, when presenting it to their classmates, commented on the debate they had when trying to establish the relationship between the concepts "team-working" and "consensus". Finally, the term they chose was "implies" indicating that teamwork "implies" or "leads" to "consensus".

Following this stage, and still during the public presentation of the map, the presenters and the listeners discuss the causal relationship, that is, whether teamwork generates consensus, or consensus, by its very existence, generates teamwork. The discussion continues about the origin of teamwork, the conditions it requires, the circumstances that favour it, and so on.

There is no doubt about the participants' interest in the session and the opportunity it has created for the in-depth definition of concepts. It also clearly manifests the value of concept maps and of the discussion they generate with a view to fostering participation, which is, as we said earlier, one of the basic goals of the subject of TW.

Finally, we want to highlight how the possibility of gathering students' comments on the regulative use of concept maps has also provided valuable information for regulating our teaching task. Here are some of the remarks we have gathered:

"I don't think I know enough of what the text is about yet, so I'll have to relate it with other things I know."

"I'm intrigued to know what the text tells me, how it tells it..."

"Now I've calmed down and I understand what it means..."

"I can see that what I've written is what I really wanted to express."

"A year ago, in a situation like this, I would've limited myself to (...): now I think I've reflected what I wanted to say."

This has enabled us to discover the perception the students have of the activities proposed and has led us to consider, on one hand, new actions regarding the planning of subjects, and on the other hand to reaffirm, in some cases, the suitability of the decisions taken.

CONCLUSIONS ON BOTH PROJECTS AND FUTURE TRENDS

In the research and innovation projects presented in these pages, we have been able to observe the importance of the role of the teacher as a guide in the handling and strategic use of concept mapping and how this process improves the quality of learning. From this it follows that, in planning the teaching process, the teacher has to take into account the skills the students need to acquire through the learning and evaluation activities they perform to learn a series of curricular contents.

This strategic use of skill learning procedures is not learned spontaneously. It must be taught, and the teaching, to be effective, must fulfil certain conditions, such as introducing the skills progressively, over a sufficient period of time, in a transversal manner, through collaborative learning and in close relation with the curricular contents, favouring dialogue and discussion of decision-making and how to use the procedures effectively in other learning situations.

The projects we present have been conducted with the aim of applying these principles, postulated in the sphere of research, to real educational situations. In the first case, the training plan carried out has allowed us to observe the differences produced in the students' learning by teaching the mere use of a procedure (concept mapping) or its regulative use, and we reach the conclusion that merely teaching a procedure does not allow students to appreciate the reasons why that procedure is beneficial. In addition to declarative and procedural knowledge of learning procedures, what is needed is conditional knowledge that enables students to decide when and why to use them.

These considerations reaffirm the responsibility we as teacher trainers have in training our students as learners and as strategic teachers. If training in regulative use is to commence in the early years of schooling (work on strategies and maps for infant education), those responsible for guiding this training, the teachers, will have to be sufficiently skilled to do so.

We would also like to conclude with a few comments on the learning procedures taught here: concept mapping and collaborative learning. As we have said, using these two strategies has improved the quality of learning and its application to different learning situations. We have also been able to observe, in consultations with the students, that they appreciate the utility of concept maps and collaborative learning for a better understanding of the basic terms of subjects.

In addition, the use of concept mapping in an interdisciplinary manner can help to avoid the compartmentalisation of knowledge and enable students to establish relationships between the contents of different subjects. On the other hand, joint activity and the progressive handing over of control involved in collaborative learning, help students to share meanings and achieve greater quality of learning.

Finally, we also want to express our support for Novak's proposal to use concept mapping as an instrument of evaluation and self-assessment, to evaluate prior knowledge and misconceptions, to analyse the changes that occur in meaning when faced with new experiences, to evaluate new relationships that are formed on the basis of new information and to assess changes in students' knowledge and how those changes have been constructed.

REFERENCES

Biggs, J. B. (1987). *Students Approaches to Learning and Studying*. Melbourne: Australian Council for Educational Research.

Biggs, J. B., & Collis, K. F. (1982). *Evaluating the quality of learning. The SOLO Taxonomy (Structure of the Observed Learning Outcome)*. New York: Academic Press.

Biott, C., & Nias, J. (2000). *Working and learning together for change*. Buckingham, UK: Open University Press.

Coll, C. (1990). *Aprendizaje escolar y construcción de conocimiento*. Barcelona: Paidós.

Coll, C., & Edwards, D. (Eds.). (1996). *Enseñanza, aprendizaje y discurso en el aula. Aproximaciones al discurso educacional*. Madrid: Fundación Infancia y Aprendizaje.

Entwistle, N. J. (2000). Approaches to studying and levels of understanding: The influences of teaching and assessment. In J.C. Smart (Ed.), *Higher Education: Handbook of Theory and Research*: *Volume XV*. (pp. 156-218). New York: Agathon Press.

Escudero, J. M. (2001). La mejora de la educación como marco de referencia para el asesoramiento pedagógico. In J. Domingo Segovia (Coord.), *Asesoramiento al centro educativo. Colaboración y cambio en la institución*. (pp.15-50) Barcelona: Octaedro.

Fullan, M. (2002). *Los nuevos significados del cambio en educación*. Madrid: Octaedro.

Guban, L. (1992). Curriculum stability and change. In P. Jackson (Ed.), *Handbook of research on curriculum* (pp. 216-247). New York: Macmillan.

Järvelä, S., & Järvenoja, H. (2007). *Socially constructed self-regulated learning in collaborative learning groups*. Paper presented at invited Symposium organised by Järvela, S. & Hadwin, A. at Earli Conference, Budapest, Hungary.

Malone, J., & Dekkers, J. (1984). The concept map as an aid to instruction in science and mathematics. *School Science and Mathematics, 84*(3), 221–231.

Mercer, N., & Coll, C. (Eds.). (1994). Teaching, learning and interaction. In P. Del Rio, A. Álvarez & J.V. Wertsch (General Eds.) *Explorations in Socio Cultural Studies* (Vol. 3). Madrid: Fundación Infancia y Aprendizaje.

Michavila, F. (2003, February). Cómo y en qué se deben formar los profesores universitarios. Communication presented at the *Congreso de la Red Estatal de Docencia Universitaria: Hacia una docencia de calidad.* Castellón, Spain: Universitat Jaume I.

Monereo, C., Clariana, M., Castelló, M., Palma, M., & Pérez Cabaní, M. L. (1994). *Estrategias de enseñanza y aprendizaje. Formación del profesorado y aplicación en la escuela.* Barcelona: Graó.

Monereo, C., & Pérez Cabaní, M. L. (1996). La incidencia de la toma de apuntes sobre el aprendizaje significativo. Un estudio en enseñanza superior. *Infancia y Aprendizaje, 19*(1), 65–86. doi:10.1174/02103709660560555

Monereo, C., & Pozo, J. I. (2003). *La universidad ante la nueva cultura educativa. Enseñar y aprender para la autonomía.* Madrid: Síntesis.

Novak, J. D. (1985). Metalearning and Metaknowledge strategies to help students learn. In L.T. West, & A.L. Pines (Eds.) *Cognitive Structure and Conceptual Change* (pp. 189-209) New York: Academic Press.

Novak, J. D., Gowin, B., & Johansen, G. T. (1983). The use of concept mapping and knowledge Vee mapping with junior high school science students. *Science Education, 67*(5), 625–645. doi:10.1002/sce.3730670511

Novak, J. D., & Symington, D. (1982). Concept mapping: a useful tool for science education. *Journal of Research in Science Teaching, 27*(10), 937–949.

Parrilla, A., & Gallego, C. (1999). Cómo pueden los profesores ayudarse entre si: diseño del trabajo comparativo. *Revista de Educación Especial, 25* (55-56).

Pérez Cabaní, M. L. (1995). Los mapas conceptuales. *Cuadernos de Pedagogía, 237*, 16–21.

Pérez Cabaní, M. L. (1996). Els mapes conceptuals. Anàlisi de les condicions de la seva utilització. *Perspectiva Escolar, 206*, 11–19.

Pérez Cabaní, M. L. (Coord.) (1997). *La enseñanza y el aprendizaje de estrategias desde el curriculum.* Barcelona: Horsori.

Pérez Cabaní, M. L. (2001). El aprendizaje escolar desde el punto de vista del alumno: los enfoques de aprendizaje. In C. Coll, J. Palacios, & A. Marchesi (Coord.) *Desarrollo psicológico y educación 2. Psicología de la educación escolar, (*pp. 285-308*).* Madrid: Alianza Editorial.

Pérez Cabaní, M. L. (2007). Aprendiendo a ser profesionales reflexivos: Aprender para enseñar a aprender. In *Enseñar a pensar: Sentando las bases para aprender a lo largo de la vida* (pp.113-135). Madrid: Ministerio de Educación y Ciencia.

Perez Cabaní, M. L., & Carretero, R. (2003). La promoción de estudiantes estratégicos a través del proceso de evaluación que proponen los profesores universitarios. In C. Monereo, & J.I. Pozo, (Eds.) *La universidad ante la nueva cultura educativa. Enseñar y aprender para la autonomía* (pp. 173-190). Madrid: Síntesis.

Pérez Cabaní, M. L., & Juandó Bosch, J. (2001). *Gestionar la información para que se comprenda*. Barcelona: Edebé.

Pressley, M., Wolosshyn, V., Lysynchuk, M. L., Martin, V., Wood, E., & Willoughby, T. (1990). A primer of research on cognitive strategy instruction: The important issues and how to address them. *Educational Psychology Review*, *2*(1), 1–58. doi:10.1007/BF01323528

Ribas, M. (2000). *Innovación educativa. Teoría, procesos y estrategias*. Madrid: Síntesis.

Salomon, G. (1998). Novel constructivist learning environments and novel technologies: Some issues to be concerned with. *Research Dialogue in Learning and Instruction*, *8*(1), 3–12. doi:10.1016/S0959-4752(98)00007-3

Scott, J., & Flanigan, E. (1996). *Achieving Consensus: Tools and Technique*. Menlo Park: Crisp Publications.

Zimmerman, B. J. (1990). Self-regulated learning and academic achievement: An overview. *Educational Psychologist*, *25*(1), 3–18. doi:10.1207/s15326985ep2501_2

KEY TERMS AND DEFINITIONS

Approaches to Learning: The learning approach refers to learning from the perspective of the student. Research developed over the last twenty years, and the controversies caused at particular moments, have led to current conception. The most recent work (see Pérez Cabaní, 2001) defines learning approaches as the purposes that guide student activity in a complex process, which includes at the same time consistency (typical actions that guide the way students act in similar tasks and contexts) and variability (as a consequence of the great influence of the specific characteristics of each situation). This concept is closely related to the learning strategy.

Assessment: Learning results cannot be converted directly into learning achieved; rather, they must be analyzed in terms of whether skills have been achieved. The temporal dimension of the assessment snapshots of different moments and its progression have to be taken into account, forming different types of demands that provide complementary information. Students, through the assessment process, must be able to transform the information into knowledge, that is, endow it with meaning in order to understand it.

Improvement: The changes proposed as part of the adaptation of European university studies to the Bologna Process imply not only a modification of the structure or duration of the studies but also a change in methodology, in culture and in the meaning of teaching and learning. The processes of change, in which the entire university community and the society of which it is a part have to participate, must be well planned in order to achieve improved higher education, with better quality and responding to the challenges faced by today's society.

Learning Strategies: The strategic concept of learning has been defined based on different theoretical models, with emphasis placed on certain of its characteristics, according to the adopted focus. A basic characteristic, widely accepted by the scientific community in current lines of research, is that in order for a learning procedure to really be considered as a strategy, it has to be used in a conscious and intentional way to achieve an objective in a specific situation under particular conditions. The strategic use of the procedures (as well as of the conceptual maps) means activating declarative knowledge (knowing what), procedural knowledge (knowing how) and conditional knowledge (knowing when and why).

Quality of Learning: This refers to the knowledge generated as a process of social construction in two ways: the construction process, by means of shared understanding and working together, and the results that emerge from the process.

Students Learning: From a constructivist conception, the activity of the student is a basic basis of learning. The Bologna Process specifies some important aspects of this activity: the European Credit Transfer System (ECTS) specifies the number of hours students must dedicate to all their different activities, with and without a teacher. Emphasis is also placed on the importance of student participation in the processes of change, on lifelong learning and on mobility and exchanges as fundamental to this process.

Teaching Methods: If we conceptualize teaching and learning as an interactive process, the teaching methods used will have a decisive influence on student learning. In order that students learn to learn, teachers must learn how to teach for them to learn. And if students cannot be replaced in the learning process, neither can teachers be substituted in their teaching function.

Chapter 17
Concept Maps and Conceptual Change in Physics

Angel Luis Pérez Rodríguez
University of Extremadura, Spain

Maria Isabel Suero López
University of Extremadura, Spain

Manuel Montanero-Fernández
University of Extremadura, Spain

Pedro J. Pardo Fernández
University of Extremadura, Spain

Manuel Montanero-Morán
University of Extremadura, Spain

ABSTRACT

The authors describe and discuss some recent applications of concept maps to physics teaching. They begin by reviewing the literature on applications of concept maps to science teaching, and argue for the usefulness of this resource in facilitating processes of conceptual change. They then describe two experiments on the collaborative use of concept maps to this end. The first was a study of how a team of teachers designed learning sequences using three-dimensional maps. In the second, concept maps were constructed and then collaboratively re-constructed by various groups of students. Finally, they discuss the preliminary results of these experiments on the processes of conceptual change, and suggest lines for further research.

CONCEPT MAPS AND CONCEPTUAL CHANGE: APPLICATIONS TO TEACHING AND LEARNING PHYSICS

From the mid 20th century onwards, researchers in various disciplines – philosophy, philology, artificial intelligence, and psychology – have developed systems for the graphical representation of declarative knowledge. The resulting diagrams consist of polygons and lines linking them to represent the underlying structure of the knowledge concerned. The polygons symbolize various kinds of nodes (objects, concepts, events, actions, etc.).

DOI: 10.4018/978-1-59904-992-2.ch017

The relationships represented by the links may be semantic (as in the classic associative networks of Quilliam), sequential (as in flowcharts), or functional (as in organigrams). A concept map is a type of diagram that represents semantic relationships between concepts in the form of a tree. The links connecting the concepts are labelled with words or phrases that indicate some property of one of the concepts. The overall diagram represents a hierarchical propositional structure of the body of knowledge.

Since their introduction by Novak in the 1960s, concept maps have grown into a powerful teaching tool with multiple applications because of their capacity to specify, design, and share knowledge. Their use has spread into many domains of learning. In science education in particular, several studies have demonstrated their usefulness as a strategy for evaluating, learning, and designing the teaching of scientific content.

Their primary use has been as a strategy for evaluation. Indeed, there is clear evidence for their effectiveness in assessing students' prior knowledge of scientific content and how it is organized (Anderson-Inman, Ditson, & Ditson, 1998; Caswell & Wendel, 1992), and the degree of understanding that students attain (Markham, Mintzes, & Jones, 1994; Novak, Gowin, & Johansen, 1983). The nature of the information that can be gathered in a map on students' learning depends, however, on the type of task asked of them. The more open procedures, such as creating a map of a single concept, provide very different information from more structured procedures, such as filling out an incomplete map or constructing one from a set list of concepts (see Ruiz-Primo, 2004).

Concept maps also constitute an interesting resource to support the learning processes of conceptual content, whether within a framework of autonomous learning activities or with the help of the teacher. Some studies, for example, have reported their usefulness for the students themselves to reconstruct strategically the knowledge they have acquired from various sources, and then to apply it to different learning tasks, i.e., to learn how to learn (Novak & Gowin, 1984). In a series of studies, Okebukola and Jegede have documented some of the main advantages of the use of concept maps as a strategy for learning science content, in particular, their positive influence on students' attitudes, strategic behaviour, and academic performance (Okebukola, 1990; Okebukola & Jegede, 1988, 1989). In this vein, McCrudden, Schraw, Lehman and Poliquin (2007) found that subjects who studied a scientific text accompanied by a diagram representing the semantic organization of the content understood it better than another group that spent the same amount of time in studying the text alone.

Concept maps have also been used as a teaching resource in contexts of joint teacher-student activities and to support students' collaborative learning at different stages of the teaching-learning process. In science education in particular, their utility has been studied as pre-organizers, i.e., as a means of presenting an initial overview of the content and connecting it with the students' prior knowledge (Montanero & Montanero, 1995). By comparing various methods of teaching a topic of science at the pre-university secondary education level, Hernández and Serio (2004) showed that the students' preparation of a concept map is especially useful as pre-organizer, whether it is presented by the teacher or the students themselves construct it, as long as the teacher explicitly helps them connect it with what has just been learnt. Other studies have also demonstrated their usefulness as a form of synthesizing the content studied during or at the end of a learning sequence (Horton, Mcconney, Gallo, Woods, Senn, & Hamelin, 1993; Pankratius, 1990).

Finally, concept maps provide a useful framework for representing knowledge in a form that can be taught or shared in a variety of contexts – scientific, educational, or professional (see Cañas, Hill, Carff, Suri, Lott, Eskridge, et al., 2004). In the educational context, the maps help to "unbundle" a teacher's expert knowledge in the design of content and learning sequences. The process of analysis that concept mapping demands can help teachers reflect on what is the best way to organize the content to be learnt, what of their previous knowledge the students need to use, the most complex conceptual relationships, etc.

This last application has not as yet given rise to sufficient empirical research. The main purpose of the present chapter in this context is therefore to analyze the potential of concept maps as a tool for collaboration in learning physics. On the one hand, we shall analyze how concept maps can enhance teamwork when teachers are designing learning sequences for the sciences. And on the other, we shall describe some collaborative learning experiments involving students preparing and revising concept maps. In both applications, we will study the potential of concept maps to facilitate the processes of conceptual change in learning physics.

FUNDAMENTALS

Misconceptions and Conceptual Change in Physics

The spontaneous conceptions that students build from their everyday experience have an extraordinary influence on their learning of scientific content. The naive ideas that students show about physical phenomena are often "primitive" and "phenomenological" (Di Sessa, 1993). They are primitive because they are based on simple and intuitive explanations, and phenomenological because they derive from the experiences that the subjects obtain from their interaction with the physical world. These consist of causal, linear and highly simplified explanations that underpin knowledge that is useful in everyday life (Claxton, 1987). Phenomena are explained in terms of a single cause which generates the observed change in the physical reality. In the official theories of science, however, there prevails an analysis of reality as a "state", defined by the interaction of systems (Montanero, Pérez, & Suero, 1995).

Many studies concur in that misconceptions condition the acquisition of knowledge in specific domains. There is no consensus, however, on whether they constitute genuine implicit theories (Lautrey & Mazens, 2004). For Di Sessa (1993), it is a case of "knowledge in pieces" rather than a logically organized body of knowledge. Vosniadou (1992) argues, however, that spontaneous knowledge develops into spontaneous theories, at two levels of depth: "framework theories" and "specific theories". Framework theories are based on certain ontological and epistemological presuppositions that define a domain (such as the presupposition of the solidity or continuity of matter). Specific theories are constrained by some framework theory. They are based on beliefs that facilitate the construction of *mental models* of specific physical phenomena (such as a collision between two bodies).

It is more difficult to change framework theories than specific theories. According to the Incompatibility Hypothesis (Chi, Slotta & de Leeuw, 1994), this difficulty is largely explained by "the existence of a mismatch or incompatibility between the categorical representation that students bring to an instructional context, and the ontological category to which the science concept truly belongs". Many students misunderstand such concepts as light or electric current because they categorize them ontologically as *matter*, when they are rather *processes* or *states* (which belong to the category "constraint-based interaction").

We should now briefly describe the hierarchical construction of its theories so as to understand how the teacher can intervene in the process of students' conceptual change.

The first rung of the hierarchy consists of the clearly identified physical objects which the learner faces. This identification must be given explicitly even when changes may occur through their possible evolution. For example, a physical object (or system) might be a pendulum consisting of a thread suspended at one end and a sphere hanging from the other which can move from one side to the other. Another simple example would be a tin rod of a certain length and thickness which can be heated and change in length and appearance.

The second step in the construction of physical explanations is a mental schematic representation, formed by certain abstractions of that physical object. This is therefore a deliberate simplification made to facilitate the idea of it in the mind. Hence, some of the features and properties of the objects will be removed from consideration, and attention will be focused on those that are really of interest. This is what we would call the physical model. Thus, in the aforementioned physical system of the pendulum, the thread is assumed to be massless and inextensible, and one is only interested in the mass of the sphere hanging from it, not its size or shape. Similarly, in the example of the tin rod, only its length and mass are considered, and one is not interested in its hardness, flexibility, etc.

It is a proven fact that the first deficiencies observed in both the formal learning of physics and in the formation of implicit theories concern the lack of clarity and explicitness in this necessary set of factual concepts (Bunge, 1978). As noted above, there must be room in the concept maps incorporating the instruction sequences for the proposal of a *factual* conflict corresponding to these initial levels of elaboration.

The third stage of the formation of physical theories consists of assigning certain states to the physical model adopted previously. These states would be determined by clearly defined magnitudes. This is what is known as a theoretical model (or theoretical system) which, moreover, is specific to each type of physical phenomenon, unlike the physical model which is applicable to different physical systems (for example, the physical model consisting of isolated particles can belong to both the theoretical system of the kinetic of the perfect gases as any other piece of the statistical physics).

In the example that we have been using of the pendulum, the states will refer to the various positions of the sphere, also including knowledge of its velocities and accelerations. In other words, the definition of the states would imply the use from the very beginning of learning of a number of concepts (such as mass, velocity, acceleration, energy,...) which are contained in those concepts supported by the initial explanations, as will be seen below. They will therefore constitute simple hierarchies of learning (from the bottom up) to be taken into account as prerequisites of knowledge in support of the physical explanations. Similarly, in the case of the tin rod, the states would be determined by values of the length, temperature...

This form of in-depth analysis of the genesis of physical theories, originating in the philosophy and epistemology of science (López Rupérez, 1990), is unfortunately taken very little into account by teachers in dealing pædagogically with the inevitable analysis of the physical models (Meléndez, Pacual, & Rosado, 1989) that underlie any explanation of the phenomena. These aspects are of particular interest when one is dealing with implicit theories, in which these phases of the construction of scientific thought are never made explicit. In sum, as was mentioned above, one needs to deal with this "modeling" of the physical explanation for factual change to be converted into true conceptual change. Concept maps again prove to be a very useful tool for unpacking the analysis of this content, as we will have the opportunity to show below.

The last step that we shall consider is the establishment of a theoretical model. This will consist of the theoretical system seen previously, together with a set of statements (laws) obtained by induction from the general facts. In the pendulum example, this statement is the equation of motion, which, given the initial data of a particular case, determines all the states of motion. Likewise, the law of linear thermal expansion would provide the way to obtain the different lengths of the tin rod as a function of temperature, and statements of the laws of melting would enable one to know what the physical state of this physical object is.

Modifying the students' implicit theories in learning physics is therefore a complex task. It requires their strongly active participation in activities that lead them to an awareness of, and a subsequent change in, their misconceptions. An educational strategy that has proved effective in this regard is to design hands-on activities that lead to factual conflict. In other words, the student experiences that his or her implicit theory about a certain physical phenomenon leads to predictions that are not fulfilled. This psychological state of uncertainty and dissatisfaction facilitates their becoming aware of and revising their particular mental model (Posner, Strike, Hewson, & Gertzog, 1982).

This empirical conflict would, however, be sterile unless it were accompanied by a genuine conceptual conflict, i.e., unless the students become aware of their ideas, verbalize them, and contrast them against alternative versions (Hewson & Thorley, 1989; Pozo, 1999). This requires a process in which physics knowledge is abstracted and made explicit which would be most difficult to foster outside systematic instruction. Indeed, such conceptual change is based on the teacher's intense support, for the students to be able to transform their mental models with which they explain physical phenomena (Rodrigo & Correa, 1999).

Collaborative Learning and Conceptual Change

Collaborative learning is based on the idea that students influence each other in their learning when they exchange knowledge and negotiate its meaning (Baker, Hansen, Joiner & Traum, 1999; Barron, 2003). There is evidence of the effect of this type of classroom task in changing mistaken preconceptions in physics. Analyzing how several pairs of students discussed the relationship between physical concepts, Roth and Roychoudhury (1992) found that the more extensive were the explanations and justifications they recorded in the discussion, the greater was the likelihood of conceptual change .

But not only how extensive the students' exchanges are is important. Above all, it is important that they reach agreement in the group. Mercer (1996) examined different types of verbal exchanges typical of collaborative learning situations. The students were not really negotiating meanings in conversations that were merely "*cumulative talk*" or "*disputational talk*". In "*exploratory talk*", however, they were critically engaged in refuting their colleagues' ideas and in reaching consensus, processes that are more conducive to learning.

Other studies have found a relationship between the quality of these communicative exchanges and the nature of the learning task. Tasks in which the students share goals and working tools encourage "positive student interdependence" (Cohen, 1994). And tasks that require externalization of specific visible products (as would be the case of a concept map) provide a space for joint work that is more useful for explicit negotiation of the students' representations, thereby facilitating the processes of conceptual change (Roth & Roychoudhury, 1992). Another important factor is that the tasks should elicit the participants' prior knowledge on the topic, and should be oriented towards the construction of consensus, either through conflict or through integration of their contributions (Fischer, Bruhn, Gräsel & Mandl, 2002).

Concept Maps and Conceptual Change

In this context, we can argue for several reasons why concept maps can help facilitate conceptual change in physics teaching. Every conceptual change is based on a gradual re-development of the theoretical model that explains certain physical phenomena. At the beginning of learning, such knowledge only allows relatively simple and unidirectional explanations of the content of physics that is to be constructed. Subsequently, the students will be able to understand and use more complex and interdependent theoretical models. As an instructional design strategy, the use of concept maps can be an invaluable tool to guide the teacher's reflection on the different levels of development of these theoretical models – from the simplest causal representations to others that are more complex and abstract.

Concept maps can also lend support to activities aimed at provoking conceptual conflict. Analyzing tutoring activities involving science content, Chi (1996) found that help from the teacher aimed at eliciting self-explanations or revision of mental models is the most effective at promoting learning in depth and changes in erroneous preconceptions. When the students ask themselves about the why of conceptual relationships, they are setting out on a process of revising the new information which often comes into conflict with a prior causal model that had until then only been implicit. In another experiment, it was found that eliciting self-explanations had a positive impact on causal understanding, even when some of those explanations were wrong and never corrected by an expert (Chi, De Leeuw, Chiu, & Lavancher, 1994). Graphical representations of scientific explanations may represent an interesting resource in this regard. The task of constructing maps can lead the students to an awareness of their own implicit causal models of the physical phenomenon being dealt with, and to an active involvement in revising their own explanations.

In accordance with the studies discussed in the previous section, collaborative construction of concept maps would seem to be potentially ideal tasks with which to generate processes of conceptual change. In these tasks, the students share working tools and generate explicit representations of their thinking which they find easy to manipulate and negotiate.

In the particular case of scientific content, Okebukola showed that the collaborative construction of maps promotes critical thought and meaningful learning (Okebukola, 1990; Okebukola & Jegede, 1988). Van Boxtel, van der Linden & Kanselaar (2000) observed that the task of constructing a map of concepts about electricity generated more discussion in which conceptual arguments and conflicts were recorded than the preparation of a poster, even though the authors note that, in this case, the results of learning were not significantly better.

APPLICATIONS OF CONCEPT MAPS TO THE DESIGN OF ELABORATION-THEORY-BASED LEARNING SEQUENCES

A key problem that all teachers have to resolve is how to organize the content so as to facilitate its learning as far as is possible. Over the past few decades, two general procedures have been developed with which to establish teaching and learning sequences.

One is Task Analysis. This decomposes and orders the skills that students are expected to master by the end of the learning process (Gagnè, 1985). The result is a hierarchy ascending from the most basic learning to more complex strategies that require good prior mastery of the preceding skills.

The other is Content Analysis. This, in contrast, focuses on identifying and organizing knowledge, especially conceptual knowledge. In this case, it is postulated that the learning sequence must proceed overall from the more general to the more specific, alternating processes of analysis and of synthesis (Ausubel, Novak, & Hanesian, 1978). Concept maps fit very well into this approach, since they allow this conceptual hierarchy to be represented graphically.

The Elaboration Theory of Reigeluth and Stein (1983, 1987) takes an alternative approach to structuring learning, conceiving of that structure in the form of a spiral. It has been found particularly useful in physics teaching (Pérez, Suero, Montanero, & Montanero, 2001). In this approach, every learning sequence has to be based on two prior analyses: reflection on the fundamental content best suited to forming the scaffold on which the students' learning can be constructed, and an analysis of the different levels of elaboration.

Curricular design begins with reflection on the type of content (conceptual, procedural, etc.) to form the scaffold on which the learning sequence will be organized. The entire content will then be organized at different levels of elaboration. A level of elaboration reflects a point in the process of constructing the learning content that has some determined degree of complexity. Each of these levels should in their turn begin with a "panoramic vision" (or epitome) of the more general content that will subsequently be dealt with in detail. The epitome synthesizes those more general ideas at the same level, in a simplified and practical form, i.e., taking what is the most inclusive and then presenting it in the most concrete and factual fashion. The student identifies them as part of a structured whole, since the teacher's explanation

Figure 1. Educational components of elaboration theory (© 2008 Pérez, Suero, Montanero-Fernández, Pardo, & Montanero-Morán)

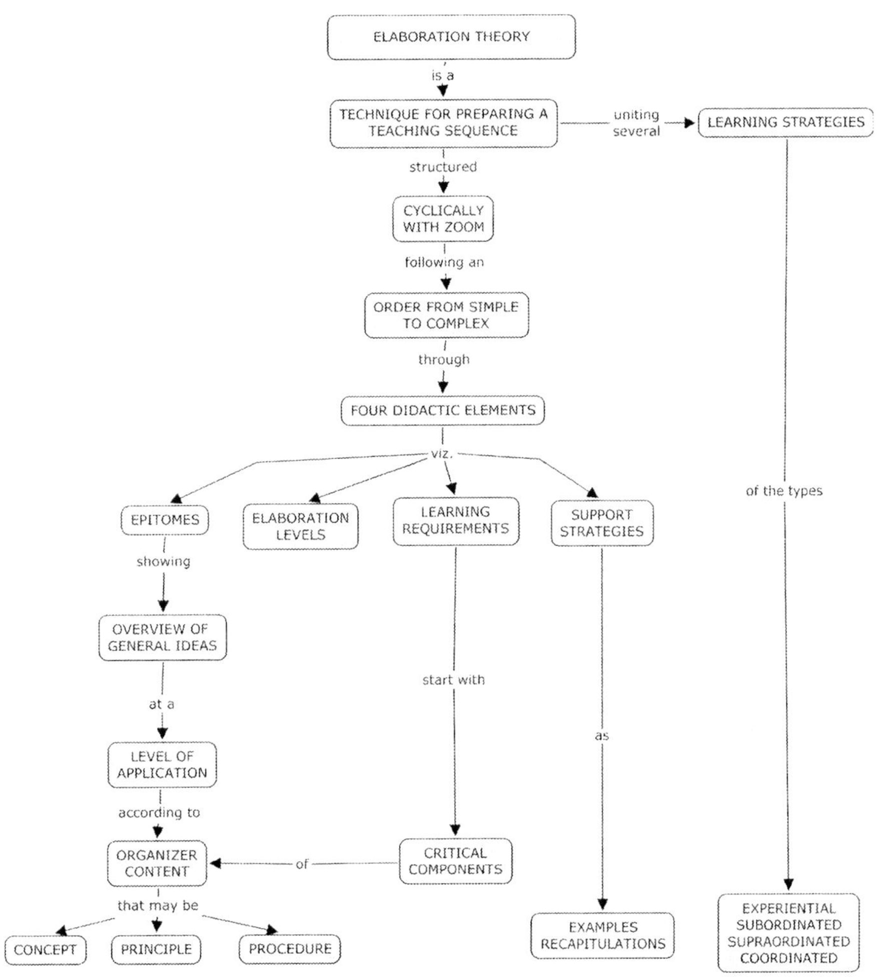

describes successive approximations to these ideas one at a time, but without exhausting them in the first presentation. This progressive "zooming" into the content therefore entails going back over the same content but in steadily greater depth.

Each epitome is in itself an item of teaching content, presented in a practical and hands-on way. Herein lies the greatest difficulty in the design process, because the student initially needs a working knowledge of the learning content to serve as the basis for further learning. At the completion of each phase of increasing depth ("elaboration"), one must insist on presenting its relationship with the overall general plan, which will thus be enriched and extended. This is the "expanded epitome". At the end of the successive phases, one will have a "final epitome" in which the basic content of the first level will now appear already reformulated, with the more complex and abstract conceptual relationships that were generated in the learning process. Figure 1 is a concept map summarizing the major educational components of elaboration theory.

Adaptation of Elaboration Theory to Physics Teaching

As we have shown elsewhere, elaboration theory offers interesting theoretical support for science teaching (Pérez et al., 2001). The importance and difficulty of the process of conceptual change in physics teaching, however, require some modifications to be made to the initial theory of Reigeluth and Stein. The most important innovation is to take physical phenomena (rather than concepts or principles) as being the organizer content in the learning sequences. To this end, a map of phenomena is elaborated on the basis of the ordered set of concepts and principles, as well as their relationships, about the content to be taught (often referred to as the *logical structure* of the content in its final stage of elaboration). This map will have to contain all the *phenomena* to be studied in a related and hierarchical form, as a *theoretical-descriptive* superordinated structure (Reigeluth, Merril, & Bunderson, 1978). We shall call it a three-dimensional expert map (Pérez, Suero, Montanero, & Montanero, 2000). Such maps define, therefore, the factual context of the content to be taught, so that the hierarchical structure of the physical phenomena provide us the leitmotif of the learning sequence, i.e., what before we called the organizer content.

This simple modification entails activities that are important for the students to gain awareness of the erroneous theories that they implicitly use to explain physical phenomena, and to be able to incorporate more complex theoretical models into their growing knowledge base. The observation of reality – in particular, of a variety of physical phenomena – is one of the foundations on which the process of generating scientific theories will be based, as we have described before. Often indeed, the implicit theories themselves were the result of this need being insufficiently satisfied in formal education. The observation, prediction, and discussion of the underlying processes in physical phenomena form the best scenario in which to foster conceptual change.

Let us consider at this point an example of the application of these principles to a thermodynamics learning sequence. The initial epitome might focus on the phenomena of expansion and change of state that bodies undergo with increasing temperature. Meaningful learning of the content of thermodynamics requires that the students' analysis of these physical phenomena will be approached from the base of a theoretical model appropriate to their pre-existing knowledge. The students are likely to begin learning with misconceptions, such as that the temperature of the change of state depends on the amount of substance or on the intensity of the heat source. The physical model that the student can understand at this stage is based on a mental representation of bodies as consisting of moving point particles, linked by some sort of bond of varying intensity. The support conceptual content (temperature, thermometer, states of matter, etc.)

should be introduced gradually, not in a "strictly scientific" way, but in terms of everyday knowledge that will subsequently be re-elaborated as hierarchies of learning (from the bottom up) as we have proposed before. (see Figure 2)

As new levels of elaboration are introduced, learning is enriched with progressively more complex explanations and theoretical models. For example, in the second epitome of a thermodynamics learning sequences, new phenomena related to heat transfer would be studied (conduction, convection, and radiation). In the third level epitome, one would now address the "ideal gases" (isochoric, isobaric, isothermal, and adiabatic processes).

The preparation of this type of learning sequence requires three types of learning activity that have a marked hands-on or experiential character.

The first consists of activities in which physical phenomena are observed in all their phases and then discussed as to their possible explanations. This basic causal explanation (BCE) should be based on the scientific model closest to the student's prior knowledge at each level of elaboration. Consider for example the observation of a phenomenon of change of state. At the first level of elaboration, the BCE that the student should deduce with the teacher's help could be summarized in the idea that, when a body's temperature increases, there is an increase in the movement of its particles to the point where the links that keep them together are weakened. A practical experiment that might help the students reach this conclusion could be to heat a container containing a piece of ice, another container with a piece of tin, and another with lead (which have different melting points) over the flame of a cigarette lighter. The question that the students should try to answer would be: "Why do these bodies melt?", or "What will have happened inside them?". The basic causal explanation that the students should try to discover from the perceptual features of the phenomenon is that on "providing the body with heat energy", and thus increasing its temperature, its particles vibrate with more energy and move away from each other, until the interaction holding the particles together gets so reduced that the body ends up "melting". (see Figure 3)

The second type of discovery learning activity is for the students to perform small experiments, in particular, to formulate hypotheses and make predictions about the variables involved in a given physical phenomenon, manipulate those variables systematically, record the results, and confirm or reject those hypotheses. Whenever possible, the activities should allow the students to recognize proportional or interdependence relationships,

Figure 2. Three-dimensional map of the initial epitome. (© 2008 Pérez, Suero, Montanero-Fernández, Pardo, & Montanero-Morán)

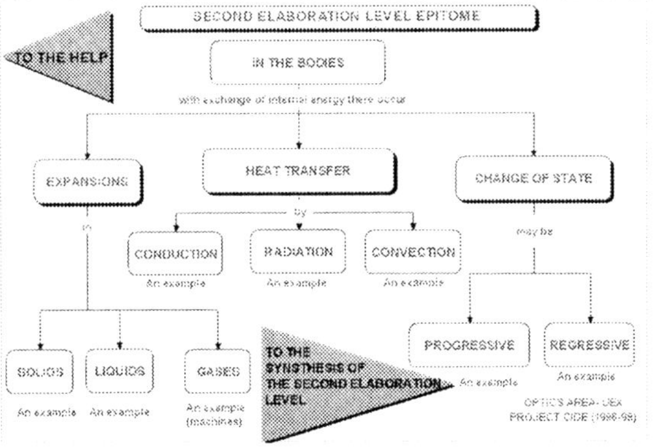

Figure 3. Map of the learning microsequence developing one of the topics of the content. (© 2008 Pérez, Suero, Montanero-Fernández, Pardo, & Montanero-Morán)

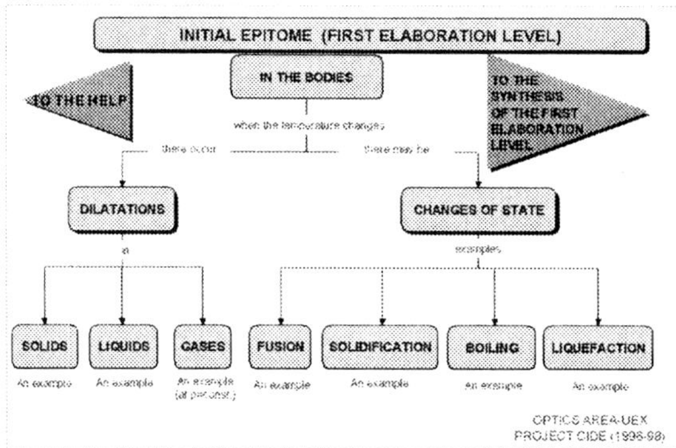

leading to their suggesting an initial idea (not necessarily mathematical) of some simple laws. Using the same example as above, since the temperature should rise with the energy supplied in heating the container, the students could be asked to discuss the following questions: Is the temperature rise always the same? Could it reach a plateau? How would that be possible? Why must we raise the temperature more for lead than for tin, and more for tin than for ice? The first three questions guide the student to an initial formulation of the law of constancy of temperature during melting. The last question would help them to conclude with another law relative to pure substances each having its own melting point. (see Figure 4)

The third type of activity should facilitate awareness of the conflict between the students' pre-existing causal theories (as yet "implicit") and the new causal explanations that have been constructed on the basis of the study of the physical phenomena. The activities in this case will be of a "pencil and paper" type, based on questions about predictions or explanations of natural phenomena related to the course content, or on the graphical representation of that content in the form of concept maps. In both cases, the teacher

Figure 4. Three-dimensional map of the second level of elaboration epitome. (© 2008 Pérez, Suero, Montanero-Fernández, Pardo, & Montanero-Morán)

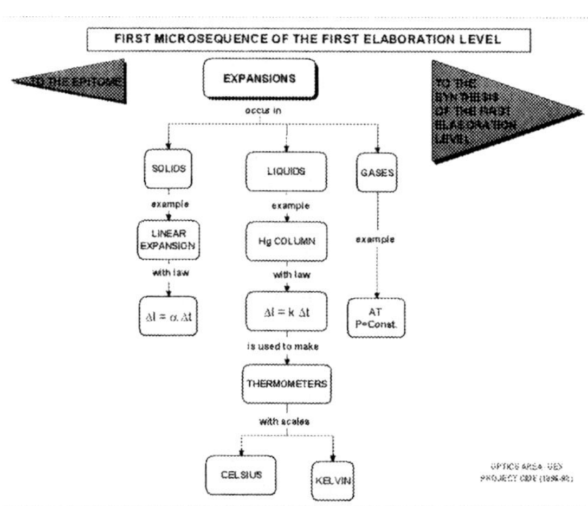

will discuss the responses and maps constructed by the students with the class as a whole, with the aim of pointing out the erroneous preconceptions that they contain and the changes undergone by the students' implicit theories during the learning process[1].

There are two major advantages for science learning and conceptual change in this adaptation of Reigeluth's theory to physics teaching. First, the design of levels of elaboration helps the students with their construction of new knowledge using a dual approach: from the simple to the complex, and from the particular to the abstract. Learning science does not consist in merely incorporating new knowledge, or replacing a number of misconceptions with others that conform the truth. Rather it is a constant re-working of the conceptual relationships that underpin a theory. At the beginning of learning, the students' knowledge only allows them to construct relatively simple and unidirectional explanations of the physics content. Subsequently, they will be able to understand and use more complex and interdependent theoretical models. Second, organizing the content around physical phenomena facilitates both hands-on learning and teaching the scientific method. The processes of conceptual change of the students' misconceptions require learning that goes from the implicit to the explicit, and from realism to perspectivism. The students must first of all become aware of their own implicit theories about the physical phenomena they are studying. They will have to put these into words, and contrast them with alternative explanations and models (Rodrigo & Correa, 1999). In this complex process, concept maps can have an important role to play, as will be shown below.

Applications of CmapTools to Elaboration Theory: Three-Dimensional Maps

There are various ways in which concept maps can be useful in the type of instructional design described above. First, they provide a visual representation of the logical structure of the knowledge that is to be taught. Second, they allow one to simulate the spiral sequence of different levels of elaboration of the content. Third, they provide a vehicle for teamwork among teachers designing their learning sequences.

The innovation that makes these goals attainable is based on adapting concept maps to a three-dimensional format. Traditionally, maps have been used to display knowledge in just two dimensions – horizontal and vertical – with the relationships of semantic subordination between concepts of different levels of inclusiveness being represented vertically, and the relationships between concepts that are at the same hierarchical level being represented horizontally. Three-dimensional maps allow one to represent a third type of content relationship: "depth", i.e., the different levels of elaboration that can be established in the instructional sequence.

The computer application CmapTools (designed by the Florida Institute for Human and Machine Cognition) makes the construction of such maps straightforward, thanks to an option that allows one to link new maps with any of the content already represented. The resulting maps contain text-labelled lines linking different items of the content (which are usually contained in a rectangular or elliptical frame). Also, for some of the content there appears a symbol indicating the connection in depth with another map.

We have used this simple technology to construct three-dimensional maps of the entire "Bachillerato" (pre-university secondary education) physics curriculum (see Pérez et al., 1999, 2000)[2]. By way of example, the following figures show some maps depicting the thermodynamics learning sequence described above.

We have found evidence that the application of this type of instructional design is more effective for science learning and conceptual change than the traditional alternatives at the Bachillerato level. The participants in our study were 5 teachers and

341 students from various secondary schools in Spain. Each teacher worked with three groups of students for approximately 15 hours. With two of these groups, the instruction sequence followed was based on an elaboration-theory design with three-dimensional maps. With the third group, it was based on the teacher's usual method. The pre- and post-trial evaluations of the students' learning consisted of three tests: a test of the comprehension of the scientific concepts studied; another with problems to identify any possible misconceptions; and a third with tasks that required the interpretation of phenomena or their application to everyday situations. Comparison of the pre- and post-trial results showed improvement in all the participating groups. Nonetheless, the improvements in all the tests were significantly greater in the groups using the elaboration theory sequences. The greatest differences were observed in the tests that evaluated conceptual changes (Montanero, Pérez, Suero, & Montanero, 2001).

APPLICATIONS OF CONCEPT MAPS TO ACTIVITIES OF COLLABORATIVE RECONSTRUCTION OF KNOWLEDGE IN PHYSICS

As was pointed out above, students can also use concept maps in certain activities of the elaboration sequence. Indeed, these maps represent an interesting resource for fostering conceptual change through collaborative learning activities. The graphical representation of the theories explaining the physical phenomena at each level of elaboration greatly facilitates discussion of possible misconceptions, as well as the students' own perception of the changes they undergo during the course of learning. Collaborative group design of concept maps also provides a context for questioning and negotiating their conceptual representations, which may be especially useful for them to revise their implicit theories.

There has been some previous work on the collaborative elaboration of concept maps in which computer resources were used to facilitate teacher-student and student-student activities. Chiu, Huang & Chang (2000) investigated group interaction processes in network-supported collaborative concept mapping. It was found that group concept mapping performance was significantly correlated to the quantity of group interaction, particularly complex cooperation interaction.

Using the application "Soups" of CmapTools[3], Cañas, Ford, Hayes, Reichherzer, Suri, Coffey, et al. (1998) carried out various experiments in which they explored the educational possibilities of collaborative learning with maps. As a student is draughting a concept map, CmapTools lets him or her publish any idea (as a minimum, two linked concepts) onto a network so that it is potentially visible to other students on that network. The images are stored in a database (the "knowledge soup") that all the students share and use in constructing their own maps[4]. Any student can question a claim made by another if he or she disagrees with it or finds it ambiguous. The other students can respond, initiating a parallel discussion on some specific content.

The collaborative elaboration of concept maps with such tools is an extremely powerful resource with which to encourage the students' active involvement in processes of meaningful learning and conceptual change. As was observed above, changing implicit theories requires activities that will lead the students to an awareness of the theory that they use to explain physical phenomena, and to contrast it with alternative versions. The construction of collaborative maps can be of particular help in attaining this important objective in physics teaching.

In a series of educational experiments (see Pérez, Suero, & Pardo, 2008), we tested the implementation of concept maps in support of processes of the collaborative reconstruction of knowledge and conceptual change in undergraduates. The experiments were carried out in four phases:

- Firstly, the students were asked to prepare individual concept maps about certain physical phenomena of optics and electricity.
- Secondly, they reviewed and proposed changes to the maps made by their classmates. One of the students took the responsibility of synthesizing the proposals of amendments to each of the individual maps. The result was what we denominated "the map in revision".
- Thirdly, the students discussed the changes proposed for each map. Each author had to justify his or her acceptance or rejection of incorporating the changes. The result of this process was each student's "revised map".
- Finally, another student draughted the "consensus map", summarizing the essential parts of all the revised maps. The maps in revision and those already revised contained links to the originals, allowing easy comparison of the differences between them. They also included notes clarifying why the amendments proposed by classmates had been accepted or rejected.

The negotiation and exchange of maps took place directly in the classrooms themselves, or over Internet by accessing either the Cmap site or the Web page to which they were being published[5]. In the following paragraphs, we shall illustrate this working strategy with an example of a map of one of the participating students concerning the content of optics relative to refraction.

As can be seen in Figure 5, the original map contains three links – to the same map in revision, to the map as subsequently revised by the author, and to the final map reached as a consensus of the entire class. Some of her classmates analyzed

Figure 5. The original map of an undergraduate student on the refraction of light. (© 2008 Pérez, Suero, Montanero-Fernández, Pardo, & Montanero-Morán)

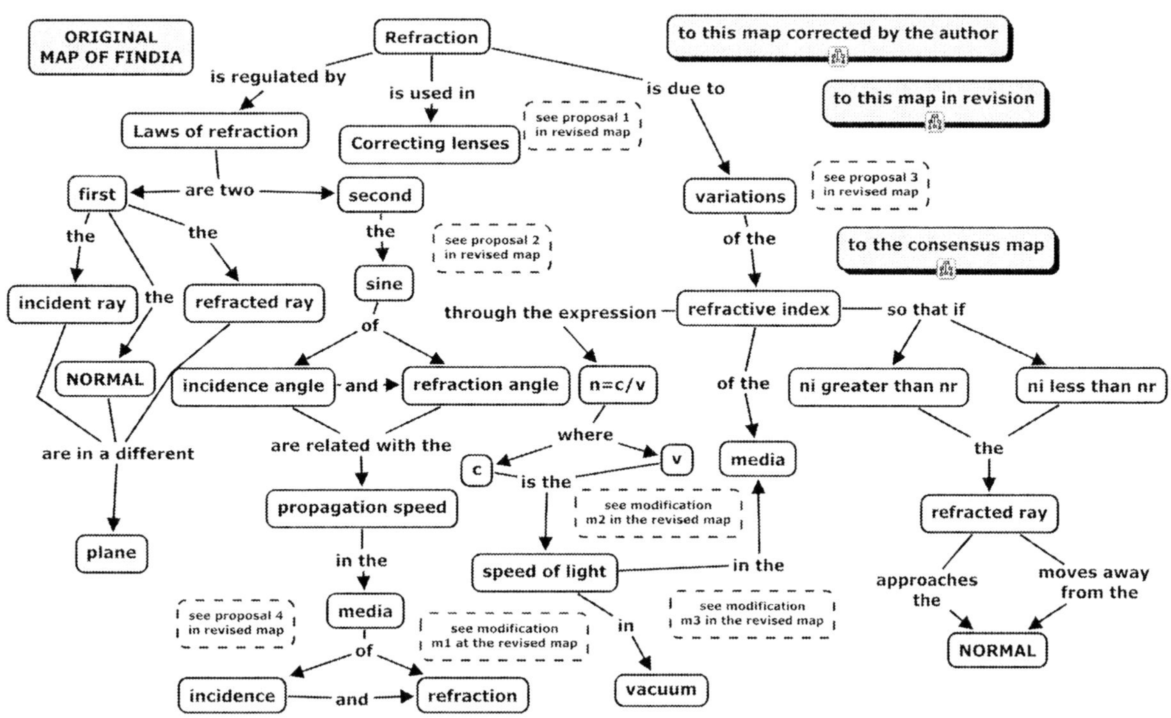

her original map, and proposed changes to some of its parts. These proposed changes are outlined with dashed lines in the figure.

Figure 6 shows the map in revision resulting from incorporating the classmates' proposals (indicated by dashed lines and in red). The figure also shows the 22 notes in which those classmates clarify or argue for their proposed changes.

Analysis of the dashed-line links and the content of the notes showed that there were essentially five types of help contributed by colleagues to facilitate the re-draughting of the original representation: clarifications, syntheses, examples, evaluations, and conflicts.

Most of the dashed lines correspond to *clarifications* (proposing new concepts or links) clarifying, complementing, or justifying an idea already expressed in the map. Other added details are given in the notes. Note 2, for example, proposes adding further information about the concept of "variation" and "refractive index" ("It is important to say that the light rays in refraction cross the surface separating media with different refractive indices"). Note 6 suggests indicating that the second law (which the "Second" concept refers to) is Snell's Law.

The focus of the help in the form of *syntheses* is on integrating information that appears disaggregated in the original map. This is the case of the proposal (in red on the right-hand side of the map) connecting $n_i > n_r$ and $n_i < n_r$ with the idea of approaching or moving away from the normal, respectively. Similarly, Note 11 suggests removing the link to which it is attached, and integrating the concepts of angles of incidence and of refraction into a single box.

Help in the form of *examples* proposes cases that are concrete instances of some idea in the original map. Thus, in the central part of the map, a student suggests considering corrective lenses as

Figure 6. Map in revision incorporating modifications to the original map of Figure 5. (© 2008 Pérez, Suero, Montanero-Fernández, Pardo, & Montanero-Morán)

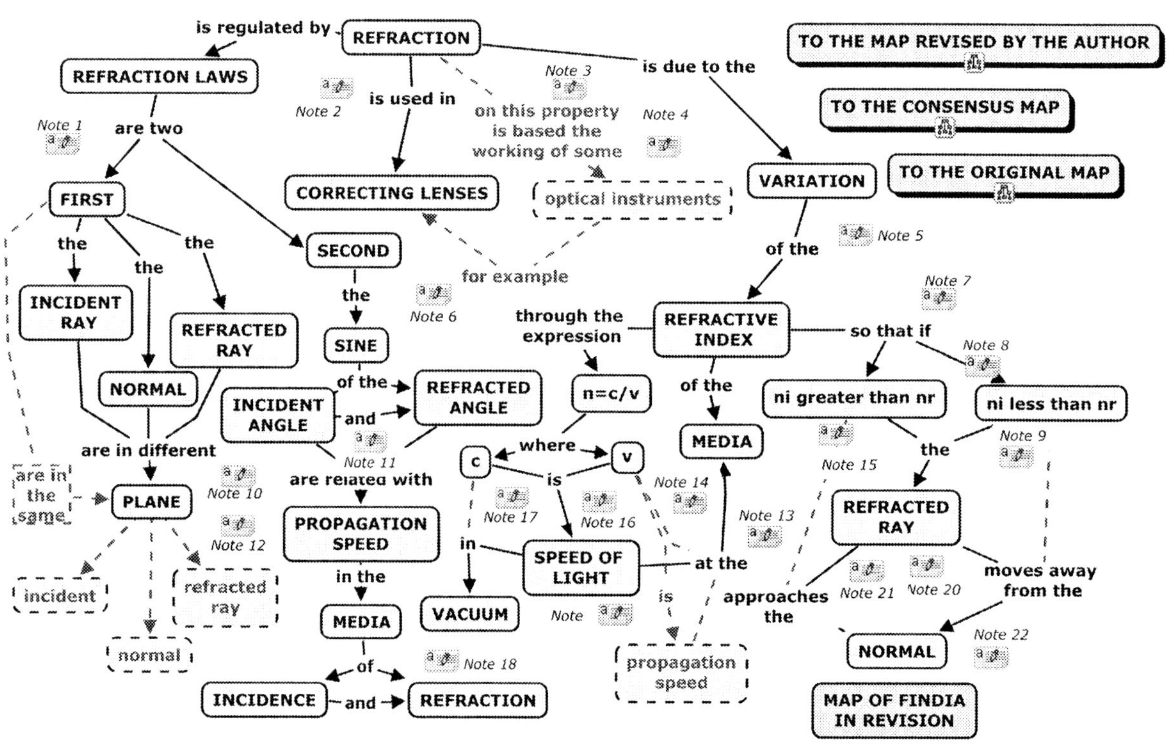

Figure 7. Map revised by the author. (© 2008 Pérez, Suero, Montanero-Fernández, Pardo, & Montanero-Morán)

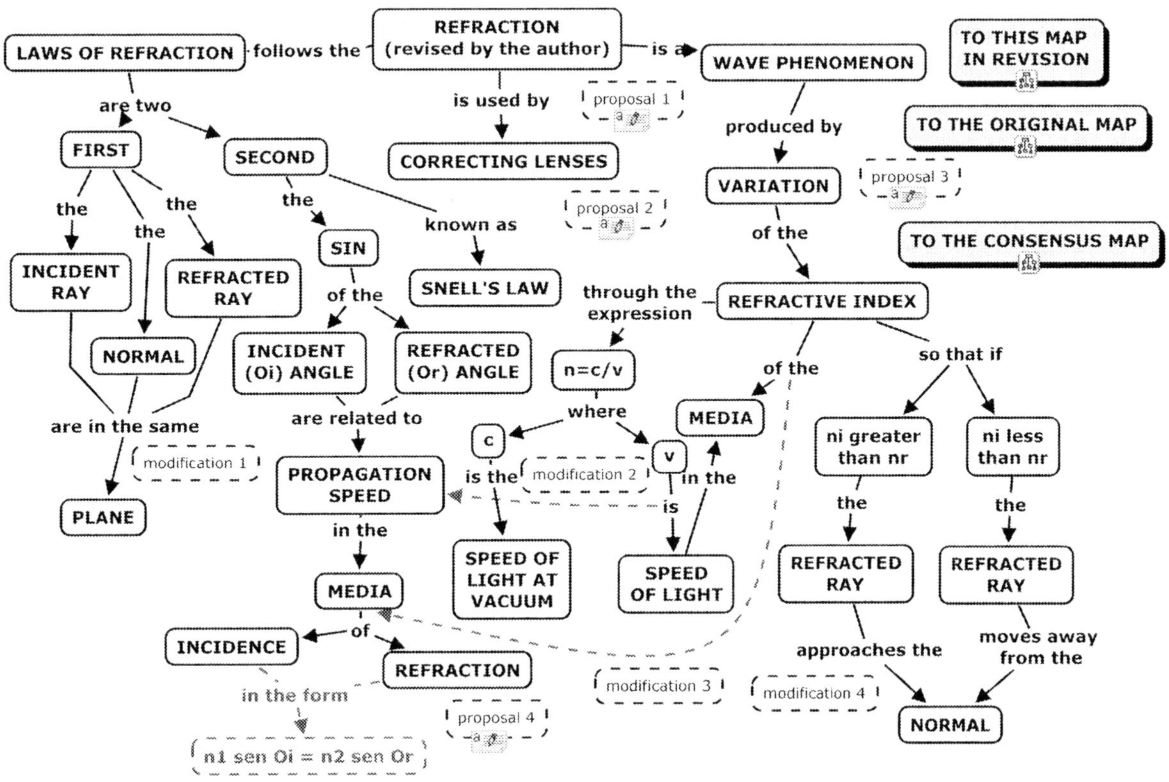

an example of an optical instrument that is based on the phenomenon of refraction.

In their *evaluations*, the other students may pose questions, or point to some incorrect or confusing idea in the original map. For example, in Note 7 a student indicates that he does not understand the relationship expressed between the indices of refraction, and that ni is greater than or less than nr. A similar problem is detected with respect to other links in Notes 13 and 22.

Finally, in the type of help that we have called "*conflicts*", other students may point to contradictions, biases, or errors. Such is the case of the new link ("are in the same" on the left-hand side of the map) proposed between concepts relating to the first law of refraction and the plane. This brings out the grave conceptual mistake the author of the map had made in confusing "plane" and "medium". Instead of "different plane", the author should have written "different media". She would thus have managed to explain that, for there to be refraction, the incident ray must be in one medium and the refracted ray in another. Often, help of this type from different students is redundant. Thus, others of her colleagues bring up this error in Notes 4 ("When does refraction take place?") and 12 ("Plane or medium?"). The instructor too can occasionally post notes that generate a conflict or help to revise relevant items of the map. In this case, for example, the instructor suggested including an explanation of cases in which the refraction has characteristics that make the medium transparent and others that make it merely translucent.

In the revised map shown in Figure 7, the author has analyzed her classmates' proposals for amendment of her original map (Figure 5). As can be seen, she accepts four amendments that greatly improve her original map. More importantly, her review process introduces new concepts and relationships (such as the idea that refraction is a wave phenomenon). All the participating students made similar modifications to their original maps, by including the suggestions of their classmates.

Finally, the students negotiated a map on the refraction of light that integrated as well as possible the set of representations that they had all constructed and re-constructed. The most viable strategy that was followed for this step began with taking the revised map that they considered to be the closest to complete. They then discussed which elements of the other maps could be added, ending with the consensus map (Figure 8).

This learning activity is a clear illustration of how certain strategies of intersubject negotiation (Wertsch, 1998) and "scaffolding" of students' cognitive representations can lead to conceptual change. We have shown that the most effective activities to promote conceptual change are those which provoke the students' own explanations and revisions of their mental models (Chi, 1996). In the maps resulting from the experiment described above, each student contributed an average of 4 inputs to each of the original maps of their colleagues (mostly clarifications, evaluations, and conflicts, in that order). Also, on average, each student integrated into their revised map more than 6 inputs from their peers. Misconceptions about the physics content were detected in more than 80% of the original maps, which, by the end of the process, the authors' of the maps themselves had identified and revised. In another experiment with the content of electricity, we analyzed the influence of this type of activity using an objective test

Figure 8. Final consensus map. (© 2008 Pérez, Suero, Montanero-Fernández, Pardo, & Montanero-Morán)

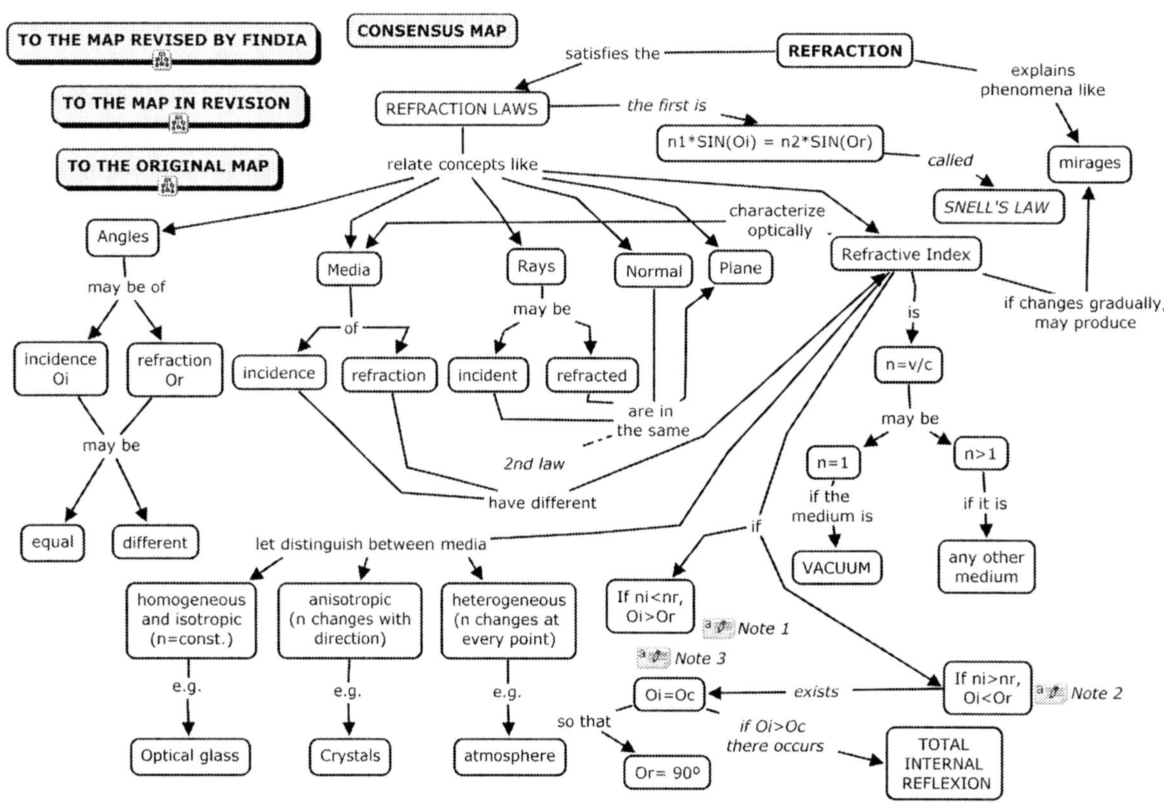

for the evaluation of misconceptions. Compared with an equivalent control group, the students who participated in the experiment responded 17% better to the test items (Pérez et al., 2008).

In sum, resources like CmapTools provide an ideal scenario for the collaboration of teacher and students in the construction of a conceptual system. The graphical representation that they provide plays a very important role in making the students' respective internal representations explicit, as well as facilitating the inclusion of help from colleagues in the group and the consequent redefinition of those representations. The task of reconstructing their maps facilitates in particular the students' active participation in recalling the necessary prior knowledge, in inferring causal relationships, and in modifying their implicit causal model of the physical phenomenon being studied.

CONCLUSIONS AND FUTURE PROSPECTS

Over the past 15 years, the Grupo Orión at the University of Extremadura has been undertaking an extensive program of educational research and innovation, focused on the application of concept maps to physics teaching. In the preceding pages, we have analyzed two specific recent applications carried out in a framework of collaborative work of teachers and students. We first described how the construction of three-dimensional maps can facilitate the collaborative design of learning sequences based on the elaboration theory of Reigeluth and Stein. We then reported on some experiments and preliminary results suggestive of the potential of this tool in support of students' collaborative learning.

We have also analyzed the reasons why the two applications can have a result on learning that is especially important in science education – conceptual change. The results of these experiments demonstrated the potential of concept maps to foster first awareness, and then the modification, of implicit theories about physical phenomena. Three-dimensional maps constitute a very useful tool to help teams of teachers discuss and design learning sequences at different levels of elaboration. Compared with other types of instructional designs, the sequences in which three-dimensional maps were used led to better learning outcomes, particularly in tests that measured the modification of misconceptions about physics content. A similar conclusion can be drawn from the experiments in which undergraduate students carried out a collaborative activity of construction and re-construction of concept maps. Our analysis of the original and revised maps showed clear evidence for the usefulness of this task in facilitating processes of conceptual change.

Another line of work with concept maps that has as yet been relatively unexplored is their use to analyze teachers' explanations in the classroom. In this sense, CmapTools can be used to represent the configuration of the verbal discourse that is made public in the classroom. In a recent study, we used various codes (types of polygons and links) to represent the classes of concept that teachers were explaining, the kinds of relationship, the order in which they were being presented, the supporting resources the teachers provided, and the degree of student participation in the construction of the ideas that were made public (Montanero, Lucero, & Méndez, 2008. Such applications will permit an analysis of the process of shared construction of knowledge as a function of various parameters – the degree of articulation of the explanation, how it is ordered, the degree of redundancy or density, the aid provided to the students to help them understand the content, etc.

REFERENCES

Anderson-Inman, L., Ditson, L. A., & Ditson, M. T. (1998). *Computer-based concept mapping: Promoting meaningful learning in science for students with disabilities. information Technology and Disabilities.* Retrieved September 22, 2008 from http://people.rit.edu/easi/easisem/itdscience.htm.

Ausubel, D. P., Novak, K. D., & Hanesian, H. (1978). *Educational Psychology*. New York: Holt, Rinehart & Winston.

Baker, M., Hansen, T., Joiner, R., & Traum, D. (1999). The role of grounding in collaborative learning tasks. In P. Dillenbourg (Eds.), *Collaborative learning: cognitive and computational approaches* (pp. 31-63). Oxford, UK: Elsevier Science Ltd.

Barron, B. (2003). When smart groups fail. *Journal of the Learning Sciences*, *12*, 307–359. doi:10.1207/S15327809JLS1203_1

Bunge, M. (1978). *Filosofía de la Física*. Buenos Aires, Venezuela: EUDEBA.

Cañas, A. J., Ford, K. M., Hayes, P. J., Reichherzer, T., Suri, N., Coffey, J., et al. (1998). *Colaboración en la Construcción de Conocimiento Mediante Mapas Conceptuales*. Proceedings of the VIII Congreso Internacional sobre Tecnología y Educación a Distancia, San José, Costa Rica, (pp. 25- 42).

Cañas, A. J., Hill, G., Carff, R., Suri, N., Lott, J., Eskridge, T., et al. (2004). CmapTools: A Knowledge Modeling and Sharing Environment. In J.A. Cañas, J.D. Novak, & F.M. González (Eds.), *Concept Maps: Theory, Methodology, Technology: Proceedings of the First International Conference on Concept Mapping* (pp. 125-133). Pamplona, Spain: Universidad Pública de Navarra. Retrieved September 2, 2008 from http://cmc.ihmc.us/papers/cmc2004-283.pdf

Caswell, R. M., & Wendel, C. T. (1992, April). Conceptual changes occurring in elementary teachers' knowledge of science concepts throughout long-term instruction. Paper presented at the *American Educational Research Association Annual Meeting*, San Francisco, CA.

Chi, M. T. H. (1996). Constructing self-explanations and scaffolded explanations in tutoring. *Applied Cognitive Psychology*, *10*, 33–49. doi:10.1002/(SICI)1099-0720(199611)10:7<33::AID-ACP436>3.0.CO;2-E

Chi, M. T. H., De Leeuw, N., Chiu, M. H., & Lavancher, C. (1994). Eliciting self-explanations improves understanding. *Cognitive Science*, *18*, 439–477.

Chi, M. T. H., Slotta, J. D., & de Leeuw, N. (1994). From things to processes: a theory of conceptual change for learning science concepts. *Learning and Instruction*, *4*, 27–43. doi:10.1016/0959-4752(94)90017-5

Chiu, C. H., Huang, C. C., & Chang, W. T. (2000). The evaluation and influence of interaction in network supported collaborative concept mapping. *Computers & Education*, *34*, 17–25. doi:10.1016/S0360-1315(99)00025-1

Claxton, G. (1987). *Vivir y aprender*. Madrid: Alianza editorial.

Cohen, E. G. (1994). Restructuring the classroom: conditions for productive small groups. *Review of Educational Research*, *64*, 1–35.

Di Sessa, A. A. (1993). Towards an epistemology of physics. *Cognition and Instruction*, *10*, 105–225. doi:10.1207/s1532690xci1002&3_2

Fischer, F., Bruhn, J., Gräsel, C., & Mandl, H. (2002). Fostering collaborative knowledge construction with visualization tools. *Learning and Instruction*, *12*, 213–232. doi:10.1016/S0959-4752(01)00005-6

Gagnè, R. (1985). *The Condition of Learning of Instruction*. New York: Holt, Rinehart & Winston.

Hernández, P., & Serio, A. (2004). ¿Cómo hacer eficaces los mapas conceptuales en la instrucción? *Infancia y Aprendizaje, 27*, 247–265. doi:10.1174/021037004323038851

Hewson, P. W., & Thorley, R. N. (1989). The conditions of conceptual change. *International Journal of Science Education, 11*, 541–553. doi:10.1080/0950069890110506

Horton, P. B., Mcconney, A. A., Gallo, M., Woods, A. L., Senn, G. J., & Hamelin, D. (1993). An investigation of the effectiveness of concept mapping as an instructional tool. *Science Education, 77*, 95–111. doi:10.1002/sce.3730770107

Lautrey, J., & Mazens, K. (2004). Is children's naive knowledge consistent? A comparison of the concepts of sound and heat. *Learning and Instruction, 14*, 399–423. doi:10.1016/j.learninstruc.2004.06.011

López Rupérez, F. (1990). Epistemología y didáctica de las ciencias: Un análisis de segundo orden. *Ensenañza de las Ciencias, 8*, 65–74.

Markham, K. M., Mintzes, J. J., & Jones, M. (1994). The concept map as a research and evaluation tool: Further evidence of validity. *Journal of Research in Science Teaching, 31*, 91–101. doi:10.1002/tea.3660310109

McCrudden, M. T., Schraw, G., Lehman, S., & Poliquin, A. (2007). The effect of causal diagrams on text learning. *Contemporary Educational Psychology, 32*(3), 367–388. doi:10.1016/j.cedpsych.2005.11.002

Meléndez, A., Pacual, I., & Rosado, L. (1989). La enseñanza de los modelos sobre la naturaleza de la luz. *Ensenañza de las Ciencias, 7*, 271–276.

Mercer, N. (1996). The quality of talk in children's collaborative activity in the classroom. *Learning and Instruction, 6*, 359–377. doi:10.1016/S0959-4752(96)00021-7

Montanero, M., Lucero, M., & Méndez, J. M. (2008). Enseñanza de contenidos históricos mediante diagramas multicausales. *Revista Española de Pedagogía, 239*(1), 27–48.

Montanero, M., & Montanero, M. (1995). *Didáctica del momento angular de una partícula*. Badajoz, Spain: ICE.

Montanero, M., Pérez, A. L., & Suero, M. I. (1995). Survey of students and teacher conceptions of action-reaction dynamics: Implicit alternative theories are manifest in the consistency of incorrect responses. *Physics Education, 30*, 277–283. doi:10.1088/0031-9120/30/5/004

Montanero, M., Pérez, A. L., Suero, M. I., & Montanero, M. (2001). Cambio conceptual y enseñanza de la Física: Aplicaciones en el marco de la Teoría de la Elaboración. *Revista de Educación, 326*, 311–332.

Novak, J. D., & Gowin, D. B. (1984). *Concept Mapping for Meaningful Learning. Learning How to Learn*. Cambridge, UK: Cambridge University Press.

Novak, J. D., Gowin, D. B., & Johansen, G. (1983). The use of concept mapping and knowledge vee mapping with junior high school science students. *Science Education, 67*, 625–645. doi:10.1002/sce.3730670511

Okebukola, P. A., & Jegede, O. J. (1988). Cognitive preference and learning mode as determinants of meaningful learning through concept mapping. *Science Education, 72*, 489–500. doi:10.1002/sce.3730720408

Okebukola, P. A., & Jegede, O. J. (1989). Students' anxiety towards and perception of difficulty of some biological concepts under the concept mapping heuristic. *Research in Science & Technological Education*, *7*, 85–92. doi:10.1080/0263514890070109

Okebukola, P. A. O. (1990). Attaining meaningful learning of concepts in genetic and ecology: An examination of the potency of the concept mapping technique. *Journal of Research in Science Teaching*, *27*, 493–504. doi:10.1002/tea.3660270508

Pankratius, W. J. (1990). Building an organized knowledge base: Concept mapping and achievement in secondary school physics. *Journal of Research in Science Teaching*, *27*, 315–333. doi:10.1002/tea.3660270404

Pérez, A. L., Suero, M. I., Montanero, M., & Montanero, M. (1999). Mapas de experto tridimensionales: Aplicaciones al diseño de secuencias instruccionales de Física. In M.E.C., *Premios Nacionales de Investigación Educativa 1998* (pp. 93-116). Salamanca, Spain: CIDE.

Pérez, A. L., Suero, M. I., Montanero, M., & Montanero, M. (2000). *Mapas de expertos tridimensionales. Aplicaciones al diseño de secuencias instruccionales de Física, basadas en la teoría de la Elaboración*. Badajoz, Spain: Junta de Extremadura.

Pérez, A. L., Suero, M. I., Montanero, M., & Montanero, M. (2001). Propuestas de innovación entorno al análisis y secuenciación de contenidos en los diseños curriculares de Física. *Bordón*, *53*, 279–286.

Pérez A.L., Suero, M.I., & Pardo, P.J. (2008) Utilización de Cmaps para mejorar los conocimientos relativos a la refracción de la Luz mediante su "Reconstrucción Colaborativa". *Óptica pura y aplicada*, *41*, 17-23.

Posner, G. J., Strike, K. A., Hewson, P. W., & Gertzog, W. A. (1982). Accomodation of scientific conception: Toward a theory of conceptual change. *Science Education*, *66*, 211–227. doi:10.1002/sce.3730660207

Pozo, J. I. (1999). Más allá del cambio conceptual: el aprendizaje de las ciencias como cambio representacional. *Ensenañza de las Ciencias*, *17*, 513–520.

Reigeluth, C. M., Merril, M. D., & Bunderson, C. V. (1978). The structure of subject matteer content and its instructional desing implications. *Instructional Science*, *7*, 107–126. doi:10.1007/BF00121929

Reigeluth, C. M., & Stein, F. S. (1983). The Elaboration Theory of Instruction. In C.M. Reigeluth (Ed.), *Instructional design theories and models: An overview of their current status* (pp. 335-381). Hillsdale, NJ: Erlbaum.

Reigeluth, C. M., & Stein, F. S. (1987). Lesson blueprints based on the Elaboration Theory of Instruction, In C.M. Reigeluth (Ed.), *Instructional theories in action. Lesson illustrating selected theories and models* (pp. 245-288). Hillsdale, NJ: Erlbaum.

Rodrigo, M. J., & Correa, N. (1999). Teorías implícitas, modelos mentales y cambio educativo. In I. Pozo & C. Monereo (Eds.), *El aprendizaje estratégico* (pp. 75-86). Madrid: Santillana.

Roth, W., & Roychoudhury, A. (1992). The social construction of scientific concepts or the concept map as conscription device and tool for social thinking in high school science. *Science Eductation*, *76*(5), 531–557. doi:10.1002/sce.3730760507

Ruiz-Primo, A. (2004, September). *Examining concept maps as an assessment tool*. Paper presented at the I International Conference on Concept Mapping, Pamplona, Spain.

Suero, M. I., Pérez, A. L., & Montanero, M. (Eds.). (2003). *Física 2*. Madrid: Santillana.

Van Boxtel, C., van der Linden, J., & Kanselaar, G. (2000). Collaborative learning tasks and the elaboration of conceptual knowledge. *Learning and Instruction*, *10*, 311–330. doi:10.1016/S0959-4752(00)00002-5

Vosniadou, S. (1992). Knowledge acquisition and conceptual change. *Applied Psychology . International Review (Steubenville, Ohio)*, *41*(4), 347–357.

Wertsch, J. (1998). *Mind as Action*. Oxford, UK: Oxford University Press.

KEY TERMS AND DEFINITIONS

Collaborative Learning: is an umbrella term for a variety of approaches in education that involve joint intellectual effort by students or students and teachers

Concept Mapping: a technique for visualizing the relationships among different concepts

Conceptual change: learning that changes an existing conception

Constructivism: a philosophical framework or theory of learning which argues humans construct meaning from current knowledge structures

Educational Psychology: is the study of how humans learn in educational settings, the effectiveness of educational interventions and more.

Elaboration Theory: An instructional design theory that argues that content to be learned should be organized from simple to complex order, while providing a meaningful context in which subsequent ideas can be integrated.

Physics Teaching: Teaching of the scientific discipline that involves mechanic, thermodynamic, electromagnetism, optics, high energy and radioactive physics.

ENDNOTES

1 The manner in which the maps can constitute a vehicle for the collaborative construction of learning in this type of activity requires a more detailed analysis, which will be specifically addressed in the next section

2 This innovation was honoured with the 2nd National Educational Research Award 1998, and has also served as the foundation for the various editions of one of the most widely used textbooks at this stage of education in Spain (Suero, Pérez, & Montanero, 2003). Much of the resulting material can be found on the Web at http://grupoorion.unex.es.

3 In particular, CmapTools contains three applications that empower the collaborative construction of maps via the Internet: Synchronous Maps, Discussion Lists, and Soups.

4 To select the most relevant ideas, the system uses filtering heuristics based on the degree of semantic interrelation with which each student has published the ideas in the "soup". (This, in turn, reinforces the students' motivation to make contributions.)

5 The results of these experiments are available deposited in our Cmap Site: "Universidad de Extremadura (España)", in the directory "Asignatura de Didáctica de la Física" and in that corresponding to the 2004/05 academic year. They can be used interactively by means of the CmapTools computer application. These maps can also be viewed on the Web at: http://grupoorion.unex.es:8001.

Chapter 18
Using Concept Maps to Assess Individuals and Teams in Collaborative Learning Environments

Tristan E. Johnson
Florida State University, USA

Dirk Ifenthanler
Albert-Ludwigs University Freiburg, Germany

Pablo N. Pirnay-Dummer
Albert-Ludwigs University Freiburg, Germany

J. Michael Spector
University of Georgia, USA

ABSTRACT

The main focus of this chapter is the use of concept mapping, broadly defined to include both graphical and textual representations, for assessment in collaborative learning contexts. Several tools developed by the authors integrate concept mapping as the primary means of assessing progress of learning in complex and problem-solving domains. This chapter presents an overview of the theoretical and empirical foundation for these assessment tools, and discusses their applicability to collaborative learning environments (CLE).

INTRODUCTION

Collaborations are formed in learning settings to meet instructional needs as well as to exploit the benefits (pedagogical, learning, and pragmatic) associated with collaborative learning (Stahl, 2006). This rationale is similar to that of organizations that rely on working collaborations to address difficult and challenging tasks (Salas & Fiore, 2004). In educational settings, collaborative learning environments (CLEs) have been used to promote participation and enhance learning. One of the main reasons for creating CLEs is to facilitate the devel-

DOI: 10.4018/978-1-59904-992-2.ch018

opment of professional skills that are promoted from group learning, such as communication, teamwork, decision making, leadership, valuing others, problem solving, negotiation, thinking creatively, and working as a member of a team (Carnevale, Gainer, & Meltzer, 1989).

Collaborative learning involves the interaction of two or more individuals and their environment with the goal of developing knowledge, skills, or attitudes. There is a need to assess the learning and interactions in order to provide feedback to the learners to improve collaborative interactions, thereby improving overall performance, and in order to evaluate the general efficacy of collaborative learning given a specific context. Studies of collaboration processes have led to improved understanding about what teams do, and how and why they do what they do (Salas & Cannon-Bowers, 2000). Another benefit in assessing collaborative learning processes is to understand the differences between highly successful collaboration process behaviors and unsuccessful process behaviors. From this understanding, a better collaboration framework can be used to inform the design, development, and deployment of collaborative learning strategies in contexts such as educational and workplace settings.

There is evidence that teams whose thinking is similar are likely to work more effectively together than teams whose thinking is not (Cannon-Bowers & Salas, 1998; Guzzo & Salas, 1995; Hackman, 1990). The degree to which a team shares similar conceptualizations is seen as a key indicator of successful overall performance (Salas & Cannon-Bowers, 2000). Developing learning activities that enhance a team's development of shared understanding has the potential to improve a team's learning and ultimately that of the individuals on the team. Methodologies that focus on a team cognition framework can provide an important view on collaborative processes and help guide the design of effective collaborative learning activities (interaction strategies) (Fiore & Salas, 2004; Langan-Fox, Anglim, & Wilson, 2004).

Several research studies have tried to establish the link between collaborative knowledge and collaborative processes. Research has shown that specific interactions, such as communication and coordination, mediate the development of team knowledge and thus mediate team performance (Mathieu, et al., 2000). Interactions among teammates coupled with shared knowledge are a predominate force in the construct of shared cognition. As teammates interact, they begin to share knowledge, thereby enabling them to create cues in a similar manner that in turn helps them make compatible decisions and take proper actions (Klimoski & Mohammed, 1994). Shared knowledge can help team members in CLE by increasing understanding about what is occurring with the learning task, develop accurate expectations about collaborative actions, and communicate efficiently in the learning environment.

Collaborative learning environments are becoming more common in educational and workplace settings in order to meet the challenges created by technological and scientific progress. Technological advances impact how we do things as well as how we think. Knowledge used to be measured by the ability to repeat or recall information in many contexts. However, knowledge is increasingly being measured by the ability to efficiently find and make effective use of information (Simon, 1996). The former approach to knowledge and assessment is appropriate for declarative knowledge – knowing facts and the definitions of concepts. However, with regard to the ability to solve problems, especially complex, dynamic and ill-structured problems, the latter kind of knowledge is also required. The challenge addressed in this paper is to develop assessment methodologies appropriate for this latter type of knowledge.

Learners not only need to gain a basic understanding of foundational topics: they also need to acquire skills associated with asking meaningful questions that will help them develop general skills and attitudes. To be successful, educational and training programs must focus on helping learners develop cognitive strategies and tools that will allow them to develop understanding and skills that will support productive thinking. These skills are essential to support the wide array of tasks such as problem solving, abstract reasoning, communication, and collaboration that learners will be required to perform to be successful in a workplace setting (Karoly & Panis, 2004).

Many individuals are trained to employ specific complex skills. However, there are cases in which the nature of certain tasks is too complicated and too great to be handled by an individual (Cooke, Salas, Kiekel, & Bell, 2004; Stout, Cannon-Bowers, Salas, & Milanovich, 1999). Learning teams are formed not only because of the benefits of having collective resources (e.g., knowledge, skills, and diverse expertise) but also because the team benefits from individual learning. Teams can provide the basis for improved performance, learning and motivation; however, developing effective teams is a challenge, and developing effective learning teams is even more challenging.

When considering activities and programs that focus on the development of knowledge and skills, it is critical to couple learning with assessment (Bransford, Brown, & Cocking, 2000). As researchers study CLEs, it is important to study their effects on individuals and teams in order to ascertain the value of specific CLE techniques and approaches. Several meta-analyses have shown that inducing subject matter oriented discussions within teams is insufficient in order to successfully facilitate purposeful collaborative learning and problem solving (e.g. Johnson et al., 1981; Qin et al., 1995). More than simple team discussions are required for effective teamwork in the context of solving complex problems; developing a shared understanding of the problem and solution approach is required. Accordingly, specific assessment techniques are needed to gather comprehensive data on the underlying processes of collaborative learning and problem solving, and consequently designing more successful instructional intervention.

In addition to measuring the products that are created by individuals and teams, there are various methods of studying the processes of individuals and teams (Johnson & O'Connor, 2008a; and Savenye, Robinson, Niemczyk, & Atkinson, 2008). A few methods use concept maps to assess individuals and teams as they engage in collaborative learning environments.

In this chapter, we examine four methodologies, including: 1) Surface, Matching, and Deep Structure (SMD) Methodology, 2) Model Inspection Trace of Concepts and Relations (MITOCAR) Methodology, 3) Dynamic Evaluation of Enhanced Problem-solving (DEEP) Methodology, and 4) Analysis Constructed Shared Mental Models (ACSMM) Methodology. These methods capture conceptual representations of individual mental models and then use analytical methods to compare the externalized mental models in order to show relationships between individual and team knowledge representations. These four methods have recently been integrated into an online assessment toolset called HIMATT (Highly Interactive Model-based Assessment Tools and Technologies). These tools include analytical comparisons that are useful in representing differences and changes in cognitive problem-solving functions.

We present the strengths and applicability of each method in order to show how these tools are used for assessment and how they can be used to facilitate further mental model research and theory development. In this chapter we describe both knowledge and conceptual elicitation methods of individuals and teams, and present several data analysis methods that compare both individual and collaborative progress over time as well as quantitative knowledge structures at a given point in time. Each of these assessment methods employs concept maps as part of the conceptual system elicitation and/or part of the analysis results.

BACKGROUND

Various fields of human science continually develop new techniques to capture key latent variables associated with cognition. One specific area of interest is concerned with internal conceptual systems (mental models and schemata). These conceptual systems are theoretical constructs of science that are not observable. Therefore, individuals have to externalize their internal constructions, and changes in these externalizations are interpreted as changes in the underlying conceptual systems. In other words, researchers can only learn about internal conceptual systems if individuals communicate or externalize their systems (Seel, 1991). A key research interest is to capture and/or create conceptual representations of these internal conceptual systems.

The research foundation for this chapter is built upon: (a) a model-facilitated framework for developing system dynamics-based interactive learning environments (Milrad, Spector, & Davidsen, 2002), (b) a tool to measure the structural and semantic similarity between concept maps (Ifenthaler, 2006), (c) a tool to develop a concept map from text and to measure structural and semantic similarities (Pirnay-Dummer, 2006), (d) a tool to elicit annotated concept maps and problem conceptualizations (Spector & Koszalka, 2004), and (e) a method for assessing team concept maps and measuring shared mental models (O'Connor & Johnson, 2004, Johnson & O'Connor, 2008b). These tools address the design and the assessment of learning in complex problem-solving domains for which collaborative learning is generally well suited.

One analytical requirement is to compare conceptual representations of one team or individual with the representations of another team or individual or even with oneself at a later time. These analytical comparisons are made to represent differences and changes in cognitive functions. As new techniques and methods are being developed, the question of creating/finding a valid and reliable measurement of change is one of the central problems of conceptual systems research (Seel, 1999). Two comparative situations merit consideration: 1) a single comparison of different teams or individuals with another team or individual, and 2) the comparison of a team or individual with themselves over time. Accordingly, the psychological and educational diagnosis of internal conceptual systems presupposes repeated measurements of these systems over the course of a given process (Ifenthaler & Seel, 2005), both for individual and team performance.

Several assessment methods focus on an analysis of team knowledge. Most of them involve techniques for elicitation of conceptual models and also offer specific strategies on how to analyze the elicited models. To measure learning in a CLE, there are three measuring techniques for assessing individuals and teams: 1) direct process measures, which are the direct actions of a team; 2) indirect process measures, which are measures at a specific point in time; and 3) indirect process measures, which are outcomes from team actions. (Johnson and O'Connor, 2008a).

Direct process measures directly capture the actions of a team. These measures are continuous, and the data are captured over time. Techniques include capturing spoken and written language, as well as actions taken during a collaborative learning activity. As the team interacts, the resultant data are in the form of a holistic collaborative data set.

Measuring collaborative learning processes indirectly involves capturing, at a specific point in time, team processes, as well as team characteristics and team products. In other words, these measures are discrete, and capture a state or condition of the team processes at a particular point in time either during or after a team activity. These techniques do not directly measure team process but capture data that are indicative and linked to specific collaborative processes. There are two subcategories of indirect process measures: holistic and collective. These measures are focused on collecting team member responses about the process and are specifically not a direct observation of the process.

There are three key techniques to eliciting knowledge related to collaborative learning processes: (1) indirect process data, (2) characteristic data, and (3) product data. Indirect collaborative process data that describe collaborative processes can include factors such as communication (verbal/non-verbal), collaborative actions, collaborative behaviors, collaborative performance, and collaborative process. Team characteristic data include factors such as collaborative knowledge, collaborative skills, collaborative efficacy, collaborative attitudes, member roles, collaborative environment, and leadership. The elicitation techniques for these types of indirect data typically include interviews, questionnaires, and conceptual modeling (Cooke, Salas, Cannon-Bowers, & Stout, 2000). Analyzing the quality of individual products can be facilitated by the use of specified criteria that make it possible to create a product rating scale that can include numerical scales, descriptive scales, or checklists.

Capturing conceptual models involves assessing individual or team understanding about a given topic. Langan-Fox, Code, and Langfield-Smith (2000) have summarized that knowledge in teams has been investigated by aggregate methods that use elicitation techniques such as cognitive interviewing, observation, card sorting, causal mapping, and pairwise ratings, along with representation techniques such as multidimensional scaling and distance ratio formulas.

Use of concept maps is one of the most common methods for assessing collaborative knowledge (Herl et al., 1999; Ifenthaler, 2006; O'Connor & Johnson, 2004; O' Neil et al., 2000). With concept mapping, alignment of individual internal conceptualizations can be calculated in terms of the proportion of nodes and links shared between one concept map and another (Rowe & Cooke, 1995). Concept maps can be used to derive patterns of change in shared conceptual models within a CLE (Johnson & O'Connor, 2008b).

In the next section, we focus on describing benefits of four assessment tools that use concept maps to assess the progress of learning in a CLE. All of the methods described are intended to minimize the influence of the modeling tool or technique, although it is impossible to totally eliminate that influence. While the talk-aloud and think-aloud protocol analysis methods can be intended to minimize the influence of the method on capturing the representations, these older methodologies are also known to introduce some bias and influence. The advantage of the tools presented in this chapter is that, unlike think/talk-aloud methods, the qualitative components have been automated and formalized. All of these methods are, therefore, scalable, and can be and have been applied to larger subject pools.

METHODOLOGY DESCRIPTIONS

The four methodologies that are described in this section include: 1) Surface, Matching, and Deep Structure (SMD) Methodology, 2) Model Inspection Trace of Concepts and Relations (MITOCAR) Methodology, 3) Dynamic Evaluation of Enhanced Problem-solving (DEEP) Methodology, and 4) Analysis Constructed Shared Mental Models (ACSMM) Methodology. While the initial need driving the development of each method was different, this section describes the methods as they relate to learning assessment. The methods employ various techniques (quantitative, qualitative) but each uses concept maps as part of their methodological tasks.

Surface, Matching, and Deep Structure (SMD) Methodology

The question of a precise assessment of mental models led to the development of the SMD Technology (Ifenthaler, 2006, 2007), which uses graphical representations or concept maps to assess individual or collaborative processes in persons solving complex and ill-structured problems at single time points or at multiple intervals over time. As a basis for the assessment of models, SMD Technology uses graphical drawings (concept maps) or natural language statements (which are then converted into a concept map) made by individuals or teams. Both the graphical drawings and natural language statements are transferred into a dataset for an automated analysis. The relational (Surface Structure), structural (Matching Structure) and semantic (Deep Structure) analysis of graphical representations or concept maps use a computer-based and automated instrument. In three phases, the SMD Technology generates three numerical indicators for the Surface (θ), Matching (μ) and Deep (δ) Structure and a standardized graphical re-representation of each individual or collaborative concept map. Additionally, various indicators derived from graph theory are applied to more precisely describe changes in individual and collaborative processes over time (e.g., Ifenthaler, Masduki, & Seel, 2008). Besides these quantitative measures, SMD generates four standardized concept-map-like representations for qualitative analysis and as ready-to-use instructional materials: 1) individual or team representation, 2) reference or expert representation, 3) similarity representation (only including semantically similar propositions between individuals/teams and experts), and 4) contrast representation (including propositions which individuals/teams and experts do not share).

The surface structure is computed as the sum of all propositions (node-link-node) of an individual or collaborative representation. θ is defined between 0 (no proposition) and n (n propositions of the representation). The matching structure is defined as the quantity of links of the shortest path between the most distant nodes of the spanning tree of the representation (see Harary, 1974; Ifenthaler et al., 2008). The complexity indicator μ is defined between 0 (no links) and n. The deep structure is computed as the semantic similarity of propositions (Tversky, 1977) between a domain-specific expert representation and the individually or collaboratively constructed representation. The semantic indicator δ is defined between 0 (no similarity) and 1 (complete similarity). The automated analysis of the SMD Technology enables researchers and instructors to give immediate feedback during the learning process.

The basic technique involved in the SMD methodology involves eliciting concept maps from individuals at multiple intervals over time as they use a specific instructional intervention. SMD is then used to determine whether the concept maps for the individuals or teams changed over time and then to combine the concept maps of students in each treatment group and compare the findings between treatment groups and expert solutions in order to determine the effect (see Ifenthaler, 2006; Ifenthaler et al., 2008). The effects are determined by four analysis phases.

Phase 1: Surface Structure Analysis—The first level of SMD Technology is made up of the Surface Structure on which a rapid and economical assessment of the number of propositions (node-link-node) is made possible. The Surface Structure θ is defined as the sum of all propositions P in an individual or team model.

$$\theta = \sum_{i=0}^{n} P_i \tag{1}$$

Phase 2: Structural Properties Analysis—The assessment of the structural properties of the externalized models is carried out on the Matching Structure. The Matching Structure μ is defined as the quantity of links L of the shortest path between the most distant nodes K of the spanning tree of an individual or team model.

$$\mu = \max_{i,j}\{d(i,j)\} \tag{2}$$

Phase 3: Deep Structure Analysis—The third SMD Technology level is defined as the Deep

Structure. This is the level on which the models are assessed in terms of their semantic structure. The Deep Structure δ is calculated as the similarity (Tversky, 1977) between an individual or shared team model M_{gr} and a domain-specific reference or expert model M_{ex}.

$$\delta = \frac{f(A \cap B)}{f(A \cap B) + \alpha \cdot f(A - B) + \beta \cdot f(B - A)} \quad (3)$$

Phase 4: Construction of Standardized Representations (representative example)—For further qualitative analysis, SMD automatically constructs four standardized representations. The individual or team representation (see Figure 1) includes all nodes and links of the previously created concept map; however, nodes and links are represented without overlaps and always have the same color and font size. The representations are automatically generated with the help of the open source graph visualization software GraphViz (see Ellson et al., 2003). Having a standardized representation at hand helps researchers or instructors identify similarities and differences of individual or team representations.

The standardized reference/expert model can be created by one or more experts, instructors, or another team involved in the same problem-solving task. The expert model is the basis for the semantic analysis of the SMD Technology and also provides a possible solution for the underlying problem-solving task (see Figure 2).

Figure 1. Standardized SMD representation of a team model

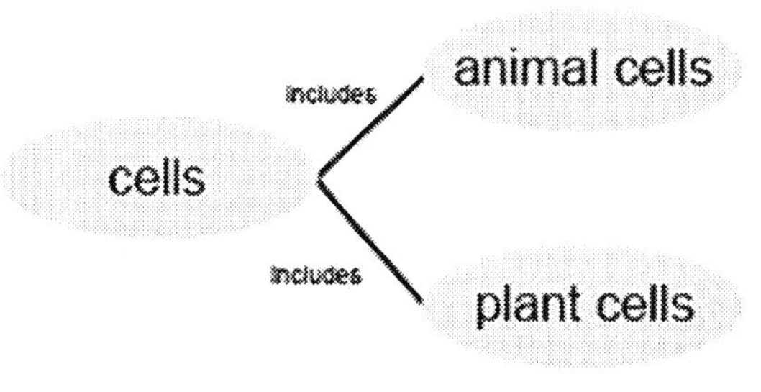

Figure 2. Standardized SMD representation of a reference/expert model

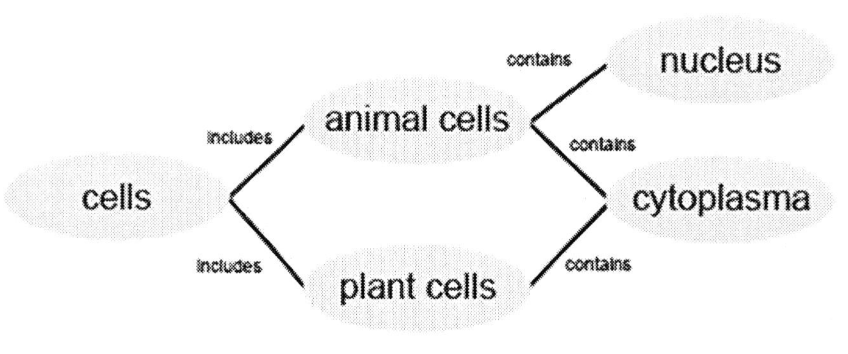

In contrast to the reference model (Figure 2), the similarity model (Figure 3) only contains semantically correct nodes and links. The similarity model can be used for individual or team self-assessment.

The contrast model includes only propositions that have no semantic similarity between the individual/team model and the reference/expert model (see Figure 4). Providing the contrast model to individuals or teams as instructional material can initiate or facilitate learning processes.

All four standardized representations are constructed "on the fly" and can be implemented in a learning environment.

Data Elicitation Representation: For SMD Cognitive structures (e.g., mental models or schemata) are elicited in the form of concept maps. The externalized models are stored pairwise (node-link-node) in a standardized list form (most concept mapping tools, (e.g., Cmap Tools, DEEP) support this function. Table 1 provides a representative example of the standardized list form that is used for the iterative calculations and detailed data analysis. The model number is a unique number for all pairwise stored propositions to be identified by the database and analysis technique.

Data Coding/Measurements: From each external representation (concept maps), SMD calculates: 1) the number of propositions in an individual model (Surface Structure) and 2) the 'diameter' of the individually constructed concept map (Matching Structure). These measures represent the structural complexity of the concept maps. The diameter and further indicators (e.g., number of cycles, number of sub-models,

Figure 3. Standardized SMD representation of a similarity model

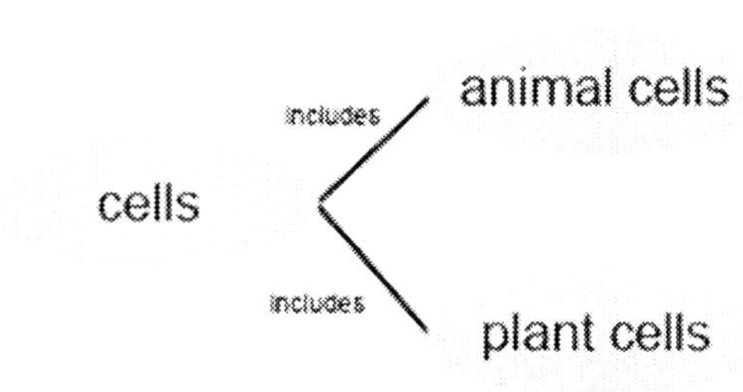

Figure 4. Standardized SMD representation of a contrast model

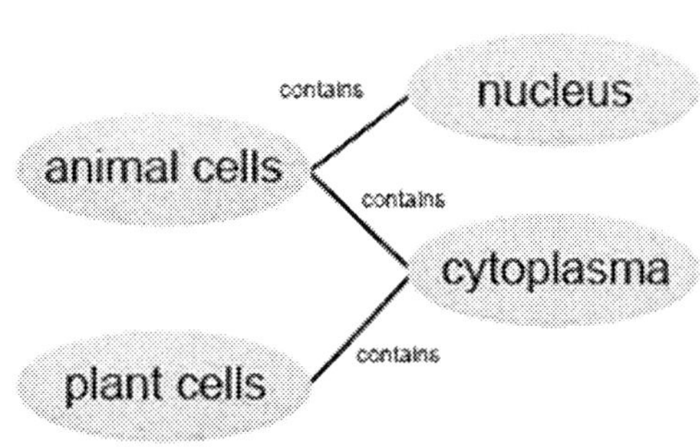

Table 1. Raw data of the contrast model of Figure 4 stored pairwise (as propositions)

Model number	*Node 1*	*Node 2*	*Link*
004	nucleus	animal cells	contains
004	cytoplasma	animal cells	contains
004	plant cells	cytoplasma	contains
...			

ruggedness) are derived from graph theory (see Harary, 1974; Ifenthaler et al., 2008). Additionally, domain-dependent semantic measures are calculated with the help of the similarity measure between individual or team models and reference models (e.g., expert representations). SMD then calculates 3) the semantic similarity of single nodes – Vertex Matching (see Pirnay-Dummer, 2006) and the semantic similarity of propositions – Deep Structure.

Data Analysis: The quantitative measures of the SMD Technology are generated automatically and stored on an SQL (structured query language) database. The core measures of the quantitative analysis include structural (number of propositions, complexity of representations) and semantic (correct nodes and propositions compared to a reference model) indicators. Further data analysis can be applied with standard statistical packages, and all stored data can also be exported into a spreadsheet. An in-depth qualitative analysis is made possible by the four standardized representations (individual/team model, expert/reference model, similarity model, and contrast model). Various validation studies have shown a high reliability and validity for the SMD Technology (e.g., Ifenthaler, 2006, 2007; Pirnay-Dummer, 2006).

Model Inspection Trace of Concepts and Relations (MITOCAR) Methodology

Model Inspection Trace of Concepts and Relations (MITOCAR) (Pirnay-Dummer, 2006, 2007) is a tool based on mental model theory (Seel, 1991). One of the goals of MITOCAR is to dig deeper into the semantics of models, especially shared models within groups of experts. This is done during a phase of assessment and an inferential phase. During the assessment phase of MITOCAR, subjects usually go through two different rounds. In the first round they only provide a number of natural phrases (usually sentences) about their specific subject matter. Before the second round, the parser extracts the most frequent concepts from the text corpus of the team and connects them to pairs of concepts.

In the second round the subjects rate how close the concepts are and how sure they are about their assessment. The participants also cluster their concepts from a random list into a list of groups – a method that is sometimes used in knowledge tracking (Janetzko, 1996). Additionally, subjects rate the plausibility of their fellow team members' source phrases.

Using the inferential modules, MITOCAR calculates a proximity vector that represents the whole model that is used to build the model representation (concept map). These data make it possible to create models, even graphical ones (Fruchterman & Reingold, 1991; Ganser & North, 1999) that can then be compared in several ways as explained below. In addition MITOCAR provides a variety of analysis measures based on graph theory and Tversky-Similarity (Tversky, 1977) to compare concept maps in general (not only the ones assessed by MITOCAR). For example, concept matching compares the use of terms between different models, and structural matching introduces an algorithm that compares concept maps in terms of structure only (providing a testing ground for hypotheses and theories about structure of expertise), and several density measures, such as the vertices-per-edge quotient γ (cf. Pirnay-Dummer, 2006).

The MITOCAR software tool uses natural language expressions as input data for model re-representation instead of using graphical drawings by the subjects. This is made possible by parsing and corpus linguistics technologies that are similar to those used to implement automated learning of concept hierarchies from text corpora to construct ontology (Maedche et al, 2002). Parsing is a process whereby text is sorted based on specific rules embedded in the parser. Whereas the semantic comparison of MITOCAR uses traditional measures of similarity (Tversky, 1977), the technology of structural comparison, which is unique to MITOCAR, can compare models from different subject domains (Pirnay-Dummer, 2006). MITOCAR is able to re-represent the models graphically (concepts and structures) on non-directed graphs, and compare them by using conceptual, structural and combined similarity measures. These measures are controlled using different statistical tests and controls for homogeneity and reliability of the team consensus models, multidimensional scaling (MDS) of the proximity vectors to test for the representation in 2D and for the comparison and tracking of model complexity.

The final outputs of MITOCAR are the graphical representations to assess a team consensus model from any subject domain and the comparison between different groups of experts. For both outputs there are automated reports that compute, present and interpret all the above-mentioned measures. The methods of MITOCAR have been applied successfully to tracking user behavior in e-learning environments (Dummer & Ifenthaler, 2005) as well as supporting a needs assessment (Pirnay-Dummer, 2006; Pirnay-Dummer, Nußbickel, in press).

Due to the modular design of MITOCAR, the assessment tools (re-representation of models by means of natural language, parsing and graph theory) can be separated from the inferential tools (comparing structures and semantics and both). This opens the MITOCAR technology for use on all kinds of model-related data.

Except of course for the interaction with the subjects, MITOCAR is a fully automated tool. Data collection, descriptors, and analysis are all handled on a web server. No manual effort is required.

The subjects write text as a number of statements about a subject domain (identification). The statements of the whole team are rated, ranked and parsed for their most frequent nouns. The nouns are presented in a randomized list for an association test (construction). Afterwards, the pairs of nouns are rated with different measures for their degree of association within a team (verification) and between teams (confrontation).

The measurements within MITOCAR are different ratings on relatedness and transition probabilities (between concepts). Models are stored and represented in a list form of pairwise concepts. The visual result is a graphical output of the model with no previous concept mapping by the subjects.

MITOCAR analyses the data to construct the model by generating a proximity vector. The data are automatically analyzed for quality (selectivity, homogeneity, representability) to make sure that MITOCAR always uses the right measures for a specific dataset.

In the first phase, a simple collection of statements on a given subject domain is collected from each subject. Participants type in 15 (the number can be pre-determined) expressions about a given topic. (Expressions can contain more than one sentence. Experts have a tendency to write more text into each expression.) The prompt is similar every time: "Please provide us with the 15 most important statements on X". Afterwards, each individual rates every expression from the team separately (peer rating). It involves three items for each expression:

1. Rate if the expressions are true
2. Rate how related the expression is to the subject matter
3. Rate how sure they are about their rating

With large sample sizes, this information can be used to structure the dataset a priori. With smaller sample sizes (usually the case in research on learning and instruction), the rating can be used as a descriptive in addition to the graphical model, e.g., to derive the "Top 10" statements from the team.

The expressions are then parsed to extract the nouns (as concepts), and the 30 most frequent terms are then used in the process.

Construction, verification, and confrontation are the three model-representation core modules within MITOCAR that are used after parsing. Each module provides a different perspective on the same set of terms, depending on the research question. After completing all three of these modules, the model can be created. The construction module can be left out for small sample sizes of less than 50 words because this module produces almost the same data as the verification module. The construction module is, however, mandatory when the research question deals directly with association. Confrontation and verification are the same methodologically. They differ only in the data they process: verification covers the within-team assessment and confrontation is for between-team assessments. For the confrontation module, the subjects deal with the set of concepts of another team in the confrontation module (rather than with the content of their own team).

Construction: In the construction phase, the subjects categorize a randomized list of the concepts into groups which can be processed into model information using Markov chains: the instrument for the construction mode is based on the knowledge-tracking technology introduced by Janetzko (1996). The subjects are asked to go through the list and click on concepts that they associate with a meaningful group. They do not have to name the groups. They are asked to keep the number of groups as small as possible while still being selective as regards content. All frequencies of pairs are calculated from the grouped list: if concept B follows concept A in a list (and no group separation marker is in between), then the pair (A, B) gets +1 added to its frequency value. The matrix of term frequencies can be transformed into basic transition probabilities that make it possible to build Markov chains from the matrix (cf. Chung, 1968). The grouped lists are then transferred to a paired-concepts matrix (just like the adjacency matrix for diameter), and every pair gets a plus one if they are neighbors. The matrix is a relative measure of the transition probability. The 30 strongest neighbors find their way into the final model representation.

Verification: In the verification module, the subjects rate each pair of terms from the selected concepts. With 30 terms, there are 435 possible combinations. Some of them may be rated repeatedly to control for reliability. Each pair is subject to three ratings by each individual of the team:

1. The item of closeness describes how closely related two concepts are rated as being by the subjects on a 5-point scale
2. For the item of contrast, the subjects rate how different two concepts are or to what extent they exclude each other (e.g., fire and water) on a 5-point scale
3. The confidence rating measures how sure the subjects are of their ratings of contrast and closeness, also on a 5-point scale. Only the higher ratings are used in subsequent data analyses.

The ratings may be combined to create a matrix of average weights. Meaningful combinations are closeness & contrast, closeness & confidence, contrast & confidence and finally closeness & contrast & confidence. Depending on the quality of the data (which is tested before re-representation), different measures may be used. For example, if the combined item has too much deviance or is inhomogeneous within a team, it will be excluded from re-representation. This is automatically tested and reported by the MITOCAR software prior to the model construction.

Confrontation: This module is exactly the same as the verification one with one important difference: in confrontation the pairs of terms come from another team (e.g., another kind of expert within the same domain, such as doctors and nurses). In general, for research questions which aim at contrasting different teams, the confrontation module may be a helpful tool.

MITOCAR uses different levels of shared model data validation. Multidimensional Scaling (MDS) is used to calculate the stress value and to represent the distance distribution of the concepts as another descriptive. The stress value from MDS displays how much the graphs' distances resemble the real distances from the data, even on only two dimensions. The methodology considers the representation of the matrix to be sufficient with 30 re-represented distances, if the error is not much higher than 30%. The error reported by the MDS occurs only in representing the distance matrix (level of association) correctly. All other information drawn within the graph fits the data matrix completely.

MITOCAR uses Within Group ANOVA tests on a pair-of-concept level to test homogeneity (levels of agreement on items). If the within-pair deviation is statistically significant, the data will be considered to be inhomogeneous. Thus, the model would no longer be considered a team consensus model. Additionally, all ratings are also correlated to check for selectivity between the measures (e.g., between closeness and contrast).

In order to construct a shared model, the data matrix which generates the model's graphical

Figure 5. Sample output graph from MITOCAR (experts on instructional design)

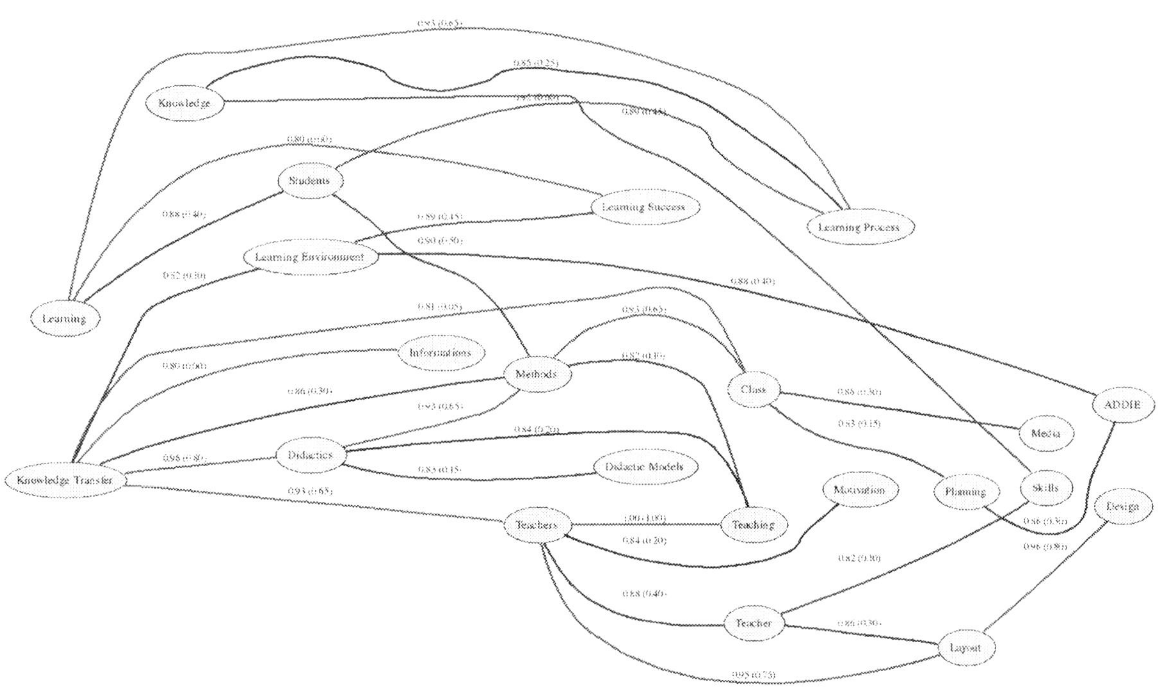

representation is referred to as the proximity vector. (see Figure 5)

The proximity vector contains the pairs and the weights that are calculated by single or combined measures (from closeness, contrast, confidence). On every link in the graph there are two numbers, one outside and one within brackets X (X.X): the first X is the normalization weight for the entire data set, and the second (X.X) is the normalized weight for just the represented graph. For example, for the numbers 0.40 (1.00), the 1.00 indicates that it is the strongest link relationship in the presented model (scale normalization) and 0.20 (0,00) says that this is the weakest link for the presented model; the 0.40 is the strength for the overall model including the parts that are not presented as well as the 0.20. The result 0.30 (0.50) … means that this is half the strength as the 1.00 and the value is 0.30. The link thus accounts for the measure of how related two concepts are, normalized on a scale of 0 to 1. The color also corresponds to the weight, ranging from red (strongest link) to blue (weakest link that made it into the graph). Usually the weakest links still have a strong weight when compared to the whole data matrix.

Dynamic Evaluation of Enhanced Problem-solving (DEEP) Methodology

The Dynamic Evaluation of Enhanced Problem-solving (DEEP) (Spector & Koszalka, 2004) methodology is based on a view of learning as becoming more expert-like (Ericsson & Smith, 1991) and more skilled in higher-order causal reasoning and problem solving (Grotzer & Perkins, 2000). A fundamental assumption is that with regard to complex task performance it is possible to predict performance and assess relative level of expertise by examining a person's conceptualization of the problem space (a problem conceptualization that suggests likely solution alternatives) for specific complex problems. These representations can be compared with prior representations or those of experts using the analytic techniques discussed with regard to MITOCAR and SMD. Moreover, these representations can be created by small teams as well as individuals, although thus far only the representations of individuals have been involved in the development and validation of the methodology.

In DEEP, learners, either individually or in small teams, are presented with a short problem scenario. They are asked to identify the ten or so most relevant factors and issues to consider in developing a solution. Respondents are then asked to describe each of these factors. Next, respondents are asked to indicate how the various factors are related and to describe the nature of each relationship. At this point, respondents will have developed what amounts to annotated causal influence diagrams, although not all of the links are necessarily causal. Some links might be correlational or procedural or even examples illustrating or elaborating another node in the structure. Respondents are then asked two final reflection questions: (1) What else would you need to know in order to actually resolve this problem situation? and (2) What assumptions have you made in responding to this problem? These reflection questions serve to resolve differences when comparing representations. Moreover, the annotated nodes and links can be further analyzed using MITOCAR since their content is primarily textual.

This method was inspired by the causal influence diagrams that system dynamicists use to elicit domain and system knowledge from experts when modeling complex, dynamic systems. One advantage of this methodology is that it is relatively simple to use and understand, minimizing the influence of the method on representing key knowledge structures.

The annotated causal representations in DEEP are compared with prior representations and with those of experts to determine the progress of learning. Three levels of analysis can be applied to these representations: surface, structural and semantic. Simple counts of nodes, links, and words per node or link constitute a simple, surface-level analysis. Determining the extent of similar nodes and links and how they were connected constitutes the structural analysis. Understanding what is said about a particular node or link constitutes the semantic analysis. Because the tool allows nodes and links to be named according to respondent preferences, it is not possible to easily separate the structural and semantic analyses. One purpose of developing DEEP was to determine the extent to which a semantic analysis would be required to determine relative level of expertise, which is the specific purpose for collecting responses to the two reflective questions. Since MITOCAR, SMD, ACSMM, and DEEP have been integrated in the tool called HIMATT (Highly Interactive Model-based Assessment Tools and Technologies), the issue of determining the similarity of specific nodes can be determined by a MITOCAR analysis within HIMATT.

A unique aspect of the DEEP methodology is that it is intended for complex problems involving causal relationships that are interrelated and that may change over time. Moreover, a variety of graphical representations (e.g., semantic networks, flowcharts, causal diagrams) can be accommodated in this methodology. The graphical representations are converted into standard causal representations (i.e., annotated causal influence diagrams). The reason for using causal representations as the basis for analysis is that these representations reflect internal relationships among factors and components (i.e., problem dynamics), and causal representations can be derived from many other graphical representations when the appropriate documentation is provided (e.g., the descriptions of individual factors). The DEEP methodology supports assessments of individual learning in problem-centered instructional modules, which can be used in evaluating problem-centered instructional programs (Baker, 1999; Herl et al., 1999). The data provide information about how well individual learners are doing in specific problem-centered modules. This enables teachers to adjust instructional scaffolding appropriately, and enables learners to adjust their learning strategies. Additionally, instructional designers are provided with information they can use to make specific modifications to the structure and sequence of various learning activities.

Variations of this methodology have been effectively demonstrated in simpler domains (Herl et al., 1999; Novak, 1998; Schvaneveldt, 1990). Those who have employed an analogous method for simpler learning tasks have relied on: (a) simple quantitative measures of similarity to expert responses (e.g., presence/absence of salient features and their location in a concept map); and (b) qualitative analysis of responses, which are notoriously time-intensive and costly and, consequently, hardly ever used when a laboratory or demonstration effort of a learning environment or instructional system scales up to full-scale implementation and deployment. DEEP within the context of the HIMATT tool set provides more sophisticated quantitative measures and the basis for more efficient qualitative analyses.

Analysis Constructed Shared Mental Model (ACSMM) Methodology

The Analysis Constructed Shared Mental Model (ACSMM) methodology (O'Connor & Johnson, 2004; Johnson & O'Connor, 2008b) was developed as a method of determining sharedness among team members. The basic technique involves eliciting concept maps from individual teammates and then using the ACSMM phases to analyze the individual concept maps, thereby creating a single concept map that represents the shared components of the team.

Through concept mapping, the similarity of mental models can be measured in terms of the proportion of nodes and links shared between one concept map (individual mental model) and another (Rowe & Cooke, 1995). By using qualitative techniques with an aggregate method of creating an ACSMM, it is possible to have a more descriptive understanding than can be attained by using quantitative techniques.

The ACSMM technique transforms individual mental models into a team sharedness map without losing the original perspective of the individual, thereby representing a more accurate representation of team sharedness. The methodology includes several phases: elicitation design and preparation, elicitation of individual team member mental models, coding of individual data, analysis of data to determine what is shared among team members, and construction of the team conceptual representation.

Phase I: Elicitation Design—Before eliciting mental models, a few critical design issues related to the elicitation of concept maps (i.e., developing mapping skills and determining mapping techniques) should be considered. It is important that the learners have an understanding of how to create a concept map. In order for an individual to create a concept map, it is reasonable to provide them with an example explaining the notion of a concept and link and a proposition explaining the relationship between concepts. Guided practice is provided on how to create a generic concept map. This practice is provided only if needed. Learners then create a very simple concept map in order to practice this skill. Then the learners receive feedback related to their general concept mapping technique. Feedback as to the quality, depth, or quality of their logic is not provided.

The next step involves considering how the maps are to be elicited. Depending on the assessment goals, concept-mapping techniques could fit along a continuum from totally unconstrained (where the learner simply creates their concept map) to fully constrained (where the learner only uses concepts in their own maps that have been provided from other concept maps). If the assessment requires a fully constrained or semi-constrained concept map, a topic analysis is performed to generate the list of related terms that will be provided to the learners to use in their concept maps. An analysis of a topic involves focusing on determining the various components of a concept and the logical relationships between concepts, if there are any. Topics can be ill-structured and allow for multiple logical arrangements. Once the components of the topic have been generated, these terms can be used to help elicit individual mental models.

Data analysis of fully constrained maps is easier to carry out because the set of terms used by each learner is the same. However, this only allows individuals to represent their concept maps based on the concepts that are provided to them, and they are not allowed to add any additional terms. This may inhibit a full elicitation of each individual learner.

Phase 2: Individually Constructed Mental Model (ICMM) Elicitation—If the learners have the skills to create concept maps, they are ready to create their ICMM. Individually, learners are asked to create their concept maps. They are provided with the materials (such as paper and pens) needed to create these maps. Creating a well-thought-out map may take approximately 30 minutes, depending on the complexity of the topic.

Phase 3: ICMM Coding—There are many components for coding concept maps. In this section we present the general framework for ACSMM coding. Refer to the work of Johnson and O'Connor (2008b) for specific details and examples of coding.

Coding of the individual maps is required in order to compare and measure a degree of sharedness in individual maps. The ACSMM method accounts for map relatedness at the concept, link, and cluster levels. Because maps are so unique, the coding strives to reduce the logical, spatial, and structural information and code them so that comparisons between maps can be made. The coding process involves documenting the explicit information on the maps as well as making assessments regarding implicit information. This assessment allows for explication of implicit relationships by considering the spatial, structural and logical information in the map. The coding process involves studying concepts, links, and clusters within each ICMM. The process of coding is much like the process of interpretation. Each map is analyzed, and then the researcher codes their interpretation in a spreadsheet or other appropriate tool. The coding factors include (1) concepts, (2) links, and (3) clusters (see Figure 6). In addition, the coding process considers coding the explicit data (for concepts and links) and coding the implicit data (for links and clusters).

Factor 1: Coding concepts. Concepts (nodes) are the terms used in the concept maps resulting from the topic analysis and/or those added by individual team members. Concept (explicit individual nodes) coding involves listing the concepts from a given map. All terms that are used in a map are coded by documenting them as concepts. If the elicitations were not constrained (i.e., if individuals could include any term they wanted), then concept coding would also involve resolving like and similar terms.

Factor 2: Coding links. Links consist of two concepts (nodes) and one connector (lines, unidirectional and bidirectional arrows). A link refers to a node-connector-node combination, indicating a relationship between the two nodes. Links can be included in a concept map in many different ways: simple explicit links, complex explicit links, complex links containing open-end connectors, and implicit links.

The descriptors that are used to describe the relationship between concepts are a key part of concept maps. Explicit links may or may not be labeled by a descriptor describing the relationship between the nodes. If a label is present, it is coded

Figure 6. Example of ICMM and coding for select items

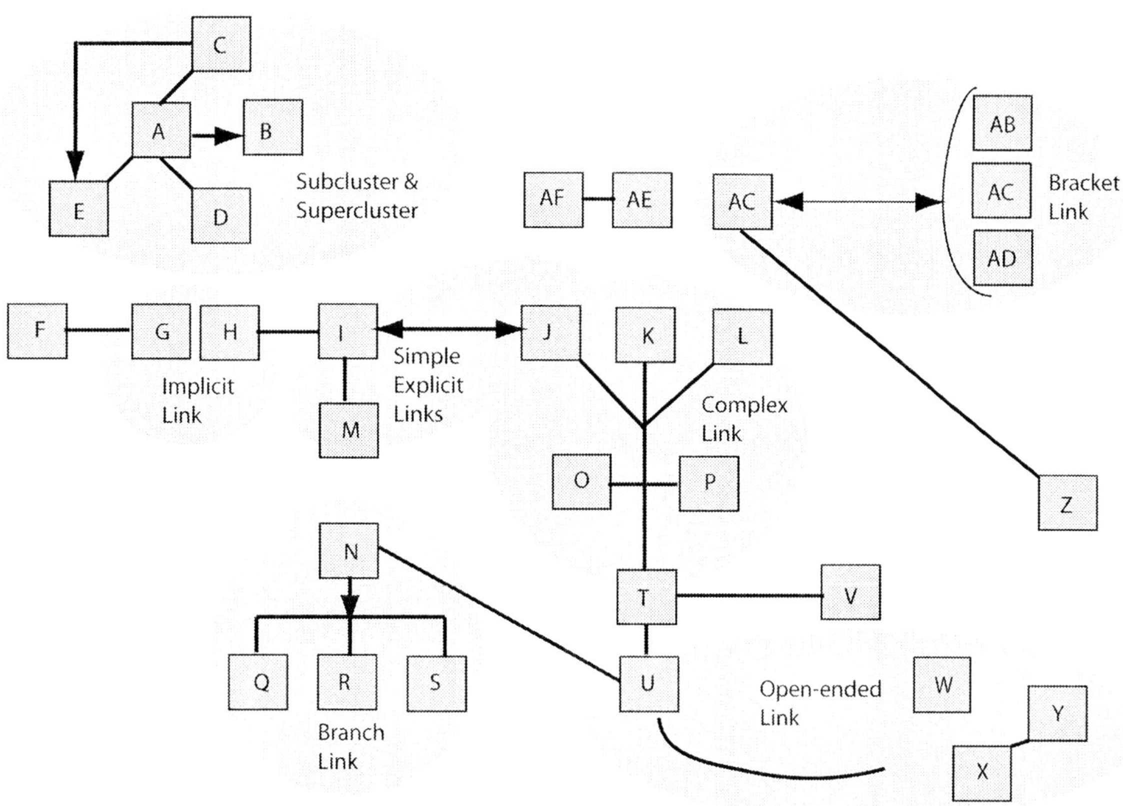

and can be included in the data analysis. While we have categorized several types of links, the most common links include simple explicit links, complex explicit links, and implicit links.

Simple explicit links are merely a connector (with or without arrowheads) between two concepts. There can be three types of simple explicit links: lines, unidirectional arrows, and bidirectional arrows.

Complex explicit links are two concepts related through two or more intersecting/shared connectors that do not intersect with another concept before establishing the link. These complex links may also include cases where multiple connectors intersect/share in a bracket, branch, or open-ended fashion.

Implicit links primarily focus on concepts that are related spatially. However, implicit links are determined not only by their spatial relationships, but also by their structural and logical relationships. Because of the complexity and difficulty of creating concept maps, there are cases in which maps seem to have meaning but this meaning is not explicit in the structure of the map. A common example is when there are three concepts (that are juxtaposed) that do not have lines connecting them. In order to make meaning of the map, the coder uses a technique for coding implicit links and clusters relying on consideration of the map's spatial orientation, structure, and semantic logic.

As a guideline for coding, before an implicit link (or cluster) is coded, the coder should check for (1) logic and spatial congruency or (2) logic and structural congruency. At least one of these guidelines must be satisfied before coding implicit clusters or links. Spatial determination of implicit links depends on the orientation of the concepts within the context of the concept map. Implicitly related concepts are generally placed more closely together compared to concepts surrounding them and other concepts within the map. In terms of structural support, implicitly related concepts have either no explicit structural support or a limited explicit structural/spatial support. Concepts are often placed adjacent (node-node) to another without the connector relationship. In the case of open-ended links, the incomplete structural support is interpreted as complete so long as there is a logical relationship between the concepts. The logic component for determining implicit links refers to the conceptual relationship among the nodes in the identified implicit link. The conceptual relationship does not have to be complete, but it does need to have good logic.

Factor 3: Coding clusters. After links are coded, maps are analyzed in search of indicators of more complex knowledge than what is represented by the combination of simple and/or complex node-connector-nodes. Clusters are minimally two connectors bridging three or more concepts. Clusters can include simple links, complex links, explicit links, implicit links, or a combination of any link types. Like implicit links, three key components must be considered when identifying clusters: spatial, structural, and semantic logic information.

Explicit clusters may contain a combination of connectors (non-directional, unidirectional, bidirectional, brackets, or branches). Explicit clusters may be identified by structural cues such as circles drawn around a group of concepts, prepositional descriptors, or other notations provided by the individual creating the ICMM. Clusters may consist of closed loop or triangulated relationships between concepts. Implicit clusters contain three or more concepts placed adjacent to one another, thus providing spatial support in addition to logical support.

Phase 4: Shared Analysis—After coding the ICMMs, the next step involves an analysis of the ICMM dataset to determine which items were shared by team members. The data tables resulting from ICMM coding are compared for similarity across team members. Start with a sharedness criterion of 50%. Then determine the shared items for that sharedness level. Depending on the sharedness level sensitivity, the level can be adjusted up or down to increase or decrease

Figure 7. AC-SMM constructed from shared data resulting from hypothetical member ICMMs

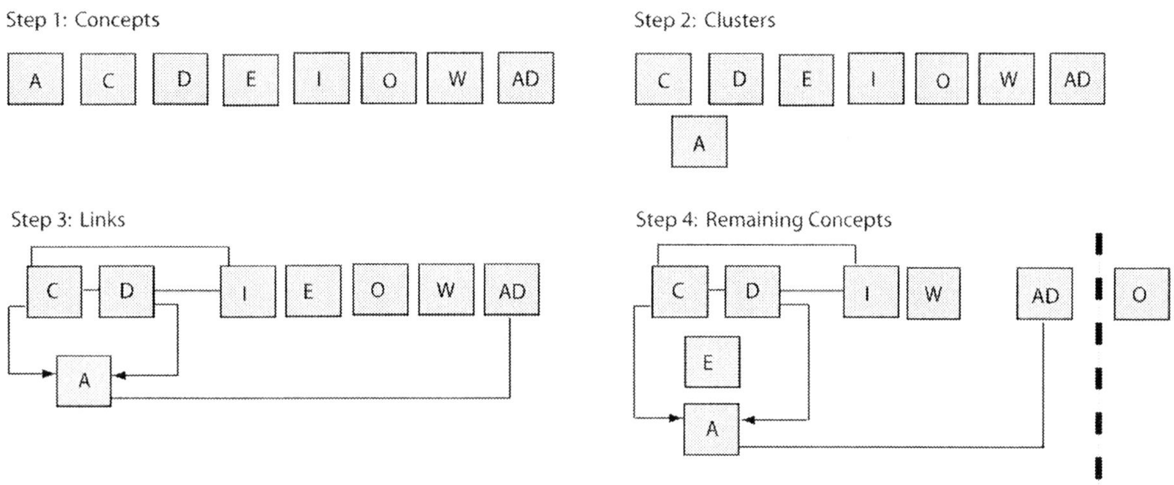

the sensitivity. The percentage or number of learners sharing each item is recorded, and all shared items are carried forward for use in constructing the representation of the team's shared mental model, the ACSMM.

Phase 5: ACSMM Construction—The analysis constructed shared mental model (ACSMM) is constructed from the shared ICMM dataset generated in Phase 4. This construction process includes the following steps: 1) List Shared Concepts, 2) Configure Shared Clusters, 3) Configure Shared Links, and 4) Configure Non-linked Concepts (see Figure 7). The ACSMM methodology is repeated for each subsequent set of concept maps. Once the sharedness among all ICMMs has been identified, the analysis constructed shared mental models can be compared by looking at change over time for a specific team, or the ACSMMs can be compared among teams to show variation of shared mental models among the learners.

FUTURE TRENDS

The manner in which concept-mapping tools are used to assess individuals and teams in collaborative learning environments continues to evolve, especially as workplace and educational organizations continue to place more emphasis on the use of learning collaboration. As more attention shifts towards collaborative learning environments, not only will this increased interest provide new opportunities for research and development of CLE, but it will also bring a more critical view of CLE, thereby exposing various systemic weaknesses as well as strengths. Future developments need to consider this new attention in the push to establish the viability and efficacy of collaborative learning environments. To establish a strong foundation for CLE claims, well-developed and empirically driven research tools (like the ones presented in this chapter) are needed to frame and carry out studies that will solidify the strategies and techniques espoused in CLE. Using these tools, we can better understand the underlying mechanisms and processes that take place within CLE, including the mental states that individuals and teams go through when collaborating.

There are several key trends that focus on using concept mapping assessment tools in CLEs.

Trend 1: Mental model comparisons—A key trend for CLEs is the exploration of the effects of interactions among individuals as well as a comparison of interaction patterns across teams.

Used as indirect measures (Johnson & O'Connor, 2008a), concept mapping assessment tools are being used to capture collaborative processes and used to compare processes across varying levels of expertise for individuals and teams. These process comparisons will enable researchers and developers to see how different individuals and collaborators interact in different CLEs. With this knowledge, process support mechanisms can be implemented to enhance the effects of CLEs.

In addition to process measures, state measures can be captured and then compared using concept mapping assessment tools. These comparisons can look at mental models of individual and teams with themselves and with an expert model. These comparisons will show developmental convergence along an expertise continuum, and will help explicate the states of individuals and teams as they engage in learning difficult and challenging tasks.

Trend 2: Learner feedback—In parallel with the comparisons trend above, concept mapping assessment tools can be used to provide feedback to learners. Such feedback can include information on how a specific collaboration is interacting and also progressing. In order to support a learning collaboration, feedback can be presented as a synthesis of information from expert collaborations related to interaction and progress. This feedback can be used as a means of coaching individuals and teams on how they are doing (progress) as well as what to do (interactions) with a target of supporting the development of a collaborative learning strategy.

Trend 3: Methodology comparison research—An important trend is that of developing an understanding of the various concept mapping assessment tools with respect to each method's strengths, unique characteristics, and collective viability. The authors have set up a series of comparative studies to address these issues. A series of pair-wise comparative studies are followed in order to detect analytical differences among the methodologies. The two points of method comparisons are focused on data conversion techniques and data analysis techniques.

Each of these methodologies has been used in specific settings; however, the work represented in these comparisons is the first project utilizing a single set of data to compare the results of these four methodologies. There are six studies involving two parts each. These studies compare the following pairs of methods: 1) ACSMM & MITOCAR, 2) SMD & MITOCAR, 3) ACSMM & SMD, 4) DEEP & SMD, 5) DEEP & MITOCAR, and 6) DEEP & ACSMM.

The first consideration is the data analysis techniques. Since each method has different techniques for data conversion prior to data analysis, the initial comparison among the methods will focus on the specific analytic techniques employed. As such, the comparisons will begin with the same dataset in order to control for the analysis phase of each method. This will involve taking a set of decomposed data from concept maps or natural language, and the analysis techniques used in each method will be employed and the output compared.

The second consideration compares the data conversion techniques. This will involve collecting data according to the method specifications and then going through the process of data conversion. The converted data will then be compared for similarities and differences.

For example, the application of the SMD-Technology in different subject domains and the comparison with other quantitative methodologies, such as MITOCAR, or qualitative methodologies, such as ACSMM and DEEP, could cross-validate the SMD-Technology (and vice versa) and give a more detailed understanding of the changes of mental models within individuals and teams.

CONCLUSION

The goal of integrating SMD, MITOCAR, DEEP, and ACSMM is to develop extensive datasets that can be used by multiple researchers and developers to carry out their work of understanding and creating CLEs to facilitate learning. Additionally, the integration of concept mapping assessment tools together will collectively evolve to include several different problem scenarios and many points of comparison, thereby helping researchers and developers understand the implications of various CLEs. In other words, this stream of research and development will provide guidance for practitioners to select the appropriate concept mapping assessment tools that can be used in collaborative learning environments.

From the integration of these tools, we hope to better qualitatively and quantitatively represent individual and team mental models, thereby facilitating greater understanding of the notion of collaborative team processes and the development of mental models. Further, by comparing teams at various points during collaboration processes, we should be able to determine how collaborative mental models change over time. With each tool, one can test both the progression/development of conceptual representations with each other from the initial state to the post state and the similarities among the individually constructed conceptual representations. Not only will this information benefit further study in individual and team dynamics, but also (if we can identify how team mental models change over time and find indicators of why they change) we should be able to develop methods for improving overall individual and team learning.

Conducting a cross validation of these assessment tools will give us the quantitative control for these assessment tools as well as qualitative control for quantitative assessment tools. With this evidence, we will have validated tools to understand the impact and value of collaborative learning environments on both individuals and teammates. With this understanding, we will be in a position to modify and adjust CLE techniques to best improve learning. This also includes the use of the tools to carry out individual and team assessment as a means of providing feedback to support optimal learning gains from a CLE.

REFERENCES

Baker, E. (1999, July). *Technology: How do we know it works?* Paper presented at the Secretary's Conference on Educational Technology, July 12-13, 1999, Washington, D.C. Retrieved from http://www.ed.gov/Technology/TechConf/1999/whitepapers/paper5.html

Bransford, J. D., Brown, A. L., & Cocking, R. R. (Eds.). (2000). How experts differ from novices. In *How people learn: Brain, mind, experience, and school* (pp. 31-50). Washington, DC: National Academy Press.

Cannon-Bowers, J. A., & Salas, E. (Eds.). (1998). *Making decisions under stress: Implications for individual and team training*. Washington, DC: American Psychological Association.

Carnevale, A., Gainer, L., & Meltzer, A. (1989). *Workplace basics: The skills employers want*. Alexandria, VA: American Society for Training and Development.

Cooke, N. J., Salas, E., Cannon-Bowers, J. A., & Stout, R. J. (2000). Measuring team knowledge. *Human Factors*, *42*(1), 151–173. doi:10.1518/001872000779656561

Cooke, N. J., Salas, E., Kiekel, P. A., & Bell, B. (2004). Advances in measuring team cognition. In E. Salas & S. M. Fiore (Eds.), *Team cognition: Understanding the factors that drive process and performance* (pp. 83-106). Washington, DC: American Psychological Association.

Doyle, J. K., Radzicki, M. J., & Trees, W. S. (1998). *Measuring Change in Mental Models of Dynamic Systems: An Exploratory Study*. Unpublished manuscript.

Dummer, P., & Ifenthaler, D. (2005). Planning and assessing navigation in model-centered learning environments. Why learners often do not follow the path laid out for them. In G. Chiazzese, M. Allegra, A. Chifari & S. Ottaviano (Eds.), *Methods and technologies for learning* (pp. 327-334). Boston: WIT Press.

Ellson, J., Gansner, E. R., Koutsofios, E., North, S. C., & Woodhull, G. (2003). *GraphViz and Dynagraph. Static and dynamic graph drawing tools*. Florham Park, NJ: AT&T Labs.

Ericsson, K. A., & Smith, J. (Eds.). (1991). *Toward a general theory of expertise: Prospects and limits*. New York: Cambridge University Press.

Fiore, S. M., & Salas, E. (2004). Why we need team cognition. In E. Salas & S. M. Fiore (Eds.), *Team cognition: Understanding the factors that drive process and performance* (pp. xi, 268). Washington, DC: American Psychological Association.

Fruchterman, T. M. J., & Reingold, E. M. (1991). Graph drawing by force-directed placement. *Software, Practice & Experience*, *21*(11), 1129–1164. doi:10.1002/spe.4380211102

Ganser, E. R., & North, S. C. (1999). An open graph visualization system and its applications to software engineering. *Software, Practice & Experience*, *00*(S1), 1–5.

Grotzer, T. A., & Perkins, D. N. (2000). *A taxonomy of causal models: The conceptual leaps between models and students' reflections on them.* Paper presented at the National Association of Research in Science Teaching (NARST), New Orleans, LA, April 28-May 1, 2000.

Guzzo, R. A., & Salas, E. (1995). *Team effectiveness and decision-making in organization.* San Francisco, CA: Jossey-Bass, Inc.

Hackman, R. A. (1990). *Groups that work (and those that don't): Creating conditions for effective team work*. San Francisco,CA. Jossey-Bass.

Harary, F. (1974). *Graphentheorie*. München, Germany: Oldenbourg.

Herl, H. E., O'Neil, H. F., Jr., Chung, G. L. W. K., Bianchi, C., Wang, S., Mayer, R., et al. (1999). *Final report for validation of problem solving measures* (CSE Technical Report 501). Los Angeles: CRESST.

Ifenthaler, D. (2006). *Diagnosis of the learning-dependent progression of mental models. Development of the SMD-Technology as a methodology for assessing individual models on relational, structural and semantic levels*. Freiburg, Germany: Universitäts-Dissertation.

Ifenthaler, D. (2007). *Relational, structural, and semantic analysis of graphical representations and concept maps.* Paper presented at the Annual Convention of the AECT, Anaheim, CA.

Ifenthaler, D., Masduki, I., & Seel, N. M. (2008). *Tracking the development of cognitive structures over time.* Paper presented at the AREA 2008, New York.

Ifenthaler, D., & Seel, N. M. (2005). The measurement of change: learning-dependent progression of mental models. *Technology, Instruction . Cognition and Learning*, *2*(4), 317–336.

Janetzko, D. (1996) *Knowledge Tracking. A Method to Analyze Cognitive Structures*. Freiburg, Germany: IIG-Berichte 2.

Johnson, D. W., Maruyama, G., Johnson, R., Nelson, D., & Skon, L. (1981). Effects of cooperative, competitive, and individualistic goal structures on achievement: A meta-analysis. *Psychological Bulletin*, *89*(1), 47–62. doi:10.1037/0033-2909.89.1.47

Johnson, T. E., & O'Connor, D. L. (2008a). Assessment of group learning processes. In M. D. Merrill, van Merriënboer, J., Spector, J. M., & Driscoll, M. (Ed.), *Handbook of research on educational communications and technology* (3 Ed., pp. 776-783). New York: Lawrence Erlbaum Associates.

Johnson, T. E., & O'Connor, D. L. (2008b). Measuring team shared understanding using analysis constructed shared mental model methodology. *Performance Improvement Quarterly*, *21*(3), 1–23.

Jonassen, D. H., Reeves, T. C., Hong, N., Harvey, D., & Peters, K. (1997). Concept mapping as cognitive learning and assessment tools. *Journal of Interactive Learning Research*, *8*(3/4), 289–308.

Karoly, L. A., & Panis, C. W. A. (2004). *The 21st century at work: Forces shaping the future workforce and workplace in the United States*. Santa Monica, CA: RAND Corporation.

Klimoski, R., & Mohammed, S. (1994). Team mental model - construct or metaphor. *Journal of Management*, *20*(2), 403–437. doi:10.1016/0149-2063(94)90021-3

Langan-Fox, J., Anglim, J., & Wilson, J. R. (2004). Mental models, team mental models, and performance: Process, development, and future directions. *Human Factors and Ergonomics in Manufacturing*, *14*(4), 331–352. doi:10.1002/hfm.20004

Langan-Fox, J., Code, S., & Langfield-Smith, K. (2000). Team mental models: Techniques, methods, and analytic approaches. *Human Factors*, *42*(2), 242–271. doi:10.1518/001872000779656534

Maedche, A., Pekar, V., & Staab, S. (2002) Ontology learning part one - on discovering taxonomic relations from the web. In *Proceedings of the Web Intelligence Conference*, (pp. 301-322). Berlin: Springer.

Mathieu, J. E., Heffner, T. S., Goodwin, G. F., Salas, E., & Cannon-Bowers, J. A. (2000). The influence of shared mental models on team process and performance. *The Journal of Applied Psychology*, *85*(2), 273–283. doi:10.1037/0021-9010.85.2.273

Milrad, M., Spector, J. M., & Davidsen, P. I. (2002). Model facilitated learning. In S. Naidu (Ed.), *Learning and Teaching with Technology: Principles and Practices* (pp. 13-27). London: Kogan Page.

Novak, J. D. (1998). *Learning, creating and using knowledge: Concept maps as facilitative tools in schools and corporations*. Mahwah, NJ: Erlbaum.

Novak, J. D., & Gowin, D. B. (1984). *Learning How to Learn*. New York: Cambridge University Press.

O'Connor, D. L., & Johnson, T. E. (2004). *Measuring team cognition: Concept mapping elicitation as a means of constructing team shared mental models in an applied setting*. First International Conference on Concept Mapping, September 14-17, 2004, Pamplona, Spain.

Pirnay-Dummer, P. (2006). *Expertise und Modellbildung: MITOCAR*. Freiburg, Germany: Universitäts-Dissertation.

Pirnay-Dummer, P. (2007). *Model Inspection Trace of Concepts and Relations. A Heuristic Approach to Language-Oriented Model Assessment*. Paper presented at the AERA 2007, Division C, TICL SIG, April 2007, Chicago.

Pirnay-Dummer, P. & Nußbickel, M. (in press). *New Ways of Finding Out What is Needed to Know. Using the latest Tools for Knowledge and Mental Model Assessment in the Processes of Needs Assessment.*

Qin, Z., Johnson, D. W., & Johnson, R. T. (1995). Cooperative versus competitive efforts and problem solving. *Review of Educational Research*, *65*(2), 129–143.

Rowe, A. L., & Cooke, N. J. (1995). Measuring mental models: Choosing the right tools for the job. *Human Resource Development Quarterly*, *6*(3), 243–255. doi:10.1002/hrdq.3920060303

Salas, E., & Cannon-Bowers, J. A. (2000). The anatomy of team training. In S. Tobias & J. D. Fletcher (Eds.), *Training & retraining: A handbook for business, industry, government, and the military* (pp. 312-335). New York: Macmillan Reference.

Salas, E., & Fiore, S. M. (2004). *Team cognition: Understanding the factors that drive process and performance*. Washington, DC: American Psychological Association.

Savenye, W., Robinson, R., Niemczyk, M., & Atkinson, R. (2008). Assessment of individual learning processes. In M. D. Merrill, van Merriënboer, J., Spector, J. M., & Driscoll, M. (Ed.), *Handbook of research on educational communications and technology* (3 Ed., pp. 776-783). New York: Lawrence Erlbaum Associates.

Schvaneveldt, R. W. (Ed.). (1990). *Pathfinder associative networks: Studies in knowledge organization*. Norwood, NJ: Ablex.

Seel, N. M. (1991). *Weltwissen und mentale Modelle*[World knowledge and mental models]. Göttingen: Hogrefe.

Seel, N. M. (1999). Educational diagnosis of mental models: Assessment problems and technology-based solutions. *Journal of Structural Learning and Intelligent Systems*, *14*(2), 153–185.

Seel, N. M., Ifenthaler, D., & Pirnay-Dummer, P. (2006). Mental Models and Problem Solving: Technological Solutions for Measurement and Assessment. In J. Strobel & D. Jonassen (Eds.) *Learning by Modelling Systems*. Mahwah, NJ: Erlbaum.

Simon, H. A. (1996). *Observations on the sciences of science learning*. Pittsburgh, PA: Department of Psychology, Carnegie Mellon University.

Spector, J. M., Dennen, V. P., & Koszalka, T. A. (2005). Causal maps, mental models and assessing acquisition of expertise. *Technology, Instruction, Cognition and Learning, 3*.

Spector, J. M., & Koszalka, T. A. (2004). *The DEEP methodology for assessing learning in complex domains* (Final report to the National Science Foundation Evaluative Research and Evaluation Capacity Building). Syracuse, NY: Syracuse University.

Stahl, G. (2006). *Group cognition: Computer support for building collaborative knowledge*. Cambridge, MA: MIT Press.

Stout, R. J., Cannon-Bowers, J. A., Salas, E., & Milanovich, D. M. (1999). Planning, shared mental models, and coordinated performance: An empirical link is established. *Human Factors*, *41*(1), 61–67. doi:10.1518/001872099779577273

Tversky, A. (1977). Features of similarity. *Psychological Review*, *84*(4), 327–352. doi:10.1037/0033-295X.84.4.327

KEY TERMS AND DEFINITIONS

Assessment of Learning: activities of measuring learning achievement, performance, outcomes, and processes with the intent to provide feedback it improve or reinforce learning.

Collaboration: two or more individuals working together to achieve a common goal.

Collaborative Learning Process: actions and interactions performed by members during the collaborative learning task.

Collective Data Collection: obtaining data first from an individual and then the data are later aggregated or manipulated into a representation of the collaboration as a whole.

Direct Process Measure: continuous elicitation of data from beginning to end of collaborative process; direct process measures involve videotaping, audio taping, direct observation or a combination of these methods.

Holistic Data Collection: obtaining data from collaboration as a whole; as this type of data collection results in a representation of the group rather than an individual, there is no need for aggregating or manipulating data.

Indirect Process Measure: measure at a specific instance during a process; often involves multiple instances; may measure processes, outcomes, products, or other factors related to collaborative processes.

List Form: data of concept maps stored pair-wise (as propositions) for analysis with various tools and technologies.

Proposition: unit of a concept map containing node – link – node.

Similarity Measure: quantity that reflects the strength of relationships between two concept maps or features of concept maps.

Section 3
The Use of Concept Mapping and Collaborative Learning at Pre-School, Primary and Secondary Schools

Chapter 19
Enhancing Autonomy, Active Inquiry and Meaning Negotiation in Preschool Concept Mapping

Gloria Gomez
Swinburne University of Technology, Australia

ABSTRACT

This chapter reports on a case study where a teacher, unfamiliar with Novak's concept mapping method, employed a personally custom designed Authoring Concept Mapping Kit for evaluating preschoolers' knowledge on big cats. The Kit provides drawing and voice-recording features which were designed under constructivist learning and user-centered design principles. In a single session these features enabled eleven five-year-olds to build a map with verbally-labeled symbols because they made their conceptual and propositional meanings explicit. With teacher guidance and despite the absence of arrows, mapping was possible: meanings were categorized, edited, revisited, retained, shared, and preserved. Such activities promoted active participation, knowledge organization and manipulation, and facilitated teacher instruction in a way that has seldom been reported for any approach currently employed in preschool concept mapping. Together these and other results from related studies showing use of arrows and hierarchy, support the assertion that with a tool of this type, preschoolers can use concept-mapping-related skills to structure knowledge individually and collaboratively.

INTRODUCTION

Underpinned by the work of researchers such as Vygotsky, Ausubel and McNamara, Novak theorized on the concept acquisition abilities of young children (two-and-a-half- to six-years-olds) and used anecdotes to explain how they acquired concepts such as "annoying", "underbrellas" [*sic*], and "grocery shopping". Through a discovery–reception learning process, involving the use of verbal language and/or concrete materials (toys, symbols), children show us how they integrate new knowledge into their cognitive structure (see chapter 4, Novak, 1998).

Our research and more recent studies by many others in countries all over the world, has shown that

DOI: 10.4018/978-1-59904-992-2.ch019

young children learn quickly how to make good concept maps...' (Novak and Wandersee, 1990, as cited in Novak, 1998, p. 31)

Novak offered theories, but has neither researched children's use of symbols when labeling concepts, nor their ability to make concept maps. I argue that the above claim is supported by his explanations of how people learn, informed comments on child development, informal observations of young children, and research results from using concept maps with early elementary age children (Gomez, c. 2008).

For more than two decades, a small but growing group of early childhood (EC) experts have reported on preschool concept mapping. Their interest has been driven by the underlying theories of concept maps, reported benefits with older children and adults, and Novak's claim that concepts can also be labeled with symbols such as + or %. Some of these experts have claimed that concept maps can help the development of learning-how-to-learn skills of preschool-age children and could be more powerful than guided conversations and drawings in the process of disclosing acquired knowledge. My doctoral study involved the investigation of an age-appropriate tool that allowed for the testing of such claims (see Gomez, 2006).

Of these EC experts, only Stice and Alvarez (1986) have reported on instructing concept mapping with Novak's template, as their preschool students were already literate. Others have reported on instructing concept mapping with an adapted template, as their preschool students were still illiterate (Badilla, 2004; Figueiredo, Lopes, Firmino, & Sousa, 2004; FOD, 2004a, 2004b; Mancinelli, Gentili, Priori, & Valitutti, 2004; Mancinelli & Guaglione, 2004; Mérida, 2001-2002, 2002; Pérez-Cabani, Falgas, Nadal, & Valenti, 1992). These teacher adaptations or symbol-based maps inspired in Novak's template present common attributes. Concepts are labeled with symbols (drawings, preset pictures or toys). When drawing on paper, concepts are still enclosed in boxes or circles. Two or more concepts are then connected with arrowhead or straight lines or woolen threads. Hierarchy is represented with visual aids: templates, color-coded threads, ordinal numbers, and/or boxes of different sizes. These options are all used to differentiate the most inclusive concepts from the more specific ones (Figueiredo, Lopes, Firmino, & Sousa, 2004; Mérida, 2002). Linking phrase meanings (e.g. have, can) are assigned to connecting lines (Pérez-Cabani, Falgas, Nadal, & Valenti, 1992) or to sections of the map (Figueiredo, Lopes, Firmino, & Sousa, 2004), but they are not explicitly instructed, as they cannot be represented with visual symbols. Only Mérida's adaptation (2001-2002;, 2002) allows for including linking phrases. Children are interviewed to elicit the meaning of a drawing, which is then teacher-annotated with handwriting next to its corresponding drawing. The annotated meaning usually stands for a proposition and the linking phrase is represented as part of it. The annotations of Mancinelli et al. (2004), usually placed in one corner of the child's drawing, represent propositions. Cross-links are not visible in any of these map adaptations. More details in the literature reviews of Birbili (2006) and Gomez (2005b).

When concept mapping with the adaptations described here, teachers' objective "appears to be the stimulation of children's critical thinking skills as well as the social interactions among peers" (Gomez, 2006, p. 33). Through verbal language and guided conversations with teachers, children establish relations between symbols (Figueiredo, Lopes, Firmino, & Sousa, 2004; Mancinelli, Gentili, Priori, & Valitutti, 2004; Mérida, 2002). The educational value of these approaches cannot be denied. One way or another, they have allowed students to develop cognitive skills relevant to concept mapping-related activities: representing and manipulating abstract concepts with symbols, organizing concepts hierarchically, and line-connecting concepts to establish relationships among them. These outcomes are relevant when considering that to date the preschool curriculum has placed little emphasis on the development of metacognitive skills, reflective and autonomous thinking (Mérida, 2001-2002). However, experts are slowly changing this trend through their investigations into the issue (Cassata & French, 2006; Cassata-Widera, 2008a, 2008b; Chacón, 2006). Discussions on a concept-based curriculum (Birbili, 2007) and research on the use of drawing as a knowledge construction tool (Brooks, 2004) have been reported in recent years.

The array of teacher adaptations currently offered has enabled the representation of some kind of map structures, but they are not concept maps as Novak envisages them. The pattern concept → linking phrase → concept, which is the simplest form of a concept map (Novak & Gowin, 1984) cannot be represented with these adaptations, as they "...do not conform to Novak's definition of concept maps and, as such, raise new problems. Unlabeled symbols 1) allow for map misinterpretations, 2) prevent the introduction of abstract concepts, and 3) do not accommodate the insertion of linking phrases" (Gomez, 2006, p. 33). Also, teacher intervention is essential to the building of these symbol-based maps, which lessens children's autonomous participation and the emergence of collaborative behavior of the kind required for learning.

This chapter reports on the development and testing of an Authoring Concept Mapping Kit ("the Kit") for five-year-olds, which was designed to address the interaction issues described above. The results of studies performed in one preschool classroom showed that collaborative learning, involving active inquiry and meaning negotiation, is possible with tools enhancing children's control over every aspect of the process of map building (e.g. the manipulation of knowledge represented with symbolic and verbal language). Autonomy at symbol labeling made their knowledge explicit, transformed the teacher into a partner, promoted peer-to-peer interactions, and generated an inclusive classroom where children at different levels (pre-literates, emergent writers, and an autistic child) worked concept mapping-related activities individually and collaboratively. Reported research contributes to improve our understanding of the issues in preschool concept mapping and the type of suitable child-friendly alternatives to Novak's template.

BACKGROUND

Coffey et al. (2003) tell us that if learning is to occur during the building of a concept map, the nature of the learner's mental interaction with the subject matter to be learned cannot be passive. A teacher or other facilitator, the learner himself or herself, a computer device or group learning, are all factors that activate learning during the building process. The learner's mental interactions with these factors promote active inquiry and organization by engaging them in a process that involves asking questions, prompting for explanation and justification, requesting clarification, embellishment, encouraging the learner to make connection among elements and formulate questions about the material, and so forth.

Group learning or collaborative learning is an umbrella term for a variety of educational approaches involving joint intellectual effort by students or students and teachers together. Its related activities vary, but most center on students' exploration (or application) of the course material, not simply on the teacher's representation or explication of it (Smith & MacGregor, 1992). Most reported experiences in preschool concept mapping include building maps collaboratively. Usually a group map is built by the teacher who employs metaphors, clinical interviews or guided conversations to stimulate students' participation (Gomez, 2005b). Some reported experiences (Badilla, 2004; Figueiredo, Lopes, Firmino, & Sousa, 2004; FOD, 2004a, 2004b; Gomez, 2005b, 2007) show that when symbol-based maps are built with pre-determined concepts, meanings, templates and pictures, children's participation can be limited and highly teacher-centered. Some identified issues are: every person (illiterate or literate) interprets symbols differently, unlabeled symbols do not help retaining meanings in memory, and instruction with pre-determined visual aids does not allow for understanding the use of hierarchy.

If children are presented with pre-determined symbol-based maps, the nature of their mental interactions with the subject matter (ideas or topic) being mapped is close to passive. These templates do not allow the owner and/or peers to effectively interact with or explore the represented knowledge, as they cannot explicitly see the conceptual meanings and how these relate to each other, which is one of the main strengths of concept maps (Novak & Gowin, 1984). Reducing the mapping process to the representation and negotiation of simple and preset symbol-based maps should not be taken as a solution for two reasons. First, children verbally disclose a lot more knowledge than what they will do with symbolic representations. Second, Novak considers meaning negotiation, of any kind, the major advantage of using concept maps:

...the same word can have significantly different meaning for each person. This is why we emphasize the constant need to negotiate meanings between teacher and learner... One reason we are enthusiastic about concept mapping as an instructional and evaluation tool is that concept maps can be enormously useful to teachers, administrators, and learners to move toward sharing the same concept meanings for the words or symbols presented. They can also be helpful to move the learner from mere representational meaning to richer conceptual meaning. (Novak, 1998, p. 38)

Promoting active inquiry and meaning negotiation of preschool children at different stages of a concept mapping activity is a critical issue to resolve. Only when children have been instructed to build self-created symbol-based maps (Mancinelli, Gentili, Priori, & Valitutti, 2004; Mérida, 2002), aspects of collaborative learning, meaning negotiation, and active participation were noticeably higher during map building. These untypical mapping experiences allowed the emergence of children's individuality at drawing representation and activity understanding. They enabled five-year-olds to construct 1) in the case of Mérida (2002) and in small groups, hierarchical symbol-based maps with abstract drawings, color-coded lines, and boxes of different sizes to differentiate more inclusive concepts from more specific ones; and 2) in the case of Mancinelli's et al. (2004) and individually, a simpler kind of symbol-based map representing the process or lifecycle of an experience e.g. the pumpkin.

Mérida (2002) reported that her students actively participated in the process of constructing maps with her approach, because they understood instruction and with practice developed the representational skills needed to make them. Once the needed skills were mastered, meaning negotiation and organization as well as peer-to-peer collaboration were visible during the building of group maps. Smith and MacGregor (1992) explain that learning is an active and constructive process, depends on rich contexts, is inherently social, allows students to talk to each other, produces an intellectual synergy of many minds coming together, and finally, allows the participation of diverse learners who bring multiple perspectives, styles, experiences, aspirations.

One issue that has been reported, and which has possibly prevented the development of effective preschool mapping approaches, is that teachers are poorly prepared on the underlying theory of concept maps, and have difficulty coupling this with child development theories and operating EC curriculum (FOD, 2004a, 2004b). These comments may have some foundation. Mérida's adaptation (2002) is strongly rooted in Ausubel's and Vygostky's theories of learning, and Mancinelli et al.'s adaptation (2004) in Novak's concept mapping theories and La Garanderie's pedagogy. Therefore, maybe these experts' child-friendly adaptations to Novak's concept map template have succeeded with preschoolers because they have a good understanding of the relevant mapping theories and how these relate to EC education.

IDENTIFIED ISSUES

Most teacher adaptations do not make children autonomous concept-mappers after a few sessions, in contrast to what usually occurs with literate people interacting with Novak's template. Building maps with pre-determined concepts and templates has shown to limit children and prevent effective scaffolding. They have 1) failed to interest children in the task, 2) been simplified to a point that no longer resembles Novak's strategy, 3) produced lack of motivation, and 4) not made clear to the children the marking of critical features that are relevant for understanding the task. Described situations show that teachers have not succeeded at building '... on what children know and extend their competencies by providing supporting structures or scaffolds for the child's performance' (Wood et al. 1976, as cited in Bransford, Brown, & Cocking, 1999, p. 92).

The adaptations of Mérida (2002) and Mancinelli et al. (2004) showed that autonomous mapping occurs when children understand the process of achieving the task. Their scaffolded designs succeeded because the children assimilated the steps for building the maps after instruction. However, they do need improvement for making the building process fully autonomous. Children should be able to label their own symbols with mastered representation skills, instead of having them teacher-annotated. This type of annotation appears to benefit the teacher only, as children are pre-literates. Symbol-based maps "do not make the children's knowledge about a particular [domain] visible because the symbolic representations are not labeled and the linking phrases are not present. As a consequence the learner [child owner] cannot see the knowledge, cannot manipulate it, and reflect upon it" autonomously (Gomez, 2005b, p. 6).

Mérida (2002) and Pérez Cabani et al. (1992) argue that it is not really the resulting map that is important, but what the mapping process involves: meaning negotiation, collaboration, and training of critical thinking skills such as clustering and hierarchy. While their reasoning is not rejected, I argue that such experiences can be made much richer and more effective if the mapping tools employed provide components that facilitate autonomous behavior and make knowledge explicit. I argue that as children become autonomous creators at every step of the map building, abstract content can be introduced (Gomez, 2006).

THE DESIGN PROCESS OF THE KIT

This section presents the theoretical and methodological reasoning underpinning the development of a child-friendly adaptation to Novak's concept maps.

Research Claim

My research claim was that preschool children would be able to make knowledge explicit with Novak's concept maps if provided with a tool that allows them to label symbolic representations with verbal inputs. The described tool should promote children's autonomy during concept manipulation and facilitate teacher instruction, which would consequently improve individual and collaborative interactions with the mapped content. The emergence of these behaviors during mapping activities with this tool would make it possible to demonstrate that the issues in preschool concept mapping are instructional, not developmental (Gomez, 2006, c. 2008).

Table 1. Guidelines and requirements

Design Guidelines		Design Requirements
Built on how children learn	—Informed→	Allow for tactile manipulation of map elements by making the features less abstract and more tangible
Support the ability to retain, organise, and self-regulate conceptual knowledge,		• Keep control and speed up the building process by presenting the mapping tasks sequentially and in a few steps • Revisit, share and preserve map elements
Support the use of multiliteracy at concept labeling		• Represent ideas or concepts with symbols or writing • Add meaning to symbolic representations, include abstract concepts and linking phrase with spoken language • Edit, review, replace individual map elements
Enable teachers to diversify instruction		• Construct low-directed and pre-constructed concept maps • Make maps adaptable to different concept mapping activities

Method

The Bridging Design Prototype ("BDP") method was used to develop the functional prototype of a concept mapping tool that could be incorporated into a preschool classroom for investigating the issues earlier described. A BDP includes features familiar to users, together with novel features the researcher chooses to incorporate after careful analysis of relevant data. It capitalizes on users' prior knowledge (embodied in techniques, theories, instruments) and recognizes their context realities (Gomez, 2007). This design method emerged as a by-product of my doctoral study and is grounded on Norman's user-centered product development process (Norman, 1999), inclusive design (Keates & Clarkson, 2003), and participatory design (Spinuzzi, 2002), as well as principles of Ausubel's and Vygostky's theories of learning.

In this chapter, I will not evaluate the BDP method or the BDP for preschool concept mapping derived from it. These evaluations are not relevant for warranting the research claims. Such analysis will need to wait for a different occasion. What is relevant to this chapter is that a functional prototype developed under the BDP method allowed me to: 1) initiate a relationship with members of a preschool community, 2) identify teachers who were willing to incorporate this design into their classroom activities, 3) observe teachers and children using the BDP while I performed observations, and finally 4) develop BDPs that fitted their environment, but also fitted my research interests.

Guidelines and Requirements

One component of the BDP method is the need for the designers to familiarize themselves with the prior knowledge of the user community. However, at the design stage of the study, I could not establish a working relationship allowing me to interact with the community in their natural setting, the preschool classroom. Therefore it was necessary for me to develop my understanding of the user community and the attendant research problem by drawing on current literature. The literature on concept mapping, child learning and development was employed to establish the gap between theory and practice in preschool concept mapping. Understanding the source of the gaps enabled the generation of design guidelines, requirements, and scaffolding features that were later employed in the development of the tool used to investigate the research claims. Table 1 shows guidelines and requirements (Gomez, c. 2008).

Two prototyping cycles were required to arrive at the version of the BDP used in the studies. It was first envisioned as a computer-based tool, as computer technologies have been often seen as a good option for augmenting the benefits of concept mapping (Coffey et al., 2003). As a result of user research, however, this idea evolved into the development of a tangible prototype that was easier to promote among the community. In Australia, where the first case study took place, mainstream preschools do not often include the use of computers in the classroom. Therefore, the BDP design focused on developing structuring features that improve some aspects of the mapping approaches developed by teachers, require little or no training, bear adaptation, and have inclusive features (Figure 1).

Authoring the Representation of Knowledge

The Kit is an example of a BDP, a tangible interface comprised of materials commonly found in the classroom (construction paper, markers, whiteboard, magnets) and one innovation, commercially available voice-recorders ("vrecorders"). A vrecorder enabled a 10-second voice message to be recorded that could be played back up to 500 times, and is small enough that could be held and operated by a preschool-age child. All these materials were re-purposed, re-designed, and/or assembled to create structuring features that scaffolded and enhanced children's control over the cognitive skills used in concept mapping: knowledge retention, organization, and sharing. Together these support and improve children's authoring at the moment of labeling and representing concepts (ideas, thoughts) with symbolic and/or verbal language. Its features can be adapted for use with individual children or groups, for building pre-constructed maps or low-directed maps (Cassata-Widera, 2008a; Gomez, 2005b, 2006, 2007).

How it Works

To build a map with the Kit, children are asked to draw ideas about a particular topic on a small piece of paper, verbally record the meaning of the drawing using the voice-recorder ("vrecorder"), and place the verbally-labeled symbol on the whiteboard. To form a verbally-labeled symbol, the child should attach the drawing to the magnetized vrecorder. To make relationships between concepts the child should use magnetized arrows or draw connections on the whiteboard with a marker. Movable hierarchical templates can be used to help children understand the concept of hierarchy (Figure 1).

Improving Interactions

Some interaction issues are resolved with the Kit. Verbally-labeled symbols representing concepts can be autonomously organized, retained, and shared. Pre-literate children can revisit them as desired by pushing the play button on the vrecorder. Similarly to Novak's concept maps, people will be able to explicitly see how concepts are related to each other because linking phrases can be verbally inserted. I was somewhat concerned whether children would be able to operate vrecorders. Making a voice-recording requires fine motor skills for performing two actions at the same time: pressing down with a pen's pointer in the tiny hole of the "REC" feature while voice-recording a message. Despite my doubts and lack of a more suitable alternative, this technology was incorporated in the studies because children's autonomy during symbol labeling could be greatly enhanced (Gomez, c. 2008).

To improve manipulation of represented knowledge, irrelevant activities were removed such as drawing boxes or circles and cutting out paper with scissors. Hand-size magnetized objects (erasable whiteboards, arrows, and paper cards)

Figure 1. Left column: graphic adaptations by the author of the mapping approach of cited teachers. Centre column: features of the Authoring Concept Mapping Kit. Right column: children using Kit features in case studies performed by the author

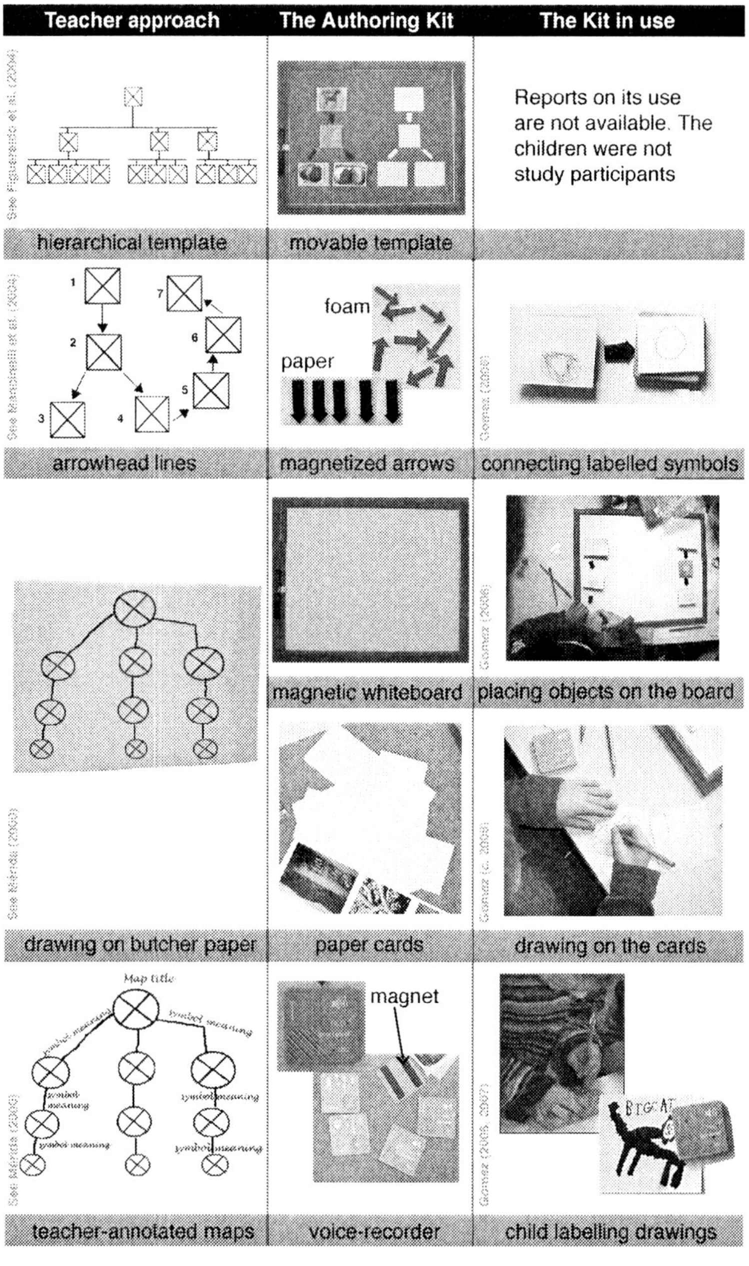

provide a kind of automation that saves building time, cognitive processing, and keeps children focused on relevant activities. Automation is a concept used in technology development (Norman, 1999) that has been loosely applied to this design process. The claim was that once irrelevant activities were automated, children would be able to focus their limited attention span (FOD, 2004a; Piernot, Felciano, Stancel, Marsh, & Yvon, 1995) in map building activities.

Extensive instruction time will not be necessary with the Kit, as children can represent meanings with multiliteracy. Drawing a concept map template is easy to achieve, Novak has reported on a six-year-old representing a simple tree hierarchy after 30 minutes of instruction (see Novak, 1998, pp. 53-54). Reducing how-to instruction allows for more time on instructing concept labeling and organization. What people really find difficult is to make a concept map that reflects what they know: organizing concepts hierarchically, differentiating general from specific concepts, and making suitable relationships among concepts (Novak, 1998; Wandersee, 1990).

Testing

Three case studies have been performed with the Kit. Children aged four-point-six to seven years have used it in one home and two classroom settings, during free-play time and teacher-led activities. All participants, unfamiliar to Novak's maps and for one time only, performed concept mapping-related activities lasting between fifty to seventy minutes. The children built maps or arrow-connected patterns with pre-determined pictures and self-created verbal labels or self-created drawings and verbal labels. Instruction of participant teachers was either not undertaken or minimal. In case study one, the teacher did not show interest in concept mapping. In case study two, the teachers were pressured by time. Despite occurring under less-than-ideal conditions, the study results so far have shown that the Kit effectively addressed the research claims, and can be easily adapted to different learning conditions and styles (Gomez, 2005b, 2006, 2007, c. 2008).

A colleague has incorporated the Kit into her doctoral research. Cassata-Widera (2008a;, 2008b) included the Kit in her study on young children's metacognitive skills with concept maps. Kit features were adapted to make pre-determined maps, and the children were instructed to use the vrecorders for only representing linking phrases. With it, a kindergarten classroom made maps for a period of eight weeks during teacher-led activities and free-play time. Prior to the study, participant teachers were given a three-hour concept mapping instruction session. She reports that teacher-guided concept mapping activities promoted three important dimensions of emergent literacy:

1) Expository language genres can be used to obtain or provide decontextualized information; 2) A relationship exists between the elements of printed or symbolic representation and words in a spoken proposition; 3) Creating text is an inherently social process, with the purposes of communicating and sharing one's thinking with others. (Cassata-Widera, 2008a, p. 1)

Cassata-Widera's results show the importance of researching suitable tools for preschool concept mapping. Her work will also allow for comparing children's metacognitive behavior and teacher instruction in a concept mapping activity, before and after the Kit. In her first study (Cassata & French, 2006), a mapping tool inspired on the work of Figueiredo et al. (2004) was employed. A meta-analysis of these two studies will focus on identifying the tool that enabled children to use self-regulation skills more independently during mapping.

The results of my study, presented in the next section, show how the Kit features transformed children into autonomous mappers, teachers into partners, and allowed for the emergence of meaningful learning, by facilitating knowledge retention, preservation and sharing.

CASE STUDY

I worked with a kindergarten classroom comprised of two teachers, and eighteen children aged between five to six. Contrary to study one (Gomez, 2006), teacher-A and teacher-B incorporated the Kit into an activity to evaluate children's knowledge on big cats, the topic of study at the time. The teachers planned to make a collaborative map during activity time (approx. 45 minutes). Because the Kit materials were not sufficient to work with the whole class, this was divided into two groups: Tigers and Leopards. Each group presented their finished map to the other group the next day at the beginning of circle time (approx. 10 minutes).

The study goal was to see how this community interpreted the Kit, and determine if the research claim was achievable in their context. Towards this end, Novak's concept maps were explained, but not imposed. The preschool director claimed her teachers knew mapping. They were familiar with webbing, a mapping strategy popular in preschool (Birbili, 2006; Helm & Katz, 2001), but unfamiliar to Novak's technique. In the pre-study meeting lasting less than thirty minutes, teacher-B was instructed on how to operate the Kit and asked, by the school's director, to instruct teacher-A who was left in the classroom taking care of the children.

The Kit was adapted for working collaboratively. A bigger magnetic whiteboard and bigger paper pieces were needed. The teachers provided markers, construction paper, and small, sturdy, square pieces of wood. The wood pieces were used as a drawing surface, as the children worked on the floor. I cut the paper into hand-size pieces and replaced paper-based arrows (study one) with foam arrows that last longer. The teachers decided not to use the removable templates (Gomez, 2005b), as they did not see a use for them.

Findings of the Leopards' session are reported: The teachers and eleven children participated in the activity. The group had pre-literates, emergent writers, and one autistic child accompanied by an aid. This session was chosen because the children participated in the map building, contrary to the children of the Tigers' session. The data analyzed involved photos of the building process, teacher interview transcripts, and the completed maps of the Tigers and the Leopards.

Before reporting and discussing the study results on collaborative learning, teachers and children interactions with the Kit features will be reported.

How Was the Kit Used?

Children's autonomy was greatly enhanced during the process of map building. All of them quickly took control over familiar activities e.g. drawing symbols on paper cards and revisiting verbally-labeled symbols by pushing vrecorders' play button. Most of them easily learnt unfamiliar ones e.g. placing objects on the whiteboard, voice-recording, peeling off sticky magnets. A few children needed help with unfamiliar activities, which was provided when requested. See Figure 2.

The Kit supported activities pivotal to concept mapping such as symbol labeling and knowledge organization. Also, it helped to reduce how-to instruction, as irrelevant time-consuming tasks such as attaching sticky magnets to the back of drawn cards or placing labeled symbols on the whiteboard did not distract children from performing relevant tasks. This result was contrary to other studies where irrelevant motor skill-related tasks occupied cognitive processing of young children such as Velcro-ing drawn cards to voice-recorders (Gomez, 2007), or using scissors to cut-out final versions of drawings that will be part of a group map built on butcher paper (Mérida, 2002).

Labeling Symbols

Verbally-labeled symbols made knowledge explicit. The participant children autonomously labeled nineteen ideas or propositions on big cats with a combination of drawing + voice, drawing + voice + emergent writing, and emergent writing + voice. The teacher encouraged them to use mastered representation skills which sped up the representation process (Figure 2). There was no need to teacher-annotate drawings for disclosing meaning or train children on repre-

Figure 2. Children interacting autonomously on different aspects of the construction of the map

sentational drawing, as it has occurred in other studies (Mancinelli, Gentili, Priori, & Valitutti, 2004; Mérida, 2002).

Verbally-labeled symbols disclosed a child's ability to represent and integrate knowledge with multiliteracy. Sometimes the verbal label explained an aspect of the concept that was not represented in the drawing. See the Leopards' map, proposition #1 (Figure 4). The verbal label is "big cats have long teeth" while the drawing is a schematic representation of a tiger. The assigned label does not represent the drawing and vice versa.

Meaning confusions were also disclosed. Some children assigned features common to all big cats to only one big cat. For instance, see propositions #6 " big cats have paws" and #10 "leopards have paws" of the Leopards' map (Figure 4). Because this was their first experience with mapping, clarification and correction of concept representation was overlooked. However, in a concept mapping activity, this issue of identifying general concepts from more specific ones should be addressed with instruction.

Children exercised self-regulation skills when planning, constructing, editing, revisiting, and sharing meanings represented with verbally-labeled symbols (Figure 3). Similarly to a previous study with the Kit (Gomez, 2006), five-year-olds self-negotiated how to draw and what to say about an idea, while disposable paper facilitated drawing editing and placement. For instance, a child was able to edit a drawing already placed on the whiteboard (Figure 2).

The teacher and I trained children on using the paper cards for drawing ideas and the voice-recorders for adding meanings to them. No one experienced problems representing ideas with drawings or emergent writing. Most children knew how to operate the voice-recorders by the end of the session. One of the children, who experienced problems, requested teacher assistant to label the drawing. On this problem teacher-B said, "...after many experiences with it, she could be fine, it was just the newness of the tool... she did not know what was going to happen, so she was a little fearful of it, but I think it depends specifically on the child... She happens to be too shy." In case study one, the same issue was reported, several four-and-a-half- to five-year-olds "had some trouble coordinating vrecording use, but persisted until achieved" (Gomez, 2007, p. 1751).

Despite the voice-recorders' shortcoming, anticipated during the design phase, children added verbal meanings to symbols and made knowledge

Figure 3. The verbally-labeled symbol "big cats hunt at night" was first verbally-labeled "cheetahs hunt at night". The child performed this process by herself

explicit with them autonomously or with teacher assistance. This study result supports my research claim and also indicates that it is worth it to invest in the design of a more suitable voice-recording tool for young children.

Making Relationships and Using Hierarchy

No instruction was provided for making relationships between verbally-labeled symbols with arrows. When asked about it, teacher-B said, "yes, we know what arrows are, they are for telling the direction they can go. The reason why we didn't use them is because I couldn't find them [the arrows]." When asked if the concept "arrow" was unfamiliar to the children, she replied:

... you can see on the playground, they are used to arrows pointing directions to things. The arrows on the playground out there show the direction that they can ride the bikes, they are used to arrows, we used them, umm,... things on the computer, sometimes there is an arrow that shows which direction a [inaudible] goes. They're definitely used to seeing arrows and have an understanding that they are a directional tool.

Observation notes and transcripts show that her students recognized the arrow symbol, knew instances of its use, liked their appearance, and noted their absence when comparing their map (Figure 4) to the Tigers' map (Figure 5). (Such comparison was established in the absence of the Tiger's map, disassembled for creating the Leopard's map.)

Figure 4. Completed map of the leopards group

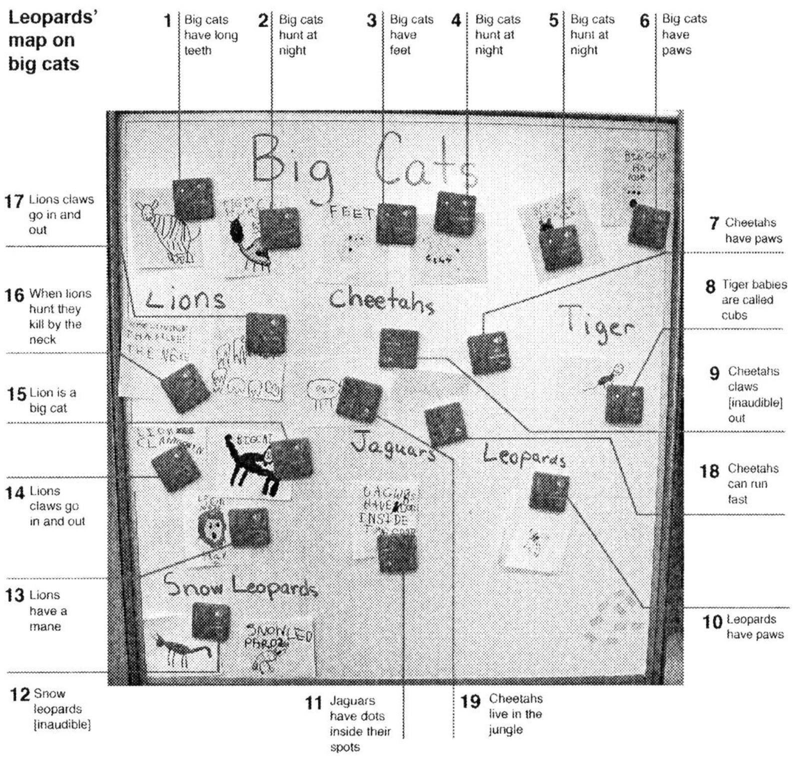

Teacher-A did not reinforce arrow instruction after I demonstrated (at her request) to the Tigers group how to use them. Contrary to teacher-B, this teacher considers that her students would never understand the use of arrow symbols. What I found was that she relates the ability to use arrows with the concept of orientation required in writing, which in her view, is only acquired until second grade. "...in their world orientation doesn't matter...they flip letters constantly... write their names backwards". See symbols represented with emergent writing in maps (Figure 4 and Figure 5).

Also teacher-A thinks using arrows is not yet needed, "they are not making those categories yet" and when it comes to making categorizations, teachers mediate the process but it is hard to know if the children understood. At the moment, and with our support, they organize knowledge using color cues, arrows are much more abstract for them". She continues:

...you can keep trying arrows and see. It helps for a teacher and even when we did our bulletin board that was over there, and we had different colors where the lion was on a certain color and the tiger was on a certain color and I guess that helped them. But I was definitely the one that said to do that...

Not instructing children in the use of arrows prevented the emergence of the pattern verbally-labeled symbol→verbally-labeled symbol. From other studies we know that children can learn to make this and lifecycle patterns, be trained to organize symbols hierarchically (Figure 6), as long as the necessary drawings and verbal skills have been mastered, or a framework for develop-

Figure 5. Modified finished map of the Tigers group

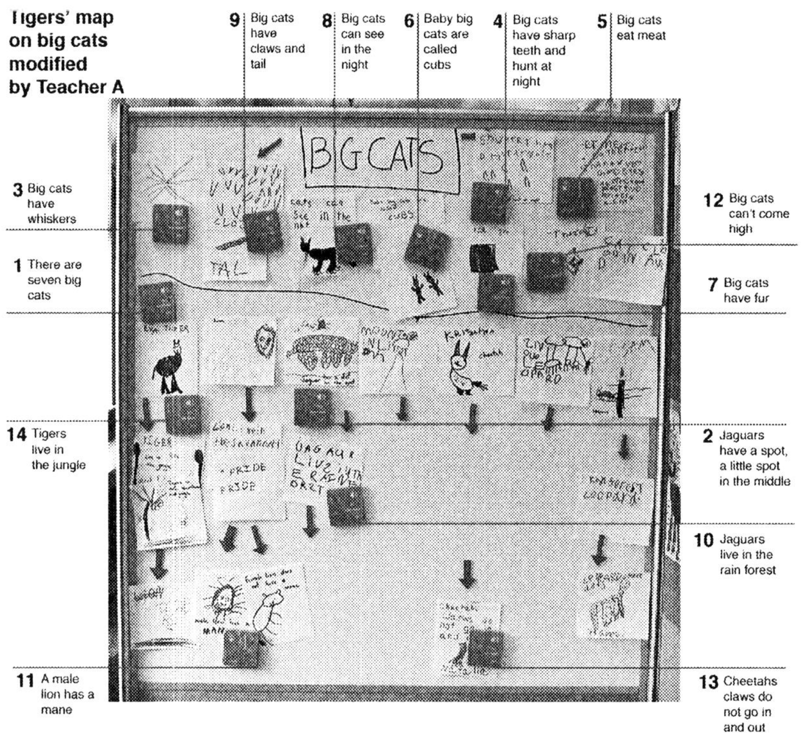

ing these skills is part of the instruction (Gomez, 2006; Mancinelli, Gentili, Priori, & Valitutti, 2004; Mérida, 2002).

Organizing knowledge hierarchically was not instructed either. However, the map built by teacher-A in the absence of her students shows that children can also use the organizational structures that she used in the map building. To build the modified Tigers' map, teacher-A organized the verbally-labeled symbol of her students using the following strategies: clustering, the verbally-labeled symbol pattern, and hierarchy (Figure 7).

When comparing these knowledge structures to the ones created by the five-year-olds in other studies (Gomez, 2006; Mancinelli, Gentili, Priori, & Valitutti, 2004; Mérida, 2002), they reveal that these children have used at least one of the strategies employed by teacher-A (Figure 6). The result of these comparisons challenges the viewpoint of teacher-A regarding categorization and of three forum teachers (FOD, 2004a, 2004b) who think that concept maps may not be appropriate for young children, due to their developmental stage and understanding.

Organizing or Mapping Knowledge into Clusters

Since arrows were not used, teacher-B organized children's verbally-labeled symbols with clustering strategy. Often used in preschool for organizing knowledge (Bjorklund & Douglas, 1997), this strategy was used to cluster the symbols into seven thematic groups: big cats, lions, cheetahs, tiger, jaguars, leopards and snow leopards (Figure 4). Mapping knowledge into clusters was possible because verbally-labeled symbols made concepts and propositions explicit to the teacher and owners.

Figure 6. Graphic adaptations by the author of knowledge structures that five-year-olds built using the mapping approaches of experts as individually referenced

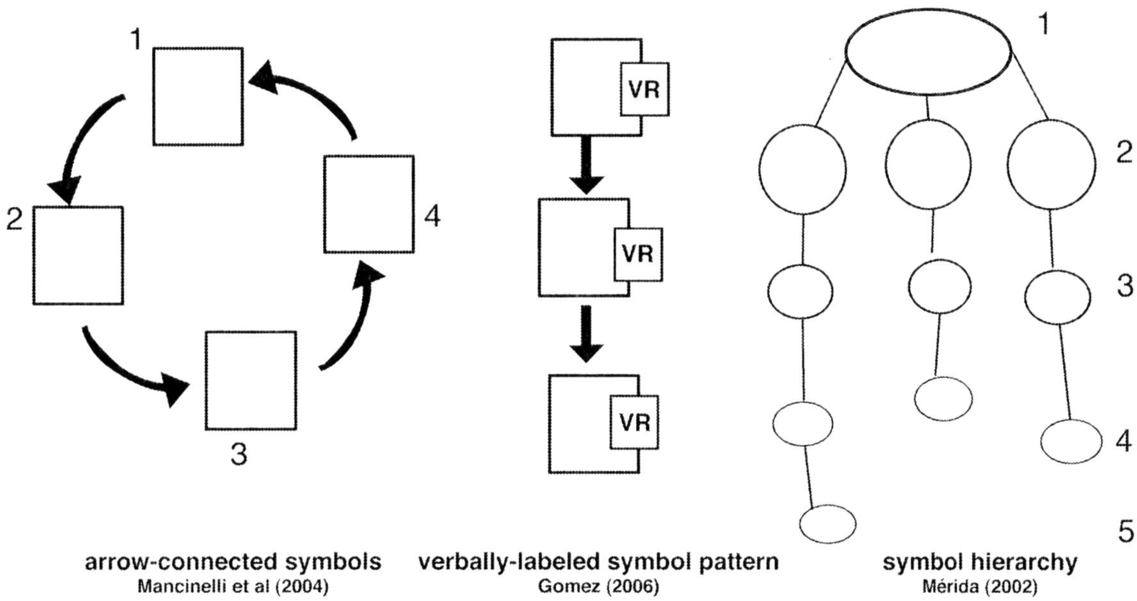

Figure 7. The knowledge structure of the Tigers' map built with the Kit (compare with figure 5). It was organized with clustering, the verbally-labeled symbol pattern, and hierarchy

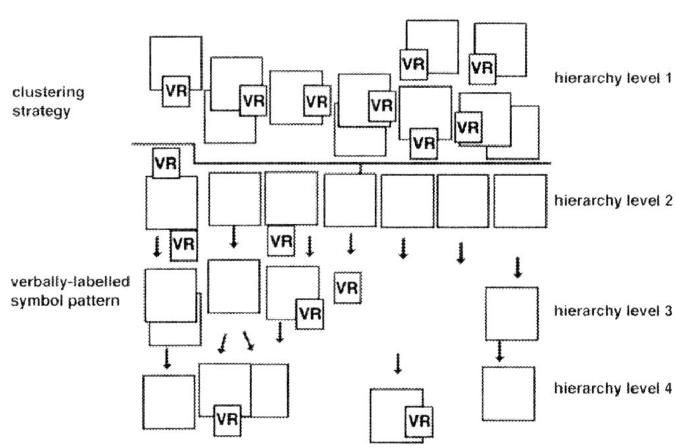

Magnets attached to paper cards and/or voice-recorders further facilitated clustering. Teacher-B could easily place, edit, explore, and re-arrange the symbols, and label the map and groupings with a red marker. Such manipulation can be compared with the flow achieved when concept mapping with a software application.

The resulting structure is a map because the knowledge produced by the children was organized logically following the rules of clustering. "Mapping is both a metaphor for connecting and overlapping knowledge structures", and also, "... the name for practical tools for mental fluency... a rich synthesis of thinking processes, mental strategies, techniques and knowledge that enable humans to investigate unknowns, show patterns of information, and then use the map to express, build and assess new knowledge" (Hyerle, 2000, p. 8). However, the Leopards' map cannot be considered a Novak's concept map because it is missing structural components: arrows, hierarchy, and concepts have not been separated from linking phrases.

In summary, the way the Kit features were employed by the study participants, to a large extent, supports my research claim. Verbally-labeled symbols facilitate mapping of knowledge because they make concepts and propositions explicit. However the kind of mapping instruction provided to children was greatly influenced by the teachers' understanding of how to use Novak's technique and their confidence in the students' abilities to represent knowledge with the Kit features. Children were instructed on independent symbol labeling, but not on making relationships with arrows or hierarchy. The time spent on instruction was not enough for the study teachers to assimilate the building rules for concept mapping with the Kit.

KNOWLEDGE SHARING

Rogoff (1982) observes that: "cognitive development depends on children's adapting and adopting the intellectual tools and skills of the larger sociocultural context, aided by other people" (p. 154). Young children's intellectual development is probably best served by opportunities to interact with adults, each other, the environment, and a variety of materials. (Katz & Chard, 1989, p. 42)

The autonomous participation of children in aspects of the map building contributed to the emergence of different collaborative modalities: teacher-led, teacher-student, peer-to-peer collaborations. These modalities show that the Kit features facilitated instruction by transforming children into active participants and teachers into partners. The nature of the mental interactions with the mapped-content produced basic elements of active inquiry and meaning negotiation. These results and their relationship to collaborative learning are presented and discussed.

Teacher-Led

Teacher-B introduced the activity, guided the idea generation process (children answered questions about ideas remembered on big cats), and mapped the verbally-labeled symbols into clusters while children observed the process: they were not consulted on how these should be organized. Also, in circle time when the Leopards presented the map to the rest of the class, teacher-B led again. She played each vrecorder, repeated aloud the meaning for everyone to listen (sometimes the recording was not loud enough for everyone in the class to hear). At the teacher's request, a child claimed ownership of the verbal label. In these sessions, teacher guided interactions and dialogue did not promote discussions among participants, as it should occur in collaborative learning activities. There was an exchange of factual information, but teacher-B certainly decided symbol placement. Barnes and colleagues' (1969, as cited in Fisher, 2002) research shows that this kind of dialogue frequently revealed less of a discussion and more of a diatribe where the teacher corrects facts throughout by asking questions and monopolizing the type of questions given.

For teachers unfamiliar to concept mapping, giving children control over knowledge organization activities is an uncommon practice. One reason can be offered for this. Mapping structures such as webbing and clustering are usually teacher-built. The earlier quoted comment of teacher-A supports this claim. However, after this experience with the Kit and after seeing children authoring symbol labeling (a concept mapping-related activity) a teacher's viewpoint can change. This occurred to teacher-B who commented:

I think the best thing is watching them teach each other.... because if you can teach it, then you understand it... if you can't teach it to somebody else then you don't know what you're doing. And watching them teaching and talk among themselves, it was nice for me to step back and let them [inaudible]. I would have liked to see how they would have put [the map] together on the [white]board, but we were pressed for time [laugh]... But it would help because there were a few readers in that group that might be able to put it into their own maps. And I would think over time... practicing it and practicing it... by this time of the year [April], if they started doing it since September, by now they might be able to map like this, completely, independently.

Despite the unsuitability of the instruction as a collaborative learning experience, some concept mapping-related activities emerged, such as identification of missing concepts, propositional similarities, and knowledge retention. Mapping knowledge by revisiting verbal labels (assigned to a verbally-labeled symbol) enabled 1) teacher-B to identify that a concept was missing from the completed map: no one represented ideas about "white lions" and 2) a child to notice a propositional similarity by pointing out that "jaguars have a spot, a spot in the middle" was also represented in the Tigers' map. Teacher-A presented this map to the Leopards during circle time and a day prior to their own activity.

This child was able to establish a propositional similarity between maps because knowledge was retained in the cognitive structure. This can be said because the child used his/her own vocabulary to disclose the proposition in question. In the Tigers' map, verbal label #2 (Figure 5), the label used was "jaguars have a spot, a little spot in the middle". In the Leopards' map, verbal label #11 (Figure 4), the label used was "jaguars have dots inside their spots". Verbal language can reveal how children assimilate concepts into their cognitive structure. Novak knew that his grandson understood the meaning of the concept of "annoy" when the child used it in the right context and correct tense (Novak, 1998). At the same time, revisiting the voice-recorded label with the teacher may have triggered the child's memory recall. Teacher-B commented that voice-recorders enabled children "to keep a record" or "...making a memory". "It is hard for them to get their thoughts down in writing because they are not able to do it yet, but the fact that they were able to record what they drew about...that was really good..."

Teachers as Partners

Teacher-student collaboration was evident when Kit features and conceptual content were explained or reviewed one-to-one at the request of a child or the teacher. Some children asked for suggestions of how to draw an idea (e.g. "How can I draw a lion's tail?") while others asked for help or instruction in operating voice-recorders (Figure 8).

Mérida (2002) has reported similar teacher-student interactions such as conversations with students regarding the representation of the concept "pueblo" –town in English. On these

Figure 8. Voice-recoding enabled meaning sharing among children and with the teacher

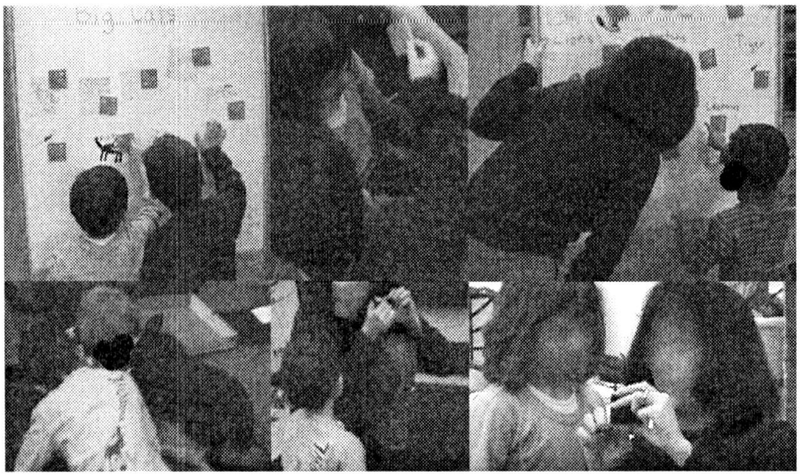

occasions, teachers have acted "less as expert transmitters of knowledge to students and more as expert designers of intellectual experiences for students -- as coaches ... of a more emergent learning process" (Belenky, Clinchy, Goldberg, & Tarule, 1985; Schön, 1983, 1987; Whipple, 1987 in Smith & MacGregor, 1992, p. 11).

In the studies where voice-recorders have been employed, facilitation changed to supporting children's own labeling process (Cassata-Widera, 2008a; Gomez, 2005b, 2006, 2007). This situation changed the class dynamic. Teacher-B commented, "the teacher who becomes a partner, it's like, what we have seen it...", and continued to explain:

It brought more life to them because it wasn't my interpreting of their words, it was their words. 'Cause usually when we do it, we put a big piece of paper on the wall, they'll dictate and I will write. I think there was more active participation on their part rather than me, them sitting talking and me writing. I think, it is very much more the preschool mentality 'cause it is active learning, it is not sitting and being passive...

The principle of active learning requires that the classroom have a stimulating environment, offering children many choices, provoking them to engage in many activities, and opportunities to represent their ideas in different media (Fraser & Gestwicki, 2002). The children found the mapping activity stimulating because the Kit features offered them choices at labeling and manipulating knowledge. Instead of "them sitting talking and me writing" as described by teacher-B in above quote, children were actively involved in meaning making, autonomously created something new, and constructed meanings product of intellectual processing, which are crucial requirements to learning, and characterize collaborative learning environments (Smith & MacGregor, 1992). High levels of engagement were also seen in case study one when children autonomously operated Kit features that enabled them to represent concepts of topics they knew (Gomez, 2006).

The Kit's features stimulated and engaged the children of the studies in autonomous, active and participatory concept mapping experiences. "The learner must want to learn" is the one condition of Ausubel's theory of meaningful learning a teacher or mentor only has indirect control: by motivating the students to choose to learn meaningfully with adequate instructional and evaluation strategies (Novak, 1998). With motivation, people are "most likely to learn something effectively when they experience success or progress that they regard as worthwhile and significantly challenging" (Crooks, 2002, p. 5).

Peer-to-Peer Collaboration

Child autonomy at labeling and manipulating symbols promoted peer-to-peer collaborations regarding feature operation or idea sharing. Teacher-B said,

...Once... I was able to step back it was really fun to watch how everyone was helping each other. Little [child's name] yesterday [inaudible] was able to go out and show everybody what you showed him... how to use the recorders. He was going and helping the other children to do that... He's very mature in that respect and... has very good skills and it was...a lot of peer teaching and they were also able to go back and forth and say hey, I already said that [inaudible], I said the same thing... they were able to compare notes with each other sort and feed off each other.

A main structuring feature of collaborative learning is that it allows student talk (Golub, 1988, as cited in Smith & MacGregor, 1992). Children discussed voice-recorded ideas among themselves and studied peers' drawings. These collaborations mainly happened during the activity (Figure 8), however after class dismissal, three children continued exploring the map, the autistic child and two pre-literate children (Figure 9).

Teachers need to ensure that classroom conversations are not one-sided and that children have genuine opportunities to say what they know and what they feel. When teachers group children for educational reasons, these groupings should be flexible and responsive to the changing needs of children. Young children naturally collaborate and cooperate when they see the need to do so, and they do so particularly when they are engaged in self-initiated activities. (Fisher, 2002, p. 111)

The analysis of the Leopards" map showed that peer-to-peer collaboration, voluntary or involuntary, influenced idea representation and discussion. Copying each other's ideas was an unavoidable result of the ability to share knowledge and working directly with peers. Teacher-B did not discourage the children from doing it: "I think you [are] all trying not to write what the other friends wrote and it's OK to write what the other friends wrote".

Inclusive Collaboration

Collaborative learning benefits diverse learners and emerged during mapping. Multiple perspectives, diverse backgrounds, learning styles were visible (Smith & MacGregor, 1992). Pre-literate children, emergent writers, and an autistic child worked on the same activity and interacted with Kit features according to mastered skills. Each child worked at own pace, developmental stage, and understanding. The active participation of an autistic child was an unexpected result, as the Kit was only designed to be inclusive of teachers and normal children in the classroom, and was never designed for children with special needs. At several professional meetings I have been asked the same question in reference to children with special needs. Also, a FOD forum teacher (2004a) requested suggestions on methods for using concept maps with children with down's syndrome, cerebral paralysis, and autism. Certainly, my study results and expert enquiries have added another dimension to this investigation that should be further explored.

Active Inquiry

Some type of active inquiry was evident when making the verbally-labeled symbols and revisiting the completed map. The below examples are not exactly the type of active inquiry comments Coffey et al (2003) referred to: the children are not inquiring for map modifications or request-

Figure 9. Pre-literate children revisiting the completed map

ing for embellishment. However, they show that the potential for more significant discussions is present, as these mapping activities promoted comments, small discussions, and propositional comparisons. Only instruction of teachers and children can promote active inquiry and participation as reported. This will be discussed later.

Some of the children's comments:

- When making verbally-labeled symbols:
  - "I will draw a tiger that has a long tail"
  - "How do you spell mane?"
  - "Everyone can hear mine" [meaning, hear my voice-recorder]
  - "Any more headphones?" [meaning, any more voice-recorders?]
  - "I need something to put it up on there" [meaning, I need a sticky magnet to put my verbally-labeled symbols on the whiteboard]. See the child at the whiteboard – top-left picture in Figure 2.
- When revisiting the content with the whole class in circle time,
  - A few children made comments such as "play it again"
  - A child commented on similarities between maps "almost everything looks the same like the Tigers"
  - A child noted arrows were not used
  - As explained earlier, a child noted and commented that the "jaguar" proposition was represented in both maps. Other children seconded this comment by repeating the phrase over and over
  - A child walked to the whiteboard and

Figure 10. Mapping and re-organizing verbally-labeled symbols - study one

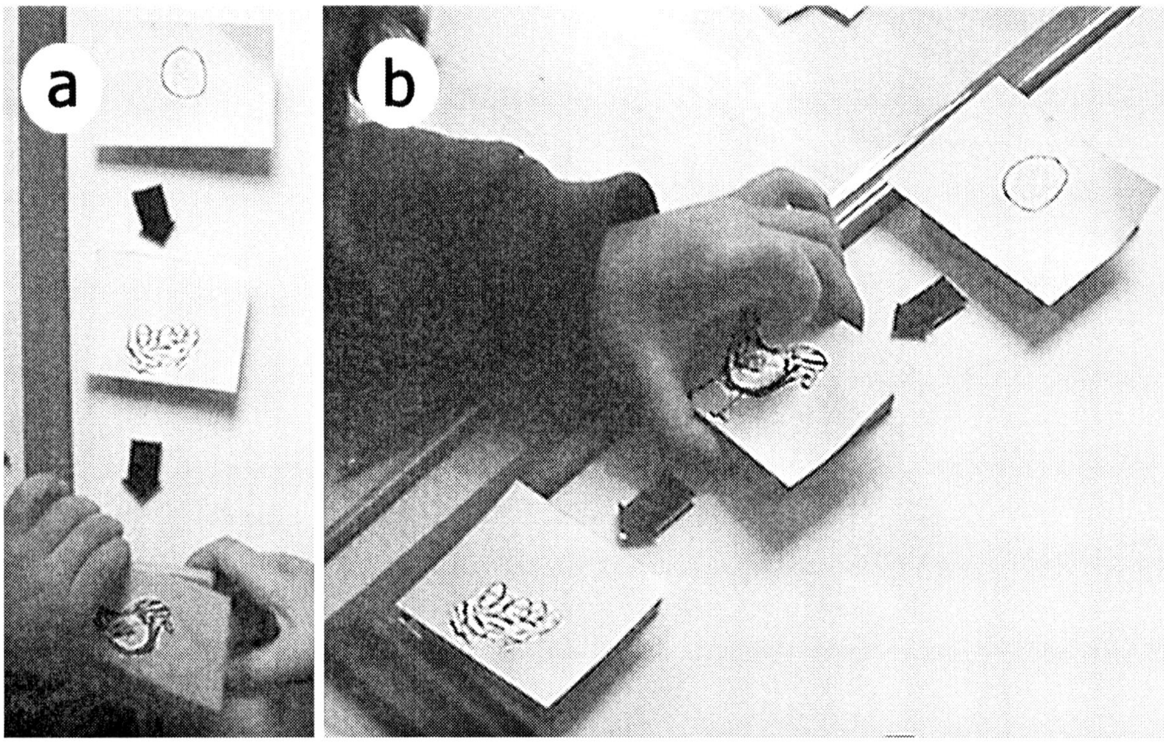

while pointing at a drawing commented, "that picture of the lion is scary to me", image with verbal label #13 in Figure 4. To this comment, the teacher responded "it's scary to you, thanks for telling us that"

- The teacher made comparisons e.g. "lots of friends had that idea", referring to the proposition "big cats hunt at night".

Contrary to the above results, a five-year-old of study one, who autonomously performed the mapping activity, used elements of active inquiry in the construction of her verbally-labeled symbol patterns (Gomez, 2006). This child prompted for clarifications, asked questions about voice-recording or symbol drawing, resulting in self-initiated changes to the representation and labeling of concepts or re-arrangement of verbally-labeled symbols. The latter activity occurred when the child realized that the sequencing represented with arrows did not match her desired thinking (Figure 10). The completed patterns also made explicit that conversations held with the teacher and/or peers influenced the representations of ideas (Gomez, c. 2008).

Meaning Negotiation

The study results support Novak's enthusiasm about concept mapping as an instructional and evaluation tool –see related quote in background section. As described earlier in this section, the knowledge represented with the Kit features promoted many opportunities for children to share conceptual meanings with teacher-B and peers. Meaning negotiation promoted concept labeling and retention, and disclosed singularities in children's thinking.

Children negotiated meanings individually when instructed to represent an idea with paper-cards and vrecorders. Figure 3 shows a child working on verbally-labeled symbol #5 of the Leopards' map (see Figure 4). The easy manipulation of these Kit features facilitated self-negotiation which is an activity present in the concept mapping process of literate learners when selecting concepts that will be included in a Novak's concept map (Novak & Gowin, 1984).

Negotiating their own or peer ideas (knowledge) by revisiting verbal labels enabled children to rehearse developing cognitive skills for remembering and reflective thinking. The comments of teacher-B, reported earlier, support this view when she said that vrecorders helped children to make a memory. DeLoache et al. (1985a, as cited in Bransford, Brown, & Cocking, 1999) reported that young children utilize less systematic and not yet well formed rehearsal strategies for keeping alive information to be remembered. The ability to revisit voice-recorded meanings allowed the participant children, especially pre-literates (Figure 9) and the autistic child, to train rehearsal strategies. Teacher-A commented on the benefits of voice-recording for the autistic child: this would be a "wonderful tool for children that have a little bit of special needs because they need that extra reinforcement… This way they can go back [and revisit voice-recorded meanings] and help themselves…"

Through the process of negotiating meanings, singularities in the children's representations surfaced that otherwise would not have been noticeable. It disclosed how children represent an idea or several aspects of the same idea using voice, drawing and/or emergent writing, and allowed the teacher and the researcher to witness their struggle when organizing the idea for representation. For instance, verbally-labeled symbols showed conceptual similarities at verbal labeling and differences at symbolic labeling. For instance, two or more children represented the same idea, but each child used different symbols, the proposition "big cats hunt at night" was represented 3 times, see #2, #4, and #5 in the Leopards' map (see Figure 4). However, if a drawing or emergent writing was undecipherable and/or unreadable, its verbal label always disclosed the child's meaning. Mancinelli et al.'s students (2004) built individual symbol-based maps of a same topic: the lifecycle of a pumpkin and the process of making papier-mâché. Each child's map revealed distinctive differences at symbolic labeling: two maps did not look the same.

DISCUSSION

Even though children were not instructed in Novak's concept mapping, the Kit features still enabled children to autonomously perform instructed concept mapping-related activities. Verbally-labeled symbols made children's knowledge explicit, prevented misinterpretation of drawings, allowed the representation of abstract symbols and incorporation of linking phrases (as part of the proposition). As a result of these, active participation, individually and collaboratively, was evident. Teacher direct intervention in the building process lessened, while support during autonomous representation and labeling increased.

The study discloses that the issues in preschool concept mapping are instructional, not developmental. Sometimes teachers are knowledgeable in the underlying theories of Novak's concept maps, but do not have suitable mapping tools for instructing children (Cassata & French, 2006; Figueiredo, Lopes, Firmino, & Sousa, 2004; FOD, 2004a, 2004b). The reverse situation also presents, as it occurred in reported study. Due to time pressures, teacher instruction was unsuitable, resulting in teachers building the mapping structures with limited student participation, as in the case of the Leopards' map, or no participation, as in the Tigers' map.

Novak (1998) states in many occasions "inadequate prior preparation or inappropriate instruction" is responsible for bad performance of the strategies for learning. Piaget's stages of development, he continues, have been used as a "convenient scapegoat" for justifying issues related to ineffective instruction (p. 47). Instruction of teachers and children is essential to re-direct activities and the cognitive abilities we have seen children employing in reported studies: knowledge retention, organization, and sharing. Teachers should be instructed in Novak's concept map and its underlying theories, which in turn, should organize instruction for their students using familiar languages. The more the Kit is incorporated into realistic concept mapping experiences, the more information will be gathered to inform the design of a suitable tool. Study results of Cassata-Widera (2008a) showed that after a three-hour teacher instruction, and a eight-week child instruction, on Novak's technique with the Kit, children were able to build maps with two, three and five concepts. The scaffolding techniques used by the teachers succeeded after constant practice and promoted the early literacy-related findings quoted in the section "testing" of this chapter.

Students' autonomous interactions with the Kit modified and challenged the study teachers' viewpoints of what they could do. Their comments revealed that having students authoring their own mapping experiences is an uncommon practice and in my view, this is why they monopolized the map building. However, this could be changed if they see students actively participating, peer teaching, and overcoming developmental limitations with practice.

Without explicit instruction, pre-literate children do not separate concepts and linking phrases, connect verbally-labeled symbols with arrows, and/or use hierarchy. Cassata-Widera (2008a), for instance, reported that "a common 'error' observed involved children recording the entire proposition into the recorder, instead of just the linking word or phrase" (p. 5). To address this issue, she explains that the teacher developed strategies for helping children understand the correspondence between spoken and symbolic element within a map. With practice and good guidance involving active inquiry and meaning negotiation, children understood that only the linking word was to be recorded. This result supports Novak's claim regarding effective instruction coupled with the right materials.

The incorporation of voice-recorders for facilitating the labeling of symbols and addition of linking phrases was pivotal for children to negotiate meanings alone, with a teacher and/or peers, as the study results have shown. The reader should be reminded that originally Novak's concept maps did not include linking phrases (Novak & Gowin, 1984), these were added later to enhance understanding (Novak & Cañas, 2006). Mapping of knowledge occurred, as the concepts were hierarchically organized, but the relationships among concepts of such mapping were only made explicit with the incorporation of linking phrases. A comparison can be drawn: the experience reports of Mérida (2002) and Mancinelli et al. (2004) have shown that the symbol-based maps that children created with their adapted templates enable them to organize knowledge with self-generated drawings. Mapping is occurring, but the child's understanding of a topic is either partially explicit or inexplicit. With the Kit we saw that verbally-labeled drawings make knowledge explicit, despite the child's drawing skills. In short, labeling with written or verbal language is needed for enhancing participation, active inquiry, and meaning negotiation in preschool concept mapping.

FUTURE TRENDS

The studies that were discussed here show that investigating issues in preschool concept mapping from a design perspective could be a fruitful approach. The Kit requires further development and systematic testing is needed to arrive at more conclusive findings regarding concept development and representation. Having colleagues incorporating the Kit into their own research is key to advancing and validating this design research. Also the active participation of an autistic child should be followed up. This unexpected outcome provides hints of the kind of tools that can succeed in inclusive education programs.

CONCLUSION

This chapter reported design research showing that authoring tools promoting the use of mastered representational skills and providing features enhancing control over developing cognitive skills can help children overcome developmental limitations when concept mapping. These results cannot be generalized, but together with studies performed by other experts provide pivotal insights into the effective introduction of Novak's concept maps in the preschool classroom as well as the instructional hurdles that need to be overcome.

Child autonomy, active inquiry and meaning negotiation are enhanced, as this kind of tool can allow for the rehearsal of mapping-related skills: knowledge organization (e.g. creating verbally-labeled symbols), retention (e.g. recalling conceptual similarities), self-regulation (e.g. planning how to represent, edit, or when to revisit a verbally-labeled symbol), and sharing (e.g. discussing ideas with peers and/or the teacher).

The ability of children to use described skills during mapping, alone, with teachers and/or peers, supports the claim that the issues in preschool concept mapping are instructional, not developmental. The instructional issues are of a representational kind, and appear when preschoolers are asked to organize conceptual knowledge with unlabeled symbols and unfamiliar representational systems such as hierarchy and arrows.

Teacher-annotated or not, symbol-based maps do not enable children to show all their potential at organizing and disclosing knowledge. Two innovations need to be incorporated in their design: one that "retains" verbal labels of any kind of concepts and propositions, and one that allows autonomous manipulation of represented knowledge. The Kit resolved these issues with voice-recorders and magnetic objects. They simplified the performance of tasks involving mental interactions with the mapped content.

With the Kit features, mapping occurred while listening to the verbal meaning accompanying a drawing, not while reading the written labels as it occurs with Novak's maps. Mapping verbally-labeled symbols enabled the identification of a missing concept, similarities among labels and maps, the analysis of represented and integrated knowledge. Verbally-labeled symbols 1) created by the children showed how knowledge can be structured with voice, drawings and/or emergent writing, and 2) organized by the teachers showed us how knowledge can be categorized thematically or hierarchically.

Tools such as the Kit could succeed in preschool, as they only slightly disrupted classroom operations. Teachers could quickly see the Kit's benefit and a way to integrate it into their activities because it was adaptable and easy-to-use by both members of the community: teachers and children.

The study teachers' prior knowledge and beliefs of children's abilities complicated instruction. However, their opinion was quickly changed after seeing their students actively participating in aspects of the map building individually and collaboratively with peers. Changing their opinion on arrow use will require instruction on Novak's concept maps and showing them examples of how other teachers have used the technique.

Finally, the reported research outcomes suggest that approaching the preschool concept mapping issues from a design perspective leads to the evolution of more appropriate tools. We should remember, Novak's concept map template was developed as the design solution to a visualization problem: an easy way to represent and manipulate the content of audio taped transcripts representing changes in students' cognitive structures about science concepts (Novak & Cañas, 2006).

ACKNOWLEDGMENT

The case study I report in this chapter was performed at the Children's School, Carnegie Mellon University while I was visiting there as a research scholar in 2006. This research was made possible by the cooperation of Prof Sharon Carver, Director of the School, the teachers Linda Hancock and Donna Perovich, together with Prof Dan Boyarski, Head of the School of Design. I will always be thankful for their generous support.

REFERENCES

Badilla, S. (2004, September 3 - 4). *Experiencia pedagógica: mapas conceptuales en preescolar.* Paper presented at the V Congreso Nacional de Educadores, San Jose, Costa Rica.

Birbili, M. (2006). Mapping knowledge: concept maps in early childhood education. *Early Childhood Research and Practice, 8*(2).

Birbili, M. (2007). Making the case for a conceptually based curriculum in early childhood education. *Early Childhood Education Journal, 35*(2), 141–147. doi:10.1007/s10643-006-0112-0

Bjorklund, D. F., & Douglas, R. N. (1997). The development of memory strategies. In N. Cowan (Ed.), *The Development of Memory in Childhood* (pp. 201-246). Hove, UK: Psychology Press.

Bransford, J., Brown, A. L., & Cocking, R. R. (Eds.). (1999). *How people learn: brain, mind, experience and school* Washington, D.C.: National Academy Press.

Brooks, M. (2004). Drawing: the social construction of knowledge. *Australian Journal of Early Childhood, 29*(2), 41–49.

Cassata, A. E., & French, L. (2006). *Using concept mapping to facilitate metacognitive control in preschool children.* Paper presented at the second international Conference on Concept Mapping, San Jose, Costa Rica.

Cassata-Widera, A. E. (2008a, September 22-25). *Concept mapping and early literacy: a promising crossroads.* Paper presented at the Third International Conference on Concept Mapping, Tallin, Estonia & Helsinki, Finland.

Cassata-Widera, A. E. (2008b). *Fostering metacognitive control skills in the kindergarten classroom using concept maps.* Paper presented at the annual meeting of the American Educational Research Association, New York, NY.

Chacón, S. (2006, September 5-8). *La pregunta pedagógica como instrumento de mediación en la elaboración de mapas conceptuales* Paper presented at the second international Conference on Concept Mapping, San Jose, Costa Rica.

Coffey, J. W., Carnot, M. J., Feltovich, P. J., Feltovich, J., Hoffman, R. R., Cañas, A. J., et al. (2003). *A summary of literature pertaining to the use of concept mapping techniques and technologies for education and performance support* (Technical Report submitted to the Chief of Naval Education and Training). Pensacola, FL: IHMC - Institute for Human and Machine Cognition.

Crooks, T. (2002, December 5-8). *Assessment, accountability and achievement - principles, possibilities and pitfalls.* Paper presented at the Twenty-fourth Annual Conference of the New Zealand Association for Research Education, Palmerston North.

Figueiredo, M., Lopes, A. S., Firmino, R., & Sousa, S. d. (2004). *"Things we know about the cow": concept mapping in a preschool setting.* Paper presented at the first international Conference on Concept Mapping, Pamplona, Spain.

Fisher, J. (2002). *Starting from the child: teaching and learning from 3 to 8* (2nd ed.). Philadelphia: Open University Press.

FOD. (2004a). Foro en línea con Doña Eleonora Badilla-Saxe en preparación del V Congreso Nacional de Educadores PRONIE ME-FOD, Preescolar, I y II Ciclos. *Foros Institucionales - Fundación Omar Dengo (FOD)* Retrieved October 15, 2004, from http://www.fod.ac.cr/Vcongreso/CronoPonencias.htm

FOD. (2004b). Foro en línea de las inquietudes iniciales de los tutores previas al foro con Doña Eleonora Badilla-Saxe. *Foros Institucionales - Fundación Omar Dengo (FOD)* Retrieved October 15, 2004, from http://www.fod.ac.cr/Vcongreso/ChatEleonora.htm

Fraser, S., & Gestwicki, C. (2002). *Authentic childhood: exploring Reggio Emilia in the classroom.* Del Mar, Canada: Thomson Learning.

Gomez, G. (2006). *An authoring concept mapping kit for the early childhood classroom.* Paper presented at the second international Conference on Concept Mapping, San Jose, Costa Rica.

Gomez, G. (2007, April 28-May 3). *A bridging design prototype for investigating concept mapping in the preschool community.* Paper presented at the CHI 2007, San Jose, California, USA.

Gomez, G. (c. 2008). *Issues in preschool concept mapping: an interaction design perspective.* Unpublished doctoral dissertation, Swinburne University of Technology, Melbourne.

Gomez, G. (2005b, July 16-19). *Young children's use of a voice-input device to transform their symbolic maps into concept maps.* Paper presented at the Our Children the Future 4 - OCTF 4, Adelaide, Australia.

Helm, J. H., & Katz, L. G. (2001). *Young investigators: the project approach in the early years.* New York: Teachers College Press.

Hyerle, D. (2000). *A field guide to using visual tools.* Alexandria, VA: ASC Publications.

Katz, L. G., & Chard, S. C. (1989). *Engaging children's minds: the project approach.* New York: Ablex Publishing Corporation.

Keates, S., & Clarkson, J. (2003). *Countering design exclusion: an introduction to inclusive design.* New York: Springer.

Mancinelli, C., Gentili, M., Priori, G., & Valitutti, G. (2004). *Concept maps in kindergarten.* Paper presented at the first international Conference on Concept Mapping, Pamplona, Spain.

Mancinelli, C., & Guaglione, G. (2004). *Le esperienze dei bambini: Indagare nella zucca - Le parole della scienza nella scuola dell'infanzia.* Retrieved 6 August 2004, from http://85.47.105.117/ces/index.htm

Mérida, R. (2002). Una nueva forma de trabajar en educación infantil: los mapas preconceptuales. *Cultura y Educación, 14*(1), 99–125. doi:10.1174/113564002317348147

Mérida, R. (2001-2002). Ayudando a organizar el pensamiento infantil: los mapas preconceptuales. *Kikiriki - Cooperación Educativa* (62-63), 76-83.

Norman, D. A. (1999). *The invisible computer: why good products can fail, the personal computer is so complex, and information appliances are the solution.* Cambridge, MA: MIT Press.

Novak, J. D. (1998). *Learning, creating and using knowledge: concept maps as facilitative tools in schools and corporations.* Mahwah, NJ: Lawrence Erlbaum Associates.

Novak, J. D., & Cañas, A. J. (2006). The origins of the concept mapping tool and the continuing evolution of the tool. *Information visualization, 5 (special issue on concept mapping)*(5), 175-184.

Novak, J. D., & Gowin, D. B. (1984). *Learning how to learn*. New York: Cambridge University Press.

Pérez-Cabani, M. L., Falgas, M., Nadal, A., & Valenti, M. (1992). Els Mapes Conceptuales: una estrategia d'aprenentatge al parvulari. *Revista Guix*.

Piernot, P. P., Felciano, R. M., Stancel, R., Marsh, J., & Yvon, M. (1995, May 07 - 11, 1995). *Designing the PenPal: blending hardware and software in a user-interface for children*. Paper presented at the SIGCHI conference on Human Factors in Computing Systems, Denver, CO.

Smith, B. L., & MacGregor, J. (1992). What is collaborative learning? In A. S. Goodsell, M. R. Maher, V. Tinto, B. L. Smith & J. MacGregor (Eds.), *Collaborative learning: a sourcebook for higher education* (pp. 10-30). University Park, PA: National center on postsecondary teaching, learning, and assessment

Spinuzzi, C. (2002). *A scandinavian challenge, US response: methodological assumptions in scandinavian and US prototyping approaches*. Paper presented at the SIGDOC 2002.

Stice, C. F., & Alvarez, M. C. (1986). *Hierarchical concept mapping: young children learning how to learn (a viable heuristic for the primary grades)*. (No. 5). Nashville, TN: Center of Excellence, Basic Skills for the Disadvantaged, Tennessee State University.

Wandersee, J. H. (1990). Concept mapping and the cartography of cognition. *Journal of Research in Science Teaching, 27*(10), 923–936.

KEY TERMS AND DEFINITIONS

Autonomous: Independent in mind or judgment; self-directed (Ref: American Heritage Dictionary, Fourth Edition)

Bridging Design Prototype or BDP Method: It is a prototyping method inspired in user-centred design and social constructivist theories of learning.

Concept Mapping-Related Skills: Under this term the author groups the following terms that Novak groups as the elements facilitating meaningful learning: knowledge retention, preservation and sharing

Inclusive Tools: Artefacts or devices that can be used by children with different levels of development, in the case of this research, emergent writers, pre-literates, and an autistic child.

Meaning Negotiation: ".... Negotiation of meaning is an internal-and-external process whereby people ascertain what mental models their interlocutors impute to particular linguistic expressions." (Ref: Allan Mark Konrad, doctoral dissertation, 1997)

Multiliteracy: this term is used to group the following representational languages: voice, drawing and emergent writing

Preschool: an educational institution for children too young for elementary school (Ref: WordNet® 3.0, © 2006 by Princeton University.)

Self-Regulation: (1) control of external behaviors, i.e. adaptation and compliance, (2) control of cognitive systems, i.e. attention control, monitored thinking, problem solving, and independent learning. (Ref: A historical examination of self-regulation: helping children now and in the future, Yesman Post, Wanda Boyer and Laura Brett, Early childhood education journal, 2006)

Chapter 20
Consensual Concept Maps in Early Childhood Education

Rosario Mérida Serrano
University of Córdoba, Spain

ABSTRACT

The study described here was designed to demonstrate the sociocognitive benefits experienced by preschool children, specifically 5 year-olds, when they participated in a collaborative activity whose central task was the production of consensual preconcept maps. This task required oral interaction throughout the process of meaning negotiation and the selection of the preconcepts required by such a task. Therefore, the objective of the author's research was twofold: (1) on one hand, to investigate the educational relevance of collaborative work to an early childhood classroom (with 5 year-old pupils), and (2) to explore the possibility of using consensual preconcept maps at this level, analysing the implications for the development of children's competence.

INTRODUCTION

In our earlier research (Mérida, 1999) we conducted a pedagogical adaptation of the concept maps devised by Novak and Gowin (1988), which in turn are based on Ausubel, Novak & Hanesian's (1983) theory of assimilation. We adapted these maps to the sociocognitive competence of 5 year-olds participating in an early childhood classroom, understood to be a social context for learning. In this study we intend to demonstrate the sociocognitive benefits experienced by young children, specifically 5 year-old children, as they participate in the production of consensual concept maps.

We rely on a concept of collaborative learning understood to be a carefully designed system of interactions that organises and leads to reciprocal influence among members of a team (Johnson and Johnson, 1989). Collaborative learning develops by means of a gradual process in which each and every participant feels mutually involved with the learning of the others, generating a positive interdependence that does not imply competition.

DOI: 10.4018/978-1-59904-992-2.ch020

Therefore, the objective of the study described here is twofold: (1) to investigate the educational relevance of collaborative work to an early childhood classroom (with 5 year-old pupils), and (2) to explore the possibility of using consensual concept maps at this level, analysing the implications for the development of children's competence.

WHAT DO WE MEAN WHEN WE REFER TO COLLABORATIVE LEARNING? SOME CONCEPTUAL CLARIFICATIONS

The terms collaborative vs. cooperative learning appear repeatedly in the literature. Although some authors (Woods, 1989) tend to collapse these terms, our own understanding is more in line with those who claim the existence of a difference between the two concepts (Edwards 1992). The latter notion maintains that collaborative learning derives from a sociocultural perspective, while cooperative learning corresponds to a Piagetian approach to constructivism. The essential difference between these two learning processes is that in the first, collaborative learning, it is the pupils who design their interaction structure and maintain control over the various decisions that will have repercussions on their learning, while in the second, cooperative learning, it is the teacher who almost entirely designs and maintains control over the interaction structures and the results to be obtained (Panitz, 1995). Therefore, each paradigm represents an extreme of the teaching-learning process, which ranges from being highly teacher-structured (cooperative) to leaving the responsibility for learning primarily with the pupil (collaborative).

Despite these important differences, the two models should be seen as part of a continuum based on constructivist epistemology as well as the theory of social interdependence proposed by Lewin (1948), which situates collaborative learning as an interdependent social activity where common goals are established with others in such a way that each person's results are affected by the actions of the others (Johnson & Johnson, 1989). In sum, we understand collaborative learning in an educational context to be a model of interactive learning that invites pupils to share efforts, talents and competencies by means of a series of transactions which permit them to achieve a shared goal together.

Is Collaborative Work Useful in Early Childhood Education?

Until very recently, the capacities of young children (below the age of seven) to really engage in collaborative work has been underestimated, perhaps due to the paucity of empirical evidence and to the effect of Piagetian developmental stage theory, which points to interactional difficulties stemming from the egocentrism of young children and their limitations in terms of adopting another's point of view and carrying out a process of cognitive decentring.

The results of some studies, such as those conducted by Azmitia and Perlmutter (1989), suggest that benefits are limited when interactions take place among preschool-aged children. They argue that children of these ages offer fewer verbal explanations than older children. In addition, Azmitia (1988) found little evidence that five year-old children learned a model-copying task by means of discussion; instead, the children tended to rely either on observational learning or on the advice of a peer who was more expert at the task. This position is supported by the contributions of Tudge and Rogoff (1995), who after reviewing research based on pre-schoolers' shared activities, concluded that "it may be that the benefits of social interaction among equals, as a specific result of discussion and coordination of perspectives, are not usually apparent until children are able to simultaneously defend their own point of view and that of another person" (p. 117).

Nevertheless, the results of other investigations seem to indicate that this position is not particularly consistent. Among the authors who point out the sociocognitive benefits derived from interactions among equals under six years of age, we find particularly interesting the contributions of Martí & Solé (1997), who insist on the fact that age is in no way a factor, "If the task is well-selected (simpler, more structured and with well-defined objectives), very young children (from two years old) are able to collaborate, and what has been referred to in the psychological literature as parallel activity (each carrying out activities without taking into account what the other does) does not necessarily arise" (p. 60).

Coll & Colomina (1990) make a similar argument when they point to studies of the level of collaboration demonstrated by two year-old children working on simple tasks, such as opening a box and taking out a doll that is inside; the children appear capable of coordinating their actions. They add that there are other studies that reveal how preschool-aged children are able to provide simple arguments, justifications and demands for clarification in a discussion carried out in the context of everyday interactions.

An interesting line of investigation in this area centres on determining to what extent small children can benefit from social interaction, primarily focusing on the sociocognitive benefits obtained by nonconservers from working with conservers (Perret-Clermont and Schubauer-Leoni, 1981). According to the results of this research, children between four and five years of age were more likely to make progress in their cognitive abilities after having been paired with other children than when they carried out the same task individually.

From our point of view, at this age (five years) children possess capacities that must be trained in interaction with their equals in order to achieve the appropriate personal development. In addition to the scientific understanding derived from research, we base our position on the need for pupils of this age to develop the following capacities:

1. **Development of oral communication abilities:** Participation in interactive contexts promotes the use of language as an instrument for expression and comprehension with respect to their equals. The use of such a linguistic code facilitates comprehension and helps develop the vocabulary as well as the syntactic structure and grammar of the language.
2. **Development of self-esteem:** Social acceptance is one of the basic foundations on which a balanced and secure identity is constructed. The child needs to construct a positive self-image by means of interactions with others.
3. **Development of social abilities:** Interactive activities offer the possibility of establishing social relations in increasingly broad social environments and of learning to progressively articulate one's own interests, points of view and contributions with those of others. As Kozulin (1994) points out, "It is necessary to facilitate in children the expansion and progressive diversification of their social relations in a climate of acceptance ad respect so that they can understand different guidelines for relations and learn to use them in each situation" (p. 39).
4. **Development of emotional capacity:** At this age the fundamental characteristics of children's personalities are being formed. For this reason, acquiring a balanced and harmonious identity requires the establishment of satisfactory affective relationships. This means learning to freely express one's own feelings and emotions to peers, with the aim of learning social self-expression without fear of rejection from others, while at the same time accepting difference and diversity in groups and the people who form them.

According to Wertsch (1988) the process of developing the representations, abilities and behaviours necessary for the fluid production of collaborative interactions may take years of work and regulation. For this reason, in order to promote interactions among children we must design situations for collaborative learning that will permit pupils to begin at the earliest ages to discuss, relate and contrast their points of view. These situations must be designed so that the emergence of conflict provokes a search for more appropriate and at the same time more consensual concepts and models of reasoning.

FROM CONCEPT MAPS TO PRECONCEPT MAPS: SOME PEDAGOGICAL ADJUSTMENTS

Preconcept maps are the result of adapting the concept maps devised by Novak and Gowin (1988) to the developmental level and the capacities of pre-school pupils. The formal elements of the technique are modified so that written concepts are transformed into preconcepts expressed at the symbolic-graphic level, written linking words become oral nexuses and the degree of hierarchical complexity is quantitatively limited and is organised by assigning an ordinal number to each level and attributing a specific colour to each (Mérida, 2001).

We speak of **preconcept maps** because learners at this age do not yet possess concepts, instead they operate mentally by means of fragmentary preconcepts, which are linked to individual idiosyncrasies and dependent on the context and experiential framework in which they are produced.

The pedagogical sequence employed for the teaching-learning of **preconcept maps** incorporates three key dimensions: (1) familiarisation, motivation and understanding of the formal elements of the technique, (2) individual and autonomous production of **preconcept maps** by pupils about various curricular topics and (3) creation of **consensual preconcept maps** that are produced by a small group of children and understood as collaborative activities. In this analysis we are focusing only on this third dimension.

In this chapter we describe research conducted through the process of producing **consensual preconcept maps** as well as through analysis of the verbal interactions expressed by the pupils as they worked. Nevertheless, we also consider it necessary to explain how the **preconcept maps** were introduced to the classroom so the reader can better understand the individual techniques for learning **preconcept maps** experienced by the pupils as the step prior to beginning the task of producing collective maps.

As participant-researcher (simultaneously assuming researcher and teacher roles), the first objective was to present the maps to the children, saying that we were going to learn a new way of drawing. As an example and motivator, we used a road map on which we found the largest population points (cities), smallest population points (small towns) and the lines that linked certain geographical points with others (roads). Based on this example we then established the relevant similarities between the road map and the concept map, associating the different-sized geographical points with concepts situated at different hierarchical levels: we associated the large population points (cities) with the inclusive preconcept (in red), and the small population points (small towns) with the preconcepts situated at the lower hierarchical levels (in yellow and blue), and these were linked together by the connectors and the linking words in the same way that the roads linked the different regions. To clarify the similarity we made a diagram on the chalkboard by means of a guided interrogation that was directed by the researcher and in which the teacher and the pupils participated. The resulting diagram was:

It was the children themselves who suggested that we codify the levels of hierarchy in order to 'not get it wrong', establishing in this way that we would use ordinal numbers and colour codes to differentiate levels from each other. The colours selected to differentiate hierarchical levels were:

Inclusive preconcept (1st and red), first hierarchical level (2nd and yellow), third hierarchical level (3rd and blue), fourth hierarchical level (4th and green)… Once the diagram of one of the arms of the map was finished, it then had to be filled with contents. To do that, the researcher proposed that the children themselves select the topic independently. The majority choice was to do a map about 'My body', arguing that 'we know a lot of things to put'. The procedure for its production was shared with the class-group. The researcher asked:

Researcher: *"How can we represent my body?"*

Researcher: *"What do we know about the body?"*

Children: *"Well, that we have a head."*

Researcher: *"And… where should we draw the head?"*

Children: *"In the yellow circle"*

Researcher: *"And… do we know anything about the head?"*

Children: *"Yes, yes… that in the head we have a mouth"*

Researcher: *Where should we draw the mouth?"*

Children: *"Under the head, in the smallest blue circle…And also put that we eat delicious spaghetti with our mouth…"*

Researcher: *"Yes, but if you want to put more drawings we have to make more circles under the blue circle"*

Children: *"OK, Miss… let's put a little green circle and let's put that about the spaghetti…"*

Researcher: *"Do you want us to read the map, although we only have an arm so far, we have to do more… OK?"*

Children: *"Let's read it, but this is read downwards, not to the side like the cards we make with the lady …"*

Researcher: *"Exactly, we're going to read it starting from the red circle and going down to the green one"*

Researcher and children: *"My body has a head, and in the head is the mouth which we use to eat delicious, delicious spaghetti…"*

Following this same procedure as described, two more arms were designed; one of these referred to the extremities and the other to the trunk of the body. Once this diagram was finished on the board the children were asked to draw it on paper. This didactic sequence based on the collective production of a preconcept map on the board, incorporating the children's contributions and then transferring it to paper was carried out over four field work sessions, and always with the participant researcher present. Afterward, once the children had interiorised the process, the graphic assistance of the map on the board was eliminated and the children were asked to proceed independently in completing their maps. The transition, moving from adult dependence and guidance to independent completion of the maps, was the most complicated part. After six field work sessions the majority of the children, following the procedure established from the beginning of the strategy-learning process, demonstrated technical ability in completing **preconcept maps** as well as correct personal readings of them. Obviously the level of achievement was not homogenous across all children, presenting significant differences in production among them which generally coincided with the ability level demonstrated by the children in other areas and/or educational tasks.

In the learning and practice sessions for map production, the classroom teacher assumed the role of collaborator. She showed willingness to participate in the research, but made her lack of expertise in the technique clear from the beginning and, as a result, negotiated with the researcher for her (the researcher) to lead the educational intervention. As a result, the classroom teacher, together with the researcher, assumed the role of process supervisors, guides and sources of confidence and security in this phase, taking on functions such as helping to resolve questions raised by the children, reorienting the errors that were detected, trying to help pupils become aware of their mistakes, propose concepts when children advanced at a particularly slow pace…in general, they occupied the pupils' Zone of Proximal Development, offering pedagogical assistance only according to their demands and needs, and as such differentiating among individuals. The aspect that presented the most difficulty was the levels of hierarchy, since the understanding that one prconcept included another, which in its turn included yet a third, and these in order, requires an extremely complex cognitive process of inclusion.

During the rest of the investigation the collaborating teacher took on the role of key informant, advising the researcher in the development of sessions and triangulating the categories and analysis of the maps as they were developed in each session. Finally, the collaborating teacher participated in an ethnographic interview to elicit her opinion about the possibilities for capacity development provided by **consensual preconcept maps** for 5 year-old children as well as her evaluation of their didactic use in early childhood classrooms.

The teacher's opinion was:

Collaborating teacher: *"I think that what the**consensual preconcept maps**are best for is to organise the preconcepts of the children. The children have diverse and disorganised understandings of a particular reality. The maps help them to, by means of an organised diagram, order their ideas, relate them and debate their different points of view. I think that the best didactic use is as a summary-diagram after having worked on a particular topic in the classroom, because that way they remember the most important aspects of the topic and they have to come to an agreement in order to find consensus in the discrepancies that they find among themselves. They are also very useful for arguing their ideas and expressing them verbally. In addition, when they read their group map they become aware of the possible mistakes they made and, independently, change the aspects that are incorrect."*

Researcher: "*Why do you think that the capacities indicated develop better with the**consensual preconcept maps**than with other didactic techniques?"*

Collaborating teacher: *"I think it's a more complete didactic strategy than, for example, other groups tasks such as making a mural together. Because they regulate their manner of participation in such a way that all children have to participate. I also find very interesting the ordering* *of their ideas, because in having to first select the preconcepts that are most relevant to a topic (you can't put all of them, only the ones that the group finds most relevant), we're helping them to differentiate between the significant and the secondary. In ordering their preconcepts from the most global or generic to the most concrete they are also exercising their capacity for mental relation, and by reading the maps they verify whether their relations are correct or not, because as you know, when two preconcepts aren't related they say that they can't read the map because they say "silly things." All of these things can only be done with the maps. The other graphic activities provide more opportunity to promote social skills, but I don't think they 'require' so much the construction of a shared understanding."*

CAN PRECONCEPT MAPS BE UTILISED AS COLLABORATIVE LEARNING STRATEGIES IN EARLY CHILDHOOD EDUCATION?

For the reasons detailed in earlier sections of this chapter, we have chosen to incorporate into the design of our work (**consensual preconcept maps**) the prerequisites for collaborative learning (Mérida, 2001, 2002 and 20032003), which can be summarised as follows: (1) group members should arrive at a consensus by means of collaboration, (2) the participation of group members should be direct and there should exist among them a commitment and willingness to carry out some shared activity, (3) active learning should develop within relationships of negotiation, discussion and agreements and not as a result of competition among group members, (4) group members should share equitably the responsibilities they assume and (5) collective production should emerge and clearly appear to be a group goal whose benefits reflect positively on all group members.

Taking into account the necessary prerequisites that must be fulfilled in a collaborative activity as well as the sociocognitive capacities characteristic of pupils at this age, we will attempt to respond to two exploratory hypotheses that can be defined as follows:

- *What pedagogical sequence is the most appropriate to employ* ***consensual preconcept maps*** *as collaborative activities in a preschool classroom?*
- *What kinds of modalities of* ***sociocognitive interactions*** *emerge among children as they collaborate in producing a consensual preconcept map?*

In order to respond to these questions, we have designed an investigation that is qualitative and ethnographic in nature and which is framed within an interpretive approach to educational research.

Working together, the classroom teacher and the researcher planned a pedagogical sequence which, as an emergent process, was modelled throughout the development of the action, that is, over the course of six work sessions conducted to carry out the process of teaching-learning of **consensual preconcept maps**. The most significant aspects of this sequence were:

1. Organisation of the classroom into work groups consisting of four or five members
2. Selection of the topic of the preconcept map through negotiation and argumentation
3. Selection of a coordinator, whose function was to incorporate the various contributions and encourage everyone to participate
4. Oral account of understandings, attitudes and experiences related to the topic selected
5. Selection of the inclusive preconcept
6. Expression, debate, negotiation and consensus around the various preconcepts comprising the map
7. Rotary-graphic representation of the preconcepts that made up the draft-map achieved by consensus of each member of the group

8. Transfer of the draft-map to a mural-map

We will describe here the activities and negotiation process experienced by one of the work groups into which the classroom was divided, specifically group 2, which was selected at random and which included both girls and boys. The oral language productions manifested by the pupils as they collaborated in the production of the **consensual preconcept maps** were video taped by the researcher and were complemented by live observations throughout the six work sessions that made up the investigation. A systematic observational methodology was employed, simultaneously using different instruments for collecting data to ensure methodological triangulation. The subsequent transcription and analysis of the data clearly illustrated the lived process of collaboration.

1. Organisation of the Classroom into Work Groups Consisting of Four or Five Members

With the help of the collaborating teacher (the teacher in the classroom in which the study took place, who took an active role in the research activity), we distributed the children into work groups with four members in each. The criterion utilised for the group distribution was to respect the child's usual physical location in the classroom, which was based on a stable place assignment over the course of the school year. The prior knowledge of peers, as well as their earlier participation in other collaborative activities, provided a positive interactive experience when taking on the challenge of producing a consensual preconcept map.

2. Selection of the Topic of the Preconcept Map Through Negotiation and Argumentation

Session 1, group 2 (initial phase)

Coordinator: *"I'm going to make it about tractors"*

Jesús: *"Yeah, tractors are OK" –responds Jesús-*

The girls look at each other and shyly (in a weak tone of voice) say:

María José: *"I don't really like this topic. I'm not really into tractors"*

Ana: *"Well, look... I, that's why I'm going with the others (referring to group 1)... because they're going to make a really cute house... of oranges!!... and I know more about those things..."*

We can see that the selection of the map content is the first element that affords dissent and the opposition of subjective points of view. In the first work sessions, the existence of different preferences relating to the selection of the map content even provoked the rupture of the group. Nevertheless, as the work sessions progressed, attitudes of resistance and premature abandonment of the collective project gave way to discussion mechanisms by which, through reasoned argumentation, children attempted to affect the opinions of the rest of the team.

3. Selection of a Coordinator, Whose Function was to Incorporate the Various Contributions and Encourage Everyone to Participate

In the first sessions certain clashes were observed as the coordinators assumed a very egocentric attitude:

Session 1, group 2 (initial phase)

Coordinator: *"Well you have to do it, because you also have to put something, because the map is from everyone... and anyway I'm the coordinator and I say what you have to do, that's why I'm the coordinator... if not, I tell the teacher..."*

In the passage transcribed here we see that the coordinator has assumed an authoritarian and individualistic role, based on coercion to impose orders rather than reasoned dialogue. In reviewing the data from this first work session, both the collaborating teacher and the researcher agreed that it was necessary to clarify and negotiate with the pupils the functions that the coordinator should carry out, as well as the process of coordinator selection itself. The children agreed that each team would vote for a child to act as coordinator. In this sense, we created a shared framework through which the basic functions of this post were regulated, assigning as fundamental attributions: (1) assume responsibility for making sure that work is divided and distributed equitably among all involved and (2) take responsibility for the quality of the group map, watching over the involvement and appropriate contribution of each participating subject.

4. Oral Account of Understandings, Attitudes and Experiences Related to the Topic Selected

Session 2, group 2

María Ángeles: *"Well I'm not going to the procession because my mother doesn't want me to."*

Francisco: *"What a pain! It's always so much fun... I always go and eat in the country with my parents and my grandparents and my neighbours and friends, and later I go on the carriages..."*

María Ángeles: *"Well, my mother tells me that the procession is silly... And that the people who ride the horses get drunk and they run over you."*

We can see that the discrepancy in meaning is more significant in the conversations about content related to the out-of-school, home-life context than in those relating to curricular content, which is characterised by considerable homogeneity. Children's diverse family experiences situate each one of the participants in a universe of various idiosyncratic meanings that, on occasion, they find difficult to move beyond.

5. Selection of the Inclusive Preconcept

Session 2, group 2

Coordinator: *"It's true, come on Francisco, we're going to start. I'm going to put the red circle and I'm putting inside the hermitage that means 'The procession.' You know why it means 'the procession?' (addressing the girls) Because that's where they take the Virgin Mary. I've been there a lot of times with my mum."*

The acquisition of the formal aspects of the technique (preconcepts, linking words and hierarchical levels) is an issue that still presents no problem at this point. Children routinely began with the development of the inclusive preconcept. We believe that this assimilation of the technical elements of the maps permitted the children to delve deeper into the debate and negotiation of the preconcepts, with the result that the process of oral interaction and consensus tended to be richer than the graphic representation produced.

6. Expression, Debate, Negotiation and Consensus Around the Various Preconcepts Comprising the Map

Session 4, group 2

(...) María Ángeles asks about some other aspect to include in the map, and Jesús tries to remember "a thing" he saw on television:

Jesús: *"Should we put what I said before? About the ones that live in Africa and Asia?"* **María Ángeles:** *"What's that?"*

Jesús: *"That's some places that are really far away and that's where the wild animals live, it's like the forest or the desert..."*

As the sessions progress we can see that the initial attitudes of conflict and cognitive resistance begin to give way to attitudes of bringing together and complementing points of view. While at the beginning of the collaborative work the group efforts were focused on reaching an agreement concerning the selection of the topic and the equitable division of work, in the later sessions we began to see the use of a collective discourse concerned with complementing, joining and finding consensus within various individual contributions to the group endeavour.

7. Rotary-Graphic Representation of the Preconcepts That Make Up the Draft-Map Achieved by Consensus of Each Member of the Group

Session 3, group 2

Jesús: *"I...now it's my turn. I can put (thinks)... ah, now I know! I can put that worms make cocoons in the silk. Yeah, because now there's a lot of cocoons that I have seen in the box. You guys haven't seen the cocoons they made? Come on, I'll show you."*

By the third group work session children had adopted a turn-taking procedure to guarantee the effective participation and involvement of all the group members. The children passed a pencil and the sheet of paper with the map to the child on the right so that he or she could continue the work. Although this resulted in an actual contribution for each child, the hierarchical relations among preconcepts was not always respected. In the following case, the coordinator and some other children in the group make suggestions and help their classmate to include a preconcept that fits the required relationship in terms of dependence. This development in the collaboration can be seen in the following passage:

Session 4, group 2

(...) Jesús receives the sheet of paper and pencil and sits thinking without knowing very well what he should put. Suddenly he says:

Jesús: *"Uhmm...I don't know what else to put... I'm going to put that elephants are bigger than ants..."*

Francisco: *"But... what are you doing? (shock). We can't put that here (indicates the yellow pre-*

concept where Africa is drawn) because then you won't be able to read the map, because it says something silly...What does Africa have to do with ants?"

Maria José: *"No, Jesús, it's not good that way. We have to think up something about Africa so that you can read our map. If not, it won't look nice..."*

We can infer that all of the children in the group perceive the consensual map as a collective product that belongs to them and for which they all feel responsible; for this reason they collaborate actively, contributing their own ideas and monitoring those that don't seem right to them. We can see that the structure of the task itself generates a very high level of interdependence among group members. In this sense, we would like to stress that the necessity to participate as an implicit requisite of the activity. At least, each child should participate in producing the graphic representation of the preconcept that was earlier negotiated and accepted by the rest of the group.

8. Transfer of the Draft-Map to a Mural-Map

Once children had produced the consensual map on their sheet of paper, they used this as a draft to transfer the map to a larger medium (a continuous cardboard sheet), increasing the size of the preconcepts to form a mural-map that would decorate the classroom. The equitable sharing of the task became a significant challenge at this point and generated considerable controversy:

Session 5, group 2

Jesús: *"Uff... I am so tired...! It's that our map is so big. You know how many drawings it has? I'm going to count them (Francisco joins in the counting). One, two three,...twelve. We have twelve. So you have to do twelve twelve and... I..." (hesitates)*

Coordinator: *"Come on! I have to do half, because it's me and you. You don't know that half is two equal things in the middle?"*

Jesús: *"Yeah,... but how many do we do?"*

Coordinator: *"Me this one, you this one (starts to assign one for each), me this one, you this one... Now I have six, and you?"*

Jesús: *"I have six, too"*

Coordinator: *"Well that's that, we each have the same, half"*

ANALYSIS OF THE ORAL INTERACTIONS IN THE PROCESS OF MEANING NEGOTIATION AND CONSENSUS BUILDING DURING PRECONCEPT MAP PRODUCTION

In order to respond to the second hypothesis, which asks whether there is an enrichment of the nature of verbal interaction used by children as they collaborate in the production of a consensual preconcept map, we will analyse the form of conversation employed by children as they carry out the proposed group activity. Our intention is to examine the modalities of sociocognitive interaction used and determine whether there is a positive development in these across work sessions, from the first to the last.

Figure 3. Group 2 producing a mural-map in session 5

Nature of the Categories

As previously mentioned, our data analysis technique is to analyse the content of the conversations maintained by the members of group 2 as they collaborated on the production of their consensual preconcept map.

We began by dividing the speech of each group member into turns according to the content of these statements, adjusting the classification system proposed by Roselli, Gimelli and Hechen (1995) to the nature of our data. In this sense our theoretical understanding is based on that described by these authors, who argue that there are three basic modalities of sociocognitive interaction: egocentric, asymmetric and egalitarian. Each of these modalities possesses a different level of interactive richness, with egocentric considered the most rudimentary and egalitarian considered the most complex. Using the authors' words, the characterisation of each modality can be briefly summarised in the following way, "The egocentric modality is characterised by isolation and low participation in the process of producing together the proposed task; the low number of interactions suggests difficulties in inter-individual articulation and an individualistic attitude. The asymmetrical or dependent modality suggests the prevalence and the dominance of one subject over another; exchange exists, but the strategy to be followed dissolves into dominant-submissive terms. The symmetrical and egalitarian modality, by contrast, is defined by a high degree of interaction within a framework of reciprocal negotiation and consensus-building through convincing; contributions from all parties are alternatively recognised" (Roselli, Gimelli & Hechen, 1995, pp. 138).

We needed a process for adapting this classification, since even though it is quite useful, it does not completely conform to the nature of our data. Once we had globally reviewed the empirical verbal data and completed a preliminary classification, we proceeded to identify the conversational turns of each team member in each group work session, employing the message as a "micro" category to distinguish among the meanings of each subject. By means of interjudgemental agreement, the collaborating teacher and the re-

Table 1. Categories of Interactive analysis

Oral Interactions of the Group
Juxtaposed Interactions
1. PRO. Propose a new idea
2. INF. Inform about a situation, feeling, or event
3. DIV. Divide up work
4. IGN. Ignore the proposal or request of a peer
5. IMP. Impose one's own opinion or ideas

searcher classified the oral statements of the group members into three different categories, which are understood to represent varying degrees of richness of sociocognitive interaction: (1) juxtaposed statements, (2) interactive statements and (3) consensual statements.

Each type of statement includes a set of specific categories which permitted us to analyse the interactive content of the pupils' different interactions. The verbal material was coded by applying Table 1.

Juxtaposed interactions are characterised by a greater degree of individualism and isolation and a low degree of adaptability to the demands and needs of the other speakers. Graphically, these statements might be represented as parallel lines, that is, they do not share a point of intersection. These statements relate to task planning and organisation of the map. We can conceptualise them as individual, unconnected contributions rather than patterns of interactive behaviours.

Interactive statements, which are considered to be qualitatively richer than the juxtaposed interactions, make an explicit reference to another subject as a fundamental part of the proposal. In other words, this second type of statement is characterised by a connection with the demands of another individual. We can say that a *transactive* discussion exists, whereby the children produce specifications about the speaker's reasoning.

Consensual interactions are considered to be the most rich and complex. Not only do speakers interact, but they complement and construct shared meanings by means of a process of linguistic argumentation and counter-argumentation. Forman (1992) describes this process as "intersubjective regulation," since meanings are negotiated through consent, creating a frame of intersubjectivity (Trevarthen, 1982).

Statistical analysis of the data

We set out to demonstrate that there was a shift from the predominant use of juxtaposed statements in the earlier sessions to a greater utilisation of consensual statements in the later sessions. To reduce the data we conducted a frequency count, obtaining the number and type of statement employed by each subject in each one of the sessions. Table 2 describes the absolute frequencies of the different sociocognitive interaction modalities utilised by the four subjects who formed the work group across the six research sessions.

The raw data was converted to continuous quantitative data (Table 3) by assigning to each category a number indicating the quality of the statement. Specifically, we assigned a 1 to juxtaposed statements, a 2 to interactive statements and a 3 to consensual statements. The score recorded in each table cell corresponds to a subject and was obtained by adding the number and type of statements of each child and dividing that sum by the total number of statements for the individual in each session.

We applied the Friedman test to determine whether the quality of **sociocognitive interactions** among participants improved over the six sessions. The results permit us to confirm the existence of a significant difference between the quality of some sessions and others, $\chi^2(5) = 12.05$, $p = .03$. Figure 1 demonstrates clearly that the quality of statements is poorer in the two earliest sessions and later improves, despite a slight decline in Session 5. Figure 2 separates the three types of statements in each group. Here we can see that the frequency of juxtaposed and interactive statements is equal in the first three sessions, while, in the following

Table 2. Raw data frequencies

Research sessions							
Subjects	**Modalities**	**1**	**2**	**3**	**4**	**5**	**6**
Francisco	*Juxtaposed*	8	15	13	10	2	13
	Interactive	9	12	13	15	8	25
	Consensual	3	4	10	12	3	18
	Session total	20	31	36	37	13	56
Jesús	*Juxtaposed*	6	10	8	8	4	10
	Interactive	9	14	5	9	8	8
	Consensual	1	2	8	10	4	13
	Session total	16	26	21	27	16	31
María José	*Juxtaposed*	4	12	4	4	6	3
	Interactive	3	10	6	6	5	7
	Consensual	4	2	6	8	2	3
	Session total	11	24	16	18	13	13
María Ángeles	*Juxtaposed*	5	8	5	3	7	1
	Interactive	2	11	6	5	3	4
	Consensual	3	1	7	1	1	2
	Session total	10	20	18	9	11	7
TOTAL		**57**	**101**	**91**	**91**	**53**	**107**

Table 3. Continuous quantitative data frequencies

Research sessions						
Subjects	**1**	**2**	**3**	**4**	**5**	**6**
Francisco	1.75	1.64	1.91	2.05	2.07	2.08
Jesús	1.68	1.69	2	2.07	2	2.09
María José	2	1.58	2.12	2.22	1.69	1.76
María Ángeles	1	1.65	2.11	1.77	1.45	2.14

three sessions, the juxtaposed statements decrease in number. At the same time, the consensual statements experience a sharp increase from the third session onward.

Although the Friedman test does not allow us to specify between which sessions the significant differences exists, the graphs clarify that the most important intersection occurs between the second and third session, in such a way that we can distinguish between two well defined phases: 1) a phase characterised by low-quality statements and 2) one characterised by medium-quality statements. (see Figure 4 and Figure 5)

Figure 1. Preconcept map about "my body" produced by Maxi (five years)

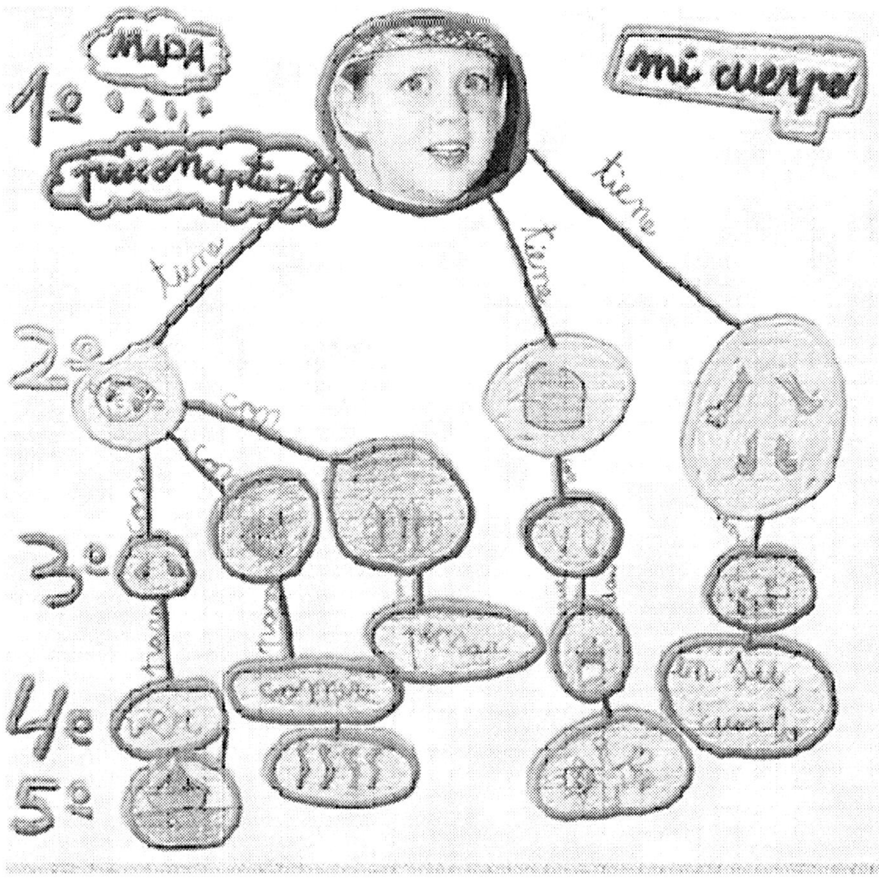

Figure 2. Preconcept map diagram

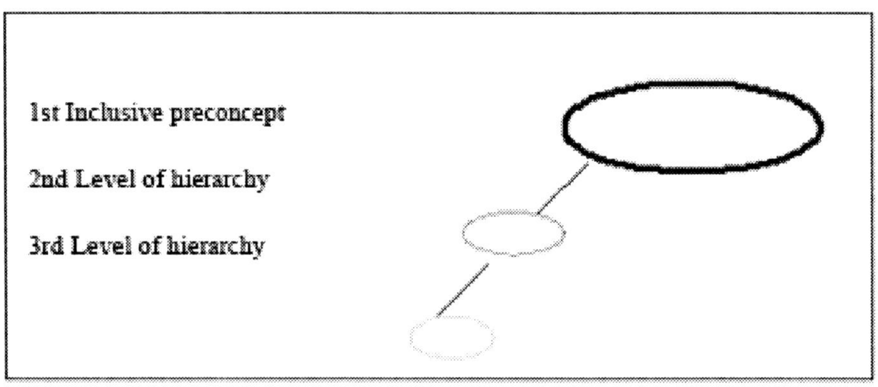

Figure 4. Global evolution of sociocognitive quality across the six sessions

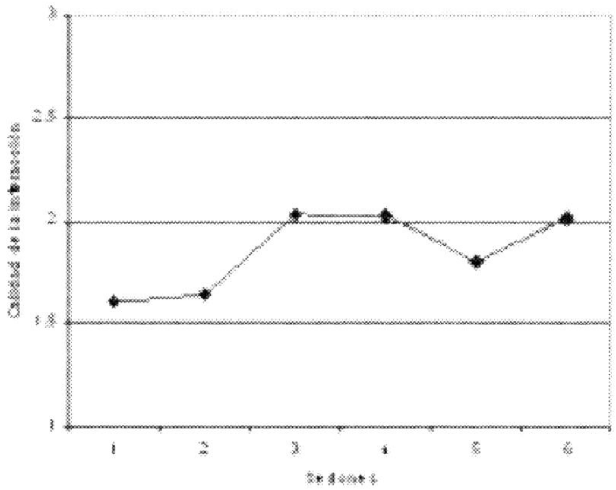

Figure 5. Evolution of the frequencies of the three statement types (juxtaposed, interactive and consensual) in the group across the six sessions

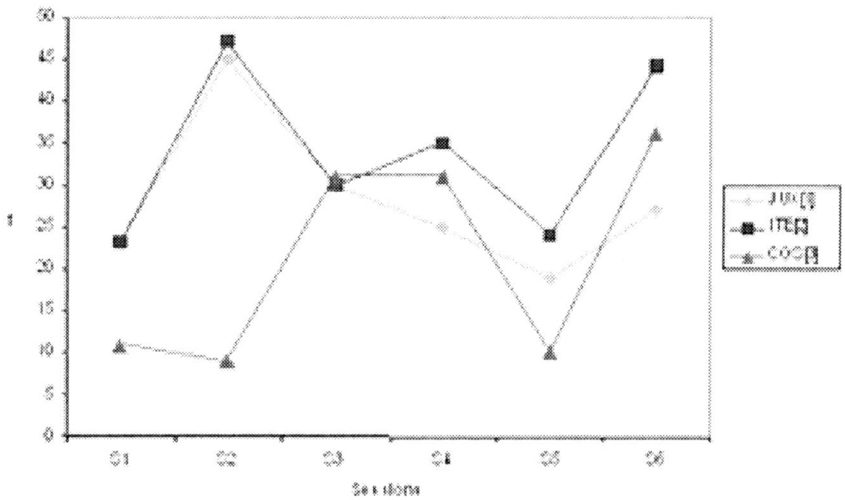

ANALYSIS OF THE SOCIOCOGNITIVE ABILITIES OF GROUP MEMBERS AND THE IMPLICATIONS FOR COLLABORATIVE TASKS

The content analysis of the verbal interaction produced by the four group members as observed over the course of the study demonstrates interesting characteristics to be taken into consideration when designing and organising future collaborative activities:

There was an excessive level of sociocognitive discrepancy among the group members investigated. The two boys in the group produced a much higher number of statements than the other two group members, both girls. These two boys possessed a higher communicative capacity and a richness of personal experiences and preconcepts available in their cognitive structure which enabled them to express proposals and ideas about any topic that arose. The girls, by contrast, presented as timid and introverted, unsure of their opinions and with little verbal ability to argue and defend their convictions with their male group mates. This situation produced dominance in terms of the boys' contributions, which increased in conjunction with the successive sociocognitive leadership incidents they experienced with respect to the girls. This dominance manifested itself in various behaviours:

1. Ignoring the Girls' Presence in the Organisation or Taking Their Own Decisions About the Task (Selecting the Map Topic)

Session 1, group 2 (Initial phase)

Francisco: *"Jesús, buddy, what are we going to do it about...? I know about war, but that's what Alvaro and the others are going to pick..."*

Jesús: *"I'm going to be the coordinator... because it's my turn...so then we can do..." (appears pensive)*

María Ángeles: *"We can do it... about..."*

Jesús: *"About what...? Say it, already... !"*

Francisco: *"So you want to be the coordinator today? So say what we're going to do it about, about the poor...or..."*

2. Disregarding the Girls' Contributions During the Task Organisation and the Selection of the Inclusive Preconcept

Session 3, group 2

Francisco: *"Let's start with the red circle like usual, right?"*

Jesús: *"Yeah...we put the fires in the red circle or no...how did that go? What was that called, what we did before...? Uhh..."*

3. Depending on the Approval of the Other Boy as a Source of Verification of Progress and Avoiding the Evaluative Capacities of the Girls

Session 3, group 2

Francisco: *"Let's see..." (expressing approval of Jesús' representation with a facial expression)*

Jesús: *(with pride) "You like that, man...? Isn't it cool...? You know how I did it...? Mixing the colours, first I put red and then I went over it with yellow..."*

4. Taking on Challenges that Involve Greater Difficulty

María Ángeles: *"Come on, we're going to start doing it"*

Jesús: *"Come on, I'll start, and I'll put the red circle here, ad inside I'll draw a little elephant... Should we put the name? Should we write it down?"*

María Ángeles: *"It doesn't matter to me. But...I don't know...do you know?"*

5. Criticising the Mistakes and Errors of the Girls

Session 6, group 2

(...) María Ángeles intervenes, saying that she has never been to Mallorca, but that her uncles work there and they've told her that one day they're going to take her to a park called "Disonaurs". As soon as María Ángeles pronounces this word the boys start laughing, looking at each other and repeating, in an exaggerated manner, "disss-oo-naurs?" María Ángeles makes an effort to repeat the word properly, trying it in a very low voice and making gestures with her mouth; nevertheless, she appears not to succeed. Francisco says to her, "say '*Di-no-saur*'" (making himself the linguistic model), but María Ángeles does not manage to repeat the word correctly.

Our study corroborates the view proposed by Perret-Clermont & Schubauer-Leoni (1981), who claim that the capabilities (personal and social) of the members of a group influence the kind of relational structure created within the group. That is, a unilateral structure is what emerges when the cognitive distance among group members is excessive. In this type of interaction, the subject who possesses a higher cognitive level ends up imposing his or her judgements about problem resolution, while the subject with a lower cognitive level becomes inhibited and conforms to the more capable peer's proposals due to an over-estimation of the value of these proposals and a lack of self-confidence.

CONCLUSION

- Working with **consensual preconcept maps** resulted in an improvement of the quality of group interactions. Over the course of the work sessions the data revealed an incremental progression in the use of the most complex sociocognitive modality of interaction, the consensual, accompanied by a reduction in the use of the more rudimentary juxtaposed modality of interaction, which were more prevalent in the earlier sessions.
- There were some imbalances in group members' interactions due to a significant discrepancy in competencies. The deliberate criterion for forming the groups proved to be an obstacle to the collaborative process, since there were excessive differences in the capabilities of various group members. For this reason the contributions of two subjects, whose sociocognitive dominance served to inhibit the other participants, tended to overpower the others. This resulted in interactions that followed a dominant-submissive pattern rather than encouraging symmetric relations.
- Working with **consensual preconcept maps** helps to establish social relationships; children learn to articulate with progressive skill their own interests, contributions and points of view with those of others. This involves developing the capacities for argument, decision-making, respect for different opinions, reconstruction of personal meanings and democratic procedures for dividing up work.
- **Consensual preconcept maps** help children utilise verbal language in ways that are increasingly appropriate for a range of different communicative situations, understand and be understood by others and regulate individual and group activity.

REFERENCES

Ausubel, B. P., Novak, J. D., & Hanesian, H. (1983). *Psicología educativa. Un punto de vista cognoscitivo.* México: Trillas.

Azmitia, M. (1988). Peer interaction and problem solving: when are two heads better than one? *Child Development, 59*, 87–96. doi:10.2307/1130391

Azmitia, M., & Perlmutter, M. (1989). Social influences on children's cognition: state of the art and future directions. In H.W. Reese (Ed.), *Advances in child development and behavior* (pp. 321-340). New York: Academic Press.

Coll, C., & Colomina, R. (1990). Interacción entre alumnos y aprendizaje escolar. In C. Coll, J. Palacios & A. Marchesi (Eds.), *Desarrollo Psicológico y Educación II. Psicología de la Educación* (pp. 435-453). Madrid: Alianza Editorial.

Edwards, D. (1992). Discurso y aprendizaje en el aula. In C. Rogers & P. Kutnick (Eds.), *Psicología Social de la escuela primaria* (pp. 63-82). Barcelona: Paidós.

Forman, E. (1992). Discourse, intersubjetivity and the development of peer collaboration: a vygotskian approach. In L. T. Winegar & J. Valsiner (Eds), *Children's development within social contexts: Metatheoretical, theoretical and methodological issues* (pp. 55-69). Hillsdale, NJ: Erlbaum.

Johnson, D., & Johnson, R. (1989). *Cooperation and competition.* Edina, MN: Interaction Book Company.

Kozulin, A. (1994). *La psicología de Vygotsky.* Madrid: Alianza Editorial.

Lewin, K. (1948). *Resolving Social Conflicts.* New York: Harper and Row Publishers.

Martí, E., & Solé, I. (1997). Conseguir un trabajo en grupo eficaz. *Cuadernos de Pedagogía, 255*, 59–64.

Mérida, R. (1999). Ayudando a organizar el pensamiento infantil: los mapas preconceptuales. *Kikirikí. Cooperación Educativa, 62/63*, 76–83.

Mérida, R. (2001). *Investigando en educación infantil: los mapas preconceptuales.* Córdoba, Spain: Servicio de Publicaciones de la Universidad de Córdoba.

Mérida, R. (2002). Una nueva forma de trabajar en educación infantil: los mapas preconceptuales. *Cultura y Educación, 14*(1), 99–123. doi:10.1174/113564002317348147

Mérida, R. (2003). Los mapas preconceptuales como estrategias de aprendizaje cooperativo en educación infantil. *Revista de Educación, 331*, 421–441.

Novak, J. D., & Gowin, D. B. (1988). *Aprendiendo a aprender*. Barcelona: Martínez Roca.

Panitz, T. (1995). Collaborative versus cooperative learning- a comparison two concepts which will help us understand the underlying nature of interactive learning. *National Teaching and Learning Forum, 6*, 24-35. Retrieved November 16, 2007, from http://home.capecod.net/~tpanitz/tedsarticles/coopdefinition.htm

Perret-Clermont, A. N., & Schubauer-Leoni, M. L. (1981). Conflict and cooperation as opportunities for learning. In W. P. Robinson (Ed.), *Communication in development* (pp. 211-243). London: Academic Press.

Roselli, N. D., Gimelli, L. E., & Hechen, M. E. (1995). Modalidades de interacción sociocognitiva en el aprendizaje de conocimientos en pareja. In P. Fernández & M. A. Melero, *La interacción social en contextos educativos*. Madrid: S. XXI.

Trevarthen, C. (1982). The primary motives for cooperative understanding. In G. Butterwood & P. Light (Eds.), *Social cognition. Studies of the development of understanding* (pp. 89-112). Chicago: The University of Chicago Press.

Wertsch, J. V. (1988). *Vygotski y la formación social de la mente*. Barcelona: Paidós.

Woods, D. (1989). Social interaction as tutoring. In M. Bornstein & J.S. Bruner (Eds.), *Interaction in human development* (pp. 132-140). Hillsdale, N.J.: LEA.

KEY TERMS AND DEFINITIONS

Collaborative Learning: Learning methodologies that promote collaboration among individuals to understand, share and expand the information that each one has about the topic.

Concept Mapping: Is a technique for visualizing the relationships among different concepts. Is a diagram showing the relationships among concepts, which are connected with labelled arrows in a downward-branching hierarchical structure. The relationship between concepts is articulated in linking phrases.

Consensual Preconcept Maps: Preconcept maps produced by a group of children (4 to 5 children) that consist of the preconcepts previously agreed upon by everyone.

Early childhood education: Phase of the Spanish educational system that serves children from 0 to 6 years of age.

Intersubjectivity: Process of interpsychological regulation that takes place within a context of social interaction among equals and where participants attempt to collaboratively co-construct solutions.

Modalities of Sociocognitive Interaction: Typology of social and cognitive relations that may be produced among subjects, normally involving two dimensions: the quantity of interactions and the symmetry of the exchanges.

Preconcept Maps: Preconcept maps are an adaptation of the concept maps devised by Novak and Gowin (1988) to the sociocognitive capacities of pre-school pupils. They are symbolic-graphic diagrams comprised of preconcepts (expressed in terms of symbols), oral connectors (relating the concepts) and hierarchical levels (coded by ordinal numbers and colours).

Social Interdependence: Process of social interaction where certain common goals are established with other people in such a way that each person's results are affected by the actions of the others.

Chapter 21
Concept Maps and Meaningful Learning

Patricia Lupion Torres
Pontifícia Universidade Católica do Paraná, Brazil

Luiza Tatiana Forte
Instituto de Ensino Superior Pequeno Príncipe, Brazil

Josiane Bortolozzi
Pontifícia Universidade Católica do Paraná, Brazil

ABSTRACT

The search for meaningful learning promotes a rethinking of the approach adopted by teachers and students to take into account innovative methodologies based on previous experience of learning objectives. This article therefore describes the fundamentals of collaborative learning and the use of concept maps with different individuals in a hospital setting to promote learning about the complexity of their particular situation and in turn achieve higher levels of quality of life for school children receiving hospital treatment. The study also aimed to involve health professionals and health educators in this process and thus gather information to develop a Learning Object.

INTRODUCTION

It would seem incoherent to say that society is today experiencing a period of transition, as it is becoming increasingly clear that transition is a continuous and accelerating process. It is a fact that teaching systems need not only to adjust to this changing reality full of new requirements, but also to be flexible so that they are always up-to-date.

In this context, the value of learning is apparent. Educational methodologies and those involved in them need to become flexible and take into account new requirements so that education reflects the latest developments and can arouse peoples' interest and promote quality of life.

Behrens (2000) defines the foundations of globalized education as collaboration, communication and creativity. These three values are obtained by exploiting educational activities that promote the development of individuals' abilities, i.e., that value talents peculiar to each individual through peer exchange and teamwork, triggering interactive processes and hence socialization.

DOI: 10.4018/978-1-59904-992-2.ch021

It becomes clear that there is a need to create an educational environment in which there are constant challenges, with interdisciplinary projects based on innovative methods. There must be a constant search for knowledge both outside the classroom and outside class time, and this must be focused on a continuing interest in learning. In other words, the aim must be to develop self-motivated learning through the pursuit of the pleasure that can be obtained from relationships that involve exchanges in order to build knowledge. If this is to be achieved, it is important for schools and teachers to make available resources that are different and that keep students' interest, as well as overcoming the limits of space and time, so that they can be used by all the students irrespective of their particular situation and thus promote social inclusion.

Various didactic and methodological resources must be shown to the teachers and students to facilitate and contribute to the development of learning and the meaningful construction of this knowledge.

According to Delors (1998), four types of learning must permeate an individual's whole life and form the pillars of knowledge: *learning to know; learning to do; learning to live together; and learning to be.*

This way of looking at education therefore presupposes a focus on continuing, extended, diversified education within a context of meaningful teaching and learning, always viewed from the perspective of citizenship.

In this evolving context, present day pedagogical practice must use all the available resources to organize educational interactions and activities and invest in the training of producers and readers by means of different paths and languages, extending their concept of writing and reading to incorporate mediation using digital technologies.

Information and communication technologies provide opportunities for individual action and the diversification and transformation of learning environments, as well as extending the traditional limits of the classroom by putting individuals in contact with the globalized world. As a result, they require new dynamics in terms of both action and interaction, as well as a new organization of time and space, as they allow synchronous and asynchronous communication.

This article described the use of Learning Objects and Concept Maps in a hospital with children and adolescents with chronic kidney failure.

The learning objects made available in virtual learning environments used in this experiment are technological resources for organizing and structuring content for use in health education.

Another issue that is addressed in this study relates to health. The World Health Organization (WHO) defines health as physical, mental and social well-being rather than just the absence of disease. It also advocates healthy public policies, the creation of support environments, the development of personal skills, the strengthening of community action and the reorientation of health services as fundamental factors in the reaffirmation of social justice and equality in health promotion.

One of the aims of "hospital psychology" is to minimize children's suffering during the time they are sick and in hospital. It takes as its starting point a holistic view of man as a bio-psycho-socio-cultural being. According to Chiattone (in CAMONN, 2001, p.15) "the need for a new approach to the provision of assistance for sick children is the product of urgent reflection on prevailing issues relating to health and disease in contemporary western society."

In this regard, the knowledge associated with pedagogy and "hospital psychology" can, together with the use of learning objects and the techniques of concept maps, be of significant help in improving the quality of life of preadolescents with chronic kidney failure and involving health professionals as educators in this process.

COLLABORATIVE LEARNING

Environments that allow learning and the exchange of information to be achieved and that promote collaboration are fundamental, particularly in special situations such as hospital education. Collaborative learning is based on a shift in the teaching focus from the teacher, or the act of teaching, to learning and interaction. The structuring of the activities, the socialization and the exchange can be helped by the use of technologies and computer networks, but technology in itself is not sufficient. For Marques (2006, p.38), "all learning activity is mediated by cultural devices that stimulate, inform and sustain the purposes for which the activity is intended." With culture as the key element of an activity, the use of new mechanisms, such as emerging technologies, allows processes to be updated, thus making new socialization paths possible.

Collaborative learning requires a planned methodology if it is to be fully achieved, producing new knowledge through interrelationships stimulated by the elements that make up an action and promoting reflective and creative environments for problem solving. For Marques (2006):

The emphasis is on promoting interactions that make the values and ways of working of different cultures emerge, and when the participants' different cultures are compared, they result in new contributions that enrich the activities being developed. (p.42)

Institutions and teachers have at their disposal countless different technological resources with which educational practice can be supplemented. The use of technology in education allows situations involving the exchange of information and ideas to be experimented with outside formal class time and supports collaborative learning functions. For this to happen, however, it is essential to know how to use the resources and to provide support so that the values of collaborative learning can be exploited.

Even in a classroom-based context, but especially in a virtual environment, hospitalized children can feel intimidated and insecure when they have their first contact because of the seemingly impersonal nature of technology. According to Harasim et al. (2005), the role of the person responsible for pedagogical mediation "is to create a warm and cozy environment that stimulates [the students] to start participating immediately and calms them with clear instructions and support structures" (p.224). In addition to these requirements, all the participants in the process must be clear about the expectations of those involved (hospitalized children and young people, parents, teachers and health professionals) and clear information about the organization of the activities should be given to them.

Teachers, pedagogical staff, psychologists and other health professionals must be able to structure the planning of activities and anticipate the inclusion of online activities to strengthen collaboration and achieve greater student involvement. For Harasim et al. (2005), some techniques should be used by teachers to stimulate active student involvement. Behrens (2000) adds a number of procedures based on Chikeribg and Ehrnmann's approach that are required in teaching practice for a collaborative learning methodology.

1. develop a friendly, welcoming atmosphere, thereby reducing the sense of insecurity felt by users, who may avoid taking a position for fear of making a mistake;
2. reduce the distance that hierarchical levels can impose, making access between people easier and making it possible for people to become acquainted with each other and exchange information.
3. show the students what objectives it is hoped to achieve, making maximum use of the available resources. It is important to encourage students with positive comments about their participation in activities and to talk to students who are not taking part in them;

4. encourage each student's stance in the discussions based on the literature. If it is worth including the texts that are referred to, it is better to summarize them and encourage discussions based on the summaries and the exchange of acquired knowledge;
5. encourage user participation by means of questions that can involve the students in the discussions;
6. finish the activities with a summing up by both the teacher and the students. It is vital that feedback and answers are provided immediately, as these represent the result of what has been achieved. Students can observe their behavior and development, allowing them to review their approach, formulate new objectives with the guidance of their teachers and thus feel more motivated to participate.
7. encouraging users to reply to comments and stimulating debates and activities that are available in the online resources is fundamental if full use is to be made of the environment; the teacher must set an example so that the students feel motivated to collaborate.
8. involve students in the assessment of the activities; get feedback from them about their expectations and opinions, their personal attitudes in relation to the group and procedures; and advise them on what areas can be improved;
9. respecting different talents and ways of learning means developing students according to their abilities. Diversity should be encouraged and multiple intelligences developed in an educational environment.

It is important to create ways of maintaining contact between students and teachers by means of different forms of communication. "There is not one unique way of communicating, and all the possibilities must be exploited" (Harasim, Teles, Turoff, & Hiltz, 2005, p.228). The teacher must be the person responsible for creating a sense of community among the students. Cooperation among students must be encouraged, thereby reducing competition and extreme individualism and allowing people to come closer to each other.

The points addressed by Harasim et al. (2005) and Behrens (2000) in relation to the procedures for the behavior of teachers are fundamental if the best is to be extracted from the functionalities of a virtual environment that well-planned and well-guided interaction can produce. These procedures are applicable in any collaborative-learning educational context, whether virtual or classroom-based.

The approach taken by teachers in a collaborative-learning methodology needs to be rethought. Although it is fundamental for teachers to plan their activities and structure their behavior, it is also important that students behave innovatively, rather than passively and independently, and be integrated in the group and actively involved in the construction of collective knowledge. Based on this, Alcântara, Siqueira, & Valaski, (2004) start from Johnson and Johnson's definition of collaborative learning given below and propose three central ideas:

a set of methods for use in groups to develop learning skills, personal knowledge and social relationships, where each member of the group is responsible both for his or her own learning and that of the rest of the group (p.172)

The first point is reflected in the procedures described by the present authors, in which the traditional educational approach is replaced by one involving group activities and collective learning whose focus is both inside and outside the classroom. This change presupposes that student involvement extends beyond existing models and places importance on continuing interest stimulated initially by the teacher, after which students must develop an awareness of their own learning process.

The second point suggests that interaction in the teaching and learning process is more important than the content in itself. This interaction favors greater closeness between teachers and students, who build knowledge collectively.

Lastly, the third idea is the responsibility of each member of the group for the learning of all the other members. This personal and collective responsibility requires that students have an independent, participative and conscious attitude; they should:

1. pay attention to their fellow students' comments and to reply to them, thus generating an exchange of opinions and providing different viewpoints;
2. ask themselves in every discussion what they really know about the subject, what their point of view is and, if necessary, research the subject so that they can make comments and thus add value to the group;
3. pay attention to all the contributions and collaborate;
4. make reading lists or quote personal examples so that they can illustrate their points with new research sources, adding their knowledge to that of the other students.

Collaborative learning promotes learning through joint action and the exchange of information between people in a process of socialization and cultural education by the group to which they belong.

Learning in this approach changes from being based on an individual perspective to involving group learning and no longer places excessive importance on individual work but involves collaboration. 'When students work together, i.e., collaboratively, they produce deeper knowledge; at the same time they cease to be independent and become interdependent' (Pallof and Pratt, in Leite, Passos, Torres, & Alcantara, 2005, p.4).

Unlike what normally happens in any group or collective exercise, where tasks are just divided, for collaboration to occur there must be interaction and discussion and each participant's opinions and values must be disclosed so that, together, knowledge can be added, new ideas reached and new knowledge constructed based on everybody's contributions.

It becomes apparent that collaborative learning requires a well-planned collective context that makes use of tools as a means of communication and interaction to facilitate group educational activities. For this to happen, however, ways of encouraging this process are required. Rather than just using new technologies, it is fundamental to review the behavior of the agents involved in the process, using as starting point a well-formulated methodology with an objective based on the construction of knowledge.

MEANINGFUL LEARNING AND CONCEPT MAPS

For David Ausubel, a learning psychologist, the most important thing in the learning process is that the learning be meaningful. In other words, the knowledge to be learnt must make some sense to the student. This happens when the new information is "anchored" in relevant concepts that already exist in the learner's cognitive structure. For Novak and Gowin (1999, p.23), to learn meaningfully, the individual relates new concepts to propositions and concepts that he or she already masters.

Dutra, Fagundes, & Cañas (2003, p.3) point out that for Novak meaningful learning requires that the student chooses to learn meaningfully and that "the material to be learnt must be conceptually clear and presented both in language and with examples that can be related to the learner's previous knowledge." The author also points out that thus "mediated by language, the learners process new meanings that must be organized progressively so that they can anchor their knowledge."

In his theory, Ausubel proposes a learning process in which there is effective communication and the student achieves meaningful learning and adopts an independent approach to knowledge construction. Within this process, the teacher can reduce the gap between theory and practice and stimulate students to develop their learning both creatively and meaningfully.

Torres and Marriott (2006, p.11) identify three principles of Ausubel's theory that guided Novak's work:

- The development of new meanings must be built on relevant preexisting concepts and propositions;
- The cognitive structure must be organized hierarchically, with more general and inclusive concepts in higher levels of the hierarchy and the more specific and less inclusive concepts below these;
- Meaningful learning ensures that the relationships between concepts become more precise and better integrated with other concepts and propositions (Novak, 2004, p.460).

Díaz Barriga and Hernández Rojas (2006, p.41) discuss meaningful learning and point out that according to this theory students relate the new information in a nonarbitrary, significant way to previously existing familiar knowledge and experience that they have in their cognitive or knowledge structures." The authors thus highlight the importance of the student having some prior knowledge and experience of the learning objectives. Otherwise, little will be achieved, even if the content matter is well prepared. Likewise, for the student to learn, teachers must be qualified, willing and motivated to teach meaningfully and also have previous knowledge and experience of the subject that they want to teach.

In this respect, the technique of concept maps developed by Joseph Novak in 1972, which has as its theoretical basis the approach advocated by Ausubel, is a resource that has great potential for the teaching/learning process. It allows the teacher to mediate the development and construction of knowledge and the student to understand and give greater meaning to what he or she is learning and so achieve meaningful learning. Maps thus constitute a tool that allows the meaningful relationships between different concepts to be represented.

Concept maps are also based on Vygotsky's theory of the zone of proximal development, particularly insofar as it relates to the construction of knowledge and the influence of the student's social interaction if the learning is to be truly meaningful. They are also based on Piaget's constructivist theory, according to which individuals construct their own knowledge and meanings based on their willingness to undertake this construction.

Concept maps are an instrument or resource to facilitate the learning of content and broaden its meaning for the person who is using them. For Soto (2004), they are a schematic resource that shows a set of conceptual meanings included in a structure of propositions. This way of presenting knowledge benefits active learning and the creative spirit at any level, as well as facilitating communication, linguistic development, reflective thought and cognition. (Torres and Marriott 2006)

Thus, concept maps constitute a very useful tool to help students reflect on their learning process, on the structure of knowledge and on its production; in other words, on metaknowledge (Novak and Gowin, 1999). When they reflect on their learning, students carry out self-assessment, which greatly benefits their personal growth and a possible reorganization of the teaching/learning process.

It should be stressed that concept maps are different from other visual forms of representation such as graphic organizers, mind maps and flowcharts in that they use connecting words or phrases, whose function is to show the relationship between the concepts. (Torres and Marriott 2006)

For Novak and Gowin, (2002 p.15) "concept maps are intended to represent meaningful relationships between concepts in the form of propositions. Propositions are two or more concept labels linked by words in a semantic unit." The same authors state that "concept maps make clear to both students and teachers the small number of key ideas they must focus on for any specific learning task" (2002 p.15). Novak (1998) also states that concepts are perceived regularities in events or objects. The concepts and propositions therefore form blocks for building knowledge in any domain.

The use of concept maps can be of significant help in developing a range of processes, of which the following are worthy of note: an understanding of what is being read, reflective thinking, thinking about meanings, spatial relationships, perceptual field integration, differentiation between the essential and the accidental, reasoning about deductive series, reasoning about inductive series and verbal comprehension. (Soto, 2004, p. 147)

Whatever use is made of concept maps, the challenge that the teacher faces is to change from a conservative paradigm to an innovative one, where the student plays an active role and the teacher takes on the role of guide or learning mediator.

RESEARCH METHODOLOGY

The present study was developed in two stages. The first involved planning the concept maps for the first module of a learning object about chronic kidney failure developed by the Center for Educational Technologies (NTE in Portuguese) in the Distance Education Directorate at PUC-PR. The second stage involved completing the concept map with 15 individuals, including preadolescents, parents, health professionals and teachers, using a stimulus question *What is CKF?* to gather information so that the learning object could be completed.

In terms of the approach to the problem, the methodology used in this study was qualitative research, which considers the existence of a dynamic relationship between the real world and the subject an indissoluble binomial between the objective world and the subjectivity of the human being.

Technically, it involves action research, in which an action and the solving of a collective problem are very closely associated.

The subjects of the research were literate preadolescents with CKF who were either being diagnosed or treated, teachers and health professionals working in a children's hospital with a highly complex environment in the city of Curitiba, Paraná state.

To obtain the desired information for the study, appropriate instruments had to be developed, and these involved concepts such as a questionnaire and systematic observation.

By combining pedagogy and psychology and using the construction of a learning object for health education about chronic kidney failure as an instrument for pedagogical mediation and educator training incorporating educational technologies, the project aimed to interfere significantly in order to improve the quality of life of preadolescents with chronic kidney failure (CKF) and to involve health professionals and teachers as health educators in this process.

The subject in question arose from the clinical experiment involving preadolescents with chronic kidney failure during psychological follow-up from the time of the diagnostic examination until they started dialytic treatment and had a kidney transplant.

It is known that although preadolescents can appreciate why they have been hospitalized, they very often have difficulty understanding what is happening to them and the explanations that are given.

It was observed that the information provided by the doctor is aimed at the adult, who is not suffering from the disease, while the preadolescent is relegated to learning about his new situation from the improvement or deterioration in his condition as well as from tests, which can be more or less invasive. There is a need to allow the preadolescent to learn about his disease and treatment from the time it is diagnosed. The more the preadolescent can master this learning and knowledge, the better chance he will have of dealing with all the situations that can arise as a result of the disease.

THE USE OF CONCEPT MAPS TO BUILD LEARNING OBJECTS

The presentation of multimedia didactic content in the learning object format proposed by the Center for Educational Technologies at PUC-PR is a complex task because of the range of different procedures and techniques required.

The content is developed taking into account the target public's profile as well as the place where it is going to be used.

The structure for the content of the learning objects defined by the NTE at PUC-PR was determined by the need to ensure that the modules could be reused and reorganized, making the learning objects easier to update and more flexible to use.

Wiley (2000, p.4-5) highlights these factors when he defines a learning object as "any digital resource that that can be reused to support learning." He also states (2002, p.4-5) that this is a traditional view "according to which learning objects can be anything from images and graphics, videos, sounds or tools to any other digital educational resource that can be used for educational purposes." Wiley also states (2002, p.4-5) that "common sense suggests that a learning object usually includes a set of such basic resources organized so as to provide the opportunity for a learning experience."

According to L'Alier (1997, in Polsani, 2003, p.8), a learning object must contain objectives; learning activities or instructional content; and practice and feedback.

This division into three parts can be a useful way of structuring the object. With regard to the first part, i.e., the objectives, these must make clear to the learner what can be learnt with this object and whether or not there is any prerequisite for this learning. The instructional content or learning activities introduce the didactic material that is essential for the learner to achieve his objectives. The practice and feedback allow the learner to check whether his performance was satisfactory and met the expectations; it is here that simulations can be used.

To develop the first module of the learning object (LO) about chronic kidney failure, the following concept map was designed. (see Figure 1 & Figure 2)

In this first construction, which has a web structure with the main concept in the center of the map, it can be seen that the map worked as a brainstorming exercise in which all the important content that is already known and that should be discussed in the construction of the object has been included. Okada, Santos, & Okada (2003, p.08) states that "knowing what one already knows about new knowledge and seeking to learn what else it is still necessary to learn to receive this new related knowledge is the correct path for learning and consolidating the new unit of knowledge that is being formed."

However, the plan used was a general one, and there remained a pressing need to specify the contents in order to establish the concepts and organize them in modules. In addition, it became clear that some content had been suppressed, and with the overall view now available after this first construction it could be seen that this content needed to be included. This map allowed the summary to be constructed and the modules that would make up this object to be organized.

Using this summary as a starting point, a concept map specifically for the content of module I could be proposed.

The summary for the first module contains the following items: objective, overview of the human body at an anatomical and functional level, how

Figure 1. General concept map of the contents of the virtual object: Part 1 (© 2007, Forte, L.T.; used with permission.)

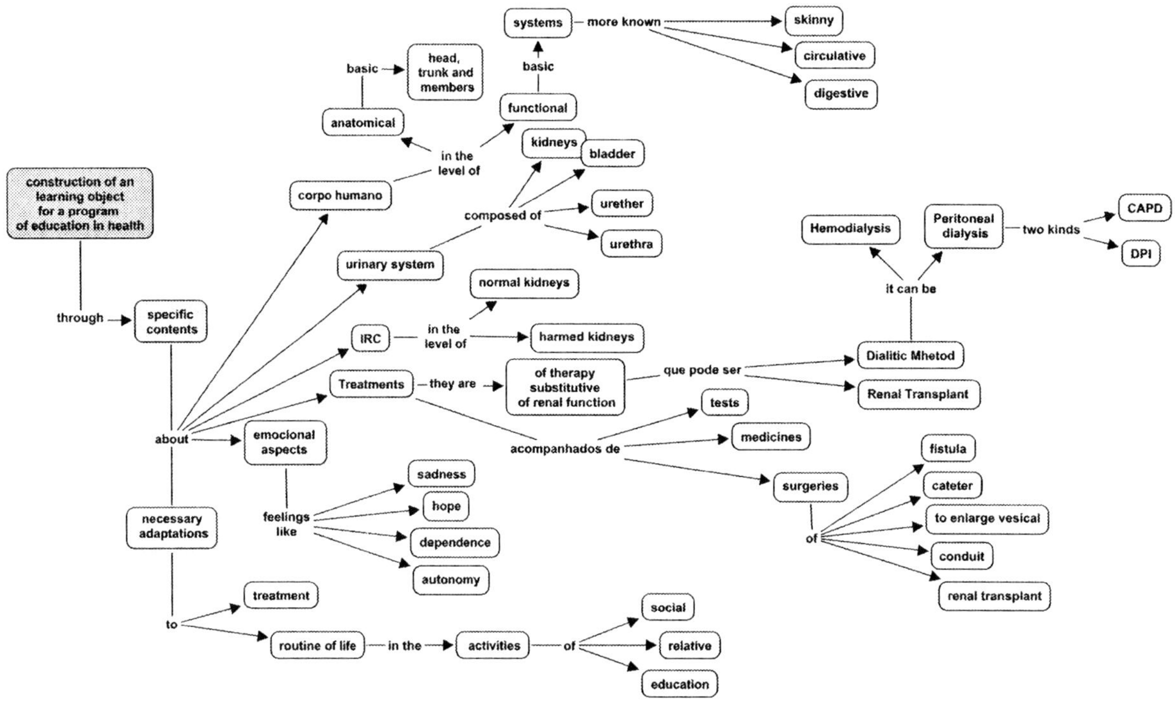

the kidneys work, healthy kidneys, damaged kidneys, chronic kidney failure, the main symptoms of ESRD, checking activity, the causes of ESRD, a checking activity (do you know what caused your IRC(ESRD)?) and conclusion. (see Figure 3)

The visual presentation of this map not only allowed new ideas to emerge and be expressed but also allowed them to be reorganized. In addition, it allowed concepts to be unified, separated and integrated as well as to be synthesized through pictures and words to articulate thinking.

CONSTRUCTION OF THE CONCEPT MAP BY THE PREADOLESCENTS

Preadolescence is a transition phase in which thinking and language become further and further removed from immediate reality and the capacity for abstraction takes over. With this acquired capacity, reflection becomes the preferred way of understanding and intervening in reality. Social relationships increase in number and intensity, and with this increase preadolescents can take into account different points of view without abandoning their own. It is a good time to encourage debates and discussions in the environments where preadolescents are usually found, whether these be school, family, social gatherings or health facilities.

In light of this, we asked some preadolescents with ESRD to make the concept map and stimulated them with the question: What is IRC (ESRD)? (see Figure 4)

In this second concept map, a clear separation can be observed between the more concrete aspects of the disease, such as tests, dialysis, medicines and not having a kidney, and the feelings arising from these connected with the preadolescent's daily routine, the feeling that it is bad, that it gets in the way of play and school and cannot be cured. When stimulated to expand further on the

Figure 2. General concept map of the contents of the virtual object: Part 2 (© 2007, Forte, L.T.; used with permission.)

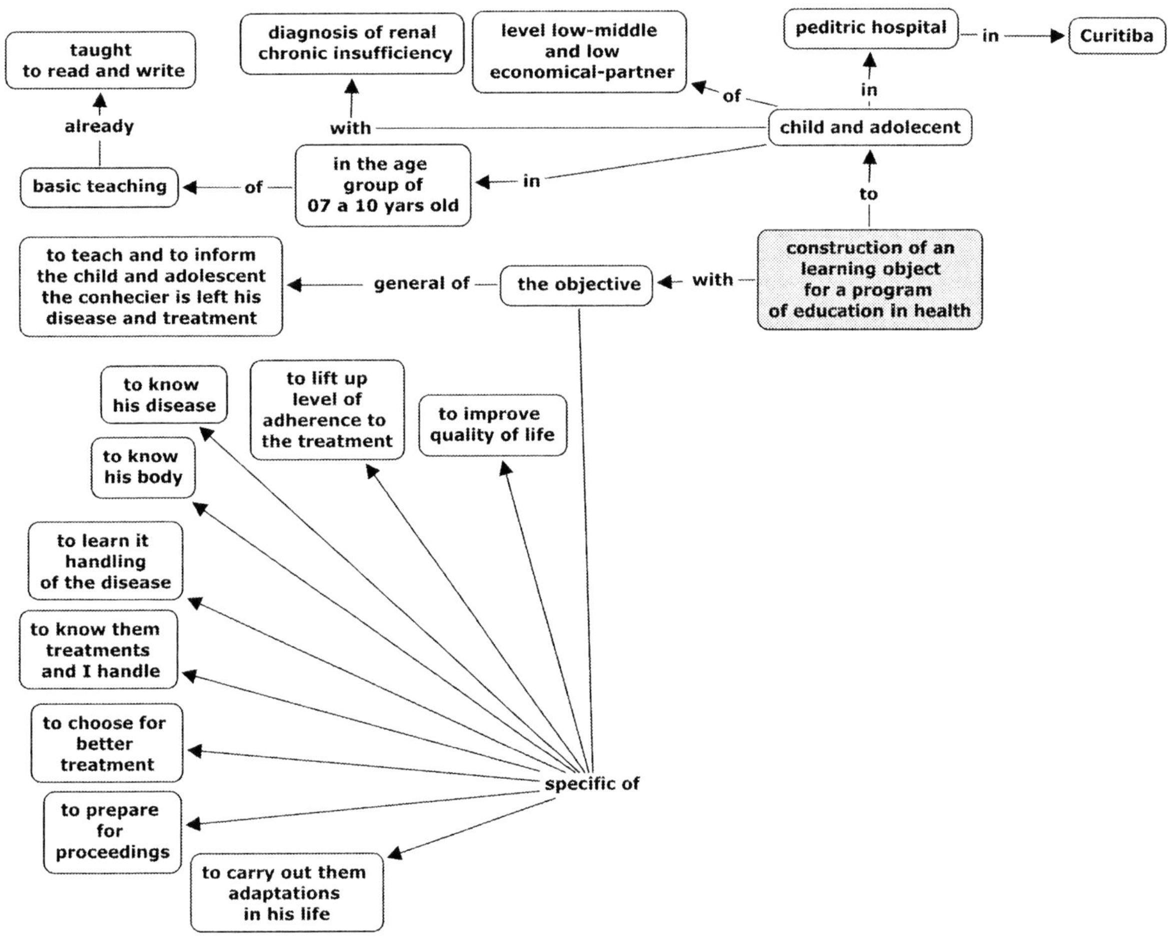

concept of a cure, the preadolescent includes the transplant and claims not to know much about this subject yet.

Human thought is built with networks and associations. We produce new knowledge in a hypertextual network; we do not think linearly. New knowledge connects to existing knowledge that has already been constructed and can be updated and even refuted. Everything depends on our production of meanings, on the meaning we attribute to concepts. (Okada and Saburo, 2003 p.09)

When it was explained to her that she did not need to *know a lot*, the preadolescent was willing to make a second map, this time about her knowledge of kidney transplants. (see Figure 5)

The third map, which was made by a 7-year-old preadolescent boy, shows more rudimentary thinking about what ESRD is. The way in which the limits imposed by the disease have been clearly delineated is particularly noticeable. (see Figure 6)

It can be observed, however, that this client makes a distinction between what he knows and what his mother knows, and notes that it is his mother who knows best. He knows the limits that are clearly identified on the map, but he does not appear to know about the possibilities of adapta-

Figure 3. Specific concept map of the contents of the module I virtual object (© 2007, Forte, L.T.; used with permission.)

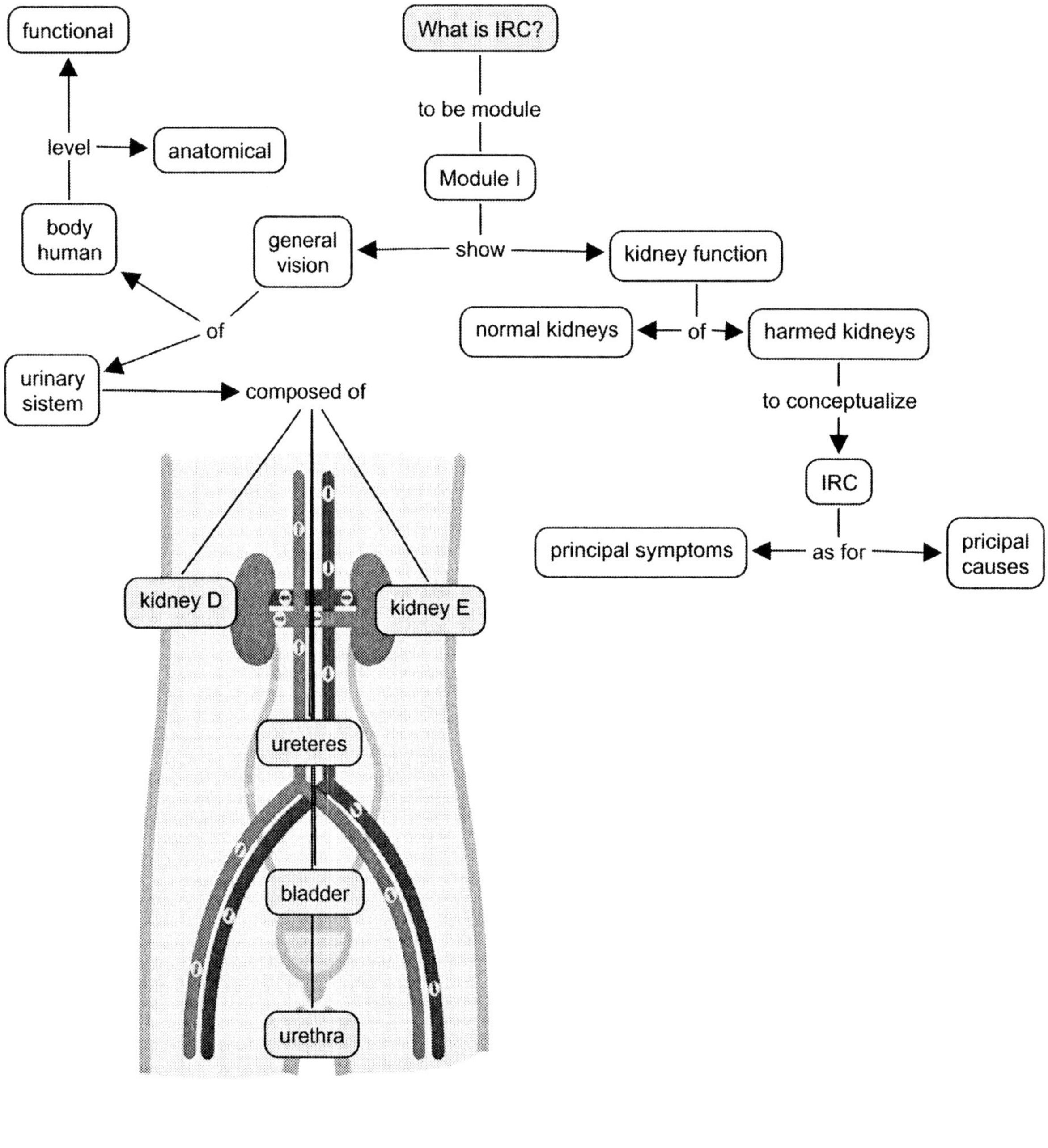

tion. That the relationship is one of dependence is made clear when he indicates that knowledge about CKF lies outside his scope. According to Okada (2000, p.4) "the schemes are constructed or reconstructed on repeated experiences and reflect the individual's current level of knowledge."

The fourth map was produced by a ten-year-old preadolescent boy. He defines ESRD not only in terms of treatment-related issues but also in terms of changes in his life. (see Figure 7)

This patient uses longer phrases in an attempt to explain why CKF is changing his life. It is as

Figure 4. Concept map 1 nine-year-old preadolescent girl (© 2007, Forte, L.T; used with permission.)

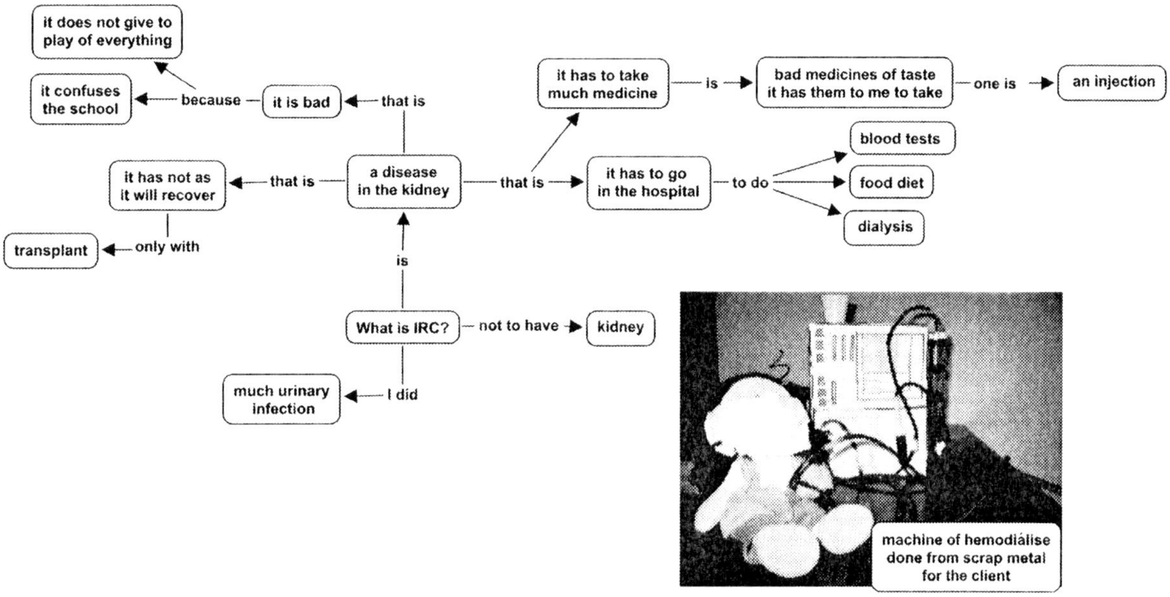

Figure 5. Concept map 2 nine-year-old preadolescent (© 2007, Forte, L.T.; used with permission.)

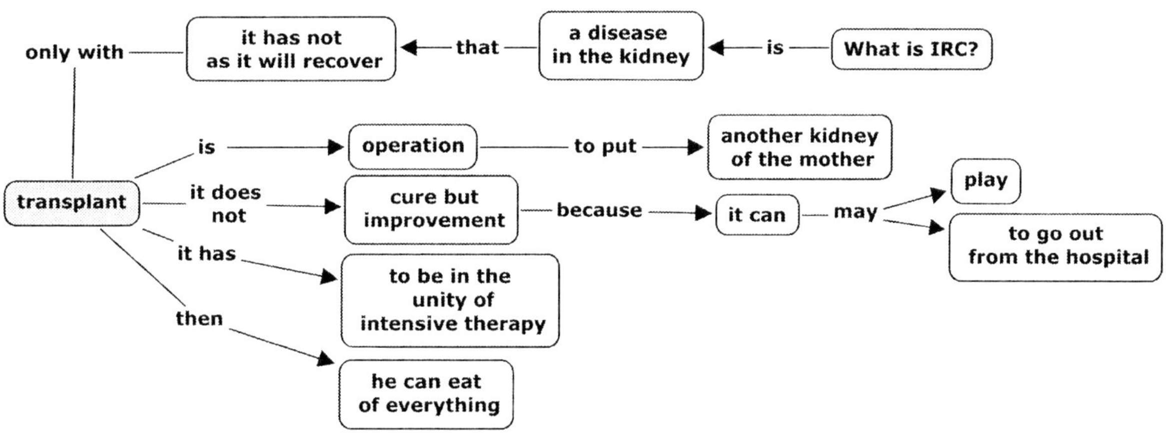

if the branch on the right of the map were his rational thinking and the branch on the left his emotional thinking, which can be externalized by the process of abstraction characteristic of this developmental phase. It can also be seen that the preadolescent is making a clear appeal to the team, namely, to take an interest in his life as a whole rather than only taking an interest in his disease. According to Wallon (1968 in Manrique, 2000, p.11) "the emotions essentially consist of systems of attitudes, each of which correspond to a particular type of situation."

The fifth map was constructed by a 10-year-old preadolescent girl. It is clear that the most important question for her is the change in the appearance of her body because of the need for a catheter in the abdominal region. This concern with her body is to be expected given her stage of development at the time. (see Figure 8)

She describes ESRD succinctly and concen-

Figure 6. Concept map seven-year-old preadolescent boy (© 2007, Forte, L.T.; used with permission.)

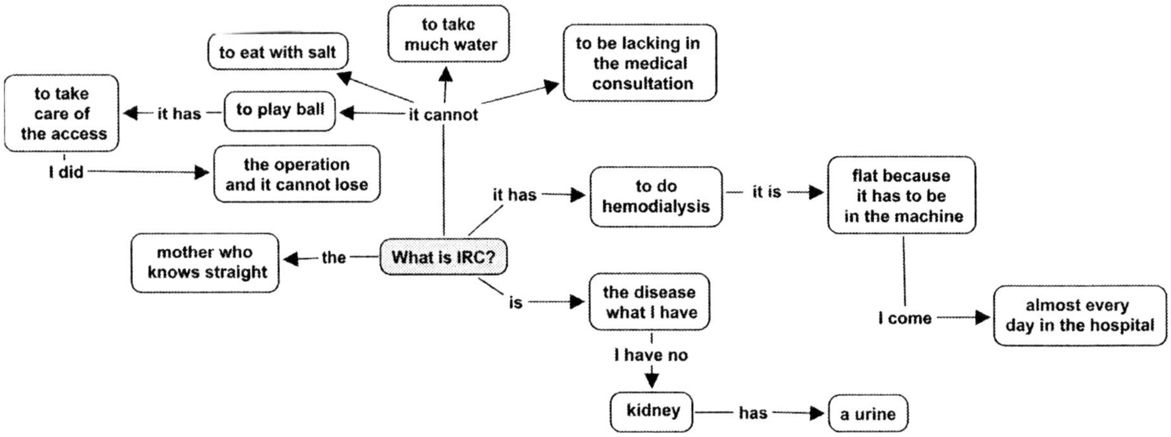

Figure 7. Concept map ten-year-old preadolescent boy (© 2007, Forte, L.T.; used with permission.)

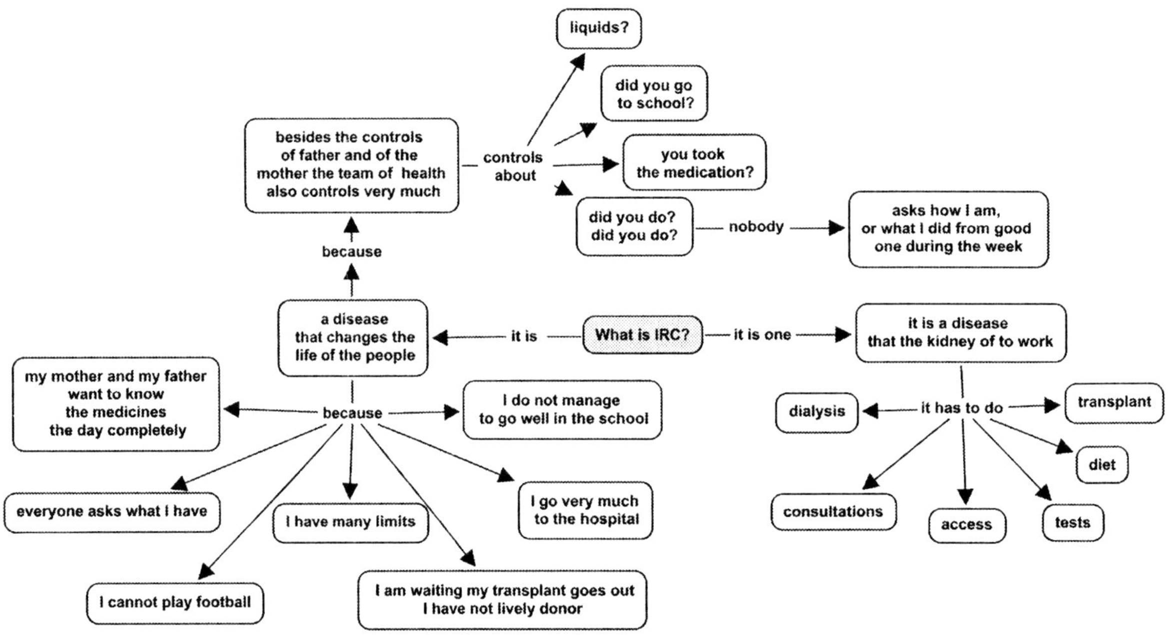

trates on describing her worries about her clothes. It makes clear her feeling of shame and lack of independence because of the time limits imposed by the therapy.

The sixth map was produced by a nine-year-old preadolescent boy. It describes the hemodialysis in more detail, with his feelings and the physical symptoms he has during the session. (see Figure 9)

On the left of the map he represents his life and places his hope in the kidney transplant. The map has been constructed in a very concise way, but with important content that is the most meaningful for him up to that time.

It should be stressed that there is no such thing as a "correct" concept map; each learner externalizes their concepts in a unique way, as a result of

Figure 8. Concept map ten-year-old preadolescent girl (© 2007, Forte, L.T.; used with permission.)

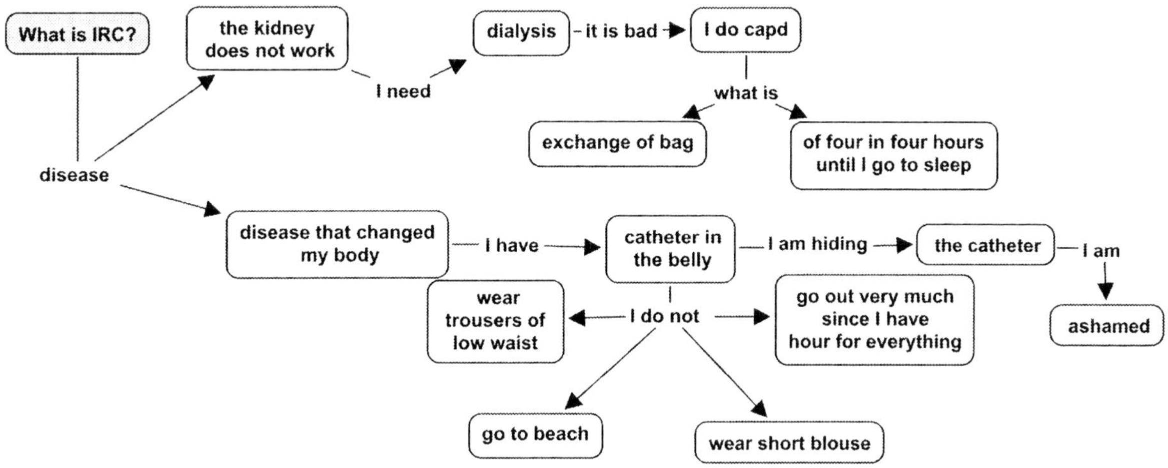

Figure 9. Concept map nine-year-old preadolescent boy (© 2007, Forte, L.T.; used with permission.)

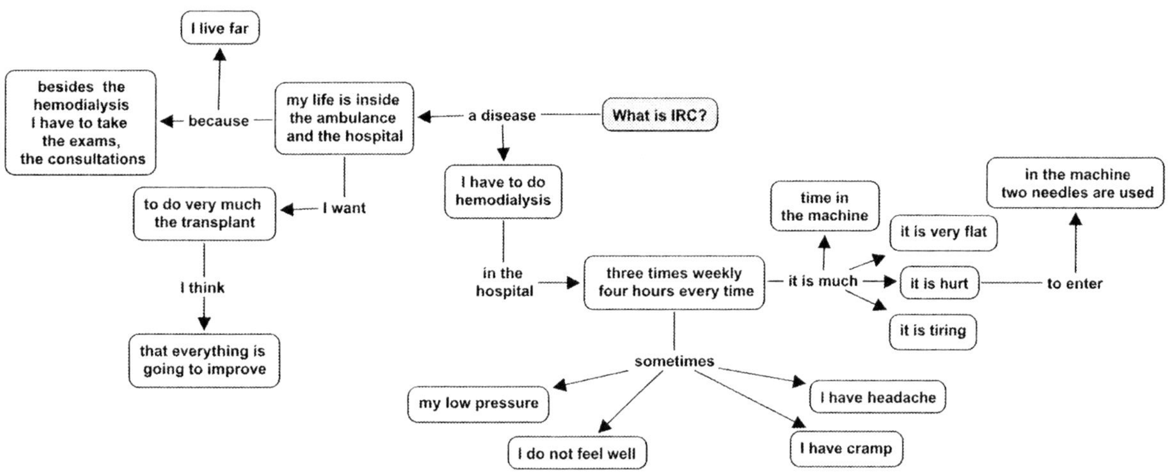

reflection and their own particular way of seeing, feeling and acting.

CONSTRUCTION OF THE CONCEPT MAP BY THE HEALTH PROFESSIONALS

The information provided by the doctors and the team of health staff is usually aimed at the adult, who is not ill but is feeling the effects of the disease, while the preadolescent is relegated to learning about his new situation from the improvement or deterioration in his condition as well as from tests, which can be more or less invasive. It is important to provide the preadolescent with knowledge about how the disease behaves and how it can be treated from the moment it is diagnosed. The more knowledgeable the preadolescent becomes and the more active he becomes in this process, the better chance he will have of dealing with all the situations that can arise as a result of the disease. It is the responsibility of the health team to orchestrate this knowledge and act as facilitators in the process.

Following are the three concept maps constructed by three health professionals who work with preadolescents with ESRD.

The first map shown below was produced by a nurse. (see Figure 10)

It is interesting that, using the stimulus question *What is ESRD?*, the map produced by the nurse encompasses the team, the family and the preadolescent, both when she describes what ESRD is and when she refers to the need for knowledge about the disease to achieve a greater quality of life. While on the one hand she defines some of the team's functions, she does not define those of the family or preadolescent.

The concepts are chosen by the subject according to their cultural representativeness based on ***typicality***, i.e., the degree to which the concepts belongs to a particular class and the extent to which the concept is representative of a class of concepts. Thus, the concepts indicate the set of characteristics to which the objects in a particular class belong for a particular social segment and allow us to distinguish them from others (Amoretti & Tarouco, 2000 p.3).

The nurse can be seen to have a holistic view of man, but it is only when the map has been constructed that she notices that she has concentrated a lot of responsibility on the team.

Figure 10. Concept map 35-year-old nurse (© 2007, Forte, L.T.; used with permission)

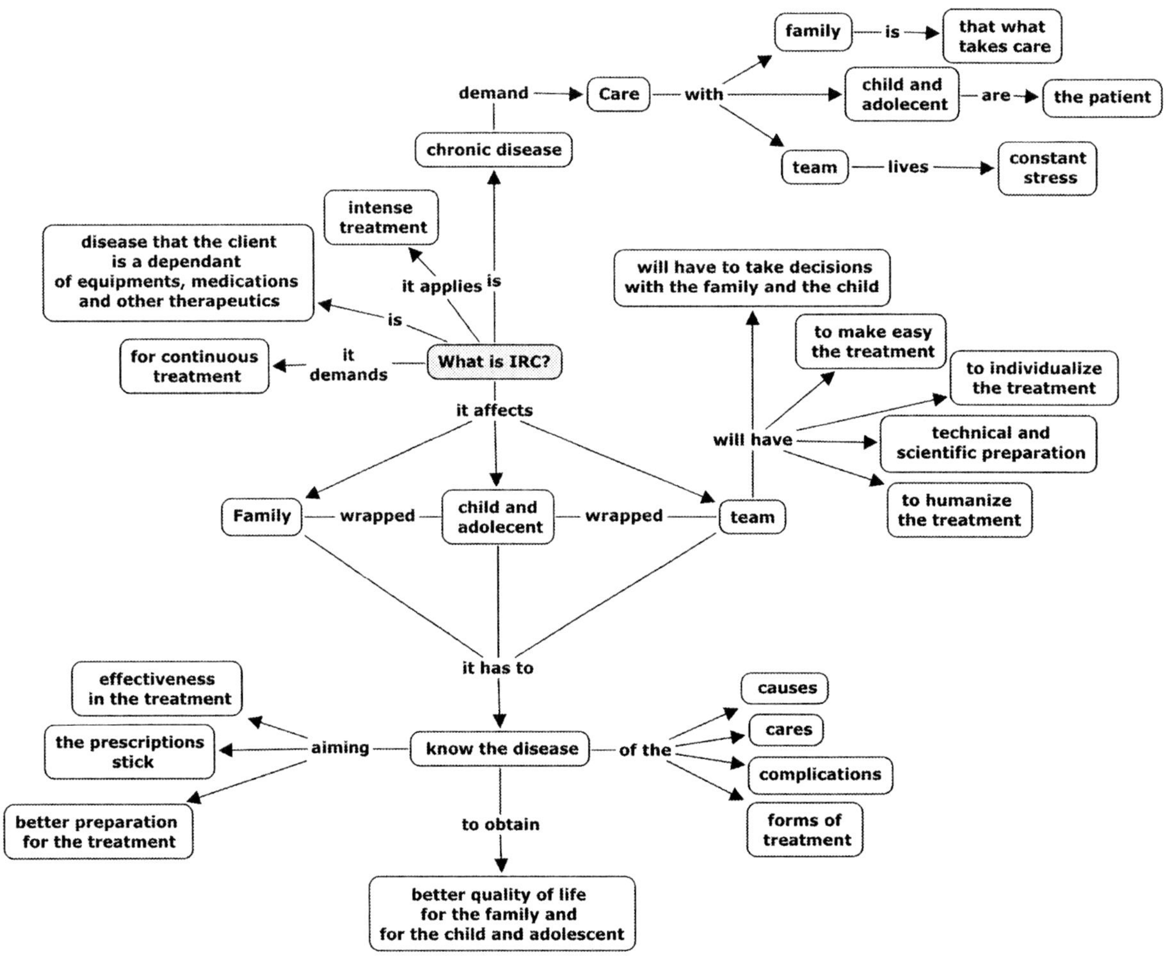

The second map was produced by a psychologist who works with these clients. (see Figure 11)

Reflecting her understanding of ESRD, the map is divided into four branches: the upper branch, which defines ESRD; the right branch, which defines the physical treatment; the lower one, which defines the symptomatology; and the one on the left, which defines the emotional consequences. In this last branch she defines, in addition to her involvement/functions, the main emotional consequences, and when she describes her involvement she includes the team, the family and the preadolescent.

The third map was constructed by a specialist in pediatric nephrology. We can see that the most important aspect for this person are the definitions of the disease and the therapies. It can also be seen that the map refers to the involvement of the family and the child but that these categories are not developed. This does not mean that she does not take the family or the child into account, but that in the context the most important thing is to master the medical issues. (see Figure 12)

Education in a hospital setting requires the support of innovative methodologies if it is to motivate and take place meaningfully. The present analysis shows the educational needs in such a setting and their receptivity to innovative methodologies, as well as the lack of attention given to hospital education, both in terms of orienting school children in hospital and health professionals working there. The data obtained using concept maps form part of the preparation of a learning object to be integrated into the EUREK@Kids project and subsequently used in hospital-based learning activities.

It must be stressed that continuing the present study will involve:

- analyzing the possibilities and limitations of education in hospital settings, taking into account social differences and differences in access to technology;
- drawing up a proposal to remodel the educational approach used to support hospitalized children and adolescents;
- proposing a methodology that will help health professionals cater for the unmet needs of hospitalized schoolchildren in order to encourage the participation and

Figure 11. Concept map 26-year-old psychologist (© 2007, Forte, L.T. ; used with permission)

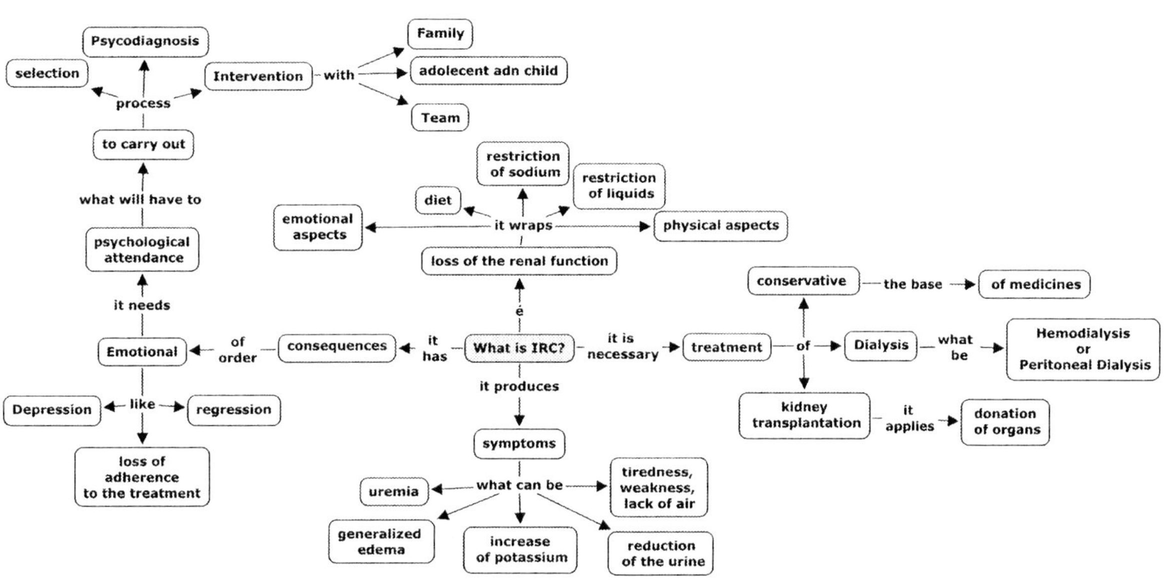

Figure 12. Concept map 34-year-old nephrologist (© 2007, Forte, L.T.; used with permission)

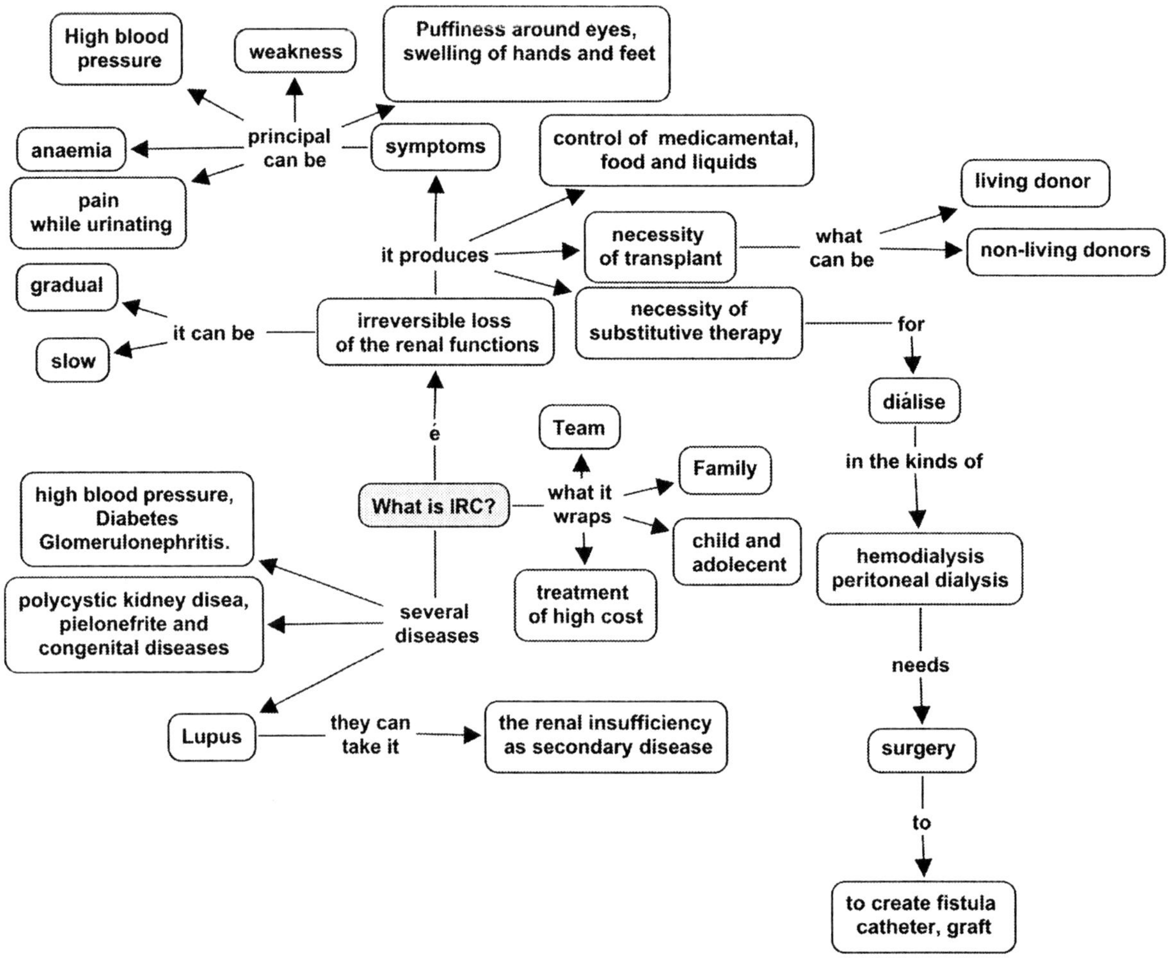

development of their individual capacities;

- validation of the learning object developed in hospitals.

SOME CONSIDERATIONS

The experience constructing maps with preadolescents and health professionals was highly profitable. Each individual organized their thinking and articulated questions of greater or lesser importance according to their way of understanding and feeling. The preadolescents, depending on their age and developmental stage, expressed concerns about their life, their leisure and also their treatment. As expected, they pinned their hopes on the kidney transplant, as they knew it was their only chance to prolong their existence. The maps produced by these preadolescents appear to show how much they know and how much they still need to learn about the disease as well as any relevant or important ways in which they need to adapt to have a better quality of life. They spontaneously incorporate their own situation and their feelings in this situation into the concept map.

The concept maps produced by the health professionals reflect their way of understanding the disease, each showing on the map the most important questions according to his or her specialty. It can be observed that there is still resistance to a more global and humanizing view, as if it were still not possible to look at, feel or perceive the human being as a whole. It is not a question of dehumanization, but of the consequence of the way that health professionals were, and are, trained to exercise their profession.

After the maps had been produced, it was decided to show those produced by the preadolescents to the health professionals once the former's permission had been obtained. The health professionals were surprised at the intensity with which the issues of feelings and quality of life appeared in the children's maps and how they, the nurse and doctor, did not attribute much importance to these when they were constructing their maps. They admitted that their maps showed the stamp of their university and the training they had received.

REFERENCES

Alcântara, P. R., Siqueira, L. M. M., & Valaski, S. (2004, maio/ago). Vivenciando a aprendizagem colaborativa em sala de aula: experiências no ensino superior. In *Revista Diálogo Educacional*. Curitiba, *4* (12) 169-188

Aldana, M. F. (2006). *O Cognitivismo de David Ausubel.* Artigo da Internet. Disponível em: Ser Professor Universitário. Retrieved October 10, 2006, from http://serprofessoruniversitario.pro.br

Amoretti, M. S. M., & Tarouco, L. M. R. (2000, September). Mapas Conceituais: modelagem colaborativa do conhecimento. *Informática na Educação: Teoria e Prática, 3* (1). Retrieved October 13, 2001, from http://pgie.ufrgs.br

Behrens, M. A. (2000) Projetos de Aprendizagem Colaborativa num Paradigma Emergente. In Moran, J. M., Masetto, M. T., & Behrens, M. A. *Novas tecnologias e mediação pedagógica.* Campinas, Spain: Papirus.

CMAPTOOLS. (n.d.). *Software para Construção de Mapas Conceituais da University of West Florida.* Retrieved September 06, 2006, from http://penta2.ufrgs.brtutindicecmap.htm

Delors,J.(1998). Relatório para a UNESCO da Comissão Internacional sobre Educação para o século XXI.

Díaz Barriga, F., & Hernández Rojas, G. (2006). *Estratégias Docentes para un Aprendizaje Signficativo.* México: McGraw Hill Interamericana.

Dutra, I., Fagundes, L., & Cañas, A. (2003). *Uma proposta de uso dos mapas conceituais para um paradigma construtivista da formação de professores a distância.* Retrieved September 06, 2006, from http://homer.nuted.edu.ufrgs.br/mapas_prof.pdf

Freire, R. S. (2007). Desenvolvendo o Pensamento Proporcional Com o Uso de um Objeto de Aprendizagem. In C. L. Prata, Nascimento, & A. C. A. (Org.). *Objetos de aprendizagem: uma proposta de recurso pedagógico.* Brazil: MEC, SEED.

Harasim, L., Teles, L., Turoff, M., & Hiltz, S. R. (2005). *Redes de Aprendizagem: Um guia para Ensino e Aprendizagem On-line.* São Paulo, Brazil: Ed. Senac São Paulo.

Leite, C. L. K., Passos, M. O., Torres, P. L., & Alcantara, P. R. (2005). *A aprendizagem Colaborativa na Educação a Distância on-line.* 12º Congresso Internacional de Educação a Distância Florianópolis, Brazil. Retrieved February 20, 2007, from http://www.abed.org.br/congresso2005/por/pdf/171tcc3.pdf.

Manrique, A. L. (2000) A afetividade explicitada. *Educação Matemática, 19.* Retrieved June 21, 2007, from http://anped.org.br/ analuciamanrique.rtf

Marques, J. C. (2006). Aprendizagem Colaborativa: Atividades de Grupo como Núcleo e Uso do Computador como Contexto. In *Psicologia Argumento*. Curitiba, *24*,(44) 37-43.

Molina, A., et al. (2006). *Potencializar a capacidade de aprender e pensar.* São Paulo, Brazil: Madras.

Moran, J. M. Masetto, Marcos T., & Behrens, M. A. (2003) *Novas Tecnologias e Mediação Pedagógica*. Campinas, Spain: Papirus, 7th edição.

Moreira, M. A. (1997). *Mapas conceituais e aprendizagem significativa.* Retrieved September 12, 2006, from http://if.ufrgs.br/~moreira/mapasport.pdf

Novak, J. D. (2004). A Science Education Research Program that led to the development of the Concept Mapping tool and a New Model for Education. In A. J. Cañas, J. D. Novak, & F. M. Gonzáles, (Eds.). *Concept Maps: Theory, Methodology, Technology. Proceedings of the First International Conference on Concept Mapping* (Vol. I). Pamplona, Spain: Universidad Pública de Navarra.

Novak, J. D.; & Gowin, D. B. (1988).*Teoría y Práctica de la Educación.*

Okada, A., Santos, E., & Okada, S. (2003). *Mapeando informação, trilhando e construindo redes de significados.* Retrieved February 20, 2007, from http://people.kmi.open.ac.uk/ale/journals

Silva, M. (2001). *Sala de Aula Interativa* (2nd Ed.). Rio de Janeiro, Brazil: Quartet, Tedesco, J. C. (2004). *Educação e Novas Tecnologias: esperança ou incerteza?* Brazil: UNESCO.

Torres, P. L., & Marriot, R. C. V. (2006). *Tecnologias educacionais e educação ambiental: uso de mapas conceituais no ensino e na aprendizagem.* Curitiba: FAEP.

Weimer, M. S. M. (2000) *Escola Cliclada*. Secretaria de Estado de Educação Mato Grosso – Brasil. Retrieved May 28, 2007, from http://mec.gov.br

Wiley, D. (2000). *The instructional use of learning objects. On-line version.* Retrieved February 20, 2007, from http://reusability.org/read

KEY TERMS AND DEFINITIONS

Collaborative Learning: Collaborative learning requires a planned methodology if it is to be fully achieved, producing new knowledge through interrelationships stimulated by the elements that make up an action and promoting reflective and creative environments for problem solving.

Concept Mapping: Concept maps are graphical tools for organizing and representing knowledge. They include concepts, usually enclosed in circles or boxes of some type, and relationships between concepts indicated by a connecting line linking two concepts.

ESRD: End Stage Renal Disease

IRC: Renal Failure

Learning Object: Is any digital resource that that can be reused to support learning. Learning objects can be anything from images and graphics, videos, sounds or tools to any other digital educational resource that can be used for educational purposes.

Meaningful Learning: The knowledge to be learnt must make some sense to the student.

Chapter 22
Concept Mapping and Formative Assessment:
Elements Supporting Literacy and Learning

Jeffrey Beaudry
University of Southern Maine, USA

Polly Wilson
University of Southern Maine, USA

ABSTRACT

From the authors observations and those of Kinchin (2001) teachers may know about concept mapping but they do not seem to use it as a consistent, effective strategy. The authors argue that the concept mapping may be better understood by using an expanded definition of traditional literacy, listening, speaking, reading and writing; to include visualizing, visual representation, and technological literacy Sinatra (1986). This ethnographic case study examines the use of concept mapping and collaborative learning strategies in the content area of marine ecology in high school science classrooms. To support students' understanding of science concept and the improvement of writing students began with a field trip to study inter-coastal zones and follow-up laboratory activities, use of digital image analysis, and collaborative group work. Key vocabulary were identified to begin concept maps, and more vocabulary was added to support multiple revisions of concept maps with concept map software, and culminated with students' writing. Concept mapping integrated with collaborative learning was used to engage students

DOI: 10.4018/978-1-59904-992-2.ch022

to construct and re-construct their understanding of a complex scientific concept, the energy cycle. The results showed that students benefited from the combination of collaborative learning and concept maps to focus their writing on key ideas, to organize their ideas, and include specific details. However, the interpretation and integration of quantitative data and laboratory results was not as consistent. Most importantly, initial concept maps and revisions provided the teacher with evidence of student learning in the form of formative assessment products, to guide teachers' focused feedback and clarify specific ideas for re-teaching, as well as students' self-assessment. The authors provide examples of concept maps and graphic organizers as formative assessment of students' knowledge, what Novak (1998) calls heuristic or "facilitative tools," and as visual representations and structures to provide flexible ways supporting learners' meaningful learning through speaking, writing and in visual forms (Sinatra, 2000; Mintzes, Wandersee, and Novak, 2004).

INTRODUCTION

Our aim is to describe the value of concept mapping and graphic organizers for learning in the context of literacy (reading, speaking, and writing, as well as visual representation and technological literacy), and the importance of understanding visual learning strategies as formative assessment. In this paper we discuss 1) the concept of literacy and visual literacy, and visual representation in particular (Sinatra, 1986), and the connection with 2) the role of formative assessment in learning (Black and Wiliam, 1998a). Research reviews on the impact of knowledge and concept mapping (Novak, 1998; Nesbit and Adesope, 2006) indicate moderate to large, positive effects, as does meta-analysis of research on formative assessment (Black and Wiliam, 1998a; Black and Wiliam, 1998b). In order for concept mapping to produce the promised effects on students it must be viewed as formative assessment of students' knowledge, what Novak (1998) calls heuristic or "facilitative tools," and "learning how to learn" (Novak and Gowin, 1984; Black, McCormick, James, and Pedder, 2006). Effective use of visual representations and structures provide flexible ways to support both teachers' and learners' quest to answer three key, instructional questions to support formative assessment: 1) Where do I want to go in my teaching and learning? 2) Where am I now? And 3) What do I need to do to get there? (Black and Wiliam, 1998a; Stiggins, Arter, Chappuis, and Chappuis, 2004) Concept maps are concrete, formative assessment products that promote dialogue and focused feedback among students, and between teacher and students (Sinatra, 2000; Mintzes, Wandersee, & Novak, 2005). Concept mapping and collaborative learning are strategies that can be "thoughtful, reflective, and focused to evoke and explore understanding" (Black and Wiliam, 1998a).

According to the meta-analysis of research by Nesbit and Adesope (2006) which included fifty-five (55), well-designed experimental studies the average effect size estimates for concept mapping were positive, and greatest for collaborative mapping strategies with mixed ability students, for students with lower verbal ability and weak domain-specific background such as sciences like biology or chemistry. In a previous meta-analysis Horton, McConney, Gallo, Woods, Senn and Hamelin (1993) estimated the effects of concept mapping to improve knowledge is 0.42, and the impact on students' attitudes and engagement is 1.57. Effect sizes are quantitative estimates of the impact on specific outcomes like achievement or attitudes, and these estimates help to summarize findings across multiple research studies. The effect size of 0.42 means that on average, experimental groups scored higher than control groups by 0.42 standard deviations. Or an effect size of 0.4 would mean that the average pupil involved in an innovation would record the same achievement as a pupil in the top 35% of those not so involved (Black and Wiliam, 1998a). Positive effect sizes of strategies such collaborative learning (effect size = 0.73), nonlinguistic representation like concept mapping (effect size = 0.75), and reinforcement and feedback (effect size = 0.80) are identified by Marzano (2003) as teacher-level factors representing effective, research-based instructional strategies. In Marzano's meta-analysis *What Works in Schools: Translating Research into Action,* nonlinguistic representations include concept maps and visual representations, the use of images and metaphors, and art forms like the use of drama, constructing models. In addition, concept mapping and visual representations can be used to support other effective strategies like identifying similarities and differences, summarizing and note-taking, and the use of advance organizers.

Several other teaching and learning models provide strong support and suggestions for the use of concept mapping and visual representation, including Trowbridge and Wandersee's model of theory-driven graphic organizers (2005), Hyerle's toolbox of visual learning tools, and the Universal Design for Learning (2002). There are three major principles in the Universal Design for Learning (UDL) framework, 1) to support recognition learning, provide multiple, flexible ways of presenting and teaching new knowledge, 2) to support strategic learning, provide multiple, flexible methods of expression and practice, and 3) to support affective learning, provide multiple, flexible options for engagement. Our approach incorporates flexible learning strategies that reflect all three of these principles, especially the use of concept mapping software for knowledge construction and organization, and the use of collaborative and individual learning to support student engagement.

Several suggestions by Trowbridge and Wandersee were relevant for this study; our procedure for concept map construction closely follows their model (2005) of a progressive map. The procedure by Trowbridge and Wandersee for constructing a concept maps focuses on individual students. A critical difference is that we suggest the use of collaborative learning as the initial step for map construction. Our interviews with students indicated strong support for the collaborative learning step. Also, Trowbridge and Wandersee suggest that expert maps be used only after students had struggled with map construction, another very helpful, useful recommendation with which we agree. Finally, we believe that for most learning goals concept maps be used as formative assessment products, and not graded or scored, a matter of specifying the purpose of the assessment. The issue in any case is that students actively construct the knowledge essential to their learning, and not depend on rote memorization (Novak and Gowin, 1984; Trowbridge and Wandersee, 2005).

In a comprehensive review Hyerle (1996) identified three types of visual learning tools: brainstorming webs, task-specific organizers, and thinking-process maps. He maintains that visual learning tools are important for designing learning that is linear and nonlinear, student-centered, developmental, interactive, and linked with technology – a constructivist environment. Hyerle explains how visual learning tools support the integration of teaching, learning and assessment, inclusive classrooms, and reflection (1996). He also pointed out that students' skills at creating and using maps develop over time, when these skills are used deliberately across the curriculum.

Our study concentrates on concept maps as formative assessment products, created and used intentionally to blur the lines that separate teaching, learning and assessment. Just as important they are developed and used in a combination of group and individual learning settings, providing multiple opportunities to benefit students' emerging understanding and teachers' need to know students' current conceptions and misconceptions of complex science concepts. The outcomes for this study are measured by students' written responses on their lab reports.

We situate our case study in a high school biology classroom with a lesson on the energy cycle. Technology components of this study consisted of both commercial software for concept mapping, Inspiration, and new software provided by a National Aeronautics and Space Administration (NASA) grant, Measuring Vegetation Health (http://mvh.sr.unh.edu) emphasizing the use of digital images for environmental monitoring. The classroom work envisioned by this grant combines field-work and outdoor experiences with analysis of remote sensing imagery (aerial and satellite images) and digital photographs to support spatial, spectral and temporal investigations of plants, trees and living things in our environment. Concept mapping and formative assessment were strategies introduced with this grant for teachers and learners to improve comprehension of complex topics and to animate thinking processes of analysis, synthesis, and creativity (Ornstein and Sinatra, 2005, p. 27).

As we highlight the potential impact of concept mapping and visual representation we have a concern for classroom teachers echoing Kinchin's (2001) observations. Concept mapping is a strategy that is familiar to many, if not most teachers, however in practice it may be used on a limited basis, or not well enough to produce positive effects on students' learning. As Novak (1998) remarked, even though concept mapping is well known, and software tools like CMAP and *Inspiration* are readily available, the use of concept mapping remains inconsistent, at best. During a series of five workshops for the NASA science grant, Measuring Vegetation Health (http://mvh.sr.unh.edu), we surveyed fifty-six teachers about their understanding and use of concept mapping in science. In self-reported responses teachers indicated that, even though they know about concept mapping, it is not a regular part of instruction (Beaudry, 2005). Kinchin pointed out that the two major barriers may be teachers' own beliefs about teaching and the underlying philosophy of the curriculum (2001). We believe that consideration of literacy and formative assessment with concept mapping and visual representation may provide a stronger rationale for re-conceptualizing teaching and learning. In this respect our views are akin to what Kinchin calls an "ecological" perspective of teaching, one in which well-prepared teachers and active, motivated students work together to create and sustain conditions in which they "communicate effectively as partners within a learning community" (2001, p. 1267)

LITERATURE REVIEW

Background: Measuring Vegetation Health

The research and development for this paper is connected primarily with Measuring Vegetation Health (http://mvh.sr.un.edu), a NASA grant focused on development and initial testing of technological tools and resources for using light and color to investigate the health of plants and trees, and the promotion of spatial learning. Sources of digital information include remote sensing imagery (aerial and satellite) and hand-held digital cameras. Digital images are analyzed using new, open-source image analysis software designed and improved by this grant, including Color Basics, Image Basics, and Image Analysis software packages.

One of the purposes of the Measuring Vegetation Health grant is to support inquiry in science classrooms with technological tools and resources. Our approach is to integrate the use of science tools and resources for inquiry with the literacy development tools like concept mapping and collaborative learning. The use of such tools relates to another new literacy, spatial literacy. According to the National Academy of Sciences (2006)

Figure 1. Literacy modes (Sinatra, 1986)

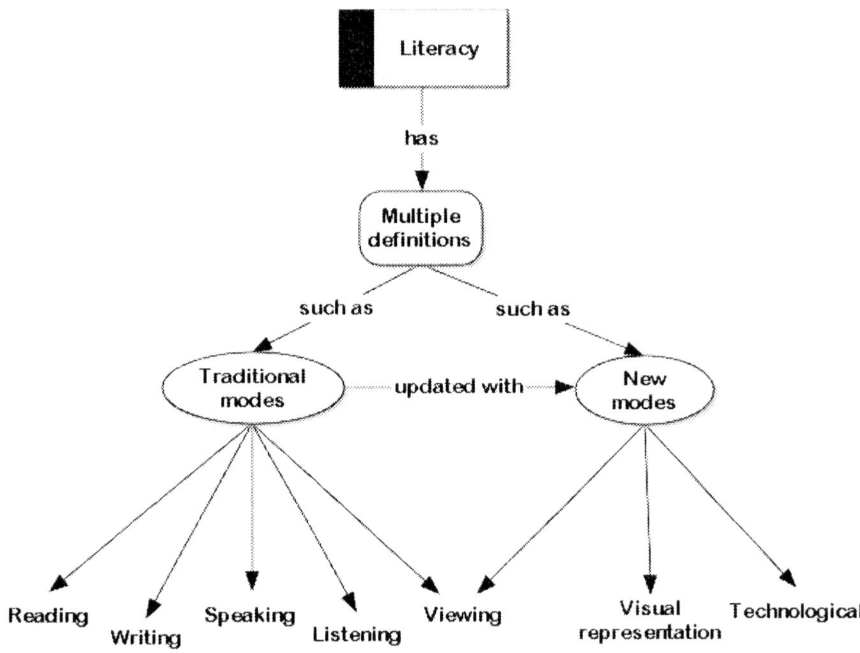

Figure 2. Visual Literacy and Relationship with Verbal and Nonverbal Literacies (Sinatra, 1986, p. 29)

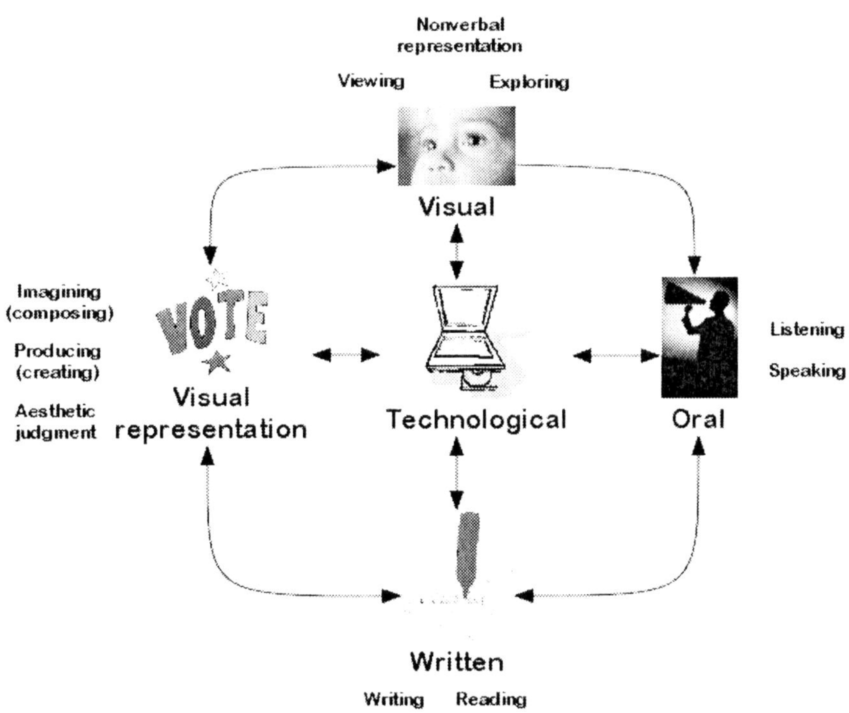

"Spatially literate students who have developed appropriate levels of spatial knowledge and skills in spatial ways of thinking and acting together with sets of spatial capabilities, have the following characteristics:

1. They have the habit of mind of thinking spatially
2. They practice spatial thinking in an informed way
3. They adopt a critical stance to spatial thinking – they can evaluate the quality of spatial data and can evaluate the validity of arguments based on spatial information. (p. 4)

This perspective fits with our view of visual literacy, and we have begun to do research on student learning with this definition as the intended outcome. For our work in the future we will try to relate spatial thinking more directly to the literacy strategies, since it has a clear connection with the expanded literacy framework (See Figure 1) and visual literacy model (See Figure 2) elaborated by Sinatra (1986) in which learners' understand how to plan, produce and evaluate visual representations.

Expanded Definition of Literacy: Reading, Writing, Listening, Speaking, Visualizing, & Visual Representation

Traditional forms of literacy include reading, writing, speaking, and listening, and Sinatra (1986) suggested expanding the model to include viewing (observing and experiencing), visual representation, and media technology (See Figure 1). Sinatra (1986) presented a model of literacy development that included four interactive phases: phase 1 - visual literacy, phase 2 - oral literacy, phase 3 - written literacy, and phase 4 – visual representation and communication (See Figure 2). Technological literacy is a rapidly evolving mode of learning involving interaction with computer hardware, software, and new information technologies. In order to "reinforce the learning of new concepts, understanding and vocabulary words" the integration of the reading, mapping and writing process is critical (Ornstein and Sinatra, 2005, p. 260; Sinatra, Beaudry, Pizzo, and Geisert, 1994). In his article, "Teaching Learners to Think, Read, and Write More Effectively in Content Subjects" Sinatra (2000) focused on three aspects of the teaching-learning construct: 1) text structure for expository writing, 2) concept mapping to clarify conceptual understanding and to represent text structure, and 3) the teaching-style shift from teacher-centered classrooms to more student participation. The review of research in the meta-analysis by Nesbit and Adesope (2006) suggests that concept mapping is a highly engaging and effective literacy development strategy, especially when students are involved directly in the construction and revision of their own visual products. Adding technology and new media environments to this model is relevant for this study because we explore a variety of software for concept mapping and spatial and spectral measurements, as well as digital cameras and images. In their framework for "technological pedagogical content knowledge" Mishra and Koehler (2006) point out that teachers need to have a deep understanding of the interaction of content, pedagogy, and technology. Teachers must have "knowledge of what makes concepts difficult or easy to learn and how technology can help redress some of the problems that students face; knowledge of students' prior knowledge and theories of epistemology; and knowledge of how technologies can be used to build on existing knowledge and to develop new epistemologies or strengthen old ones" (pp. 1028-1029). For our study literacy and technological pedagogical content knowledge (TPCK) make important contributions. Literacy elaborates ways of knowing or epistemologies, especially viewing and visual representation, and TCPK offers a framework for

Figure 3. The National Reading Panel (2000) Summary of Essential Elements

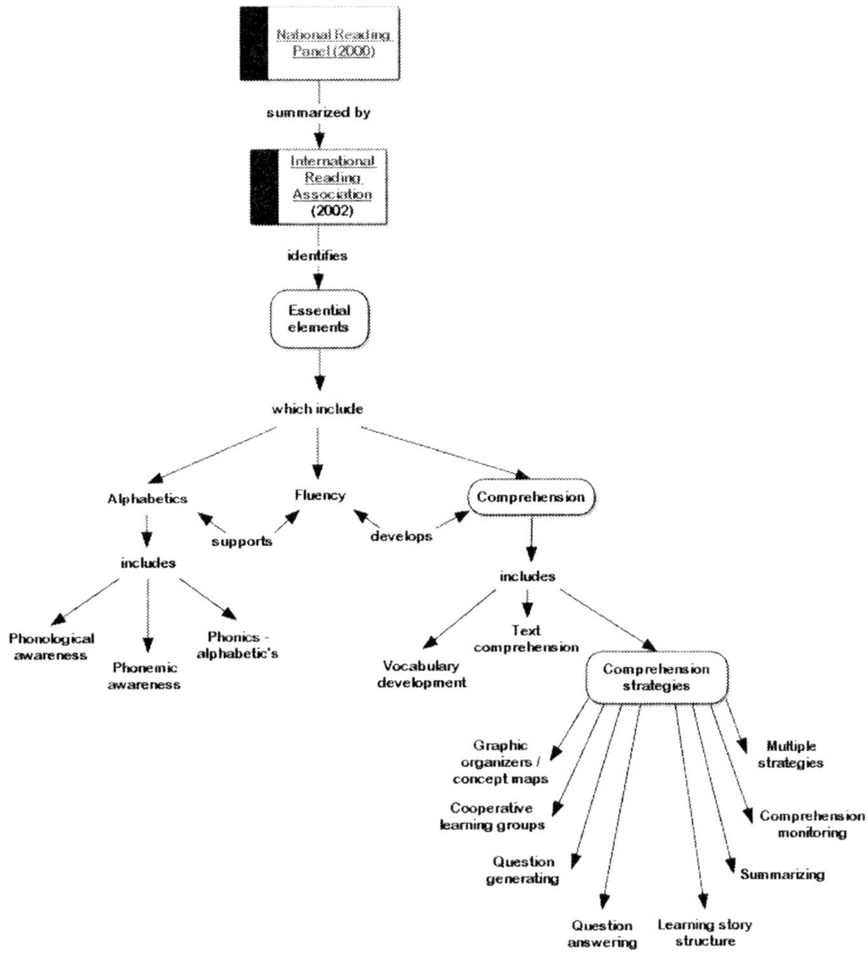

planning and designing learning, teaching, and assessment with technology tools.

Our perspective combines and extends the current recommendations for practice found in the USA's national reports on literacy by the National Reading Panel and Writing Next. First, the National Reading Panel (2000) emphasizes the use of concept mapping to improve comprehension by using graphic organizers and concept maps. (see Figure 3) Second, the recommendations to improve writing in Writing Next (2006) emphasize writing in the content areas, use of summarization, prewriting, the study of models, and specific product goals. (See Figure 4) The elements of writing are arranged according to the magnitude of the effect size, which are the results of the meta-analysis in the Writing Next report. Effect sizes represent a summary statistic of the impact of a strategy or element on specified, quantitative outcome. The largest effect size is the use of multiple strategies with an effect size of 0.82, the use of summarization with an effect size of 0.82, collaboration with an effect size of 0.75 and students understanding specific product goals in advance of the assignment with an effect size of 0.70.

Concept mapping and related visual learning approaches engage students "in discovery, in creation, and in acts of synthesis" which are essential to learning relationships among words and concepts (Sinatra, 1986, p. 27). What is striking

Figure 4. The Essential Elements of Writing (Writing Next, 2006)

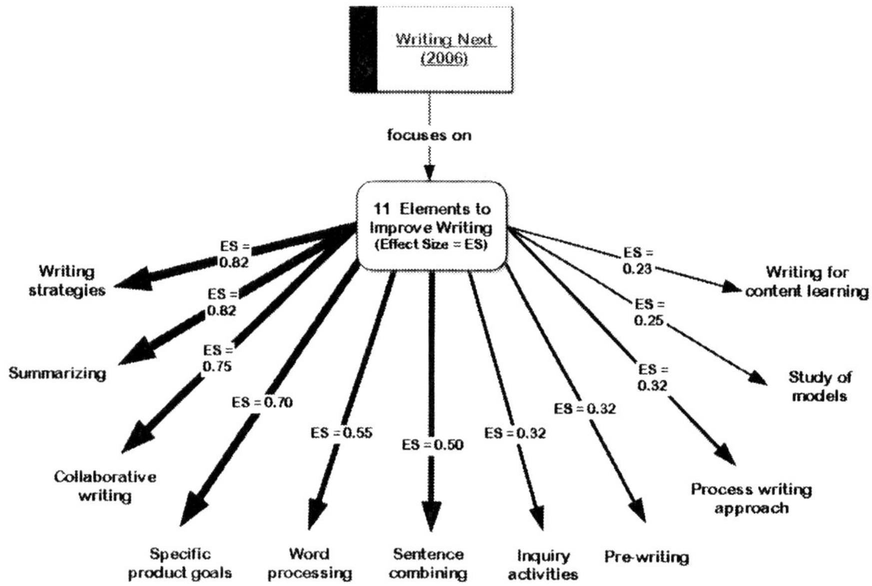

about the two models of literacy is that concept mapping is cited explicitly as a strategy for reading comprehension, but it is not specifically cited as a strategy for creating successful writers. Also, do not overlook the presence of collaborative group work in both reading and writing, and we suggest to concept mapping.

The use of mapping as an organizing framework for student-writers fits with pre-writing strategies, and according to Sinatra (2000) applying and practicing the use of organizational, concept maps "works for students at all grade levels and may benefit at-risk, low-achieving students to a high degree" (p. 272). Our suggestion to teachers would be to infuse visual literacy strategies like concept mapping to the list of literacy strategies explicitly across content and from early childhood to adolescent years and beyond. Concept mapping must be in the repertoire of teaching and learning strategies for strengthening students' reading comprehension, and as a means for scaffolding students' writing so that they can make the connection between reading and writing when they encounter new, complex ideas and difficult text. "Concept maps are to informational, expository text what story maps are to narrative literature," as they are used to generate and organize concepts (Ornstein and Sinatra, 2005, p. 369). In addition, concept maps and graphic organizers are products that facilitate interactions of multiple forms of literacy, bringing thinking, writing, visualizing together with listening and speaking.

Finally, our approach embeds the expanded model of literacy in the strands of scientific proficiency identified by the National Research Council (2007) in *Taking Science to School*. In this framework for science there are four broad learning goals, "1) know, use and interpret scientific explanations of the natural world, 2) generate and evaluate scientific evidence and explanations, 3) understand the nature and development of scientific knowledge, and 4) participate productively in scientific practices and discourse" (Duschl, Schweingruber, and Shouse, 2007, p. 36). The field study of inter-coastal tidal zones is a key instructional strategy in our study because it provides a concrete experience in a complex, natural setting for all students to view and explore. Furthermore, it is an image and point of reference for the teacher and students to support classroom investigations, interpretation of data, and expression in speaking, visual representation and writing. All of the strategies contribute to the development of yet another form of literacy, science literacy.

Formative Assessment Links Learning with Final Products

The purpose of this study is to investigate the role of formative assessment to focus learners' attention on three questions: 1) Where am I now? 2) Where do I want to go? and, 3) What do I need to get there? In an age of accountability we struggle to find the balance between accountability through summative assessment like standardized testing and grading, and formative assessment, the need to support the learning of students and teachers. There are seven strategies associated with formative assessment suggested by Stiggins Arter, Chappuis, & Chappuis (2004).

- Where am I going?
    1. Provide a clear and understandable version of the learning target.
    2. Use examples and models of strong and weak work.
- Where am I now?
    3. Offer regular, descriptive feedback.
    4. Teach students to self-assess and set goals.
- What do I need to do to get there? How can I close the gap?
    5. Design lessons to focus on one aspect of the quality at a time.
    6. Teach students focused revision.
    7. Engage students in self-reflection, and let them keep track of and share their learning. (p. 42)

One of the key findings in the research review by Black and Wiliam (1998a) is that the quality of feedback from teachers and/or other students can be a very positive component to student learning. Comments in writing and orally carry specific guidance and messages to support students' growth, whereas grades and marks tend to interrupt learning. Dweck observed that the shift caused by grading which emphasizes the fixed mindset, only ability counts, and de-emphasizes the growth mindset, I can do this if I keep trying, is most pronounced for students struggling with basic comprehension, understanding and performance (2005). The purpose of formative assessment is to keep students focused on self-efficacy and learning, especially through peer- and self-assessment, and that frequent self-assessment of specific learning strategies and processes produces more positive effects than instruction focusing on grading or performance.

The primary purpose of formative assessment is the promotion of student learning. "That is, the first priority in its design and practice is to serve the purpose of promoting students' learning. This distinguishes it from any assessment designed primarily to serve the purpose of determining accountability, ranking, or competence" (Black and Wiliam, 2004, p. 21). Formative assessment has some lasting effects on classroom roles and responsibilities by: 1) developing the capacity for a variety of feedback sources (teacher-to-student, and student-student and student-to-teacher), 2) active involvement of students in their own learning, 3) adjustment of teaching to account for the results of assessment, 4) understanding of the impact of assessment on students' motivation and self-esteem, and 5) by being explicit about self-assessment and how to improve learning (Black and Wiliam, 2004, pp. 4-5). Effective feedback, from teacher to students, and student to student, and students to teacher, is vital for effective formative assessment. Concept maps are flexible tools for allowing students to reveal the active construction of new understanding by generating and answering their own questions, to make novice errors, to support students' collaborative dialogue, and to focus teacher-expert feedback on students' individual formative assessment products. Misconceptions can be revealed to the teacher in an efficient and precise manner, with a quick look. During this process students learn how to make focused revision, and teachers gain another source of data for students to triangulate with other sources like interviews, students' writing, other performance skills, and tests and quizzes.

Formative assessment is not a trendy term meant to circumvent grades or tests. Novak and Gowin's initial work focused on cognitive development of students, but also incorporated the concept of formative assessment within Ausubel's model of "meaningful learning." Rather it is a key attribute of effective teaching and learning, as he/she learns how to use productive, cognitive and meta-cognitive strategies. Emphasizing formative assessment allows for more purposeful and sustained interactions between teacher-experts and student-novices to learn new ideas and to consolidate understanding of prior knowledge. Examples of cognitive strategies are concept mapping, summarizing, de-coding and self-questioning. Meta-cognition, the student's own regulation of cognition, might include a learner's identification of areas of misunderstanding, detecting errors, and re-writing sentences to improve sentence variety.

PURPOSE OF THE RESEARCH

Our purpose was to explore how a high school science teacher and her students used concept maps to develop understanding of the inter-tidal system, a progressively complex science concept. We conducted an ethnographic case study, from two perspectives, that of a teacher researcher and a university-based participant observer. The study focused on concept mapping, formative assessment and scientific inquiry in high school science in one science classroom in a medium-size high school with approximately 1, 200 students in a city in northern New England.

High school science courses introduce difficult subject matter to students who may or may not have sufficient prior knowledge for successful assimilation of the content. The problem begins in middle school, where typical science textbooks present the same complex and abstract concepts that are later taught in high school. The result is that high school students often profess to know concepts that in reality they understand incorrectly if at all. The problem is exacerbated by the fact that many science concepts are encased in totally unfamiliar language or in language that is used in a new way.

Typically, students learn concepts through direct vocabulary-based instruction and are assessed with selected response tests. This use of summative assessments like grades does little to help students integrate new knowledge or to provide opportunity to clear up prior misconceptions (Novak and Canas, 2006). Even in science classes that are rich in inquiry studies, hands on activities, and constructivist pedagogy, students cannot engage in meaningful learning if they are building on partial understandings or misconceptions. Concept maps are useful tools to help students construct, describe and revise their basic understanding of concepts. In constructing maps, students can correctly link new information to what they already know and expand learning while teachers can quickly and easily see what their students know and adjust their teaching accordingly.

We have found that in high school science classes, students struggle to comprehend the interconnectedness and cyclical patterns central to biological and ecological concepts. As the literature review demonstrates, concept mapping strategies go well beyond memorization because they engage students in thinking about hierarchies and specific connections. However, while concept mapping is a powerful strategy it must be taught and learned. We have noticed that when asked to map their understanding of a topic, students tend to create images that are linear. They typically draw one concept that leads to another like an arrow passing through several targets lined up one behind each other. Another commonly drawn map shows concepts that radiate from a central idea, where each concept stands alone in its connection to the center. We view these visual representations as evidence of the naïve and underdeveloped concepts and lack of exposure that characterize novice learners as they reveal their understanding of complex concepts with the instructional strategy of concept mapping.

Setting for the Research

As indicated, the research occurred in one high school science class and was collaboratively conducted by a teacher researcher from inside her own classroom and a university researcher who functioned as a participant observer. We chose to focus on the use of concept maps in the study of matter and energy moving through ecosystems. Earlier classes had identified this topic as being most problematic. This evaluation was substantiated by the teacher researcher who noted that student work and assessments reflected consistent misunderstandings. Energy moving through ecosystems is a complicated topic that relies on basic understanding of the difference between energy and matter and the biological processes of photosynthesis, cellular respiration and decomposition. It also incorporates knowledge of the elements important to building the basic molecules of organic matter. In an ecosystem, matter is cycled from inorganic reservoirs to organic reservoirs and energy is transferred from one form to another eventually dissipating in less usable forms of energy such as heat. Students often are confused by the difference between energy and matter and puzzled by the fact that the same matter held in the tissues of living things can be found in the air or water. Although these concepts are represented in familiar ecological food webs, the science behind them is abstract and difficult to grasp. The study began with the research question: could individually mapping these concepts using a software program and formative assessment strategies improve understanding of the complex concepts comprising ecosystems?

We wanted to move students beyond the typical rudimentary visual representations of prior classes. Typically, concept mapping is introduced as a collaborative learning experience in a strategy we called "glue, paper, scissors," of "GPS." Groups of students worked together with a list of 10 to 12 vocabulary words from a selected topic. They cut out and arranged the words together to describe and build their understanding of the topic. Each team described its map in a brief presentation to the teacher, who provided opportunities for students to ask questions and to expand and rethink their maps. When maps represented the ideas correctly, or close to accurately, students glued the words into place and posted the maps for whole class discussion. The teacher listened in on lively, focused conversations as the students worked out their ideas together. All groups completed their first map, an indicator that students were engaged. The maps and conversations offered an informative peek into the cognitive journey on which students had embarked; each was engaged in building his/her own understanding of complex scientific concepts and processes.

For the purpose of this study, we refined the GPS activity and also introduced the innovation of concept mapping software to follow the GPS activity with the hope these modifications would yield new and more sophisticated student products, visual representations, and writing. The strategies combining literacy, concept mapping, and collaborative learning are summarized in the Technology and Teaching Progression (See Appendix B). A complete narrative of the implementation of the innovation and its outcomes is presented in the Results Achieved section of this chapter.

RESEARCH METHODOLOGY

The research relied on qualitative methods of sampling, data collection and data analysis. Both authors implemented all phases of this study as a collaborative research team, one as the teacher researcher and the other as the participant observer. The multiplicity of roles is characteristic of an ethnographic design, one in which the immersion of the researcher in the setting is both empowering and transformative as well as conflicted and ambiguous (Smith, 1979; Harklau and Norwood, 2005). The ethnographic design provided triangulation of data including 1) long-term participant-observation, 2) collection of artifacts, i.e., concept maps and written laboratory assignments, and 3) interviews, both informal and formal.

Sampling

The classroom of middle- to low-achieving students was a purposive sample; it was selected because we wanted to work with students struggling to understand science concepts and who were challenged in communicating their understanding of science concepts in writing. There were a total of eighteen (18) students in the classroom. All of the students were observed over the course of the semester; for the in-depth interviewing at the end of the semester six students, four females and two males were selected as a focal group to represent a range of accomplishment on the written conclusion. For a focal group to study we deliberately selected students who represented the entire range of writing achievement, two high-, two medium, and two low-achievers.

Data Collection

The study was conducted as an ethnographic study that took place over a 12-month period and culminated with intensive data collection in the final three months of the study. Over the period of a year, we worked as a collaborative team, one as teacher-researcher and the other as a participant-observer. This approach allowed us to combine "emic" and "etic" perspectives, as pioneered by Smith (1979). There were at least six classroom observations, beginning with the field trip to collect specimens and make observations at the inter-coastal tidal zone. Field notes were made after each visit, which constituted "thick description" of the classroom interactions. The participant-observer provided technical assistance in taking digital photographs and assisting students with the basics of concept mapping software.

Interviews were a key source of information throughout the study. We met as a collaborative research team before and after each lesson. We talked about each lesson to determine specific roles for each person, and conducted a de-briefing interview after each lesson. The most intensive interviewing took place at the end of the semester with a group of six students. After the six students completed their writing assignment each of them was interviewed. A semi-structured interview approach was used with the following questions and prompts:

- What is your background in biology and marine ecology? Have you taken biology already?
- What is your prior experience with concept mapping? Have you used this strategy in prior courses?
- Describe the value of the collaborative mapping activity as a starter for the concept mapping.
- How did the concept mapping software help or hinder your understanding of the energy cycle? Were there any problems using the concept mapping software?
- How did you use the concept map for your writing assignment? How did you use your own map and the expert map produced by the teacher?

Each of the students was interviewed with his or her concept map in front of them to assist the interview as recommended by Novak and Gowin (1984). Interviews were designed to last 10-15 minutes to accommodate student schedules.

Document analysis was used to accomplish the ultimate goal of the study, examination of the concept maps and evaluation of the written communication of the students. The concept maps were collected and used as the basis for interviews after the lessons. The students' written summaries were scored with a rubric by the teacher-researcher. In addition, the teacher-researcher wrote numerous reflections during the study. For example, teacher reflections were written about the instructional sequence, and the use of the concept mapping software.

Finally, as another check for data credibility, an expert science educator, with prior experience as a teacher, a curriculum designer, and a researcher, conducted a final, debriefing interview at the end of the semester with the teacher-researcher and the participant-observer. The purpose of the interview was to validate the classroom observations of the participant-observer. The results of these interviews verified that the classroom activities focused on development of science literacy, and students used image analysis and concept mapping software as part of the investigation and the write-up of the investigation.

Data Analysis

Data analysis for the study relied on a constant comparative approach, which considered all of the data sources: written field notes, samples of student maps and written essays, student interviews, reflective teacher journal. Tools of analysis included coding of data, member checks, analytic memos, and checks for reliability by an outside reader.

RESULTS ACHIEVED

This section provides a narrative of the implementation of the refined version of the GPS activity and the incorporation of the innovation of computer based concept mapping. The narrative describes and reflects on the sequence of instruction and results attained in terms of student products, teacher reflections, and student interview data.

Sequence of Instruction and Results

The sequence of presenting concepts began as students were introduced to the topic of matter and energy movement in ecosystems through primary production and nutrient cycles. The lab component involved growing populations of fresh water algae in various solutions of nutrients and collecting data on growth, pH (carbon dioxide in the water) and dissolved oxygen. The summative assessment goal was for students to articulate in a written lab conclusion, their understanding of how matter was cycled from the nutrient solutions in the water into the bodies of the algae and back into the water as oxygen or carbon dioxide. While the algae were growing (or not) students were asked to describe their knowledge by manipulating 22 terms or words into the familiar cooperative paper maps. At the point where the instructions previously ended, students continued their work in the computer lab where each pair transferred their ideas to the concept mapping program.

The software was new to most students, but everyone picked up the basics in five minutes or less. Constructing a map in collaborative groups with paper cut-outs was a critical first step, a step we named "glue-paper-scissors" or "GPS." This provided continuity from instruction and separated the construction of the initial map from learning the Inspiration software. It was surprising to see that students did not quickly reproduce their paper maps and declared they were finished. Instead, they puzzled over connections, considering how they could represent their ideas more clearly and discovering components of the software that make the maps more visually interesting. Hands were popping up to ask for work to be checked and for feedback. At the end of the block, every student had a his/her own, unique formative assessment product, which could be the basis for feedback and improvement. Upon careful review the maps still needed work to clearly and correctly represent the concepts. A few changes to a rubric for concept maps (See Appendix A) and it was posted online and introduced during the next class. Students used the rubric to evaluate and comment on the maps created the previous day then the students found their original mapping partner and returned to the lab to act on the comments. The resulting maps and written comments from students pointed to increased understanding of how matter is cycled in ecosystems. The answers to one question on a post-mapping survey were troubling. When asked if their map described a cycle or a chain of events, the most common answer was yes. Teacher evaluation of the maps revealed a distinct difference; the information was organized, but usually not as a cycle.

The next step asked students to expand their understanding by relating specific details of marine ecosystems to the mapped concepts. We began with assignment criteria and a rubric but skipped the introductory paper maps. Students were not presented with a list of words or concepts, just an outline of information they needed to include. Confusion resulted. In order to begin the work, the terms to be used in the maps had to be identified and categorized. Students did not know where to begin. They needed to consider how the variety of pieces of information might be grouped into organizational categories in order to understand where to begin. After clarification and briefly modeling the process, everyone worked individually and again the level of engagement was surprising. The students were engrossed in solving puzzles with a tool that made it easy to fix, revise and act on feedback. As they worked with the concept mapping strategy their level of comfort with the software also evolved. Soon many were embellishing their work by incorporating images and arranging concept categories with similar shapes and colors. The maps were varied, but the software had a leveling effect on the finished, written products. More students produced higher quality written products than previous science investigations. In particular students whose work is often messy and less finished than the work of their peers, created equally polished products. Most importantly, more immediate and focused feedback was given to the based on how each student described their knowledge and that feedback could be instantly acted upon to improve the map. After class, more teacher review, more comments on the maps and the next day, students made final revisions to their work. Each student printed two copies, one to hand in and one to use as a guide when articulating their knowledge in the lab conclusions. The concept maps became tools for an even more challenging assignment, writing about science content.

Teacher Researcher's Pedagogical Reflection

A progression of knowledge is evident when results of the two mapping exercises are compared (See Figure 5 and Appendix B). Although some connecting lines may need an arrow reversed, most of the maps grew less confusing and more clear and concise even though information was added. Circular patterns were evident and concepts were organized more coherently. Students could apply and articulate the larger concepts from their maps in a lab conclusion. As a summative assessment, the algae lab, started at the beginning of the unit, required students to explain their knowledge of how primary producers, and the biological processes they engage in, contribute to the cycling of matter in an ecosystem. The writing assignment, the lab conclusion consists of three parts, 1) a restating of the purpose of the lab, 2) a paragraph describing the students' knowledge of the ecological concepts demonstrated in the lab and 3) a paragraph in which specific data and results from the investigation are cited to support and provide examples of the concepts at work. These three criteria were used to score students' essays. Students were able to demonstrate an understanding of the concepts

Figure 5. First map and Revised map: Before and after teaching and focused feedback

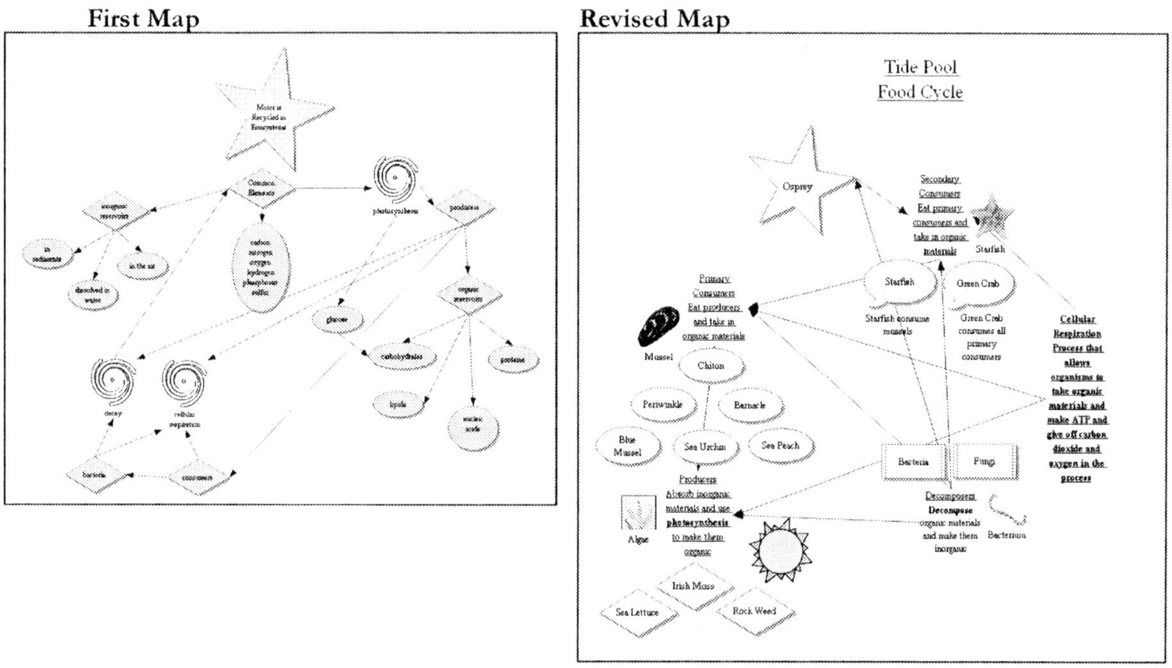

with more accuracy and authority than students in the past classes that did not work with concept mapping software. However, students still did not connect the lab data and results with their writing. They were unable to show how the data from the lab investigations is used to support and explain the concepts of energy

Students' Interview Data

We followed up the writings with semi-structured interviews with a focal group of selected students. After the lesson was completed six students representing the range of achievement from high to low were interviewed. There were two boys and four girls; one of the boys speaks English as a second language, now referred to in Maine public education as an English Language Learner (ELL). Each student had his/her concept map during the interview as a reference resource as recommended by Novak and Gowin (1984). The results of the interviews were analyzed according to the interview questions: background in biology and marine ecology, familiarity with concept mapping as an instructional strategy, value of collaborative, group work, ease of use of concept mapping software to understand key concepts, and use of concept map for writing. The results are presented below.

All students had taken other high school science courses, and four out of six had taken a general biology course with Ms. Wilson. That meant that four out of six were familiar with her teaching approaches including concept mapping, and using Inspiration software to create a concept map to acquire and consolidate understanding, and use it to scaffold his/her writing. Students responded to questions in the following areas: initial collaborative learning, going from collaborative to individual maps, writing from maps, and transfer of mapping strategies to other courses and content areas.

All of the students in the Marine Ecology course strongly underscored the importance of initiating concept maps in a collaborative, group setting. The students were accustomed to group work, which made the challenge of constructing maps possible. As one girl stated, "the groups gave me the chance to see if I was wrong, and if I was questioning something." Another talked about group work as a supportive setting for argumentation, "Agreement was good, but controversy was good. I could remember the arguments over the maps and the resolution." Group work also allowed students to express their frustration with the content and the mapping activity, as they struggled to create a clear map. Working as a group to start up the mapping activity gave students a chance to offer feedback as peers, in a learning community permitting academic and emotional learning to be shared with peers, too. The ELL student spoke about the importance of his vocabulary development; when he worked collaboratively he found out that algae were living organisms and not "dirt", as he had believed. Other themes from the interviews were consistent but did not match the clear message, the benefits of collaborative work for the initial steps in concept mapping.

As students went from the paper "cut outs" in the "GPS" activity to the software, Inspiration, all but one student completed the assignment to construct his/her own map of the cycles. The student who continued to struggle had difficulties because she wanted to make a hierarchical map, and not a cycle. A key struggle for students was making too many connections, a common problem with map construction (Heinze-Fry, 1992). Some students used the visual-graphic tools like shape, color and line weight to improve and to differentiate concepts. Only one student mentioned that using connecting or linking words was important to improving his understanding. The lack of linking words indicates an area for improvement in the instruction of map making.

Did concept mapping help students write in the content area? All of the students agreed that the mapping positively affected their writing about cycles, and was a very engaging strategy for learning. In essence, the mapping served as a pre-writing activity, and as a means to summarize the science inquiry activities. In addition, for the ELL student the vocabulary was defined and organized into a structure. Another student noticed the similarity between her map and an outline, "but," she said, "with my map I could experiment with ideas." Overall, students demonstrated sustained effort in thinking about and communicating ideas orally in writing about the topic with their concept maps.

Do students incorporate mapping as a strategy for other purposes? Several students experienced using concept mapping with other teachers, however, none had used mapping software. None of them mentioned the use of concept mapping for pre-writing or to plan the organization of his/her writing for other courses or assignments. However, the ELL student described how he used mapping to take notes and improve his understanding of events and people in his modern history course.

Finally, we asked about how to improve the concept mapping activity in Marine Ecology. Students realized that they did not have enough experience with concept mapping to design and produce a map of the cycles in one lesson, a skill that was missing or undeveloped in previous grades. The concept maps provided students with a reasoning tool to understand the difference between a hierarchical and a cycle map. Finally, they wanted to understand how to manage the links between concepts more meaningfully. "I wanted to draw lines, but there were too many lines and connections," a comment made out of frustration by many of those interviewed. Overall, students suggested that the total number of concepts in the concept mapping activity be trimmed because the map became too complicated; with more concepts came the propensity for adding unnecessary links.

FUTURE TRENDS

The biggest trend for the future is the continued use and investigation of concept mapping, collaborative learning, and formative assessment by the teacher researcher in her classroom. Collaboratively, we continue to refine the process, both the GPS activity and the computer-based concept mapping. The GPS activity now begins with fewer words at the start. It is more carefully monitored to show how a hierarchy of concepts is visually represented and linking words are a requirement, as recommended by Novak and Gowin (1984) and Novak and Canas (2006).

We will continue to explore the use of concept mapping software. Concept mapping software tools are not so different from hand-drawn maps, but the transfer to computer software allows for more flexibility in the re-drawing of maps, more tools for the development of multi-layered concept maps, the sharing of maps electronically, and the creation of concept maps for group presentations. There is more capacity to make complex links, and to use of shapes and colors to represent similarities and differences of concepts. Other concept mapping software such as CMAP tools represent new means of collaboration, also opportunities for research.

Future research can use the instructional sequences we have identified for more controlled experimental research. In order to validate our ethnographic research findings other quantitative research designs can be with computer software concept mapping strategies as the educational treatment. For example, research by Stull and Meyer (2007) utilizes an experimental design to investigate the impact of learner-generated versus expert-generated graphic organizers on students' reading comprehension of difficult text in a college biology course. Our approach would emphasize the comparison of concept mapping and collaborative learning with other strategies.

Under the broad concept of strategies for teaching and learning there are numerous questions concerning concept mapping and collaborative learning to investigate. For example, how should teachers use both their own expert concept maps and student-generated concept maps in different learning contexts? How much training do teachers need to become expert producers and users of concept maps? How does peer assessment affect the development of concept mapping skills in individual students? The Universal Design for Learning (UDL) framework suggests that teachers provide opportunities to learn that are differentiated to match the needs of learners. Are there preferences for different visual representations such as concept maps and outlines? How is learning, especially meta-cognition, affected by these preferences? How do the strategies of concept mapping and collaborative learning provide teachers with the necessary tools for a differentiated learning environment? By utilizing concept maps as formative assessment we provided students with the opportunity to produce different visual representations of their understanding, but did not investigate individual representations from the UDL. We see the future trends for concept mapping and collaborative learning as positive, generative and beneficial for teaching and learning.

CONCLUSIONS AND RECOMMENDATIONS FOR VISUAL LITERACY AND CONCEPT MAPPING, COLLABORATIVE LEARNING AND FORMATIVE ASSESSMENT

In this section we return to the framework of the seven strategies for formative assessment (Stiggins et al., 2004) as an organizing framework to summarize the connections with concept mapping and collaborative learning. We believe that meaningful learning means that teachers use a combination of formative assessment strategies to address the three major questions:

- Where am I going?
  1. Provide a clear and understandable version of the learning target.
  2. Use examples and models of strong and weak work.
- Where am I now?
  3. Offer regular, descriptive feedback.
  4. Teach students to self-assess and set goals.
- What do I need to do to get there? How can I close the gap?
  5. Design lessons to focus on one aspect of the quality at a time.
  6. Teach students focused revision.
  7. Engage students in self-reflection, and let them keep track of and share their learning. (p. 42)

Clear learning targets are essential to establish and hold steady for students to succeed, and for effective teaching and assessment. The teacher researcher provided oral and written descriptions of the writing assignment at multiple points in the lesson. The concept mapping activity is a strategy to clarify conceptual understanding. Collaborative learning groups provided students with the forum for peer assessment to clarify their understanding of the content, the energy cycle. As students said, using their own words created argumentation and emotional discourse, essential to correct misconceptions and sharpen definitions of key vocabulary as they co-constructed the initial maps in the GPS activity.

To provide the clear target for all students, and to share a high quality exemplar the teacher researcher drew her expert map as an exemplar to represent the arrangement of the key concepts, and used it for a lesson to summarize the topic. These strategies helped focus on the question "where am I going?"

The initial concept map drawn by students was a visual representation of their current understanding of the topic, energy flow in an ecosystem. It established the baseline for his/her understanding of "where am I now?" With challenging content it is difficult for students to get a picture that is a personal representation of "where am I now" but that is just what the concept map did. Each concept map was a depiction of a student's private universe, a visual representation that showed each stage of understanding, and marked progress along the continuum from novice to accomplished novice and for some, to expert. The progression of learning assisted students to assimilate and reconstruct their understanding of the energy cycle in an ecosystem (Mintzes, Wandersee, and Novak, 2005). Concept maps were a new tool for self-assessment, as well as a tangible product for the teacher researcher to assess formatively and provide descriptive feedback. The teacher researcher collected and examined the concept maps with each step in construction, and used the concept maps to re-set the learning targets. The collaborative learning strategy promoted peer assessment in which students provided feedback to each other. Students continued to look at each other's maps and ask questions. Students were provided with a rubric as a self-assessment guide to improve the construction of their concept map. The maps were not graded, a critical adjustment in students' expectations of the role of feedback. In teaching this difficult concept the teacher had to monitor students' struggle to comprehend, to persist in "meaningful learning" and prevent this from becoming a frustrating and dissatisfying learning experience.

There were several key strategies that provided the teacher and students with meaningful answers to the question "how do I get there?" One thing that was clear from the beginning the teacher researcher wanted students to become better writers. That decision helped students focus on specific aspects of quality writing. The drawing and re-drawing of the concept maps assisted each student in making focused revisions even before actual writing took place. As mentioned the drawing and re-drawing of the concept maps the benefited both the students and the teacher researcher. Revising and improving concept maps was effortful and intentional for students, and helped each student to struggle with the complexity of the big picture and multiple levels of detail. Finally, students appreciated the opportunities to talk about their ideas with their peers at the early stages of concept mapping, and throughout the lesson. The impact of collaboration and peer assessment led us to an alternative name for the GPS activity, "group-paper-scissors." In addition, the computer software component is a powerful, flexible tool, but by itself technology is not a formative assessment strategy without the inclusion of collaborative, group process, and other strategies. This lesson represents a teacher involved in using her "technology pedagogical content knowledge" for the specific purpose of selecting appropriate technology tools to support students' learning (Mishra and Koehler, 2006).

Based on our research we have a number of points for consideration. Collaborative learning, highlighted in both reading and writing research as an effective strategy, was found to be extremely important as a strategy used in combination with concept mapping for formative assessment and communications for high school science students. From our interviews and sustained observations in the high school classroom we developed several themes:

- Immerse students in authentic learning experiences (field trips, labs, digital photography, image analysis) to support rich literacy and language development.
- Use as many of the seven formative assessment strategies as needed to support students' learning.
- To improve the ecology of teaching from a literacy perspective, use multiple strategies to distribute the roles and responsibilities for teaching, learning and assessing between the teacher and students.
- Use concept maps and graphic organizers as formative assessment – no grades or numbers (i.e., the rubric was a formative assessment tool) only grades with specific feedback for improvement, if necessary.
- Use concept maps and graphic organizers as pre-writing and text structure organizers, as well as for comprehension.
- Begin concept maps as collaborative group work (formative assessment with peers), "group-paper-scissors" (GPS), and transition to individual map making.
- Use technology tools purposefully for concept mapping to facilitate teachers' understanding of students' current knowledge, to design feedback, assist in focused revision of concept maps, and provide a flexible medium for revision.
- Expert, teacher-made maps may help students struggling to construct his/her own map to clarify concepts.
- Reflect on teaching before, during and after lessons, and make changes and adjustments based on evidence and feedback, not just grades and test scores.

Numerous sources have underscored the essential role of concept mapping to support and differentiate learning to accommodate all preferences. Grapho-motor skills are a part of spatial learning skills that figure prominently in Levine's neuro-developmental systems. In the Universal Design for Learning, concept mapping creates a more inclusive learning environment by insuring the opportunity for students to learn using alternative, but equivalent strategies like webs and outlines. Webs and concept maps that show cycles and ecosystems are challenging topics that benefit from nonlinguistic (visual) representation of important information (Marzano, 2003).

We believe that the neglect of concept mapping may begin with limited understanding of the vital role of visual representation in literacy development. Edward Tufte suggested, "Many believe that graphical displays should divert and entertain those in the audience who find the words in the text too difficult" (1983, p. 80). Are visuals to be dismissed as mere decoration and less rigorous than text? And, will our obsession with grades and scores on all student work create a dilemma for teachers to minimize formative assessment? To promote student success teachers must resolve the balance between summative and formative assessment. The connection of visual representational thinking to content areas like science has resulted in documents like *Science Literacy by the American Association for the Advancement of Science* (AAAS). While the atlas of concepts and standards provides excellent expert models for teachers, it may remove teachers who are not comfortable with or masters of visual-graphic representations from learning about and doing his/her own concept mapping. The point is, even as finished products these teacher-expert maps need continuous elaboration and adaptation to allow for the growth of knowledge and for teaching and learning.

In terms of literacy development, teachers and students are now concentrating on the need to strengthen students' content-area reading, and, we would add, content-area writing and visual representation. The multiple meanings of literacy serves as a reminder of the complexity of choices facing teachers and learners, and the need for integrative, constructive approaches that empower us to "learn how to learn" (Novak & Gowin, 1984; Black, McCormick, James, & Pedder, 2006); or, as Kinchin pointed out, the ecology of teaching and learning (2001). In the literacy model we have discussed how peer and group learning are collaborative approaches embedded in a model of literacy promoting meaningful communications–teacher to student, student to student and student to teacher. A deep understanding of literacy as it pertains to reading, writing, speaking, visualizing, visual representation, and technology, and of formative assessment can support rigorous and engaging, productive and sustainable teaching, learning and assessment.

REFERENCES

Ausubel, D. (1968). *Educational psychology: A cognitive view.* New York: Holt, Rinehart & Winston.

Beaudry, J. (2005). *Measuring vegetation health: Year 2 evaluation report.* Gorham, ME: Center for Policy Analysis, Research and Evaluation.

Beaudry, J. (July, 2007). *Concept mapping: Picture pathways for formative assessment.* Paper presented at the summer meeting of the Educational Testing Service on "Assessment For Learning" Portland, OR.

Black, P., McCormick, R., James, M., & Pedder, D. (2006). Learning how to learn and assessment for learning. *Research Papers in Education, 21*(2), 119–132. doi:10.1080/02671520600615612

Black, P., & Wiliam, D. (1998a). Inside the black box. *Phi Delta Kappan, 80*(2), 9–21.

Black, P., & Wiliam, D. (1998b). Assessment and classroom learning . *Assessment in Education, 5*(1), 7–85. doi:10.1080/0969595980050102

Duschl, R., Schweingruber, H., & Shouse, A. (Eds.). (2007). *Taking Science to School.* Washington, D.C.: National Academy Press.

Dweck, C. (2006). *Mindset: The new psychology of success.* New York: Random House.

Dweck, C. (2007). The perils and promises of praise. *Educational Leadership, 65*(2), 34–39.

Graham, S., & Perin, D. (2007). *Writing next: Effective strategies to improve the writing of adolescents in middle and high school. A report to the Carnegie Commission of New York.* Washington, DC: Alliance for Excellent Education.

Grotzer, T. (2002). *Causal Patterns in Ecosystems: Lessons to Infuse into Ecosystems Units to Enable Deeper Understanding: Understandings of Consequence Project, President and Fellows of Harvard College.* Retrieved from http://www.pz.harvard.edu/ucp/curriculum/ecosystems/index.htm

Harklau, L., & Norwood, R. (2005). Negotiating researcher roles in ethnographic program evaluation: A postmodern lens. *Anthropology & Education Quarterly, 36*(3), 278–288. doi:10.1525/aeq.2005.36.3.278

Heinze-Fry, J. (1992). Using student self-assessment of biology concepts in an introductory biology course. *Journal of College Science Teaching, 22*(1), 39–43.

Horton, P., McConney, A., Gallo, M., Woods, A., Senn, G., & Hamelin, D. (1993). An investigation of the effectiveness of concept mapping as an instructional tool. *Science Education, 77*, 95–111. doi:10.1002/sce.3730770107

Hyerle, D. (1996). *Visual tools for constructing knowledge.* Alexandria, VA: Association for Supervision and Curriculum Development.

Inman-Andersen, L., & Ditson, L. (1998). Computer-based Concept Mapping: Promoting Meaningful Learning in Science for Students with Disabilities. *Information Technology and Disabilities,* (ITDVO5N1-2 Article 2, June).

Kinchin, I. (2001). If concept mapping is so helpful to learning biology, why aren't we all doing it? *International Journal of Science Education, 23*(12), 1257–1269. doi:10.1080/09500690010025058

Kinchin, I., Hay, D., & Adams, A. (2000). How a qualitative approach to concept map analysis can be used to aid learning by illustrating patterns of conceptual development. *Educational Research 42*(1), pp.

Levine, M. (2001). *A mind at a time.* New York: Simon & Schuster.

Markow, P. G., & Lonning, R. A. (1998). Usefulness of concept maps in college chemistry laboratories: Students' perceptions and effects on achievement. *Journal of Research in Science Teaching, 35*(9), 1015-1029. doi:10.1002/(SICI)1098-2736(199811)35:9<1015::AID-TEA4>3.0.CO;2-G

Marzano, R. (2003). *What works in schools: Translating Research into Action.* Reston, VA: Association for Supervision and Curriculum Development.

Mintzes, J., Wandersee, J., & Novak, J. (2005). *Teaching science for understanding.* London: Elsevier Academic Press.

Mishra, P., & Koehler, M. (2006). Technological pedagogical content knowledge: A framework for teacher knowledge. *Teachers College Record, 108*(6), 1017–1054. doi:10.1111/j.1467-9620.2006.00684.x

National Academy of Sciences. (2006). *Learning to think spatially.* Washington, D.C.: The National Academies Press.

National Reading Panel. (2000). *Teaching children to read: An evidence-based assessment of the scientific literature on reading and its implications for reading instruction.* http://www.nichd.nih.gov/publications/nrp/smallbook.cfm

Nesbit, J., & Adesope, O. (2006). Learning with concept and knowledge maps: A meta-analysis. *Review of Educational Research, 76*(3), 413–448. doi:10.3102/00346543076003413

Novak, J. (1990). Concept maps and Vee diagrams: two metacognitive tools to facilitate meaningful learning. *Instructional Science, 19*(1), 29-52. doi:10.1007/BF00377984

Novak, J. (1998). *Learning, creating, and using knowledge: concept maps as facilitative tools in schools and corporations.* Mahwah, NJ: Lawrence Erlbaum Associates.

Novak, J., & Canas, A. (2006). *The Theory Underlying Concept Maps and How to Construct Them.* (, Technical Report IHMC CmapTools 2006-01). Pensacola, FL: Florida Institute for Human and Machine Cognition.

Novak, J., & Gowin, D. (1984). *Learning how to learn*. New York: Cambridge University Press.

Ornstein, A., & Sinatra, R. (2005). *K-8 instructional methods: A literacy perspective.* Boston: Pearson Education.

Rice, D., Ryan, J., & Samson, S. (1998). Using concept maps to assess student learning in the science classroom: Must different methods compete? *Journal of Research in Science Teaching, 35*(9).

Sinatra, R. (1986). *Visual literacy connections to thinking, reading, and writing*. Springfield, IL: Charles Thomas Press.

Sinatra, R. (2000). Teaching learners to think, read, and write more effectively in content subjects. *Clearing House (Menasha, Wis.), 73*(5), 266–273.

Sinatra, R., Beaudry, J., Pizzo, J., & Geisert, G. (1994). Using computer-based semantic mapping, reading, and writing approach with at-risk fourth graders. *Journal of Computing in Childhood Education, 5*(11), 93–112.

Smith, L. (1979). An evolving logic of participant observation, educational ethnography, and other case studies. In L. Shulman, (Ed.) *Review of Research in Education*. Chicago: Peacock Press.

Stiggins, R., Arter, J., Chappuis, J., & Chappuis, S. (2004). *Classroom Assessment for Student Learning*. Portland, OR: Assessment Training Institute.

Stull, A., & Meyer, R. (2007). Learning by doing versus learning by viewing: Experimental comparisons of learner-generated versus author-provided graphic organizers. *Journal of Educational Psychology, 99*(4), 808–820. doi:10.1037/0022-0663.99.4.808

Trowbridge, J., & Wandersee, J. (2005). Theory-driven graphic organizers. In J. Mintzes, J. Wandersee, & J. Novak, (Eds.) *Teaching Science for Understanding*. London: Elsevier Academic Press.

Tufte, E. (1983). *The visual display of quantitative information*. Cheshire, CT: Graphics Press.

Tufte, E. (1990). *Envisioning information.* Cheshire, CT: Graphics Press.

KEY TERMS AND DEFINITIONS

Concept Mapping: A strategy in which verbal and visual thinking are integrated and displayed in a way distinctive from but traditional writing. Concept maps are constructed by the individual learner, in collaboration with other learners, or created and shared by teacher-experts. Key concepts are arranged in space, not in a linear, grammatically structured form, and may be accompanied by icons and graphic images. Novak and Gowin definition of concept maps specified maps be constructed in a hierarchical format, in which big ideas are linked with verbs or key words to supporting concepts and details. The National Reading Panel (2000) issued a definition of literacy for reading in which concept mapping is included as a key strategy to assist learners in the comprehension of text.

Formative Assessment: Formative assessment represents the use of evidence and data to inform teaching and learning. The seven strategies for formative assessment (Stiggins et al., 2004) aim to improve students' self-assessment, and rely on teachers' feedback and peer assessment and self-assessment. Three essential questions help to organize the concept: Where am I going? Where am I now? What do I need to do to get there? Formative assessment is another concept with multiple meanings, but can be understood as "encompassing all those activities undertaken by teachers, and/or by their students, which provide information to be used as feedback to modify teaching and learning activities in which they are engaged" (Black and Wiliam, 1998, p. 8). The important distinction here is between summative, grades or test scores, and formative assessment, evidence collected and used to modify instruction and learning.

Literacy: "Literacy is a normative statement of what members of a culture should know and be able to do with that knowledge. The Workforce Investment Act of 1998 (Public Law 105-220) stated that '...the term literacy means an individual's ability to read, write, and speak in English, compute, and solve problems, at levels of proficiency necessary to function on the job, in the family of the individual, and in society' (Title II, Section 203, Number 12) (*Learning to Think Spatially*, National Research Council, 2006, p. 4). The definition of literacy is extended to include visual literacy. (See Visual Literacy and Figures 1 - 4).

Technological Literacy: The newest phase of literacy development is technology, and the field of technology is evolving very quickly. Technology is cited as a major strategy for reading and writing literacy in national reports, and is the unifying theme of the model for designing instruction, technological pedagogical content knowledge (TCPK). The model includes computer hardware and software, and all types of digital equipment like digital cameras, recording devices, laboratory equipment, and multimedia. Technological literacy is an essential component supporting the accessibility of learners to information as described by the Universal Design for Learning.

Universal Design for Learning: Universal design for learning (UDL) is a framework for providing alternative learning strategies for learners. According to the Center for Applied Special Technology (CAST) "requires 1) multiple means of representation, to give learners various ways of acquiring information and knowledge, 2) multiple means of expression, to provide learners alternatives for demonstrating what they know, and 3) multiple means of engagement, to tap into learners' interests, offer appropriate challenges, and increase motivation."

Visual Literacy: The first stage of literacy development for learners with all five senses, visual literacy, is followed by language and then written literacy (Sinatra, 1986). (See Figure 1) Humans with sight actively view and seek information as hearing develops. It is the responsibility of educators to create learning environments for all learners to develop cognitive skills. "Language, then, becomes the natural extension of symbolic thought, and symbolic thoughts help form the mental schemata of a visually literate person" (Sinatra, 1986, p. 11).

APPENDIX A: TABLE 1

CATEGORY	Exemplary	Proficient	Developing
Content	Reflects essential information; information is correct and logically arranged; concepts succinctly presented;	Reflects most of the essential information correctly; is generally logically arranged; concepts presented without too many excess words	Contains extraneous information, information is incorrect or is missing; is not logically arranged.
Arrangement of Concepts	Main concept easily identified; sub-concepts branch appropriately from main idea	Main concept easily identified; most sub-concepts branch from main idea.	Main concept not clearly identified; sub-concepts don't consistently branch from main idea.
Links and Linking Lines	Linking lines connect related terms/point in correct direction; linking words accurately describe relationship between concepts.	Most linking lines connect properly; most linking words accurately describe the relationship between concepts.	Linking lines not always pointing in correct direction; linking words don't clarify relationships.
Graphics	Graphics used appropriately; greatly enhance the topic and aid in comprehension; are clear, crisp and well situated on the page.	Graphics used appropriately most of the time; most graphics selected enhance the topic, are of good quality, and are situated in logical places on the page.	Graphics used inappropriately and excessively; graphics poorly selected and don't enhance the topic; some graphics are blurry and ill-placed.
Design	Clean design; high visual appeal; four or fewer symbol shapes and text fonts; fits page without a lot of scrolling; color used effectively for emphasis. Text easy to read.	Design is fairly clean, with a few exceptions; diagram has visual appeal; four or fewer symbol shapes and text fonts; fits page well; uses color effectively most of time. Text easy to read.	Cluttered design; too many shapes and fonts, low in visual appeal; requires a lot of scrolling to view entire diagram; choice of colors lacks visual appeal and impedes comprehension. Text difficult to read.

APPENDIX B: TABLE 2

Date	Teaching – Learning Activity	Technology Connection
Aug 21	Planning meeting PW JB for two independent studies	How can concept mapping and MVH software tools be used?
Sept 11	Planning meeting PW-JB to study macro-algae in marine ecology	Use MVH software tools
Oct 13	Visit to beach with two Marine Ecology classes	Data collection with digital cameras
Oct 19	Talked about light as energy and what makes up light. In-class digital photos of macro-algae for analysis with MVH software. Flash and available light. Trans-illumination lighting.	Data collection with digital cameras
Oct 25	In the computer lab to use image analysis software. Used Image Basics (Pixel View component) with group photo of the class. Played Color Basics (Color Matching) Games to understand the difference between color separation and color intensity, percent reflectance and absorption	Use MVH software tools for color analysis; experience color concepts
Oct 26	In the computer lab, use of MVH Image to analyze macro-algae color values of red, green and blue, and the absorption/percent reflectance	Use of MVH software tools for color analysis; collect color reflectance data
Oct 30	Adjusted plans for implementation, more concept mapping to assist comprehension of difficult concepts.	
Oct 31	In-class experiment to grow phyto-plankton. Close-up digital photos of treatment and control test tubes. Lighting indirect with backlight.	Use of MVH software tools for color analysis; collect color reflectance data

Date	Teaching – Learning Activity	Technology Connection
Nov 2	Concept mapping activity. First step in drawing the system or cycle of events. Began activity with discussion of big ideas. Used a list of 12 words that characterized the marine eco-system, students cut out pieces of paper and arranged them into a map. Transferred the original paper map to Inspiration concept mapping software in the computer lab. Resulted in Concept Map #1	Group-paper-scissors (GPS) activity, followed by use of Inspiration concept mapping software
Nov 7	Concept Map #2 – create a "Matter Web", with 12 organisms and processes. Students were asked to represent three levels of consumers, with producers and decomposers. Written reflection about concept mapping.	Individual students use Inspiration concept mapping software
Nov 9	Teacher-expert Map and Revision of Concept Map #2 – Based on the results of the reflective writing the teacher provides her own expert map and asks students to identify components of her map, and asks students to revise their maps, in particular to include the energy cycle, not just the components.	Teacher constructs her own expert map with Inspiration and individual students use Inspiration concept mapping software
Nov 13	Students use MVH software to calculate color of digital photos of phyto-plankton at the conclusion of the experiment.	Use of MVH software tools for color analysis; collect color reflectance data
Nov 16	Students write an in-class assignment about the system including the energy cycle.	Students write essays in class by hand.

Compilation of References

Aarnio, H. (1999). *Dialogia etsimässä* [Searching for dialogue]. Doctoral dissertation, Acta Universitatis Tamperensis, (p. 676).

Aarnio, H., & Enqvist, J. (2001, June 25-30). Dialogic knowledge construction as the crucial issue in network-based learning in vocational education. In C. Montgomerie, & J. Viteli, (Eds.), *Proceedings of ED-MEDIA 2001: World Conference on Educational Multimedia, Hypermedia & Telecommunications* (pp. 1-6). Tampere, Finland: AACE.

Adey, P., Csapo, B., Demitriou, A., Hautamäki, J., & Shayer, M. (2007). Can we be intelligent about intelligence? Why education needs the concept of plastic general ability? *Educational Research Review*, *2*(2), 75–97. doi:10.1016/j.edurev.2007.05.001

Adriessen, J., Suthers, D., & Baker, M. (2003). *Arguing to Learn: Cronfronting Cognitions in Computer-Supported Collaborative Learning Environments*. London: Kluwer Academic Publishers.

Åhlberg, M. (1993). *Concept maps, Vee diagrams and rhetorical argumentation (RA) analysis: Three educational theory-based tools to facilitate meaningful learning*. Paper presented at The Third International Seminar on Misconceptions in Science and Mathematics, August 1- 5, 1993, Cornell University. Retrieved June 20, 2008 from http://www.scribd.com/doc/2868664/Ahlberg-Cornell-University-August-1-5-1993</eref>

Åhlberg, M. (1998). Education for sustainability, good environment and good life. In Åhlberg, M. & Leal Filho, W. (Eds.) *Environmental education for sustainability: good environment, good life*. (pp. 25-43). Frankfurt, Germany: Peter Lang.

Åhlberg, M. (2004).Varieties of concept mapping. In A. Cañas, J. Novak, & F. Gonzales, (Eds.), *Concept Maps: Theory, Methodology, Technology. Proceedings of the First International Conference on Concept mapping (CMC 2004)*, Pamplona, Spain, Sept 14 – 17, (Vol. 2, pp. 25-28). Retrieved January 26, 2008 from http://cmc.ihmc.us/papers/cmc2004-206.pdf

Åhlberg, M. (2004). Concept mapping for sustainable development. In Canas, A., Novak, J., Gonzales, F. (Eds.) *Concept Maps: Theory, Methodology, Technology. Proceedings of the First International Conference on Concept mapping (CMC 2004)*, Pamplona, Spain, September 14-17, (Vol. 1, pp. 39 – 44). Retrieved January 26, 2008 from http://cmc.ihmc.us/papers/cmc2004-233.pdf

Åhlberg, M. (2005). Integrating education for sustainable development. In W. Leal Filho (Ed.), *Handbook of Sustainability Research*, (pp. 477-504). Frankfurt, Germany: Peter Lang.

Åhlberg, M., & Ahoranta, V. (2002). Two improved educational theory based tools to monitor and promote quality of geographical education and learning. *International Research in Geographical and Environmental Education, 11*(2), 119–137. Retrieved June 20, 2008 from http://www.multilingual-matters.net/irgee/011/0119/irgee0110119.pdf

Åhlberg, M., & Kaivola, T. (2006). Approaches to learning and teaching environmental and geographical education for sustainability. In Lee, J. & Williams, M. (Eds.) *Environmental and Geographical Education for Sustainability* (pp. 79 – 93). New York: Nova.

Åhlberg, M., Äänismaa, P., & Dillon, P. (2005). Education for sustainable living: Integrating theory, practice, design and development. *Scandinavian Journal of Educational Research*, *49*(2), 167–186. doi:10.1080/00313830500048923

Åhlberg, M., Kaasinen, A., Kaivola, T., & Houtsonen, L. (2002). Collaborative kowledge bilding to pomote in-service teacher training on environmental education: First year results. *Journal of Information Technology for Teacher Education*, *10*(3), 227–238.

Åhlberg, M., Turja, L., & Robinson, J. (2003). *Educational research and development to promote sustainable development in the city of Helsinki*: Helping the accessible Helsinki Programme 2001 - 2011 to achieve its goals. International Journal of Environment and Sustainable Development, 2(2), 197–209. doi:10.1504/IJESD.2003.003321

Alcântara, P. R., Siqueira, L. M. M., & Valaski, S. (2004, maio/ago). Vivenciando a aprendizagem colaborativa em sala de aula: experiências no ensino superior. In *Revista Diálogo Educacional*. Curitiba, *4* (12) 169-188

Aldana, M. F. (2006). *O Cognitivismo de David Ausubel*. Artigo da Internet. Disponível em: Ser Professor Universitário. Retrieved October 10, 2006, from http://serprofessoruniversitario.pro.br

Allison, E. H., & Ellis, F. (2001). The livelihoods approach and management of small-scale fisheries. *Marine Policy*, *5*(5): 377–388. doi:10.1016/S0308-597X(01)00023-9

Alpert, S. (2004). Flexibility of Expressiveness: State of the Practice. In P. Kommers (Ed.), *Cognitive Support for Learning: Imagining the Unknown* (pp. 253-268). Amsterdam: IOS Press.

Amer, A. A. (1994). The effect of knowledge-map and underlining training on the reading comprehension of scientific texts. *English for Specific Purposes*, *13*, 35–45. doi:10.1016/0889-4906(94)90023-X

Amoretti, M. S. M., & Tarouco, L. M. R. (2000, September). Mapas Conceituais: modelagem colaborativa do conhecimento. *Informática na Educação: Teoria e Prática*, *3* (1). Retrieved October 13, 2001, from http://pgie.ufrgs.br

Anderson, K., & Witfelt, C. (2003). Educational game design: Bridging the gab between computer-based learning and experimental learning environments. In *Conference Proceedings of the 6th International Conference on Computer-Based Learning in Science* (Vol.1, pp. 402-413).

Anderson, L. W., & Krathwohl, D. R. (Eds.). (2001). *A taxonomy for learning, teaching and assessing: A revision of Bloom's taxonomy of educational objectives*. New York: Addison, Wesley, Longman.

Anderson, T. (2004). Towards a theory of online learning. In T. Anderson & F. Elloum (Eds.), *Theory and practice of online learning* (pp. 33-60). Edmonton, Canada: Athabasca University Press.

Anderson, T. (2007). *Reducing the Loneliness of Distant Learner Using Social Software*. Open and Distance Learning Conference. Retrieved on January, 17, 2007 from < http://www2.open.ac.uk/r06/conference/TerryAndersonKeynoteCambridge2007.pdf >

Anderson, T., Rourke, L., Garrison, R., & Archer, W. (2001). Assessing teaching presence in a computer conferencing context. Journal of Asynchronous Learning Networks, 5(2), 1–17.

Anderson-Inman, L., & Ditson, L. (1999). Computer-based concept mapping: A tool for negotiating meaning. Learning and Leading with Technology, 26(8), 7–13.

Anderson-Inman, L., Ditson, L. A., & Ditson, M. T. (1998). *Computer-based concept mapping: Promoting meaningful learning in science for students with disabilities. information Technology and Disabilities*. Retrieved September 22, 2008 from http://people.rit.edu/easi/easisem/itdscience.htm.

Anohina, A., Graudina, V., & Grundspenkis, J. (2007). Using concept maps in adaptive knowledge assessment. In G. Magyar, G. Knapp, W. Wojtkowski, W. G. Wojtkowski, & J. Zupancic (Eds.), *Advances in Information Systems Development New Methods and Practice for the Networked Society*. New York: Springer.

Ardoíno, J. (2003). *Para uma pedagogia socialista*. Brazil: Editora Plano.

Ashby, J. (2003). Uniting Science and Participation in the Process of Innovation – Research *for* Development. In B. Pound, S. Snapp, C. McDougall, and A. Braun (Eds.) *Managing Natural Resources for Sustainable Livelihoods: Uniting Science and Participation*. London: Earthscan.

Ausubel, D. P. (1963). *The psychology of meaningful verbal learning*. New York: Grune & Stratton.

Ausubel, D. P. (1968). *Educational psychology: A cognitive view*. New York: Holt, Rinehart & Winston.

Ausubel, D. P. (2000). *The acquisition and retention of knowledge: a cognitive view*. Dordrecht, the Netherlands: Kluwer Academic Publishers.

Ausubel, D. P., Novak, K. D., & Hanesian, H. (1978). *Educational Psychology*. New York: Holt, Rinehart & Winston.

Avouris, N. M., Dimitracopoulou, A., & Komis, V. (2003). On analysis of collaborative problem solving: An object-oriented approach. *Computers in Human Behavior, 19*(2), 147–167. doi:10.1016/S0747-5632(02)00056-0

Azmitia, M. (1988). Peer interaction and problem solving: when are two heads better than one? Child Development, 59, 87–96. doi:10.2307/1130391

Azmitia, M., & Perlmutter, M. (1989). Social influences on children's cognition: state of the art and future directions. In H.W. Reese (Ed.), *Advances in child development and behavior* (pp. 321-340). New York: Academic Press.

Baddeley, A. D. (2000). Short term and working memory. In E. Tulving & F. I. M. Craik (Eds.). *The Oxford handbook of memory* (pp. 77-92). New York: Oxford University Press.

Badilla, S. (2004, September 3 - 4). *Experiencia pedagógica: mapas conceptuales en preescolar*. Paper presented at the V Congreso Nacional de Educadores, San Jose, Costa Rica.

Bahr, G. S., & Dansereau, D. F. (2001). Bilingual knowledge maps (BiK-Maps) in second-language vocabulary learning. Journal of Experimental Education, 70, 5–24.

Bahr, G. S., & Dansereau, D. F. (2004). Bilingual knowledge (BIK-) maps: Study strategy effects. In A. J. Cañas, J. D. Novak, & F. M. Gonzáles (Eds.), *Concept Maps: Theory, Methodology, Technology. Proceedings of the First International Conference on Concept Mapping* (Vol. 1). Pamplona, Spain: Universidad Pública de Navarra.

Baker, E. (1999, July). *Technology: How do we know it works?* Paper presented at the Secretary's Conference on Educational Technology, July 12-13, 1999, Washington, D.C. Retrieved from http://www.ed.gov/Technology/TechConf/1999/whitepapers/paper5.html

Baker, M. (2003). Computer-mediated argumentative interactions for the co-elaboration of scientific notions. In J. Andriessen, M. Baker, & D. Suthers (Eds.), *Arguing to Learn: Confronting Cognitions in Computer-Supported Collaborative Learning environments* Amsterdam, the Netherlands: Kluwer Academic Publishers.

Baker, M., Hansen, T., Joiner, R., & Traum, D. (1999). The role of grounding in collaborative learning tasks. In P. Dillenbourg (Eds.), *Collaborative learning: cognitive and computational approaches* (pp. 31-63). Oxford, UK: Elsevier Science Ltd.

Baroody, A. J., & Bartels, B. H. (2000). Using concept maps to link mathematical ideas. Mathematics Teaching in the Middle School, 5(9), 604–609.

Barron, B. (2003). When smart groups fail. Journal of the Learning Sciences, 12, 307–359. doi:10.1207/S15327809JLS1203_1

Barton, D., & Hamilton, M. (2005). Literacy, reification and the dynamics of social interaction. In D. Barton, & K. Tusting, (Eds.) *Beyond communities of practice: language, power and social context*. Cambridge, UK: Cambridge University Press

Basque, J. (2004). Le transfert d'apprentissage: Qu'en disent les contextualistes? In M. Frenay & A. Presseau (Eds.), *Le transfert des apprentissages: Comprendre pour mieux intervenir*. Québec, Canada: Presses de l'Université Laval.

Basque, J., & Lavoie, M. (2006). Collaborative concept mapping in education: Major research trends. In A. J. Cañas & J. D. Novak (Eds.), *Concept maps: Theory, methodology, technology: Proceedings of the Second International Conference on Concept Mapping,* (Vol. 1, pp. 79-86). San Jose, Costa Rica.

Basque, J., & Pudelko, B. (2002). *La (co)construction des cartes de connaissances dans le contexte du téléapprentissage: quelques pistes de recherche.* Paper presented at the Atelier CIRTA (Centre Interuniversitaire de Recherche sur le Télé-apprentissage), Télé-université, Montréal, Canada.

Basque, J., & Pudelko, B. (2003). Using a concept mapping software as a knowledge construction tool in a graduate online course. In D. Lassner & C. McNaught (Eds.), *Proceedings of ED-MEDIA 2003, Wold Conference on Educational Multimedia, Hypermedia & Telecommunications, Honolulu, June 23-28, 2003* (pp. 2268-2264). Norfolk, VA: AACE.

Basque, J., & Pudelko, B. (2004). The effect of collaborative knowledge modeling at a distance on performance and on learning. In A. J. Canas, J. D. Novak & F. M. Gonzalez (Eds.), *Proceedings of the First International Conference on Concept Mapping (CMC 2004), Pamplona, September 14-17,* (Vol. 1, pp. 67-74). Pamplona, Spain: Universidad Publica de Navarra.

Basque, J., Dao, K., & Contamines, J. (2005. June 27-July 2). Participating virtually in a scientific conference: A collaborative e-learning scenario for authentic learning. In P. Kommers & G. Richards (Eds.) *Higher Education, Proceedings of ED-MEDIA 2005*, Montreal, (pp. 1953-1960).

Basque, J., Dufresne, A., Léonard, M., Lundgren-Cayrol, K., Paquette, G., & Prom Tep, S. (2002). *Vers un modèle générique de système d'assistance pour le téléapprentissage.* Paper presented at the 70e Congrès de l'ACFAS, 13-17 mai 2002, Université Laval, Québec, Canada.

Basque, J., Imbeault, C., Pudelko, B., & Léonard, M. (2004). Collaborative knowledge modeling between experts and novices: A strategy to support transfer of expertise in an organization. In A. J. Canas, J. D. Novak & F. M. Gonzalez (Eds.), *Proceedings of the First International Conference on Concept Mapping (CMC 2004), Pamplona, September 14-17, vol. 1* (pp. 75-81). Pamplona, Spain: Universidad Publica de Navarra.

Basque, J., Rocheleau, J., Paquette, G., & Paquin, C. (1998). An Object-Oriented Model of a Computer-Enriched High School. In T. Ottmann & I. Tomek (Eds.), *Proceedings of ED-MEDIA/ED-TELECOM 98.* Charlottesville, VA: Association for the Advancement of Computing in Education.

Bastos, H. M. (1998). A escrita no ensino de uma língua estrangeira: reflexão e prática. In V. L. M. Paiva (Ed.) *Ensino de Língua Inglesa: reflexões e expressões.* Campinas, Spain: Pontes.

Beaudry, J. (2005). *Measuring vegetation health: Year 2 evaluation report.* Gorham, ME: Center for Policy Analysis, Research and Evaluation.

Beaudry, J. (July, 2007). *Concept mapping: Picture pathways for formative assessment.* Paper presented at the summer meeting of the Educational Testing Service on "Assessment For Learning" Portland, OR.

Behrens, M. A. (2000). Projetos de aprendizagem colaborativa num paradigma emergente. In J. M. Moran, M. T. Masetto & M. A. Behrens (Eds.), *Novas Tecnologias e Mediação Pedagógica* (pp. 67-132). Campinas, Spain: Papirus.

Bell, D. (1999). *The Coming of Post-Industrial Society: A Venture in Social Forecasting.* Jackson, TN: Basic Books.

Bell, P., Davis, E. A., & Linn, M. C. (1995). *The Knowledge Integration Environment: Theory and design.* Proceedings of the Computer Supported Collaborative Learning Conference (CSCL '95: Bloomington, IN), (pp. 14-21). Mahwah, NJ: Lawrence Erlbaum Associates.

Bennett, C., Howe, C., & Truswell, E. (2002). *Small group teaching and learning in psychology.* Retrieved February 15, 2007, from http://www.psychology.heacademy.ac.uk/docs/pdf/p20030116_smallgroupsreport.pdf

Bennett, S. (2003). Supporting collaborative project teams using computer-based technologies. In T. S. Roberts (Ed.), *Online collaborative learning: theory and practice* (pp. 1-26). Hershey, PA, USA: Information Science Publishing.

Berkes, F., Colding, J., & Folke, C. (2003). Introduction. In F. Berkes, J. Colding, and C. Folke, (eds.) *Navigating Social-Ecological Systems: building resilience for complexity and change*. Cambridge, UK: Cambridge University Press.

Biggs, J. B. (1987). *Students Approaches to Learning and Studying*. Melbourne: Australian Council for Educational Research.

Biggs, J. B., & Collis, K. F. (1982). *Evaluating the quality of learning. The SOLO Taxonomy (Structure of the Observed Learning Outcome)*. New York: Academic Press.

Biggs, J., & Tang, C. (2007). *Teaching for quality learning at university*, (3rd ed.). Maidenhead, UK: Open University Press.

Biott, C., & Nias, J. (2000). *Working and learning together for change*. Buckingham, UK: Open University Press.

Birbili, M. (2006). Mapping knowledge: concept maps in early childhood education. *Early Childhood Research and Practice, 8*(2).

Birbili, M. (2007). Making the case for a conceptually based curriculum in early childhood education. Early Childhood Education Journal, 35(2), 141–147. doi:10.1007/s10643-006-0112-0

Birch, D., & McDonald, J. (2005). *Students' perceptions of a virtual team assessment item (VTAI): An Australian case study*. Paper presented to The Open and Distance Learning Association of Australia (ODLAA) Breaking Down Boundaries Conference, Adelaide. Retrieved February 15, 2007 from http://www.odlaa.org/events/2005conf/ref/ODLAA2005Birch-McDonald.pdf

Bjorklund, D. F., & Douglas, R. N. (1997). The development of memory strategies. In N. Cowan (Ed.), *The Development of Memory in Childhood* (pp. 201-246). Hove, UK: Psychology Press.

Black, P., & Wiliam, D. (1998). Inside the black box. *Phi Delta Kappan, 80*(2), 9 21.

Black, P., & Wiliam, D. (1998). Assessment and classroom learning. *Assessment in Education, 5*(1), 7–85. doi:10.1080/0969595980050102

Black, P., McCormick, R., James, M., & Pedder, D. (2006). Learning how to learn and assessment for learning. *Research Papers in Education, 21*(2), 119–132. doi:10.1080/02671520600615612

Blankenship, J., & Dansereau, D. F. (2000). The effect of animated node-link displays on information recall. Journal of Experimental Education, 68, 293–308.

Blaxter, L., Hughes, C., & Tight, M. (2001). *How to Research*. Buckingham, UK: Open University Press.

Bloom, B., Englehart, M., Furst, E., Hill, W., & Krathwohl, D. (1956). *Taxonomy of educational objectives: The classification of educational goals. Handbook I: Cognitive domain*. New York, Toronto: Longmans, Green.

Bochniak, R. (1998). *Questionar o conhecimento. A interdisciplinariedade na escola e fora dela.* (2nd Ed). São Paulo, Brazil: Loyola.

Boyd, R. D., & Myers, J. G. (1988). Transformative education. International Journal of Lifelong Education, 7, 261–284. doi:10.1080/0260137880070403

Bransford, J. D., Brown, A. L., & Cocking, R. R. (Eds.). (2000). *How People Learn: Brain, Mind, Experience, and School*. Washington, DC: National Academy Press.

Brewer, J., & Hunter, A. (2006). *Foundations of multimethod research*. Thousand Oaks, CA: SAGE.

Brooks, M. (2004). Drawing: the social construction of knowledge. Australian Journal of Early Childhood, 29(2), 41–49.

Brown, D. H. (1994). *Teaching by principles: An interactive approach to language pedagogy*. Englewood Cliff, NJ: Prentice Hall, Regents.

Brown, D. S. (2003). High school biology: A group approach to concept mapping. *The American Biology Teacher, 65*, 192–197. doi:10.1662/0002-7685(2003)065[0192:HSBAGA]2.0.CO;2

Bunge, M. (1978). *Filosofía de la Física*. Buenos Aires, Venezuela: EUDEBA.

Burnard, P. (2006). Reflecting on the creativity agenda in education. *Cambridge Journal of Education, 36*(3), 313–318. doi:10.1080/03057640600865801

Cabral, A. R. Y., Zeve, C. M. D., Nicolao, M., & Amoretti, M. S. M. (2004). Use of conceptual maps in distance learning courses. In A. J. Cañas, J. D. Novak, & F. M. González (Eds.), *Concept maps: Theory, methodology, technology: Proceedings of the First International Conference on Concept Mapping,* (Vol. 2, pp. 111-114). Pamplona, Spain.

Calderón, M. D., Agüera, E., & Alfageme, M. B. (2004). Los maps conceptuales, herramienta hipertextual para el trabajo colaborativo y desarrollo de habilidades comunicativas y docentes. In A. J. Cañas, J. D. Novak, & F. M. González (Eds.), *Concept maps: Theory, methodology, technology. Proceedings of the First International Conference on Concept Mapping* (Vol. 2, pp. 115-118). Pamplona, Spain.

Cañas, A. J., & Novak, J. D. (2006). Re-examining the foundations for effective use of concept maps. In A. J. Cañas & J. D. Novak (Eds.), *Proceedings of the Second International Conference on Concept Mapping,* (Vol. 1, pp. 247-255). San Jose, Costa Rica: Universidad de Costa Rica.

Cañas, A. J., Coffey, J. W., Carnot, M. J., Feltovich, P., Hoffman, R. R., Feltovich, J., & Novak, J. D. (2003). *A summary of literature pertaining to the use of concept mapping techniques and technologies for education and performance support.* Report from The Institute for Human and Machine Cognition.

Cañas, A. J., Ford, K. M., Hayes, P. J., Reichherzer, T., Suri, N., Coffey, J., et al. (1998). *Colaboración en la Construcción de Conocimiento Mediante Mapas Conceptuales.* Proceedings of the VIII Congreso Internacional sobre Tecnología y Educación a Distancia, San José, Costa Rica, (pp. 25- 42).

Cañas, A. J., Ford, K. M., Novak, J. D., Hayes, P., Reichherzer, T. R., & Suri, N. (2001). Using concept maps with technology to enhance collaborative learning in Latin America. Science Teacher (Normal, Ill.), 68, 49–51.

Cañas, A. J., Ford, K. M., Novak, J. D., Hayes, P., Reichherzer, T. R., & Suri, N. (2001). Using concept maps with technology to enhance collaborative learning in Latin America. *Science Teacher* (Normal, Ill.), *68*, 49–51.

Cañas, A. J., Hill, G., & Lott, J. (2003). *Support for Constructing Knowledge Models in CmapTools,* (Technical Report CmapTools 93-03). Institute for Human and Machine Cognition.

Cañas, A. J., Hill, G., Carff, R., & Suri, N. (2003). *CmapTools: A knowledge modeling and sharing toolkit.* (Technical Report IHMC CmapTools 93-01). Institute for Human and Machine Cognition.

Cañas, A. J., Hill, G., Carff, R., Suri, N., Lott, J., Eskridge, T., et al. IHMC (2004). CmapTools: A knowledge modelling and sharing environment. In A. J. Cañas, J. D. Novak & F. M. Gonzáles (Eds.), *Concept Maps: Theory, Methodology, Technology, Proceedings of the First International Conference on Concept Mapping* (Vol. 1, pp. 125-133). Pamplona, Spain: Universidad Pública de Navarra.

Cañas, A., & Novak, J. (2008). Concept mapping using Cmap tools to enhance meaningful learning. In Okada, A., Buckingham Shum, S. & Sherborne, T. (Eds.) *Knowledge Cartography: software tools and mapping techniques.* London: Springer.

Cañas, A., Valerio, A., Lalinde-Pulido, J., Carvalho, M., & Arguedas, M. (2003, Octobre 2003). *Using WordNet for word sense disambiguation to support concept map construction.* Paper presented at the SPIRE 2003 - 10th international symposium on string processing and information retrieval, Manaus, Brésil.

Cannon-Bowers, J. A., & Salas, E. (Eds.). (1998). *Making decisions under stress: Implications for individual and team training.* Washington, DC: American Psychological Association.

Carnevale, A., Gainer, L., & Meltzer, A. (1989). *Workplace basics: The skills employers want.* Alexandria, VA: American Society for Training and Development.

Cassata, A. E., & French, L. (2006). *Using concept mapping to facilitate metacognitive control in preschool children.* Paper presented at the second international Conference on Concept Mapping, San Jose, Costa Rica.

Cassata-Widera, A. E. (2008, September 22-25). *Concept mapping and early literacy: a promising crossroads.* Paper presented at the Third International Conference on Concept Mapping, Tallin, Estonia & Helsinki, Finland.

Cassata-Widera, A. E. (2008). *Fostering metacognitive control skills in the kindergarten classroom using concept maps.* Paper presented at the annual meeting of the American Educational Research Association, New York, NY.

Caswell, R. M., & Wendel, C. T. (1992, April). Conceptual changes occurring in elementary teachers' knowledge of science concepts throughout long-term instruction. Paper presented at the *American Educational Research Association Annual Meeting*, San Francisco, CA.

Chacón, S. (2006, September 5-8). *La pregunta pedagógica como instrumento de mediación en la elaboración de mapas conceptuales* Paper presented at the second international Conference on Concept Mapping, San Jose, Costa Rica.

Chaka, C. (2008). Revisiting knowledge management and learning organisation: Factoring in multiple intelligences and cognitive flexibility theory. In A. Koohang, K, Harman & J. Britz (Eds.), *Knowledge management: Theoretical foundations* (pp. 191-222). Santa Rosa, CA: Informing Science Press.

Chang, K., Sung, Y., & Chen, I. (2002). The effect of concept mapping to enhance text comprehension and summarization. Journal of Experimental Education, 71, 5–23.

Chang, K.-E., Sung, Y.-T., & Lee, C.-L. (2003). Web-based collaborative inquiry learning. Journal of Computer Assisted Learning, 19, 56–59. doi:10.1046/j.0266-4909.2003.00006.x

Chang, W. (1994). The effects of using concept mapping to supplement class notes on the biology test scores of seventh-grade students in Taiwan, R.O.C. (Doctoral dissertation, The University of Iowa). *Dissertation Abstracts International, 55,* 3800.*

Chen, C., & Czerwinski, M. (1997). Spatial ability and visual navigation: An empirical study. *The New Review for Hypertext and Multimedia, 3,* 40-66. http://research.microsoft.com/~marycz/nrhm.htm

Chi, M. T. H. (1996). Constructing self-explanations and scaffolded explanations in tutoring. Applied Cognitive Psychology, 10, 33–49. doi:10.1002/(SICI)1099-0720(199611)10:7<33::AID-ACP436>3.0.CO;2-E

Chi, M. T. H., De Leeuw, N., Chiu, M. H., & Lavancher, C. (1994). Eliciting self-explanations improves understanding. Cognitive Science, 18, 439–477.

Chi, M. T. H., Slotta, J. D., & de Leeuw, N. (1994). From things to processes: a theory of conceptual change for learning science concepts. Learning and Instruction, 4, 27–43. doi:10.1016/0959-4752(94)90017-5

Chickering, A. W., & Ehrmann, S. C. (1996). Implementing the seven principles: Technology as lever. AAHE Bulletin, 49(2), 3–6.

Chiu, C. H. (2003). Exploring how primary school students function in computer supported collaborative learning. International Journal of Continuing Engineering Education and Lifelong Learning, 13, 258–267. doi:10.1504/IJCEELL.2003.003276

Chiu, C. H., Huang, C. C., & Chang, W. T. (2000). The evaluation and influence of interaction in network supported collaborative concept mapping. Computers & Education, 34, 17–25. doi:10.1016/S0360-1315(99)00025-1

Chiu, C.-H. (2004). Evaluating system-based strategies for managing conflict in collaborative concept mapping. Journal of Computer Assisted Learning, 20, 124–132. doi:10.1111/j.1365-2729.2004.00072.x

Chularut, P., & DeBacker, T. K. (2004). The influence of concept mapping on achievement, self-regulation, and self-efficacy in students of English as a second language. Contemporary Educational Psychology, 29, 248–263.* doi:10.1016/j.cedpsych.2003.09.001

Chung, G. K., O'Neil, H. F. J., & Herl, H. E. (1999). The use of computer-based collaborative knowledge mapping to measure team processes and team outcomes. Computers in Human Behavior, 15(3-4), 463–494. doi:10.1016/S0747-5632(99)00032-1

Cilliers, P. (2000). *Complexity and postmodernism – understanding complex systems*. London: Routledge.

Clark, W. (2007). Mapping the syllabus: multimodal digital mapping tools. In *Reflecting education: Fascinating cultural artefacts: multimodal (concept) mapping in teaching and learning* (pp. 43-60). Retrieved November 3, from www.reflectingeducation.net/index.php/reflecting

Claxton, G. (1987). *Vivir y aprender*. Madrid: Alianza editorial.

Cliburn, J. W., Jr. (1986). An Ausubelian approach to instruction: The use of concept maps as advance organizers in a junior college anatomy and physiology class. (Doctoral Dissertation, University of Southern Mississippi). *Dissertation Abstracts International, 47*, 852A.

CMAPTOOLS. (n.d.). *Software para Construção de Mapas Conceituais da University of West Florida.* Retrieved September 06, 2006, from http://penta2.ufrgs.brtutindicecmap.htm

Coffey, J. W., Carnot, M. J., Feltovich, P. J., Feltovich, J., Hoffman, R. R., Cañas, A. J., et al. (2003). *A summary of literature pertaining to the use of concept mapping techniques and technologies for education and performance support* (Technical Report submitted to the Chief of Naval Education and Training). Pensacola, FL: IHMC - Institute for Human and Machine Cognition.

Cohen, E. G. (1994). Restructuring the classroom: conditions for productive small groups. *Review of Educational Research, 64*, 1–35.

Cohen, J. (1988). *Statistical power analysis for the behavioral sciences* (2nd Ed.). Hillsdale, NJ: Lawrence Erlbaum Associates.

Coll, C. (1990). *Aprendizaje escolar y construcción de conocimiento.* Barcelona: Paidós.

Coll, C., & Colomina, R. (1990). Interacción entre alumnos y aprendizaje escolar. In C. Coll, J. Palacios & A. Marchesi (Eds.), *Desarrollo Psicológico y Educación II. Psicología de la Educación* (pp. 435-453). Madrid: Alianza Editorial.

Coll, C., & Edwards, D. (Eds.). (1996). *Enseñanza, aprendizaje y discurso en el aula. Aproximaciones al discurso educacional.* Madrid: Fundación Infancia y Aprendizaje.

Condivi, A. (1999). *The life of Michelangelo.* University Park, PA: Pennsylvania State University Press.

Conklin, J. (2006). *Dialogue Mapping: Building Shared Understanding of Wicked Problems.* Chicester, UK: John Wiley.

Cook, L. K., & Mayer, R. E. (1988). Teaching readers about the structure of scientific text. *Journal of Educational Psychology, 80*, 448–456. doi:10.1037/0022-0663.80.4.448

Cooke, N. J., Salas, E., Cannon-Bowers, J. A., & Stout, R. J. (2000). Measuring team knowledge. *Human Factors, 42*(1), 151–173. doi:10.1518/001872000779656561

Cooke, N. J., Salas, E., Kiekel, P. A., & Bell, B. (2004). Advances in measuring team cognition. In E. Salas & S. M. Fiore (Eds.), *Team cognition: Understanding the factors that drive process and performance* (pp. 83-106). Washington, DC: American Psychological Association.

Cooper, H. M., & Hedges, L. V. (1994). *The Handbook of Research Synthesis.* New York: The Russell Sage Foundation.

Cord, B. (2001). *Internet et pédagogie: état des lieux.* Retrieved from http://wwwadm.admp6.jussieu.fr/fp/uaginternetetp/definition_travail_collaboratif.htm

Crooks, T. (2002, December 5-8). *Assessment, accountability and achievement - principles, possibilities and pitfalls.* Paper presented at the Twenty-fourth Annual Conference of the New Zealand Association for Research Education, Palmerston North.

Cuthell, J. P. (2002). A learning community: A community of learners. Journal of Interactive Learning Research, 13(1/2), 169–188.

Cuthell, J. P. (2005). *What does it take to be active? Teacher participation in online communities.* Geneva, Switzerland: Inderscience.

Czerniak, C. M., & Haney, J. J. (1998). The effect of collaborative concept mapping on elementary preservice teachers' anxiety, efficacy, and achievement in physical science. *Journal of Science Teacher Education, 9,* 303–320. doi:10.1023/A:1009431400397

D'Incao, F., & Reis, E. G. (2002). Community-based management and technical advice in Patos Lagoon estuary (Brazil). *Ocean and Coastal Management, 45*(8): 531–539. doi:10.1016/S0964-5691(02)00084-4

da Costa, J. V. J., da Rocha, F. E. L., & Faveo, E. L. (2004). Linking phrases in concept maps: A study of the nature of inclusivity. In A. J. Canas, J. D. Novak & F. M. Gonzalez (Eds.), *Concept maps: Theory, Methodology, Technology - Proceedings of the First International Conference on Concept Mapping* (Vol. 1, pp. 2M-249). Pamplona, Spain: Universidad Publica de Navarra.

Daly, C., & Pachler, N. (2007). Learning with others in mind. In J. Pickering, C. Daly & N. Pachler (Eds.) *New Designs for Teachers' Professional Learning.* London: Bedford Way Papers.

Dansereau, D. F. (2005). Node-link mapping principles for visualizing knowledge and information. In S. O. Tergan & T. Keller (Eds.). *Knowledge and information visualization: Searching for synergies.* (LNCS 3426, pp. 61-81). Heidelberg, Germany: Springer-Verlag.

Darmofal, D. L., Soderholm, D. H., & Brodeur, D. R. (2002). *Using concept maps and concept questions to enhance conceptual understanding.* Retrieved March 11, 2008, from http://www.cdio.org/papers/concept_maps_fie.pdf

Davis, N. E. (2008 in press). How may teacher learning be promoted for educational renewal with IT? In J. Voogt & G. Knezek (Eds.), *International handbook of information technology in education.* Amsterdam: Kluwer Press.

Davis, N. E., Preston, C., & Sahinl, I. (2008 in press). Theoretical and evaluation frameworks to inform technology-related professional development for teachers, tested with evidence from a national study of ICT professional development for teachers. *British Educational Research Journal.*

Davis, N., Preston, C., & Sahin, I. (2008 in press). ICT teacher training impacts multiple ecologies: evidence from a national initiative. *British Journal of Education Technology (BJET).*

De Beni, R., Magnussen, S., Cornoldi, C., Larsson, M., & Rønnberg, J. (2007). Memory experts. Visual learning, wine tasting, orienteering and speech reading. In S. Magnussen & T. Helstrup (Eds.) *Everyday memory* (pp. 201-227). Hove, UK: Psychology Press.

De Masi, D. (2000). *Ozio creativo. Conversazione con Maria Serena Palieri.* Italy: Rizzoli.

De Masi, D. (2003). *Creatività individuale e di gruppo.* Italy: Rizzoli.

de Menezes, L. C. (2000). Ensinar ciências no próximo século. In E. W. Hamburguer & C. Matos (Eds.), *O desafio de ensinar ciências no século XXI* (pp. 48-54). São Paulo, Brazil: Edusp.

De Simone, C. (2007). Applications of concept mapping. College Teaching, 55(1), 33–36. doi:10.3200/CTCH.55.1.33-36

DeHaan, R. L. (2005). The impending revolution in undergraduate science education. Journal of Science Education and Technology, 14(2), 253–269. doi:10.1007/s10956-005-4425-3

Delors, J. (1998). Relatório para a UNESCO da Comissão Internacional sobre *Educação para o século* XXI.

Delors, J. (1998). *Report to UNESCO of the International Commission on Education for the Twenty-first Century.* Retrieved March 22, 2003 from http://www.unesco.org/delors/

Derbentseva, N., Safayeni, F., & Cañas, A. J. (2004). *Experiments on the Effects of Map Structure and Concept Quantification during Concept Map Construction.* Proceedings of the First International Conference on Concept Mapping. Pamplona, Spain, 2004.

Derbentseva, N., Safayeni, F., & Cañas, A. J. (2007). Experiments on dynamic thinking. *Journal of Research in Science Teaching, 44*(3), 448-465. doi:10.1002/tea.20153

DfES. (2003). *Towards a Unified e-learning strategy: consultation document.* Retrieved from http://www.dfes.gov.uk/elearningstrategy/strategy.stm.

Di Sessa, A. A. (1993). Towards an epistemology of physics. Cognition and Instruction, 10, 105–225. doi:10.1207/s1532690xci1002&3_2

Dias, R. (1996). A Incorporação de estratégias ao ensino-aprendizagem de leitura em inglês como língua estrangeira. In V. L. M. de O. Paiva (Ed.) *Ensino de Língua Inglesa: Reflexões e Experiências* (pp. 137–160). Campinas, Spain: Pontes Editores.

Díaz Barriga, F., & Hernández Rojas, G. (2006). *Estratégias Docentes para un Aprendizaje Signficativo.* México: McGraw Hill Interamericana.

Dillenbourg, P. (1999). What do you mean by collaborative learning? In P. Dillenbourg (Ed.) *Collaborative-learning: Cognitive and computational approaches.* Oxford, UK: Elsevier.

Dillenbourg, P. (2000). Learning in the new millennium: Building education strategies for schools. In *Workshop on Virtual Learning Environments.* Retrieved July, 2008 from http://tecfa,unige.ch/tecfa/publicat/dil-papers-2/Dil.7.5.18.pdf

Dodge, M., & Kitchin, R. (2001). *Mapping cyberspace.* London: Routledge.

Doise, W., & Mugny, G. (1984). *The social development of the intellect.* Oxford, UK: Pergamon Press.

Donnelly, J. F. (2004). Humanizing science education. Science Education, 88(5), 762–784. doi:10.1002/sce.20004

Doré, S., & Basque, J. (2002). *Why not apply an engineering methodology when creating courses?* Paper presented at the American Society for Engineering Education Annual Conference & Exposition (ASEE), Montréal, June 2002.

Doyle, J. K., Radzicki, M. J., & Trees, W. S. (1998). *Measuring Change in Mental Models of Dynamic Systems: An Exploratory Study.*Unpublished manuscript.

Dummer, P., & Ifenthaler, D. (2005). Planning and assessing navigation in model-centered learning environments. Why learners often do not follow the path laid out for them. In G. Chiazzese, M. Allegra, A. Chifari & S. Ottaviano (Eds.), *Methods and technologies for learning* (pp. 327-334). Boston: WIT Press.

Duschl, R., Schweingruber, H., & Shouse, A. (Eds.). (2007). *Taking Science to School.* Washington, D.C.: National Academy Press.

Dutra, I., Fagundes, L., & Cañas, A. (2003). *Uma proposta de uso dos mapas conceituais para um paradigma construtivista da formação de professores a distância.* Retrieved September 06, 2006, from http://homer.nuted.edu.ufrgs.br/ mapas_prof.pdf

Dweck, C. (2006). *Mindset: The new psychology of success.* New York: Random House.

Dweck, C. (2007). The perils and promises of praise. *Educational Leadership, 65*(2), 34–39.

Edelson, D. C. (2001). Learning-for-use: A framework for integrating content and process learning in the design of inquiry activities. *Journal of Research in Science Teaching, 38,* 355–385. doi:10.1002/1098-2736(200103)38:3<355::AID-TEA1010>3.0.CO;2-M

Edson, E. (1997). *Mapping time and space: how medieval mapmakers viewed their world.* London: The British Library.

Edwards, D. (1992). Discurso y aprendizaje en el aula. In C. Rogers & P. Kutnick (Eds.), *Psicología Social de la escuela primaria* (pp. 63-82). Barcelona: Paidós.

Elliott, J. (1991). *Action Research for Educational Change.* Buckingham, UK: Open University Press.

Ellson, J., Gansner, E. R., Koutsofios, E., North, S. C., & Woodhull, G. (2003). *GraphViz and Dynagraph. Static and dynamic graph drawing tools*. Florham Park, NJ: AT&T Labs.

Engeström, Y., Miettinen, R., & Punamäki, R.-L. (1999). *Perspectives on Activity Theory*. Cambridge, UK: Cambridge University Press.

Entwistle, N. J. (2000). Approaches to studying and levels of understanding: The influences of teaching and assessment. In J.C. Smart (Ed.), *Higher Education: Handbook of Theory and Research: Volume XV*. (pp. 156-218). New York: Agathon Press.

Ericsson, K. A., & Smith, J. (Eds.). (1991). *Toward a general theory of expertise: Prospects and limits*. New York: Cambridge University Press.

Escudero, J. M. (2001). La mejora de la educación como marco de referencia para el asesoramiento pedagógico. In J. Domingo Segovia (Coord.), *Asesoramiento al centro educativo. Colaboración y cambio en la institución*. (pp.15-50) Barcelona: Octaedro.

Esiobu, G. O., & Soyibo, K. (1995). Effects of concept and vee mappings under three learning modes on students, cognitive achievement in ecology and genetics. *Journal of Research in Science Teaching*, *32*, 971–995.* doi:10.1002/tea.3660320908

Facione, N., & Facione, P. (1997). *Critical thinking assessment in nursing education: An aggregate data analysis*. Millbrae, CA: California Academic Press.

Faletti, J., & Fisher, K. M. (1996). The information in relations in biology, or the unexamined relation is not worth having. In K. M. Fisher & M. R. Kibby (Eds.), *Knowledge acquisition, organization, and use in biology* (pp. 182-205). Berlin, Germany: Springer.

Figueiredo, M., Lopes, A. S., Firmino, R., & Sousa, S. d. (2004). *"Things we know about the cow": concept mapping in a preschool setting*. Paper presented at the first international Conference on Concept Mapping, Pamplona, Spain.

Fiore, S. M., & Salas, E. (2004). Why we need team cognition. In E. Salas & S. M. Fiore (Eds.), *Team cognition: Understanding the factors that drive process and performance* (pp. xi, 268). Washington, DC: American Psychological Association.

Fischer, F., Bruhn, J., Gräsel, C., & Mandl, H. (2002). Fostering collaborative knowledge construction with visualization tools. *Learning and Instruction*, *12*(2), 213–232. doi:10.1016/S0959-4752(01)00005-6

Fisher, J. (2002). *Starting from the child: teaching and learning from 3 to 8* (2nd ed.). Philadelphia: Open University Press.

Fisher, K. (1990). Semantic networking: The new kid on the block. *Journal of Research in Science Teaching*, 27(10), 1001–1018.

Fisher, K. M. (2000). SemNet® Semantic Networking. In K. M. Fisher, J. H. Wandersee & D. E. Moody (Eds.), *Mapping biology knowledge* (pp. 143-165). Dordrecht, the Netherlands: Kluwer Academic Publishers.

FOD. (2004). Foro en línea con Doña Eleonora Badilla-Saxe en preparación del V Congreso Nacional de Educadores PRONIE ME-FOD, Preescolar, I y II Ciclos. *Foros Institucionales - Fundación Omar Dengo (FOD)* Retrieved October 15, 2004, from http://www.fod.ac.cr/Vcongreso/CronoPonencias.htm

FOD. (2004). Foro en línea de las inquietudes iniciales de los tutores previas al foro con Doña Eleonora Badilla-Saxe. *Foros Institucionales - Fundación Omar Dengo (FOD)* Retrieved October 15, 2004, from http://www.fod.ac.cr/Vcongreso/ChatEleonora.htm

Fonteyn, M. (2007). Concept mapping: An easy teaching strategy that contributes to understanding and may improve critical thinking. The Journal of Nursing Education, 46(5), 199–200.

Forman, E. (1992). Discourse, intersubjetivity and the development of peer collaboration: a vygotskian approach. In L. T. Winegar & J. Valsiner (Eds), *Children's development within social contexts: Metatheoretical,*

theoretical and methodological issues (pp. 55-69). Hillsdale, NJ: Erlbaum.

Fourez, G. (1997). Scientific and technological literacy as a social practice. Social Studies of Science, 27, 903–936. doi:10.1177/030631297027006003

Fraser, S., & Gestwicki, C. (2002). *Authentic childhood: exploring Reggio Emilia in the classroom*. Del Mar, Canada: Thomson Learning.

Freire, P. (1967). *Educação como prática da liberdade*. Rio de Janeiro, Brazil: Paz e Terra.

Freire, P. (1979). *Conscientização: Teoria e prática da libertação*. São Paulo, Brazil: Moraes.

Freire, P. (1996). *Pedagogia da autonomia: Saberes necessários à prática educativa*. São Paulo, Brazil: Paz e Terra.

Freire, P. (2000). *Pedagogy of the Oppressed* (30th Ed.). New York: Continuum International Publishing Group.

Freire, P. (2001). *Pedagogy of Freedom: Ethics, Democracy and Civic Courage* (2nd Ed.). Lanham, MA: Rowman & Littlefield.

Freire, P. (2005). *Education for Critical Consciousness*. New York: Continuum International Publishing Group.

Freire, R. S. (2007). Desenvolvendo o Pensamento Proporcional Com o Uso de um Objeto de Aprendizagem. In C. L. Prata, Nascimento, & A. C. A. (Org.). *Objetos de aprendizagem: uma proposta de recurso pedagógico*. Brazil: MEC, SEED.

Freitas, L. V., & Freitas, C. V. (2002). *Aprendizagem Colaborativa*. Porto, Portugal: Edições Asa.

Friedman, T. L. (2007). *The World Is Flat [Updated and Expanded]: A Brief History of the Twenty-first Century*. New York: Picador.

Fruchterman, T. M. J., & Reingold, E. M. (1991). Graph drawing by force-directed placement. Software, Practice & Experience, 21(11), 1129–1164. doi:10.1002/spe.4380211102

Fullan, M. (2002). *Los nuevos significados del cambio en educación*. Madrid: Octaedro.

Fung, Y. H. (2004). Collaborative online learning: interaction patterns and limiting factors. *Open Learning, 19*(2), 135-149. doi:10.1080/0268051042000224743

Gagnè, R. (1985). *The Condition of Learning of Instruction*. New York: Holt, Rinehart & Winston.

Ganser, E. R., & North, S. C. (1999). An open graph visualization system and its applications to software engineering. *Software, Practice & Experience, 00*(S1), 1–5.

Gao, H., Shen, E., Losh, S. C., & Turner, J. (2007). A review of studies on collaborative concept mapping: What have we learned about the technique and what is next? Journal of Interactive Learning Research, 18, 479–492.

Gardner, H. (2003). *Multiple intelligences after twenty years*. Retrieved November 12, 2006, from http://pzweb.harvard.edu/Pls/HG_MI_after_20_years.pdf

Gardner, H. (2006). *Five Minds for the Future*. Boston: Harvard Business School Publishing.

Garrison, D. R. (2003). Cognitive presence for effective asynchronous online learning: The role of reflective inquiry, self-direction and metacognition. In J. Bourne & J. C. Moore (Eds.), *Elements of quality online education: Practice and direction*, (pp. 47-58). Needham, MA: Sloan-C, Sloan Center for Online Education.

Garrison, D. R., Anderson, T., & Archer, W. (2000). Critical inquiry in a text-based environment: computer conferencing in higher education. *The Internet and Higher Education, 2*(2-3), 87–105. doi:10.1016/S1096-7516(00)00016-6

Garrison, D. R., Anderson, T., & Archer, W. (2001). Critical thinking, cognitive presence and computer conferencing in distance education. *American Journal of Distance Education, 15*(1), 7–23.

Gates, B. (1995). *The Road Ahead*. New York: Viking.

Georghiades, P. (2004). From the general to the situated: three decades of metacognition. *Interna-*

tional Journal of Science Education, 26(3), 365–383. doi:10.1080/0950069032000119401

Giddens, A. (1984). *The constitution of society: an outline of the theory of structuration*. Cambridge, UK: Polity Press.

Gil, A. C. (2002). *Como elaborar projetos de pesquisa* (4th ed.). São Paulo Brazil: Atlas.

Gilly, M. (1988). *La fonction sociale des outils cognitifs*. Montréal, Canada: CIRADE.

Gokhale, A. A. (1995). *Collaborative Learning enhances critical thinking*. Retrieved May 16, 2008 from http://scholar.lib.vt.edu/ejournals/JTE/jte-v7n1/gokhale.jte-v7n1.html

Gomes, P. (2003). Internet e universidade: o EUREKA da PUCPR, 35.000 usuários depois. In C. Maia (Ed.), *Experiências em educação a distância no Brasil: Reflexões atuais, em tempo real*. São Paulo, Brazil: Anhembi Morumbi.

Gomes, P. (2003). O MATICE da PUCPR: Uso da Internet no Ensino de arquitetura. In *CREAD: Congresso de Educação a Distância* Mercosul 2003, 7., 2003, Florianópolis. Aprendizagem e Trabalho: Políticas e Tecnologias, Anais, Florianópolis.

Gomez, G. (2005, July 16-19). *Young children's use of a voice-input device to transform their symbolic maps into concept maps*. Paper presented at the Our Children the Future 4 - OCTF 4, Adelaide, Australia.

Gomez, G. (2006). *An authoring concept mapping kit for the early childhood classroom*. Paper presented at the second international Conference on Concept Mapping, San Jose, Costa Rica.

Gomez, G. (2007, April 28-May 3). *A bridging design prototype for investigating concept mapping in the preschool community*. Paper presented at the CHI 2007, San Jose, California, USA.

Gomez, G. (2008). *Issues in preschool concept mapping: an interaction design perspective*. Unpublished doctoral dissertation, Swinburne University of Technology, Melbourne.

Gouli, E., Gogoulou, A., & Grigoriadou, M. (2003). *A coherent and integrated framework using concept maps for various educational assessment functions*. Retrieved March 11, 2008, from http://jite.org/documents/Vol2/v2p215-240-23.pdf

Gowin, D. B., & Alvarez, M. (2005). *The Art of Educating with V Diagrams* (pp. 19-21). New York: Cambridge University Press.

Graham, S., & Perin, D. (2007). *Writing next: Effective strategies to improve the writing of adolescents in middle and high school. A report to the Carnegie Commission of New York*. Washington, DC: Alliance for Excellent Education.

Grotzer, T. (2002). *Causal Patterns in Ecosystems: Lessons to Infuse into Ecosystems Units to Enable Deeper Understanding: Understandings of Consequence Project, President and Fellows of Harvard College*. Retrieved from http://www.pz.harvard.edu/ucp/curriculum/ecosystems/index.htm

Grotzer, T. A., & Perkins, D. N. (2000). *A taxonomy of causal models: The conceptual leaps between models and students' reflections on them*. Paper presented at the National Association of Research in Science Teaching (NARST), New Orleans, LA, April 28-May 1, 2000.

Guastello, E. F., Beasley, M., & Sinatra, R. C. (2000). Concept mapping effects on science content comprehension of low-achieving inner-city seventh graders. Remedial and Special Education, 21, 356–365.* doi:10.1177/074193250002100605

Guban, L. (1992). Curriculum stability and change. In P. Jackson (Ed.), *Handbook of research on curriculum* (pp. 216-247). New York: Macmillan.

Gunawardena, C. N., & Zittle, F. J. (1997). Social presence as a predictor of satisfaction within a computer-mediated conferencing environment. American Journal of Distance Education, 11(3), 8–26.

Guzzo, R. A., & Salas, E. (1995). *Team effectiveness and decision-making in organization*. San Francisco, CA: Jossey-Bass, Inc.

Hackman, R. A. (1990). *Groups that work (and those that don't): Creating conditions for effective team work*. San Francisco,CA. Jossey-Bass.

Hall, R. H., Dansereau, D. F., & Skaggs, L. P. (1992). Knowledge maps and the presentation of related information domains. *Journal of Experimental Education, 61*, 5–18.

Harary, F. (1974). *Graphentheorie*. München, Germany: Oldenbourg.

Harasim, L., Teles, L., Turoff, M., & Hiltz, S. R. (2005). *Redes de Aprendizagem: Um guia para Ensino e Aprendizagem On-line*. São Paulo, Brazil: Ed. Senac São Paulo.

Harklau, L., & Norwood, R. (2005). Negotiating researcher roles in ethnographic program evaluation: A postmodern lens. *Anthropology & Education Quarterly, 36*(3), 278–288. doi:10.1525/aeq.2005.36.3.278

Harley, J. B., & Woodward, D. (1987).Cartography in prehistoric, ancient and medieval Europe and the Mediterranean. *The history of cartography* (Vol.1). Chicago: Chicago University Press.

Harrison, C., Comber, C., et al. (2002). *ImpaCT2 The Impact of Information and Communication Technologies on Pupil Learning and Attainment*. London: Becta for DfES.

Haugaard, M. (2003). Reflections on seven ways of creating power. *European Journal of Social Theory, 6*(1), 87–113. doi:10.1177/1368431003006001562

Hedges, L. V., & Olkin, I. (1985). *Statistical methods for meta-analysis*. San Diego, CA: Academic Press.

Heinze-Fry, J. (1992). Using student self-assessment of biology concepts in an introductory biology course. *Journal of College Science Teaching, 22*(1), 39–43.

Helm, J. H., & Katz, L. G. (2001). *Young investigators: the project approach in the early years*. New York: Teachers College Press.

Henao Cálad, M. (2004). Experiencia con el Uso de Mapas Conceptuales como Estrategia de Enseñanza en un Curso de Ingeniería del Conocimiento. In A. J. Cañas, J. D. Novak & F. M. Gonzáles (Eds.), *Concept Maps: Theory, Methodology, Technology, Proceedings of the First International Conference on Concept Mapping* (Vol. I, pp. 325-332). Pamplona, Spain: Universidad Pública de Navarra.

Henri, F. (2002). Formation à distance et téléconférence assistée par ordinateur: Interativité, quasi-interativité, ou monologue? *CADE: Journal of Distance Education, 7*(1), 1–4.

Herl, H. E. (1996). Construct validation of an approach to modeling cognitive structure of U.S. history knowledge. *The Journal of Educational Research, 89*, 206–218.

Herl, H. E., O'Neil, H. F., Jr., Chung, G. L. W. K., Bianchi, C., Wang, S., Mayer, R., et al. (1999). *Final report for validation of problem solving measures* (CSE Technical Report 501). Los Angeles: CRESST.

Hernández, P., & Serio, A. (2004). ¿Cómo hacer eficaces los mapas conceptuales en la instrucción? Infancia y Aprendizaje, 27, 247–265. doi:10.1174/021037004323038851

Hewson, P. W., & Thorley, R. N. (1989). The conditions of conceptual change. International Journal of Science Education, 11, 541–553. doi:10.1080/0950069890110506

Hibberd, R., Jones, A., & Morris, E. (2002). *The use of Concept Mapping as a Means to Promote and Assess Knowledge Acquisition*, (CALRG Report No. 202).

Hilbert, T. S., & Renkl, A. (2007). Concept mapping as a follow-up strategy to learning from texts: what characterizes good and poor mappers? *Learning and Instruction, 12*, 213–232.

Hill, L. H. (2004). *Concept mapping in a pharmacy communications course to encourage meaningful student learning*. Retrieved March 19, 2008, from http://www.ajpe.org/ aj6805/aj6805109.pdf

Hinkle, D. E., Wiersma, W., & Jurs, S. G. (2003). *Applied statistics for the behavioral sciences* (5th Ed.). New York: Houghton-Mifflen.

Hmelo-Silver, C. E. (2004). Problem-based learning: What and how do students learn? *Educational*

Psychology Review, 16, 235–266. doi:10.1023/B:EDPR.0000034022.16470.f3

Hmelo-Silver, C., Duncan, R., & Chinn, C. (2007). Scaffolding and Achievement in Problem-Based and Inquiry Learning. *Educational Psychologist, 42*(2), 99–107. Mahwah, NJ: Lawrence Erlbaum Associates, Inc. http://www.cogtech.usc.edu/publications/hmelo_ep07.pdf

Hobsbawn, E. (1996). *The Age of Extremes: A History of the World, 1914-1991*. New York: Vintage.

Hoffman, J. (2004). *Avaliação: mito e desafio: Uma perspectiva construtivista*. Porto Alegre, Brazil: Mediação.

Hoffman, J. (2004). *Avaliação mediadora: Uma prática em construção da pré-escola à universidade*. Porto Alegre, Brazil: Mediação.

Hoffman, J. (2004). *Avaliar para promover: As setas do caminho*. Porto Alegre, Brazil: Mediação.

Holbrook, J., & Rannikmae, M. (2007). The nature of science education for enhancing scientific literacy. International *Journal of Science Education, 28*, 1–16.

Holley, C. D., & Dansereau, D. F. (1984a). Networking: The technique and the empirical evidence. In C. D. Holley & D. F. Dansereau (Eds.), *Spatial learning strategies: Techniques, applications and related issues* (pp. 81-108). Orlando, FL: Academic Press.

Holley, C. D., & Dansereau, D. F. (1984b). *Spatial Learning Strategies. Techniques, applications, and related issues*. New York: Academic Press.

Holmes, B., Tangney, B., & Fitzgibbon, A. (2001). Communal Constructivism: Students constructing learning for as well as with others. In *Proceedings of the 12th International Conference of the Society for Information Technology & Teacher Education* (SITE 2001), Charlottesville, VA, Association for the Advancement of Computing in Education.

Holmevik, J. R., & Haynes, C. (1998). *High wired. On the design, use and theory of educational MOOs*. Ann Arbor, MI: University of Michigan Press.

Horton, P. B., Mcconney, A. A., Gallo, M., Woods, A. L., Senn, G. J., & Hamelin, D. (1993). An investigation of the effectiveness of concept mapping as an instructional tool. *Science Education, 77*, 95–111. doi:10.1002/sce.3730770107

Howell Richardson, C., & Preston, C. (2007). Multimodal concept mapping in teaching and learning: editorial: Fascinating cultural artefacts: multimodal concept mapping in teaching and learning. *Reflecting Education*. Retrieved from www.reflectingeducation.net/index.php/reflecting

Howell Richardson, C., & Preston, C. (Eds.). (2005). *Special Inaugural Edition: E-learning*. London: Reflecting Education.

Hung, D. W. L., & Nichani, M. (2001). Constructivism and e-learning: Balancing between the individual and social levels of cognition. *Educational Technology, 41*(2), 40–44.

Hyerle, D. (1996). *Visual tools for constructing knowledge*. Alexandria, VA: Association for Supervision and Curriculum Development.

Hyerle, D. (2000). *A field guide to using visual tools*. Alexandria, VA: ASC Publications.

Ifenthaler, D. (2006). *Diagnosis of the learning-dependent progression of mental models. Development of the SMD-Technology as a methodology for assessing individual models on relational, structural and semantic levels*. Freiburg, Germany: Universitäts-Dissertation.

Ifenthaler, D. (2007). *Relational, structural, and semantic analysis of graphical representations and concept maps*. Paper presented at the Annual Convention of the AECT, Anaheim, CA.

Ifenthaler, D., & Seel, N. M. (2005). The measurement of change: learning-dependent progression of mental models. *Technology, Instruction*. Cognition and Learning, 2(4), 317–336.

Ifenthaler, D., Masduki, I., & Seel, N. M. (2008). *Tracking the development of cognitive structures over time*. Paper presented at the AREA 2008, New York.

Infante-Malachias, M. E., & Correia, P. R. M. (2007). Problemas complejos en el mundo post-industrial. *Novedades Educativas, 203*, 29–33.

Inman-Andersen, L., & Ditson, L. (1998). Computer-based Concept Mapping: Promoting Meaningful Learning in Science for Students with Disabilities. *Information Technology and Disabilities,* (ITDVO5N1-2 Article 2, June).

Janetzko, D. (1996) *Knowledge Tracking. A Method to Analyze Cognitive Structures*. Freiburg, Germany: IIG-Berichte 2.

Järvelä, S., & Järvenoja, H. (2007). *Socially constructed self-regulated learning in collaborative learning groups*. Paper presented at invited Symposium organised by Järvela, S. & Hadwin, A. at Earli Conference, Budapest, Hungary.

Jewitt, C. (2003). *A multimodal framework for computer mediated learning: The ReShaping of curriculum knowledge and learning culture, language and communication.* London: Institute of Education, University of London.

Jo, I.-H. (2001). The effects of different concept mapping techniques on college-level writing. *Educational Technology International, 3*(1).

Johnson, D. W., & Johnson, R. T. (1989). *Leading the cooperative school.* Edina, MN: Interaction.

Johnson, D. W., & Johnson, R. T. (1991). *Learning together and alone. Cooperative, Competitive, and Individualistic Learning,* (4th Ed.). Englewood Cliffs, NJ: Allyn & Bacon.

Johnson, D. W., & Johnson, R. T. (2007). *Cooperative learning and moral education.* Retrieved March 10, 2008, from http://www.co-operation.org/pages/Volume%2022(1)%20Internet.doc

Johnson, D. W., Johnson, R. T., & Holubec, E. J. (1994). *The Nuts and Bolts of Cooperative Learning*. Edina, MN: Interaction Book Company.

Johnson, D. W., Maruyama, G., Johnson, R., Nelson, D., & Skon, L. (1981). Effects of cooperative, competitive, and individualistic goal structures on achievement: A meta-analysis. *Psychological Bulletin, 89*(1), 47–62. doi:10.1037/0033-2909.89.1.47

Johnson, D., & Johnson, R. (1989). *Cooperation and competition.* Edina, MN: Interaction Book Company.

Johnson, T. E., & O'Connor, D. L. (2008). Assessment of group learning processes. In M. D. Merrill, van Merriënboer, J., Spector, J. M., & Driscoll, M. (Ed.), *Handbook of research on educational communications and technology* (3 Ed., pp. 776-783). New York: Lawrence Erlbaum Associates.

Johnson, T. E., & O'Connor, D. L. (2008). Measuring team shared understanding using analysis constructed shared mental model methodology. *Performance Improvement Quarterly, 21*(3), 1–23.

Jolly, A. B. (1998). The effectiveness of learning with concept mapping on the science problem solving of sixth grade children (Doctoral dissertation, Indiana University of Pennsylvania, 1998). *Dissertation Abstracts International, 59,* 3356. *

Jonassen, D. H. (1992). Semantic networking as cognitive tools. In P. Kommers, D. H. Jonassen & J. T. Mayes (Eds.), *Cognitive Tools for Learning* (Vol. F81, pp. 19-21). Berlin: Springer Verlag.

Jonassen, D. H. (2000). *Computers as mindtools for schools: Engaging critical thinking* (2 Ed.). Upper Saddle River, NJ: Prentice Hall.

Jonassen, D. H., Beissner, K., & Yacci, M. A. (1993). *Structural knowledge: Techniques for representing, assessing, and acquiring structural knowledge.* Hillsdale, NJ: Lawrence Erlbaum Associates.

Jonassen, D. H., Reeves, T. C., Hong, N., Harvey, D., & Peters, K. (1997). Concept mapping as cognitive learning and assessment tools. Journal of Interactive Learning Research, 8(3/4), 289–308.

Jones, A., & Issroff, K. (2005). Learning technologies: affective and social issues in computer-supported collaborative learning. Computers & Education, 44(4), 395–408. doi:10.1016/j.compedu.2004.04.004

Jones, S. (2002). Social constructionism and the environment: through the quagmire. *Global Environmental Change, 12*, 247–251. doi:10.1016/S0959-3780(02)00062-6

Joritz-Nakagawa, J. (2004). *Spencer Kagan's cooperative learning structures*. Retrieved March 10, 2008, from http://www.jalt.org/pansig/PGL2/HTML/Nakagawa.htm

Kalikoski, D. C., & Satterfield, T. (2004). On crafting a fisheries co-management arrangement in the estuary of Patos Lagoon (Brazil): opportunities and challenges faced through implementation. *Marine Policy, 28*(6): 503–522. doi:10.1016/j.marpol.2003.12.001

Kapur, M., & Kinzer, C. (2007). The effect of problem type on interactional activity, inequity, and group performance in a synchronous computer-supported collaborative environment. *Educational Technology Research and Development, 55*(5), 439–459. doi:10.1007/s11423-007-9045-6

Karoly, L. A., & Panis, C. W. A. (2004). *The 21st century at work: Forces shaping the future workforce and workplace in the United States*. Santa Monica, CA: RAND Corporation.

Katz, L. G., & Chard, S. C. (1989). *Engaging children's minds: the project approach*. New York: Ablex Publishing Corporation.

Keates, S., & Clarkson, J. (2003). *Countering design exclusion: an introduction to inclusive design*. New York: Springer.

Keller, T., Tergan, S.-O., & Coffey, J. (2006). *Concept maps as a "knowledge and information awareness" tool for supporting collaborative problem solving in distributed groups*. Retrieved March 18, 2008, from http://cmc.ihmc.us/cmc2006Papers/cmc2006-p.125.pdf

Khamesan, A., & Hammond, N. (2004). *Synchronous collaborative concept mapping via ICT: Learning effectiveness and personal and interpersonal awareness*. Retrieved March 18, 2008, from http://cmc.ihmc.us/papers/cmc2004-212.pdf

Kharatmal, M., & Nagarjuna, G. (2006). A proposal to refine concept mapping for effective science learning. In A. J. Canas & J. D. Novak (Eds.), *Concept Maps: Theory, Methodology, Technology - Proceedings of the Second International Conference on Concept Mapping* (Vol. 1, pp. 151-158). San José, Costa Rica: Universidad.

Kim, B., Yang, C., & Tsai, I. (2005). Review of computer-mediated collaborative concept mapping: implications for future research. *Proceedings of the 2005 Conference on Computer Support for Collaborative Learning: Learning 2005: The next 10 years!* (pp. 291-295).

Kim, B., Yang, C.-c., & I-Chun, T. (2005). *Review of computer-mediated collaborative concept mapping: Implication for future research*. Paper presented at the 10th international conference on Computer-Supported Collaborative Learning (CSCL), Tapei, Taïwan, May 30 to June 4, 2005.

Kinchin, I. (2001). If concept mapping is so helpful to learning biology, why aren't we all doing it? *International Journal of Science Education, 23*(12), 1257–1269. doi:10.1080/09500690010025058

Kinchin, I., Hay, D., & Adams, A. (2000). How a qualitative approach to concept map analysis can be used to aid learning by illustrating patterns of conceptual development. *Educational Research 42*(1), pp.

Kirschner, P., Buckingham Shum, S., & Carr, C. (Orgs) (2003). *Visualizing Argumentation: Software Tools for Collaborative and Educational Sense-making*. London: Springer-Verlag.

Kirschner, P., Strijbos, J. W., & Kreijns, K. (2005). Designing integrated collaborative e-learning. In W. Jochems, J. van Merrienboer, & R. Koper (Eds.), *Integrated E-Learning: Implications for Pedagogy, Technology & Organization*. New York: RoutledgeFalmer.

Klein, J. (2002). Deforestation in the Madagascar Highlands – Established 'truth' and scientific uncertainty. GeoJournal, 56, 191–199. doi:10.1023/A:1025187422687

Klimoski, R., & Mohammed, S. (1994). Team mental model - construct or metaphor. Journal of Management, 20(2), 403–437. doi:10.1016/0149-2063(94)90021-3

Komis, V., Ergazaki, M., & Zogza, V. (2007). Comparing computer-supported dynamic modeling and 'paper & pencil' concept mapping tehnique in students' collaborative activity. *Computers & Education, 49*, 991–1017. doi:10.1016/j.compedu.2005.12.007

Kommers, P. A. M., Jonassen, D. H., & Mayes, J. T. (Eds.). (1992). *Cognitive tools for learning.* Berlin: Springer-Verlag.

Kozulin, A. (1994). *La psicología de Vygotsky.* Madrid: Alianza Editorial.

Kress, G., & Van Leeuwen, T. (2001). *Multimodal Discourse: The modes and media of contemporary communication.* Arnold, UK: Hodder Headline Group.

Kress, G., & Van Leeuwen, T. (2007). *Reading Images: The Grammar of Visual Design* (2nd Ed.). London: Routledge.

Laampere, M., Matsak, E., & Kippar, J. (2006). Integrating a concept mapping tool into a virtual learning environment: Pedagogical and technological challenges. In A. J. Cañas & J. D. Novak (Eds.), *Concept maps: Theory, methodology, technology: Proceedings of the Second International Conference on Concept Mapping,* (Vol. 1, pp. 280-287). San Jose, Costa Rica.

Laight, D. W. (2004). Attitudes to concept maps as a teaching/learning activity in undergraduate health professional education: influence of preferred learning style. Medical Teacher, 26(3), 229–233. doi:10.1080/0142159042000192064

Laister, J., & Kober, S. (2005). *Social aspects of collaborative learning in virtual learning environments.* Retrieved July, 2008 from http://comma.doc.ic.ac.uk/inverse/papers/patras/19.htm

Lajoie, S.P., & Derry, S.J. (1993). *Computers as cognitive tools.*

Lambiotte, J. G., & Dansereau, D. F. (1992). Effects of knowledge maps and prior knowledge on recall of science lecture content. *Journal of Experimental Education, 60*, 189–201.

Lambiotte, J. G., Dansereau, D. F., Cross, D. R., & Reynolds, S. B. (1989). Multirelational semantic maps. *Educational Psychology Review, 1*, 331–367. doi:10.1007/BF01320098

Lambiotte, J. G., Skaggs, L. P., & Dansereau, D. F. (1993). Learning from lectures: Effects of knowledge maps and cooperative review strategies. *Applied Cognitive Psychology, 7*, 483–497.* doi:10.1002/acp.2350070604

Lane, C. (2000). *Implementing multiple intelligences and learning styles in distributed learning/IMS projects.* Retrieved May 04, 2007, from http://www.tecweb.org/styles/ imslsindl.pdf

Langan-Fox, J., Anglim, J., & Wilson, J. R. (2004). Mental models, team mental models, and performance: Process, development, and future directions. *Human Factors and Ergonomics in Manufacturing, 14*(4), 331–352. doi:10.1002/hfm.20004

Langan-Fox, J., Code, S., & Langfield-Smith, K. (2000). Team mental models: Techniques, methods, and analytic approaches. *Human Factors, 42*(2), 242–271. doi:10.1518/001872000779656534

Lanzing, J. (1997). *What is concept mapping?* Retrieved January 17, 2008, from http://users.edte.utwente.nl/lanzing/cm_home.htm

Laurea. (2006). *Physiotherapy Education Curriculum.* Espoo, Finland: Laurea University of Applied Sciences.

Laurillard, D. (2002). *Rethinking University Teaching: A Framework for the Effective Use of Learning Technologies.* London: RoutledgeFalmer.

Lautrey, J., & Mazens, K. (2004). Is children's naive knowledge consistent? A comparison of the concepts of sound and heat. *Learning and Instruction, 14*, 399–423. doi:10.1016/j.learninstruc.2004.06.011

Lawes, S. (2003). What, when, how and why? Theory and foreign language teaching. *Language Learning Journal, 28*(1): 22–29. doi:10.1080/09571730385200171

Lazar, D. (2004). Selected issues in the philosophy of social sciences. In C. Seale, (ed.) *Researching society and culture.* London: Sage.

Leask, M., & Younie, S. (2001). Building On-Line Communities for Teachers: Ideas Emerging from Research. In M. Leask (Ed.), *Issues in Teaching Using ICT.* London, Routledge.

Leask, M., & Younie, S. (2002). Communal constructivist theory: ICT pedagogy & internationalisation of the curriculum. *Journal for IT for Teacher Education, 10*(1&2), 117–134.

Leite, C. L. K., Passos, M. O., Torres, P. L., & Alcantara, P. R. (2005). *A aprendizagem Colaborativa na Educação a Distância on-line.* 12° Congresso Internacional de Educação a Distância Florianópolis, Brazil. Retrieved February 20, 2007, from http://www.abed.org.br/congresso2005/por/pdf/171tcc3.pdf.

Leite, C., & Fernandes, P. (2002). *Avaliação das aprendizagens dos alunos.* Porto, Portugal: Asa.

Lemos, A. (2002). *Cibercultura: Tecnologia e vida social na cultura contemporânea.* Porto Alegre, Brazil: Sulina.

Lemos, A. (2002). *Cultura das redes: Ciberensaios para o século XXI.* Salvador, Brazil: EDUFBA.

Leontiev, A. (1974). The problem of activity in psychology. *Social Psychology, 13*(2), 4–33.

Levin, B., He, Y., & Robbins, H. (2006). Comparative analysis of preservice teachers' reflective thinking in synchronous vs asynchronous online case discussions. *Journal of Technology and Teacher Education, 14*(3), 439–460.

Levine, M. (2001). *A mind at a time.* New York: Simon & Schuster.

Lewin, K. (1948). *Resolving Social Conflicts.* New York: Harper and Row Publishers.

Lima, M., Koehler, M. J., & Spiro, R. J. (2004). Collaborative interactivity and integrated thinking in Brazilian business schools using cognitive flexibility hypertexts: The Panteon project. *J. Educational Computing Research, 31*(4), 371-406. Retrieved March 30, 2007, from http://mkoehler.educ.msu.edu/OtherPages/Koehler_Pubs/Lima_Koehler_Spiro_ 2004.pdf

Lipponen, L. (2002). *Exploring foundations for computer-supported collaborative learning.* Retrieved October 12, 2006 from http://newmedia.colorado.edu/cscl/31.html

Lipsey, M. W., & Wilson, D. B. (2001). *Practical meta-analysis.* Applied Social Research Methods Series (Vol. 49). Thousand Oaks, CA: SAGE Publications.

Llewellyn, D. (2005). *Teaching High School Science Through Inquiry: A Case Study Approach.* NSTA.

Long, N. (2001). *Development sociology: actor perspectives.* London: Routledge.

López Rupérez, F. (1990). Epistemología y didáctica de las ciencias: Un análisis de segundo orden. *Ensenañza de las Ciencias, 8*, 65–74.

Lowry, P. B., Curtis, A., & Lowry, M. R. (2004). Building a taxonomy and nomenclature of collaborative writing to improve interdisciplinary research and practice. *Journal of Business Communication, 41*(1), 66–99. doi:10.1177/0021943603259363

Luckie, D. B., Harrison, S. H., & Ebert-May, D. (2004). Introduction to C-TOOLS: Concept mapping tools for online learning. In A. J. Cañas, J. D. Novak, & F. M. González (Eds.), *Concept maps: Theory, methodology, technology. Proceedings of the First International Conference on Concept Mapping,* (Vol. 2, pp. 261-264). Pamplona, Spain.

Lüdke, M., & André, M. E. D. (1986). *A. Pesquisa em educação: abordagens qualitativas.* São Paulo, Brazil: EPU.

Maas, J. D., & Leauby, B. A. (2005). *Concept mapping: Exploring its value as a meaningful learning tool in accounting education.* Retrieved March 03, 2008, from http://gpae.bryant.edu/~gpae/vol2/04-038%20Concept%20Mapping.pdf

MacNeil, M. S. (2007). Concept mapping as a means of course evaluation. *The Journal of Nursing Education, 46*(5), 232–234.

Maedche, A., Pekar, V., & Staab, S. (2002) Ontology learning part one - on discovering taxonomic relations from the web. In *Proceedings of the Web Intelligence Conference*, (pp. 301-322). Berlin: Springer.

Maine, R. A., Cam, B., & Davis-Case, D. (1996). *Participatory Analysis, Monitoring and Evaluation for Fishing Communities: a manual* (FAO Fisheries Technical Paper. No. 364). Rome: FAO.

Malone, J., & Dekkers, J. (1984). The concept map as an aid to instruction in science and mathematics. *School Science and Mathematics*, *84*(3), 221–231.

Mancinelli, C., & Guaglione, G. (2004). *Le esperienze dei bambini: Indagare nella zucca - Le parole della scienza nella scuola dell'infanzia.* Retrieved 6 August 2004, from http://85.47.105.117/ces/index.htm

Mancinelli, C., Gentili, M., Priori, G., & Valitutti, G. (2004). *Concept maps in kindergarten.* Paper presented at the first international Conference on Concept Mapping, Pamplona, Spain.

Manrique, A. L. (2000) A afetividade explicitada. *Educação Matemática, 19.* Retrieved June 21, 2007, from http://anped.org.br/ analuciamanrique.rtf

Markham, K. M., Mintzes, J. J., & Jones, M. G. (1994). The concept map as a research and evaluation tool: Further evidence of validity. *Journal of Research in Science Teaching*, *31*(1), 91–101. doi:10.1002/tea.3660310109

Markow, P. G., & Lonning, R. A. (1998). Usefulness of concept maps in college chemistry laboratories: Students' perceptions and effects on achievement. *Journal of Research in Science Teaching*, *35*(9), 1015-1029. doi:10.1002/(SICI)1098-2736(199811)35:9<1015::AID-TEA4>3.0.CO;2-G

Marques, J. C. (2006). Aprendizagem Colaborativa: Atividades de Grupo como Núcleo e Uso do Computador como Contexto. In *Psicologia Argumento.* Curitiba, *24*,(44) 37-43.

Marriott, R. C. V. (2004). *Do LOLA—Laboratório on-line de aprendizagem ao LAPLI—Laboratório de Aprendizagem de Línguas: Uma proposta metodológica para o ensino semi-presencial em ambiente virtual.* Masters in Education, Pontifícia Universidade Católica do Paraná (PUCPR), Brazil.

Marriott, R. C. V., & Torres, P. (2008). Enhancing collaborative and meaningful learning using concept mapping. In A. Okada, S. Buckingham Shum, & T. Sherborne (Eds.), *Knowledge Cartography: Software Tools and Mapping Techniques.* London: Springer.

Marriott, R. C. V., & Torres, P. L. (2006). *Tecnologias educacionais e educação ambiental: uso de mapas conceituais no ensino e na aprendizagem.* Curitiba: FAEP.

Marshal, C., & Rossman, G. B. (1999). *Designing qualitative research.* London: Sage.

Martí, E., & Solé, I. (1997). Conseguir un trabajo en grupo eficaz. *Cuadernos de Pedagogía, 255*, 59–64.

Marzano, R. (2003). *What works in schools: Translating Research into Action.* Reston, VA: Association for Supervision and Curriculum Development.

Mason, J. (1996). *Qualitative researching.* London: Sage.

Mathieu, J. E., Heffner, T. S., Goodwin, G. F., Salas, E., & Cannon-Bowers, J. A. (2000). The influence of shared mental models on team process and performance. *The Journal of Applied Psychology*, *85*(2), 273–283. doi:10.1037/0021-9010.85.2.273

Matthews, H. (2002). *Hybrid courses and the future of teaching and learning at UC.* Retrieved April 30, 2003 from http://www.uctltc.org/news/2002/12/matthews.html

Matthews, R. S., Cooper, J. L., & Hawkes, P. (2004). Building bridges between cooperative and collaborative learning. Change, 27, 35–40.

Maturana, H. M., & Varela, F. (1998). *Tree of knowledge: the biological roots of human understanding.* Boston: Shambhala.

Mavers, D. (2004, June). *Observations on ImpaCT2 concept mapping.* MirandaNet seminar, Institute of Education, University of London, London.

Mavers, D. (2004). *Multimodal design: the semiotic resources of children's graphic representation.* Doctoral dissertation, Culture, Language and Communications, Institute of Education, University of London, London.

Mavers, D. (2007). Investigating how children make meaning in multimodal maps. Reflecting Education: Fascinating cultural artefacts: multimodal (concept) mapping in teaching and learning. *Reflecting Education* (pp. 24-28). Retrieved November 3, from www.reflectingeducation.net/index.php/reflecting

Mavers, M., Somekh, B., & Rosterick. (2002). Interpreting the externalised images of pupils' conceptions of ICT: methods for the analysis of concept maps. *Computers & Education, 38*, 187–207. doi:10.1016/S0360-1315(01)00074-4

Mayer, R. (2003). *Learning and Instruction.* Englewood Cliffs: NJ: Merrill Prentice Hall.

Mayer, R. E. (2002). Rote versus meaningful learning. *Theory into Practice, 41*(4), 226–232. doi:10.1207/s15430421tip4104_4

Mayer, R. E., & Anderson, R. B. (1991). Animations need narrations: An experimental test of a dual-coding hypothesis. *Journal of Educational Psychology, 83*, 484–490. doi:10.1037/0022-0663.83.4.484

McCagg, E. C., & Dansereau, D. F. (1991). A convergent paradigm for examining knowledge mapping as a learning strategy. The Journal of Educational Research, 84, 317–324.

McConnell, D. (2000). *Implementing Computer Supported Cooperative Learning* (2nd ed.) London: Kogan Page.

McCrudden, M. T., Schraw, G., Lehman, S., & Poliquin, A. (2007). The effect of causal diagrams on text learning. Contemporary Educational Psychology, 32(3), 367–388. doi:10.1016/j.cedpsych.2005.11.002

McDaniel, E., Roth, B., & Miller, M. (2005). *Concept mapping as a tool for curriculum design.* Retrieved March 19, 2008 from http://proceedings.informingscience.org/InSITE2005/142f49McDa.pdf

McTighe, J. (1992). Graphic Organizers: collaborative links to better thinking. In N. Davidson, & T. Worsham (Eds.)*Enhancing thinking through cooperative learning.* New York: Teachers College, Columbia University Press.

Medland, M. D. (2007). Tools for knowledge analysis, synthesis, and sharing. Journal of Science Education and Technology, 16(2), 119–153. doi:10.1007/s10956-006-9019-1

Meléndez, A., Pacual, I., & Rosado, L. (1989). La enseñanza de los modelos sobre la naturaleza de la luz. Ensenañza de las Ciencias, 7, 271–276.

Mendes, E., Mosley, N., & Counsell, S. (2001). *The cognitive flexibility theory: An approach for teaching hypermedia engineering.* Retrieved March 30, 2007, from http://www.cs.auckland.ac.nz/~emilia/publications/ITICSE2001.pdf

Mercer, N. (1996). The quality of talk in children's collaborative activity in the classroom. Learning and Instruction, 6, 359–377. doi:10.1016/S0959-4752(96)00021-7

Mercer, N., & Coll, C. (Eds.). (1994). Teaching, learning and interaction. In P. Del Rio, A. Álvarez & J.V. Wertsch (General Eds.) *Explorations in Socio Cultural Studies* (Vol. 3). Madrid: Fundación Infancia y Aprendizaje.

Mérida, R. (1999). Ayudando a organizar el pensamiento infantil: los mapas preconceptuales. *Kikirikí.* Cooperación Educativa, 62/63, 76–83.

Mérida, R. (2001). *Investigando en educación infantil: los mapas preconceptuales.* Córdoba, Spain: Servicio de Publicaciones de la Universidad de Córdoba.

Mérida, R. (2001-2002). Ayudando a organizar el pensamiento infantil: los mapas preconceptuales. *Kikiriki - Cooperación Educativa* (62-63), 76-83.

Mérida, R. (2002). Una nueva forma de trabajar en educación infantil: los mapas preconceptuales. *Cultura y Educación, 14*(1), 99–125. doi:10.1174/113564002317348147

Mérida, R. (2003). Los mapas preconceptuales como estrategias de aprendizaje cooperativo en educación infantil. *Revista de Educación, 331*, 421–441.

Merrill, M. D. (1994). *Principles of instructional design*. Englewood Cliffs, NJ: Educational Technology Publications.

Mezirow, J. (1991). *Transformative dimensions of adult learning*. San Francisco, CA: Jossey-Bass.

Michaelsen, L. K., Knight, A. B., & Fink, L. D. (Eds.). (2004). *Team-based learning: A transformative use of small groups in college teaching*. Sterling, VA: Stylus Publishing, LLC.

Michaelsen, L. K., Parmelee, D. X., McMahon, K. K., & Levine, R. E. (2007). *Team-Based Learning for Health Professions Education*. Herndon, VA: Stylus Press.

Michavila, F. (2003, February). Cómo y en qué se deben formar los profesores universitarios. Communication presented at the *Congreso de la Red Estatal de Docencia Universitaria: Hacia una docencia de calidad*. Castellón, Spain: Universitat Jaume I.

Milam, J. H., Santo, S. A., & Heaton, L. A. (n.d.). *Concept maps for web-based applications: ERIC technical report*. Retrieved February 07, 2008, from http://www.highered.org/docs/milam-conceptmaps.PDF

Milrad, M., Spector, J. M., & Davidsen, P. I. (2002). Model facilitated learning. In S. Naidu (Ed.), *Learning and Teaching with Technology: Principles and Practices* (pp. 13-27). London: Kogan Page.

Mintzes, J. J., Wandersee, J. H., & Novak, J. D. (1998). *Teaching science for understanding: a human constructivist view*. San Diego, CA: Academic Press.

Mishra, P., & Koehler, M. (2006). Technological pedagogical content knowledge: A framework for teacher knowledge. *Teachers College Record*, *108*(6), 1017–1054. doi:10.1111/j.1467-9620.2006.00684.x

Molina, A., Ontoria, A., & Gómez, J. P. R. (2006). *Potencializar a capacidade de aprender e pensar*. São Paulo, Brazil: Madras.

Monereo, C., & Pérez Cabaní, M. L. (1996). La incidencia de la toma de apuntes sobre el aprendizaje significativo. Un estudio en enseñanza superior. *Infancia y Aprendizaje*, *19*(1), 65–86. doi:10.1174/02103709660560555

Monereo, C., & Pozo, J. I. (2003). *La universidad ante la nueva cultura educativa. Enseñar y aprender para la autonomía*. Madrid: Síntesis.

Monereo, C., Clariana, M., Castelló, M., Palma, M., & Pérez Cabaní, M. L. (1994). *Estrategias de enseñanza y aprendizaje. Formación del profesorado y aplicación en la escuela*. Barcelona: Graó.

Moni, R. W., & Moni, K. B. (2008). *Student perceptions and use of an assessment rubric for a group concept map in physiology*. Retrieved March 19, 2008, from http://advan. physiology.org/cgi/reprint/32/1/47

Montanero, M., & Montanero, M. (1995). *Didáctica del momento angular de una partícula*. Badajoz, Spain: ICE.

Montanero, M., Lucero, M., & Méndez, J. M. (2008). Enseñanza de contenidos históricos mediante diagramas multicausales. *Revista Española de Pedagogía*, *239*(1), 27–48.

Montanero, M., Pérez, A. L., & Suero, M. I. (1995). Survey of students and teacher conceptions of action-reaction dynamics: Implicit alternative theories are manifest in the consistency of incorrect responses. *Physics Education*, *30*, 277–283. doi:10.1088/0031-9120/30/5/004

Montanero, M., Pérez, A. L., Suero, M. I., & Montanero, M. (2001). Cambio conceptual y enseñanza de la Física: Aplicaciones en el marco de la Teoría de la Elaboración. *Revista de Educación*, *326*, 311–332.

Moran, J. M. Masetto, Marcos T., & Behrens, M. A. (2003) *Novas Tecnologias e Mediação Pedagógica*. Campinas, Spain: Papirus, 7th edição.

Moreira, M. A. (1997). *Mapas conceituais e aprendizagem significativa*. Retrieved September 12, 2006, from http://if.ufrgs.br/~moreira/mapasport.pdf

Moreira, M. A. (1999). *Teorias de aprendizagem*. São Paulo, Brazil: EPU.

Moreira, M. A., & Masini, E. F. S. (1982). *A aprendizagem significativa: A teoria de David Ausubel*. São Paulo, Brazil: Moraes.

Morin, E. (1999). *Ciência com consciência.* Rio de Janeiro, Brazil: Bertrand Brasil.

Morin, E. (2005). *Introducao ao pensamento complexo.* Porto Alegre, Brazil: Sulina.

Murphy, P. (2002). *The hybrid strategy: Blending face-to-face with virtual instruction to improve large lecture courses.* Retrieved April 30, 2003 from http://www.uctltc.org/news/2002/12/feature.html

Murray, J. T., & Murray, M. J. (1996). *Computing Crisis: A Millennium Date Conversion Plan.* New York: McGraw-Hill.

Myers, I. B., McCaulley, M. H., Quenk, N. L., & Hammer, A. L. (Eds.). (1998). *MBTI manual: A guide to the development and use of the Myers-Briggs Type Indicator* (3rd Ed.). Palo Alto, CA: Consulting Psychologists Press, Inc.

Nanjappa, A., & Grant, M. (2003). Constructing on constructivism: The role of technology. *Electronic Journal for the Integration of Technology in Education, 2*(1). Retrieved March 15, 2008, from http://ejite.isu.edu/Volume2No1/nanjappa.htm

National Academy of Sciences. (2006). *Learning to think spatially.* Washington, D.C.: The National Academies Press.

National Reading Panel. (2000). *Teaching children to read: An evidence-based assessment of the scientific literature on reading and its implications for reading instruction.* http://www.nichd.nih.gov/publications/nrp/smallbook.cfm

Nelson, D. B. (2007). *Academic concept mapping (ACM): A critical thinking tool in academic advising for improving academic performance in college freshmen.* Retrieved February 07, 2008, from http://etd.lsu.edu/docs/available/etd-06262007-190651/ unrestricted/DorothyNelsonDissertation.pdf

Nesbit, J. C., & Adesope, O. O. (2006). Learning with concept and knowledge maps: A meta-analysis. Review of Educational Research, 76(3), 413–448. doi:10.3102/00346543076003413

Nesbit, J., & Adesope, O. (2006). Learning with concept and knowledge maps: A meta-analysis. *Review of Educational Research, 76*(3), 413–448. doi:10.3102/00346543076003413

Newell, K. M. (2003). Whose values, which ethics? Science education and the civil society. *Australian Science Teachers Journal, 49*(2), 6–11.

Newman, L. (2005). Descartes' Epistemology. *The Stanford Encyclopedia of Philosophy (Summer 2005 Edition),* E. N. Zalta (Ed.). Retrieved September 23, 2008, from http://plato.stanford.edu/archives/sum2005/entries/descartes-epistemology/

Ng, W. (2008). Virtual teamwork: Students learning about ethics in an online environment. *Journal of Research in Science & Technological Education, 26*(1), 13–29. doi:10.1080/02635140701847421

Ng, W., & Nicholas, H. (2007). Conceptualising the use of online technologies for gifted secondary students. *Journal of Gifted Education, 29*(3), 190–196.

Ng, W., Nicholas, H., & Williams, A. (in press). How experiences in schools influence pre-service teachers' beliefs about teaching efficacy and good teaching. *Teaching and Teacher Education Journal.*

Nicholls, S. M. (2001). *Aspectos pedagógicos e metodológicos do ensino de inglês.* Maceió, Brazil: EDUFAL.

Nicoll, G., Francisco, J., & Nakhleh, M. (2001). An investigation of the value of using concept maps in general chemistry. *Journal of Chemical Education, 78,* 1111–1117.

Norman, D. A. (1991). Cognitive artefacts. In J. M. Carroll (Ed.), *Designing interaction: Psychology at the human-computer interface* (pp. 17-38). Cambridge, UK: Cambridge University Press.

Norman, D. A. (1999). *The invisible computer: why good products can fail, the personal computer is so complex, and information appliances are the solution.* Cambridge, MA: MIT Press.

Norman, D. A., & Rumelhart, D. E. (1981). The LNR approach to human information processing. *Cognition, 10*, 235–240. doi:10.1016/0010-0277(81)90051-2

Novak, A. D. (1994). The effects of cooperative learning with concept mapping in a preservice teacher education course, (Doctoral dissertation, Indiana University). *Dissertation Abstracts International, 56,* 895.

Novak, J. D. (1985). Metalearning and Metaknowledge strategies to help students learn. In L.T. West, & A.L. Pines (Eds.) *Cognitive Structure and Conceptual Change* (pp. 189-209) New York: Academic Press.

Novak, J. D. (1990). Concept mapping: a useful tool for science education. *Journal of Research in Science Teaching, 27*(10), 937–949.

Novak, J. D. (1990). Concept maps and Vee diagrams: two metacognitive tools to facilitate meaningful learning. *Instructional Science, 19,* 29–52. doi:10.1007/BF00377984

Novak, J. D. (1998). *Learning, creating and using knowledge: Concept maps as facilitative tools in schools and corporations*. Mahwah, NJ: Erlbaum.

Novak, J. D. (2002). Meaningful learning: The essential factor for conceptual change in limited or inappropriate propositional hierarchies leading to empowerment of learners. *Science Education, 86*, 548–571. doi:10.1002/sce.10032

Novak, J. D. (2004). A Science Education Research Program that led to the development of the Concept Mapping tool and a New Model for Education. In A. J. Cañas, J. D. Novak, & F. M. Gonzáles, (Eds.). *Concept Maps: Theory, Methodology, Technology. Proceedings of the First International Conference on Concept Mapping* (Vol. 1). Pamplona, Spain: Universidad Pública de Navarra.

Novak, J. D., & Cañas, A. J. (2004). Building on New Constructivist Ideas and CmapTools to Create a New Model for Education. In A. J. Cañas, J. D. Novak & F. M. Gonzáles (Eds.), *Concept Maps: Theory, Methodology, Technology, Proceedings of the First International Conference on Concept Mapping* (Vol. I, pp. 469-476). Pamplona, Spain: Universidad Pública de Navarra.

Novak, J. D., & Cañas, A. J. (2006). The Origins of the Concept Mapping Tool and the Continuing Evolution of the Tool. *Information Visualization Journal, 5*(3), 175-184. Retrieved April 2, 2008, from http://cmap.ihmc.us/Publications/ResearchPapers/OriginsOfConceptMappingTool.pdf

Novak, J. D., & Canãs, A. J. (2008). *The theory underlying concept maps and how to construct and use them.* Retrieved March 17, 2008, from http://cmap.ihmc.us/publications/ResearchPapers/TheoryUnderlyingConceptMaps.pdf

Novak, J. D., & Gowin, D. B. (1984). *Learning how to learn*. Cambridge, UK: Cambridge University Press.

Novak, J. D., & Musonda, D. (1991). A twelve-year longitudinal study of science concept learning. *American Educational Research Journal, 28*, 117–153.

Novak, J. D., & Symington, D. (1982). Concept mapping: a useful tool for science education. *Journal of Research in Science Teaching, 27*(10), 937–949.

Novak, J. D., Gowin, B., & Johansen, G. T. (1983). The use of concept mapping and knowledge Vee mapping with junior high school science students. *Science Education, 67*(5), 625–645. doi:10.1002/sce.3730670511

Novak, J. D.; & Gowin, D. B. (1988).*Teoría y Práctica de la Educación.*

O'Connor, D. L., Johnson, T. E., & Khalil, M. K. (2004, September). *Measuring team cognition: Concept mapping elicitation as a means of constructing team shared mental models in an applied setting.* Presented at the First International Conference on Concept Mapping, Pamplona, Spain.

O'Donnell, A. M., Dansereau, D. F., & Hall, R. H. (2002). Knowledge maps as scaffolds for cognitive processing. *Educational Psychology Review, 14*(1), 71–86. doi:10.1023/A:1013132527007

O'Donnell, A. M., Hmelo-Silver, C., & Erkens, G. (2006). *Collaborative learning, reasoning, and technology.* Mahwah, NJ: Lawrence Erlbaum.

Officeport (2006). *A version of Bloom's Taxonomy.* Retrieved from http://www.officeport.com/edu/blooms.htm

Okada, A. (2004). Cartografia cognitiva: novos desafios e possibilidades. In A. Okada, E. Santos, and F. Almeida, (Eds.), *Curso online: Uso de software na pesquisa qualitativa.* PUS-SP, COGEAE. Retrieved March 2004 from http://cogeae.dialdata.com.br/soft/520/1/1/modulos/texto2.php

Okada, A. (2004). *Software Tools and Mapping Techniques for Academic Research - Online Course.* Retrieved on January, 17, 2007 from <http://cursosonline.cogeae.pucsp.br/index2.php?wcurso=EDP>

Okada, A. (2005). The Collective Building of Knowledge in Collaborative Learning Environments. In T. Roberts, (Org.) *Computer-Supported Collaborative Learning in Higher Education.* (Vol. 1, pp. 70-99). Hershey, PA: Idea Group Publishing.

Okada, A. (2006). *Cartography for Inquiry: epistemological and communicational interfaces to map knowledge in academic projects.* Doctoral Thesis, São Paulo PUC-SP University & The Open University – Knowledge Media Institute OU-UK.

Okada, A., & Buckingham Shum, S. (2006). *Knowledge Mapping With Compendium in Academic Research and Online Education.* 22nd ICDE World Conference [www.icde22.org.br], Rio de Janeiro, Brazil.

Okada, A., & Connolly, T. (2008). Designing Open Educational Resources through Knowledge Maps to Enhance Meaningful Learning. *Common Ground: The International Journal of Learning, 15*(7), 209-220.

Okada, A., & Zeiliger (2003). The building of knowledge through virtual maps in collaborative learning environments. In *Proceedings of EdMedia Hawaii USA*, (pp. 1625-1628).

Okada, A., Buckingham Shum, S., & Sherborne, T. (2008, in press). *Knowledge Cartography.* London: Springer.

Okada, A., Santos, E., & Okada, S. (2003). *Mapeando informação, trilhando e construindo redes de significados.* Retrieved February 20, 2007, from http://people.kmi.open.ac.uk/alc/journals

Okebukola, P. A. (1990). Attaining meaningful learning of concepts in genetics and ecology: An examination of the potency of the concept-mapping technique. *Journal of Research in Science Teaching, 27*, 493–504.* doi:10.1002/tea.3660270508

Okebukola, P. A. (1992). Concept Mapping with a cooperative learning flavor. The American Biology Teacher, 54, 218–221.*

Okebukola, P. A., & Jegede, O. J. (1988). Cognitive preference and learning-mode as determinants of meaningful learning through concept mapping. *Science Education, 72*, 489–500. doi:10.1002/sce.3730720408

Okebukola, P. A., & Jegede, O. J. (1989). Students' anxiety towards and perception of difficulty of some biological concepts under the concept mapping heuristic. *Research in Science & Technological Education, 7*, 85–92. doi:10.1080/0263514890070109

Olson, D. R. (1985). Computers as tools of the intellect. Educational Researcher, 14(5), 5–8. doi:10.2307/1174200

Ornstein, A., & Sinatra, R. (2005). *K-8 instructional methods: A literacy perspective.* Boston: Pearson Education.

Paas, F., Renkl, A., & Sweller, J. (2003). Cognitive load theory and instructional design: Recent developments. Educational Psychologist, 38, 1–4. doi:10.1207/S15326985EP3801_1

Paivio, A. (1986). *Mental representations: A dual coding approach.* Oxford, UK: Oxford University Press.

Paivio, A. (1991). Dual coding theory: Retrospect and current status. Canadian Journal of Psychology, 45, 255–287. doi:10.1037/h0084295

Panitz, T. (1995). Collaborative versus cooperative learning- a comparison two concepts which will help us understand the underlying nature of interactive learning. *National Teaching and Learning Forum, 6*, 24-35. Retrieved November 16, 2007, from http://home.capecod.net/~tpanitz/tedsarticles/coopdefinition.htm

Pankratius, W. J. (1987). Building an organized knowledge base: Concept mapping and achievement in secondary school physics (Doctoral dissertation, Georgia State University College of Education). *Dissertation Abstracts International, 49*, 474.*

Pankratius, W. J. (1990). Building an organized knowledge base: Concept mapping and achievement in secondary school physics. *Journal of Research in Science Teaching*, 27, 315–333. doi:10.1002/tea.3660270404

Paquette, G. (2002). *Modélisation des connaissances et des compétences*. Sainte-Foy, Canada: Presses de l'Université du Québec.

Paquette, G. (2003). *Instructional Engineering in Networked Environments*. San Francisco, CA: Pfeiffer/Wiley.

Parrilla, A., & Gallego, C. (1999). Cómo pueden los profesores ayudarse entre si: diseño del trabajo comparativo. *Revista de Educación Especial, 25* (55-56).

Patterson, M. E., Dansereau, D. F., & Newbern, D. (1992). Effects of communication aids and strategies on cooperative teaching. *Journal of Educational Psychology, 84*, 453–461.* doi:10.1037/0022-0663.84.4.453

Patterson, M. E., Dansereau, D. F., & Wiegmann, D. A. (1993). Receiving information during a cooperative episode: Effects of communication aids and verbal ability. *Learning and Individual Differences, 5*, 1–11.* doi:10.1016/1041-6080(93)90022-K

Paul, R. W. (1992). Critical Thinking: what, why and how. In C.A. Barnes (Ed.) *Critical thinking: educational imperative*. San Francisco, CA: Jossey-Bass.

Pea, R. D. (1993). Practices of distributed intelligence and designs for education. In G. Salomon (Ed.), *Distributed cognitions: Psychological and educational considerations* (pp. 47-87). Cambridge, UK: Cambridge University Press.

Pelley, J. W. (2005). *Concept Mapping: A Tool For Teaching Integrative Thinking*. Presented in the International Association of Medical Science Educators webcast series: Learner Centered Education. Retrieved from http://www.iamse.org/development/2005/was_2005_spring.htm.

Pelley, J. W. (2006, September). *Effect of concept mapping on Myers-Briggs personality types*. Presented at the Second International Conference on Concept Mapping, San Jose, Costa Rica.

Pelley, J. W., & Dalley, B. K. (1997). *SuccessTypes for medical students: A program for improving academic performance*. Lubbock, TX: Texas Tech University Division of Extended Learning.

Pelley, J.W., & Steele, M. (2002, April). *Teaching tools for medical knowledge competency: Concept mapping*. Presented at meeting of the Society for Academic Continuing Medical Education, Charleston, SC.

Pérez A.L., Suero, M.I., & Pardo, P.J. (2008) Utilización de Cmaps para mejorar los conocimientos relativos a la refracción de la Luz mediante su "Reconstrucción Colaborativa". *Óptica pura y aplicada, 41*, 17-23.

Pérez Cabaní, M. L. (1995). Los mapas conceptuales. Cuadernos de Pedagogía, 237, 16–21.

Pérez Cabaní, M. L. (1996). Els mapes conceptuals. Anàlisi de les condicions de la seva utilització. Perspectiva Escolar, 206, 11–19.

Pérez Cabaní, M. L. (2001). El aprendizaje escolar desde el punto de vista del alumno: los enfoques de aprendizaje. In C. Coll, J. Palacios, & A. Marchesi (Coord.) *Desarrollo psicológico y educación 2. Psicología de la educación escolar, (*pp. 285-308*)*. Madrid: Alianza Editorial.

Pérez Cabaní, M. L. (2007). Aprendiendo a ser profesionales reflexivos: Aprender para enseñar a aprender. In *Enseñar a pensar: Sentando las bases para aprender a lo largo de la vida* (pp.113-135). Madrid: Ministerio de Educación y Ciencia.

Pérez Cabaní, M. L. (Coord.) (1997). *La enseñanza y el aprendizaje de estrategias desde el curriculum*. Barcelona: Horsori.

Perez Cabaní, M. L., & Carretero, R. (2003). La promoción de estudiantes estratégicos a través del proceso de evaluación que proponen los profesores universitarios. In C. Monereo, & J.I. Pozo, (Eds.) *La universidad ante la nueva cultura educativa. Enseñar y aprender para la autonomía* (pp. 173-190). Madrid: Síntesis.

Pérez Cabaní, M. L., & Juandó Bosch, J. (2001). *Gestionar la información para que se comprenda.* Barcelona: Edebé.

Pérez, A. L., Suero, M. I., Montanero, M., & Montanero, M. (1999). Mapas de experto tridimensionales: Aplicaciones al diseño de secuencias instruccionales de Física. In M.E.C., *Premios Nacionales de Investigación Educativa 1998* (pp. 93-116). Salamanca, Spain: CIDE.

Pérez-Cabani, M. L., Falgas, M., Nadal, A., & Valenti, M. (1992). Els Mapes Conceptuales: una estrategia d'aprenentatge al parvulari. *Revista Guix.*

Perret-Clermont, A. N., & Schubauer-Leoni, M. L. (1981). Conflict and cooperation as opportunities for learning. In W. P. Robinson (Ed.), *Communication in development* (pp. 211-243). London: Academic Press.

Perry, N. E., & Winne, P. H. (2006). Learning from learning kits: gStudy traces of students' self-regulated engagements with computerized content. Educational Psychology Review, 18, 211–228. doi:10.1007/s10648-006-9014-3

Phillips, D. (Ed.). (2000). *Constructivism in education: Opinions and second opinions on controversial issues.* Chicago: University of Chicago Press.

Phillips, D. C. (1995). The good, the bad, and the ugly: The many faces of constructivism. Educational Researcher, 24(7), 5–12.

Piernot, P. P., Felciano, R. M., Stancel, R., Marsh, J., & Yvon, M. (1995, May 07 - 11, 1995). *Designing the PenPal: blending hardware and software in a user-interface for children.* Paper presented at the SIGCHI conference on Human Factors in Computing Systems, Denver, CO.

Piirainen, A., & Sarekoski, I. (2008). Family care giver's CaringTv concept. In Arja Piirainen & Irma Sarekoski (Eds), Customer-driven CaringTV® concept for elderly family care givers living at home. Final report of the Coping at Home research. *Laurea Publication A 62.* 32- 67

Piirainen, A., Julin, M., & Immonen-Orpana, P. (2007). Curriculum process in Physiotherapy. In Kallioinen, O. (Ed.) The Competence-Based curriculum at Laurea. *Laurea Publications B 25,* 147- 155.

Pilkington, R. (2007). Collaborative and problem-based learning [Unpublished notes]. *Learning and ICT, Session 8.*

Pintrich, P. R. (2004). A conceptual framework for assessing motivation and self-regulated learning in college students. Journal of Educational Psychology Review, 16(4), 385–407.

Pintrich, P. R., & Sinatra, G. M. (2003). Future Directions for Theory and Research on Intentional Conceptual Change. In G. M. Sinatra & P. R. Pintrich (Eds.) *Intentional Conceptual Change* (pp. 429-443). Mahwah, NJ: Lawrence Erlbaum Associates.

Pirnay-Dummer, P. & Nußbickel, M. (in press). *New Ways of Finding Out What is Needed to Know. Using the latest Tools for Knowledge and Mental Model Assessment in the Processes of Needs Assessment.*

Pirnay-Dummer, P. (2006). *Expertise und Modellbildung: MITOCAR.* Freiburg, Germany: Universitäts-Dissertation.

Pirnay-Dummer, P. (2007). *Model Inspection Trace of Concepts and Relations. A Heuristic Approach to Language-Oriented Model Assessment.* Paper presented at the AERA 2007, Division C, TICL SIG, April 2007, Chicago.

Plotnick, E. (1997). *Concept Mapping: A graphical system for understanding the relationship between concepts.* (ERIC document, EDO-IR-97-05).

Posner, G. J., Strike, K. A., Hewson, P. W., & Gertzog, W. A. (1982). Accomodation of scientific conception: Toward a theory of conceptual change. Science Education, 66, 211–227. doi:10.1002/sce.3730660207

Pozo, J. I. (1999). Más allá del cambio conceptual: el aprendizaje de las ciencias como cambio representacional. Ensenañza de las Ciencias, 17, 513–520.

Prater, D. L., & Terry, C. A. (1988). Effects of mapping strategies on reading comprehension and writing performance. Reading Psychology, 9, 101–120.* doi:10.1080/0270271880090203

Pressley, M., Wolosshyn, V., Lysynchuk, M. L., Martin, V., Wood, E., & Willoughby, T. (1990). A primer of research on cognitive strategy instruction: The important issues and how to address them. Educational Psychology Review, 2(1), 1–58. doi:10.1007/BF01323528

Prestera, G., & Moller, L. (2001). Facilitating Asynchronous Distance Learning: Exploiting opportunities for knowledge building in asynchronous distance learning environments. *Proceedings of the Mid-South Instructional Technology Conference, Middle Tennessee State University.* Retrieved January 24, 2008 from http://www.mtsu.edu/~itconf/proceed01/3.html

Preston, C. (2004). *Learning to use ICT in Classrooms: teachers' and trainers' perspectives: An evaluation of the English NOF ICT teacher training programme 1999-2003.* London: MirandaNet and the Teacher Training Agency. Retrieved from www.mirandanet.ac.uk/tta.

Preston, C. (2006). Etopia: Mapping interactive learning in an e-community context. *MirandaNet Braided Learning e-journal.* Retrieved from www.mirandanet.ac.uk

Preston, C. (2007). Braided Learning: Promoting active professionals in education. In C. Haythornthwaite (Ed.), *New International Theories and Models Of and For Online Learning.* Chicago: First Monday.

Preston, C. (2007). Social networking between professionals: What is the point? In *Self-regulated Learning in Technology Enhanced Learning Environments: Individual Learning Individual Learning and Communities of Learners.* Amsterdam: Springer Verlag.

Preston, C. (2008). Braided Learning: an emerging practice observed in e-communities of practice. International Journal of Web Based Communities, 4(2), 220-243. doi:10.1504/IJWBC.2008.017674

Preston, C., & Cuthell, J. (2005). Teaching in ICT-rich environments: Using e-learning to create a knowledge base for 21st century teachers. In M. Leask & N. Pachler, (Eds.), *Learning to teach using ICT in the Secondary School,* (2nd Ed.). London: Routledge.

Preston, C., & Cuthell, J. (2007). Mapping Inspiration. *Braided Learning e-Journal.* Retrieved from www.mirandanet.ac.uk/ejournal.

Preston, C., & Danby, M. (2004). *An evaluation of ICT skills training for educators*: An internal report, Business Link Hereford.

Preston, C., & Holmes, B. (2002). Capturing the Online Knowledge, Building of Educator: ICTS, Authorship and Living Design. ITTE Conference, Dublin, Ireland, ITTE

Pudelko, B. (2006). *Étude microgénétique des médiations épistémiques d'un outil informatisé de représentation graphique des connaissances au cours d'une activité de compréhension de texte: Propositions pour une approche instrumentale étendue des médiations des outils cognitifs dans l'apprentissage.* Doctoral thesis, Université Paris 8. Paris: Bibliothèque de Paris.

Punch, K. (2005). *Introduction to social research: quantitative and qualitative approaches.* London: Sage.

Qin, Z., Johnson, D. W., & Johnson, R. T. (1995). Cooperative versus competitive efforts and problem solving. *Review of Educational Research, 65*(2), 129–143.

Quillian, M. R. (1967). Word concepts: A theory and simulation of some basic semantic capabilities. *Behavioral Science, 12,* 410–430. doi:10.1002/bs.3830120511

Rábago, A. R. (2004). La construction de mapas conceptuales en educacion a distancia. In A. J. Cañas, J. D. Novak, & F. M. González (Eds.), *Concept maps: Theory, methodology, technology. Proceedings of the First International Conference on Concept Mapping,* (Vol. 2, pp. 309-312). Pamplona, Spain.

Raij, K. (2007). Learning by Developing. *Laurea Publications A, 58.*

Raimes, A. (2002). Ten steps in planning a writing course and training teachers of writing. In J. C. Richards & W. A. Renandya (Eds.), *Methodology in language teaching: An anthology of current practice* (pp. 306-314). Cambridge, UK: Cambridge University Press.

Ralston, J., & Cook, D. (2007). Collaboration, ICT and mind mapping. *Reflecting Education,* (pp. 61-73). Retrieved November 3 from www.reflectingeducation.net/index.php/reflecting

Ramalingan, B., & Jones, H. (2008). *Exploring the science of complexity: Ideas and implications for development and humanitarian efforts* (Working Paper 285). London: Overseas Development Institute.

Rauhala, P. (2007). Foreword. In O. Kallioinen, (Ed.) The Competence-Based curriculum at Laurea. *Laurea Publications B* 25, 5.

Rauhala, P. (2007). The Competence –Based Core Curriculum as Laurea's Choice. In O. Kallioinen, (Ed.) The Competence-Based curriculum at Laurea. *Laurea Publications B* 25, 11-12.

Reader, W. R., & Hammond, N. (1994). A comparison of structured and unstructured knowledge mapping tools in psychology teaching. In *Proceedings of CiP 94*. New York.

Reigeluth, C. M. (1999). *Instructional-design Theories and Models*. (pp. 284-286). Mahwah, NJ: Lawrence Erlbaum Associates.

Reigeluth, C. M., & Stein, F. S. (1983). The Elaboration Theory of Instruction. In C.M. Reigeluth (Ed.), *Instructional design theories and models: An overview of their current status* (pp. 335-381). Hillsdale, NJ: Erlbaum.

Reigeluth, C. M., & Stein, F. S. (1987). Lesson blueprints based on the Elaboration Theory of Instruction, In C.M. Reigeluth (Ed.), *Instructional theories in action. Lesson illustrating selected theories and models* (pp. 245-288). Hillsdale, NJ: Erlbaum.

Reigeluth, C. M., Merril, M. D., & Bunderson, C. V. (1978). The structure of subject matteer content and its instructional desing implications. Instructional Science, 7, 107–126. doi:10.1007/BF00121929

Resta, P. (2002). *Information and communications technologies in teacher education: A planning guide*. Paris: UNESCO. Retrieved from http://unescobkk.org/education/ICT or http://unesdoc.unesco.org/images/0012/001295/129533e.pdf

Rewey, K. L., Dansereau, D. F., Skaggs, L. P., Hall, R. H., & Pitre, U. (1989). Effects of scripted cooperation and knowledge maps on the processing of technical material. Journal of Educational Psychology, 81, 604–609.* doi:10.1037/0022-0663.81.4.604

Rewey, K., Dansereau, D., Dees, S., Skaggs, L., & Pitre, U. (1992). Scripted cooperation and knowledge map supplements: Effects on the recall of biological and statistical information. Journal of Experimental Education, 60, 93–107.*

Reynolds, S. B., Patterson, M. E., Skaggs, L. P., & Dansereau, D. F. (1991). Knowledge hypermap and cooperative learning. Computers & Education, 16, 167–173.* doi:10.1016/0360-1315(91)90023-K

Ribas, M. (2000). *Innovación educativa. Teoría, procesos y estrategias*. Madrid: Síntesis.

Rice, D. C., Ryan, J. M., & Samson, S. M. (1998). Using concept maps to assess student learning in the science classroom: Must different methods compete? *Journal of Research in Science Teaching, 35,* 1103–1127. doi:10.1002/(SICI)1098-2736(199812)35:10<1103::AID-TEA4>3.0.CO;2-P

Ricoeur, P. (1974). *The Conflict of Interpretations: Essays in Hermeneutics*. Chicago: Northwestern University Press.

Riley, N. (2005). Methods for evaluating critical learning using online discussion forums. *Technology, Pedagogy and Education, 15*(1), 61–76.

Riley, N. (2007). Investigating the impact of ICT-based multimodal mapping in developing effective learning dialogues. *Reflecting Education,* 74-97. Retrieved November 3 from www.reflectingeducation.net/index.php/reflecting

Riley, N., & Alhberg, M. (2004). Investigating the use of ICT (information and communication technologies)-based concept mapping techniques on creativity in literacy tasks. Journal of Computer Assisted Learning, 20, 244–256. doi:10.1111/j.1365-2729.2004.00090.x

Rodrigo, M. J., & Correa, N. (1999). Teorías implícitas, modelos mentales y cambio educativo. In I. Pozo & C. Monereo (Eds.), *El aprendizaje estratégico* (pp. 75-86). Madrid: Santillana.

Rodriguez, A. J. (2006). *Using concept maps as a tool for curriculum planning and student assessment in culturally diverse schools*. Retrieved March 19, 2008, from http://cmc.ihmc.us/cmc2006Papers/cmc2006-p166.pdf

Rojas-Drummond, S., & Anzures-Tapia, A. (2007). Oracy, Literacy, Conceptual maps and ICT as mediators of the social construction of knowledge amongst peers. *Reflecting Education*. Retrieved from www.reflectingeducation.net/index.php/reflecting

Romizowski, A. J. (1999). Designing instructional systems: Decision making in course planning and curriculum design. Stylus Pub, LLC.

Roschelle, J. (1994). Designing for cognitive communication: Epistemic fidelity or mediating collaborative inquiry? *The Arachnet Electronic Journal on Virtual Culture* [On-line serial], *2*(2). Retrieved from ftp://ftp.lib.ncsu.edu/pub/stacks/aejvc/aejvc-v2n02-roschelle-designing

Roschelle, J., Patton, C., & Tatar, D. (2007). Designing networked handheld devices to enhance school learning. In M. Zelkowitz (Ed.), *Advances in Computers, 70*, 1-60.

Roselli, N. D., Gimelli, L. E., & Hechen, M. E. (1995). Modalidades de interacción sociocognitiva en el aprendizaje de conocimientos en pareja. In P. Fernández & M. A. Melero, *La interacción social en contextos educativos*. Madrid: S. XXI.

Roth, W. M., & Roychoudhury, A. (1993). The concept map as a tool for the collaborative construction of knowledge: A microanalysis of high school physics students. *Journal of Research in Science Teaching*, *30*, 503–534. doi:10.1002/tea.3660300508

Roth, W. M., & Roychoudhury, A. (1994). Science discourse through collaborative concept-mapping: New perspectives for the teacher. *International Journal of Science Education*, *16*(4), 437–455. doi:10.1080/0950069940160405

Roth, W., & Roychoudhury, A. (1992). The social construction of scientific concepts or the concept map as conscription device and tool for social thinking in high school science. *Science Eductation*, *76*(5), 531–557. doi:10.1002/sce.3730760507

Rowe, A. L., & Cooke, N. J. (1995). Measuring mental models: Choosing the right tools for the job. Human Resource Development Quarterly, 6(3), 243–255. doi:10.1002/hrdq.3920060303

Ruiz-Primo, A. (2004, September). *Examining concept maps as an assessment tool*. Paper presented at the I International Conference on Concept Mapping, Pamplona, Spain.

Ruiz-Primo, M. A., & Shavelson, R. J. (1996). Problems and issues in the use of concept maps in science assessment. Journal of Research in Science Teaching, 33(6), 569–600. doi:10.1002/(SICI)1098-2736(199608)33:6<569::AID-TEA1>3.0.CO;2-M

Rumelhart, D. E. (1980). Schemata: The building blocks of cognition. In R. J. Spiro, B. C. Bruce & W. F. Brewer (Eds.), *Theoretical issues in reading comprehension*. Hillsdale, NJ: Erlbaum.

Ryve, A. (2004). Can Collaborative Concept Mapping Create Mathematically Productive Discourses? Educational Studies in Mathematics, 56(3), 157–177. doi:10.1023/B:EDUC.0000040395.17555.c2

Safayeni, F., Derbentseva, N., & Cañas, A. (2005). A theoretical note on concepts and the need for cyclic concept maps. Journal of Research in Science Teaching, 42(7), 741–766. doi:10.1002/tea.20074

Salas, E., & Cannon-Bowers, J. A. (2000). The anatomy of team training. In S. Tobias & J. D. Fletcher (Eds.), *Training & retraining: A handbook for business, industry, government, and the military* (pp. 312-335). New York: Macmillan Reference.

Salas, E., & Fiore, S. M. (2004). *Team cognition: Understanding the factors that drive process and performance*. Washington, DC: American Psychological Association.

Salomon, G. (1993). No distribution without individuals' cognition: a dynamic interactional view. In Salomon, G. (Ed.). *Distributed Cognition: Psychological and educational perspectives.* Cambridge, UK: Cambridge University Press.

Salomon, G. (1998). Novel constructivist learning environments and novel technologies: Some issues to be concerned with. Research Dialogue in Learning and Instruction, 8(1), 3–12. doi:10.1016/S0959-4752(98)00007-3

Santos, E. (2003). Dos. Articulação de saberes na educação *online*: por uma rede interdisciplinar e interativa de conhecimentos em ambientes virtuais de aprendizagem. In M. Silva, (org.). *Educação Online* (pp. 217-272). São Paulo, Brazil: Loyola.

Santos, E. (2003). Portfólio e cartografia cognitiva: dispositivos e interfaces para a prática da avaliação formativa em educação online. In E. Santos, & M. Silva, (Eds.) *Avaliação da aprendizagem em educação online* (pp. 316-331). São Paulo, Brazil: Loyola

Santos, E. (2005). *Educação Online: cibercultura e pesquisa-formação na prática docente.* Doctoral thesis, Faculdade de Educação. Universidade Federal da Bahia - FACED/UFBA, Salvador, Brazil.

Saussure, F. (1916). *Course in general linguistics.* W. Baskin (Trans.). New York: Philosophical Library.

Savenye, W., Robinson, R., Niemczyk, M., & Atkinson, R. (2008). Assessment of individual learning processes. In M. D. Merrill, van Merriënboer, J., Spector, J. M., & Driscoll, M. (Ed.), *Handbook of research on educational communications and technology* (3 Ed., pp. 776-783). New York: Lawrence Erlbaum Associates.

Sawyer, R. K. (2006). Educating for innovation. Thinking Skills and Creativity, 1, 41–48. doi:10.1016/j.tsc.2005.08.001

Schlemmer, E. (2005). Ambiente virtual de aprendizagem (AVA): Uma proposta para a sociedade em rede na cultura da aprendizagem. In B. Valentini, Soares & M. S. Eliana (Eds.), *Ambientes virtuais: Compartilhando idéias e construindo cenários.* Caxias do Sul.

Schvaneveldt, R. W. (Ed.). (1990). *Pathfinder associative networks: Studies in knowledge organization.* Norwood, NJ: Ablex.

Schwartz, M. S., & Sadler, P. M. (2007). Empowerment in science curriculum development: a microdevelopment approach. International Journal of Science Education, 29(8), 987–1017. doi:10.1080/09500690600931053

Scott, D. M., Taylor, L. L., & Pelley, J. W. (2006, September). *The use of concept mapping in integrative learning with allied health profession students.* Presented at the Second International Conference on Concept Mapping, San Jose, Costa Rica.

Scott, J., & Flanigan, E. (1996). *Achieving Consensus: Tools and Technique.* Menlo Park: Crisp Publications.

Seale, C. (2004). *Researching society and culture.* London: Sage.

Seel, N. M. (1991). *Weltwissen und mentale Modelle* [World knowledge and mental models]. Göttingen: Hogrefe.

Seel, N. M. (1999). Educational diagnosis of mental models: Assessment problems and technology-based solutions. *Journal of Structural Learning and Intelligent Systems, 14*(2), 153–185.

Seel, N. M., Ifenthaler, D., & Pirnay-Dummer, P. (2006). Mental Models and Problem Solving: Technological Solutions for Measurement and Assessment. In J. Strobel & D. Jonassen (Eds.) *Learning by Modelling Systems.* Mahwah, NJ: Erlbaum.

Semple, A. (2000). Learning theories and their influence on the development and use of educational technologies. Australian Science Teachers'Journal, 46(3), 21–28.

Sharples, M., Taylor, J., & Vavoula, G. N. (2007). A theory of learning for the mobile age. In R. Andrews & C Haythornthwaite (Eds.), *The SAGE Handbook of E-learning Research* (pp. 221-247). London: Sage.

Shor, I., & Freire, P. (1986). *Medo e ousadia: O cotidiano do professor.* Rio de Janeiro, Brazil: Paz e Terra.

Shuyska, J. A. (2007). *The CONTEXT tool – Going beyond the narrative.* Retrieved March 25, 2008, from http://www.reflectingeducation.net/index.php/reflecting/article/viewFile/44/47

Silva, M. (2001). *Sala de Aula Interativa* (2nd Ed.). Rio de Janeiro, Brazil: Quartet, Tedesco, J. C. (2004). *Educação e Novas Tecnologias: esperança ou incerteza?* Brazil: UNESCO.

Silva, M. (2005). *Educatión interactiva: enseñança y aprendizaje presencial y on-line.* Barcelona, Spain: Gedisa.

Silva, M. (2006). *Sala de Aula Interatia* (4th ed.). Rio de Janeiro, Brazil: Quartet.

Silva, M. (Ed.). (2003). *Educação on-line.* São Paulo, Brazil: Loyola.

Silva, M., & Santos, E. (2007). A Pedagogia da Transmissão e a Sala de Aula Interativa. In P. L. Torres (Ed.), *Algumas Vias para Entretecer o Pensar e o Agir.* Curitiba, Brazil: SENAR-PR.

Silva, M., & Santos, E. (Eds.). (2006). *Avaliação da aprendizagem em educação online.* São Paulo, Brazil: Loyola.

Simon, H. A. (1996). *Observations on the sciences of science learning.* Pittsburgh, PA: Department of Psychology, Carnegie Mellon University.

Sinatra, R. (1986). *Visual literacy connections to thinking, reading, and writing.* Springfield, IL: Charles Thomas Press.

Sinatra, R. (2000). Teaching learners to think, read, and write more effectively in content subjects. Clearing House (Menasha, Wis.), 73(5), 266–273.

Sinatra, R., Beaudry, J., Pizzo, J., & Geisert, G. (1994). Using computer-based semantic mapping, reading, and writing approach with at-risk fourth graders. Journal of Computing in Childhood Education, 5(11), 93–112.

Siqueira, E. (2005). *2015: como viveremos.* São Paulo, Brazil: Saraiva.

Sizmur, S., & Osborne, J. (1997). Learning processes and collaborative concept mapping. *International Journal of Science Education, 19*(10), 1117–1135. doi:10.1080/0950069970191002

Slavin, R. E. (1989). Cooperative learning and student achievement. In R. E. Slavin (Ed.), *School and classroom organization* (pp. 129-151). Hillsdale, NJ: Erlbaum.

Slavin, R. E. (1990). *Cooperative learning: Theory, research and practice.* Englewood Cliffs, NJ: Prentice Hall.

Slavin, R., & Cheung, A. (2005). A synthesis of research on language of reading instruction for English language learners. Review of Educational Research, 75, 247–284. doi:10.3102/00346543075002247

Smith, B. E. (1992). Linking theory and practice in teaching basic nursing skills. The Journal of Nursing Education, 31, 16–23.*

Smith, B. L., & MacGregor, J. (1992). What is collaborative learning? In A. S. Goodsell, M. R. Maher, V. Tinto, B. L. Smith & J. MacGregor (Eds.), *Collaborative learning: a sourcebook for higher education* (pp. 10-30). University Park, PA: National center on postsecondary teaching, learning, and assessment

Smith, L. (1979). An evolving logic of participant observation, educational ethnography, and other case studies. In L. Shulman, (Ed.) *Review of Research in Education.* Chicago: Peacock Press.

Somekh, B. (2005). *Action Research: A methodology for change and development.* Milton Keynes, UK: Open University Press.

Somekh, B. (2007). *Pedagogy and Learning with ICT: Researching the art of innovation.* New York: Routledge.

Spector, J. M., & Koszalka, T. A. (2004). *The DEEP methodology for assessing learning in complex domains* (Final report to the National Science Foundation Evaluative Research and Evaluation Capacity Building). Syracuse, NY: Syracuse University.

Spector, J. M., Dennen, V. P., & Koszalka, T. A. (2005). Causal maps, mental models and assessing acquisition of expertise. *Technology, Instruction, Cognition and Learning, 3.*

Spinuzzi, C. (2002). *A scandinavian challenge, US response: methodological assumptions in scandinavian and US prototyping approaches*. Paper presented at the SIGDOC 2002.

Spiro, R. J., Feltovich, P. L., Jacobson, M. L., & Coulson, R. L. (1992). Cognitive flexibility, constructivism, and hypertext: Random access instruction for advanced knowledge acquisition in ill-structured domains. In T. M. Duffy & D. H. Jonassen (Eds.), *Constructivism and the technology of instruction: A conversation* (pp. 57-76). Hillsdale, NJ: Lawrence Erlbaum Associates

Stahl, G. (2006). *Group cognition: Computer support for building collaborative knowledge*. Cambridge, MA: MIT Press.

Stahl, G., Koschmann, T., & Suthers, D. (2006). Computer-supported collaborative learning. In R. K. Sawyer (Ed.), *Cambridge handbook of the learning sciences* (pp. 409- 426). New York: Cambridge University Press.

Star, S. L., & Griesemer, J. R. (1989). Institutional ecology, 'translations' and boundary objects: Amateurs and professionals in Berkeley's museum of vertebrate zoology, 1907-39. *Social Studies of Science*, *19*, 387–M0.

Steketee, C. (2006). Modelling ICT integration in teacher education courses using distributed cognition as a framework. Australasian Journal of Educational Technology, 22(1), 126–144.

Steven, J. (2001). *Cultura da interface: como o computador transforma nossa maneira de criar e comunicar*. Rio de Janeiro, Brazil: Jorge Zahar.

Stice, C. F., & Alvarez, M. C. (1986). *Hierarchical concept mapping: young children learning how to learn (a viable heuristic for the primary grades)*. (No. 5). Nashville, TN: Center of Excellence, Basic Skills for the Disadvantaged, Tennessee State University.

Stiggins, R., Arter, J., Chappuis, J., & Chappuis, S. (2004). *Classroom Assessment for Student Learning*. Portland, OR: Assessment Training Institute.

Stout, R. J., Cannon-Bowers, J. A., Salas, E., & Milanovich, D. M. (1999). Planning, shared mental models, and coordinated performance: An empirical link is established. Human Factors, 41(1), 61–67. doi:10.1518/001872099779577273

Stoyanov, S., & Kommers, P. (2006). *WWW-intensive concept mapping for metacognition in solving ill-structured problems*. Retrieved January 17, 2008, from http://www.ou.nl/Docs/Expertise/OTE/Publicaties/slavi%20stoyanov/10%20Stoyanov1. pdf

Stoyanova, N., & Kommers, P. (2002). Concept mapping as a medium of shared cognition in computer supported collaborative problem solving. Journal of Interactive Learning Research, 13, 111–133.

Stull, A., & Meyer, R. (2007). Learning by doing versus learning by viewing: Experimental comparisons of learner-generated versus author-provided graphic organizers. Journal of Educational Psychology, 99(4), 808–820. doi:10.1037/0022-0663.99.4.808

Suero, M. I., Pérez, A. L., & Montanero, M. (Eds.). (2003). *Física 2*. Madrid: Santillana.

Suthers, D. (1999). *The Effects of Representational Bias on the Collaborative Inquiry*. Paper presented at the HCI.

Suthers, D. (2006). Technology affordances for intersubjective meaning making: A research agenda for CSCL. Computer-Supported Collaborative Learning, 1, 315–337. doi:10.1007/s11412-006-9660-y

Suthers, D. D. (2001). Toward a systematic study of representional guidance for collaborative discourse. Journal of Universal Computer Science, 7(3).

Suthers, D. D. (2003). Representational Guidance for Collaborative Inquiry. In J. Andriessen, M. Baker & D. D. Suthers (Eds.). *Arguing to Learn: Confronting Cognitions in Computer-Supported Collaborative Learning Environments*. Dordrecht, the Netherlands: Kluwer.

Suthers, D. D., Girardeau, L. E., & Hundhausen, C. D. (2002). *The roles of representation in online collaborations*. Paper presented at the AERA 2002, New Orleans, LA.

Suthers, D. D., Toth, E. E., & Weiner, A. (1997, 14 octobre). *An integrated approach to implementing collaborative inquiry in the classroom.* Paper presented at the Computer support of collaborative learning, Toronto, Canada.

Sweller, J., Van Merrienboer, J., & Paas, F. (1998). Cognitive architecture and instructional design. Educational Psychology Review, 10, 251–296. doi:10.1023/A:1022193728205

Tarouco, L., Geller, M., & Medina, R. (2006). Cmap as a communication tool to promote meaningful learning. In A. J. Cañas & J. D. Novak (Eds.), *Concept maps: Theory, methodology, technology: Proceedings of the Second International Conference on Concept Mapping,* (Vol. 2, pp. 44-49). San Jose, Costa Rica.

Telebinezhad, M. R. (2007). The Effect of Explicit Teaching of Concept Mapping in Expository Writing on EFL Students' Self-regulation. *The Linguistics Journal 2*(1). Retrieved July 25, 2007 from http://www.linguistics-journal.com/April_2007_mrt&gmn.php

Tennyson, R. D., & Rasch, M. (1988). Linking cognitive learning theory to instructional prescriptions. Instructional Science, 17(4), 369–385. doi:10.1007/BF00056222

Thomas, G. P. (2006). Metacognition and science education: pushing foward from a solid foundation. Journal of Research in Science Teaching, 36(1-2), 1–6.

Tifi, A., & Lombardi, A. (2006). *WWMaps, a community on education through collaborative concept mapping.* Retrieved March 07, 2008, from http://cmc.ihmc.us/cmc2006Papers/cmc2006-p.92.pdf

Tompkins, E. L. & Adger, W. N. (2004). Does adaptive management of natural resources enhance resilience to climate change? *Ecology and Society,* (9).

Torres, P. L. (2002). *Laboratório on-line de aprendizagem: Uma proposta crítica de aprendizagem colaborativa para a educação.* Doctoral dissertation, Universidade Federal de Santa Catarina (UFSC), Brazil.

Torres, P. L. (2007). A Trama do Conhecimento. In P. L. Torres (Ed.), *Alguns Fios para Entretecer o Pensar e o Agir.* Curitiba, Brazil: SENAR-PR.

Torres, P. L. (2007). Sugerindo Caminhos para Explorar a Trama do Conhecimento no Programa Agrinho. In P. L. Torres (Ed.), *Alguns Fios para Entretecer o Pensar e o Agir.* Curitiba, Brazil: SENAR-PR.

Torres, P. L., & Marriot, R. C. V. (2006). *Tecnologias educacionais e educação ambiental: uso de mapas conceituais no ensino e na aprendizagem.* Curitiba: FAEP.

Torres, P. L., & Marriott, R. C. V. (2004). The contribution of concept maps to LOLA: The online learning lab. *Proceedings of the First International Conference on Concept Mapping, Spain,* (pp. 603-610).

Touraine, A. (2007). *New Paradigm for Understanding Today's World.* Cambridge, UK: Polity Press.

Trevarthen, C. (1982). The primary motives for cooperative understanding. In G. Butterwood & P. Light (Eds.), *Social cognition. Studies of the development of understanding* (pp. 89-112). Chicago: The University of Chicago Press.

Trowbridge, J., & Wandersee, J. (2005). Theory-driven graphic organizers. In J. Mintzes, J. Wandersee, & J. Novak, (Eds.) *Teaching Science for Understanding.* London: Elsevier Academic Press.

Tuckman, B. W. (1965). Developmental sequence in small groups. Psychological Bulletin, 63, 384–399. doi:10.1037/h0022100

Tufte, E. (1983). *The visual display of quantitative information.* Cheshire, CT: Graphics Press.

Tufte, E. (1990). *Envisioning information.* Cheshire, CT: Graphics Press.

Tversky, A. (1977). Features of similarity. Psychological Review, 84(4), 327–352. doi:10.1037/0033-295X.84.4.327

UNESCO. (2005). *Towards Knowledge Societies: UNESCO World Report.* Paris, France: UNESCO Publishing.

Van Boxtel, C. (2000). *Collaborative concept learning - Collaborative learning tasks, student interaction and the learning of physics concepts*. Retrieved March 17, 2008, from http://edu.fss.uu.nl/medewerkers.gk/files/ Boxtel_dissertation.pdf

van Boxtel, C., van der Linden, J., & Kanselaar, G. (1997). Collaborative Construction of Conceptual Understanding: Interaction Processes and Learning Outcomes Emerging From a Concept Mapping and a Poster Task. *Journal of Interactive Learning Research*, *8*(3/4), 3F–361.

Van Boxtel, C., van der Linden, J., & Kanselaar, G. (2000). Collaborative learning tasks and the elaboration of conceptual knowledge. *Learning and Instruction*, *10*, 311–330. doi:10.1016/S0959-4752(00)00002-5

van Boxtel, C., van der Linden, J., Roelofs, E., & Erkens, G. (2002). Collaborative concept mapping: Provoking and supporting meaningful discourse. *Theory into Practice*, *41*, 40–46. doi:10.1207/s15430421tip4101_7

Van Gelder, T. J. (2002). Argument Mapping with Reason!Able. *The American Philosophical Association Newsletter on Philosophy and Computers*, 85-90.

Veerman, A. (2003). Constructive discussions through electronic dialog. In: Andriessen, Baker & Suthers (Eds.) *Arguing to Learn Confronting Cognitions in Computer-Supported Collaborative Learning Environments*. Dordrecht, the Netherlands: Kluwer.

Vernooy, R., & McDougall, S. (2003). Principles for good Practice in Participatory Research: Reflection on Lessons from the Field. In B. Pound, S. Snapp, C. McDougall, & A. Braun (Eds.) *Managing Natural Resources for Sustainable Livelihoods: Uniting Science and Participation*. London: Earthscan.

Vilela, R. B., Austrilino, L., & Costa, A. C. (2004). *Using concept maps for collaborative curriculum development*. Retrieved March 19, 2008, from http://cmc.ihmc.us/ papers/ cmc2004-239.pdf

Vosniadou, S. (1992). Knowledge acquisition and conceptual change. *Applied Psychology*. International Review (Steubenville, Ohio), 41(4), 347–357.

Vosniadou, S., & Kollias, V. (2003). Using Collaborative, Computer-Supported, Model Building to Promote Conceptual Change in Science. In E. De Corte, L. Verschaffel, N. Entwistel and J. Van Merrienboer (Eds.) *Powerful learning environments: Unravelling basic components and dimensions* (pp. 181–196). Oxford, UK: Elsevier Press.

Vygotsky, L. S. (1978). *Mind in society: The development of higher psychological process*. Cambridge, MA: Harvard University Press.

Wandersee, J. (1992). *A standard format for concept maps*. Paper presented at the National Convention of the National Science Teachers Association, Boston.

Wandersee, J. H. (1990). Concept mapping and the cartography of cognition. Journal of Research in Science Teaching, 27(10), 923–936.

Warschauer, M. (1986). Computer-assisted language learning: An introduction. In Fotos S. (Ed.), *Multimedia language teaching*. Tokyo: Logos International. Retrieved July 01, 2004 from http://www.ict4lt.org/en/ warschauer.htm

WCPT. (2003). *European Physiotherapy Benchmark Statement*, Barcelona, Spain. Retrieved from http://www. fysiot.ee/dok/01.pdf

Weimer, M. S. M. (2000) *Escola Cliclada*. Secretaria de Estado de Educação Mato Grosso – Brasil. Retrieved May 28, 2007, from http://mec.gov.br

Wenger, E. (1998). *Communities of practice: learning, meaning, and identity*. Cambridge, UK: Cambridge University Press.

Wenger, E. (2004). *Learning for a small planet: A research agenda*. Retrieved from www.ewenger.com/research.

Wenger, E., McDermott, R., & Snyder. (2002). *Cultivating Communities of Practice: A Guide to Managing Knowledge*. Boston: Harvard Business School Press.

Wenger, R., & Lave. (1998). *Communities of Practice: Learning, Meaning and Identity*. Cambridge, UK: Cambridge University Press.

Wertsch, J. (1998). *Mind as Action*. Oxford, UK: Oxford University Press.

Wertsch, J. V. (1988). *Vygotski y la formación social de la mente*. Barcelona: Paidós.

West, C. K., Farmer, J. A., & Wolff, P. M. (1991). *Instructional design: Implications from cognitive science*. Englewood Cliffs, NJ: Prentice Hall.

West, D. C., Pomeroy, J. R., Park, J. K., Gerstenberger, E. A., & Sandoval, J. (2000). Critical thinking in graduate education: A role for concept mapping assessment? Journal of the American Medical Association, 284(9), 1105–1110. doi:10.1001/jama.284.9.1105

Whyte, W. F. (1991). *Participatory action research*. Newbury Park, CA: Sage Publications.

Widdowson, H. G. (1991). *O Ensino de Línguas para a Comunicação*. J. C. P. de Almeida Filho (Trans.). Campinas, Spain: Pontes.

Wiley, D. (2000). *The instructional use of learning objects. On-line version*. Retrieved February 20, 2007, from http://reusability.org/read

Willinsky, J. (2006). *The Access Principle - The Case for Open Access to Research and Scholarship*. Cambridge: MIT Press.

Willis, J. (1996). *A Framework for Task-based Learning*. London: Longman

Wilson, J. A., Acheson, J. M., Metcalfe, M., & Kleban, P. (1994). Chaos, complexity and community management of fisheries. Marine Policy, 184(4), 291–305. doi:10.1016/0308-597X(94)90044-2

Winne, P. H., Nesbit, J. C., Kumar, V., Hadwin, A. F., Lajoie, S. P., Azevedo, R. A., & Perry, N. E. (2006). Supporting self-regulated learning with gStudy software: The Learning Kit Project. *Technology, Instruction*. Cognition and Learning, 3, 105–113.

Woods, D. (1989). Social interaction as tutoring. In M. Bornstein & J.S. Bruner (Eds.), *Interaction in human development* (pp. 132-140). Hillsdale, N.J.: LEA.

Yin, R. (2003). *Case study research. Design and methods* (3rd Ed.). London: SAGE.

Yin, R. B. (2003). *Case study research - Design and methods* (3rd Ed.). London: Sage.

Young, J. (2002, March 22). 'Hybrid' Teaching Seeks to End the Divide Between Traditional and Online Instruction by blending approaches, colleges hope to save money and meet students' needs. *The Chronicle of Higher education Information Technology*. Retrieved April 30, 2003 from http://chronicle.com/free/v48/i28/28a03301.htm

Zeichner, K. (2008). *Keynote: The Third Space: where teachers and academics meet*. British Education Research Association (BERA) Annual Conference, Edinburgh, BERA.

Zeichner, K., & Liston, D. P. (2006). Teaching Students to Reflect. In D. Hartley & M. Whitehead, (Eds.), *Teacher Education: Major Themes in Education*. London: Routledge.

Zimmerman, B. J. (1990). Self-regulated learning and academic achievement: An overview. Educational Psychologist, 25(1), 3–18. doi:10.1207/s15326985ep2501_2

Zull, J. (2002). *The art of changing the brain*. Sterling, VA: Stylus Publishing, LLC.

About the Contributors

Patricia Lupion Torres has been a professor at PUCPR (Pontifícia Universidade Católica do Paraná/ Brazil) since 1982. She was head of the Education Department from 1994 to 1998 and again from 2003 to 2005. She was the coordinator of research and assessment of the Distance Learning Department from 2002 to 2003. She was also a lecture at UFSC (Universidade Federal de Santa Catarina/Brazil) on the Media and Knowledge Masters Course from 1999 to 2002. A pedagogue, she is a specialist in psycho-pedagogy and in sociological theories and has a master in education from PUCPR. She took her doctorate on media and knowledge at UFSC. She was on the board of directors of PUCWEB from 2003 to 2005. She manages projects on the use of virtual learning environments for distance learning in higher education, teaches the masters and doctorate in education courses at PUCPR, and is the director of distance learning at the same institution.

Rita de Cássia Veiga Marriot, BA, MED, is a member of the academic staff in the Department of Hispanic Studies and at the Centre for Modern Languages at the University of Birmingham UK, where she is a Portuguese language tutor. She has lectured at the Pontifícia Universidade Católica do Paraná (PUCPR) and at Universidade Federal do Paraná (UFPR) in Brazil in subjects including EFL, ESP, collaborative learning online, translation studies, new technologies in education in the Department of Languages and the Department of Education at graduate and postgraduate levels, and has also given teacher training courses related to computer assisted language learning at CELIN (UFPR's Language Centre). She is currently a doctoral student in e-learning and language acquisition in the School of Education at the University of Birmingham UK.

* * *

Olusola O. Adesope is a doctoral student at the Faculty of Education, Simon Fraser University. His research interests include a broad range of issues related to teaching and learning with multimedia resources and cognitive tools. His current work focuses on learning with interactive concept maps and diagrams and application of self-regulated learning models to enhance learning with multimedia resources. He is also interested in meta-analytic approach for aggregating results of empirical studies.

Mauri Åhlberg, PhD is Full Professor of Biology and Sustainability Education at University of Helsinki. He has been visiting scholar in many foreign universities, e.g. at Cornell University, Toronto University, Griffith University, University of Exeter. Åhlberg is visiting professor of Exeter University. He has published widely, mainly in Finnish, but also in English, homepage: http://www.helsinki.fi/people/mauri.ahlberg. His fields of research include tools for monitoring and promoting learning, e.g. improved concepts and improved Vee heuristics, individual and social learning, collaborative knowledge building, digital learning environments, like CmapTools and NatureGate® (http://www.naturegate.net)

Maria Julia Baldor is a doctoral student in the urban education doctoral program at the University of Wisconsin-Milwaukee. She is interested in exploring students' critical thinking processes and investigating factors that retention of minority students in higher education.

Josianne Basque, PhD in psychology, is professor in educational technology at Tele-universite, Montreal, a French Canadian distance university. She designed online courses in the fields of learning and cognitive science, educational technology and instructional design. She is also a researcher at the LICEF Research Center, dedicated to research in the field of cognitive informatics and learning environments and located at Tele-universite. Her current research interests include collaborative knowledge modeling applied to learning and to transfer of expertise in educational and workplace settings, competency-based learning and higher education pedagogy.

Jeff Beaudry is an associate professor in the educational leadership program, College of Education and Human Development, at the University of Southern Maine. His specialty areas are classroom assessment, visual learning, instructional technology, and research methods. He has co-authored on numerous research articles on semantic mapping, visual learning strategies and writing with Dr. Richard Sinatra, and is now working on continuing to do research in the area of visual and technological literacy. He is currently interested in student involvement in classroom assessment, and visual strategies for learning quantitative data analysis.

Josiane Maria Bortolozzi, master of education (MEd) from *Pontifícia Universidade Católica do Paraná* in 2007 whose dissertation's title was *The conception of virtual learning environments for hospitalized students*, which was inserted in the CNPq-sponsored research project called *Eurek@Kids*. Bachelor of graphic design from *Pontifícia Universidade Católica do Paraná* in 2003 and technician in industrial design *by Centro Federal de Educação Tecnológica do Paraná* in 2000. Professional experience in graphic design since 1998. Development of endomarketing projects at *Doma Design* in Curitiba/PR since 2004, creation of corporative identities, printed communication, conception and planning of internal campaigns, team management, and design projects.

Josep Juandó Bosch studied for his undergraduate degree in primary school teaching (1975) at the Universitat Autònoma de Barcelona. In the same university he obtained his graduate degree in pedagogy (1980) at the same university. Josep worked as a teacher (1975-1986) in a primary school in Anglès, where he was director for two academic years. Between 1986 and 2004 he worked as an educational administrator of the Catalan Government in Girona, managing teaching support services for primary and secondary schools. In 1999 he started teaching at the University of Girona as an associate professor. At present he works full time at the university, as a lecturer in teacher training and the Deputy to the Rector for teaching support.

Maria Luisa Pérez Cabaní, PhD in psychology, 1993. Lecturer in educational psychology at the University of Girona, where she has been director of the psychology department, vice-rector for teaching and academic policy, and is presently vice-rector for international policy. Her research is in the field of teaching and learning strategies, learning assessment, the professional identity of educators, and teacher training. She is the main researcher of the GRHCS-46 research team http://psicologia.udg.es/GRecerca/GRHCS46/index.htm and a member of SINTE http://www.sinte.es/joomla_cast/index.php?lang=en, an interuniversity research team. She has participated in several scientific meetings and has written several books, chapters and articles in which she has exposed her research results and conclusions.

Chaka Chaka is a senior lecturer in the Department of English at Walter Sisulu University (Eastern Cape, South Africa). His research interests include: collaborative learning (CL); concept mapping; computer-mediated communication (CMC); electronic learning (e-learning); computer assisted language learning (CALL); mobile learning (m-learning); mobile assisted language learning (MALL); Web 2.0 learning/Mobile Web 2.0 learning; Web 3.0/Mobile Web 3.0 learning; Semantic Web/Mobile Semantic Web; knowledge management (KM); and learning organisation (LO).

Simone C. O. Conceição is an associate professor for the University of Wisconsin-Milwaukee School of Education Department of Administrative Leadership and holds a PhD in adult learning and distance education from the University of Wisconsin-Madison and a master's degree in administration and development of adult and continuing education programs from the University of Wisconsin-Milwaukee. Her research interests include adult learning, distance education, impact of technology on teaching and learning, instructional design, learning objects, and staff development. Dr. Conceição was born in Brazil and has lived in the United States for over 19 years. Her diverse background brings an international perspective to the fields of education and training. She has researched and identified many aspects of good practice in online environments, and she is an expert in helping instructors and trainers understand web-based technology tools, software, and design processes.

Paulo Rogério Miranda Correia studies the effects of methodological and didactic strategies on the classroom's organizational features. Dr. Correia believes that changes in education involve transform classrooms and schools into learning organizations. His current research focuses on ICTs and concept mapping to foster creativity, thinking skills and collaboration among people during the learning process. Dr. Correia received his BSc in Chemistry in 1998 and his PhD in Analytical Chemistry in 2004 at São Paulo University (Universidade de São Paulo). Since 2005 he is a full-time professor and researcher at School of Arts, Science and Humanities (Escola de Artes, Ciências e Humanidades). Dr Correia is the head of the Group of Interdisciplinary Research on Science Teaching (GruPIEC, Grupo de Pesquisas Interdisciplinares em Ensino de Ciências).

Carrie Ann Desnoyers is a doctoral student in the urban education doctoral program at the University of Wisconsin-Milwaukee. She has a master's degree in adult and continuing education from the University of Wisconsin-Milwaukee. Her research focuses on adult and continuing education and using technology with adult learners.

Pedro José Pardo Fernández is an assistant professor at the University of Extremadura (Uex) in Spain. He received a BS in physical sciences from the University of Extremadura in 1998, and MS and PhD in physics at the same university in 2000 and 2004, respectively. He has been involved in several projects relating to the detection and influence of visual anomalies in students. His research interest include color vision, optical instrumentation, neural networks, computer networks and physics education.

Luiza Tatiana Forte had been graduate in psychology at Catolic University of Paraná. At present she is director of extension of the Faculties Pequeno Principe and Coordenadora of Posgraduation, teacher of the disciplines of psychology and moment integrador for courses of pharmacy and biomedicine. pupil of the master's degree in education for the Catolic University of Paraná, under the direction of the teacher Patricia Lupion Torres, working in his dissertation with the technique of concept map for the construction of learning object. She uses the technique of concept map also in the teaching with teachers and pupils.

Gloria Gomez is an experienced designer in the area of education and learning. Her doctoral dissertation researched issues in preschool concept mapping from an interaction design perspective. This research is representative of the cross-disciplinary nature of her interests: design methods for user research, knowledge representation, inclusive education and social change, and educational technologies promoting learning-how-to-learn skills in people of all ages. A former member of the internationally recognized IHMC Cmap Tools project, and the award winning proyecto Ludomatica, she is currently working on a teaching grant of the Occupational and Aviation Medicine Unit at the University of Otago in New Zealand. Gloria has published a number of scholary articles on how design can promote educational innovation. Her research career has been funded with awards granted by Colciencias, the Commonwealth Government of Australia, and Swinburne University of Technology.

Ria Hanewald is on the state council of Information and Communication Technology in Education of Victoria (ICTEV) and co-editor of *ICT in Education Journal*. She has written numerous articles and book chapters about educational technology. An educator for more than two decades, Ria has taught at early childhood, primary, secondary, tertiary and adult education levels in Europe, Australia and Asia. Ria worked as a curriculum consultant in using technology for teaching and learning, with an interest in conducting research on the use of Web 2.0 technologies. Dr Hanewald is a research fellow at The University of Melbourne in the Graduate School of Education.

Denis Hellebrandt's research focuses on understanding how people perceive change and uncertainty, and how they develop livelihood strategies in dynamic settings. Dr. Hellebrandt looks at both environmental and social change in the context of small-scale fisheries, particularly in fishery systems in southern Brazil. He applies mostly qualitative research techniques, with a special focus on participatory methods and concept mapping. His current work is the result of a trajectory of almost ten years, in which he has been doing research and has been actively involved with the fisherfolk in the Patos Lagoon estuary (Rio Grande-RS, Brazil). During this period Dr. Hellebrandt was able to carry out different projects independently, usually as member of research teams, but also as a professional consultant to university-led projects and as a "pro-bono" consultant giving support to grass-roots fisherfolk organisations.

Dirk Ifenthaler is assistant professor at the Department of Educational Science at the Albert-Ludwigs-University of Freiburg, Germany. Dr. Ifenthaler's research interests focus on the learning-dependent progression of mental models, problem solving, decision making, situational awareness, emotions, and instructional science. He developed an automated and computer-based methodology for the analysis of graphical and natural language representations (SMD Technology). Additionally, he developed components of course management software and an educational simulation game (SeSIM).

Päivi Immonen-Orpana (MSc. Health Care) has been working as a senior lecturer in Laurea University of Applied Sciences since 1999 and nowadays in Laurea's Well Life Center in Finland. She is a doctoral student since 2006 in University of Helsinki, Finland, The Faculty of Behavioural Sciences, and Department of Applied Sciences of Education. She is an active member of Professor Mauri Åhlberg's research group of Education for Sustainable Development.

Maria Elena Infante-Malachias studies pre and in-service pedagogical formation of science teachers from Paulo Freire's humanistic perspective. Dr. Infante-Malachias considers interdisciplinarity as the key element for devising new educational paradigms for post-industrial society. Her current research focuses on narratives and argumentation in science classes to promote metacognition and metalearning. Dr. Infante-Malachias received her BSc in biology in 1989 and her PhD in molecular genetics in 1999 at Unicamp State University (Universidade Estadual de Campinas). Since 2005 she is a full-time professor and researcher at School of Arts, Science and Humanities (Escola de Artes, Ciências e Humanidades). Dr Infante-Malachias is the head of the Group of Interdisciplinary Research on Science Teaching (GruPIEC, Grupo de Pesquisas Interdisciplinares em Ensino de Ciências).

Tristan E. Johnson is the director for the International Center for Learning, Education and Performance Systems (ICLEPS) in the Learning Systems Institute at Florida State University. He is also on the faculty of the Instructional Systems Program in the Department of Educational Psychology and Learning Systems in the College of Education. He has several years of experience that includes the design, development, and implementation of learning and performance technology tools. His research focuses on team cognition and it's links to team performance. He is involved in the development of innovative methodologies to measure individual and team performance and shared mental models. He has been involved in validating the use of cognitive maps to measure individual knowledge. Based on these individually constructed maps, qualitative and quantitative analysis methods are used to measure team members shared understanding. This state of understanding is then compared to team performance to see if and how these two measures are linked. Tristan has also studied the link between instructional, learning, and gaming strategies, specifically measuring the use of technology tools with and without deliberate practice.

Antonietta Lombardi is a teacher, since 32 years in a primary school, currently at the first Circolo Didattico in Novi Ligure Piemonte (Italy) where she applies concept mapping as a relevant tool for educative mediation and students' metacognition, and develops activity of inquiry in science education in primary school. She is co-founder of World Wide Maps community, devoted to collaborative concept mapping among informal groups of students from different countries and languages and a member of IHMC-Cmap Tools cmappers community, as a contributor to the three conferences on concept mapping and as Italian promoter of Cmap Tools.

María Isabel Suero López is an associate professor at the University of Extremadura (Uex). She received a BS in chemical sciences and a PhD in the field of spectroscopy from UEx in 1975 and 1982, respectively. She has been involved in several projects relating to physics education, detection of misconceptions and collaborative learning. She also has an interest in color vision in challenged and normal observers, adaptative behavior for color-blind observers, and spectroscopy, as well as physics education.

Dawndra Meers-Scott, OTR, is in the program director for the Master of Occupational Therapy Program at the Texas Tech University Health Sciences Center (TTUHSC) in Lubbock, Texas. She has been at TTUHSC for over seven years. Her interests are in concept mapping as a teaching strategy and Team-Based Learning. The work for her chapter was based on her dissertation, *Concept Mapping to Promote Integrative Learning in Master of Athletic Training and Occupational Therapy Students.*

Manuel Montanero-Fernández received a BA in education and psychology from Catholic University of Salamanca, and a PhD in education from University of Extremadura. He is senior lecturer in the Faculty of Education (University of Extremadura). His research interests are text and discourse comprehension, classroom interaction and science education. Along the last years he has working on testing the effects of graphic representation of knowledge to enhance the student's learning.

Manuel Montanero-Morán received PhD degree in physics from University of Extremadura, Spain in 1994 and he was principal and professor of physics at 'San José' College in Villafranca de los Barros (Spain) during 39 years. His research interests include physics education.

John C. Nesbit is an associate professor in the Faculty of Education at Simon Fraser University who teaches and conducts research in the areas of multimedia learning and cognition, self-regulated learning, evaluation of learning resources and argumentation in education. He has authored or co-authored over 30 peer-reviewed journal articles and book chapters.

Wan Ng is senior lecturer in the areas of science & technology education and gifted education in the Faculty of Education (Bundoora) at La Trobe University, Australia. Her major research interest lies in the use of innovative and integrated curriculum in learning, and the use of technology, including mobile technology, in education at both school and tertiary levels.

Alexandra Okada is a postdoctoral researcher in knowledge mapping for open content initiative at the Knowledge Media Institute, the Open University UK. She is also a visiting lecturer at Getúlio Vargas University in Brazil and at University of the Arts London. She holds a BSc in computer science, a MBA in knowledge management and marketing, a MA and PhD in education. Her interest focuses on how knowledge maps can be used to facilitate research, teaching and learning. She has published widely in the areas of knowledge mapping, online learning and new technologies. Two of her latest books are *Knowledge Cartography: Mapping Techniques and Software Tools* published by Springer USA and *Cognitive Cartography: Knowledge Maps for Research, Learning and Teaching* published in KCM Brazil.

John W. Pelley is in the Department of Cell Biology and Biochemistry at the Texas Tech University Health Sciences Center School of Medicine in Lubbock, Texas. He joined the founding faculty of the Texas Tech University School of Medicine in 1972 and during the past 36 years he has served for as acting chair for the biochemistry department, admissions dean, and curriculum dean. His interests are in brain-based learning, complexity/chaos science, and team-based learning. For the past 10 years he has been a regular speaker to students, advisors, and faculty at medical schools across the country.

Pablo N. Pirnay-Dummer is assistant professor at the Department of Educational Science at the Albert-Ludwigs-University of Freiburg, Germany. His research and publications are located in the area of cognition, learning and technology where his practical focus is on model based knowledge management. He developed, implemented and validated the language oriented model assessment methodology MITOCAR (Model Inspection Trace of Concepts and Relations) which is built to assess, analyze and compare individual and group models of expertise and was a key developer of the Highly Integrated Model Assessment Technology and Tools (HIMATT) assessment software. Pirnay-Dummer also developed the web based training software Empirix, including new approaches of automated evaluation and automated tasks synthesis algorithms.

Christina J. Preston has always been committed to the potential of digital technologies as a catalyst for change in teaching and learning. She founded the international MirandaNet Fellowship in 1992 at the Institute of Education, London University (IOE). The 3,000 members of this 'community of practice', who are teachers, teachers educators and researchers, seek pedagogical change through innovative continuing professional development (CPD) programmes based on the principles of action research and social interaction. She also specialises in the design of multimodal accreditation as well as the building professional e-communities of practice and industry/education partnership.

Béatrice Pudelko, PhD in cognitive psychology, is associate researcher at LICEF Research Center, dedicated to research in the fields of cognitive informatics and learning environments and located at Tele-universite, Montreal. In her doctoral research, which she completed in 2006, she used the Vygotskian framework to study the epistemic mediations of a graphical knowledge representation tool used by university students during a learning-from-text activity. Her current research interests are related to knowledge modeling for learning and for knowledge elicitation, to the development of cognitive skills and to artifact-mediated learning activities. She is also a tutor in an online course on cognitive science and learning offered at Tele-universite.

Angel Luis Pérez Rodríguez is an associate professor at the University of Extremadura (Uex). In 1976, he received a BS in chemical sciences and in 1981 a PhD on the subject of electronic devices for teaching, both from UEx. He has been involved in several projects relating to physics education, detection of misconceptions, concepts mapping and collaborative learning. His research interests include collaborative reconstruction of knowledge, physics education, color vision in challenged and normal observers, adaptative behavior for color blind observers.

Edméa Santos holds a bachelor's degree and a PhD in education. She is a lecturer at the Faculty of Education of the State University of Rio de Janeiro, where she works in the Distance Learning Department. E-mail: mea2@uol.com.br; Website: www.docenciaonline.pro.br/moodle .

Rosario Mérida Serrano has worked as a teacher of early childhood education for nine years, conducting field investigations on the implementation of the concept maps in children between 4 and 5 years. Later she graduated in pedagogy, and began to work in the Department of Education at the University of Cordoba (Spain), where she obtained the degree of doctor in 2000. The central subject of her doctoral thesis is research the didactic adaptation of these maps to the pupil's sociocognitive competences. She has published different articles in relation to early childhood education in national and international reviews.

Marco Silva holds a bachelor's degree in sociology and a PhD in education. He is a lecturer in the postgraduate program in education at Estació de Sá University, in the state of Rio de Janeiro, and at the State University of Rio de Janeiro (UERJ), where he works in the distance learning department. E-mail: marcoparangole@uol.com.br; Website: www.saladeaulainterativa.pro.br.

J. Michael Spector is a research professor at the Learning and Performance Laboratory at the University of Georgia. Formerly, he was associate director of the Learning Systems Institute, professor of instructional systems, and principal investigator for the International Center for Learning, Education and Performance Systems at Florida State University (2004-2008). He is a distinguished graduate of the United States Air Force Academy and earned a PhD in philosophy from The University of Texas at Austin. His recent research is in the areas of intelligent support for instructional design, system dynamics based learning environments, assessing learning in complex domains, distance learning, and technology integration in education. Dr. Spector served on the International Board of Standards for Training, Performance and Instruction (IBSTPI) as executive vice president; he is on the executive committee of the IEEE Learning Technology Task Force, and he is president-elect of the Association for Educational and Communications Technology (AECT). He is the editor of the development section of *ETR&D* and serves on numerous other editorial boards. He co-edited the third edition of the *Handbook of Research on Educational Communications and Technology* and has more than 100 journal articles, book chapters and books to his credit.

LesLee Taylor, ATC, LAT is the program director for the Master of Athletic Training (MAT) program at the Texas Tech University Health Sciences Center (TTUHSC) in Lubbock, Texas. She joined the TTUHSC faculty in 2000 to develop the MAT program. Her interests are in concept mapping, Myers-Briggs Type Indicator use in teaching, and team-based learning. She has employed the methods described in this chapter in her own classroom and teaching experiences.

Alfredo Tifi is a chemistry teacher, since 22 years in secondary school, currently at Technical Institute "E. Divini" in San Severino Marche (Italy) where he applies concept mapping as a relevant tool for educative mediation and students' metacognition, and where he get inspiration for constructivist educative research, with a strong concern also for inquiry-based science education in primary school and laboratory problem solving in chemistry. He is co-author of several textbooks on general chemistry and co-founder of World Wide Maps community, devoted to collaborative concept mapping among informal groups of students from different countries and languages and a member of IHMC-CmapTools cmappers community, as a contributor to the three conferences on concept mapping and as Italian promoter of CmapTools, responsible for the Italian translation of the software and of the online help (and other tutorials) for the use of CmpTools, and charged for several online and in-presence courses about Cmaps.

Polly Wilson teaches biology and marine ecology at Deering High School in Portland, Maine. She is a mentor and supervisor for pre-service teachers and a contributor to *What's Working in High Schools*, edited by Anne Davies, PhD and Kathy Busick, PhD.

Index